Physics of Cancer, Volume 1
(Second Edition)

Interplay between tumor biology, inflammation and cell mechanics

Biophysical Society series

About the series

The Biophysical Society and IOP Publishing have forged a new publishing partnership in biophysics, bringing the world-leading expertise and domain knowledge of the Biophysical Society into the rapidly developing IOP ebooks program.

The program publishes textbooks, monographs, reviews, and handbooks covering all areas of biophysics research, applications, education, methods, computational tools, and techniques. Subjects of the collection will include: bioenergetics; bioengineering; biological fluorescence; biopolymers *in vivo*; cryo-electron microscopy; exocytosis and endocytosis; intrinsically disordered proteins; mechanobiology; membrane biophysics; membrane structure and assembly; molecular biophysics; motility and cytoskeleton; nanoscale biophysics; and permeation and transport.

Physics of Cancer, Volume 1
(Second Edition)

Interplay between tumor biology, inflammation and cell mechanics

Claudia Tanja Mierke
University of Leipzig

IOP Publishing, Bristol, UK

ISBN 978-0-7503-1753-5 (ebook)
ISBN 978-0-7503-1751-1 (print)
ISBN 978-0-7503-1752-8 (mobi)

DOI 10.1088/978-0-7503-1753-5

Version: 20181001

IOP Expanding Physics
ISSN 2053-2563 (online)
ISSN 2054-7315 (print)

British Library Cataloguing-in-Publication Data: A catalogue record for this book is available from the British Library.

Published by IOP Publishing, wholly owned by The Institute of Physics, London

IOP Publishing, Temple Circus, Temple Way, Bristol, BS1 6HG, UK

US Office: IOP Publishing, Inc., 190 North Independence Mall West, Suite 601, Philadelphia, PA 19106, USA

This book was written with the support of Thomas M L Mierke.

Contents

Part II The role of the mechanical properties of cancer cells in cellular invasion

Preface

This book is about the novel and promising field of the physics of cancer, which has become the focus of many biophysical research groups all over the world. The book aims to present several biophysical approaches used in cancer research. The major findings that contribute significantly to the field are presented from a biophysical point of view. The intended readership includes everyone interested in cancer research, especially those readers studying biological physics, physics or tumor biology. The book is suitable for upper undergraduate, graduate, doctoral and postdoctoral students, senior scientists, lecturers, principal investigators and professors. I wrote this book because I was kindly invited to by IOP Publishing and because there is currently no such book available in any format. Many parts of it are supported by figures that I have drawn myself or designed. I hope that the book will promote understanding of why the physics of cancer is so important for cancer research. All types of readers, including students and researchers, can use the book to be guided through the physics of cancer. Every chapter is a complete part, with references and further reading suggestions. The book also has a glossary that will help the reader to follow the book easily. I hope that this book will help to establish the physics of cancer as an essential part of cancer research.

Preface for second edition

Since the appearance of the first edition of *Physics of Cancer*, I have received many suggestions and really helpful comments. Therefore, I decided to write this second edition of the book, which was offered to me by IOP Publishing. In January 2018 I began writing this second edition in the evenings and at the weekends, again without any support of the University of Leipzig. In particular, I have completely revised the second edition in terms of language, an increase in the number of figures, the inclusion of four novel chapters and seven partly new chapters, and each original chapter has been fully updated. The novel chapters are

- Actin filaments during matrix invasion
- Microtubules during migration and matrix invasion
- Nuclear deformability during migration and matrix invasion
- The active role of the tumor stroma in regulating cell invasion

Thus, the size of the book has grown so much that it has proven best to split it into two volumes. The two volumes now contain six parts. I hope that the broader scope of the two volumes of the second edition of *Physics of Cancer* will attracted more readers from even more research fields.

Claudia Tanja Mierke
Machern, May 2018

Acknowledgments

I thank Thomas M L Mierke for searching, editing and formatting of the references for all chapters and for helping me to keep in the time schedule for writing this book.

Author biography

Claudia Tanja Mierke

Claudia Tanja Mierke studied biology at TU Braunschweig in Germany and received her doctoral degree from the Medical School of Hannover in March 2001; her thesis concerned human endothelial and mast cell cell–cell interactions. Her postdoctoral research was performed in different institutes of the University of Erlangen-Nuremberg, and covered a number of fields: cancer research; inflammation of endothelial cells and molecular cancer research; and biophysical cancer research. In 2012 she habilitated in biophysics. Her scientific results have been presented at many international conferences and published in leading international peer-reviewed journals in this field. Since 2010, she has worked as a professor at the University of Leipzig and she has published many review articles and book chapters on the subject of physical-driven cancer research. As a professor of soft matter physics and biological physics, her research focus is on biophysical cancer research, including cell motility and transmigration through endothelial vessel linings.

With a background in biology and molecular oncology, Professor Mierke is now head of the Biological Physics Division at the University of Leipzig, where she regularly teaches molecular and cell biology, biophysics and soft matter physics to physicists, and is concerned with various research areas that contribute to our understanding of the physical aspects of cancer.

Part I

Introduction to tumor biology from a biophysical point of view

Part 1 of this book helps readers get into the field of physical cancer research by first introducing them to the broad field of tumor biology. Moreover, the book explains important aspects of classical tumor biology from a biophysical point of view and hence provides the basis for understanding the following three parts of this book. Whenever possible, the book criticizes the restrictively tumor-biological viewpoint available in many review articles on classical cancer research and includes the missing biomechanical aspects and principles. Finally, the book contains state-of-the-art physical cancer research and integrates it into the classical field of tumor biology by addressing physical concepts and principles as well as hypotheses and theories.

IOP Publishing

Physics of Cancer, Volume 1
Interplay between tumor biology, inflammation and cell mechanics
Claudia Tanja Mierke

Chapter 1

Initiation of a neoplasm or tumor

Summary

Chapter 1 presents the classical, tumor-biological viewpoint on the initiation of a tumor, and its further growth and proliferation, and points out the importance of the process of neoangiogenesis for reduced hypoxia in solid primary tumors, and their enhanced tumor growth and malignancy. A normal cell undergoes genetic mutations and epigenetic alterations and transforms into a malignant and aggressive cancer cell. The process of malignant cancer progression impacts cancer cell behavior and mechanical properties and supports the propagation of the main steps of cancer metastases in detail. The malignant transformation is associated with an accumulating loss of tissue homeostasis and enhanced perturbations of the entire structure and architecture of tissues causing the functional phenotypes of cancer cells, such as highly invasive cancer cells. The motility of cancer cells constitutes the focus, in particular their ability to transmigrate through barriers such as basement membranes and the endothelial-cell layers of vessels. Classical movement or sorting theories discuss and represent the importance of cancer cell properties for providing cellular motility or stalled behavior in cooperative networks such as tissues or spheroids. The hallmarks of cancer are presented from a biophysical point of view and the still missing mechanical aspect is addressed and included as a novel hallmark. Finally, mechanical properties profoundly affect the migration and invasion of cancer cells, and hence represent a critical aspect for the establishment of an altered mechanical phenotype in a distinct cancer type. The mechanical aspects are revisited throughout the later chapters of this book and hence chapter 1 builds the basis for understanding these chapters.

1.1 Initiation of a neoplasm, tumor growth and neoangiogenesis

What promotes the initiation of a neoplasm? What evokes tumor growth and thus malignant progression of tumors? Why can a nonvascularized tumor only grow to a restricted tumor size and not grow further? Why is vascularization of a tumor so

doi:10.1088/978-0-7503-1753-5ch1

important for its survival and malignancy? Why is tumor angiogenesis called neoangiogenesis? All these questions will be answered in the following sections of chapter 1.

1.1.1 Initiation of a neoplasm and tumor growth

The onset of a neoplasia starts as a complex scenario in a complicated process consisting of multiple steps that basically involve alterations in proto-oncogenes and tumor suppressor genes. In particular, proto-oncogenes are activated, while tumor suppressor genes are inactivated (Knudson 1971). Proto-oncogenes consist of a group of genes transforming normal cells to cancerous cells when they are mutated (Adamson 1987, Weinstein and Joe 2006). Typically, mutations in proto-oncogenes stay dominant; the mutated proto-oncogene is called an oncogene. Proto-oncogenes encode proteins related to cell division stimulation, inhibition of cell differentiation and reduction of cell death (Weinstein and Joe 2006). These processes promote normal human development and help to maintain tissues and organs. However, oncogenes regulate the elevated production of these proteins, causing increased cell division, suppression of cell differentiation and omission of cell death (figure 1.1). All these effects create up the phenotypes of cancer cells. For this reason, oncogenes are regarded as a potential molecular target for anticancer drugs.

Five to six independent mutational events usually contribute to the formation of human solid tumors, whereas three to four mutational events involving different genes are sufficient to cause leukemias in humans (Thomas *et al* 2007). In animals, carcinogenesis can be induced by chemical mutagens such as 7, 12-dimethylbenzanthracene (DMBA) (Balmain and Brown 1988) and a simultaneously administered chemical promotor to stimulate the growth of mutated cells such as 12-O-tetradecanoylphorbol-13-acetate (TBA), which belongs to the group of phorbol

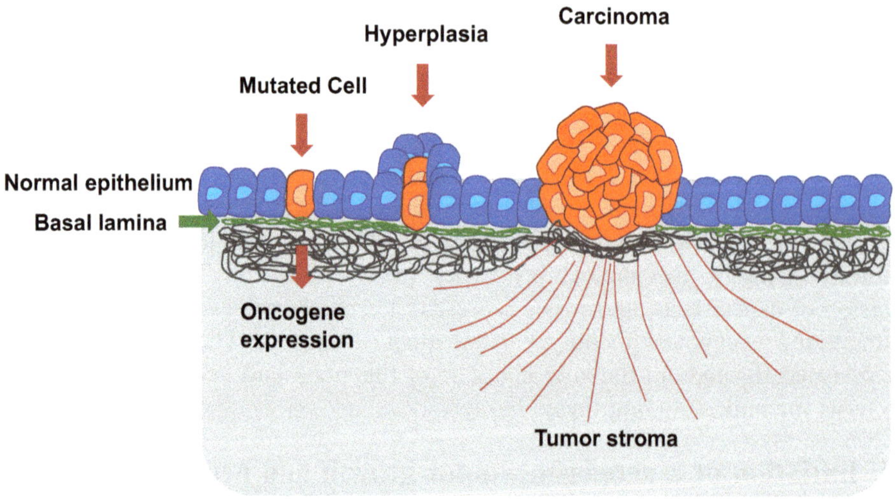

Figure 1.1. Initiation of a carcinoma. A mutation of a cell leads to the expression of an oncogene, which causes a hyperplasia. The hyperplasia develops to a carcinoma, which alters the surrounding microenvironment that is then termed the tumor stroma.

esters (Slaga 1983). First, precancerous papillomas are formed over a period of months and then they progress to skin carcinomas. DMBA may lead to a mutation in codon 61 of the H-ras oncogene, but still the growth needs to be stimulated by TPA (Slaga 1983, Balmain and Brown 1988). Chemical carcinogenesis in animals such as mice is used to model skin carcinogenesis, which is divided into three phases: initiation, promotion and progression (Moolgavkar and Knudson 1981). In addition to chemical mutagenesis, virus-induced mutagenesis is also able to cause carcinogenesis. In particular, a single oncogene is able to facilitate tumor formation by infection with some rapidly transforming retroviruses, such as Respiratory Syncytical Virus (RSV) (Temin 1988). However, the virus may also carry two different oncogenes that work together in order to cause a neoplastic phenotype (Temin 1988). An example is an avian erythroblastosis virus carrying the erbA and erbB oncogenes (Damm *et al* 1987). Transformation studies using nonimmortalized cell lines showed cooperation between the myc protein in the nucleus and the ras protein associated with the cytoplasmic-membrane site (Wang *et al* 2011) in transforming rat embryo fibroblasts. The cooperation between SV40 large T product and the mutated H-ras gene is necessary to transform 'normal' human epithelial cells and fibroblast cells, if these cells constitutively express the catalytic subunit of the telomerase enzyme (Wang *et al* 2011). Thus, these cells display a complex pattern leading to the neoplastic transformation of human cells. Taken together, the interplay between two different types of oncogenes (nuclear and cytoplasmic) has been demonstrated several times but is not strictly necessary for the malignant transformation of cells. For example, single myc oncogene expression leads to multiple genetic alterations that facilitate tumor formation. This results in increased incidence of clonal neoplasia and subsequently primary tumors (Wang *et al* 2011). Other events are necessary for neoplasia and subsequently tumor formation.

The onset and progression of human neoplasia is associated with the activation of oncogenes and the inactivation or complete loss of tumor suppressor genes. Due to the high variability of tumor types, the mechanisms of oncogene activation are highly variable. The common feature is that the activation of oncogenes leads to genetic alterations of cellular proto-oncogenes. This is generally associated with an advantage in cellular growth. Three genetic mechanisms for the activation of proto-oncogenes in human neoplasms are possible: mutations, gene amplifications and chromosome rearrangements (figure 1.2). All of these lead to alterations of the proto-oncogene structure or to an increase in the expression of the proto-oncogene. Given the multistep nature of the neoplasia process, we expect that more than one mechanism accounts for the formation of human tumors through alteration of the numbers of cancer-associated genes. For the whole expression of the neoplastic phenotype and the total capacity to metastasize, a combination of the activation of proto-oncogenes and the inactivation or loss of tumor suppressor genes is necessary (Adamson 1987, Weinstein and Joe 2006). In particular, first, the mutations are in critical regulatory domains or regions of the gene and mediate structural alterations (figure 1.2). Examples are retroviral oncogenes that have deletions contributing to their activation (Temin 1988). There can also be substitutions (called point mutations), which means only a single amino acid is altered within the protein

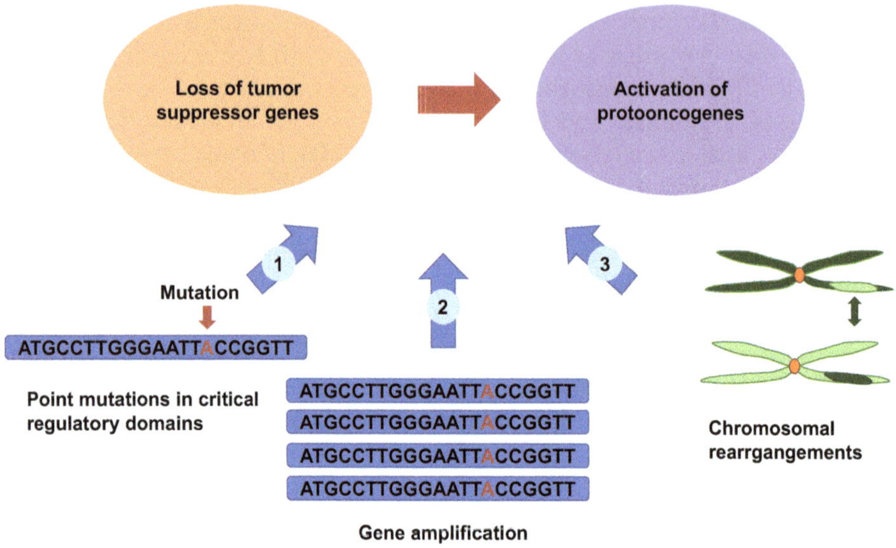

Figure 1.2. Three genetic mechanisms leading to the activation of proto-oncogenes and a loss of tumor suppressor genes: gene mutation, gene amplification and chromosome rearrangement.

(Temin 1988). Second, gene amplification is increased by expansion of the copy numbers of a single gene, providing resistance to growth-inhibiting drugs (figure 1.2) (Temin 1988). Third, chromosomal rearrangements can occur, as frequently observed in hematologic malignancies, soft-tissue sarcomas and certain solid tumors (figure 1.2) (Temin 1988). In the Burkitt lymphoma, chromosomal translocations are often observed and less pronounced chromosomal inversions have been detected (Dalla-Favera *et al* 1982). In the latter case, chromosomal breakpoints between two genes support these inversions. This may then lead on the one hand to transcriptional activation of certain proto-oncogenes and on the other hand to the fusion of genes, for example in chronic myelogenous leukemia (CML) (Lozzio and Lozzio 1975).

Although there is much variability in the pathways for the initiation and progression of tumors in humans, numerous studies of different types of malignancy have revealed the multistep character of human cancer. All the above-mentioned mechanisms for initiating a neoplasm and promoting tumor progression may affect the mechanical properties of cancer cells and subsequently alter their physiological function in a certain microenvironment, for example they may become more motile compared to nontransformed 'normal' cells.

1.1.2 The primary tumor changes from normoxia to hypoxia

The transformation of normal cells to cancer cells is additionally guided by alterations of the oxygen concentration in the primary solid tumor. During the excessive growth of a primary solid tumor such as a carcinoma, a local environment for the constituent cancer cells is created that is defined by the rapid establishment of an oxygen gradient, which has the lowest oxygen levels at the tumor's center of mass and increases towards the outermost tumor regions (figure 1.3). These dramatic alterations of the tumor's growth conditions evoke the adaption of cancer cells to these environmental changes. One adaption of the cells is the reprogramming of the energy metabolism, which has hence been regarded as a new hallmark of cancer (Hanahan and Weinberg 2011). Inside the primary tumor mass, cancer cells sense altered conditions and display in turn enhanced or decreased activation and expression of molecules. Moreover, the deregulated proliferation leads to chronic and uncontrolled growth of cancer cells that is essential for ongoing neoplastic diseases such as cancer. As the proliferation of cancer cells is increased, the energy metabolism requires an adaption in order to increase the production of elevated levels of substrates for cell growth and division that serve as cellular food. Under physiological conditions, normal, healthy cells grow and proliferate by processing primarily glucose under aerobic conditions, which leads first to the production of pyruvate through the glycolysis cycle that operates in the cytoplasm of the cells and second the production of carbon dioxide in the mitochondria. Conversely, cells of the interior of the tumor mass need to grow under anaerobic conditions that are used by prokaryotic cells such as bacteria to perform fermentation processes such as anaerobic glycolysis.

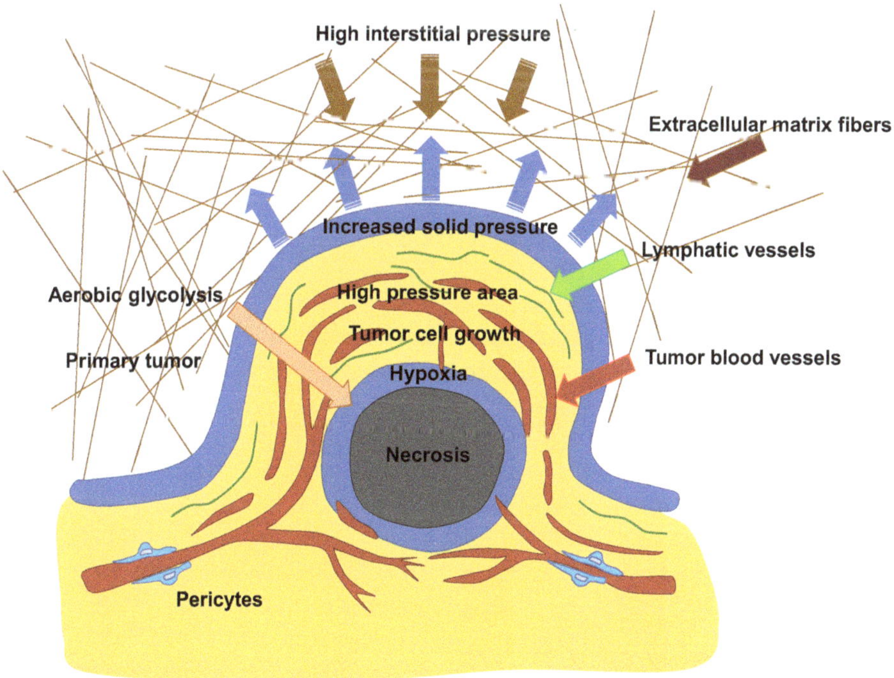

Figure 1.3. Primary tumors change from normoxia to hypoxia. Inside the tumor the oxygen content is low (hypoxia conditions). Aerobic glycolysis is performed by the cancer cells. The tumor growth is high and causes a high pressure area inside the tumor, which in turn exerts an increased solid pressure towards the tumor microenvironment. Inside the tumor are tumor blood vessels (red) and lymphatic vessels (green).

Cancer cells of the primary tumor mass still prefer the process of glycolysis for energy generation, whereas pyruvate degradation in the oxygen-requiring mitochondria is prevented. The cancer cell's energy metabolism is characterized by its restriction to purely glucose metabolism via glycolysis (figure 1.3), which was initially described by Otto Warburg and termed aerobic glycolysis (Warburg 1930, 1956a, 1956b). The transition of cancer cells' energy metabolism to undergo aerobic glycolysis is established and is now known as a novel hallmark of cancer, termed the reprogramming of energy metabolism. The transition seems to be at first glance unexpected, as the cancer cells have chosen to perform energy metabolism with an approximately 18-fold lower efficiency of ATP production compared to ATP generated by an oxidative phosphorylation in the mitochondria that is based on the citrate acid cycle. How can the cancer cells compensate for the lower efficiency of glycolysis-based ATP generation? In fact, cancer cells can even out the inefficiency by up-regulation of glucose transporters such as GLUT1, which increases glucose import into the cytoplasm (DeBerardinis *et al* 2008, Hsu and Sabatini 2008, Jones and Thompson 2009). Based on the increased uptake of glucose, elevated intracellular glucose levels are found in numerous human cancer types. The glucose uptake can be measured via the uptake of a radiolabeled analog of glucose as a reporter, such as 18F-fluorodeoxyglucose (FDG), using positron emission tomography (PET) (Haberkorn *et al* 1991, Vesselle *et al* 2000).

The glycolytic energy generation is associated with enhanced levels of activated oncogenes such as Ras and myc and connected to mutations in tumor suppressor genes such as TP53 (DeBerardinis *et al* 2008, Jones and Thompson 2009). The alterations in tumor suppressor genes of cancer cells favor those mutations providing increased capacity for cell proliferation, suppression of cytostatic controls and overcome apoptosis, with each of the three representing hallmarks of the malignant progression of cancer (Hanahan and Weinberg 2011). Under hypoxic conditions the dependence on glycolysis is even more pronouncedly promoted and thus largely enhanced in various tumors (figure 1.3). The hypoxia response causes, through a multifactorial signaling cascade, an increase in the amount of glucose transporters and many enzymes of the glycolysis cycle (DeBerardinis *et al* 2008, Jones and Thompson 2009, Semenza 2010a). Subsequently, the Ras oncogene and hypoxia both individually increase the levels of the HIF1α and HIF2α transcriptions factors, which then enhance the glycolytic production of ATP (Kroemer and Pouyssegur 2008, Semenza 2010a, 2010b). To date, it is still not yet known why cancer cells switch their energy metabolism from mitochondrial oxidative phosphorylation dependent ATP energy production to the rather inefficient glycolytic ATP energy production. A hypothesis states that the increased glycolysis leads to glycolytic intermediates, which can act themselves in various biosynthetic signaling pathways such as the generation of nucleosides and amino acids (Potter 1958, Vander Heiden *et al* 2009). In turn, the generation of these products induces the biosynthesis of biomacromolecules and entire organelles that are essential for building new cells. The Warburg-dependent metabolism seems to be utilized in diverse fast proliferating tissues such as embryonic tissues and manifests its role in large-scale and fast biosynthesis programs that are active in cells undergoing increased proliferation.

In distinct tumors, two subpopulations of cancer cells are identified that are distinguishable in their generation of energy. First, one subpopulation uses the Warburg-dependent metabolism (termed the Warburg effect) that is identified by increased secretion of lactate and, second, the other subpopulation employs lactate as a primary substrate for main energy generation by utilizing specific parts of the citrate acid cycle (Semenza 2008, Feron 2009, Kennedy and Dewhirst 2010). Both subpopulations act together in symbiosis, as the first subpopulation of hypoxic cancer cells requires glucose for energy generation and produces lactate that is secreted as waste into the surrounding environment. Lactate is endocytosed by the second subpopulation to utilize it for the highly efficient oxygenated energy production. This symbiotic mechanism still needs to be identified for different cancer types, however, it is not a unique feature of primary tumor growth, as muscle cells utilize the same mechanism for energy production (Semenza 2008, Feron 2009, Kennedy and Dewhirst 2010). Oxygenation in solid tumors ranges from normoxia to hypoxia, which is altered by continuous and varying fluctuations in time and space (Hardee *et al* 2009) that is manifested in the disordered nature of the tumor-neovasculature depending on assembly and disassembly as well as its dynamically changing architecture. Altered energy metabolism is a characteristic feature of cancer cells and has been established as another basic and potential core hallmark of cancer, as aerobic glycolysis can be seen as another phenotype of cancer cells evoked by oncogenes facilitating cell proliferation (figure 1.3) (Hanahan and Weinberg 2011). Support for determining energy metabolism as a hallmark of cancer is based on the finding that activating (gain-of-function) mutations in enzymes of the citrate acid cycle, such as isocitrate dehydrogenase 1/2 (IDH), are found in gliomas and several other human tumors (Yen *et al* 2010). Mutations of IDH can be identified within cells and selected as distinct cell clones, which still possess a reproducibly altered energy metabolism. However, it is not clear whether the core hallmark of an altered energy metabolism is independent of other core hallmarks such as decreased genome stability, increased angiogenesis and invasion, which can additionally lead to elevated oxygenation and increased stability of HIF-1 transcription factors (Reitman and Yan 2010). Therefore, the hallmark of altered energy metabolism is regarded as an emerging hallmark, which continues to demonstrate its importance in cancer progression, but there remains the question of whether it is independent of other core hallmarks.

1.1.2.1 Heterogeneity in primary solid tumors

Within solid tumors exist enormous heterogeneity that is reflected in the easily detectable heterogeneity of circulating tumor cells (CTCs) and subsequently in the emergence of multiple spreads of single new tumor seeds (figure 1.4). One representative example for the heterogeneity of the primary tumor mass is prostate cancer, which contains a heterogeneous tumorigenic cell mass of different cancer cell clones (Andreoiu and Cheng 2010, Cooper *et al* 2015). The metastases formation of prostate cancer cells is usually based on polyclonal seeding of several diverse cancer cell clones (figure 1.4) (Gundem *et al* 2015). However, there exist opposite approaches that propose development of metastases through conserved so-called

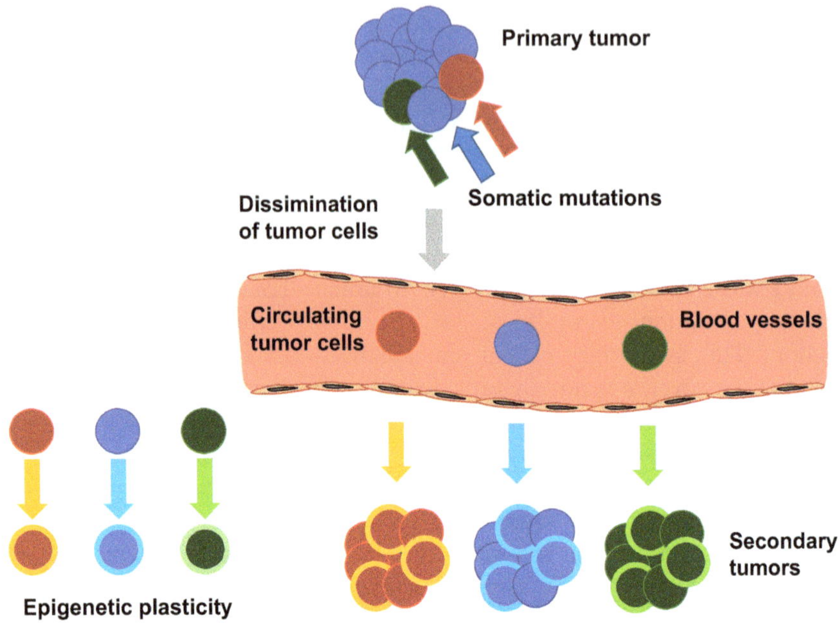

Figure 1.4. Somatic mutations occur and induce the development of a primary tumor. Tumor cells migrate out of the tumor into blood vessels (circulating tumor cells), which can undergo several further mutations. Hence the tumor cells represent many subsets of populations exhibiting different properties, which is termed epigenetic plasticity.

driver lesions in targeted organs or tissues (Kumar *et al* 2016). The identification of driver lesions is promising in selecting specific treatments based on predicted molecular vulnerabilities in the patient for an immediate and positive treatment response. These personalized treatments are based on the finding that the susceptibility of the individuals to specific therapeutics is based on malignancies with a specific alteration, whereas all other patients will not respond to the treatment. The existence of a broad diversity in genomic aberrations and treatment responses of equal histological cancer classifications favors the personalized treatment approach (Perou *et al* 2000, Tannock *et al* 2004, Frattini *et al* 2013, Sato *et al* 2013, Verhaak *et al* 2013). The strategies of individualized treatments are based on distinct molecular biomarkers. As there is a large genetic instability in cancer cells, the monoclonality of these cancer cells may be lost upon adapting a genetic heterogeneity during the progression of cancer (Zhang *et al* 2016). The heterogeneity can exist firstly between the primary tumor and secondary metastases, secondly among various metastatic lesions and thirdly within the primary or secondary solid tumors (figure 1.4). Since the heterogeneity of tumors is manifested in diverse protein functions, the diagnosis and treatment of the tumors must be precise and efficient. To determine and monitor tumor heterogeneity, CTCs are easily analyzed to identify the mutations and alterations in gene expression for each individual patient (Zhang *et al* 2016).

1.1.2.2 Development of hypoxia

The hypoxic nature of solid primary tumors is a major feature of multiple epithelially originated tumors (Wilson and Hay 2011) and subsequently a consequence of inefficient and dysfunctional vascularization of the tumors (figure 1.5). In addition, tumor endothelial vessels are often leaky and even consist of cancer cells within the endothelial-cell vessel lining that mimic endothelial-cell function and morphology by adapting their gene expression pattern such as VE-cadherin and PECAM-1 expressions, which all contribute to the chaotic and heterogeneous organization of the vessels in tumors and their nearby microenvironment (figure 1.5). In line with this, the branching of the vessels is irregularly distributed and the vessels contain abnormal shunting, resulting in heterogeneous blood flow (figure 1.5) (Carmeliet and Jain 2011), which then manifests endogenous hypoxia gradients within the entire solid tumor that contain extremely low partial pressure levels of oxygen (pO_2) in distinct regions of the primary tumor mass (Li *et al* 2007). In succession, the primary tumor can adapt its metabolism to the reduced oxygen levels, which creates a cell-layered metabolite gradient in the tumor tissue. In summary, the vascular heterogeneity of the vessels causes extracellular oxygen gradients (or pO_2), which alter the intravascular pO_2 and subsequently affect the metabolic consumption rate of cancer cells and stromal cells within the entire tissue including tumor stroma. Finally, the combination of all events causes pO_2 and metabolite gradients inside the tissue, which have a major impact on the behavior of cancer cells.

1.1.2.3 Alterations of the 'normal' pH gradient in primary solid tumors

Under physiological and pathological conditions, living cells and tissues regulate their production or chemical conversion of acids or bases through chemical reactions (Roos and Boron 1981). As cells perform respiration and thereby produce CO_2 and lactic acid, they are producing an acidic intracellular microenvironment, which decreases the intracellular pH (pH_i) over time, if the environmental fluid is not

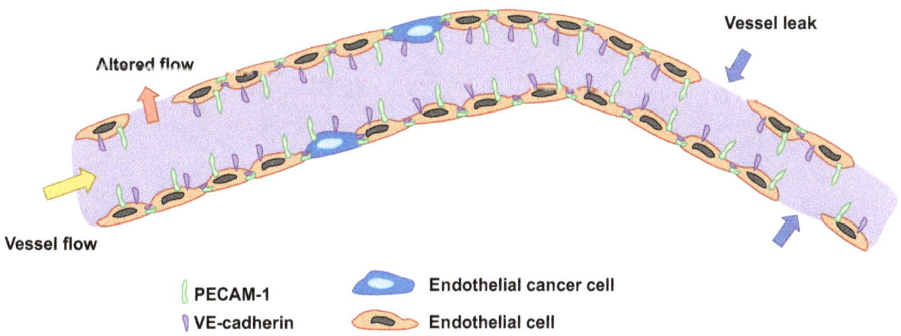

Figure 1.5. Leaky tumor blood vessels. Inside of the primary tumor blood vessels are formed by tumor endothelial cells and cancer cells mimicking endothelial cells by expressing endothelial marker proteins such as VE-cadherin and PECAM-1. Many vessels are leaky, which means there are some endothelial cells missing in the vessel lining, which alters the blood flow.

exchanged. In contrast to a pure solution, living matter such as cells and tissues is divided into compartments by semi-permeable lipid-bilayer membranes. Due to the semi-permeable nature of these membranes, the passage of selected molecules helps to establish pH differences between the compartments (figure 1.6). The strategy of cells for the regulation of pH_i is based on a balance between the internal production of acids (or bases) and an equal efflux of acids (or bases) across the cell's membrane into the surrounding environment, which represents a corrective flux (figure 1.6). However, human tumors preferentially grow around blood vessels up to an outermost distance of 200 μm. Beyond this distance the cancer cells become necrotic, which has been revealed by histological analysis of tumor samples (figure 1.5) (Thomlinson and Gray 1955). Due to the enhanced metabolism of cancer cells and long diffusion distances, a gradient of O_2 tension is established across the tissue (Vaupel *et al* 1989). Low (below 1%) O_2 tension is present in tissues of patients with increased metastasis and hence poor survival rate (Hockel *et al* 2001), which indicates hypoxia as a hallmark of malignant cancer progression. Cancer cells require the ability to sense and respond to microenvironmental factors, which in turn guide the somatic evolution of cancer cells. Under hypoxia, HIF acts as a transducing molecule between the O_2 tension gradient and cellular effects. The identification of pH sensors responsible for cancer disease progression has revealed to be difficult, as it is not easy to distinguish between physiological pH sensors from the numerous proteins simply binding H^+ ions (Schonichen *et al* 2013).

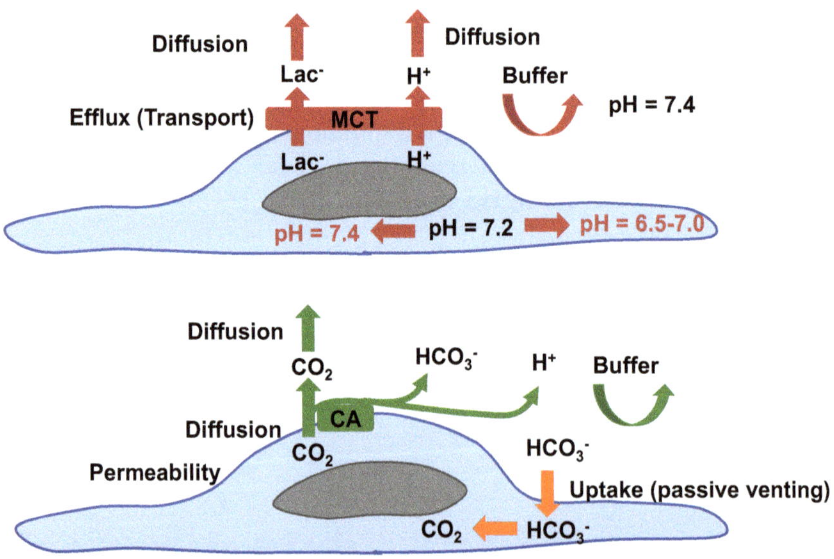

Figure 1.6. (Top) Semi-permeable membrane and venting of metabolically produced acids and the regulation of intracellular pH. H^+ efflux across the cell membrane is driven by H^+-monocarboxylate transport (MCT) activity. The pH of a normal cell is 7.2 (extracellular 7.4) and can be either 7.4 or 6.5 to 7.0 in cancer cells. The diffusion of lactate and H^+ is required for MCT activity. (Bottom) CO_2 can diffuse through the cell membrane. CO_2 hydration can be spontaneous and enhanced by exofacial carbonic anhydrase (CA) enzymes. To regulate the intracellular pH, cells can passively vent HCO_3^- to maintain a neutral pH.

Due to hypoxia, the biological view on cancer has changed (Harris 2002) and numerous hypoxia-inducible genes facilitated by the hypoxia-inducible factors (HIFs) such as HIF1α (Wang *et al* 1995) have been revealed that regulate the switch of the citrate acid cycle-based metabolism to a glycolytic metabolism, which is termed the Warburg effect (Warburg 1930). Tumor hypoxia has been established as a research topic and reveals promising knowledge for understanding etiology, providing more precise diagnosis and novel treatment developments (Harris 2002, Semenza 2003). In addition to other tumor environmental factors, extracellular acidity has manifested as a hallmark of cancer (Wike-Hooley *et al* 1984, Tannock and Rotin 1989, Gillies *et al* 1994). In contrast to the low extracellular pH (pH_e), the intracellular compartment has been detected to be alkaline (Griffiths *et al* 1981). Unlike in solid tumors, this specific transmembrane H^+ distribution, in which the extracellular space is acidic and the intracellular space is alkaline, is highly unusual for healthy tissues. How can solid tumors produce low pH_e within certain limits and manage at the same time to provide an alkaline pH_i? How does the specific transmembrane pH distribution account for malignant cancer progression?

In healthy differentiated epithelial cells, is the intracellular pH (pH_i of 7.2) that is slightly lower than the extracellular pH (pH_e of 7.4). Inside a primary tumor the pH can change dramatically depending on the pO_2 content and the access of metabolites: a normal pH gradient can be altered to a higher pH_i of 7.4 (Webb *et al* 2011) or a lower pH_e of 6.5–7.0 (figure 1.6) (Stuewe *et al* 2007, Gallagher *et al* 2008, Hashim *et al* 2011). In particular, alterations of the pH within cancer cells are due to enhanced expression and/or activity of ion pumps and ion transporters such as H^+ ATPases (Martinez-Zaguilan *et al* 1993), Na^+–H^+ exchanger NHE1 of the SLC9A family (McLean *et al* 2000) and monocarboxylate–H^+ efflux cotransporters MCT1 and MCT4 or the SLC16A family (Pinheiro *et al* 2008) that control the efflux of H^+, causing the high pH_e. During hypoxia, based on the increased activity of the carbonic anhydrases CAIX and CAXII, the hydration of extracellular CO_2 to HCO_3^- and H^+ is elevated (Chiche *et al* 2009, Swietach *et al* 2009). The reduced blood flow velocity and the decreased oxygen levels at the early stages of solid primary tumors seem to cause the switch from a citrate acid cycle-driven to a glycolysis-driven metabolism (Gatenby and Gillies 2004). Based on the glycolytic metabolism there exists an enhanced flux of carbons around the primary tumor that cause an acidification of the extracellular pH, which is enhanced in luminal structures (Barathova *et al* 2008). There are pO_2, pH and metabolite gradients around primary tumors, which in alter the behavior of cancer cells and thus cause the broad heterogeneity within tumors.

When cells sense an alkaline pH_i, they proliferate (Pouyssegur *et al* 1982), enter the cell cycle for cell division (Turchi *et al* 2000, Putney and Barber 2003), differentiate (Wang *et al* 1997), migrate (Klein *et al* 2000, Lagana *et al* 2000), impair apoptosis (McConkey and Orrenius 1996) and experience rearrangements of the chromosomes, including additions and deletions, termed clastogenesis (Morita *et al* 1992), and subsequently support the malignant transformation of cancers (Gillies *et al* 1990, Reshkin *et al* 2000) including cancer metastasis. Due to the complexity of these diverse processes, the observed pH_i sensitivity of tumors seems to require numerous H^+-binding molecular switches. The sensing of the pH is not restricted to the cytoplasm, since H^+ sensing G-protein-coupled receptors (Ludwig *et al* 2003) and H^+ sensing ion channels (Waldmann *et al* 1997, Glitsch 2011) enable cells to adapt to pH changes of the tumor environment. During the progression of cancer, cells accumulate genetic alterations through positive selection by their microenvironment (Nowell 1976, Fang *et al* 2008, Gillies *et al* 2012). The extracellular acidity is a complex phenomenon regulated by diffusion distance, buffering, metabolic rate and membrane transport, represents a major selection pressure and thereby has become a prominent cancer hallmark, as several mutations are found in genes or gene regulators for H^+/H^+-equivalent transporters, pH sensors or molecules acting in acid-yielding metabolic processes (Fang *et al* 2008). In particular, poor prognosis for cancer patients has been associated with low pH_e, which shows that acidity characterizes a distinct cell population possessing appropriate pH sensing mechanisms and a regulatory apparatus to facilitate drug resistance such as the drug doxorubicin (Gerweck *et al* 1999, Raghunand *et al* 1999). Cancer cells and possibly even tumor stromal cells (Koukourakis *et al* 2006) can control the pH_e by altering the direction and rate of pH-change for adaption of the selection pressure for optimized progression of cancer disease. Moreover, the higher acid yield of the glycolysis per produced ATP compared to mitochondrial respiration provides a possible explanation for the preferred Warburg effect in cancer disease (Gatenby and Gillies 2004). In conclusion, the plasticity of the pH_i regulation and the large variety of protein pH sensitivity mechanisms enable cancer cells to utilize pH as a selection pressure, whereas normal host cells cannot adapt to such a microenvironment due to their genetic stability. Hence, the pH may serve as a possible target for cancer therapy.

1.1.3 Neoangiogenesis

Neoangiogenesis occurs in a tumor of a certain size and enhances its malignancy state. This process of neoangiogenesis in a tumor is called tumor angiogenesis (Chaplin *et al* 1999). A tumor without a vascular system or nearby blood vessels will stop growing. This indicates that the process of angiogenesis is necessary for a tumor to overcome the restriction in size to a maximal diameter of 1–2 mm (Folkman *et al* 1963). This means that a tumor without angiogenesis is of small size and hence not malignant. Finally, the tumor will probably cause no damage to surrounding or hosting organs. In particular, tumor angiogenesis is the proliferation of endothelial cells lining blood vessels that break through the tumor and grow to novel vessels in order to supply the cancer cells of the inner tumor mass with nutrients and oxygen (figure 1.5) (Gimbrone *et al* 1972, 1974). The vessels can also serve to remove waste products, such as toxic substances that reduce tumor proliferation (Folkman *et al* 1963).

The onset of tumor angiogenesis begins with a cancerous tumor consisting of cells that secrete molecules to their surrounding microenvironment of 'normal' host tissue, such as the extracellular matrix of connective tissue. The signaling molecules activate genes in the cells of the microenvironment to produce proteins promoting or inducing the growth of new blood vessels (Gimbrone *et al* 1974). The gradients of these signaling molecules are most commonly directed around the primary tumor and thus the new vessel expands in the direction of the tumor in order to grow and migrate into it (Gimbrone *et al* 1974). This behavior indicates that tumor growth needs angiogenesis.

A pioneering experiment was performed to test the importance of angiogenesis, in which a cancerous tumor from an animal was removed and a 'normal' healthy organ from an animal of the same strain not carrying a tumor was isolated. Some cancer cells from the isolated tumor were injected into the healthy organ and cultured in a glass chamber containing a nutrient solution that was pumped into the organ to keep it healthy. After one or two weeks only small tumors had formed, no larger than 1–2 mm in diameter and without any connection to the organ's vascular system (Gimbrone *et al* 1974). This result indicated that without angiogenesis the tumor stops growing at an early stage and at a diameter of 1–2 mm.

The process of angiogenesis is regulated by the amount of activating and inhibitory proteins. The number of inhibitors is normally significantly higher than the number of activators, leading to the inhibition of vessel growth (Otrock *et al* 2007). In the case of blood vessel injury or organ growth, the quantity of angiogenesis activators increases, whereas the number of inhibitory proteins decreases, leading to neoangiogenesis by the division of the vascular endothelial-cell lining of 'older' blood vessels. The outgrowth of these endothelial cells is the onset of new blood vessel formation (Otrock *et al* 2007). The walls of blood vessels consist of vascular endothelial cells that normally do not divide, but if they do, they perform this on average every three years upon stimulation through angiogenesis (Otrock *et al* 2007). Angiogenesis does not just occur in tumors, it may occur normally during developmental stages and growth; if it occurs within a developing embryo, it needs to build up a primitive network of capillaries, veins and arteries, and this is called vasculogenesis (Patan 2004). At later stages angiogenesis can remodel this network by creating new blood vessels and capillaries that build up the circulatory system.

A basic experiment was conducted in order to discover which molecules are involved in inducing angiogenesis and what their origin is. Are these molecules provided by the cancer cells of the neoplasm or the primary tumors themselves, or are they rather produced from the surrounding tissue microenvironment? The experiment was performed by implanting cancer cells in a chamber with a membrane that served as a permeable border for molecules, but not for cells, which could not pass through (Gimbrone *et al* 1973). The result was that angiogenesis was induced in the regions surrounding the implant, indicating that small activating molecules exerted from the cancer cells passed through the membrane and induced angiogenesis in the local microenvironment (Gimbrone *et al* 1973). Key players in the angiogenesis process are vascular endothelial growth factor (VEGF) and basic fibroblast growth factor (bFGF), which are produced and secreted from cancer cells within the primary tumor in the local microenvironment. Endothelial cells possess receptors on their cell surface for these two molecules that bind and induce a signaling cascade in order to transmit a signal in the endothelial cell's nucleus (Bergers *et al* 2003, Mierke *et al* 2011). In the nucleus several genes are now transcribed, which are necessary to facilitate new endothelial-cell growth and subsequently new vessel formation. In particular, the activation of endothelial cells through VEGF and bFGF induces several consecutive steps for building new blood vessels (Gimbrone *et al* 1973, Patan 2004). The first event is the production of matrix metalloproteinases (MMPs), which can degrade the surrounding extracellular matrix when released into the microenvironment (Patan 2004). The degradation of the extracellular matrix leads to the existence of a space where endothelial cells can migrate to and divide in order to form hollow tubes and finally mature blood vessels (Patan 2004).

Other inhibitors of angiogenesis exist, such as angiostatin, endostatin, interferons and tissue inhibitor of metalloproteinases (TIMPs) such as TIMP-1, -2 and -3 (Ribatti 2009). These substances are very promising for inhibiting tumor growth. Unfortunately, only endostatin reveals such therapeutic effects on cancer growth as the primary tumor disappears after several rounds of treatment (Eder *et al* 2002). In addition, no resistance effect to endostatin treatment was observed in mice after repeated treatments. Moreover, it has been suggested that these inhibitors are able to reduce the speed of cancer metastasis. Whether this hypothesis holds true has been analyzed by injecting different types of mouse cancer cells under the skin of mice. The cells were grown for up to two weeks and then the primary tumor was removed by surgery. The mice were monitored for weeks to assess whether they developed secondary tumors (metastases). Normally, the mice form up to fifty visible tumors that spread in the lungs even before the primary tumor resection, whereas mice treated with angiostatin displayed on average only 2–3 tumors, indicating an approximately twenty-fold reduction in the spreading rate (metastasis) of the cancer cells from the primary tumor (Kirsch *et al* 1998, Bergers *et al* 1999). Why do certain metastases remain dormant for years? One possible answer is that no angiogenesis has occurred and thus no further tumor growth has evolved, since blood vessels are missing (Gimbrone *et al* 1974). An explanation for this result may be that certain primary tumors secrete angiostatin into the blood fluid, inhibiting blood vessel growth throughout the whole body in other tissues. Then, the preliminary tiny tumors are no longer visible and cannot grow into secondary tumors, unless the primary tumor is removed and the angiostatin is no longer released into the blood fluid.

In addition to the above-mentioned role of TIMP-1 as an inhibitor of pro-tumorigenic matrix metalloproteinases, TIMP-1 has recently been reported as a pro-metastatic factor that is strongly associated with poor prognosis in many cancer types (McCarthy *et al* 1999, Kuvaja *et al* 2007, Cui *et al* 2015). There seems to be a disparity between this finding and the inhibitory function of TIMP-1, but this new function of TIMP-1 is independent of and additional to its inhibitory function. TIMP-1 can signal as a molecule regulating cancer progression. In particular, it has been found that in lung adenocarcinoma cells an increase of exogenous and endogenous TIMP-1 up-regulates microRNA-210 (miR-210) by using an CD63/PI3K/AKT/HIF-1-dependent signal transduction pathway (Wang *et al* 2014, Cui *et al* 2015). This miR-210 belongs to the short RNAs that regulate the expression levels of other genes and is strongly linked to the hypoxia pathway; to be more specific, TIMP-1 induced P110/P85 PI3K-signaling and the phosphorylation of

AKT. This then induces an increase of hypoxia-inducible factor-1α (HIF-1α) protein levels, together with an increase of HIF-1-regulated mRNA expression, and subsequently the up-regulation of the microRNA miR-210 is facilitated. If TIMP-1 is overexpressed in cancer cells, miR-210 accumulates in exosomes *in vitro* and *in vivo* (Cui *et al* 2015). In turn, these exosomes induce tube formation activity in human endothelial cells of the umbilical vein (HUVECs), as indicated by the enhanced angiogenesis activity in A549L-derived tumor xenografts (Cui *et al* 2015). In summary, TIMP-1 has a new pro-tumorigenic signaling function that may explain why elevated TIMP-1 levels are found in lung cancer patients and why these are associated with poor prognosis for the patients.

The network of proteases, their inhibitors, and effector molecules is balanced and is a determinant of tissue homeostasis. For example, imbalances of this network and the tissue homeostasis caused by elevated levels of the host tissue inhibitor TIMP-1 increase the susceptibility of target organs to support metastasis by activation of the hepatocyte growth factor (HGF) pathway (Schelter *et al* 2011). In addition, up-regulated expression of HIF-1α is associated with cancer progression and has been found to induce HGF-signaling through the up-regulation of the HGF-receptor Met via a canonical means of stress induction, for example lack of oxygen (Hellman *et al* 2002). However, it has long been supposed that there is a connection between TIMP-1, HIF-1α and HGF-signaling in the promotion of metastasis. Indeed, it has been reported that HIF-1α and HIF-1-signaling were enhanced during the liver metastasis of L-CI.5s T-lymphoma cells in syngeneic (genetically identical, or sufficiently identical and immunologically compatible to permit transplantation) DBA/2 mice overexpressing TIMP-1 (Schelter *et al* 2011). Moreover, the addition of recombinant TIMP-1 to L-CI.5s cells *in vitro* induced HIF-1α and HIF-1-signaling. In line with this, the knock-down of HIF-1α within L-CI.5s cells did not induce HIF-1α and HIF-1-signaling and thus the cells were not invasive. *In vivo* experiments showed that HIF-1α knock-down pronouncedly impaired Met receptor expression and Met receptor phosphorylation, reducing liver metastasis (Lee *et al* 2008). Moreover, the HGF-dependent TIMP-1 induced phosphorylation of Met and thus the increased invasiveness *in vitro* were facilitated by HIF-1α (Comito *et al* 2011). Finally, increased levels of TIMP-1 in the local microenvironment of cancer cells caused metastasis by inducing HIF-1α-dependent HGF-signaling. The finding that there is a connection between the protease inhibitor TIMP-1 and the stress-related factor HIF-1α is novel and impacts on the tissue homeostasis regulating cancer metastasis (Schelter *et al* 2011).

The hypothesis that interfering with the process of angiogenesis restricts tumor growth was further supported by genetic studies of mice lacking the two genes, Id1 and Id3. The absence of these two genes inhibits angiogenesis. Angiogenesis-deficient mutant mice were injected with mouse breast cancer cells and observed for tumor growth. There was indeed only a short period of tumor growth and even then the whole tumor vanished completely after several weeks, resulting in the mice becoming healthy again (Li *et al* 2004). However, if cancer cells of the lung are injected into these angiogenesis-deficient mutant mice, the results are slightly different. In particular, the lung cancer cells develop slow growing tumors in these mice. Moreover, these tumors do not to spread to other organs and thus do not form metastases, resulting in a prolonged lifetime for these tumor-bearing mice compared to normal mice carrying tumors (Li *et al* 2004).

As these experiments were very promising, the following question was raised. Can the inhibition of angiogenesis slow down or prevent the growth and spread of a tumor even in humans? To answer this question many angiogenesis inhibitors belonging to different categories have been tested to cure cancer patients. Among these inhibitors are those inhibiting endothelial cells in a direct manner, whereas other inhibitors block the angiogenesis signaling cascade or abolish the endothelial cell's ability to degrade the surrounding extracellular matrix (Lee *et al* 2011, El-Kenawi and El-Remessy 2013). In more detail, one class of inhibitors, for example endostatin for angiogenesis, contains molecules that directly inhibit the endothelial cell's growth. Another inhibitory drug is combretastatin A4, which induces apoptosis (programmed cell death) specifically in endothelial cells, whereas other drugs can interact with cell-surface receptors such as integrins and subsequently destroy selectively proliferating endothelial cells (figure 1.7) (Ding *et al* 2011, Wu *et al* 2015). A second group of angiogenesis inhibitors is composed of molecules that interfere with steps in the angiogenesis signaling cascade of humans, for example anti-VEGF antibodies inhibiting the binding of growth factors to VEGF receptors (figure 1.7). One anti-VEGF monoclonal antibody, bevacizumab (Avastin), has been demonstrated to impair tumor growth and thus extend the survival of cancer patients. A third agent is interferon-alpha (IFNα), which can counteract the production of basic fibroblast growth factor (bFGF) and vascular endothelial growth factor (VEGF) and thus impair the initiation of the growth-factor-driven signal transduction cascade (figure 1.7) (Frey *et al* 2011).

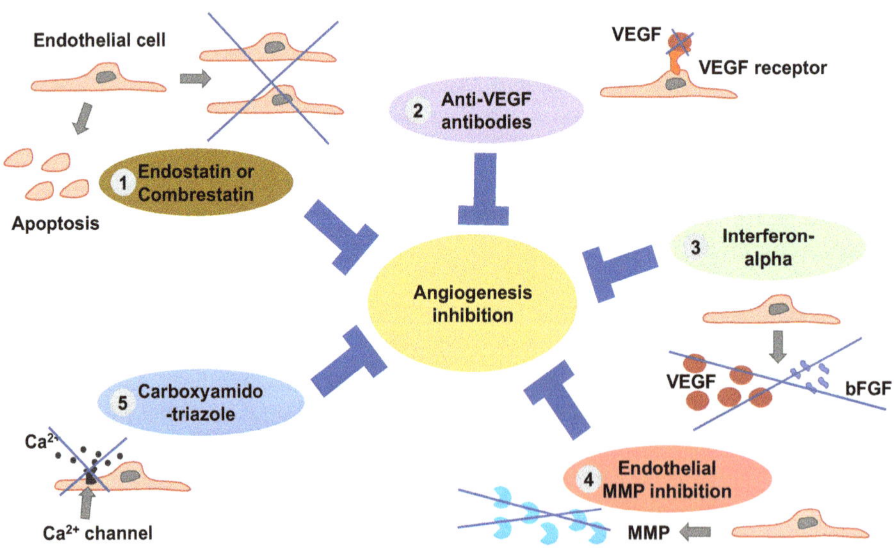

Figure 1.7. Five groups of angiogenesis inhibitors.

The fourth group of angiogenesis inhibitors contains substances that are directed against the endothelial produced MMPs (figure 1.7) (enzymes that initiate the breakdown of the local microenvironment). Since the breakdown of the surrounding extracellular matrix is required for the migration of endothelial cells into surrounding tissues and the endothelial-cell proliferation for the outgrowth of new blood vessels, inhibitory drugs targeting endothelial MMPs have impaired angiogenesis and hence tumor growth and malignant tumor progression.

A fifth group of drugs is being investigated intensively for inhibition of angiogenesis and subsequent tumor growth; these drugs are either nonspecific or not clearly understood, for example carboxyamido-triazole (CAI), which is an inhibitor of nonvoltage-gated calcium channels, works by inhibiting the calcium ion influx into all kinds of cells, including endothelial cells (figure 1.7). As this restriction of calcium uptake specifically suppresses the growth of endothelial cells, it is expected that such a general mechanism can also affect other cell types and many other cellular processes. What is still not under discussion? The mechanical impact of endothelial cells on the mechanical properties of cancer cells and their function. This is described in more detail under transmigration, a process in which cancer cells can migrate through an endothelial-cell monolayer in order to migrate into or out of blood or lymph vessels.

What role do the biomechanical properties of the primary tumor play in tumor angiogenesis?

In addition to the aforementioned growth factors and cytokines regulating neo-angiogensis, the mechanical properties of a primary tumor can also facilitate endothelial vessel growth within the tumor to increase the size of the primary tumor and cause the malignant progression of cancer. The tumor stiffness and the high interstitial pressure within the primary tumor blocks the diffusion of metabolites (Jain 1987). In turn, the tumor can regulate cellular angiogenesis in order to induce neoangiogenesis within the primary tumor. In particular, the local vascular system in the tumor microenvironment is induced by the outgrowth of new capillaries from pre-existing vessels to grow in the primary tumor in order to supply the proliferating cancer cells with metabolites (Ruddell *et al* 2014). Even within the tumor, the maturation and remodeling of new microvessels needs the perfect coordination of many diverse processes in the microvasculature (figure 1.8) (Klagsbrun and Moses 1999). For the induction of new blood vessel sprouts, the pericytes located around endothelial vessel linings must be moved from the branching vessel. Then, the endothelial-cell basement membrane and extracellular matrix surrounding the blood vessels must be degraded and restructured by MMPs (figure 1.8) (Kraeling *et al* 1999). In the next step, the new extracellular matrix is synthesized and secreted by neighboring stromal cells (Lu *et al* 2012). This newly designed extracellular matrix, together with several soluble growth factors, evokes the migration and proliferation of neighboring endothelial cells (figure 1.8). In the last step, endothelial cells build up a monolayer that results in a tube-like structure. Next, mural cells such as pericytes wrapping microvessels and smooth muscle cells wrapping large vessels are recruited to the nonluminal surface of the novel endothelial-cell lining (figure 1.8). The remaining uncovered vessels regress, showing that the process of angiogenesis is

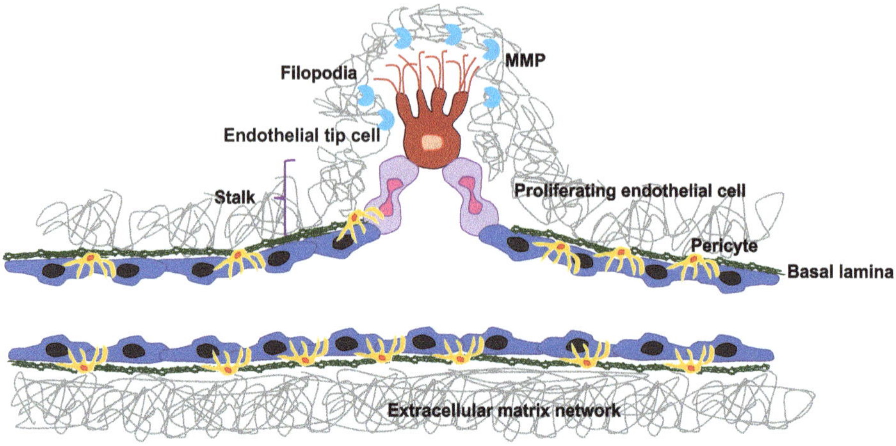

Figure 1.8. New sprouting of blood vessels. An endothelial cell becomes activated and forms a tip cell of a new vessel branch. The extracellular matrix needs to be degraded as well as the basement membrane. Additionally, the pericytes need to be removed. The neighboring endothelial cells proliferate and form a stalk which grows as a new vessel sprout into the tissue.

highly ordered and strongly regulated, as quiescent mature endothelial cells within the endothelial-cell lining of vessels need to divide and branch out of an existing vessel by omitting excessive endothelial growth. Thus, cell–cell and cell–matrix adhesions of endothelial cells are crucial for normal survival within vessels and for tumor neoangiogenesis.

In tumors, the 'tumor' endothelium is dysregulated regarding hypoxia and chronic growth-factor stimulation (such as vascular endothelial growth factor (VEGF)): tumor blood vessels possess irregular vessel diameters, are fragile, leaky and the blood flow is abnormal, suggesting that the tumor endothelium regulates tumor growth and metastasis. In particular, chaotic networks of the endothelium lacking the normal hierarchical arrangement of artery–arteriole–capillary have been found (Warren *et al* 1978, Konerding *et al* 1999). This leads to a poor stability of tumor endothelial vessels, together with lower numbers of pericytes, which are even less tightly attached to tumor endothelial cells than they are to normal endothelial cells (Baluk *et al* 2005). Vessel stability has a major impact on the blood flow and its directionality, which can be measured even at single-capillary resolution in primary tumors (Kamoun *et al* 2010). The blood vessel density increases during early tumor formation, but decreases in larger tumors. The tumor vessel system itself can serve as a marker for malignant tumor progression as the poor quality of tumor vessels with irregular pericytes on the vessel wall and multiple basement membranes are associated with cancer metastasis (McDonald and Choyke 2003, Nagy *et al* 2009).

Tumor endothelial cells are identified and isolated using cell-surface markers such as PECAM-1 (CD31) (Hida *et al* 2004), Intracellular Cell Adhesion Molecule 2 (ICAM-2) (Dudley *et al* 2008) and CD146 (St Croix *et al* 2000), which are all still markers of endothelial cells from all vascular beds (capillary, venous, arterial and lymphatic). Among the isolated tumor endothelial cells are the normal endothelial cells of the host organ. Defective, discontinuous endothelial monolayers have been detected in tumors where the gaps within the endothelial layer are filled with cancer cells mimicking endothelial cells as they express VE-cadherin (Maniotis *et al* 1999). Tumor endothelial cells have a high turnover rate and are more motile than normal endothelial cells located in healthy tissues. All this may contribute to the numerous gap formations in tumor endothelia. Moreover, even the intercellular cell–cell junctions are loosely formed, as well as the basement membranes, which results in large holes in the monolayers. These holes are assumed to be the entry sites for cancer cells to migrate through the human body and subsequently to metastasize at targeted sites. How these abnormalities of the tumor endothelial blood vessels are caused is not yet understood. Tumors are in fact dysfunctional organs: metabolic pathways are corrupted, cancer cells withdraw their microenvironment of nutrients by secreting toxic waste products such as lactate into their local tumor microenvironment, and tumors render local microenvironments areas of nonperfusion, leading to hypoxia. In this 'abnormal' microenvironment it is likely that abnormal endothelial vessels grow within the tumor (Merlo *et al* 2006). In line with this, most primary tumors produce high concentrations of vascular endothelial growth factor-A (VEGF-A), which serves as a potent vasodilator that is able to cause fluid leakage and high interstitial pressures, abnormal branching morphogenesis of the endothelium as well as small gaps in the endothelial monolayer (Nagy *et al* 2009). VEGF stimulates the endothelium, leading to a breakdown of the entire endothelial barrier function. Within a tumor, the vessels are squeezed and compressed by surrounding cancer cells, which causes external mechanical tension, strain and may finally alter the blood flow, leading to abnormal endothelial walls (Padera *et al* 2004, De Val and Black 2009). What is the function of these abnormal endothelial cells within the tumor? Do abnormal endothelial cells promote tumor growth and cancer progression to support metastasis? The interaction between tumor endothelial cells and other cell types such as leukocytes may indeed be altered by the endothelial-cell abnormalities within the tumor vasculature. In more detail, special adhesion molecules may be decreased on the cell surface of the tumor endothelial cells, helping primary tumors to escape immune surveillance due to impaired crosstalk between T-lymphocytes and the endothelial vessel linings (Griffioen *et al* 1996, Dirkx *et al* 2006). This hypothesis can be further supported by a study showing that the penetration and efficacy of primed T-lymphocytes used for tumor immunotherapy was increased by the addition of proinflammatory cytokines, up-regulating ICAM-1 and VCAM-1 on the tumor endothelial cells (Garbi *et al* 2004). More support is provided by the detection of leukocytes at the periphery of tumor vessels that secret factors such as endothelial-cell growth factors and cytokines required for tumor angiogenesis (Dudley *et al* 2010). In addition, it has been demonstrated that proinflammatory cells such as macrophages facilitate metastasis (Qian and Pollard 2010).

1.2 Malignant progression of cancer (metastasis)

The worst outcome for a cancer patient is that the cancer has formed metastases. The process of metastasis involves many consecutive steps as well as some optional side steps or side pathways. Basically, each neoplasm has in principle a possibility of metastasizing, but not all of them do. The potential of cancer metastasis is related to the special microenvironment, the aggressiveness of a special subtype of cancer cell within the primary tumor and the start of medication, for example resection of the primary tumor before metastasis has occurred and subsequently some chemo- or radiotherapy. Nonetheless, there are still cases in which the metastasis has started and it finally leads to the death of the cancer patient, although the patient has undergone chemo- or radiotherapy. Indeed, metastasis is the main cause of cancer deaths and no major progress has been made to reduce the number of cancer-related deaths.

Thus, many cancer research projects are needed on this particular subject that start from an alternative prospective and from another viewpoint than that of the classical one of tumor biologists. Here, the novel viewpoint of a biophysicist investigating cancer research should be helpful to overcome classical barriers and to reach novel ground. Thus, a new field called simply ‘physics of cancer’ has become increasingly important in cancer research and many researchers are interested in this novel viewpoint. In the following, the metastasis cascade is described briefly from a biophysical viewpoint.

What kinds of mechanisms are present in the malignant progression of cancer?
Based on the biological and molecular mechanisms that cause primary tumors, a clear picture can be drawn, but for the mechanisms causing the subsequent invasion and metastasis no clear picture arises and many aspects remain elusive. Research to generate conceptual outlines for cancer malignancies and their causes has begun. However, the new physics of cancer field has arrived to try to fill the gap and bring another dimension to cancer research in order to understand why certain cancers become malignant and metastasize and others do not. There is a classical understanding of how metastatic dissemination of cancer cells occurs, called the invasion-metastasis cascade (Fidler 2003). This cascade involves the local invasion of primary cancer cells into the vascular system, such as the blood or lymph vessels (intravasation), their transport through the whole body via the vessel flow, their accumulation in microvessels of distant tissues, their possible transmigration through the vessel lining (extravasation), their invasion of the parenchyma of targeted tissue and their building of micrometastatic cell clusters that may eventually grow to macroscopic metastases. This process is called colonization and is the starting point for the worst outcome of cancer (figure 1.9).

What determines the ‘early determination’ of a primary tumor?
A major point is the timing of the individual cancer cell’s acquisition of the capacity to invade and finally metastasize. Can every cancer cell within a primary tumor reach the ability to become malignant? Or does only a small subpopulation of cancer cells in this tumor gain the ability to invade the surrounding tissue and metastasize? What about the majority of cancer cells within a primary tumor, do they have an identical ability to invade and metastasize, or different abilities? If the ability to invade and metastasize is identical, it is suggested that the acquisition of this ability occurs early in tumor development. In contrast, if only a small subset of cancer cells

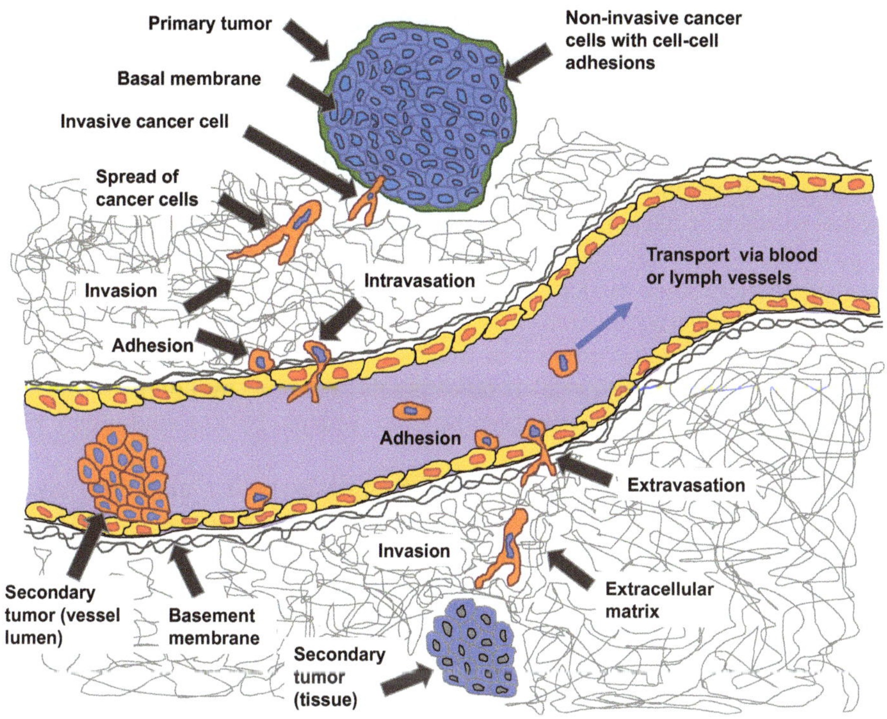

Figure 1.9. Cancer metastasis cascade.

acquires this ability to invade and metastasize, they must have obtained this in a later phase of the primary tumor formation and thus cancer progression.

How can this issue be addressed in an experiment? One possible approach is to analyze whether there is a subpopulation that can be found by clonal expansion favoring successive mutation and after several selection cycles we will obtain subclones. An alternative way of obtaining subclones is selection for a special marker, such as a cell–matrix or cell–cell surface receptor through cell sorting. Another alternative is sorting for special mechanical properties, such as the deformability of cancer cells that may support tumor pathogenesis through the selection of an aggressive cancer cell phenotype. A big problem of the model is: why should an invasive or metastatic phenotype be an advantage for a cell confined within the primary tumor?

Indeed, it seems to have no advantage for the primary tumor at first glance. This suggests that the occurrence of metastasis is not a phenotype of the primary tumor selected by the formation of the primary tumor (Bernards and Weinberg 2002). Instead, there seems to be another phenotype that may arise as an inadvertent consequence of the acquisition of alleles providing an advantage for the growth of the primary tumor mass. This finding means that the alleles behave pleiotropically by encoding a selected phenotype of the primary tumor (growth and survival of the initial tumor) and an unselected phenotype (malignant tumor progression, including invasiveness and metastatic behavior). However, there may be another possibility, that the selection of highly invasive and metastatic cancer cells arises from the microenvironment that is the tumor itself or the surrounding extracellular matrix. In particular, the selection process may be driven by cellular mechanical properties that are altered within the cells of the primary tumor (Huang and Ingber 2005) and also of the surrounding microenvironment (Lokody 2014). This view is supported by the complexity of the cellular invasiveness and the metastatic cascade that completes the early steps of the primary tumor formation. In line with this, the question may be raised as to whether the metastatic traits require several accumulated mutations that complete the variety of genetic lesions causing the primary tumor formation. However, there is some evidence that the metastatic dissemination of cancer cells does not rely on the acquisition of additional genetic lesions beyond those of the primary tumor. There exist mutations that are drivers of biological phenotypes and mutations that are passengers, solely reflecting the increased mutability of the cancer cell's genome and thus representing random genetic background noise of the primary tumor (Wood *et al* 2007).

In the following, the proposition that metastasis-specific mutations acquired during malignant cancer progression have a role is argued against. The first argument is that the prognosis of primary tumors, including their ability to metastasize, can be determined by analyzing the gene expression profile of the tumors. In particular, the gene expression profiles are altered by acquired somatic mutations and promotor methylation events that a normal cell has not acquired during its differentiation. In order to analyze the transcription patterns of primary breast carcinomas using gene expression arrays, it was possible to predict which primary tumors would undergo a malignant progression and which would stay nonmalignant (van de Vijver *et al* 2002, Fan *et al* 2006). However, this prediction is not always possible and includes some errors. If this method for predicting the malignant tumor progression works, it implies that the mechanisms regulating cancer metastasis require a majority of neoplastic cells in a primary tumor to display an altered expression profile. This is in contrast to the finding that only a small number of neoplastic cells in the primary tumor have the ability to metastasize, suggesting that additional parameters may drive malignant cancer progression, and finally metastasis, such as mechanical alterations to these metastatic cells. Moreover, the metastatic spread of cancers seems to be determined in the early phase of cancer progression in order to be expressed in the majority of cells in the primary tumor (Bernards and Weinberg 2002).

A second argument for the early determination of the cells' ability to show a metastatic spread in the primary tumor is that only the early passage of human mammary epithelial cells cultured in a defined culture medium and transfected with a defined set of genes, such as the SV40 virus early region (specifying the large T and small t oncoproteins), the hTERT gene (encoding the catalytic subunit of the telomerase holoenzyme) and a Ras oncogene developed metastasis, whereas the same cells cultured in a different medium did not. In particular, these two cell culture populations exhibit different gene expression patterns and upon transformation with one of the three genes (see above) developed histopathologically distinct tumors: the first is a squamous cell carcinoma and the second is an invasive ductal adenocarcinoma of the breast (Ince *et al* 2007).

These results support the idea that the differentiation program of the normal cell of origin is a strong determinant for the behavior of cancer cells upon transformation. The expression patterns of the two transformed human mammary epithelial cell types are closely related to those of their respective normal precursors and display relatively minimal similarity to one another. This may be a reason why one type of tumor, the invasive ductal carcinoma, metastasized to the lungs, whereas another type, the squamous cell carcinoma, did not metastasize (Ince *et al* 2007). Both types of tumor had undergone the same set of experimentally introduced genes representing somatic mutations, but they still had different metastasizing abilities. This finding shows that the differentiation state of the normal cell of origin plays a major role in determining the metastatic spread of a tumor. In summary, the properties of the normal cell of origin that is the progenitor of all the neoplastic cells of a primary tumor determine whether its offspring will have the ability to metastasize.

Another example is normal human melanocytes that have been transformed with the same oncogenes. The transformed melanocytes (the model of spontaneous melanomas) built up primary tumors with a large number of metastases in different targeted organs (Gupta *et al* 2005). Thus, the differentiation program of the normal cell of origin has a strong impact on the probability of metastasis occurring.

A third argument for the lack of metastasis-specific genes is that the same set of genetic lesions (mutations) in the genomes of primary cancer cells is generally found in the genomes of their derived metastases (Jones *et al* 2008).

Thus, the driving force for metastasis is not based on specific genetic lesions that are evolved during the multistep formation of a primary tumor. Instead, the dissemination of cancer cells from a primary tumor occurs as a side-effect of the primary tumor formation and does not seem to be a property achieved during the multistep process of primary tumor formation.

How is the invasiveness determined and how does metastasis arise?

It has been established that a complex invasion-metastasis cascade is necessary. What drives metastasis, if there are no additional mutations required beyond those necessary for the formation of the primary tumor? A classical tumor-biological answer would be that cancer cells utilize complex biological programs used by normal, healthy cells and organismic physiology. In many steps of the morphogenesis of normal cells the epithelial–mesenchymal transition (EMT) plays an important role (Thiery 2002) (figure 1.10). In more detail, several distinct morphogenetic steps involve the local migration of epithelial cells as well as their migration to distant sites during embryonic development. However, normal epithelial cells are incapable of these translocations, evoked by active movement, as they only move laterally in the epithelial plane while maintaining adhesion to the underlying basement membrane or basal lamina. Under certain conditions, there is a special case when these epithelial cells acquire active movement and invade the extracellular matrix: the shedding by epithelial cells and the switch to mesenchymal properties. In summary, EMTs play an important role during embryogenesis during gastrulation and emigration of cells to distinct targeted sites within the embryo (Thiery 2003). EMTs are induced by a number of transcription factors (TFs) that are transiently

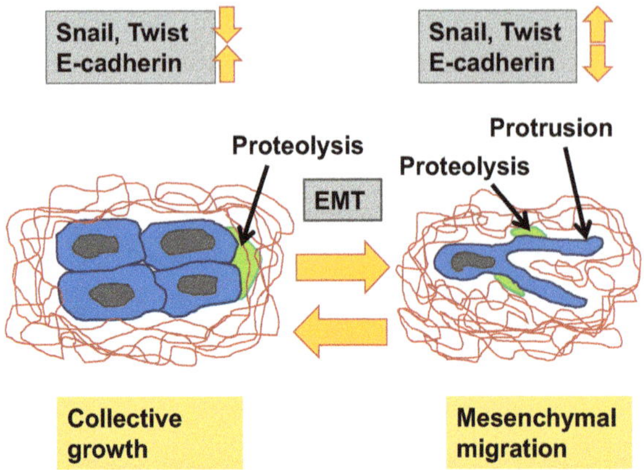

Figure 1.10. Epithelial to mesenchymal transition (EMT). Individual and undifferentiated malignant cancer cells and/or small clusters of malignant cancer cells that are frequently located in the tumor stroma and near the invasive front of the primary tumor.

active during special stages of embryogenesis and at specific sites within the embryo (Batlle *et al* 2000, Yang *et al* 2004, Gumbiner 2005, Hartwell *et al* 2006, Thiery and Sleeman 2006, Peinado *et al* 2007). The down-regulation of the TF expressions leads to the reversion of EMT, the so-called MET, so that the cells are now back in the epithelial state (the ground state), whereas the mesenchymal state seems to be the activated state. As EMTs occur at specific sites of the embryo, they need to be induced via the local microenvironment, for example neighboring cells and extracellular matrix-protein signals, suggesting that EMT-inducing TFs may be regulated by the neoplastic cell's local extracellular-induced signaling. However, these descriptions do not show the relevance of EMT during cancer progression, but two aspects of cancer cells support the relevance of EMT induced by TFs. First, many phenotypes of embryogenic cells are imitated by aggressive and invasive cancer cells. Second, numerous embryonic TFs, such as Slug, Snail, Twist, Goosecoid, SIP-1, FOXC2 and ZEB1, which regulate EMTs during embryogenesis, are also detected in human cancer cells and their expression correlates with the aggressiveness of these cancer cells (figure 1.10).

Moreover, most of the TFs were identified during the investigation of the development of model organisms, such as *Xenopus* and *Drosophila*. As these results are transferable to humans, these TFs must be strongly conserved in the genomes of distantly related animals, indicating their critical roles in the embryogenesis of diverse organisms. All these observations lead to the suggestion that TFs enable carcinoma cells to obtain highly malignant properties, such as cell invasiveness, resistance to apoptosis and secretion of proteases that degrade extracellular matrix confinement. Furthermore, the EMT and hence the expression of these TFs is not restricted to initial embryonic development, it also plays a prominent role in wound healing processes in adults and re-epithelialization processes (Savagner *et al* 2005).

What role does the induction of EMT-inducing TF expressions play?
As described above, various EMT-inducing TFs act during embryogenesis upon certain signals from nearby cells. Thus, it seems to be the case that the same type of signal (for example, Wnts and Hedgehogs, members of the transforming growth factor beta family, as well as the ligands of tyrosine kinase receptors) affects various carcinoma cells during malignant cancer progression. In particular, none of these ligands has the capacity to trigger EMT alone, but the combination is able to induce EMT of cancer cells. The exact rules are still under intensive investigation. It has been suggested that during cancer progression these signals for EMT are provided by the tumor-associated stroma containing mesenchymal cells. These stromal cells originate from the stroma of the tissue in which the tumor grows or are directed from the bone marrow, which generates many distinct types of mesenchymal progenitor cells and releases them into the circulation so that they are available for local recruitment by carcinoma cells (Direkze and Alison 2006). Indeed, these cells are then located in the tumor-associated stroma and subsequently differentiate into a variety of mesenchymal cell types, such as myofibroblasts and endothelial cells. Additionally, there might be EMT-inducing signals that are not released by the tumor stroma of early stage tumors. During tumor progression the tumor stroma grows and is activated (reactive) in a similar way to tissues exposed to active wound healing processes or chronic inflammatory processes. What role do the mechanical properties of the tumor stroma play? In most biological and medical studies this question is simply omitted and in the physics of cancer field this topic is still under investigation.

In addition to EMT induction, the tumor stroma may have another impact on the behavior of cancer cells outside of the primary tumor, further enhancing cancer cell motility and guiding cancer cells towards their targeted tissue (called secondary sites). On their way to the targeted tissue cancer cells migrate through other vastly different nontumor-associated tissues. At these secondary sites, the stroma is not altered compared to the primary tumor associated stroma through permanent stimulation by cancer cells and may hence not have these activated states that induce further EMT. Thus, EMT will be reversed and the cancer cells reassemble to a neoplasm that has similar properties to the primary tumor. However, it has been suggested that the absence of mesenchymal phenotypes in cancer metastasis disproves the idea that cancer cells must undergo an EMT to disseminate from primary tumors (Rhim *et al* 2012). This is in contrast to the argument that the reversibility of the EMT may have finished, so that the cancer cells returned to their ground state (of zero motility) in these secondary tumors. As heterotypic signals can induce EMT in carcinoma cells, these cells do not need to undergo additional mutations in order to become highly aggressive and invasive. Hence, primary carcinoma cells are able to perform EMT if the extracellular signals for EMT are recognized. What variable factors define whether cancer cells within a primary tumor will undergo an EMT? Is it really the appropriate mix of heterotypic signals that must be detected by cancer cells and that is essential for their EMT? It has been suggested that the differentiation program of the normal cell of origin represents one critical determinant of this responsiveness to the different signals regulating EMT as this program is supposed to set the stage for the malignant progression of cancers. Accumulated somatic mutations and promotor methylations during primary tumor progression also favor the responsiveness to EMT-inducing signals. However, the exact mechanism of the EMT induction and reversion is still not well understood, although many factors have been identified as regulating EMT. What role do mechanical properties play in inducing EMT? There is currently no answer, as this is still under intensive investigation. As these malignant features are manifested due to signals from activated stroma, they are not thought to be the objects of selection leading to primary tumor formation. In particular, the somatically generated alleles provide the responsiveness to these signals, whereas the alleles selected during primary tumor formation directly specify malignant features. Thus, the expression of highly malignant features occurs as an accidental consequence of the initial actions of alleles that are 'unrelated' to the phenotypes of cell invasion and cancer metastasis.

When carcinoma cells undergo an EMT, they adopt mesenchymal phenotypes, invade the 'activated' surrounding tumor stroma and then move into adjacent normal tissues outside the primary tumor borders during metastasis; they are nearly indistinguishable from normal mesenchymal cells. This raises the question of whether the EMT is solely an artifact. There are at least two arguments against this. The first argument is that most carcinoma cells undergo EMT incompletely, for example E-cadherin and cytokeratins are down-regulated and mesenchymal markers such as N-cadherin, vimentin and fibronectin are up-regulated (Schramm 2014). The coexistence of epithelial and mesenchymal markers may provide evidence for this hypothesis. The second argument is that carcinoma cells, which have undergone EMT, express certain markers not expressed by 'true' mesenchymal cells and may be detectable within the normal tissue as aggregates (Van Aarsen *et al* 2008).

The impact of the EMT and invasion during the progression of the metastasis cascade

How do these pleiotropically acting EMT-inducing TFs enable cancer cells to succeed in invasion and the metastasis cascade? These TFs also provide increased resistance to apoptotic cell death, cell movement, secretion of matrix degrading enzymes and tissue invasiveness (Jiang *et al* 2014). Can a single TF enable the metastatic cascade? Can this disseminated cancer cell survive in the new microenvironment, where it does not fit in and to which it is not adapted? The success rate for building up a secondary tumor is relatively low. One reason may be the poor adaptation of the cancer cell to its target site. Thus, the growth from micrometastasis to a macroscopic metastasis and hence colonization is a rare event as only one out of thousands succeeds. It has emerged that the colonization is not a problem, it is rather the increased resistance to apoptosis that is critical for the metastasis cascade. Are the steps of the metastatic cascade governed by one of the TFs? If this holds true, all the multiple steps of the metastatic cascade would appear to be quite simple when regulated by one TF or a small group of them. However, the regulation of these TFs would then be highly critical in order to keep them under controlled tissue homeostasis.

Is there a special type of permanent EMT?

Cancer cells have been found locked in the mesenchymal state, not displaying any plasticity (Gregory *et al* 2011). This result indicates that the reversion to an epithelial phenotype is no longer possible. How may this irreversible EMT be caused? Possible explanations for this are: genetic or other biochemical alterations, or altered mechanical properties of cells due to their microenvironment or stimulation. In recent decades it has turned out that cell-surface protein E-cadherin has a prominent role in mediating cell–cell adherence junctions between neighboring epithelial cells (Tania *et al* 2014). This role has determined E-cadherin as a typical, canonical epithelial marker. In line with this, the promoter of the E-cadherin encoding gene exhibits binding sites for several EMT-inducing TFs that are able to repress E-cadherin transcription (Cano *et al* 2000, Comijn *et al* 2001, Bolos *et al* 2003). E-cadherin repression seems to be one of the main functions of these TFs. However, via an unknown mechanism cytokeratin expression is down-regulated, whereas various mesenchymal genes are up-regulated or induced (Tania *et al* 2014). Taken together, E-cadherin expression is a main target of the regulation by EMT-inducing TFs.

By contrast, certain tumors show alterations in E-cadherin expression, containing point mutations or deletions in the E-cadherin gene that lead to a production of truncated or unstable proteins (Kanai *et al* 1994). Another associated change of functional E-cadherin is that the cytoplasmic proteins that usually serve to physically link E-cadherin to the actomyosin cytoskeleton are translocated (Kam and Quaranta 2009) and hence rapidly degraded (Gerlach *et al* 2014). An exception is β-catenin, this molecule can survive by escaping phosphorylation by the non-phosphorylated glycogen synthase kinase-3β (GSK-3β) and hence proteosomic degradation (Gerlach *et al* 2014). The GSK-3β is hyperphosphorylated by Akt kinases and thus is inactive. The free β-catenin can localize to the nucleus when it is associated with a T-cell factor group of TFs and together with other signals triggers a large number of downstream target genes that are mostly involved in the regulation of the EMT switch (Onder *et al* 2008). In particular, gene expression analysis has been shown that the loss of E-cadherin leads to the induction of multiple transcription factors, such as Twist. Twist in turn is necessary for the loss of E-cadherin and finally induces metastasis, indicating that the loss of E-cadherin in primary tumors facilitates metastatic dissemination through transcriptional and functional alterations (Onder *et al* 2008). However, in principle there is a permanent EMT as cancer cell lines exist that are able to stay in their mesenchymal phenotype.

Is there an imperfect or partial EMT?

There exists a lot of evidence that the classical EMT is imperfect, as the cells are only partially transformed in several cancer types. In order to answer this question, the concept of tumor budding needs to be addressed, as it may represent a distinct type of imperfect or partial EMT (figure 1.10) (Grigore *et al* 2016).

The concept of tumor budding

A phenomenon called tumor budding has been observed in various cancers. The histological phenomenon of tumor budding is characterized by individual and undifferentiated malignant cancer cells and/or small clusters of malignant cancer cells that are found frequently in the tumor stroma and mostly near the invasive front of the tumor (figure 1.11) (Van Wyk *et al* 2015, De Smedt *et al* 2016, Koelzer *et al* 2016). As the tumor budding phenomenon has been correlated with poor cancer outcomes, it has been proposed to mirror the EMT (Almangush *et al* 2016). The budding phenomenon of tumors has been initially observed in stomach cancer (Imai 1960) and refined in various other cancer types such as tongue, larynx, breast, stomach, colon, rectum, cervix and colorectal cancer (CRC) (Morodomi *et al* 1989, Hase *et al* 1993). Before the tumor budding can be used routinely as a marker for cancer progression, it requires clear definition, standardized assessment and applicability and clinical significance to distinct cancer types (Grigore *et al* 2016). Similar to the tumor budding phenomenon, a phenomenon has been found in experimentally induced colon cancers in mice that have been treated with dimethylhydrazine-dihydrochloride (Gabbert *et al* 1985). The invasive front of the tumor is characterized by a dramatic disorganization of its architecture that is characterized by a loss of glandular aspect in differentiated carcinomas and a loss of trabecular aspect for undifferentiated carcinomas and associated with cellular dedifferentiation. Finally, these alterations select isolated cancer cells with a uniform phenotype that is independent of main tumor mass differentiation (figure 1.11). Is the tumor bud level correlated with imperfect or partial EMT, as indicated by the connection with tumor proliferation, quiescence and stemness? As coexpression of epithelial and mesenchymal marker genes in many tumor buds can be readily observed, can tumor budding represent the hybrid epithelial/mesenchymal phenotype that performs collective cell migration?

Indeed, the tumor bud phenotype possess features that are associated with cellular motility, such as the decrease of cell junctions and impaired basement membrane, the presence of cytoplasmic microfilaments and pseudopodia, and altered cellular shapes. All of which lead to the hypothesis that the entire tumor budding process at the invasive front facilitates the migration of cancer cells out of the primary tumor mass and their invasion of surrounding tissues (Gabbert *et al* 1985). In line with this, in surgical resections of colon cancer, dedifferentiated malignant cancer cells are detected that migrate far beyond the invasive front as either individual single cells or

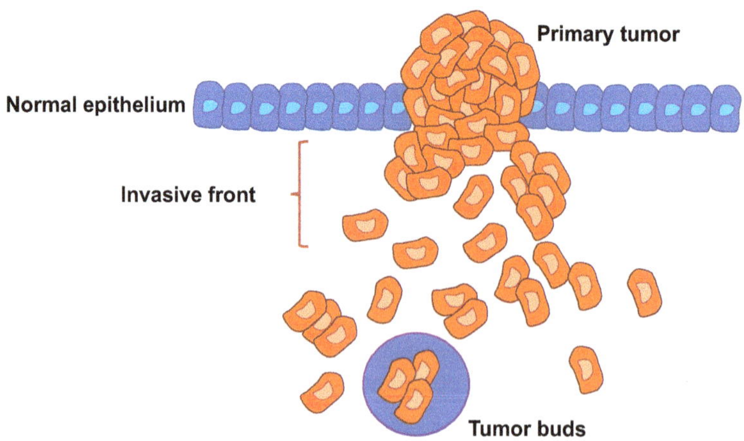

Figure 1.11. Tumor budding.

collective cell clusters. In particular, single cells exert at their migration front cytoplasmatic perturberances similar to lamellipodial protrusions, whereas a cluster of cells possess one large cytoplasmatic flap at their leading edge. These features characterize cellular motility (Carr *et al* 1986).

Tumor budding implies that the cluster of neoplastic cells is still connected to the primary tumor, which is not true for 2D histological tissue sections (Amato *et al* 1994). However, in 3D reconstructions of serial tissue sections from various cancer types, it has been revealed that tumor budding is not a static process and hence it seems to be rather dynamic as the tumor exerts multiple fingerlike projections (figure 1.11) (Carr *et al* 1986, Bronsert *et al* 2014). These projections consist of numerous cells, which can at a later time point be disconnected from the primary tumor mass or rupture into small cell clusters. Hence, while 2D sections imply the disconnection between a tumor bud and the primary tumor, 3D reconstructions draw a clear picture of a cocktail of growing tumor buds still connected to the tumor mass and others are clearly detached from the primary tumor.

How is tumor budding defined, identified and quantified within tumor tissue?
The general principles of tumor buds are discussed in a CRC model, as it is the first cancer type in which tumor budding has been systematically investigated. However, the methodology for assessing tumor budding is not standardized and hence differs widely in several aspects. First, the definition of the budding structure is not unique, as different cut-off values exist. Second, the identification of budding structures depends on the staining method such as hematoxylin and eosin (H&E) staining or pan-cytokeratin antibody immunostaining. Third, the final identification of budding structures is performed by exclusion of similar, but not bud-like structures. Fourth, regions are defined, such as assessment across the entire tumor or in the regions containing maximal budding structures. Fifth, instrument parameters such as magnification and field-of-view size vary. Sixth, the quantification of tumor budding can either use descriptive and subjective patterns or semi-quantitative determination of relative size, such as the ratio between the width of the tumor budding region and the width of the invasive front and quantitatively counting of the buds. Seventh, the settings for the critical prognostic values vary. Eighth, the application of the methods to primary lesions of tumors also vary.

However, sometimes in routine H&E tissue section staining it is difficult to distinguish tumor budding from similar bud-like structures. These difficulties are present under distinct conditions such as: abundant inflammatory regions at the invasive front containing activated lymphocytes, macrophages or dendritic cells; ample stromal reaction at the invasive front that can build artificial stromal buds; fragmented tumor glands caused by abundant inflammatory infiltrate that appear as bud-like structures, and tumor budding seems to be inversely correlated with the degree of inflammation (Max *et al* 2016); bud-like artifacts based on retraction around fragmented tumor glands; and tumor tissue fragments surrounded by an ample mucinous extracellular matrix—these are also similar to buds. The first two points can be easily addressed by performing a cytokeratin immunostaining. The latter three points should be excluded from the scoring, as it is not yet precisely defined (Mitrovic *et al* 2012). The other points should be solved by precisely defined parameters and improved data aquisition.

Imperfect or partial EMT

The EMT transition represents a multistep cellular phenotype switch that can be dynamically reversed. In particular, epithelial cells reduce their intercellular adherence junction dependent cell–cell adhesion and become migratory and invasive, similar to mesenchymal dependent cells (figure 1.10) (Kalluri and Weinberg 2009). The EMT and the reverse process, the mesenchymal-epithelial transition (MET), are both transitions occurring under physiological conditions such as embryonic development, wound healing and tissue repair. Under pathological conditions, the EMT activation is a hallmark of malignant cancer progression such as metastasis (Kalluri and Weinberg 2009, Micalizzi *et al* 2010, Nieto 2013). The classical EMT is characterized by a decrease of cell-surface E-cadherin and an increase of mesenchymal markers such as vimentin, N-cadherin, α-smooth muscle actin or fibronectin, which is associated with the reduction of the internal apical-basal polarization of epithelial cells. Since E-cadherin sequesters β-catenin at cell–cell adherence junctions, a decrease in E-cadherins and subsequently a decrease in membranous levels of β-catenin evokes the nuclear translocation of β-catenin and subsequently the activation of the Wnt/β-catenin signal transduction pathway during EMT progression (Schmalhofer *et al* 2009). Increased amounts of one or more of the EMT-promoting TFs, such as ZEB1, ZEB2, SNAI1 (Snail), SNAI2 (Slug), TWIST, GSC and FOXC2, usually accompany the progression of the EMT (De Craene and Berx 2013).

The EMT and MET transitions are not usually a full transition as a switch between purely epithelial and purely mesenchymal phenotypes (Nieto 2013), it is rather a multiple state process that can range from purely epithelial to purely mesenchymal through one or even more intermediate and hence imperfect phenotypes. These not yet fully defined intermediate phenotypes are also referred to as intermediate, incomplete, imperfect, metastable or partial EMT as well as hybrid epithelial/mesenchymal or similar phenotypes (Leroy and Mostov 2007, Lundgren *et al* 2009, Savagner 2010, Eades *et al* 2011, Futterman *et al* 2011, Jordan *et al* 2011, Morbini *et al* 2011, Bednarz-Knoll *et al* 2012, Jung *et al* 2012, Huang *et al* 2013, Lu *et al* 2013, Tian *et al* 2013, Bastos *et al* 2014, Sampson *et al* 2014, Umbreit *et al* 2014, Zhang *et al* 2014, Hong *et al* 2015, Jia *et al* 2015, Ribeiro and Paredes 2015, Savagner 2015, Schliekelman *et al* 2015, Steinway *et al* 2015).

When partial EMT is reported, experiments have been performed at a cell population level, which are compared to experiments at a single-cell level. Thereby it needs to be answered whether the partial EMT represents first, a mixed cell population containing epithelial and mesenchymal cells, second, a pure cell population consisting of a single type of hybrid epithelial/mesenchymal cells (E/M) or, third, a mixed cell population containing various different cell types such as epithelial, mesenchymal and hybrid E/M cells. Additionally, several theoretical models of EMT propose a hybrid E/M phenotype at a single-cell level (Lu *et al* 2013, Steinway *et al* 2015, Tian *et al* 2013, Hong *et al* 2015, Jia *et al* 2015). In line with this, fluorescence activated cell sorting of certain cell lines have identified a subpopulation of cells exhibiting both epithelial and mesenchymal markers (Grosse-Wilde *et al* 2015, Hong *et al* 2015). In addition, lung cancer cells can stably exhibit a hybrid E/M phenotype at a single-cell level over multiple culture passages (Jolly *et al* 2016). Cells exhibiting a hybrid E/M phenotype can provide adhesion (an epithelial trait) and migration (a mesenchymal trait) at the same time and hence perform collective cell migration. Hence, cells migrate collectively during the process of wound healing (Arnoux *et al* 2005), the mammary morphogenesis (Watanabe *et al* 2014) and during cancer as clusters of circulating tumor cells (Sarioglu *et al* 2015) and therefore manifest the hybrid E/M phenotype and encompass tumor buds for a collection of cells.

How can EMT be identified from imperfect or partial EMT? The magnitude of alterations in the epithelial and mesenchymal marker levels are suitable for distinguishing the type of EMT. The perfect EMT is driven by a complete loss of cell-surface expressed E-cadherin, which is associated with a so-called cadherin switch from E-cadherin to N-cadherin. In contrast, the coexpression of an epithelial marker (usually E-cadherin) and a mesenchymal marker within a single cell is regarded as a reliable procedure to characterize a hybrid E/M phenotype. In lung cancer cell lines such as A549, LT73 and H460, the coexpression of SNAI2 and E-cadherin have been detected and characterize different subpopulations such as epithelial, hybrid E/M and mesenchymal cell populations (Andriani *et al* 2015). A breast cancer cell line HMLER has been sorted through the expression of both an epithelial surface marker CD24 and a mesenchymal stem-cell marker CD44 and hence displays the cells have the $CD24^{+}/CD44^{+}$ phenotype (Grosse-Wilde *et al* 2015). The mathematical remodeling predicts that the coexpression of ZEB and E-cadherin in cells produces a hybrid E/M phenotype (Lu *et al* 2013). In line with this, H1975 lung cancer cells have been revealed to coexpress E-cadherin and vimentin stably over months (Jolly *et al* 2016) and metastatic brain tumors have been found to coexpress E-cadherin and vimentin within single cells (Jeevan *et al* 2016). Despite the unclear stage of the coexpression within a single cell or a population of cells, relative levels of E-cadherin and vimentin at a population level have been used to group cells into epithelial, partial EMT and mesenchymal types (Park *et al* 2008, Huang *et al* 2013, Schliekelman *et al* 2015). However, it remains elusive whether these groups contain the hybrid E/M phenotype.

Connection between tumor budding and imperfect or partial EMT

The phenomenon of tumor budding is mostly detected at the invasive leading edge of the primary tumor, as there is increased detachment of distinct cancer cells from the tumor mass (Lugli *et al* 2009, Van Wyk *et al* 2015, De Smedt *et al* 2016, Koelzer *et al* 2016). Indeed, the cell-surface localization of E-cadherin is decreased in numerous tumor buds such as in buds of CRC (Bronsert *et al* 2014), esophageal cancer (Niwa *et al* 2014), pancreatic ductal adenocarcinoma (Masugi *et al* 2010, Kohler *et al* 2015), head and neck cancer (Wang *et al* 2011, Almangush *et al* 2014, Attramadal *et al* 2015), lung cancer (Yamaguchi *et al* 2010) and invasive ductal breast cancer (Liang *et al* 2013). Indeed, decreased expression of E-cadherin on the cell surface is a hallmark of EMT (Mitrovic *et al* 2012), hence tumor buds are suggested to be EMT-like (Liang *et al* 2013). Do tumor buds mirror full or partial EMT?

There is rarely evidence connecting tumor budding and EMT such as the purely mesenchymal phenotype. One weak piece of evidence has been found in the connection of tumor budding with EMT in CRC ahead of the invasive front, as cells exerted protrusions similar to lamellipodia (Carr *et al* 1986). In esophageal squamous cell cancer high tumor budding event occurrence has been correlated with EMT (Niwa *et al* 2014). In line with this, in pancreatic ductal adenocarcinoma, tumor budding is associated with both decreased E-cadherin and increased vimentin amounts, which seems to be the morphological feature of EMT (Masugi *et al* 2010). However, there may still be some staining artifacts, which have not ruled out precisely.

However, there is most commonly an association of tumor budding with partial EMT. In pancreatic ductal adenocarcinoma, the cell-surface expression of E-cadherin and membrane localization of β-catenin in tumor buds are decreased at a cell population level compared to a cell population in the tumor center. In contrast, the amount of vimentin within tumor buds did seem to correlate inversely with the amount of E-cadherin (Kohler *et al* 2015). Moreover, only a few tumors display vimentin expression, indicating partial EMT occurrences in pancreatic ductal adenocarcinomas (Kohler *et al* 2015). However, in invasive ductal breast cancer and tongue squamous cell carcinoma, the tumor bud cells displayed decreased E-cadherin on their cell surface and higher cytoplasmic vimentin levels than tumor cells of the center (Wang *et al* 2011, Liang *et al* 2013) indicating that high budding is associated with reduced membrane E-cadherin levels and increased cytoplasmic vimentin levels. In addition, in various tumor types such as CRC, liver metastases of CRC, pancreatic ductal adenocarcinoma, lung adenocarcinoma and invasive breast ductal cancer, no single-cell migration events have been detected, which either means single migrating cells are not present or the event is very rare (0.000%–0.003% of the total amount of cancer cells) (Bronsert *et al* 2014). Therefore, cancer cell invasion may be driven apart from single-cell migration also by collective migrating cells, which in turn requires the persistence of intercellular epithelial adherence junctions. Moreover, these findings suggest that not all cells switch to purely mesenchymal cells supporting imperfect or partial EMT. Other EMT-related markers further support the association of tumor budding and partial EMT. In tumor buds of lung cancer, reduced levels of membranous β-catenin have been observed (Taira *et al* 2012), whereas in CRC and oral squamous cell carcinomas, tumor buds possess decreased levels of miR-200 family members (Paterson *et al* 2013, Jensen *et al* 2015) that are efficient inhibitors of ZEB1/2 (Gregory *et al* 2008, Gregory *et al* 2011, Huang *et al* 2015). Tumor buds also show increased levels of the extracellular matrix glycoprotein laminin-5γ2 (Yamaguchi *et al* 2010, Almangush *et al* 2014, Okado *et al* 2015), which is expressed at the leading edge (Kainulainen *et al* 1998) of collectively migrating cells during the wound healing processes (Micalizzi *et al* 2010), and hence supports additionally the connection with partial EMT (Watanabe *et al* 2014). On the one hand, reduced levels of epithelial markers support the connection of tumor budding with perfect EMT and on the other hand increased amounts of mesenchymal markers support the connection of tumor budding with imperfect or partial EMT.

Tumor buds of breast carcinoma and oral squamous cell carcinoma possess higher vimentin (Liang *et al* 2013) and fibronectin levels (Jensen *et al* 2015), respectively. Solely a fraction of oral squamous cell carcinoma performs a full 'cadherin switch' from E-cadherin to N-cadherin (Attramadal *et al* 2015). Moreover, tumor buds of lung cancer display increased ZEB1 expression (Brabletz and Brabletz 2010, Lu *et al* 2013, Jolly *et al* 2015) compared to cancer cells of the main tumor mass (Taira *et al* 2012). In pancreatic ductal adenocarcinoma tumor buds, ZEB1 is significantly increased, whereas β-catenin is still not increased in the nucleus (Galván *et al* 2015). Tumor buds of oral squamous cell carcinoma possess increased amounts of ZEB1 proteins compared to cancer cells of the tumor mass center. However, these levels are still lower than those in adjacent stromal cells (Jensen *et al* 2015). This detected trimodal distribution of ZEB is further supported by the hypothesis that the miR-200/ZE double negative feedback loop reveals the three phenotypes such as epithelial (high miR-200 and low ZEB), mesenchymal (low miR-200 and high ZEB), and hybrid E/M (intermediate miR-200 and intermediate ZEB) (figure 1.12 (Lu *et al* 2013)). In summary, all these studies propose a mesenchymal phenotype for tumor buds, but contrary the results indicate a partial EMT phenotype. To resolve the question of whether tumor bud cells exhibit a mesenchymal or an imperfect or partial EMT phenotype, more quantitative

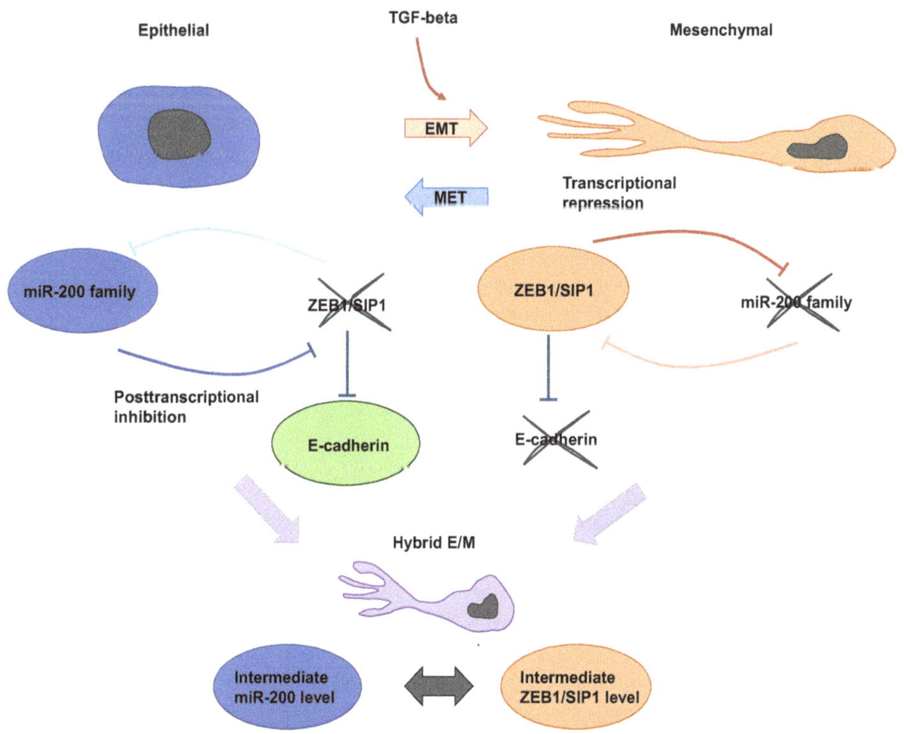

Figure 1.12. Regulation of EMT by ZEB1 and miR-200. The three phenotypes, epithelial (high miR-200 and low ZEB), mesenchymal (low miR-200 and high ZEB) and hybrid E/M (intermediate miR-200 and intermediate ZEB), of cancer cells can be adapted.

measurements are required. A main question is yet to be answered: is it possible to detect a 'real' single-cell hybrid E/M phenotype in tumor bud cells?

In esophageal squamous cell cancer, the primary tumors with increased tumor budding displayed in tumor bud cells decreased cell-surface expressed E-cadherin and increased cytoplasmic vimentin using immunostaining, while primary tumors with low budding displayed increased cell-surface expressed E-cadherin and nondetectable vimentin in their cytoplasms (Niwa *et al* 2014). In pancreatic ductal adenocarcinoma, increased tumor budding is associated with decreased E-cadherin and increased vimentin levels, but still E-cadherin is stained positively in the majority of the cells (Masugi *et al* 2010), which provides support for a hybrid E/M phenotype.

Tumor buds of oral squamous cell carcinoma contain reduced cell-surface E-cadherin and elevated cytoplasmic vimentin levels that have been detected at the cell population and the single-cell level (Attramadal *et al* 2015). Additionally, in various tumor types such as CRC, liver metastases of CRC, pancreatic ductal adenocarcinoma, lung adenocarcinoma and invasive breast ductal cancer, the tumor bud cells exhibit reduced cell polarity, reduced levels of E-cadherin (decreased cell surface membrane localization compared to increased cytoplasmic E-cadherin) and increased levels of nuclear ZEB1 within single cells (Bronsert *et al* 2014). As single-cell migration is hardly detectable and the hallmarks of EMT, such as spindle-like shape, reduced E-cadherin and increased ZEB1, levels are usually not present even among tumor bud cells, hence all these results may just point out the collective migration of cell clusters that display a hybrid E/M phenotype indicating imperfect or partial EMT and no full mesenchymal phenotype (Bronsert *et al* 2014). Do molecular events exist that drive the process of tumor budding? Is there an explanation for the proposed connections of EMT with tumor budding? Indeed, several EMT initiating pathways such as TGF-β and Wnt have been shown to be active in tumor budding processes (De Craene and Berx 2013, Dawson *et al* 2014, Dawson and Lugli 2015, Goodall 2011, Jensen *et al* 2015). What about the association of tumor budding at metastatic sites?

Proliferation and stem cells in tumor budding connects to imperfect or partial EMT
In CRC, tumor buds possess decreased nuclear Ki-67 levels indicating decreased proliferation and increased nuclear p16INK4, cyclin D1 and β-catenin levels (Palmqvist *et al* 2000, Jung *et al* 2001, Jass *et al* 2003) in comparison to the levels within the main tumor mass (Palmqvist *et al* 2000, Jung *et al* 2001). Since the these markers correlate at a cell population and a single-cell level (based on serial sections) in contrast to cells of the main primary tumor mass exhibiting opposite levels (Jung *et al* 2001), it has been proposed that an increase in p16INK4 levels leads to sequestering of Cdk4 in the cytosol, which then enables nuclear cyclin D1 to bind Cdk2 preventing thereby the binding of cyclins A and E to Cdk2 and finally resulting in decreased tumor bud proliferation (Jass *et al* 2003). Cells in the tumor buds are mainly less- or nonproliferating cells (Rubio 2008). In line with this, in breast cancer, the tumor bud cells possess decreased Ki-67 levels as compared to the main tumor mass cells (Liang *et al* 2013). In lung cancer, tumor bud cells contain decreased nuclear amounts of geminin (a key cell-cycle regulator) compared to cells of the

main tumor mass, which represents an alternative explanation for the co-occurrence of decreased E-cadherin and β-catenin amounts under nonproliferative conditions in cells undergoing budding (Taira *et al* 2012). Similarly, tumor bud cells of CRC display decreased levels of the apoptosis markers caspase-3 and M30 (Dawson *et al* 2014) as well as decreased levels of the proliferative marker Ki-67 (Brabletz *et al* 2001). These results indicate that tumor buds are nonproliferating and hence quiescent, which additionally contribute the possible involvement of EMT in tumor budding. There are at least three supportive points. First, the 'go-or-grow' (proliferation or migration) hypothesis is confirmed (Hatzikirou *et al* 2012), as it has been demonstrated in a developmental context (Matus *et al* 2015), in which the invasive phenotype needs a cell-cycle arrest. Second, the induction of EMT is able to induce cell-cycle inhibition and acquires resistance to apoptosis (Sanitarias *et al* 2002, Vega *et al* 2004, Robson *et al* 2006, Lovisa *et al* 2015) and, third, the overexpression of miR-200 facilitating MET can restore the process of apoptosis and anoikis, which is a cell-death phenomenon triggered by the detachment from the extracellular matrix (Radisky 2011). This issue needs further in-depth investigation.

Tumor buds contain increased levels of stem-cell surface markers such as CD133 (Lugli *et al* 2012) and aldehyde dehydrogenase 1 (Luo *et al* 2012), indicating that cancer stem-cell like features are associated with tumor budding. Tumor budding cells expressing aldehyde dehydrogenase 1 are directly accounted for aggressiveness and lower overall survival rate (Lugli *et al* 2009). Theoretical approaches (Jolly *et al* 2014, 2015) and experimental approaches (Swetha *et al* 2011, Goldman *et al* 2015, Grosse-Wilde *et al* 2015) addressing the interplay between EMT and cellular stemness have both reported that cells displaying a partial EMT phenotype can be more stem-like compared to either purely epithelial or purely mesenchymal cells. Hence, the connection between tumor budding and imperfect or partial EMT is further strengthen. Therefore, tumor buds at the invasive edge of a primary tumor are termed migrating cancer stem cells (Brabletz *et al* 2005).

1.2.1 Spreading of cancer cells and collective cell behavior

The onset of malignant progression starts with the spreading of cancer cells from the primary tumor into the surrounding microenvironment. The principles behind this phenomenon are not yet clear. Several questions remain unanswered. How do these malignant and highly aggressive cancer cells manage to migrate out of the primary tumor where intercellular junctions usually exist between adjacent cancer cells? What does the differential adhesion hypothesis contribute to the understanding of how certain cancer cells migrate out of the primary tumor? Do these special cancer cells walk out by a mechanism called jamming? Do these 'highly aggressive' cancer cells possess different mechanical properties, such as cellular deformability?

Differential adhesion hypothesis (DAH)

The differential adhesion hypothesis (DAH) was initially postulated by Foty and Steinberg (figure 1.13) (Steinberg 1970, 1978, Foty and Steinberg 2004) and provides a physical explanation for the spontaneous liquid-like tissue segregation, mutual

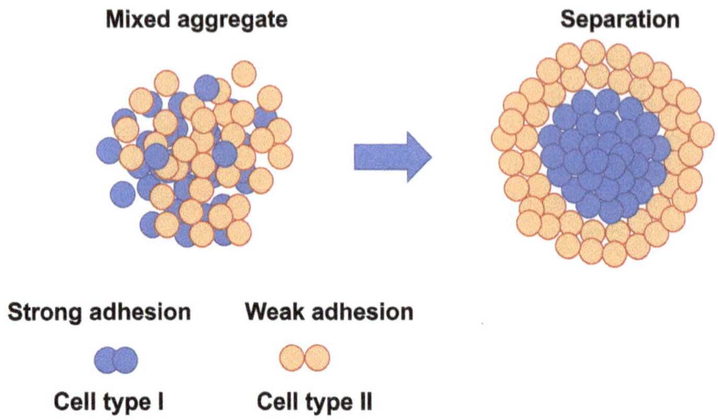

Figure 1.13. Differential adhesion hypothesis (DAH) for the segregation of two cell populations that have been mixed together.

envelopment and sorting-out behaviors of embryonic tissues and embryonic cells. The key idea behind this hypothesis is that the aggregation of cells underlies equivalent thermodynamic principles as the behavior of drops of a liquid. Thereby the degree of cell adhesion between neighboring cells is seen similar to the degree of molecule-to-molecule contact and hence the cell aggregates behave as if they have contractile surfaces. An alternative explanation is that the cell aggregates possess contractile surfaces based on increased actomyosin contraction of the cell-surface parts that are not subject to cell–cell adhesion (Harris 1976). The DAH has been successfully tested by Steinberg using seven different cell lines that have been chemically dissociated and pair-wise cocultured in all possible combinations (figure 1.13). The DAH theory predicts that if cell line one sorted out in the interior of the coculture spheroid with cell line two and cell line two sorted out in the interior of the coculture spheroid with cell line three, cell line one will sort out in the interior of the coculture spheroid with cell line three. The sorting out of cells is based on the cell's adhesiveness, which means that the more adhesive cells are in the interior of the coculture spheroid. In particular, cell line one is more adhesive than two and two is more adhesive than three. This behavior is termed transitivity (figure 1.13) and is a partial order relation, as not absolute values of adhesiveness are determined.

Another key experiment for the confirmation of DAH has been the application of a centrifugal force or the application of pressure towards cell aggregates (Phillips *et al* 1977, Phillips and Steinberg 1978). The stronger the resistance of the cells in the aggregate is to flattening, the more the cells are sorted out in the interior of the coculture spheroid experiments. Cell rounding is associated with active contraction and hence cell aggregate flattening can be seen as work against adhesion.

The DAH was reported previously to depend upon tissue affinities (Holtfreter 1939, Townes and Holtfreter 1955). In particular, the DAH explains the segregation, envelopment and sorting-out behaviors and the rounding-up of irregular embryonic tissue fragments as rearrangements of cells that seek to decrease the cell population's adhesive-free energy, whereas the overall cell–cell bonding events increase. This thermodynamic hypothesis has been formulated based on experiments comparing the behavior of cell populations during the cell sorting process and mutual tissue spreading, with expectations based upon each of the hypotheses, but it cannot explain these processes (Steinberg 1962a, 1962b, 1962c, 1963, 1964, 1970). However, a main question here is whether thermodynamics can be applied to living cells. Up to now, only the DAH makes correct predictions. Analysis of embryonic tissues capable of morphogenetic behavior revealed that these tissues could be characterized macroscopically as elasticoviscous liquids whose elemental components are motile, mutually adhesive cells. In tissues, cells first quickly stretch, but then cells gradually slide by one another. Ultrastructural and mechanical studies of rearranging cell aggregates were able to confirm these findings (Gordon *et al* 1972, Phillips *et al* 1977, Phillips and Steinberg 1978, Steinberg and Poole 1982, Forgacs *et al* 1998).

Extension of DAH by surface tension

Multiple cell–cell adhesion molecules have been identified such as N-CAM and L-CAM (Edelman *et al* 1963) as well as various cadherins (Takeichi 1988), which reduce the importance of the amount of one single cell–cell adhesion molecule for providing cell–cell adhesion strength differences. In addition to the amount of distinct cell–cell adhesion molecules, they may differ in the cell–cell adhesion strength, which is another variable for the DAH. The explanation for the resistance to flattening was that distinct cadherins facilitate the binding of cells of the same type and the amount of the contractile strength causes the interior or more outside locations of the cells within the aggregate, indicating that contractile strength has been measured. These finding have been reproduced by computer simulations (Brodland 2002) and are termed the differential interfacial tension hypothesis (DITH) (figure 1.14).

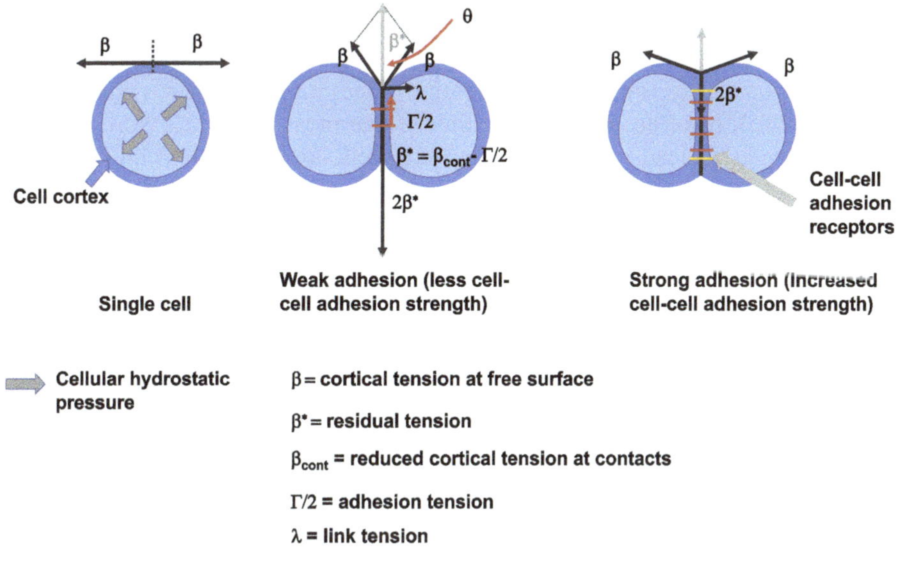

Figure 1.14. Extension of DAH by surface tension, termed the differential interfacial tension hypothesis (DITH). Single-cell tension (left) and weakly (middle) and strongly adherent cell pairs (right) at force equilibrium.

The relative surface tensions of two immiscible liquids determine which liquid will envelop the other. The DAH concludes that the mutual spreading ability of tissues (Phillips and Davis 1978, Davis 1984, Foty *et al* 1994, 1996, Davis *et al* 1997) is specified by their relative surface tensions (Steinberg 1970), which was proved by newly developed tissue surface tensiometers. Thus, using these surface tensiometers it has been shown that in every mutually adhesive tissue pair tested, the tissue with lower surface tension always envelops that with higher surface tension. Moreover, it has been reported that this behavior is independent of the identities of the adhesion molecules used by the interacting cells (Foty *et al* 1994, 1996, Duguay *et al* 2003). What has to be confirmed by experiment is the postulation that the tissue surface tensions underlying mutual tissue segregation, spreading and cell sorting are obtained solely from the intensities of the cell–cell adhesions building up these tissues. Indeed, it has been suggested that cells sort out because of a different cadherin expression level, for example N-cadherin (Foty and Steinberg 2005).

Using nonadhesive mouse cell lines (L-cells) for insertion of multiple copies of the P-cadherin cDNA, it has been shown that the mixing of two cell clones with different copy numbers of the P-cadherin cDNA sorted out the cell clone with the most copies to the interior of the aggregate (Steinberg and Takeichi 1994). However, the contractility strength has unfortunately not been considered and, thus, the presented confirmation of the DAH is still not undisputable. From the biological point of view, the differences in the cDNA copies in the transfected cells do not need to correlate with the cell-surface expression of P-cadherin on the cell's surface.

There is still a fundamental question to be answered: where is the directional information that initiates and coordinates the orientation of cell-layer movements in embryonic development stored? In zebrafish embryos, the pulling forces of cell–cell adhesions have been determined using atomic force microscopy (AFM) and compared to the forces required for the cell aggregate flattening (Puech *et al* 2005, Krieg *et al* 2008). Thus, the self-sorting ability of early embryonic cells is not facilitated exclusively by pure differential adhesion, it rather employs other processes, which have been shown by direct force measurements. Moreover, these force measurements lead to the suggestion that cellular cortical tension plays a role in cell sorting during restructuring processes, however, the epithelial cell wrapping dominates the cell sorting of enveloped mesenchymal cells (Green 2008). As the DAH is still under debate, another theory, the differential surface contraction (DSC) model has been proposed, in which the cortical tension (force generated inside the cells parallel to the cell's membrane surface) and not the cell–cell adhesion drives the process of cell sorting (figure 1.15) (Harris 1976).

Both the DAH and the DSC theories agree that cell sorting is based on the same forces facilitating developmental processes such as gastrulation, neurulation and other physiological rearrangements during embryonic movement. In particular, both theories support the idea that the majority of cancer cells differ from normal equivalents in the same properties that drive the sorting out of the cells to interior or exterior locations in a cell aggregate such as a primary tumor (Danowski and Harris 1988, Foty and Steinberg 2004).

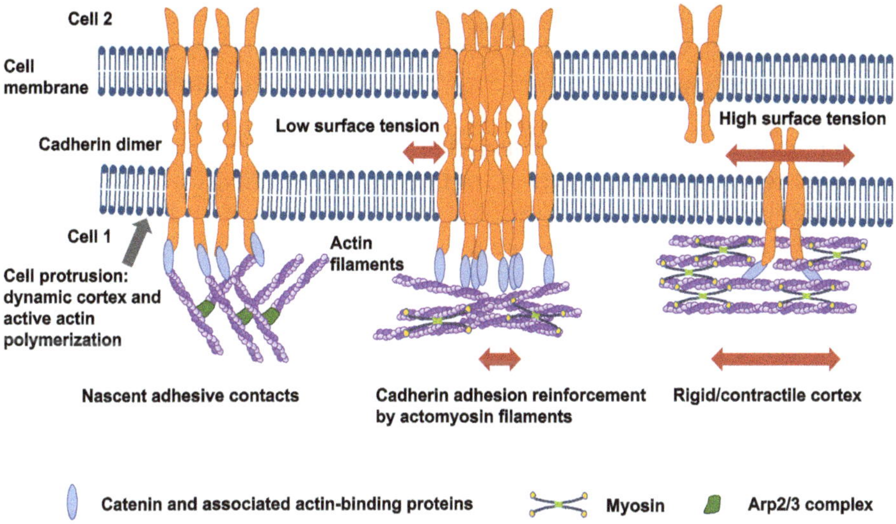

Figure 1.15. Cell surface tension depends on the cell cortex, cadherin adhesion and subsequently actomyosin contraction. The actin cytoskeleton underneath the cell membrane can adopt two major configurations: first, a dynamically polymerizing network typical of expanding protrusions or, second, a rigid system containing bundles of contractile actomyosin fibers. The protrusive activity favors the formation of cadherin contacts. Cadherin–catenin complexes become associated with diverse actin regulatory proteins such as Arp2/3 that induces actin dynamics and protrusive activity (dynamic feedback loop) (left). However, strong cortical contractility is antagonistic to cell adhesion which impairs focal adhesion formation and destabilizes or disrupts existing focal adhesions (right). Actomyosin structures are associated with maturing adhesions that anchor cadherin clusters and then reinforce adhesion (middle).

Cells are not passive material, which is pulled around by rearrangements such as the spatial enlargement of cell–cell adhesions that ignores the cell–cell adhesions' strength. It is a misconception that only conservative and reversible forces, which do not continuously expand energy, can enable cells to gravitate to consistent end results through exhibiting multiple series of intermediate geometries. Cells of the body can manage to crawl actively, which seems similar to amoebae movement. The cells pull themselves towards novel adhesions and leave older adhesions of the rear behind. The cells are not passively transported or pulled around by forces, such as an attraction to substrates such as glass, plastic or extracellular matrix proteins, as was predicted a long time ago (Weiss 1961, Carter 1967, 1970). The supporters of 'differential adhesion' do not take into account that the calculations of the 'reversible work of adhesion' are not valid when the cells are actively migrating and are not passively pulled around by enhanced cell–cell adhesions.

The DAH theory was revisited by Foty and Steinberg (2005) and some points proved to be still right. First, cell sorting is driven by combinations of physical properties of distinct cells. Second, the same properties of cells drive the gastrulation, neurulation and rearrangements of organs within the body. Third, one quantitatively variable causes during a sorting out process the final location of cell types inside or outside of a coculture aggregate and exerts a force towards the cells that then exhibit a rounded shape and resistance to flattening. However, the actomyosin contractility plays a role in this sorting out process, whereas there is no regulatory role for the cell–cell adhesiveness. Anatomical shapes and arrangements are able to evoke the counterbalance of forces. However, there exists a misunderstanding of thermodynamics here, as cellular forces can be regarded as 'reversible works of adhesion'. It seems to be true that counterbalance is involved. However, active forces can also be counterbalanced either against each other or against so-called conservative and reversible forces such as elastic stretching evoked forces. Do these counterbalanced forces really exist? If yes, how can they measured and are they reversible? The mechanical stability of organs has been investigated and a model, based on tensegrity structures, has been presented and refined (Ingber 1997, Levin 2002, Beloussov 2012). In order to achieve geometric stability, it is adequate to have pairs of opposed forces (of equal strength) that are counterbalanced, only when a distinct shape exists. All imbalances require to be set back towards a stably balanced shape. This phenomenon is called 'shape homeostasis' (Andersen *et al* 2009). As active feedback controls are usually over-correct, which may be partly a hyper-restoration phenomenon, it evokes necessary embryonic cell rearrangements (Beloussov 2012).

Extension of DAH by surface tension to embryogenesis-dependent cell sorting
As the DAH has limitations for describing cellular behavior such as cell sorting, the DAH is extended by surface tension (Amack and Manning 2012). Embryogenesis needs a proper cell sorting process and compartmentalization of various, distinct cell types. However, the DAH seems not be suited for all cancer cell lines such as the human MDA-MB-231 breast cancer cells sorting out in a coculture spheroid with the human MCF-10A mammary epithelial cell line (Pawlizak *et al* 2015). An explanation for the observed non-DAH sorting out behavior of MDA-MB-231 cells is yet to be provided. Physics-based theories have explained *in vitro* studies of cell sorting and tissue surface tension. However, this approach has been extended, as adhesive molecules may even act as signaling molecules facilitating the local reorganization of the actomyosin cytoskeleton. Moreover, cells at the boundary of a cell cluster beginning as identical cells undergo mechanical polarization. The extension of physical models by mechanical polarization enables us to explain why *in vitro* cell sorting experiments differ from *in vivo* situations and may help to solve differences in the magnitudes of tissue surface tensions. However, new studies are needed to solve the issue of how mechanical polarization and tissue boundaries are related.

How are cells guided in their motion inside a developing embryo? How do cells know how to behave when they reach their targeted position? A programmed chemical signal (a so-called morphogen) can guide cells, whereas under other conditions cells self-organize. The cell sorting and tissue layering behaviors are similarly stable. When two different types of dissociated embryonic tissues are homogeneously mixed, they start spontaneously to sort out into layers. What regulates the cell sorting process in tissues?

Application of the differential adhesion hypothesis to cells
Partially, this behavior can be answered by the DAH, which proposes that each type of embryonic tissue possesses its own individual and unique tissue surface tension

(TST) guiding the sorting out of tissues similar to the fluid surface tension regulating how immiscible liquids demix and hence differences in TST are based on different cell–cell adhesion (figure 1.15) (Steinberg 1963). As intercellular interactions cannot really be investigated inside a living embryo, DAH has been investigated primarily using *in vitro* approaches. In order to investigate the intercellular forces within cellular aggregates *in vitro*, a surface tension-like quantity, termed effective surface tension, can be determined, which provides a precise explanation for the sorting behavior of the cells even in tissues and colonies, where cells alter their mechanical phenotype actively at locally distinct positions. There remain important questions that have not yet been clearly addressed. What is the impact of single-cell properties such as cell adhesion and membrane-associated cortical tension on the entire TST? What relation exists between *in vitro* analyses and *in vivo* tissue layering and compartmentalization during embryonic development? An approach to answer the first question seems to be, at first glance, simple. Cells within a tissue can be treated similarly to molecules within a fluid. In a fluid, the surface tension is first correlated with the depth of the interaction potential between the molecules, which is termed 'adhesion energy', and second with the interaction area between the molecules. Hence, for the cellular scenario, the TST is approximately equal to the cell adhesion energy per interaction area. The adhesion energy can also be seen as the cell's mechanical energy that depends on cortical tension, adhesion expression and elasticity and contributes to the mechanical polarization at the boundaries. The mechanical energy of cells can exactly explain the cell shapes of clusters or spheroids in *in vitro* cell culture assays and even in epithelial layers *in vivo*, where fluctuations and cell divisions occur (Staple *et al* 2010, Maître *et al* 2012). Hence, it can be hypothesized that TST is the mechanical energy that needs to be spent by moving cells to the boundary in order to increase tissue surface area. Moreover, it has been reported that when cells at the boundary are mechanically polarized, the overall mechanical effect on the entire tissue is of the same amount as if they had additionally cell–cell adherence junctions compared to cells of the interior (Manning *et al* 2010).

As cell–cell adhesion in tissues is pronouncedly facilitated by cadherin cell-surface receptors, it has been suggested that the TST is equal to the energy required to rupture a single cadherin bond multiplied by the total number of cadherin bonds per unit interaction area. Indeed, the TST has been found to be proportional to the surface cadherin amounts (Foty and Steinberg 2005). In contrast, the measured magnitude of TST in cell aggregates is approximately 1×10^{-3} N m^{-1} and hence is four orders of magnitude lower than the predicted adhesion energy per unit area of approximately 1×10^{-7} N m^{-1}, as the rupture energy of a single cadherin bond is approximately 10 kT (or 10^{-19} J) and ten cadherins are located within a square micron of a typical cell membrane (Evans and Yeung 1989, Youssef *et al* 2011, Maître *et al* 2012). Why is there such a large difference? Based on *in vitro* measurements, it is hypothesized that the cell sorting process correlates more strongly with the magnitude of the cortical tension of a single cell, which is generated by active actomyosin contractility rather than purely cell adhesion (Harris 1976, Brodland 2002, Krieg *et al* 2008, Youssef *et al* 2011). Hence the membrane cannot be treated separately from the underlying cytoskeleton, such as the actomyosin cytoskeleton. When including these points, the TST is proportional to a so-called 'effective adhesion'. Moreover, the TST seems to be driven by boundary mechanical polarization rather than by the mechanical energy of adhesive bonds. In addition, when the mechanical polarization is indeed regulated by cadherin signal transduction, it may explain why TST depends on the amount of surface cadherins and correlates with the actomyosin-dependent cortical tension of cells.

What is the mechanical polarization of boundary cells?
The discrepancies may rely on the crude assumption that all cells in an undifferentiated tissue are nearly identical and hence a single cell reflects the properties of the entire tissue. These discrepancies can be resolved by assuming that cells at the tissue boundary are different compared to cells of the interior tissue regions, which is based on the hypothesis that boundary cells actively alter their mechanical properties and hence become mechanically polarized (Trepat *et al* 2009, Liu *et al* 2010). Using a traction force imbalance method (TFIM), the precise amount of forces that a cell transmits through cell–cell adherence junctions within cell clusters or monolayers of adherent cells on 2D substrates without confinements, can be analyzed (Maruthamuthu *et al* 2011). Dissimilar to laser ablation (Hutson *et al* 2003) or micropipette aspiration (Maître *et al* 2012) approaches, TFIM does not allow any assumptions about cellular mechanical properties. Instead, TFIM quantifies the mechanical polarization in individual cell pairs or triplets (Maruthamuthu *et al* 2011). As hypothesized, the cells are able to arrange their adhesive and cytoskeletal structural and functional network in a manner that the tension forces and actin network density are increased underneath external interfaces of doublets or triplets of cells compared to cells embedded within internal interfaces.

In larger coherent populations of initially fully identical cells on flat 2D surfaces, they can be described by a so-called effective surface tension (Mertz *et al* 2012). Similarly, the traction forces are primarily localized in close proximity to the boundary of the entire cell population and hence boundary cells are mechanically polarized. In line with this, cells on the surface of 3D zebrafish embryonic explants exhibit increased apical-basal actin polarization despite the expression of epithelial markers at later time points and their ability to become subsequently differentiated (Krens *et al* 2011). Thus, various cell types in different microenvironments can sense their localization at a boundary and respond to this cue by inducing mechanical polarization. What are the driving factors of the mechanical polarization of boundary cells? Cadherins are supposed to act as signaling molecules that regulate nearby actomyosin dynamics and facilitate cell polarization. Moreover, cadherin-driven cell–cell adhesion induces the Rho family guanosine triphosphatases to impair the actomyosin contractility nearby the cell–cell contact interface, whereas at the boundary no actomyosin contractility decrease occurs (Yamada and Nelson 2007). In larger cell aggregates, the cortical tension is supposed to be impaired at the interface of all cell–cell contacts of cells located inside a confluent tissue monolayer. In contrast boundary cells display impaired cortical tension only along internal cell–cell interfaces that drives their mechanical polarization (Mertz *et al* 2012). In addition to the small amount of mechanical energy in single and small cadherin bonds, cadherin signaling seems to be a major factor in driving boundary polarization.

Does mechanical polarization exist in in vivo systems?
Is the tissue layering in an embryo also driven by the mechanical polarization of the boundary cells? Indeed, boundary polarization seems to present *in vivo*, as it has been shown in *Drosophila* embryos that the actomyosin cytoskeletal network is crucial in the maintenance of the compartment boundaries (Monier *et al* 2011). In particular, clonal cells exhibiting an increased adhesion form an actomyosin cable at the clone interface (Laplante and Nilson 2006). Can the mechanisms of the mechanical polarization between cells be modeled in *in vitro* cell culture systems that are essential in the formation of tissue boundaries in embryos? Indeed, tissue intercalation and compartmentalization *in vivo* only partly correlate with *in vitro* cell sorting or TST measurements (Ninomiya *et al* 2012). In more detail, *in vitro* sorting seems to be driven by short time-scale interactions between external parts of the adhesion molecules, whereas *in vivo* the intercalation is facilitated by long time-scale spatial cadherin reorganization and cortical tensions. Indeed, some *in vivo* studies in the *Xenopus* embryo are based on dense, intercalating tissues, in which the cell–cell adherence junctions are mature and stable over a long time-scale. In *in vivo* processes within zebrafish, cells are not so densely packed and hence can exchange their neighbors quickly similar to germ-layer progenitor cells during the shield stage (Schoetz *et al* 2008, Arboleda-Estudillo *et al* 2010) and hence these short time-scale interactions seem to be crucial for *in vitro* cell sorting processes.

Indeed, the boundary mechanical polarization may explain the *in vivo* results and unite results obtained in *in vivo* and *in vitro* systems. When the surface tension is regulated by the long time-scale boundary polarization, the cadherin density on the cell surface may not affect the surface tension in the absence of internal cytoplasmic domains interacting with the actin cytoskeleton. It can be hypothesized that *in vivo* at cell–cell interfaces, the individual cells are not mechanically polarized, although individual tissue boundaries can polarize in *in vitro* cell culture fluids. Hence, this behavior may provide an explanation for the discrepancy between differential TST and cell intercalation. Cells *in vivo* are within structural networks composed of an extracellular matrix and interact with them, which possibly affects the boundary polarization and subsequently the cell sorting.

In order to refine the model and hypothesis, new techniques and experiments need to be thought out to analyze the cytoskeletal organization of cells within tissue boundaries, quantify interfacial tensions and forces *in vivo* and reveal signaling pathways involved in mechanical polarization.

The phenomenon of mechanical polarization needs to be demonstrated in 3D tissues that are formed by identical and undifferentiated cells similar to the 2D experiments. In more detail, the actin and myosin reorganization at the boundary of a cell aggregate needs to be investigated using fluorescent labeling. Similarly, the mechanical polarization at the boundaries between two tissue types *in vivo* and *in vitro* should be analyzed. Are the actin and myosin concentrations higher underneath the cell surface at the interface of two adjacent tissue types? If yes, is the interaction between two cells less pronounced at a two-tissue interface displaying mechanical polarization? What are the differences between the actomyosin reorganization at two-tissue interfaces compared to the tissue–cell culture fluid interfaces?

In order to answer these questions, experimental approaches are required to exactly quantify interfacial tensions within two different tissues. The laser ablation method can be utilized to analyze tensions in 2D epithelial layers, whereas the tissue interfaces are distinct in 3D tissues, which still cannot be analyzed. TFIM is an alternative approach for the quantification of interfacial tensions between two tissue types and helps to identify whether these tensions are based on actomyosin or cadherin dynamics. Other approaches for interfacial tension analysis involve molecular force sensors (Grashoff *et al* 2010) or membrane fluctuation analyses (Betz *et al* 2009, Tambe *et al* 2009).

Can extracellular signals trigger the mechanical polarization of boundary cells? Several signaling pathways have been shown to be involved in tissue boundary formation *in vivo* (Rodriguez and Basler 1997, Xu *et al* 1999, Rohani *et al* 2011). The Hedgehog signaling pathway has been shown to facilitate the myosin activity and trigger expression of cadherins at tissue boundaries in *Drosophila*. Can the alterations of Hedgehog signaling pathway remodel the actomyosin accumulation during the formation of the boundary? What are the downstream effects of the mechanical polarization? The polarized cytoskeletal activity requires alterations in cellular morphology and enables cells to undergo differentiation, when located at tissue boundaries.

The limits of physical theories for collective phenomena and the formation of patterns are that even the localization and transport of adhesion molecules affects these processes *in vivo* (Ninomiya *et al* 2012). Hence a physical model needs to be created that includes the mobility effect of adhesive molecules on cellular shapes and/or forces and the inverse. In particular, these models must also address a broader range of mechanical effects at boundaries including the differences in cell protrusivity and even the presence of oriented cell divisions. However, an interpolation between short time-scale mechanical interactions, which are present in tissues with highly mobile cells, and long time-scale interactions, which play a role in dense tissues with mature contacts. Thus, collective mechanical interactions seem to be a mechanism for strong cellular movements and actomyosin reorganization at cell boundaries are hypothesized to contribute to these mechanical interactions. However, the upstream signaling pathways of the actomyosin reorganization may reveal more insight into the tight regulation during embryonic development or the malignant progression of cancer.

Cell jamming

Cell jamming has been observed during studies of inert soft condensed matter. In particular, spontaneous intermittent fluctuations, dynamic heterogeneity, cooperativity, force chains and kinetic arrest are the hallmarks of the glass transition supposed to be associated with jamming (figure 1.16) (Liu *et al* 1995, Liu and Nagel 1998, 2010, Bi *et al* 2011, Garrahan 2011, Trappe *et al* 2001, Vitelli and van Hecke 2011). Although jamming remains debatable and not very well understood, the

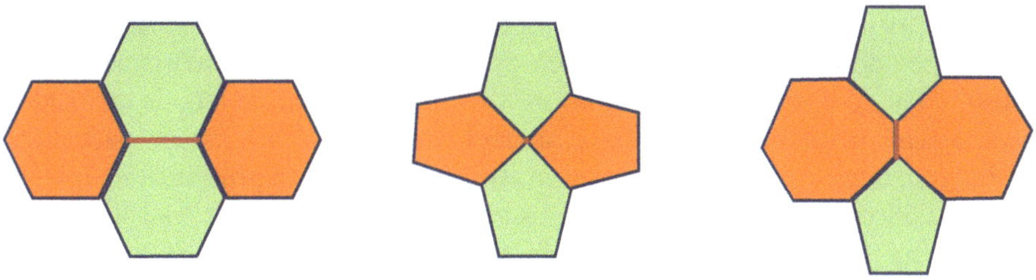

Figure 1.16. Cell jamming transition. The contractile energy associated with the cell–cell boundary acts to decrease the cell–cell contact perimeter, whereas the adhesive energy associated with the cell–cell boundary acts to increase the cell–cell contact perimeter. The sum of these opposite effects is the net line tension that spans along cell–cell junctions.

concept has become a focus of much research as it tries to unite understanding of a broad range of soft matter forms, such as foams, pastes, colloids, slurries and suspensions, which can flow in some situations but jam in others.

The same hallmarks can be found in the dynamics of cell monolayers. These dynamics conform quantitatively to the Avramov–Milchev equation, which describes the rate of structural rearrangements (Angelini *et al* 2011) and shows the growing length- and time-scales that are quantified using the more rigorous four-point susceptibility (Berthier *et al* 2005, Tambe *et al* 2011). As inert and living condensed systems' dynamics are constrained by many of the same physical factors, the assertion of cell jamming may be suitable. In more detail, looking at the basic unit such as a living cell, a foam bubble or a colloidal particle, they all include volume exclusion (two particles cannot occupy the same space at the same time), volume (size) (Zhou *et al* 2009), deformability (Mattsson *et al* 2009), mutual crowding, mutual caging (Segre *et al* 2001, Schall *et al* 2007), mutual adhesion or repulsion (Trappe *et al* 2001) and evoked mechanical deformation, such as stretch or shear (Wyss *et al* 2007, Trepat *et al* 2007, Krishnan *et al* 2009, Oliver *et al* 2010). In a cell monolayer, the cells move more freely as their size, crowding, stiffness or mutual adhesion decreases, or as their motile forces or imposed stretch become greater. However, as adhesion or crowding increases, or as motile forces decrease, cellular rearrangements might slow, cooperativity will increase, and the monolayer will be topologically frozen and all cells will be caged by their neighboring cells (figure 1.16) (Ladoux 2009, Angelini *et al* 2010, Wilk *et al* 2014, Tambe *et al* 2011, Angelini *et al* 2011). It seems that the jamming hypothesis can unify the effects of diverse biological factors that have been considered to be separate and independent of each other.

There are still major open questions in cell biology: when and why is a cell inside a confluent epithelial layer quiescent and fixed in place, and when and why does it instead adopt the ability to become motile and migrate over large distances in a cooperative and collective manner, such as multicellular packs, swirls, streaks and clusters? The collective cell migration of the epithelial layers is critical for physiological processes such as development and wound healing or pathological processes such as chronic inflammation and cancer cell migration and invasion. How the collective migration of cell clusters is driven is still not well understood. Collective scenarios are important in physiological processes such as wound repair after tissue injury, embryonic development such as gastrulation, epiboly as well as morphogenesis and in pathological processes such as cancer cell invasion through the healthy tissue environment (Fischer *et al* 2015, Munjal *et al* 2015, Zheng *et al* 2015, Pattabiraman *et al* 2016). However, an overall physical description for understanding collective cellular events is not yet available.

Most conceptual frameworks are focusing on the individual and separated role of each underlying factor or a few of them and hence may not be able to describe the behavior of a cellular collective that covers strong and complex physical and molecular interactions. To close this gap, the concept of cell jamming needs to be addressed, which predicts that solely a small set of factors regulate a cellular collective that jams and hence becomes rigid, similar to a solid, or unjams and hence flows similar to a fluid. The two gateways to cellular migration, the unjamming transition (UJT) and the EMT share at first glance superficial similarities, but at a closer look the similarities seem to vanish (figure 1.17).

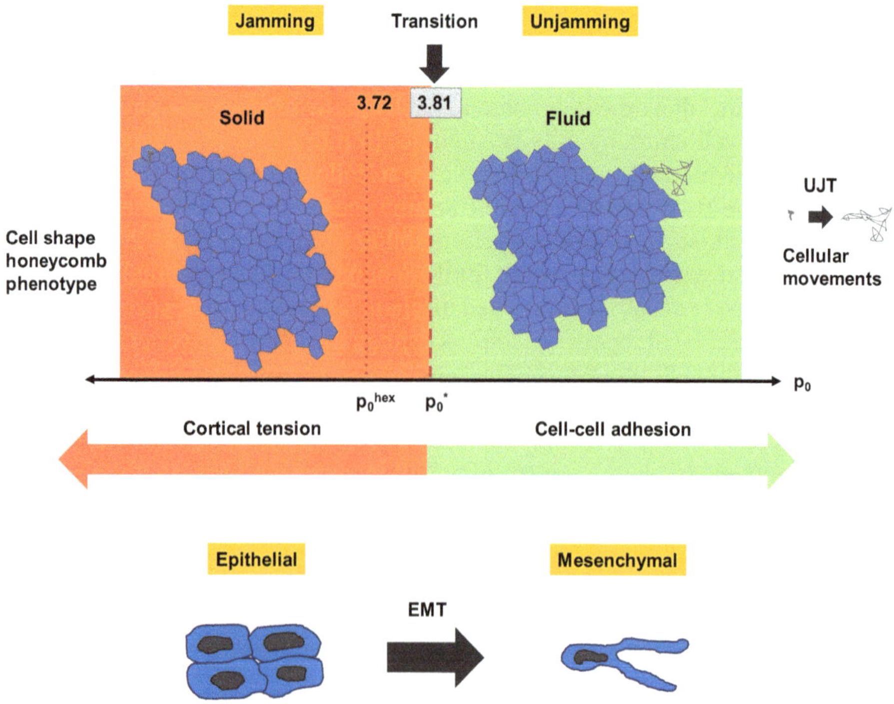

Figure 1.17. Comparison of unjamming to jamming transition and epithelial to mesenchymal transition.

Is cell jamming still speculative in cancer cell invasion? Epithelial cells form the lining of the inner and outer surfaces of all organs and body cavities and hence build a strong barrier towards the exterior environment to protect, separate, sense, transport, secrete and absorb for the entire organ. Under quiescent conditions, each epithelial cell fulfills physiological functions at a fixed spatial localization and hence is tightly connected to the underlying basement membrane and tightly bound to its nearest neighbors. Even under disturbance by an immediate neighbor cell, a cell still will not move far. However, the same cell manages to mobilize and migrate over large distances, similar to developmental processes. An epithelial cell within a confluent layer can migrate, but not as an individual cell, it rather is part of a coordinated and cooperative cellular collective (Holmes 1914, Arboleda-Estudillo *et al* 2010, Das *et al* 2015, Haeger *et al* 2015). Within these collective migration processes, the underlying physical and molecular factors may largely vary. As the interactions among these factors are strong and highly complex, the collective cellular migration is not yet clearly understood. Does a physical integrative framework exist to cover all these factors? The epithelial layer performs collective cellular behavior, which seems to be similar to the transition between the fluid-like and solid-like phases of matter (Szabó *et al* 2006, Farhadifar *et al* 2007, Angelini *et al* 2011, Vicsek and Zafeiris 2012, Kim and Hilgenfeldt 2015). A focus lies on the jamming to unjamming transition. Indeed, the epithelial cellular collective has the capability for switch-like alterations, as in certain circumstances they are solid-like and virtually frozen and in others they are fluid-like and highly mobile (Fredberg 2014, Kim and Hilgenfeldt 2015). By only modest alterations in its physical properties, the collective achieves a large range of adaptability and physiological scope.

The two collective phases: the fluid-like versus solid-like phase

How can an epithelial layer become solid-like and be in a jammed phase? Without any gaps or voids within the cell layer, each cell is surrounded by neighboring cells and hence physically caged by those neighbors. However, the strength of such a cage has a finite limit and represents an effective energy barrier. This energy barrier needs to be overcome when a cell escapes its cage and, thereby, successfully undergoes a cellular rearrangement with its neighboring cells. When the barrier is so strong that it is rarely overcome, the cell is effectively fixed in its position and hence trapped in its cage. Hence, this behavior leads to a so-called glassy solid, which has been identified to correspond to stable tissues consisting of homeostatic epithelial cell packing (Sadati *et al* 2013). When the energy barrier is overcome by increased cellular propulsive forces or the energy barrier is reduced or even abolished, the cell escapes its cage more easily. In various cellular collectives, the characteristic time and length scales for these cellular rearrangements have been determined (Angelini *et al* 2011, Nnetu *et al* 2013, Garcia *et al* 2015, Park *et al* 2015b). These events can promote cellular rearrangements among neighbors and the entire collective in order to unjam and flow (Bi *et al* 2015, Park *et al* 2015b, Bi *et al* 2016). The unjamming of a confluent cell layer may be helpful to migrate into unfilled space to heal a wound, during embryonic development, morphogenesis or the invasion of malignant cancer cells into a healthy tissue (Serra-Picamal *et al* 2012, Haeger *et al* 2014, Bazellières *et al* 2015, Das *et al* 2014, 2015, Fischer *et al* 2015, Zheng *et al* 2015). However, cell jamming and unjamming events are connected to cellular shapes and their alterations.

Cell shape alterations towards adoption of a honeycomb phenotype
There exists a similarity between the structure of an epithelial cellular collective and a soap foam (Thompson 1917), which leads to the simplest view of the epithelial cell shape in a cellular collective, the 'honeycomb conjecture' (figure 1.17). Hence it has been suggested that any partition of the plane into cells with equal area must possess a net cellular perimeter, which may be achieved by exhibiting a regular, hexagonal honeycomb tiling (Hales 2001). Compared to the hexagonal honeycomb structure, all other configurations tiling the plane into cells of equal areas, such as triangles, squares, parallelograms, necessarily possess a greater net perimeter. In addition, if the cell perimeter is based on lowest energy levels, the honeycomb configuration represents the optimal packing at minimal energy cost. For this configuration, the tiling of the cell plane is regular and hence ordered, the ratio of cell perimeter to the square root of cell area results in a nondimensional index of cell shape close to 3.72 (figure 1.17).

How are cell shape and cell proliferation related?
Indeed, epithelial sheets try to adapt such a regular honeycomb configuration as the mature *Drosophila* imaginal disc. Ongoing cellular divisions and apoptosis in a cell layer provide a strong disordering influence on the shape and packing of the cells. These events systematically alter through cellular arrangements the honeycomb packing to a polydisperse packing with irregular polygons spanning four-sided to nine-sided cells, but still keep preferentially a hexagonal shape (Gibson *et al* 2006, Farhadifar *et al* 2007). Cell division orientates along the long axis of the interphase cell (Hertwig 1893), which facilitates both stress relaxation and isotropic growth without the transduction of local mechanical signals (Wyatt *et al* 2015). Tricellular junctions in the *Drosophila* epithelium represent a polarity cue for geometry and mechanical stress that is provided by the dynein-associated protein Mud (Bosveld *et al* 2016). Not all epithelial events need to be restricted to one single plane such as in the presumptive enveloping layer on the zebrafish embryonic surface, in which cell divisions that are oriented in-plane versus out-of-plane evoke alterations of the epithelial cell shape and are log-normal distributed (Xiong *et al* 2014). In the airway epithelium, overcrowding based on proliferation and migration facilitates Piezo1-dependent extrusion and spreading of living cells out of the initial cell plane (Eisenhoffer *et al* 2012). Hence the regular honeycomb structure of tissues seems to be rather unusual, when the cells proliferate and undergo apoptosis, as tissues then usually display a polydispersity of cellular shapes and their location is not confined within one single plane.

How are cell shape, cell–cell adhesion and cell sorting connected?
In order to explain collective cell migration and cell transitions between fluid-like and solid-like phases, the cell within the cellular layer can be modeled in several ways, such as a self-propelled particle, as a member of a swarm and even as a distinct locus on a fixed grid interacting energetically with other nearby loci (termed the

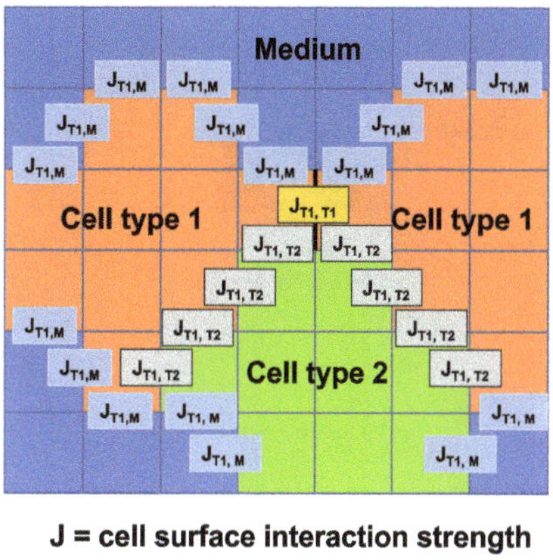

Figure 1.18. Cell-surface interaction and cellular Potts model. Each of the three cells fill several lattice sites and interact with one another or the medium. The strength J of the interaction depends on the two cell types T1 and T2 and on the cell medium (the microenvironment).

cellular Potts model) (figure 1.18) (Graner and Glazier 1992, Glazier and Graner 1993, Henkes *et al* 2011, Vicsek and Zafeiris 2012, Basan *et al* 2013, Garcia *et al* 2015). Regarding the rearrangements of cells within a tissue, each model revealed some major insights, however, none of these models can predict the physical determinants of the cell shape and its relationship to the cell jamming phenomenon. Morphogenetic movements such as invagination, evagination and layer spreading seem to be based on alterations in cell–cell adhesions (Townes and Holtfreter 1955). The DAH has been used to explain the cell sorting process during tissue morphogenesis (Steinberg 1970, 2007). In order to explain the associated alterations of the cell shape and cellular rearrangements of cell neighbors within a tissue without the formation of gaps, epithelial layer dynamics can be mathematically modeled using a continuous and complete tiling of the cell-layer plane (Sulsky *et al* 1984). In this model, each polygonal cell is based on mutual cell–cell adhesion energies that are presented as a mutual surface tension. The equilibrium relies on a force balance at each vertex, at which point the cell–cell adherence junctions are formed. These theoretical approaches are termed vertex models (figure 1.19) (Alt *et al* 2017).

A unified physical description of epithelial mechanics is provided by a theory combining physical forces within a cell layer with cellular motions, which in turn exert these forces (Sulsky *et al* 1984). A unifying description has been suggested to be helpful, but has yet only been provided theoretically (Tambe *et al* 2011). Cellular stresses within a continuous cell layer and their distributions have been determined, such as the distribution of traction stresses exerted by each single cell on its substrate and of both normal and tangential stresses exerted by each cell toward its neighbors across their cell–cell adherence junctions (Trepat *et al* 2009, Tambe *et al* 2011, Trepat and Fredberg 2011, Kim *et al* 2013, Bazellières *et al* 2015). However, only limited systematic connections between cellular stresses and cellular motions have been identified.

Far beyond any boundary, each cell within a collective seems to migrate along a trajectory, in which the shear stresses exerted to neighboring cells across mutual

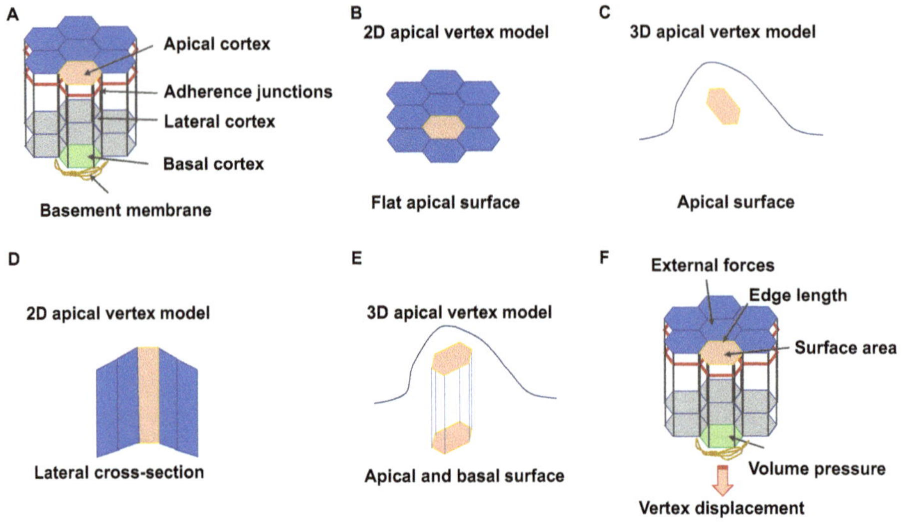

Figure 1.19. 2D and 3D vertex models for tissue mechanical properties. (A) Schematic drawing of an epithelial tissue. Cytoskeletal components generate forces inside cells and to the basement membrane. In apical vertex models in 2D (B) and in 3D (C) the epithelial cells are represented by the shape of their apical surfaces (polygons). (D) In 2D lateral vertex models, the cells are represented by their lateral cross-sections. (E) In 3D vertex models the tissue is represented by its apical and basal geometry. (F) Four parameters such as external forces, edge length, surface area and volume pressure regulate the vertex displacement. The tissue is in mechanical equilibrium when the forces acting on all vertices is zero.

cell–cell junctions are minimized (termed plithotaxis), whereas cells in close proximity to a cell-free void, exert a traction stress on their substrate that is aligned toward the void (termed kenotaxis) (Tambe *et al* 2011, Trepat and Fredberg 2011, Kim *et al* 2013).

How are cell shape, cell jamming and the vertex model related?
In the vertex model, each cell in the plane possesses a preferred value of its cell–cell contact perimeter (termed the perimeter set point, P_0) and departures from this set point require a so-called spring-like energy penalty (Brodland 2002, Farhadifar *et al* 2007). It may be useful to separate the perimenter into two competing contributions (Brodland 2002, Farhadifar *et al* 2007, Maître *et al* 2012, Fletcher *et al* 2014). First, the contractile forces are associated with the cell–cell adherence junctions acting to decrease the perimeter of the cell–cell adherence junction. Second, the adhesion between cells at their cell–cell adherence junctions acts to enhance the perimeter. Indeed, the two effects are opposite and competitive. The net effects on the cell perimeter, such as the junctional contraction and inversely the junctional adhesion, can be presented as an aggregate line tension with either positive or negative values. When contraction is dominant, the net line tension exhibits positive levels, whereas when adhesion is dominant, the net line tension exhibits negative levels (Brodland 2002, Farhadifar *et al* 2007). Each cell seems to have a preferred value of its projected area, termed the area set point A_0 and changes from its set point can necessitate a so-called spring-like energy payment. This payment was originally attributed to elastic

deformations of the cellular body (so-called height elasticity), whereas it in fact includes alterations in active cellular tension that persists without viscoelastic relaxation over long time-scales (Vincent *et al* 2015) and also increases with increasing cell density (Zimmermann *et al* 2016). So far it has not been seen that adhesion of the cell basal site to the cell substrate need to be addressed additionally. For each cell in a continuous layer, all these factors yield a total energy that is given by:

$$E = K_A\,(A - A_0)^2 + K_P(P - P_0)^2. \tag{1.1}$$

The pre-factors K_A and K_P present spring-like coefficients for either alterations in area (cell projected area) or perimeter. The theoretical analysis of these systems revealed that fluid–solid transitions together with characteristic alterations of the cellular shape can be observed (Farhadifar *et al* 2007, Kim and Hilgenfeldt 2015). Moreover, a formal analysis has been performed and hence established the existence of critical behavior that is correlated with a jamming transition between a fluid-like and a solid-like phase (Bi *et al* 2015, 2016, Park *et al* 2015b). In particular, three parameters facilitating the fluid to solid transition have been revealed and combined in a unified jamming phase diagram (figure 1.20). The first parameter is $p_0 = P_0/\sqrt{A_0}$, which is important for the following three reasons. First, p_0 is a dimensionless factor and can be regarded as a simple index for the preferred cell shape. Second, p_0 is additionally a material property of the cell, as it depends on the ratio between cell–cell adhesive stress and cellular cortical tension (Farhadifar *et al* 2007, Bi *et al* 2015). Third, there exists a discrete or critical value of p_0, termed $p_0{}^* = 3.81$, at which a transition between the fluid-like phase and the solid-like phase of the cellular collective can be observed (Bi *et al* 2015, 2016, Park *et al* 2015b):

$p_0 < 3.81$	fluid-like
$p_0 = p_0{}^* = 3.81$	critical point and jamming transition
$p_0 > 3.81$	solid-like

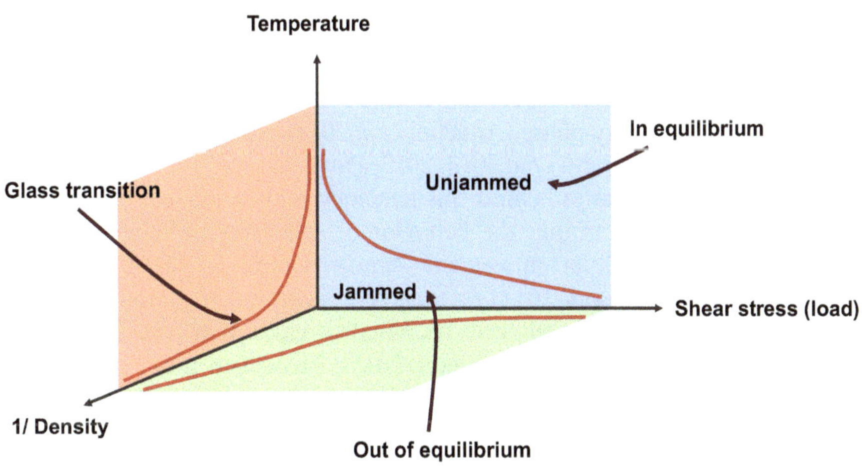

Figure 1.20. Jamming phase diagram.

In a cell, p_0 can be altered due to changes in cell–cell adhesion or cortical tension or a combination of both and also various underlying factors. However, the theory predicts a preserved transition, when p_0 is approaching the numerical value $p_0^* \approx 3.81$. The preferred cell shape p_0 may be comparable to or may differ from the actually attained cell shape. The attained cell shape can be described by a shape index $q = P/\sqrt{A}$. It has been predicted by the jamming theory that when $p_0 < 3.81$, the cell layer becomes solid-like and jammed (Bi *et al* 2015). Then each cell is trapped in a cell shape that is different from its preferred shape with $q \approx 3.81$. However, when $p_0 > 3.81$, the cell layer becomes fluid-like and unjammed and hence each cell is now free to adapt its preferred shape and subsequently q is equal to p_0. The solid-to-fluid transition is performed when cells 'wet' (or adherently contact) each another more than they pull on each another and vice versa. The second parameter within the jamming phase diagram addresses cellular propulsion. These propulsive forces are exerted toward the substrate by the cell layer, fluctuate in space and in time, and are highly cooperative (Trepat *et al* 2009, Angelini *et al* 2011, Tambe *et al* 2011). In order to refine the theory the so-called self-propelled Voronoi model (SPV model) has been revealed and used to predict a motility-facilitated unjamming transition (Bi *et al* 2016). In particular, the Voronoi model predicts that propulsive forces can become sufficiently large to overcome energy barriers and subsequently unjam the layer (the SPV model is addressed in more detail below). Moreover, the SPV model shows that when motility-facilitated unjamming occurs, the shape factor q indeed exceeds the critical value close to 3.81. The third and last parameter addresses the persistence of propulsive forces and how the persistence increases the propulsive force-dependent effects. The propulsive forces are usually directionally persistent and hence have a higher impact than those forces that are random in their direction. In conclusion, this model describes that the jamming transition is based on three natural parameters such as the preferred cellular shape (p_0), propulsion and persistence, and organizes them into a unifying jamming phase diagram. Indeed, the heterogeneity within the cell layer is based on various effects and is proposed to impact the transition in terms of sharpness and can also lead to other considerations such as the development of a perspective for the airway epithelium, which is not yet a mainstream view.

On the one hand, there is agreement that the cellular collective will evolve an amorphous solid-like glassy phase (Angelini *et al* 2011, Garrahan 2011, Tambe *et al* 2011, Garcia *et al* 2015, Nnetu *et al* 2012, 2013, Pawlizak *et al* 2015), whereas on the other hand, the mechanism(s) behind this tendency remain unclear. In addition to the mechanism of cell jamming, the behavior of cells may be based on contact-facilitated inhibition of cellular movement (Zimmermann *et al* 2016), enhanced cell packing density (Henkes *et al* 2011) or even cell packing density independent, but still dependent on elevated cell–cell frictional stresses that are based on the maturation of adhesive bonds (Garcia *et al* 2015). Moreover, it has been proposed that within the maturing cellular layer, the frictional stresses emerge at the levels of cell–cell adhesion and cell–substrate adhesion, and they couple to velocities by restricting the velocity correlation length. The theory is based on the formation and breaking of adhesive bond linkages, however, it does not include nonfrictional stresses or the effect of the competition between adhesive forces and cortical tension. Finally, a perspective into the mechanism underlying cellular jamming is provided that is complementary to the jamming transition. There remain open questions: What are the dominant mechanisms of collective cellular migration? What impact does cell jamming behavior have on collective cellular migration?

Self-propelled particle (SPP) model of a persistent random walk

As the jamming process occurs also in inert and nonbiological matter, such as particulate systems including granular materials, polymers, colloidal suspensions and foams, when their packing density is increased over a distinct critical value and glass transitions are observed when the fluid is cooled below a distinct transition temperature. These phenomena are described by the jamming phase diagrams (Liu *et al* 2010, Trappe *et al* 2001). Hence self-propelled particle (SPP) models have been utilized to describe dense biological tissues (Belmonte *et al* 2008, Henkes *et al* 2011, Sepulveda *et al* 2013, Garcia *et al* 2015, Soumya *et al* 2015, Szabo *et al* 2006). These SPP models are similar to those for inert particulate matter, as cells are described as disks or spheres interacting with an isotropic soft repulsive potential. However, unlike Brownian particles in a thermal bath, these self-propelled particles perform persistent random walks. SPP models typically undergo a glass transition from a diffusive fluid state to an arrested sub-diffusive solid state, which is driven by firstly the strength of self-propulsion (Henkes *et al* 2011, Ni *et al* 2013, Garcia *et al* 2015) and secondly the packing density φ (Henkes *et al* 2011, Fily *et al* 2012, 2014, Ni *et al* 2013, Berthier 2014). Similar to thermal systems, a jamming transition is observed at a critical packing density φ_G, but is altered by the persistence time of the random walks (Henkes *et al* 2011, Ni *et al* 2013, Berthier 2014, Fily *et al* 2014). During various biological processes, a tissue still remains at confluence (packing fraction equal to unity), while it is able to reversely switch between a liquid-like and a solid-like state. In wound healing processes, cells can collectively organize to build a so-called moving sheet without altering their packing density (Kim *et al* 2013), and similarly during vertebrate embryogenesis, mesendoderm tissues exhibit a more fluid-like behavior than ectoderm tissues, while both tissues still display a packing fraction equal to unity (Ben-Isaac *et al* 2015).

It has been reported that the well-studied vertex model for 2D confluent tissues (Nagai and Honda 2001, Farhadifar *et al* 2007, Hufnagel *et al* 2007, Manning *et al* 2010, Staple *et al* 2010, Bi *et al* 2016,) undergoes a rigidity transition in the limit of zero cell motility (Bi *et al* 2015). Interestingly, the rigidity of the tissue vanishes at a critical balance between the cortical tension and the cell–cell adhesion. The rigidity transition is driven sensitively by the cell shapes, which are well defined in the vertex model. However, the vertex models are useful, but have their limitations in not incorporating the cell motility.

Self-propelled Voronoi (SPV) model

The SPV model represents a hybrid between the vertex model and the SPP model (figure 1.21) (Bi *et al* 2016) and is suitable to describe the cell jamming or unjamming behavior of cells. Cellular motion within dense tissues regulates several biological

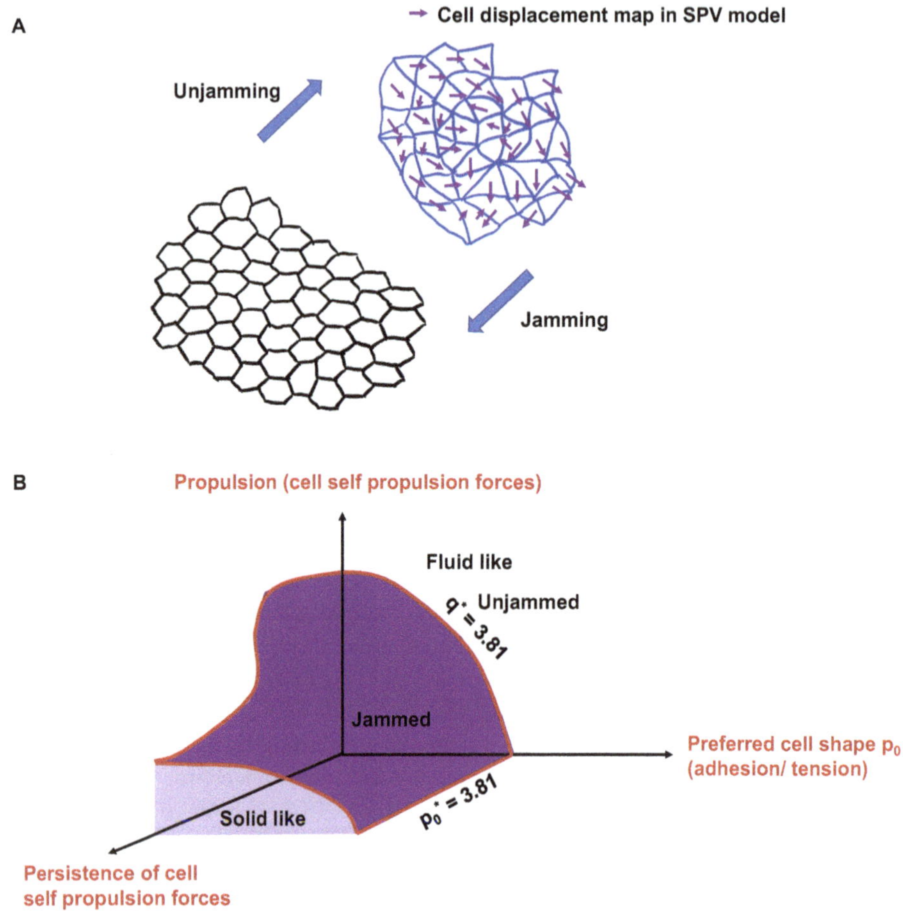

Figure 1.21. Self-propelled Voronoi (SPV) model. (A) The cell displacement map in the SPV model for a fluid close to glass transition over a certain time corresponding to a distinct structural relaxation. (B) Jamming phase diagram adapted to cells. The unjamming to jamming transition depends on three parameters such as propulsion, persistence of propulsion and preferred cell shape.

processes such as embryonic development and malignant progression of cancer and these tissues are supposed to exhibit collective glassy behavior. For quantitative predictions about glass transitions in tissues, the self-propelled Voronoi (SPV) model is applied, which simultaneously addresses polarized cell motility and multibody cell–cell interactions within a confluent tissue excluding cell-free gaps. Moreover, this model predicts a jamming transition from a solid-like to a fluid-like phase, which is regulated by three parameters: the migration speed of single cells, the persistence time of single-cell migration tracks and a target shape index defining the competition between cell–cell adhesion and cortical tension (Doyle *et al* 2013). Contrary to traditional particulate glasses, a structural order parameter can be measured that identifies the entire jamming surface as a function of model parameters. Indeed, a continuum soft glassy rheology model seems to predict the transition precisely in the limit of small persistence times, whereas it still fails in the limit of large persistence times. All of which helps us to reveal the nature of the collective solid-to-liquid transitions that are seen in embryonic development and malignant progression of cancer assumed to driven by 'tissue' EMT.

Can a continuum model be applied for glass transitions in tissues?
The continuum hydrodynamic equations of motion have been revealed by coarse graining SPP models in the dilute limit, however, no continuum model for dense active matter near the glass transition exists. A simple trap (Monthus and Bouchaud 1996) or soft glassy rheology (SGR) (figure 1.22) (Sollich 1998) model seems to be a suitable continuum approximation for the phase behavior in the large D_r Brownian regime, whereas it fails in the small D_r limit. For large D_r, the particle behaves like a Brownian particle with an effective temperature $T_{\text{eff}} = v_0^2/2\mu D_r$ (Fily and Marchetti 2012). Moreover, this mapping becomes exact when $D_r \rightarrow \infty$ at fixed 'effective inertia' $(\mu D_r)^{-1}$ (Fily *et al* 2014). In particular, similar to granular systems (Abate and Durian 2008, Cecconi *et al* 2003), the effective temperature in SPP is predominantly based on kinetic effects. Hence, the temperature seems to scale quadratically with the velocity:

$$T_{\text{eff}} \propto c v_0^2. \tag{1.2}$$

From a physical point of-view, this effective temperature provides the amount of energy that individual cells can utilize to vibrate within their cage (or synonymously trap). How can the trap depths (or energy barriers) between metastable states be determined? In the Brownian regime with large D_r, there exists no cellular dynamical mechanism to rearrange collectively and hence the rearrangements are proposed to be small and local.

The statistics of energy barriers have been analyzed for local rearrangements using the vertex model at equilibrium conditions (Bi *et al* 2014). Using the 2D vertex

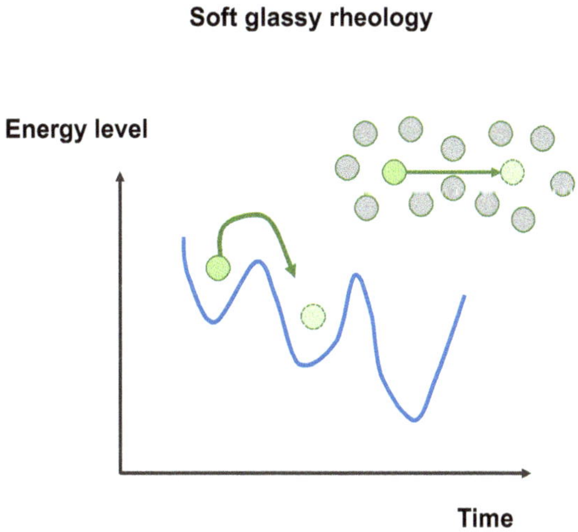

Figure 1.22. Soft glassy rheology (SGR) model. Schematic representation of the SGR model that describes the glassy and weak power law behavior of soft disordered materials such as cells or foams.

model, local rearrangements have been found to occur via so-called T1 transitions (Weaire and Hutzler 1999). Using a trap model (Monthus and Bouchaud 1996) or the SGR framework (Sollich 1998), these statistics have been employed to calculate an analytic prediction for the glass transition temperature T_g as function of p_0 without any fit parameters (Bi *et al* 2016). Does the SGR prediction for the glass transition still functions for the SPV model with large D_r? Hence, data points of glassy states from the SPV model are compared with the glass transition T_g line (Bi *et al* 2014). Indeed, it has been demonstrated by one fitting parameter c (the proportionality constant) that the SPV data for $D_r = 10^{-3}$ are in agreement with the SGR prediction, as $T_{\text{eff}} \sim v_0^2$ and the glass transition line scales as $T_g \sim p_0^* - p_0$ for $p_0 \ll p_0^*$ and $D_r \rightarrow 0$, the glass transition line scales as $v_0 \sim (p_0^* - p_0)^{0.5}$ within these limits. Why is the effective temperature SGR model suitable? Similar to SPP models for spherical active Brownian colloids, the angular dynamics of each single cell can be treated independently of cell–cell interactions and the angular dynamics of other neighboring cells. Additionally, an alignment interaction connecting the angular and translational dynamics seems to alter this behavior. Finally, an SGR/trap model prediction has been precisely analyzed in a glassy system, as dissimilarly to common glass models, all trap depths can be listed for local transition paths within the vertex model. However, for small D_r values, collective normal modes are regulating cell displacements and thus energy barriers for localized T1 transitions are likely to be irrelevant in this regime. There exists a deviation (L2-norm) between glass transition lines in the SPV model and T1-based SGR prediction as a function of D_r, as indeed SRG cannot work in the small D_r limit. How can the theory for collective modes be altered to include small D_r values?

It has been shown with a minimal model that confluent tissues with cell motility display glassy dynamics at constant density. This model provides a quantitative prediction for how the fluid (unjamming) to solid (jamming) transition in biological tissues depends on parameters such as cell migration speed, the persistence time associated with directed cell movement and cell mechanical properties, which are dependent on adhesion and cortical tension. The cell shape index is as a structural order parameter a simple parameter that indicates jamming transition. Moreover, a simple analytic model, which is based on localized T1 rearrangements, can precisely predict jamming transition in the limit of large D_r. However, the prediction fails in the limit of small D_r, as collective normal modes regulate the instantaneous particle displacements. This model provides various experimentally verifiable predictions regarding cell shape and tissue mechanics: The order parameter $q = 3.81$ represents a structural signature for the glass transition even in tissues, where significant cell motility or dynamics occur, which indeed has been found in epithelial lung tissue (Park *et al* 2015b). What about other tissues? In proliferating MDCK monolayers (Deforet *et al* 2014) and the convergent extension in fruit fly development (Etournay *et al* 2015), similar shapes have been determined supporting the order parameter. Where does the order parameter breakdown? In the limit of decreasing cell motility, the shape and pressure fluctuations are abolished when the jamming transition is approached from the solid side and are still zero in the fluid. A finite motility v_0 still evokes such fluctuations in the fluid phase, as confirmed by cellular stresses and pressure calculations in the SPV model (Yang *et al* 2014). Cell motility and alterations in cellular stress and pressure within cell monolayers need to be investigated at least in 2D motility assays near the unjamming transition. Moreover, it needs to be assumed that the local velocity of the monolayer is very small above the transition, which can be modeled by particle image velocimetry (PIV). The proliferation of cells causes an increase in cell density in confluent tissues and hence needs to be addressed, which is associated with a decrease in individual cell motility v_0, through contact inhibition of movement. If proliferation is dominant and alterations of the ratio between A_0 and P_0 are negligible, proliferation is supposed to facilitate the unjamming to jamming transition (Deforet *et al* 2014). However, this needs to be refined precisely. In tissues with low v_0 values at all times (Puliafito *et al* 2012), the SPV model predicts that the proliferation can cause both jamming or unjamming, which relies on the orientation of cell divisions decreasing or increasing the cell shape anisotropy (Bi *et al* 2016).

The spatial correlations and fluctuations of the cell displacement field such as swirl sizes (Poujade *et al* 2007, Angelini *et al* 2010, 2011, Petitjean *et al* 2010, Garcia *et al* 2015) are predicted to increase when a tissue is close to the glass transition from the fluid side. In line with this, based on a particle-based model this prediction has been verified for a single cell type (Garcia *et al* 2015). The SPV model tracks cell centers and hence has two degrees of freedom for each cell and in the limit of zero cell motility it performs the same rigidity transition as the vertex model that has two degrees of freedom per vertex (Bi *et al* 2016). An 'active vertex model' has been tested, in which active motile forces are included in the vertex model vertices and thereby it still shows a robust glass transition defined by the well-known shape order parameter q. Hence a deeper universality has been suggested and needs to be tested. Why do confluent models, such as vertex and SPV models and particle-based models such as Lennard-Jones glasses and SPP models, lead to differences?

In SPP models it has been hypothesized that the location of the glass transition packing density ϕG at zero motility (which is the density with dynamics ceasing in the limit of $v_0 \rightarrow 0$) determines the value of noise, D_r (Fily and Marchetti 2012, Berthier 2014). Moreover, the jamming and glass transition are not regulated by the same critical point in nonactive systems (Ikeda *et al* 2012, Olsson and Teitel 2013), which is not true for the SPV model, as the glass transition point p_0* shifts with D_r at finite values of v_0, in the limit of nearly zero motility, all glass transition lines merge into a single point in the limit $v_0 \rightarrow 0$ that is $p_0* = 3.81$. Hence SPV models seem to be more appropriate for various biological tissues. SPP models are based on two-body interactions depending only on particle center positions, whereas instead both SPV and vertex models naturally incorporate contractility, which is a key property of living cells, and include the inherently multibody nature of cell–cell forces dependent on shape deformations. Dissimilar to equilibrium vertex models, SPV models include cell motility and they are easier to simulate in 3D.

A confluent cell can be modeled as a Voronoi tessellation of the plane (Li and Sun 2015). An important difference of the Voronoi model is that cell–cell adhesion included through a potential which is quadratic in the distance between cell centers (Li and Sun 2015) is similar to particle models. Stronger cell–cell adhesion seems to be associated with stiffening of the tissue, which is a common feature for particle-based models. Can the Voronoi tessellation be used in active systems? Contrary in SPV models, adhesion is included through coupling of shape energy to cell perimeter. In particular, enhanced cell–cell adhesion (or reduced cortical tension) causes a larger value of p_0 and renders the tissue to be more compliable and hence softer. Other shape-based models of confluent tissue dynamics will lead similarly to the glass transition such as the cellular Potts model (Szabo *et al* 2010, Kabla 2012, Durand and Guesnet 2016) and the modified SPP model (Garcia *et al* 2015), when the cell's motile force is decreased below a certain threshold, the motion of cells transit from diffusive to sub-diffusive motion, which is similar to the glass transition in the SPV model by decreasing the value of v_0.

In the vertex model, cell volume is usually assumed to be fixed, which is appropriate in developmental systems such as *Drosophila* (Gelbart *et al* 2012, Kasza *et al* 2014) and zebrafish (Manning *et al* 2014), whereas epithelial tissue cells can display volume fluctuations (Zehnder *et al* 2015a, 2015b). Thus, volume fluctuations need to be incorporated in the vertex model or the SPV model, as these alterations in the cell volume represent another source of active shape fluctuation and hence may lead to jamming or unjamming of the tissue locally and pronouncedly shift the overall location of the rigidity and glass transitions. In the SPV model, cell polarity proposed to be regulated by simple rotational white noise (Bi *et al* 2016). Even more complex mechanisms for cell polarity may be included such as external chemical or mechanical cues by coupling v_0 and $\hat{n}_i$ to chemoattractant or mechanical gradients and allowing waves or other pattern formation mechanisms to interact with the jamming transition (Fredberg and Trepat 2012). Similarly, simple alignment rules (the so-called Vicsek model) may evoke collective flocking modes regulating glassy dynamics (Vicsek *et al* 1995). As mentioned before, the inclusion of cell–cell friction that regulates collective dynamics in particle-based tissue models seems to be a suitable extension of the SPV model (Garcia *et al* 2015). The SPV model includes viscous frictional coupling between the cell to the 2D substrate and cell–cell adhesion and thereby enters as a negative line tension on interfaces (Bi *et al* 2016). A frictional force between cells proportional to the length of the edge shared two cells can be integrated into the SPV model in analogy to particulate glasses, where these local friction forces can alter the location of the jamming or glass transition as well as the nature of spatial correlations within a glass (Henkes *et al* 2010, Silbert 2010). How are the unjamming transition captured by the SPV model and the EMT related in predicting the cell spread out of a primary solid tumor mass into the surrounding healthy tissue environment? Indeed, the EMT involves pronounced alterations in cell–cell adhesion and cytoskeletal composition, which are inducing alterations in cell shape and cell motility. Hence, it is supposed that the spread of cells from the primary tumor mass is not simply driven by an enzymatic breakdown of the basement membrane. It seems to be the case that the spread of the cancer cells from the primary tumor mass is facilitated through distinct alterations of mechanical properties of both individual cancer cells and the surrounding tumor stroma (Wong *et al* 2014). Subsequently, the collective unjamming revealed the initial step towards essential mechanical alterations for cancer cell spreading out of primary solid tumors.

As primary tumors and cancer cell lines are known to be heterogeneous regarding their genomic and proteomic phenotype, it has been shown for cancer cell lines that there exists a mechanically heterogeneous phenotype containing subpopulations of stiff and soft cells with various degrees of active contractility (Mierke *et al* 2011a, Mierke *et al* 2011b, Mierke 2013a). The jamming phase diagram predicts that soft cells exhibit mesenchymal phenotypes, seem to have higher values of p_0 and hence are supposed to unjam and migrate to primary tumor boundaries more effectively than their stiff cells. Therefore, the analysis of the effects of tissue heterogeneity on the distributions of cellular rigidity and migration speeds seems to be useful for defining novel tumor invasion strategies and subsequently cancer metastasis.

Jamming behavior in healthy and asthmatic airway epithelium

The ideas of the jamming concept can be revisited for asthmatic airway epithelium. How can this jamming framework and the jamming phase diagram help to reveal the pathogenesis of asthma? In particular, asthma involves epithelial injury repair responses, which appear mostly aberrant and dysregulated (Lambrecht and Hammad 2012). Among these dysregulated processes are nuclear factor κB (NF-κB), sonic hedgehog (shh) and Wnt, or even growth-factor signaling pathways (Knight *et al* 2004, Holgate 2008). During fetal lung development, all these factors contribute to the activation of the epithelial–mesenchymal trophic unit (EMTU), which consist of opposing layers of epithelial and mesenchymal cells (Evans *et al* 1999). In asthma disease, the EMTU becomes reactivated and seems to be partly responsible for progressive diminishing of lung function and structural remodeling of the lung wall, promoting goblet-cell hyperplasia, thickening of the basement membrane, sub-epithelial angiogenesis and airway smooth muscle hypertrophy as well as hyperplasia (Roche *et al* 1989, Huber and Koessler 1922, Lazaar and Panettier 2003, Homer and Elias 2005, Durrani *et al* 2011). Dysregulations of these processes are evoked by a cascade of immune and inflammatory events such as type 2 inflammation. Hence most current asthma based on type 2 inflammation therapies may not be suitable. In particular, although inflammation might be a major disease modifier in asthma, it may not be the core abnormality (Fahy 2015), which may be structural abnormalities within the lung cell, such as the airway smooth muscle cell or the airway epithelial cell. In line with this, it has been found that chronic exposure to low doses of endotoxin even protects against allergic asthma by induction of the protein A20 in airway epithelial cells, which indeed shows an essential and key role of the airway epithelium in the pathogenesis of asthma (Schuijs *et al* 2015). Additionally, the airway epithelium is important in aberrant airway remodeling by purely mechanical events such as bronchoconstriction, in which the airway epithelium is compressed and hence buckled into a rosette pattern. Indeed, these compressive stresses have the capacity to induce a cascade of mechanotransduction processes regulating the airway remodeling in the absence of 'classical' airway inflammation (Tschumperlin *et al* 2004, Grainge *et al* 2011, Park *et al* 2015a).

Could cell jamming be an unappreciated factor? If yes, is the cell jamming behavior in some innate or in some systemic ways different in cell layers from nonasthmatic and asthmatic donors? Indeed, the dynamic and structural signatures of the jamming transition have been determined in human bronchial epithelial cells (HBECs) from nonasthmatic and asthmatic donors (Park *et al* 2015b). The differentiation and repair processes that occur in the maturing or injured airway epithelium *in vivo* are modeled using *in vitro* cell culture models with an air–liquid interface (ALI) to mimic well-differentiated pseudostratified HBEC layers (Whitcutt *et al* 1988, Dvorak *et al* 2011). As these HBEC layers mature and differentiate, the speed maps of regional cellular motions show a transition from a mobile fluid-like and unjammed phase to an immobile solid-like and jammed phase. The jamming transition has been observed in cells from nonasthmatic donors between 6 and 8 days of the ALI culture, whereas the jamming transition in cells from asthmatic donors is delayed until day 14 (Dvorak *et al* 2011). When the mature and fully jammed nonasthmatic HBEC layer is mechanically compressed (similarly as in bronchoconstriction) during asthma, the layer is immediately unjammed and rejams very slowly over 36–48 h. Hence, in cells derived from nonasthmatic and asthmatic donors, the jammed phase corresponds to a mature and quiescent layer with full barrier function (Park *et al* 2015b). The unjammed phase is related to the immature and active layer, as well as the mature layer under but mechanical compression and the delayed transition to jamming in asthmatic donor cells. Indeed, the vertex model propose that the measured cell shape parameter q approaches $p_0^* \approx 3.81$, when the cell layer is close to jamming. In HBEC layers from nonasthmatic donors, the median q reaches the value of p_0^* on day 8, when the layer jams, whereas in asthmatic donors, p_0^* is reached on day 14, when the layer jams. In addition, there exists a substantial degree of variability in the q between individual cells, but the origin of the variability is elusive, but can reflect cell-to-cell variability or dynamical heterogeneity. Subsequently, in the jammed layer that gets unjammed by compression, q increases substantially in response to compression, when it unjams (Park *et al* 2015b). Finally, these results confirm the prediction of a structural signature of the layer jamming and provide an explanation for the delay of the jamming in asthmatic cells.

Are EMT and UJT both real?

The EMT is the change of a polarized epithelial cell within a cell layer to a mesenchymal phenotype that is associated with increased production of extracellular matrix, resistance to apoptosis, enhanced migratory capacity and increased invasiveness into tissues (Hay 1995). The reverse process is termed mesenchymal-to-epithelial transition (MET) (Kalluri and Weinberg 2009). Biological markers of EMT are extensively studied, such as the enhanced activity of transcriptional factors including TWIST, SNAI and ZEB families, the decrease of epithelial cell junction proteins such as E-cadherin and the induction of mesenchymal proteins such as vimentin (Thiery *et al* 2009). As the mesenchymal phenotype is required for the initiation of cell migration and invasion of epithelial cells, the EMT has been associated with invasion of carcinoma cells and cancer metastases (Pattabiraman *et al* 2016). Hence, it has been proposed that the EMT is dispensable for carcinoma cell invasion and metastasis (Fischer *et al* 2015, Zheng *et al* 2015). In more detail, in genetically engineered mouse models of pancreatic ductal adenocarcinoma, a loss of either Twist1 or Snail does not affect the progression of cancer nor metastasis formation in the lung and liver (Zheng *et al* 2015). The overexpression of miR-200 impairing Zebl and Zeb2 expressions, and thus restricting cells to maintain the epithelial phenotype, does not affect metastasis in lungs within murine models (Fischer *et al* 2015). In conclusion, EMT increases chemoresistance, whereas it has no effect on cellular motility or the metastatic potential. In addition, overexpression of Twist1 promotes the dissemination of breast cancer cells and thereby needs the expression of E-cadherin (Shamir *et al* 2014). Hence, cancer cells can adapt migratory dynamics and invasion, the EMT seems not to be required. Hence the EMT and the unjamming transition (UJT) are seen as alternative ways for cell migration, although both start from a quiescent, polarized and epithelial state. As the morphological and molecular requirements for the EMT are high, those for the UJT may be less high and hence are preferred. However, whether and to what extent the UJT is independent of the EMT or partial EMT still remains to be fully understood (Tam and Weinberg 2013).

Critical view on UJT

Does the concept of cell jamming and unjamming help us? For airway epithelium in asthma patients, the time-scale (days or weeks) needed to jam the maturing HBEC layer *in vitro* seem to correlate with the period of recovery from an asthmatic exacerbation *in vivo* (Petheram *et al* 1979, Jenkins *et al* 1981). A mechanistic role for rejamming in the recovery process or for unjamming (or delayed jamming) in an asthmatic exacerbation cannot be provided. Hence, it is still not clear whether rejamming, unjamming or delayed jamming can be precipitating factors. Some questions remain open. Are the jamming dynamics observed in HBECs *in vitro* (Park *et al* 2015b) similarly present in the bronchial epithelial layer *in vivo*? If yes, do the alterations in jamming dynamics between HBECs in asthmatic versus nonasthmatic patients drive asthma pathogenesis by defining a core hallmark of abnormality in asthma (Fahy 2015) or are they simply an effect? The jammed phase describes a quiescent, stable and homeostatic state of mature healthy epithelium, in which the cell layer is not motile and hence solid-like, whereas the unjammed phase is a dynamic state of injured epithelium, in which the layer is motile and hence fluid-like.

In carcinomas, most metastases are based on circulating multicellular tumor cell clusters (CTCs) and retain most often an epithelial phenotype during metastases formation (Aceto *et al* 2014, Fischer *et al* 2015, Zheng *et al* 2015). The transbarrier crossing of CTCs and their migration from the primary solid tumor to the distant metastatic niches seems to require the maintenance of an epithelial phenotype (Cheung and Ewald 2016). The existence of epithelial phenotype throughout all stages of the metastasis abolishes the concept of EMT. What is the underlying alternative physical process for the initiation and support of migration? Cell jamming has been found *in vivo* using murine cancer models (Haeger *et al* 2014). Can the three following (sub)hypotheses be experimentally confirmed?

First, at the primary tumor site, the migration of a leader cell and following cells through surrounding solid tissues is facilitated by the transition of the cells from a solid-like jammed phase to a fluid-like unjammed phase. Second, when the vascular system is entered, CTCs are formed and are exposed to destabilizing mechanical effects, such as blood-borne shear stresses, and thereby the CTCs seem to stabilize themselves by undergoing a re-transition to a solid-like jammed phase. Third, when the CTCs found their targeted metastatic site, they unjam (Au *et al* 2016) to induce collective cellular invasion into noncancerous solid tissue. The process of metastasis has been hypothesized to utilize EMT and its inverse MET and hence this hypothesis can be compared with the alternative hypothesis that metastasis requires that events are driven by less drastic transitions between solid-like (jammed) and fluid-like (unjammed) phases of the cell layer, during which strong collectivity and an epithelial phenotype are maintained. It is therefore interesting that counterintuitively, the phenomenon of enhanced cell–cell adhesion induces unjamming and vice versa.

Each axis of the jamming phase diagram is a determinant of the jamming transition, whereas it is simultaneously altered by multiple underlying molecular and physical factors, which in turn may alter other axes. Based on this crosstalk and even associated pleiotropic effects, it can be proposed that there is no single key regulatory molecule for cell jamming or unjamming. In particular, all axes of the jamming phase diagram need to be taken into account. Additionally, the effects of friction between cells and the contact-driven inhibition of locomotion within a jamming phase diagram have yet to be ruled out precisely (Garcia *et al* 2015, Zimmermann *et al* 2016). In addition to bioinformatics approaches, how cell jamming in epithelial cell layers is controlled needs to be figured out and hence key molecules are required to be identified. Moreover, we need to determine the extent to which the quiescent epithelial layer is jammed and whether the unjamming of this layer and hence the collective migration of the cells are prerequisites for physiological processes such as pattern formation, tissue remodeling and wound repair. Under these processes, an individual cell of the unjammed collective is regulated cooperatively through chemical and mechanical cues from its cell neighbors. During pathophysiological processes such as aberrant repair of the asthmatic airway wall or malignant progression of cancer such as metastasis, unjamming may be seen as a malfunction, such as the breakdown of cooperative agreements that individual cells adopt upon their initial assembling into a multicellular collective during organogenesises. From an evolutionary and biological point of-view, it has been supposed that these agreements must be conserved and also deeply rooted in generic physical processes expressed by inert soft matter and only utilized by earliest multicellular living soft matter (Newman 2012). Among these generic physical processes is the jamming transition between the fluid-like and solid-like phases of cellular collectives (Garrahan 2011, Zhou *et al* 2013, Fredberg 2014). Finally, when cell jamming as a biological mechanism has early evolutionary roots, the dynamics of cell signal transduction and cell jamming seem to be intertwined in a manner such that both need to be considered for understanding the entire mechanism.

The jamming phase diagram

As jamming phase diagrams are known from inert soft matter (Liu *et al* 1995, Trappe *et al* 2001), these effects may also play a role in the living systems within a supposed jamming phase diagram (figure 1.21). In particular, we have the first axis for cellular crowding (expressing the inverse as the reciprocal of cellular density with infinite density in the origin). On the second axis we have the cell–cell adhesion (expressed as a reciprocal with infinitely sticky cells mapped to the origin). On the third axis we display the effects of cell motile forces. Additional axes may be possible, for example for imposed stretch or shear loading (Trepat *et al* 2007, Krishnan *et al* 2009), cellular volume (Zhou *et al* 2009), cellular stiffness (Mattsson *et al* 2009) and substrate stiffness (Krishnan *et al* 2011, Angelini *et al* 2010), but often they are not available. In a phase diagram with multidimensional space, the origin, and regions near the origin, are jammed and rearrangements are not possible as each cell is fully caged (confined) by its neighbors, or even glued to its neighbors, or it cannot generate the necessary driving motile force. However, away from the origin, especially along certain trajectories, structural rearrangements are possible with increasing probability. Unjamming can take place if cells are stretched, undergo apoptosis or are extruded from the monolayer; these all lead to a decrease in cell density (Eisenhoffer *et al* 2012). In line with this, if adhesive interactions are weak enough, or if the stretch becomes large enough in order to rupture cell–cell adhesions, the unjamming mode of the cellular system is preferred. Unjamming is present if motile forces are strong enough to pull individual cells away and dissociate from neighboring cells to reach a loose and disaggregated system.

An example is MCF-10A human breast cancer cells that overexpress ErbB2, which promotes proliferation and cell crowding, pushing the system towards a glassy and jammed state, whereas the overexpression of the potential tumor-associated and L1-CAM interacting protein 14-3-3ζ degrades cell–cell junctions, moving the system out of the jamming state towards a fluidized and unjammed state. Another example is hepatocyte growth factor (HGF) facilitated scattering of MDCK cells leading to the disruption of cadherin-dependent cell–cell adherence junctions depending upon integrin adhesion and the phosphorylation of the myosin regulatory light chain (MRLC) (de Rooij *et al* 2005). All these findings fit perfectly into the jamming phase diagram.

Uniting biology and physics

The binary alternatives of jammed and unjammed states are not the only possibilities. In inert systems the jamming phase diagram includes fragile intermediate states with possibly no less relevance to the biology of the monolayer (Bi *et al* 2011, Vitelli and van Hecke 2011). The jamming hypothesis states that specific events at the molecular scale necessarily modulate and respond to cooperative heterogeneities caused by jamming at a much larger scale of organization, but specific events at the molecular scale can never by themselves explain these cooperative large-scale events. In this physical view the specific molecular events are ignored or they belong to an integrative framework that was unanticipated and overwritten.

This leads to novel questions that have not been asked before. Does an epithelial monolayer build a solid-like aggregated sheet displaying excellent barrier function with little possibility of cell invasion or escape due to the phenomenon that the constituent cells are jammed (Eisenhoffer *et al* 2012)? Do certain specific cell populations become fluid-like and hence permissive of paracellular leakage, transformation, cell escape or cell invasion due to the phenomenon that they become unjammed? Do the formation of pattern and wound healing processes require that the cells are in an unjammed state (Serra-Picamal *et al* 2012)? If the answer is yes, what represents the critical physical threshold? What are the signaling events at the level of gene expression and signaling and the resulting physical changes that favor or hinder the jamming of cells? One possibility may be force-dependent thresholds and novel pathways that regulate cellular polarization (Prager-Khoutorsky *et al* 2011), but can these thresholds and pathways work in collective processes? However, trials targeting adhesion molecules in order to decrease tumor progression have been reported to be ineffective, indicating that migration events are somehow reprogrammed to maintain invasiveness through morphological and functional dedifferentiation (Friedl and Wolf 2003, Friedl *et al* 2004, Rice 2012). Does jamming allow certain cancer cell subpopulations to unjam, awake from dormancy and thus evolve so as to maintain invasiveness by selection for adhesive interaction, compressive stress and cyclic deformation? Each of these complex questions cannot be answered solely by physics or biology. One possibility seems to be to combine these approaches to answer these important questions and reach new horizons in cancer research by incorporating the physics of cancer.

1.2.2 Single-cell migration of cancer cells into the microenvironment

The motility of cells in three-dimensional (3D) extracellular matrices is a prerequisite for tissue assembly and regeneration, immune cell trafficking and diseases such as cancer. In more detail, the process of metastasis depends on the migration of single cancer cells that migrate out of the primary tumor. The migration of cancer cells through the extracellular matrix of connective tissue is a cyclic process comprising multiple steps, including: (1) the actin polymerization-dependent protrusion of pseudopods at the leading edge; (2) the integrin-mediated adhesion to the extracellular matrix; (3) the contact-dependent degradation of the extracellular matrix evoked by its cleavage through cell-surface proteases; (4) the actomyosin-facilitated contraction of the cell's body, increasing longitudinal tension; and (5) the retraction of the cell's rear part followed by the translocation of the whole cell body (figure 1.23) (Doyle *et al* 2013). All these steps describe only one specific mode of migration that can be chosen by metastatic cancer cells: the protrusive mode (Maruthamuthu and Gardel 2014). However, there are still other modes of cancer cell migration, such as the blebbing mode (Laser-Azogui *et al* 2014). Which mode of migration is favored by a certain type of cancer cell and how the choice of mode is altered by the specific microenvironmental conditions are still under discussion. In addition, there is an intermediate lobopodial migration mode, so far only detected for fibroblasts, and it has been suggested that cancer cells also choose this under certain circumstances. Hence, further investigation is required to determine whether cancer cells are able to use this fibroblastoid lobopodial mode.

The ability of cancers to metastasize depends on the cell's ability to migrate to and invade connective tissue, adhere, and possibly transmigrate through a barrier such as basal membrane and the endothelium. However, what determines a special mode of invasion and how the appearance or the switch between the different modes is regulated is not yet understood. In more detail, the invasion mode is supposed to play an important role for the regulation of the basement membrane or endothelial barrier-crossing transmigration potential of cancer cells and has a major impact on their invasion speed.

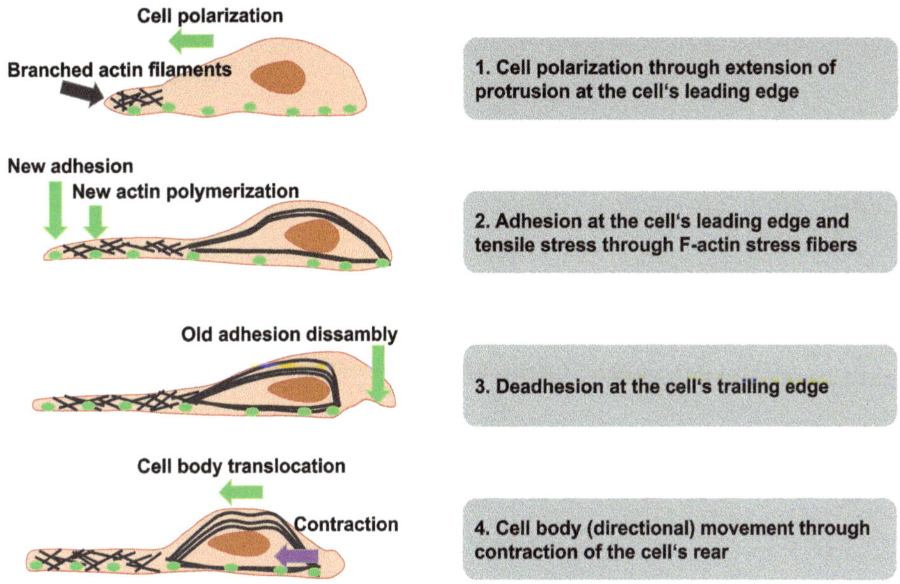

Figure 1.23. 2D migration model.

How powerfully a migrating and invading cell overcomes the different obstacles found in dense 3D matrices depends strongly on its mechanical properties and how it is able to generate and transmit its protrusive forces. Hence forces and material properties determine which mode of invasive cell migration is favored for a special cancer cell type or cancer cell subpopulation. Indeed, it has been established that cancer cells with certain mechanical properties such as contractile force transmission and generation are able to invade 3D extracellular matrices more efficiently than less contractile cancer cells (Mierke *et al* 2008a, Mierke *et al* 2011). Nonetheless, what kind of migration is chosen when cells try to squeeze through narrower spaces, such as vascular endothelial-cell walls? In this special case of transmigration blebbing motion may be more favorable. Additionally, in preliminary experiments it has been found that the stiffness of the plasma membrane drastically softens in primary human mamma and human cervix carcinoma cells, favoring the blebbing process of cell migration (figure 1.24). Thus, one may hypothesize that the mechanical properties and the type of force generation of cancer cells determine their invasion mode and may also regulate the switch between the different migration modes. The following questions remain unanswered. What are the major mechanisms that regulate the invasion mode of cancer cells? What role do microenvironmental properties such as the mechanics and structure of the extracellular matrix play regarding the migration mode of cancer cells? In order to investigate this, one needs to dissect the crosstalk between the invasion modes and the environmental confinements such as mesh or pore size, stiffness of the entire cell, the plasma membrane and the extracellular matrix, and the proteomics of the cellular adhesion machinery as well as the extracellular protein composition.

To understand the interaction between invading cancer cells and their micro-environmental confinement and why certain cancer cells use a particular invasion mode such as blebbing or protrusive modes, the following major questions should be

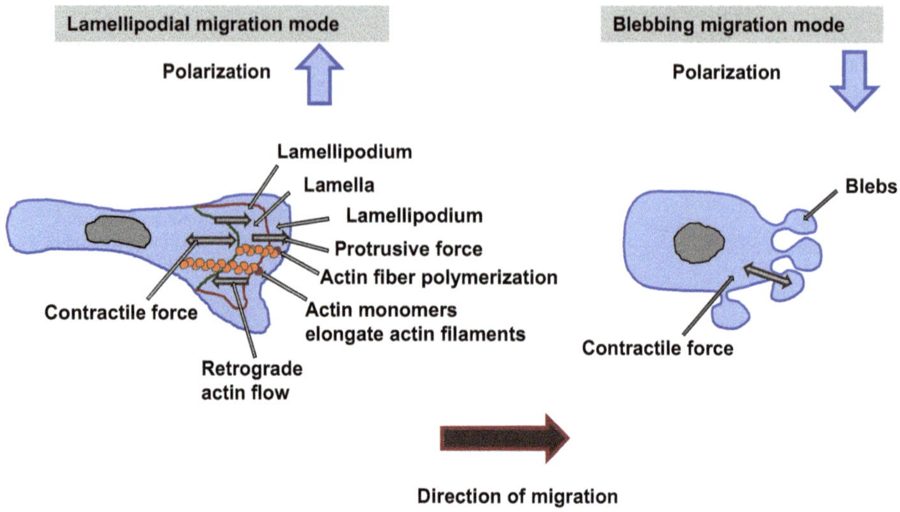

Figure 1.24. Modes of migration in 3D.

answered. What roles do the influence of cytoskeletal stiffness and cell contractility play regarding the invasive cell motility? To what extent do these factors favor either the protrusive or blebbing-based migration modes of cancer cells? How does the stiffness of the plasma membrane impact on the cellular invasive behavior, such as protrusive or blebbing-based motion? How strongly do adhesive cancer cells differ from weakly adhesive cells with respect to their preferred invasive motility modes? Is the blebbing mode of invasion preferred by small mesh sizes of the microenvironmental confinement, such as the connective tissue matrix scaffold, whereas the protrusive mode is supported by large mesh sizes? Can the blebbing or protrusive modes of cancer cells support transmigration through barriers such as vascular endothelial-cell linings and basal membranes?

Knowing the answers will contribute significantly to the understanding of how cancer cells utilize a certain invasion mode in order to migrate through the 3D microenvironment and what role the material properties of cancer cells and their microenvironments play. Moreover, these data will help to reveal the respective contributions of the mechanical properties of cancer cells and their microenvironments in supporting the invasive behavior of epithelial-derived carcinomas, in particular metastasis.

Malignant cancer progression involves the process of metastasis, which makes it a systemic disease that leads to death. The complex process of metastasis is composed of many steps, which follow a linear propagation. The metastatic cascade can be delayed by stopping at special steps in order to start again after some relapse time in which the aggressive cancer cells are dormant. How this phenomenon is regulated or induced is not yet well understood and thus needs further investigation. The metastatic cascade begins with the spreading of cancer cells from the primary tumor (dissemination), which migrate into the surrounding tumor microenvironment that is locally 'transformed' by the primary tumor (Bizzarri and Cucina 2014). Then these cancer cells transmigrate into blood or lymph vessels (intravasation) through the basal membrane and endothelial barriers, they are transported through the vessel flow, adhere to the endothelial-cell lining, grow and build up a secondary tumor (Al-Mehdi *et al* 2000). Thereby the new tumor is initiated either in the blood or lymph vessel, or cancer cells possibly transmigrate through the endothelial vessel lining (extravasation) into the extracellular matrix of the connective tissue. In the latter case, these cancer cells then migrate further into the targeted tissue, divide and form a secondary tumor, which means that the primary tumor has metastasized.

During the last two decades, the typical migration modes, such as mesenchymal and amoeboid motion, have been investigated (Friedl and Wolf 2003, Wolf *et al* 2003a, Taddei *et al* 2014). To date, these migration modes have not been clearly defined. Although these migration modes are reported to be a mechanistically well-described concept, they are solely a morphological description of the migration mode rather than an exactly defined and hence distinguishable invasion mode (Laemmermann and Sixt 2009). This needs to be clarified in order to compare the migration modes and to define their occurrence and regulation. In particular, we will analyze whether a cancer cell prefers a certain mode of invasion and which mechanical phenotype defines this mode exactly and makes it distinguishable from the other invasion modes. The term amoeboid is not precisely defined, as migrating cells with roundish shape are categorized as amoeboid without analyzing the mechanistic aspects of their cytoskeletal remodeling dynamics and their mechanical forces. In line with this, a definition of amoeboid migration has been based on the cell morphology, adhesiveness and proteolytic remodeling of the microenvironment for the interstitial migration of leukocytes (Wolf *et al* 2003b, Sabeh *et al* 2009). Proteolytic degradation is linked to the amoeboid migration mode depending on cellular deformability (Sabeh *et al* 2004, Wolf *et al* 2007, Rowe and Weiss 2009). In addition, proteolytic degradation is also associated with interstitial invasion of mesenchymal cells into collagen-rich and hence dense 3D extracellular matrices, and these mesenchymal cells migrate by using proteolytic and non-proteolytic degradation through narrow constrictions. Important questions remain unanswered. How do cellular forces such as contractile or protrusive forces facilitate the switch between the different invasion modes? How does the definition of protrusive or blebbing-based invasion modes fit into this scenario?

In previous studies the mechanical analysis was centered solely on actomyosin mechanics, whereas regulatory aspects of the actomyosin networks or the mechanical impact of other cytoskeletal elements such as actin-crosslinkers, or intermediate filaments such as vimentin are still elusive (Brown *et al* 2001, Laevsky and Knecht 2003, Tooley *et al* 2008, Xu *et al* 2005). In particular, a connection between the actomyosin cytoskeleton and the intermediate cytoskeleton has been proposed, but the proteins mediating this linkage have not yet been discovered (Seltmann *et al* 2013). However, understanding the regulatory function of these cytoskeletal factors in well-defined migration models will be necessary and will encourage understanding of the diverse mechanics involved in the different migration modes. There are still many open questions that should be answered. Which types of physical invasion strategies can cancer cells utilize and are they indeed derived from leukemias or fibroblasts?

How do cancer cells facilitate the switch between different invasion modes, such as the protrusive and blebbing-based modes? How fast is this switch and is it permanent? How are these two 'novel' invasion modes related to the classical epithelial–mesenchymal transition that supports the invasive behavior of cancer cells?

Despite these open questions regarding the cellular migration of cancer cells, a general conceptual model has been developed for how a lamellipodium can support protrusive cell motility. In more detail, actin filaments polymerizing below the leading edge of the cell membrane generate a pushing force (like a thermal ratchet) towards the cell membrane required for the formation of cellular protrusions. The surface tension of the cell membrane is able to oppose the free anterograde expansion (outward) of the actin network and thus actin filaments are then pushed back into the cytoplasm of the cell, which is detectable as a retrograde (inward) flow of actin (Ponti *et al* 2004, Gardel *et al* 2008, Mierke 2014). In more detail, cell–matrix adhesion receptors such as integrins connect the internal cytoskeleton to the external extracellular matrix through focal adhesions containing focal adhesion proteins such as vinculin (Mierke *et al* 2008b; Mierke *et al* 2010), focal adhesion kinase (Mierke 2013b), paxillin (Schaller 2001) and talin (Ziegler *et al* 2008, Das *et al* 2014). These adhesive contacts ensure that the retrograde-directed forces, which are enforced by actomyosin contraction, are transformed into outward locomotion of the cell's body in the migration direction (clutch hypothesis) (Vicente-Manzanares *et al* 2009). Taken together, the basic concept of the lamellipodial motion relies on cellular mechanics. In particular, the actin polymerization facilitates the formation of membrane protrusions. This particular migration mode of cancer cells is defined as the protrusive mode. Apart from these lamellipodium-supported outward forces, filopodia-like extensions are required for cell invasion, which are suggested to be more force sensing rather than force generating, as well as invadosomes, which are needed for 3D invasion of tissue barriers (Ridley 2011). Lamellipodial-driven motion has predominately been investigated on two-dimensional (2D) substrates (figure 1.24). Recent studies have found that many aspects differ significantly in three dimensions, supposing that the results obtained in 2D assays should be confirmed in 3D assays (Mierke *et al* 2010, Meyer *et al* 2012, Mierke 2013b).

One alternative way for cancer cells to migrate without protrusive forces is migration through membrane blebbing. Blebs are cellular spherical extensions that are instrumental for cell migration in developmental processes as well as diseases like cancer. These cellular blebs are anterior cellular extensions and devoid of actin filaments. For a long time, they have only been considered to be exclusively a hallmark of apoptosis (Mills *et al* 1998). In the last decade, they have become a hallmark for a special migration mode (Charras *et al* 2006, Charras and Paluch 2008, Paluch and Raz 2013). The phenomenon of blebbing has been seen in 2D (Yoshida and Soldati 2006) and 3D migration assays (Charras and Paluch 2008). In particular, the intracellular hydrostatic pressure generated by actomyosin contraction causes (1) the rupture of the actin cortex (Tinevez *et al* 2009) and/or (2) the rupture of the focal adhesion protein mediated linkage between the actin cytoskeleton and the cell membrane (Charras *et al* 2006). These two proposed mechanisms depend on the cellular mechanics. In particular, if the membrane loses its mechanical anchor to the extracellular matrix, the intracellular pressure facilitates the formation of a membrane bleb (Charras and Paluch 2008). These mechanisms are difficult to determine, as they may act in combination. The blebbing lifecycle is divided into three phases: initiation, growth and retraction. The membrane bleb grows until a new actin cortex is reassembled, which then may also contract, repeating the whole blebbing cycle (Charras and Paluch 2008). The site of the bleb initiation during migration has not yet been discovered. The cellular blebbing strategy of cells is physiologically relevant in motility, in which the cells migrate by a directed and persistent motion (Blaser *et al* 2006). Moreover, membrane blebbing may alter the surface tension of the cell membrane and subsequently the mechanical properties of the entire cell.

However, the precise mechanisms when cancer cells use either the protrusive or the blebbing-based invasion mode and how cancer cells can switch between these two are not clear. Furthermore, what impact does the microenvironmental confinement have on the determination of the invasion mode? What we know is that in tissue and cell cultures, several cells are able to switch between blebbing and protrusive motility due to microenvironmental confinements, in response to genetic alterations or pharmacological drugs (Laemmermann and Sixt 2009, Diz-Munoz *et al* 2010, Poincloux *et al* 2011). A clear definition of the individual conditions for the switch between these two invasion modes has not yet been postulated.

However, the key points for analysis of the blebbing migration compared to the protrusive mode are the effect of the cell–matrix adhesion, matrix geometry, matrix mechanics and the regulation of the signaling processes between the front and the rear of a motile cell. For both motility modes, the blebbing and the protrusive mode of migration depend on stabilized cell–cell or cell–matrix adhesions (Renkawitz and Sixt 2010). The cell adhesion provides stability and has a broad physiological importance, but seems to be significantly diminished in fast-migrating amoeboid cells such as immune cells, which are able to migrate through the tissue in the blebbing and in the protrusive migration mode (called synonymously lamellipodial mode) (Laemmermann and Sixt 2009). For example, the slow mesenchymal movement of cells through the extracellular matrix, which relies completely on focalized cell–matrix adhesions, is a solely protrusive migration mode.

Another important feature of cellular motility is the dimensionality of the surrounding microenvironment in which the cells will migrate; this is supposed to have a broad impact on the motility of cells (Mierke *et al* 2010, Rubashkin *et al* 2014), and in particular, the blebbing and protrusive migration modes may occur on 2D substrates and in 3D microenvironments. A main difference that has been reported between 2D and 3D motility assays is that 3D, but not 2D, microenvironments support cellular motility and invasion under minimal adhesion forces and thus decreased adhesion strength (Friedl and Wolf 2010). Moreover, how cellular polarity supports cellular motility in 2D and 3D migration systems and whether cellular polarity is exclusively associated with the protrusive migration mode are not yet known.

In addition, it has been found that cells migrate by forming cylindrical-shaped lobopodia with protrusions containing multiple tiny blebs at their tips, which might be an intermediate invasion mode between the blebbing and protrusive modes (figure 1.24) (Petrie *et al* 2012). Until now, this lobopodial mode has only been described for fibroblasts, whether cancer cells are also able to use this intermediate invasion mode has not been determined in a more precise way. The classical method to analyze cell migration was to use 2D migration systems. On these planar substrates the cells form flat lamellipodia, whereas when seeded into 3D collagen matrices of noncross-linked bovine collagen type I these cells exert some kind of 3D lamellipodia, the so-called invadopodia. In several reports invadopodia have been associated with the proteolytic degradation of the extracellular matrix (Destaing *et al* 2011, Linder *et al* 2011). However, if the collagen fibers are crosslinked, the cells start to migrate further in a lobopodial mode, in which they sense the mechanical properties of the surrounding microenvironment, such as the elastic properties of the matrix's scaffold. In particular, if the 3D matrix stiffness is low then the cells utilize a lobopodial mode of invasion, whereas the actomyosin contraction causes stress hardening of the extracellular matrix that may automatically activate the blebbing mode, where cells need to squeeze through the pores of the extracellular matrix. Finally, it can be summarized that the stiffness of the local microenvironment is not the major driving factor in lobopodial-based migration. It seems that it is rather the shape of the stress–strain curve that is the driving factor, as lobopodia are only formed in linearly elastic microenvironments such as the skin and the cell-derived matrix, and not in noncross-linked 3D collagen matrices showing strain stiffening, where the formation of lamellipodia is promoted (protrusive invasion mode) (Petrie *et al* 2012). These findings are in contrast to the hypothesis that matrix stiffness increases lamellipodia/invadopodia-driven cancer cell invasion (Levental *et al* 2009, Alcaraz *et al* 2011, Pathak and Kumar 2013, Mouw *et al* 2014). There are still open questions, for instance regarding the stabilized functional polarity of cells that perform persistent directional migration, and how cells manage to sense linear elasticity and distinguish it from pure strain stiffening. How is the extracellular proteolysis of the matrix confinement connected or associated with lobopodial migration? Do cells migrating in a lobopodial mode secrete extracellular matrix proteins such as fibronectin to remodel their microenvironment? Can cells other than fibroblasts, for example cancer cells, use this lobopodial migration?

However, matrix geometry has been suggested as a means to determine the invasion strategy of cancer cells (Tozluoglu *et al* 2013). In addition, matrix geometry, surface tension and cell–cell coupling through adherence junctions may play a major role in selecting the optimal and efficient migration mode under specific constraints (Maître *et al* 2012, Campinho *et al* 2013).

Another mode of migration is the degradation of the extracellular matrix

Another important mode of migration is the one in which the cancer cells degrade the surrounding extracellular matrix in order to migrate deeper into it (figure 1.25). What we have discussed up to now is that cancer invasion is a multistep process, which regulates the interplay between cancer cells and the underlying mechanotransduction signaling processes towards the surrounding extracellular matrix. This results in morphologically well-defined cancer invasion strategies. As refined by the microenvironment, in mesenchymal tumors such as melanoma and fibrosarcoma,

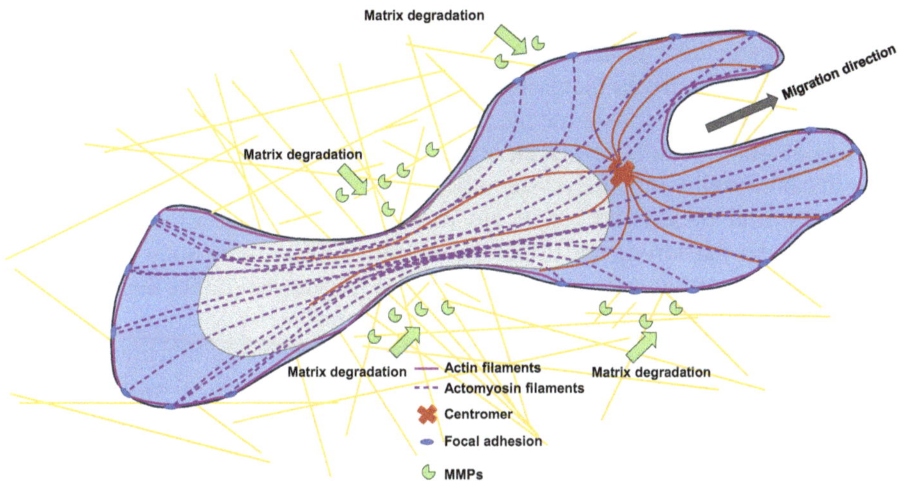

Figure 1.25. Matrix degradation.

either single-cell or collective invasion modes can be detected. What regulates such plasticity of mesenchymal invasion programs on the mechanical and molecular levels is still not well studied. Do tissue properties regulate the switch of invasion modes? To investigate this issue, spheroids of MV3 melanoma and HT1080 fibrosarcoma cells are embedded and cultured into 3D collagen fiber matrices of varying density and stiffness and analyzed for their migration type. In addition, the migration efficacy is analyzed with the MMP-dependent collagen degradation enabled or pharmacologically inhibited. It has been suggested that with increasing density of the collagen matrices, and dependent on the ability to proteolytically break down collagen fibers and provide track clearance, but independent of matrix stiffness, cells are able to switch from a single-cell to a collective mode of invasion (Haeger *et al* 2014). The conversion from single cell to collective invasion included the gain of cell-to-cell adherence junctions, supracellular polarization and possibly joint guidance along the pre-defined cellular migration tracks. In general, the network density of the extracellular matrix regulates the type of invasion mode used by mesenchymal cancer cells. It has been observed that fibrillar, high porosity extracellular matrices enable the dissemination of single cells, whereas dense extracellular matrices induce cell–cell interactions, a leader-to-follower cell behavior leading to collective migration with obligatory protease-degradable migration. Taken together, these results show that the plasticity of cancer invasion modes in response to the extracellular matrix properties such as matrix porosity and confinement plays an important role in providing the efficient invasion strategy, thereby recapitulating invasion patterns of mesenchymal tumors *in vivo*. In particular, the conversion from single to collective cell invasion due to increasing extracellular matrix confinement favors the concept of cell jamming as a guiding principle for the migration of certain cancer cell types such as melanoma and fibrosarcoma into dense connective tissue.

Even when migrating as a collective cluster of motile cells, each individual cell undergoing the cellular collective senses signals or gradients in order to mobilize physical forces. These forces themselves regulate local cellular motions from which the collective cellular migrations subsequently arise. However, these forces cannot explain spontaneous noisy fluctuations (even large ones). In particular, the implicit assumption is linear causality in which systematic local motions, on average, are caused by local forces, and these local forces are in turn evoked by local signals. A novel hypothesis states that strong and dominant mechanical events are not solely local. This indicates that the mechanical causality cascade is not so linear, and the fluctuations may not be seen as noise, they are rather an essential feature of this mechanism. Fluctuations and nonlocal cooperative events defining the cellular collective motion are described by the process of cell jamming. Cellular jamming can even unite several factors, including cellular crowding, force transmission, cadherin-dependent cell–cell adhesion, integrin-dependent cell–substrate adhesion, myosin-dependent motile forces and contractility, actin-dependent deformability, proliferation, compression and cell stretch, which have been reported, but still mostly act separately and independently.

1.2.3 Distinct features of the collective migration phenotype of cancer cells

During the process of cancer metastasis, the migration of single cells out of the primary tumor plays a role. However, in the malignant progression of cancer, cancer cells can also migrate collectively into the surrounding tissue microenvironment of primary tumors that causes the infiltration of the nearby tissue or organ with transformed cancer cells. When does collective migration and not individual single-cell migration play a role? What are the differences between collective and single migration of cells? The chemical composition and physical properties of the microenvironment regulate proliferation, differentiation and apoptosis of cells in the long term. In order to become motile and migrate through a 3D extracellular matrix environment, the cells need to adapt immediately to the physical properties of the environment. The interactions between cells and the extracellular matrix are facilitated by focal adhesions through the clustering and activation of integrins by fibrillary extracellular matrix proteins. As the cells in collective interact with their neighboring cells, intercellular adhesive structures such as tight junctions, cadherin-based adherence junctions and desmosomes. Indeed, cell–matrix and cell–cell adhesions facilitate mechanosensing the transduction of mechanical cues into biochemical signals (outside–in-signaling) and in turn the transmission of intracellular generated forces to the surrounding extracellular matrix (inside–out-signaling). Migrating cells sense through these adhesive structures the mechanical properties of their local environment, hence adapt in turn their own mechanical properties and exert forces needed for their movements in confinements representing a bidirectional response. The collective migration of cells is partly highly similar to the migration of single cells, but due to the connections with neighboring cells in a cluster, tissue sheet, spheroid or tumor it need to be taken into account that cell–cell adhesion strength and mechano-coupling between migrating cells are critical determents for migration speed and efficiency (de Pascalis and Etienne-Manneville 2017).

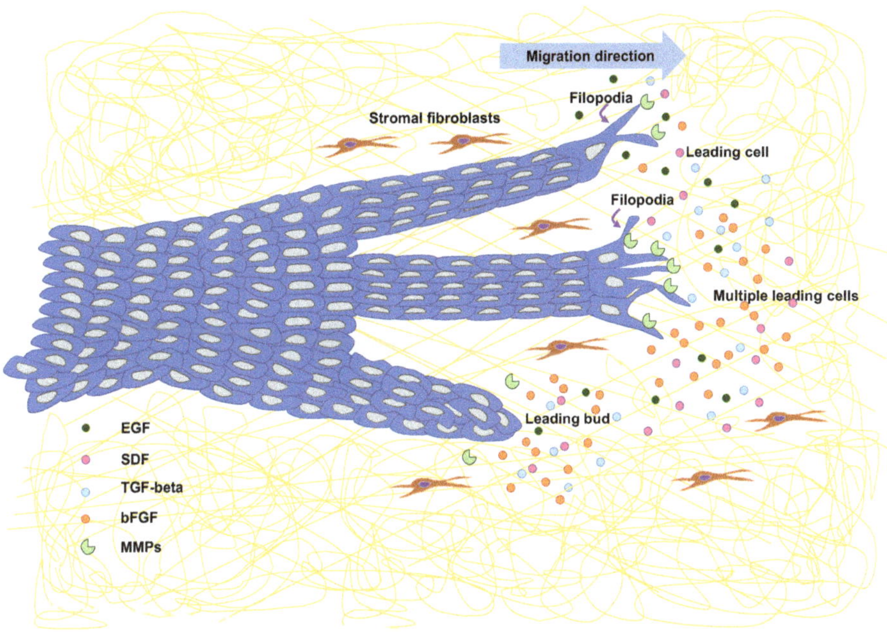

Figure 1.26. Collective migration. The three main types of collective invasion are cell cluster migration with one leader cell, multiple leader cells or tumor bud formation.

The collective migration is performed under physiological conditions such as developmental processes, shaping of tissues and wound healing processes. In particular, collective migration is the coordinated movement of cell groups, cell sheets or streams. Under pathological conditions, collective migration is crucial for the malignant progression of various tumors (Friedl *et al* 2004). Similar to the migration of single cells, collectively migrating cells can adapt to mechanical alterations. In addition, the collective migration is dependent on several parameters such as crowding (cell density), cohesion (adhesion strength) and constraints such as boundary conditions evoked by the extracellular matrix environment (Doxzen *et al* 2013). Finally, the cells sense and adapt to environmental cues including neighboring cells during the collective migration (figure 1.26).

Mechanical properties of collectively moving cellular groups
Similar to single-cell migration, collectively migrating cell groups sense the biochemical and physical properties of their microenvironment. Migrating collectives cannot still not be treated as a group of independent and single cells moving with the same speed and directional persistence. Moreover, the collective migratory behavior leads to an efficient migration that acquires distinct features (Mayor and Etienne-Manneville 2016). Based on the communication between neighboring migrating cells and the directional persistence of the movement of each individual cell depends on its immediate neighbors (Vicsek *et al* 1995, Szabo *et al* 2006). When cells migrate collectively, they are mechanically and chemically connected via cell–cell adherence

junctions and through the actin cytoskeleton (Mayor and Etienne-Manneville 2016). Hence, the cells affect the behavior of neighboring cells and alter the supracellular front–rear polarity of the migrating collective. Subsequently, a hierarchy inside the group has emerged by selecting a group of leaders sensing the mechanical and chemical cues of the environment and initiating collective cell migration. Leader cells alter following cells through the phenomenon of mechanical coupling. As cells within the migrating group additionally impact each other, there may occur the phenomenon of contact inhibition of locomotion (termed collective guidance) which will not be addressed here further (Theveneau *et al* 2013, Scarpa *et al* 2015, Scarpa and Mayor 2016, Zimmermann *et al* 2016). How can cells collectively adjust their forces? How can cells sense and transduce the mechanical properties of their neighbors? These questions still remain unanswered.

How are the forces transmitted in migrating collectives?
Using micropillars, the forces in collectively migrating Madin–Darby canine kidney (MDCK) cells have been determined. In particular, the tractions (with an average force of 5 nN) are mainly exerted at the cell front and perpendicular to edge of the monolayer. Thus, the migration is based on the pulling of the leader cells onto the substrate and not based on pushing of the follower cells (du Roure *et al* 2005). The strongest traction forces are applied at the leading edge, but in epithelial cell and endothelial monolayers, traction forces are still observed hundreds of micrometers within the monolayer (Raupach *et al* 2007, Trepat *et al* 2009, Tambe *et al* 2011, Serra-Picamal *et al* 2012). In contrast, using a silicone wrinkling, temperature-sensitive substrate it has been demonstrated that tractions are rather limited to the first row of leader cells in MDCK cell monolayers (Yokoyama *et al* 2016). However, cells far away from the leader cells exert smaller forces than the leader cells, whereas the tractions within the monolayer are heterogeneous as the cells are not synchronized their cell-cycle stage and hence tractions fluctuate continuously over time (Serra–Picamal *et al* 2012). Although leader cells provide biochemical and mechanical cues to following cells, cells inside the monolayer can even slow down, migrate in different directions or create swirls (Petitjean *et al* 2010, Vedula *et al* 2012, Reffay *et al* 2014). The distribution of forces within a cell monolayer is dynamic and fluctuating based on alterations in cell–substrate adhesion and contractility (Ng *et al* 2014). These local variations may promote the polarization of cells and the exertion of tractions in other directions than those of the global movement during the entire migration time. Subsequently, tractions need to be regulated by migration speed and also other local parameters such as cellular polarity (Notbohm *et al* 2016). During collective cell migration the cell layer is under global tensile stress and hence forces are transmitted at large scale, which is due to the unbalanced distribution of tractions between leader and following cells (Trepat *et al* 2009, Tambe *et al* 2011, Serra-Picamal *et al* 2012). Although the leaders and the following cells act against each other, the entire monolayer still migrates forward as the leader cells exert the strongest forces. The leader cell forces are in the range of up to 100 nN, which is still not sufficient to pull the entire monolayer. Hence, a mechanical 'X-wave' propagates continuously from the leader cell front to the back of the monolayer in order to

deliver information on the mechanics and polarity towards the following cells during collective migration (Serra-Picamal *et al* 2012). The coordinated behavior of the following cells is required to align their traction forces with the general velocity of the entire cell group (Basan *et al* 2013).

Using monolayer stress microscopy (MSM) the distribution of stresses (σ) within in the monolayer are heterogeneous, with large regions of tensile (positive) stresses and regions of weak compressive (negative) stresses, which fluctuate over time. In more detail, stresses are defined by their components such as the shear stress (σxy and σyx) tangent to the surface and the normal stress (σxx and σyy) perpendicular to the surface. In biological systems, these stresses correspond to the tangential and perpendicular forces transmitted to cell–cell junctions by neighboring cells. During collective cell migration, cells within the monolayer migrate in the direction that possess the shear stress minimum and the normal stress maximum. The tendency of collective migrating cells within a monolayer to migrate along the orientation of maximal principal stress is termed plithotaxis (Tambe *et al* 2011, Trepat and Fredberg 2011). Thereby, leader cells induce traction forces on following cells and shear stress on neighboring cells, which transforms initially local forces into coordinated and polarized traction forces promoting plitothaxis (Zaritsky *et al* 2015). The tumor suppressor and regulator of the Hippo pathway merlin is suggested to be crucial for plithotaxis of epithelial cells through Rac1 alterations (Das *et al* 2015). When leader cells start their migration, Rac1 is activated at the cell's front. The leader cells pull on the following cells to facilitate the release of merlin from adherence junctions and subsequently induce the polarized activation of Rac1 in the following cells. The transmission of forces within a group of cells is mainly facilitated by adherence junctions (Tambe *et al* 2011, Trepat and Fredberg 2011). To investigate the mechanical role of adherence junctions between neighboring cells migrating collectively on extracellular matrix proteins, cells can be plated as doublets, which represents the smallest possible group on adhesive micropatterns that is clearly not a model for collective cell migration. In particular, a pair of endothelial cells can withstand forces at cell–cell junctions of approximately 100–120 nN perpendicular to the cell–cell contact (Liu *et al* 2010, Maruthamuthu *et al* 2011). In addition, different adherence junction proteins may alter the physical parameters regulating the monolayer kinematics and/or forces. In MCF-10A cells, P-cadherin and E-cadherin display altered responses to mechanical stress using magnetic tweezers. In particular, E-cadherin facilitates the adaption of the cells to an extracellular force through the activation of a mechanotransduction pathway via vinculin, whereas P-cadherin cannot reinforce adherence junctions. These two proteins are competing for the same mechanotransduction pathway, as P-cadherin can indeed rescue the loss of E-cadherin phenotype (Bazellières *et al* 2015). However, the specific role of each cadherin in providing intercellular stresses during collective migration is not yet resolved. The diverse expression of cadherins during EMT or tumor invasion may enable cells to migrate and invade collectively through a cadherin-dependent regulation of forces (Friedl and Mayor 2017). In a carcinoma model, cancer associated fibroblasts (CAFs) can pull cancer cells to induce collective invasion, due to a heterophilic interaction between the N-cadherin of CAFs and

E-cadherin of cancer cells, which then actively reacts to forces and induces the polarization of CAFs (Labernadie *et al* 2017).

1.2.4 Transendothelial migration of cancer cells

What impact do transmigrating and invasive cancer cells have on the biomechanical properties of the endothelial-cell lining?

The role of endothelial cells in the regulation of cancer cell invasiveness into 3D extracellular matrices is not yet fully understood. Moreover, the coordination and regulation of cancer cell transendothelial migration may be a complex scenario, which is not yet fully characterized. In numerous or almost all previous studies, the endothelium acts as a barrier against the migration and invasion of cancer cells (Al-Mehdi *et al* 2000, Zijlstra *et al* 2008). In more detail, the endothelium decreases pronouncedly the invasion of cancer cells and subsequently metastasis (Van Sluis *et al* 2009). Contrary to these studies, several recent reports propose a novel paradigm in which endothelial cells regulate the invasiveness of special cancer cells by increasing their dissemination through vessels (Kedrin *et al* 2008) or even by increasing the invasiveness of cancer cells (Mierke *et al* 2008a). Although special cell adhesion molecules have been identified to facilitate tumor–endothelial-cell interactions and subsequently support metastasis formation, the role of endothelial cells' mechanical properties during the transmigration and invasion of cancer cells is not well understood. However, it has been supposed that the altered mechanical properties of endothelial cells can support one of the two main functions of the endothelium in cancer metastasis, regulation of tumor growth and metastasis. First, endothelial cells act as a barrier inhibiting transmigration and, second, they serve as an enhancer for cancer cell invasion. How this behavior of endothelial cells is actually regulated is not yet known. Moreover, the conditions that determine whether the endothelium acts as an enhancer or a repressor of cancer cell transmigration are still elusive. A main biochemical pathway of the tumor–endothelial interaction has so far been reported to involve cell adhesion receptors and integrins such as platelet endothelial-cell adhesion molecule-1 (PECAM-1) and $\alpha v\beta 3$ integrins (figure 1.27) (Voura *et al* 2000). As integrins facilitate a linkage between the extracellular matrix and the cell's actomyosin cytoskeleton (Neff *et al* 1982, Damsky *et al* 1985, Riveline *et al* 2001), the outside–inside connection is facilitated through the mechano-coupling focal adhesion and cytoskeletal adaptor protein vinculin (Mierke *et al* 2008b). In particular, this connection determines the amount of cellular counter-forces that maintain the cell shape, cell morphology and cell stiffness (Rape *et al* 2011). Finally, an overall biomechanical approach investigating the endothelial barrier breakdown in the presence of cocultured invasive cancer cells is still elusive. Microrheological measurements, such as magnetic tweezer microrheology, have been adequate for determining the endothelial cell's mechanical properties, such as cellular stiffness during coculture with invasive or noninvasive cancer cells compared to mono-cultured endothelial cells (Mierke 2011). Indeed, the endothelial cell's stiffness is influenced during the coculture by cancer cells, depending on their aggressiveness. In particular, highly invasive human breast cancer cells

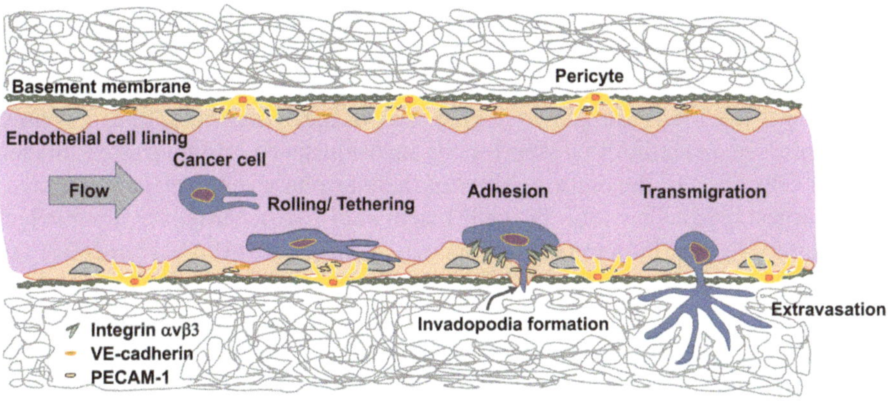

Figure 1.27. Transendothelial migration. A cancer cell in the blood vessel is transported through the vessel flow, rolls on the endothelium (tethering), attaches to it and exerts an invadopodium through the endothelial-cell layer and the basement membrane into the extracellular matrix. The cancer cell interacts via PECAM-1 and integrin $\alpha v\beta a3$ interactions with the neighboring endothelial cells. The interendothelial cell–cell adhesion is mainly facilitated through hemophilic PECAM-1 or hemophilic VE-cadherin interactions.

influence the mechanical properties of cocultured microvascular endothelial cells by lowering the stiffness of endothelial cells, whereas noninvasive cancer cells have no effect on endothelial-cell stiffness (Mierke 2011). Whether a direct cell–cell contact between endothelial cells and cancer cells is necessary is still under intensive investigation. In addition, the nanoscale particle-tracking method is used for diffusion measurements of actomyosin cytoskeletal-bound beads serving as intracellular markers for structural changes of the intercellular cytoskeletal scaffold. As expected, this method has turned out to be suitable to analyze actomyosin-driven cytoskeletal remodeling dynamics (Mierke 2011). Thus, the cytoskeletal remodeling dynamics of endothelial cells are reported to increase in coculture with highly invasive cancer cells, when the coculture is performed by direct cell–cell contact, whereas they are unchanged by coculture with noninvasive cancer cells (Mierke 2011). Finally, these results demonstrate that highly invasive breast cancer cells, but not less invasive breast cancer cells, actively alter the biomechanical properties of cocultured endothelial cells. Thus, these findings may provide an explanation for the dramatic breakdown of the endothelial barrier function of vessel walls.

What impact do the biomechanical properties of cancer cells have in intravasation? During the entry into and the exit from the blood or lymph vessels, cancer cells experience pronounced morphological shape changes, which are regulated by contractile forces transmission and the generation or cytoskeletal remodeling dynamics, which empower them to migrate through endothelial cell–cell adhesions. The cytoplasm of a cell is regarded as a highly complex system that behaves like an elastic material such as a spring at high deformation rates, whereas at low deformation rates it acts like a viscous material such as honey that exhibits a yield stress (Wirtz 2009). In particular, cellular elasticity mirrors the ability of the cytoplasm to restructure and remodel upon the application of external stresses such as forces. In turn, the cellular viscosity describes the capability of the cytoplasm to flow upon external shear stresses. Thus, cancer cells possess viscoelastic properties prepared to react to all kinds of mechanical stimulation from their local micro-environment. The MMP-facilitated degradation of the extracellular matrix is only to a certain degree a rate-limiting step for the invasion speed and migration efficiency of cancer cells, instead the deformability of the interphase cell nucleus is decisive, as it is the largest organelle within the cell (Friedl *et al* 2010). Moreover, the nucleus is at least ten-fold stiffer than the soft cytoplasm of cancer cells (Dahl *et al* 2004, Tseng *et al* 2004). In particular, the elasticity of the cellular nucleus depends on the mechanical properties of the nuclear lamina underneath the nuclear envelope consisting of intermediate filaments (Dahl *et al* 2004), the chromatin organization (Gerlitz and Bustin 2010) and special linker proteins connecting the nucleus and the cytoskeleton via LINC complexes (Crisp *et al* 2006, Lee *et al* 2007, Hale *et al* 2008, Stewart-Hutchinson *et al* 2008). The precise nucleus position within the cell body is important for many cellular processes such as cell motility and invasion through narrow pores of the extracellular matrix. LINC complexes consist of special proteins that span over the nuclear envelope and mediate mechanical connections between the nuclear lamina and the cytoskeleton (Crisp *et al* 2006). In more detail, these tight connections between the nucleus and the cytoplasm are facilitated by interactions between SUN domain-containing proteins such as SUN1 and SUN2 and Klarsicht homology (KASH) domain-containing proteins at the outer nuclear membrane such as nesprin 2 and nesprin 3, which are able to bind actin directly as well as indirectly (Starr *et al* 2001, Starr and Han 2003, Technau and Roth 2008, Lei *et al* 2009). Subsequently, the loss of LINC complex components leads to nuclear shape defects, a softening of the nucleus and the cytoplasm (Lammerding *et al* 2004) and has a major impact on the cellular motility on 2D substrates (Lee *et al* 2007, Hale *et al* 2008). Although mutations of nesprins and lamins A/C have been found in breast cancer (Wood *et al* 2007), their role in the 3D motility of cancer cells is not yet well understood.

In line with this, the mechanical properties have been shown to be softer in cancer cells compared to normal, healthy cells and thus are associated with an increased metastatic potential (Guck *et al* 2005, Bloom *et al* 2008, Roesel *et al* 2008). The softer cytoplasm of aggressive cancer cells can be explained as a loss of cytoskeletal organization. The softening of the cytoskeleton should be further investigated in *in vivo* models for cancer metastasis and in a 3D extracellular matrix invasion assay. The development and major improvement of novel biophysical methods such as nanoscale particle tracking, magnetic tweezer microrheology, traction force measurements in 3D, optical cell stretching and twisting microrheology will help to develop stiffness measurements that can be carried out directly in animal models or 3D environments (Bausch *et al* 1999, Guck *et al* 2005, Raupach *et al* 2007, Wirtz 2009, Legant *et al* 2010). The knowledge of the values of these mechanical properties may be used as a biophysical diagnostic marker of cancer and the determination of the metastatic potential (Cross *et al* 2007). Currently, there is no clue as to why aggressive cancer cells should be softer than nontransformed healthy cells. One hypothesis is that the invasion of cancer cells through 3D extracellular matrices and transendothelial migration needs precisely regulated mechanical properties. If the cancer cells are too stiff to deform themselves or too soft to generate and transmit forces to their microenvironment, they may no longer be able to squeeze through the pores of highly crosslinked collagen fibers of the connective tissue. It has been suggested that the mechanical properties of invasive cancer cells may depend on the current cell-cycle phase, the impact of neighboring cells, the regulatory role of the connective tissue's physical properties, cytokine or chemokine concentration around the cell, growth-factor stimulation, receptor ligands and receptor shedding enzymes, membrane surface tension, and membrane domains such as lipid rafts and their mobility. Indeed, single-cell measurements have consistently revealed that individual cancer cells display a wide range of mechanical properties. These findings indicate that only a certain subtype of cancer cells is able to migrate and to intravasate into the endothelial-cell lining of blood or lymph vessels. These special invasive cancer cells possess mechanical properties that facilitate cancer cell motility in connective tissue and intravasation into blood or lymph vessels. These invasive properties of cancer cells as well as their mechanical properties are stably associated even over a huge number of passages in cell culture, after injection into animals and re-isolation from animal tumor models. How can physical properties of cancer cells such as stiffness be passed on from generation to generation? When these physical properties are based on genomic mutations, they can be altered by the addition of pharmacological drugs in order to regulate the proteins affecting cellular mechanics, invasion and trans-endothelial migration. In more detail, cancer cells are supposed to vary their mechanical properties according to their progression level in the metastatic process.

Moreover, it is suggested that the optimal mechanical properties for the invasion of cancer cells into the surrounding tumor stroma near the primary tumor are different from those for the majority of stationary cancer cells within the solid primary tumor. In particular, these special mechanical properties seem to be essential for the cancer cells to overcome the steric restrictions of 3D extracellular matrices such as the extracellular matrix or endothelial vessel walls. Hence, the mechanical properties of cancer cells need to be adopted during the metastatic process in order to react to the altered environmental properties. Cancer cells must overcome microenvironmental restrictions to reach the next step of the metastatic process. These alterations in the mechanical properties of cancer cells seem to be regulated by biochemical gradients (Tseng *et al* 2004), interstitial flow rates (Swartz and Fleury 2007) and neighboring cells such as the endothelial cells lining the vessel wall (Mierke 2011).

The impact of shear stress on endothelial lining and transmigrating cells

During their travel through the blood or lymphoid system, cancer cells experience external dynamic forces, immunological stimulation and collisions with other cell types, such as red blood cells, and interact with the endothelial lining of the vessels. All of which may have an impact on the survival of motile cancer cells and the cancer cells' capacity to metastasize in targeted organs. Those circulating cancer cells that have been able to withstand fluid shear stress and immunosurveillance can adhere to the endothelial-cell monolayer of the vessel, transmigrate and invade distant organs and then enter targeted tissues. However, only a small and specialized subtype fraction of the cancer cells survives and generates metastases, whereas the majority of the cancer cells die or remain dormant within the vessels or tissues (Fidler *et al* 2002).

Upon entry into the circulatory system (intravasation), the migration path and cell fate of a cancer cell is influenced by a number of physical and mechanical parameters: the flow rate and the diameter of the blood vessels as well as the interplay between shear flow and intercellular adhesion leads to the abolition of cancer cell movement in larger vessels. In more detail, the shear stress arises between adjacent layers of the blood fluid (of certain viscosity) moving at different velocities: the velocity of a liquid in a cylinder (the blood vessel) is maximal in the center, but zero at the border of the cylinder. The total shear stress is the product of fluid's viscosity and shear rate. The blood fluid behaves as a Newtonian fluid at shear rates higher than 100 1 s^{-1}, with a shear stress that increases linearly with the shear rate. In contrast to cylinders, the maximum shear stress is experienced at the vessel wall. The shear flow influences the translational and rotational motion of cancer cells and regulates the orientation of cancer cells as well as the speed of receptor–ligand interactions during cell adhesion. What we do not yet know is whether the shear flow induces the deformation of cancer cells or even affects the magnitude of adhesion. Moreover, the impact of the shear flow on the viability and proliferation rates of cancer cells is not yet well understood.

The extravasation of cancer cells from blood or lymph vessels

If a cancer cell within the vessel lumen wants to exit, it must first adhere to the endothelial blood vessel lining. There are two proposed mechanisms of cancer cell trapping: physical arrest (called occlusion) and the cell adhesion. Which of these two mechanisms is preferred depends on the actual diameter of the blood vessel.

Physical arrest means that a cancer cell intravasates into a vessel whose diameter is less than that of the cancer cell's diameter. Thus, the cancer cell arrests within the vessel by mechanical trapping (physical occlusion). Circulating cancer cells of epithelial origin are normally larger than 10 μm in size, hence physical occlusion only plays a role in small vessels, such as capillaries smaller than 10 μm. In line with this, cancer cells have been found at branches in blood vessels of the brain. In particular, cancer cells' extravasation through the endothelium and finally their metastasis has been analyzed *in vivo* by intravital microscopy using a mouse model (Kienast *et al* 2010).

In the cell adhesion mechanism, the extravasation of a cancer cell from a large blood vessel possessing a vessel diameter that is larger than that of the cancer cell initially requires the adhesion of the cancer cell to the endothelial vessel wall through intercellular cell–cell adhesion. The probability of getting caught in a large vessel is the collision frequency between membrane-bound receptors on the surface of the cancer cells and endothelial ligands exposed on the surface as well as the residence time, which depends on the shear force and adhesive forces between the circulating cancer cell and the endothelial vessel lining (Zhu *et al* 2008).

During their journey within the vessels, cancer cells are exposed to translational and tangential velocity. In particular, the translational velocity of a cell traveling through the vessel is always larger than the surface tangential velocity, which leads to a rolling (slipping) motion relative to the endothelial vessel wall. The slipping motion increases the binding rate (depending on the rotation rate of the cancer cell) between a single receptor on a cancer cells in the blood stream and ligands on the surface of endothelial cells lining the vessel wall (Chang and Hammer 1999). The total adhesion strength is defined by the tensile strength of the particular receptor–ligand connection and the number of receptor–ligand interactions involved, whereas possible cooperative effects are excluded (Niessen and Gumbiner 2002, Duguay *et al* 2003). What can help to reveal the cooperative effects? The further development and invention of biophysical methods for measuring the mechanical properties of ligand–receptor bonds may help to distinguish between single-molecule affinity and multimolecular avidity exactly (Marshall *et al* 2003, Huang *et al* 2010). In particular, the kinetic and mechanical properties, such as the tensile strength of a single receptor–ligand connection, regulate whether a bond will form at a certain shear stress and subsequently lead to cell adhesion. The initiation of a receptor–ligand-dependent adhesion of cancer cells to the endothelium is facilitated by selectins under shear stress and requires a sufficient fast binding rate, sufficient tensile strength in order to resist the dispersive hydrodynamic force and a slow rupture rate of receptor–ligand bonds. Thus, this all helps to promote the formation of additional receptor–ligand interactions. As integrins have slower binding rates than selectins, they are only able to bind after selectin–ligand mediated cancer cell binding. Finally, these integrin receptors have cooperative effects: for example, integrin clustering leads to the formation of multiple bonds and hence to strong adhesion of cancer cells (Hynes 2002).

Cancer cells are supposed to escape immune surveillance through association with platelets when migrating through the blood vessels (McCarty *et al* 2000). This association may additionally provide guidance for the extravasation step of cancer cells (McCarty *et al* 2000). Nonetheless, the role of these platelets in promoting cancer metastasis is still not clearly understood and thus needs further investigation through using an established mouse model (Gasic *et al* 1968, Camerer *et al* 2004). There are still open questions. Are cancer cells masked by platelets in the blood stream in order to be protected from immune response reactions such as the clearance by phagocytosis? Can platelets support the cancer cell's adhesion to the endothelial vessel lining? Does the secretion of VEGF by platelets alter the vascular hyperpermeability and subsequently the extravasation of cancer cells? What is the reason for having leukocytes in close neighborhood to metastatic and highly aggressive cancer cells? Why do transmigrating cancer cells use similar mechanisms as transmigrating leukocytes, such as neutrophils? What regulates the specificity of cancer cells to metastasize exclusively in targeted organs and not in other sites (Weiss 2000, Fidler 2003, Steeg 2006, Jacob *et al* 2007)? There are currently at least three hypotheses. The first is the one called the seed and soil hypothesis, which states that a cancer cell will only metastasize to a tissue site where the microenvironment is suitable and favors cancer cell survival and secondary tumor growth (Paget 1989). The second is called the mechanical hypothesis and is based on structural restrictions; this states that the process of cancer metastasis occurs at sites related to the directional pattern of blood flow, as the blood fluid flows from most organs to the heart, then to the lungs by the venous system and back via the arterial system

(Weiss 2000). The third is called the stiffness hypothesis, which states that cancer metastasis occurs only at sites where the mechanical properties of the targeted tissue are similar or even equal to the mechanical conditions of the primary tumor.

Does the endothelial cell's actomyosin cytoskeleton serve as a migration grid for transmigrating cancer cells?

In the classical view, the endothelial-cell monolayer of vessels acts as a barrier and as such it is a key rate-limiting step for the transmigration, invasion and metastasis of invasive cancer cells (Zijlstra *et al* 2008). The endothelial vessels wall is commonly presented as a strong tissue barrier towards the dissemination of cancer cells (Wittchen *et al* 2005). However, recent reports demonstrate a novel role for the endothelial vessel wall, in which the endothelial-cell lining increases the invasiveness of special cancer cells. In the first report, breast cancer cells display enhanced dispersion through hematogeneous dissemination in close neighborhood to blood vessels (Kedrin *et al* 2008). In the second report, the invasiveness of certain cancer cell lines is endothelial-cell-dependent and is even enhanced for special cancer cell types (Mierke *et al* 2008a). Nevertheless, the process of cancer cell invasion has been investigated in numerous research studies, but the molecular and mechanical mechanisms of cancer cell transendothelial migration are not yet well understood.

The intravasation of cancer cells in the vessels includes as a physical process the interaction of at least three cell types: the special invasive cancer cell, the nearby macrophage and the opposing endothelial cell. This interaction will engage the mechano- and biochemical-transduction cytoskeletal properties of these three neighboring cells. To reveal whether signals from cancer cells stimulate endothelial cells, a 3D coculture assay with a collagen fiber matrix can be used in which the real-time intra-endothelial signaling events evoked by invasive cancer cells or macrophages are analyzed and compared to mono-cultured endothelial cells (Khuon *et al* 2010). In more detail, this assay involves the assembly of a vasculature network in a 3D collagen matrix using endothelial cells expressing a fluorescent resonant energy-transfer-based biosensor that reports the activity of myosin light chain kinase (MLCK) inside the endothelial cell in real time (Chew *et al* 2002). As expected, endothelial cells sensed their environment, such as the 3D collagen fiber matrix, mechanically. Indeed, the 3D microenvironment induces lumen formation in endothelial cells, which then display basal–apical polarity in the proper orientation supported by α4 laminin deposition. As suggested, invasive cancer cells are able to alter the MLCK-mediated actomyosin function in the surrounding underlying endothelium. Subsequently, cancer cells are able to transmigrate through the endothelial-cell monolayer in at least two ways: by a transmigration mode using a transcellular path (through neighboring endothelial cells) or by a transmigration mode using a paracellular (though endothelial cell–cell junctions) path (Khuon *et al* 2010). When cancer cells use a transcellular mode to invade, they trigger MLCK activation inside the endothelial cell, which correlates with enhanced local phosphorylation of the myosin-II regulatory light chain (RLC) and endothelial myosin contraction. This has been functionally investigated using endothelial cells expressing an RLC mutant that cannot be phosphorylated; thus, the transcellular modes of cancer cell intravasation are reduced. Finally, the transcellular transmigration mode is indeed used and can be characterized: (1) invasive cancer cells undergo transcellular migration; (2) cancer cells undergoing the transcellular transmigration mode induce transient and local MLCK activation as well as myosin contraction of adjacent endothelial cells at the transmigration site and subsequently tissue invasion; and (3) the transcellular transmigration path depends on the phosphorylation of myosin-II RLC. In summary, these findings demonstrate that the endothelium plays an active rather than a passive role in supporting cancer cell intravasation.

1.2.5 Secondary tumor in targeted tissues

If a secondary tumor occurs in the progression of cancer disease, the primary tumor has metastasized in a targeted tissue. How the tissue specificity is determined is still elusive and under debate. It is still commonly held that the hypothesis formulated by Stephen Paget in 1989 holds true, in which the process of metastasis depends on a crosstalk between special cancer cells (the seeds) and the specific microenvironments of organs (the soil) (Paget 1989). In established reports it is stated that the metastatic potential of a cancer cell depends on its interactions with the homeostatic factors that facilitate primary tumor-cell growth, survival, angiogenesis, invasion and metastasis. What major findings are most likely to answer the question of why certain cancer cells metastasize in a special target organ? It is suggested that the mechanical properties of the primary tumor tissue are similar to those of the targeted tissue (Kumar and Weaver 2009). In order to see how these mechanical properties fit into the process of metastatic cancer progression, we need to discuss the classical and established hallmarks of cancer.

1.3 Hallmarks of cancer

In the last 25 years, many medical and biological aspects of the field of classical tumor biology have been investigated and much effort has been put into this field of cancer research. Thus, it is not surprising that in 2000 six 'classical' hallmarks of cancer were proposed in order to describe the major steps of the malignant cancer progression cascade process in a more detailed and precise mode (Hanahan and Weinberg 2000): (1) sustaining proliferative signaling, (2) evading growth suppressors, (3) activating invasion and metastasis, (4) enabling replicative immortality, (5) inducing angiogenesis and (6) resisting cell death (figure 1.28). After defining these six hallmarks, the following question may arise: how many of these regulatory circuits must a cancer-candidate cell deregulate to be finally cancerous? However, this question cannot be fully answered yet, as the six proposed hallmarks describe only the first set of regulatory circuits and have been revealed to be insufficient to describe all possible types of cancer. Thus, this first set of hallmarks has to be refined. Indeed, 11 years later the same two researchers who proposed the first set of six hallmarks of cancer defined the next two novel, additional hallmarks to improve the principles for classification of malignant cancer progression (Hanahan and Weinberg 2011): (1) the utilization of abnormal metabolic pathways and (2) the repression of the immune system. The addition of the seventh and eighth hallmarks

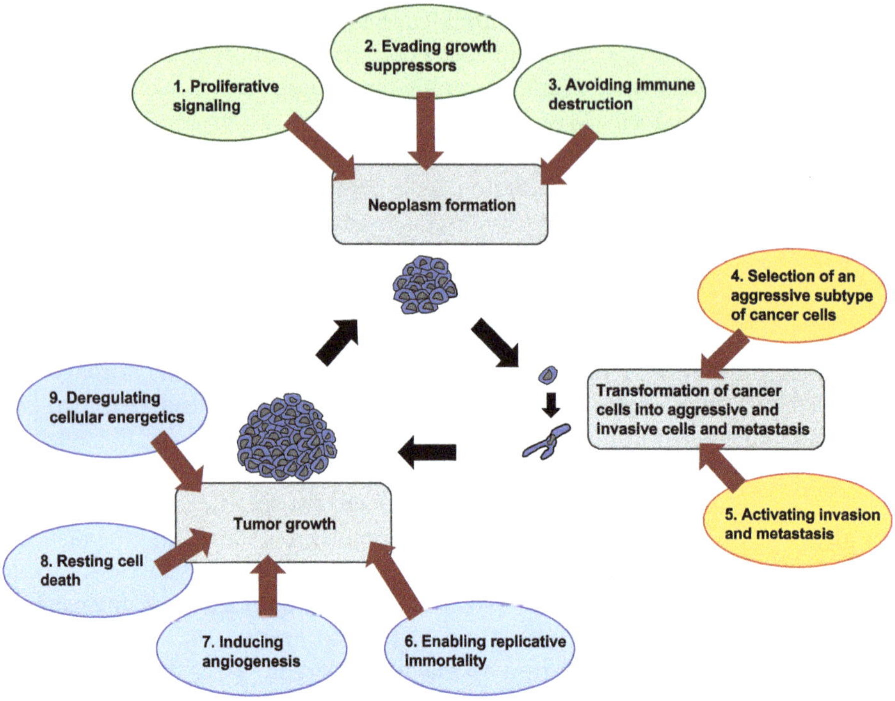

Figure 1.28. Hallmarks of cancer.

of cancer to their list showed that the immune system was viewed as being more important in malignant cancer progression.

Many molecules relevant for the motility and invasion of cancer cells, such as $\alpha6\beta4$, $\alpha v\beta3$, $\alpha v\beta5$, $\alpha5\beta1$, E-cadherin, N-cadherin, Notch1-4 receptors, CXCR2 and CXCR4, have been reported to play a regulatory role in cancer disease (Gong *et al* 1997, Bauer *et al* 2007, Mierke *et al* 2008a, 2008b, 2008c, Sawada *et al* 2008, Gilcrease *et al* 2009, Ricono *et al* 2009, Teicher and Fricker 2010). Despite all of these major findings based on novel molecular or biochemical approaches such as genomics and proteomics, these novel approaches were not able to fundamentally change patients' survival rates in the field of cancer research. After the initial enthusiastic reception, it has been seen that these molecular biological technology-based approaches have not yet reached their goals, although they have contributed to the discovery of significant insights into cancer biology, cancer diagnosis and prognosis. The detailed classification of tumors due to numerous novel marker proteins for special cancer types and the detailed mapping of specific human cancer types has begun. Nonetheless, a major criticism of these novel approaches is the high variation among the gene and protein expression levels during the progression of cancer, possibly depending on the cancer disease stage, but also varying between cancer patients, possibly due to altered environmental conditions. With this knowledge, it is not yet well understood how these novel biological approaches contribute in a regulatory way to the progression of cancer and to what extent. In these genomic-

and proteomic-based analysis methods the spatial localization of the proteins in special compartments such as lipid rafts (Runz *et al* 2008), their state of activation such as phosphorylation and assembly, and finally their lifetime, turnover rate, modification rate (for glycosylation and recycling, for example) are still not included (Veiga *et al* 1997, Garcia *et al* 1998, Caswell *et al* 2008, Gu *et al* 2011, Liu *et al* 2011).

Current classical biological and biochemical approaches to reveal insights into cancer biology have not yet covered the whole complexity of cancer disease and have so far failed, in particular, to give more appreciation to the malignant progression of cancer. Cancer research has adapted classical physical approaches. Moreover, novel biophysical analysis methods have been developed in order to be suitable for use in the field of cancer research. To date, these promising new directions for physical-based cancer research have significantly changed the field and have demolished the classical biological and biochemical view on cancer disease. In addition, cancer disease is closely associated with inflammatory processes and hence the physical view has changed and needs to be adapted to inflammatory diseases. In addition to solid epithelial originated tumors, 'soft' leukemias will also be addressed in order to compare them with the results obtained for epithelial cancer cells.

As there is still at least one hallmark missing, including the mechanical properties of cancer cells, this book will focus on the impact of the physical properties of cancer cells, their physical microenvironment such as the extracellular matrix, and on neighboring cells such as endothelial cells, cancer-associated macrophages and fibroblasts. These processes or regulatory circuits are presented here from a biophysical point of view and thus break down the classical and well-known eight-hallmarks-based view of cancer.

Do physical aspects affect the classical hallmarks of cancer?

All the eight postulated hallmarks of cancer miss the aspect of the mechanical properties of cancer cells and their local microenvironment, indicating that the common view on cancer is not from a physical viewpoint. Here, the physical view on cancer disease is addressed and is the focus of discussion.

There is agreement that physical aspects in cancer research can no longer be excluded and hence must be adapted into the current understanding of cancer research. In particular, this means that a ninth hallmark, which incorporates physics into classical cancer research, is needed to fully reveal the complex regulatory scenario of cancer metastasis. Indeed, this ninth hallmark states that the primary tumor and the tumor microenvironment can alter the survival conditions and cellular properties of a special type of cancer cells, which then subsequently supports the selection of an aggressive (highly invasive) subtype of cancer cell. Moreover, this aggressive subtype of cancer cells is supposed to decrease cell–cell adhesions to neighboring cells, affecting the mechanical properties of neighboring and interacting cells, migrate through the tumor boundary of the primary tumor, including the tumor-surrounding basement membrane, and invade the local tumor stroma.

In summary, each of these steps affects either the mechanical properties of the migrating cancer cells, their microenvironment, the neighboring cells representing a stable barrier and the interacting endothelial cells lining blood or lymph vessels.

Indeed, this novel ninth hallmark can be included after the hallmark of avoiding immune destruction and before the hallmark of activating invasion and metastasis (Mierke 2013b). In more detail, the novel ninth hallmark, the selection of an aggressive subtype of cancer cells, may have the ability to reduce cell–cell adhesions, possibly even increase cell–matrix adhesions and adapt the mechanical properties of cancer cells, facilitating cancer cell transmigration through the basement membrane and migration into the connective tissue. Moreover, these nine hallmarks can then be put into three groups: the neoplasm formation group (including hallmarks 1–3), the transformation of cancer cells into aggressive and invasive cells group (including hallmarks 4–5) and the tumor growth group (including hallmarks 6–9) (Mierke 2013b). Finally, when re-investigating the eight classical hallmarks, the physical aspect should be included in each of them to refine this particular hallmark.

1.4 The impact of the mechanical properties of cancer cells on their migration

During the last decade, many studies have suggested that the mechanical properties of cancer cells are a key player in facilitating cancer cell migration. Novel biophysical methods have been developed and further improved in order to measure the mechanical properties of cancer cells precisely.

The biomechanical properties of cancer cells determine their own aggressiveness
What determines the cellular motility in 2D and 3D migratory environments is still elusive. Nonetheless, the capacity of cancer cells to migrate and invade 3D connective tissues depends on certain mechanical parameters determining the overall migration speed through a dense 3D extracellular matrix (with a typical pore size of 1–2 μm). Among these parameters are (1) cell adhesion and de-adhesion dynamics, such as turnover of focal adhesions and adhesion strength; (2) cytoskeletal remodeling dynamics, cellular fluidity and cellular stiffness; (3) extracellular matrix remodeling and enzymatic degradation of the extracellular matrix; and (4) protrusive (contractile) force generation (Friedl and Brocker 2000, Webb *et al* 2004, Mierke *et al* 2008c). The exact values do not restrict the migratory potential of cancer cells, it is rather the balance between the parameters that is crucial for the efficiency of cancer cell invasion, invasion speed and invasion depths into 3D extracellular matrices (Zaman *et al* 2006). These parameters can be altered and shifted towards a single parameter, whereas they are still all important for regulating the invasiveness and invasion efficiency of the cancer cells. An imbalance between these parameters may switch the mode of cell invasion from epithelial to mesenchymal, or even to the amoeboid mode with or without the presence of traction forces. Taken together, the invasion strategy is determined by the utilization of these parameters. How cancer cells keep these parameters precisely balanced, and how the microenvironment, such as growth factors, cytokines, chemokines, matrix-protein composition, structure and concentration, and matrix mechanical stiffness, impacts cell invasiveness is still elusive or the subject of controversial discussion (Steeg 2006). There are at least two mechanisms. The first is that invasive cancer cells digest the

dense 3D extracellular matrix through the secretion of enzymes such as MMPs (Wolf *et al* 2003a, 2007, Friedl and Wolf 2009). The second mechanism is that sheddases exposed on the cell surface such as the alpha-secretases ADAM-10 and ADAM-17 cut cell–cell adhesion molecules such as NOTCH receptors (Micchelli and Blair 1999), ephrins or E-cadherins of cancer cells or facilitate signaling, leading to nuclear translocation and the induction of gene expression, and finally induce cancer cell invasiveness (Brou *et al* 2000, Itoh *et al* 2008, Li *et al* 2008, Bozkulak and Weinmaster 2009, Riedle *et al* 2009, Singh *et al* 2009, van Tetering *et al* 2009). Additionally, ADAM-17 cuts pro-TNF-α exposed on the cell surface and releases mature TNF-α into the microenvironment (Black *et al* 1997). In addition to proteolysis of the extracellular matrix or shedding of membrane receptors, another important parameter for regulating the invasion speed and determining the invasion depth of cancer cells in dense 3D extracellular matrices is the mechanical ability to generate and transmit contractile forces (Mierke *et al* 2008a, Roesel *et al* 2008). Novel biophysical methods measuring contractile forces in 3D collagen or fibrin matrices have been reported (Bloom *et al* 2008, Legant *et al* 2010, Koch *et al* 2012, Gjorevski and Nelson 2012). Thus, the invasion of cancer cells is analyzed by taking z-stack images. In some assays, matrix embedded beads may serve as markers, whereas in others the collagen fiber structure is the marker of the cellular force displacement evoked by invasive cancer cells. The tracking of collagen fibers seems to be more complicated, but also more reliable and it reduces the effect of marker phagocytosis as only a minor number of collagen fibers is cut by enzymes and internalized compared to the embedded beads in close neighborhood to invasive cancer cells. One problem is that bead internalization (bead phagocytosis) may affect the mechanical properties of cancer cells and consequently may alter their invasiveness into dense 3D extracellular matrices.

However, major players in the regulation of cancer cell invasion into dense 3D extracellular matrices are the cell–matrix adhesion receptors such as integrins and their associated focal adhesion molecules such as vinculin and focal adhesion kinase (FAK). Integrins are key players in the regulation of the invasiveness of cancer cells, because they regulate cell–matrix adhesion and de-adhesion, adhesion strength and the generation or transmission of contractile forces upon outside–in (via ligand) or inside–out (via integral signaling through growth factors, etc) signaling, stimulation of integrin receptors and cytoskeletal remodeling dynamics (Giancotti 2000, Discher *et al* 2005). The integrins are a family of cell–matrix adhesion receptors and are composed of two noncovalently linked α (one out of 18) and β (one out of 8) subunits. In addition, they regulate transmembrane connections between the cell's actomyosin cytoskeleton and the extracellular microenvironment, such as an extracellular matrix scaffold. This connection is mainly mediated by the focal adhesion protein vinculin, which can act as a mechano-coupling and mechano-regulating protein (Mierke 2009). The focal adhesions of cancer cells fulfill two functions in cell invasiveness: first, they transmit contractile forces to the extra-cellular matrix microenvironment and second, they anchor the cancer cells to its substrate to withstand de-adhesion and apoptosis (programmed cell death) (Loftus and Liddington 1997, Palecek *et al* 1997, Zaman *et al* 2006). The composition of focal adhesions may vary significantly, depending on the integrin type and the integrin activation state. In addition, the integrin signal transduction through the α integrin subunit is not uniform: the activation of $\alpha4$ and $\alpha9$ integrin subunits leads to decreased cell spreading, but the activation of $\alpha1$, $\alpha3$, $\alpha5$, $\alpha6$, αv and αIIb integrin subunits even enhances cell spreading (Horwitz and Parsons 1999, Rolli *et al* 2003, Rout *et al* 2004, Schmid *et al* 2004, Pawar *et al* 2007). However, the function of vinculin in the focal adhesions is supposed to vary due to the integrin type and activation state. In more detail, integrins are able to modulate the function of other integrins, for example the $\alpha5\beta1$ integrin can regulate the $\alpha v\beta3$ integrin facilitated focal adhesions and migration on extracellular matrices (Ly *et al* 2003). In line with this, the expression of the $\beta1$ integrin subunit has been reported to support cancer cell migration in 3D matrigel and 3D extracellular matrices (Mierke *et al* 2011).

References and further reading

Abate A R and Durian D J 2008 Effective temperatures and activated dynamics for a two-dimensional air-driven granular system on two approaches to jamming *Phys. Rev. Lett.* **101** 245701

Aceto N, Bardia A and Miyamoto D T *et al* 2014 Circulating tumor cell clusters are oligoclonal precursors of breast cancer metastasis *Cell* **158** 1110–22

Adamson E D 1987 Oncogenes in development *Development* **99** 449–71

Al-Mehdi A B, Tozawa K, Fisher A B, Shientag L, Lee A and Muschel R J 2000 Intravascular origin of metastasis from the proliferation of endothelium-attached tumor cells: a new model for metastasis *Nat. Med.* **6** 100–2

Alcaraz J, Mori H, Ghajar C M, Brownfield D, Galgoczy R and Bissell M J 2011 Collective epithelial cell invasion overcomes mechanical barriers of collagenous extracellular matrix by a narrow tube-like geometry and MMP14-dependent local softening *Integr. Biol. (Camb.)* **3** 1153–66

Almangush A, Karhunen M, Hautaniemi S, Salo T and Leivo I 2016 Prognostic value of tumour budding in oesophageal cancer: a meta-analysis *Histopathology* **68** 173–82

Almangush A, Salo T, Hagström J and Leivo I 2014 Tumour budding in head and neck squamous cell carcinoma—a systematic review *Histopathology* **65** 587–94

Alt S, Ganguly P and Salbreux G 2017 Vertex models: from cell mechanics to tissue morphogenesis *Phil. Trans. R. Soc.* B **372** 20150520

Amack J D and Manning M L 2012 Knowing the boundaries: extending the differential adhesion hypothesis in embryonic cell sorting *Science* **338** 212–5

Amato A, Scucchi L and Pescatori M 1994 Tumor budding and recurrence of colorectal cancer *Dis. Colon Rectum* **37** 514–5

Andersen T, Newman R and Otter T 2009 Shape homeostasis in virtual embryos *Artif. Life* **15** 161–83

Andreoiu M and Cheng L 2010 Multifocal prostate cancer: biologic, prognostic, and therapeutic implications *Hum. Pathol.* **41** 781–93

Andriani F, Bertolini G and Facchinetti F *et al* 2015 Conversion to stem-cell state in response to microenvironmental cues is regulated by balance between epithelial and mesenchymal features in lung cancer cells *Mol. Oncol.* **10** 253–71

Angelini T E, Hannezo E, Trepat X, Fredberg J J and Weitz D A 2010 Cell migration driven by cooperative substrate deformation patterns *Phys. Rev. Lett.* **104** 168104

Angelini T E, Hannezo E, Trepat X, Marquez M, Fredberg J J and Weitz D A 2011 Glass-like dynamics of collective cell migration *Proc. Natl Acad. Sci. USA* **108** 4714–9

Angelini T, Hannezo E, Trepat X, Fredberg J and Weitz D 2010 Cell migration driven by cooperative substrate deformation patterns *Phys. Rev. Lett.* **104** 168104

Arboleda-Estudillo Y, Krieg M, Stühmer J, Licata N A, Muller D J and Heisenberg C P 2010 Movement directionality in collective migration of germ layer progenitors *Curr. Biol.* **20** 161–9

Arnoux V, Côme C, Kusewitt D F, Hudson L G and Savagner P 2005 Cutaneous wound reepithelialization *Rise and Fall of Epithelial Phenotype* (New York: Springer), pp 111–34

Attramadal C G, Kumar S, Boysen M E, Dhakal H P, Nesland J M and Bryne M 2015 Tumor budding, EMT and cancer stem cells in T1–2/N0 oral squamous cell carcinomas *Anticancer Res.* **35** 6111–20

Au S H, Storey B D and Moore *et al* 2016 Clusters of circulating tumor cells traverse capillary-sized vessels *Proc. Natl Acad. Sci. USA* **113** 4947–52

Balmain A and Brown K 1988 Oncogene activation in chemical carcinogenesis *Adv. Cancer Res.* **51** 147–82

Baluk P, Hashizume H and McDonald D M 2005 Cellular abnormalities of blood vessels as targets in cancer *Curr. Opin. Genet. Dev.* **15** 102–11

Barathova M, Takacova M and Holotnakova T 2008 Alternative splicing variant of the hypoxia marker carbonic anhydrase IX expressed independently of hypoxia and tumour phenotype *Br. J. Cancer* **98** 129–36

Berthier L, Biroli G, Bouchaud J P, Cipelletti L, El Masri D, L'Hôte D, Ladieu F and Pierno M 2005 Direct experimental evidence of a growing length scale accompanying the glass transition *Science* **310** 1797–800

Basan M, Elgeti J, Hannezo E, Rappel W J and Levine H 2013 Alignment of cellular motility forces with tissue flow as a mechanism for efficient wound healing *Proc. Natl Acad. Sci USA* **110** 2452–9

Bastos L G D R, de Marcondes P G, de-Freitas-Junior J C M, Leve F, Mencalha A L, de Souza W F, de Araujo W M, Tanaka M N, Abdelhay E S F W and Morgado-Díaz J A 2014 Progeny from irradiated colorectal cancer cells acquire an EMT-like phenotype and activate Wnt/β-catenin pathway *J. Cell. Biochem.* **115** 2175–87

Batlle E, Sancho E, Francí C, Domínguez D, Monfar M, Baulida J and García De Herreros A 2000 The transcription factor snail is a repressor of E-cadherin gene expression in epithelial tumor cells *Nat. Cell Biol.* **2** 84–9

Bauer K, Mierke C and Behrens J 2007 Expression profiling reveals genes associated with transendothelial migration of tumor cells: a functional role for $\alpha v\beta 3$ integrin *Int. J. Cancer* **121** 1910–8

Bausch A R, Möller W and Sackmann E 1999 Measurement of local viscoelasticity and forces in living cells by magnetic tweezers *Biophys. J.* **76** 573–9

Bazellières E, Conte V, Elosegui-Artola A, Serra-Picamal X, Bintanel-Morcillo M, Roca-Cusachs P, Muñoz J J, Sales-Pardo M, Guimerà R and Trepat X 2015 Control of cell-cell forces and collective cell dynamics by the intercellular adhesome *Nat. Cell Biol.* **17** 409–20

Bednarz-Knoll N, Alix-Panabières C and Pantel K 2012 plasticity of disseminating cancer cells in patients with epithelial malignancies *Cancer Metastasis Rev.* **31** 673–87

Belmonte J M, Thomas G L, Brunnet L G, de Almeida R M C and Chaté H 2008 Self-propelled particle model for cell-sorting phenomena *Phys. Rev. Lett.* **100** 248702

Beloussov L V 2012 Morphogensis as a macroscopic self-organizing process *BioSystems* **109** 262–79

Ben-Isaac E, Fodor É, Visco P, van Wijland F and Gov Nir S 2015 Modeling the dynamics of a tracer particle in an elastic active gel *Phys. Rev.* E **92** 012716

Bergers G, Javaherian K, Lo K M, Folkman J and Hanahan D 1999 Effects of angiogenesis inhibitors on multistage carcinogenesis in mice *Science* **284** 808–12

Bergers G, Song S, Meyer-Morse N, Bergsland E and Hanahan D 2003 Benefits of targeting both pericytes and endothelial cells in the tumor vasculature with kinase inhibitors *J. Clin. Investig.* **111** 1287–95

Bernards R and Weinberg R A 2002 Metastasis genes: a progression puzzle *Nature* **418** 823

Berthier L 2014 Nonequilibrium glassy dynamics of self-propelled hard disks *Phys. Rev. Lett.* **112** 220602

Betz T, Lenz M and Sykes J F 2009 ATP-dependent mechanics of red blood cells *Proc. Natl Acad. Sci. USA* **106** 15320–5

Bi D, Lopez J H, Schwarz J M and Manning M L 2015 A density-independent rigidity transition in biological tissues *Nat. Phys.* **11** 1074–9

Bi D, Lopez J H, Schwarz J M and Manning M L 2014 Energy barriers and cell migration in densely packed tissues *Soft Matter.* **10** 1885–90

Bi D, Yang X, Marchetti M C and Manning M L 2016 Motility-driven glass and jamming transitions in biological tissues *Phys. Rev.* X **6** 021011

Bi D, Zhang J, Chakraborty B and Behringer R P 2011 Jamming by shear *Nature* **480** 355–8

Bizzarri M and Cucina A 2014 Tumor and the microenvironment: a chance to reframe the paradigm of carcinogenesis? *Biomed. Res. Int.* 934038

Black R A, Rauch C T, Kozlosky C, Peschon J J, Slack J L and Wolfson M F 1997 A metalloproteinase disintegrin that releases tumor-necrosis factor-alpha from cells *Nature* **385** 729–33

Blaser H, Reichman-Fried M, Castanon I, Dumstrei K, Marlow F L, Kawakami K, Solnica-Krezel L, Heisenberg C P and Raz E 2006 Migration of zebrafish primordial germ cells: a role for myosin contraction and cytoplasmic flow *Dev. Cell.* **11** 613–27

Bloom R J, George J P, Celedon A, Su S X and Wirtz D 2008 Mapping local matrix remodeling induced by a migrating tumor cell using 3D multiple-particle tracking *Biophys. J.* **95** 4077–88

Bolos V, Peinado H, Pérez-Moreno M A, Fraga M F, Esteller M and Cano A 2003 The transcription factor slug represses E-cadherin expression and induces epithelial to mesenchymal transitions: a comparison with Snail and E47 repressors *J. Cell Sci.* **116** 499–511

Bosveld F, Markova O and Guirao B *et al* 2016 Epithelial tricellular junctions act as interphase cell shape sensors to orient mitosis *Nature* **530** 495–8

Bozkulak E C and Weinmaster G 2009 Selective use of ADAM10 and ADAM17 in activation of notch signaling *Mol. Cell Biol.* **29** 5679–95

Brabletz S and Brabletz T 2010 The ZEB/miR-200 feedback loop—a motor of cellular plasticity in development and cancer? *EMBO Rep.* **11** 670–7

Brabletz T, Jung A, Reu S, Porzner M, Hlubek F, Kunz-Schughart L A, Knuechel R and Kirchner T 2001 Variable β-catenin expression in colorectal cancers indicates tumor progression driven by the tumor environment *Proc. Nat. Acad. Sci. USA* **98** 10356–61

Brabletz T, Jung A, Spaderna S, Hlubek F and Kirchner T 2005 Migrating cancer stem cells—an integrated concept of malignant tumour progression *Nat. Rev. Cancer* **5** 744–9

Brodland G W 2002 The differential interfacial tension hypothesis (DITH): a comprehensive theory for the self-rearrangement of embryonic cells and tissues *J. Biomech. Eng.* **124** 188–97

Bronsert P, Enderle-Ammour K and Bader M *et al* 2014 Cancer cell invasion and EMT marker expression—a three-dimensional study of the human cancer-host interface *J. Pathol.* **234** 410–22

Brou C, Logeat F, Gupta N, Bessia C, LeBail O, Doedens J R, Cumano A, Roux P, Black R A and Israel A 2000 A novel proteolytic cleavage involved in Notch signaling: the role of the disintegrin-metalloprotease TACE *Mol. Cell* **5** 207–16

Brown M J, Hallam J A, Colucci-Guyon E and Shaw S 2001 Rigidity of circulating lymphocytes is primarily conferred by vimentin intermediate filaments *J. Immunol.* **166** 6640–6

Camerer E, Qazi A A, Duong D N, Cornelissen I, Advincula R and Coughlin S R 2004 Platelets, protease-activated receptors and fibrinogen in hematogenous metastasis *Blood* **104** 397–401

Campinho P, Behrndt M, Ranft J, Risler T, Minc N and Heisenberg C P 2013 Tension-oriented cell divisions limit anisotropic tissue tension in epithelial spreading during zebrafish epiboly *Nat. Cell Biol.* **15** 1405–14

Cano A, Pérez-Moreno M A, Rodrigo I, Locascio A, Blanco M J, del Barrio M G, Portillo F and Nieto M A 2000 The transcription factor snail controls epithelial-mesenchymal transitions by repressing E-cadherin expression *Nat. Cell Biol.* **2** 76–83

Carmeliet P and Jain R K 2011 Principles and mechanisms of vessel normalization for cancer and other angiogenic diseases *Nat. Rev. Drug Discov.* **10** 417–27

Carr I, Levy M and Watson P 1986 The invasive edge: invasion in colorectal cancer *Clin. Exp. Metastasis* **4** 129–39

Carter S B 1967 Haptotaxis and the mechanism of cell motility *Nature* **213** 256–60

Carter S B 1970 Cell movement and cell spreading: a passive or an active process? *Nature* **225** 858–9

Caswell P T, Chan M, Lindsay A J, McCaffrey M W, Boettiger D and Norman J C 2008 Rab-coupling protein coordinates recycling of $\alpha5\beta1$ integrin and EGFR1 to promote cell migration in 3D microenvironments *J. Cell Biol.* **183** 143–55

Cecconi F, Puglisi A, Marconi U, Marini B and Vulpiani A 2003 Noise activated granular dynamics *Phys. Rev. Lett.* **90** 064301

Chang K C and Hammer D A 1999 The forward rate of binding of surface-tethered reactants: effect of relative motion between two surfaces *Biophys. J.* **76** 1280–92

Chaplin D J, Presta M and Denekamp J 1999 Evidence for characteristic vascular patterns in solid tumors: quantitative studies using corrosion casts *Br. J. Cancer* **80** 724–32

Charras G and Paluch E 2008 Blebs lead the way: how to migrate without lamellipodia *Nat. Rev. Mol. Cell Biol.* **9** 730–6

Charras G T, Hu C K, Coughlin M and Mitchison T J 2006 Reassembly of contractile actin cortex in cell blebs *J. Cell Biol.* **175** 477–90

Cheung K J and Ewald A J 2016 A collective route to metastasis: seeding by tumor cell clusters *Science* **352** 167–69

Chew T L, Wolf W A, Gallagher P J, Matsumura F and Chisholm R L 2002 A fluorescent resonant energy transfer-based biosensor reveals transient and regional myosin light chain kinase activation in lamella and cleavage furrows *J. Cell Biol.* **156** 543–53

Chiche J, Ilc K, Laferriere J, Trottier E, Dayan F, Mazure N M, Brahimi-Horn M C and Pouyssegur J 2009 Hypoxia-inducible carbonic anhydrase IX and XII promote tumor cell growth by counteracting acidosis through the regulation of the intracellular pH *Cancer Res.* **69** 358–68

Comijn J, Berx G, Vermassen P, Verschueren K, van Grunsven L, Bruyneel E, Mareel M, Huylebroeck D and van Roy F 2001 The two-handed E-box-binding zinc finger protein SIP1 downregulates E-cadherin and induces invasion *Mol. Cell* **7** 1267–78

Comito G, Calvani M, Giannoni E, Bianchini F, Calorini L, Torre E, Migliore C, Giordano S and Chiarugi P 2011 HIF-1α stabilization by mitochondrial ROS promotes Met-dependent invasive growth and vasculogenic mimicry in melanoma cells *Free Radic. Biol. Med.* **51** 893–904

Cooper C S, Eeles R and Wedge D C *et al* 2015 Analysis of the genetic phylogeny of multifocal prostate cancer identifies multiple independent clonal expansions in neoplastic and morphologically normal prostate tissue *Nat. Genet.* **47** 367–72

Crisp M, Liu Q, Roux K, Rattner J B, Shanahan C, Burke B, Stahl P D and Hodzic D 2006 Coupling of the nucleus and cytoplasm: role of the LINC complex *J. Cell Biol.* **172** 41–53

Cross S E, Jin Y S, Rao J and Gimzewski J K 2007 Nanomechanical analysis of cells from cancer patients *Nat. Nanotechnol.* **2** 780–3

Cui H *et al* 2015 Tissue inhibitor of metalloproteinases-1 induces a pro-tumorigenic increase of miR-210 in lung adenocarcinoma cells and their exosomes *Oncogene* **34** 3640–50

Dahl K N, Kahn S M, Wilson K L and Discher D E 2004 The nuclear envelope lamina network has elasticity and a compressibility limit suggestive of a molecular shock absorber *J. Cell Sci.* **117** 4779–86

Dalla-Favera R, Bregni M, Erikson J, Patterson D, Gallo R C and Croce C M 1982 Human c-myc onc gene is located on the region of chromosome 8 that is translocated in Burkitt lymphoma cells *PNAS* **79** 7824–7

Damm K, Beug H, Graf T and Vennström B 1987 A single point mutation in erbA restores the erythroid transforming potential of a mutant avian erythroblastosis virus (AEV) defective in both erbA and erbB oncogenes *EMBO J.* **6** 375–82

Damsky C H, Knudsen K A, Bradley D, Buck C A and Horwitz A F 1985 Distribution of the cell substratum attachment (CSAT) antigen on myogenic and fibroblastic cells in culture *J. Cell Biol.* **100** 1528–39

Danowski B A and Harris A K 1988 Changes in fibroblast contractility, morphology, and adhesion in response to a phorbol ester tumor promoter *Exp. Cell Res.* **177** 47–59

Das M, Subbayya Ithychanda S, Qin J and Plow E F 2014 Mechanisms of talin- dependent integrin signaling and crosstalk *Biochim. Biophys. Acta* **1838** 579–88

Das T, Safferling K, Rausch S, Grabe N, Boehm H and Spatz J P 2015 A molecular mechanotransduction pathway regulates collective migration of epithelial cells *Nat. Cell Biol.* **17** 276–87

Davis G S 1984 Migration-directing liquid properties of embryonic amphibian tissues *Am. Zool* **24** 649–55

Davis G S, Phillips H M and Steinberg M S 1997 Germ-layer surface tensions and 'tissue affinities' in Rana pipiens gastrulae: quantitative measurements *Dev. Biol.* **192** 630–44

Dawson H and Lugli A 2015 Molecular and pathogenetic aspects of tumor budding in colorectal cancer *Front. Med.* **2** 11

Dawson H, Koelzer V H, Karamitopoulou E, Economou M, Hammer C, Muller D E, Lugli A and Zlobec I 2014 The apoptotic and proliferation rate of tumour budding cells in colorectal cancer outlines a heterogeneous population of cells with various impacts on clinical outcome *Histopathology* **64** 577–84

de Craene B and Berx G 2013 Regulatory networks defining EMT during cancer initiation and progression *Nat. Rev. Cancer* **13** 97–110

de Pascalis C and Etienne-Manneville S 2017 Single and collective cell migration: the mechanics of adhesions *Mol. Biol. Cell* **28** 1833–46

de Rooij J, Kerstens A, Danuser G, Schwartz M A and Waterman-Storer C M 2005 Integrin-dependent actomyosin contraction regulates epithelial cell scattering *J. Cell Biol.* **171** 153–64

de Smedt L, Palmans S and Sagaert X 2016 Tumour budding in colorectal cancer: What do we know and what can we do? *Virchows Arch.* **468** 397–408

de Val S and Black B L 2009 Transcriptional control of endothelial cell development *Dev. Cell* **16** 180–95

DeBerardinis R J, Lum J J, Hatzivassiliou G and Thompson C B 2008 The biology of cancer: metabolic reprogramming fuels cell growth and proliferation *Cell Metab.* **7** 11–20

Deforet M, Hakim V, Yevick H G, Duclos G and Silberzan P 2014 Emergence of collective modes and tri-dimensional structures from epithelial confinement *Nat. Commun.* **5** 3747

Destaing O, Block M R, Planus E and Albiges-Rizo C 2011 Invadosome regulation by adhesion signaling *Curr. Opin. Cell Biol.* **23** 597–606

Ding X, Zhang Z, Li S and Wang A 2011 Combretastatin A4 phosphate induces programmed cell death in vascular endothelial cells *Oncol. Res.* **19** 303–9

Direkze N C and Alison M R 2006 Bone marrow and tumor stroma: an intimate relationship *Hematol. Oncol.* **24** 189–95

Dirkx A E *et al* 2006 Antiangiogenesis therapy can overcome endothelial cell anergy and promote leukocyte-endothelium interactions and infiltration in tumors *FASEB J.* **20** 621–30

Discher D E, Janmey P and Wang Y L 2005 Tissue cells feel and respond to the stiffness of their substrate *Science* **310** 1139–43

Diz-Munoz A, Krieg M, Bergert M, Ibarlucea-Benitez I, Muller D J, Paluch E and Heisenberg C P 2010 Control of directed cell migration *in vivo* by membrane-to-cortex attachment *PLoS Biol.* **8** e1000544

Doxzen K, Vedula S R K, Leong M C, Hirata H, Gov N S, Kabla A J, Ladoux B and Lim C T 2013 Guidance of cell migration by substrate dimension *Biophys. J.* **104** 313–21

Doyle A D, Petrie R J, Kutys M L and Yamada K M 2013 Dimensions in cell migration *Curr. Opin. Cell Biol.* **25** 642–9

du Roure O, Saez A, Buguin A, Austin R H, Chavrier P, Silberzan P and Ladoux B 2005 Force mapping in epithelial cell migration *Proc. Natl Acad. Sci. USA* **102** 2390–5

Dudley A C, Khan Z A, Shih S-C, Kang S-Y, Zwaans B M M, Bischoff J and Klagsbrun M 2008 Calcification of multipotent prostate tumor endothelium *Cancer Cell* **14** 201–11

Dudley A C, Udagawa T, Melero-Martin J M, Shih S C, Curatolo A, Moses M A and Klagsbrun M 2010 Bone marrow is a reservoir for pro-angiogenic myelomonocytic cells but not endothelial cells in spontaneous tumors Blood **116** 3367–71

Duguay D, Foty R A and Steinberg M S 2003 Cadherin-mediated cell adhesion and tissue segregation: qualitative and quantitative determinants *Dev. Biol.* **253** 309–23

Durand M and Guesnet E 2016 An efficient cellular Potts model algorithm that forbids cell fragmentation *Comput. Phys. Commun.* **208** 54–63
Durrani S R, Viswanathan R K and Busse W W 2011 What effect does asthma treatment have on airway remodeling? Current perspectives *J. Allergy Clin. Immunol.* **128** 439–48
Dvorak A, Tilley A E, Shaykhiev R, Wang R and Crystal R G 2011 Do airway epithelium air–liquid cultures represent the in vivo airway epithelium transcriptome? *Am. J. Respir. Cell Mol. Biol.* **44** 465–73
Eades G, Yao Y, Yang M, Zhang Y, Chumsri S and Zhou Q 2011 miR-200a regulates SIRT1 expression and epithelial to mesenchymal transition (EMT)-like transformation in mammary epithelial cells *J. Biol. Chem.* **286** 25992–6002
Edelman G M, Gallin W J, Delouvée A, Cunningham B A and Thierry J P 1963 Early epochal maps of two different cell adhesion *molecules Proc. Natl. Acad. Sci. USA* **80** 4384–88
Eder J P Jr, Supko J G and Clark J W *et al* 2002 Phase I clinical trial of recombinant human endostatin administered as a short intravenous infusion repeated daily *J. Clin. Oncol.* **20** 3772–84
Eisenhoffer G T, Loftus P D, Yoshigi M, Otsuna H, Chien C B, Morcos P A and Rosenblatt J 2012 Crowding induces live cell extrusion to maintain homeostatic cell numbers in epithelia *Nature* **484** 546–9
El-Kenawi A E and El-Remessy A B 2013 Angiogenesis inhibitors in cancer therapy: mechanistic perspective on classification and treatment rationales *Br. J. Pharmacol.* **170** 712–29
Etournay R *et al* 2015 Interplay of cell dynamics and epithelial tension during morphogenesis of the *Drosophila* pupal wing *eLife* **4** e07090
Evans E and Yeung A 1989 Apparent viscosity and cortical tension of blood granulocytes determined by micropipet aspiration *Biophys. J.* **56** 151–60
Evans M J, Van Winkle L S, Fanucchi M V and Plopper C G 1999 The attenuated fibroblast sheath of the respiratory tract epithelial-mesenchymal trophic unit *Am. J. Respir. Cell Mol. Biol.* **21** 655–7
Fahy J V 2015 Type 2 inflammation in asthma—present in most, absent in many *Nat. Rev. Immunol.* **15** 57–65
Fan C, Oh D S, Wessels L, Weigelt B, Nuyten D S, Nobel A B, van't Veer L J and Perou C M 2006 Concordance among gene-expression-based predictors of breast cancer *N. Engl. J. Med.* **355** 560–9
Fang J S, Gillies R D and Gatenby R A 2008 Adaptation to hypoxia and acidosis in carcinogenesis and tumor progression *Semin. Cancer Biol.* **18** 330–7
Farhadifar R, Roeper J C, Aigouy B, Eaton S and Julicher F 2007 The influence of cell mechanics, cell–cell interactions, and proliferation on epithelial packing *Curr. Biol.* **17** 2095–104
Feron O 2009 Pyruvate into lactate and back: from the Warburg effect to symbiotic energy fuel exchange in cancer cells *Radiother. Oncol.* **92** 329–33
Fidler I J 2003 The pathogenesis of cancer metastasis: the 'seed and soil' hypothesis revisited *Nat. Rev. Cancer* **3** 453–8
Fidler I J, Yano S, Zhang R D, Fujimaki T and Bucana C D 2002 The seed and soil hypothesis: vascularization and brain metastases *Lancet Oncol.* **3** 53–7
Fily Y and Marchetti M C 2012 Athermal phase separation of self-propelled particles with no alignment *Phys Rev Lett.* **108** 235702
Fily Y, Henkes S and Marchetti M C 2014 Freezing and phase separation of self-propelled disks *Soft Matter* **10** 2132–40

Fischer K R, Durrans A and Lee S *et al* 2015 Epithelial-to-mesenchymal transition is not required for lung metastasis but contributes to chemoresistance *Nature* **527** 472–6
Fletcher A G, Osterfield M, Baker R E and Shvartsman S Y 2014 Vertex models of epithelial morphogenesis *Biophys. J.* **106** 2291–304
Folkman M J, Long D M and Becker F F 1963 Growth and metastasis of tumor in organ culture *Cancer* **16** 453–67
Forgacs G, Foty R A, Shafrir Y and Steinberg M S 1998 Viscoelastic properties of living embryonic tissues: a quantitative study *Biophys. J.* **74** 2227–34
Foty R A and Steinberg M S 2004 Cadherin-mediated cell–cell adhesion and tissue segregation in relation to malignancy *Int. J. Dev. Biol.* **48** 397–409
Foty R A and Steinberg M S 2005 The differential adhesion hypothesis: a direct evaluation *Dev. Biol.* **278** 255–63
Foty R A, Forgacs G, Pfleger C M and Steinberg M S 1994 Liquid properties of embryonic tissues: measurement of interfacial tensions *Phys. Rev. Lett.* **72** 2298–301
Foty R A, Pfleger C M, Forgacs G and Steinberg M S 1996 Surface tensions of embryonic tissues predict their mutual envelopment behavior *Development* **122** 1611–20
Frattini V, Trifonov V and Chan J M *et al* 2013 The integrated landscape of driver genomic alterations in glioblastoma *Nat. Genet.* **45** 1141–49
Fredberg J and Trepat X 2012 Mechanical waves during tissue expansion *Nat. Phys.* **8** 628–34
Fredberg J J 2014 Power steering, power brakes, and jamming: evolution of collective cell–cell interactions *Physiology* **29** 218–19
Frey K, Zivanovic A, Schwager K and Neri D 2011 Antibody-based targeting of interferon-β to the tumor neovasculature: a critical evaluation *Integr. Biol. (Camb.)* **3** 468–78
Friedl P and Brocker E B 2000 The biology of cell locomotion within three-dimensional extracellular matrix *Cell. Mol. Life Sci.* **57** 41–64
Friedl P and Mayor R 2017 Tuning collective cell migration by cell-cell junction regulation *Cold Spring Harb. Perspect. Biol.* **9** 1–17
Friedl P and Wolf K 2003 Tumor-cell invasion and migration: diversity and escape mechanisms *Nat. Rev. Cancer* **3** 362–74
Friedl P and Wolf K 2009 Proteolytic interstitial cell migration: a five-step process *Cancer Metas. Rev.* **28** 129–35
Friedl P and Wolf K 2010 Plasticity of cell migration: a multiscale tuning model *J. Cell Biol.* **188** 11–9
Friedl P, Hegerfeldt Y and Tusch M 2004 Collective cell migration in morphogenesis and cancer *Int. J. Dev. Biol.* **48** 441–9
Friedl P, Wolf K and Lammerding J 2010 Nuclear mechanics during cell migration *Curr. Opin. Cell Biol.* **23** 1–10
Futterman M A, García A J and Zamir E A 2011 Evidence for partial epithelial-to-mesenchymal transition (pEMT) and recruitment of motile blastoderm edge cells during avian epiboly *Dev. Dyn.* **240** 1502–11
Gabbert H, Wagner R, Moll R and Gerharz C D 1985 Tumor dedifferentiation: an important step in tumor invasion *Clin. Exp. Metas.* **3** 257–79
Gallagher F A, Kettunen M I and Day S E *et al* 2008 Magnetic resonance imaging of pH *in vivo* using hyperpolarized 13C-labelled bicarbonate *Nature* **453** 940–43

Galván J A, Zlobec I, Wartenberg M, Lugli A, Gloor B, Perren A and Karamitopoulou E 2015 Expression of E-cadherin repressors SNAIL, ZEB1 and ZEB2 by tumour and stromal cells influences tumour-budding phenotype and suggests heterogeneity of stromal cells in pancreatic cancer *Br. J. Cancer* **112** 1944–50

Garbi N, Arnold B, Gordon S, Hämmerling G J and Ganss R 2004 CpG motifs as proinflammatory factors render autochthonous tumors permissive for infiltration and destruction *J. Immunol.* **172** 5861–9

Garcia A J, Huber F and Boettiger D 1998 Force required to break a5b1 integrin-fibronectin bonds in intact adherent cells is sensitive to integrin activation state *J. Biol. Chem.* **273** 10988–93

Garcia S, Hannezo E, Elgeti J, Joanny J F, Silberzan P and Gov N S 2015 Physics of active jamming during collective cellular motion in a monolayer *Proc. Natl Acad. Sci. USA* **112** 15314–9

Gardel M L, Sabass B, Ji L, Danuser G, Schwarz U S and Waterman C M 2008 Traction stress in focal adhesions correlates biphasically with actin retrograde flow speed *J. Cell Biol.* **183** 999–1005

Garrahan J P 2011 Dynamic heterogeneity comes to life *Proc. Natl Acad. Sci. USA* **108** 4701–2

Gasic G J, Gasic T B and Stewart C C 1968 Antimetastatic effects associated with platelet reduction *Proc. Natl Acad. Sci. USA* **61** 46–52

Gatenby R A and Gillies R J 2004 Why do cancers have high aerobic glycolysis? *Nat. Rev. Cancer* **4** 891–9

Gelbart M A, He B H, Martin A C, Thiberge S Y, Wieschaus E F and Kaschube M 2012 Volume conservation principle involved in cell lengthening and nucleus movement during tissue morphogenesis *Proc. Natl Acad. Sci. USA* **109** 19298–303

Gerlach J P, Emmink B L, Nojima H, Kranenburg O and Maurice M M 2014 Wnt signalling induces accumulation of phosphorylated β-catenin in two distinct cytosolic complexes *Open Biol.* **4** 140120

Gerlitz G and Bustin M 2010 Efficient cell migration requires global chromatin condensation *J. Cell Sci.* **123** 2207–217

Gerweck L E, Kozin S V and Stocks S J 1999 The pH partition theory predicts the accumulation and toxicity of doxorubicin in normal and low-pH-adapted cells *Br. J. Cancer* **79** 838–42

Giancotti F G 2000 Complexity and specificity of integrin signalling *Nat. Cell Biol.* **2** E13–4

Gibson M C, Patel A B, Nagpal R and Perrimon N 2006 The emergence of geometric order in proliferating metazoan epithelia *Nature* **442** 1038–41

Gilcrease M Z, Zhou X, Lu X, Woodward W A, Hall B E and Morrissey P J 2009 $\alpha 6\beta 4$ integrin crosslinking induces EGFR clustering and promotes EGF-mediated Rho activation in breast cancer *J. Exp. Clin. Cancer Res.* **28** 67

Gillies R J, Liu Z and Bhujwalla Z 1994 31P-MRS measurements of extracellular pH of tumors using 3-aminopropylphosphonate *Am. J. Physiol.* **267** C195–203

Gillies R J, Martinez-Zaguilan R, Martinez G M, Serrano R and Perona R 1990 Tumorigenic 3T3 cells maintain an alkaline intracellular pH under physiological conditions *Proc. Natl Acad. Sci. USA* **87** 7414–8

Gillies R J, Verduzco D and Gatenby R A 2012 Evolutionary dynamics of carcinogenesis and why targeted therapy does not work *Nat. Rev. Cancer* **12** 487–93

Gimbrone M A, Cotran R S and Folkman J 1974 Tumor growth and neovascularization: an experimental model using rabbit cornea *J. Natl Cancer Inst.* **52** 413–27

Gimbrone M A, Leapman S B, Cotran R S and Folkman J 1972 Tumor dormancy *in vivo* by prevention of neovascularization *J. Exp. Med.* **136** 261–76

Gimbrone M A, Leapman S, Cotran R S and Folkman J 1973 Tumor angiogenesis: iris neovascularization at a distance from experimental intraocular tumors *J. Natl Cancer Inst.* **50** 219–28

Gjorevski N and Nelson C M 2012 Mapping of mechanical strains and stresses around quiescent engineered three-dimensional epithelial tissues *Biophys. J.* **103** 152–62

Glazier J A and Graner F 1993 Simulation of the differential adhesion driven rearrangement of biological cells *Phys. Rev.* E **47** 2128–54

Glitsch M 2011 Protons and Ca^{2+}: ionic allies in tumor progression? *Physiology (Bethesda)* **26** 252–65

Goldman A, Majumder B, Dhawan A, Ravi S, Goldman D, Kohandel M, Majumder P K and Sengupta S 2015 Temporally sequenced anticancer drugs overcome adaptive resistance by targeting a vulnerable chemotherapy-induced phenotypic transition *Nat. Commun.* **6** 6139

Gong J, Wang D, Sun L, Zborowska E, Willson J K and Brattain M G 1997 Role of $\alpha 5\beta 1$ integrin in determining malignant properties of colon carcinoma cells *Cell Growth Differ.* **8** 83–90

Goodall G J 2011 An autocrine TGF-beta/ZEB/miR-200 signaling network regulates establishment and maintenance of epithelial-mesenchymal transition *Mol. Biol. Cell* **22** 1686–98

Gordon R, Goel N S, Steinberg M S and Wiseman L L 1972 A rheological mechanism sufficient to explain the kinetics of cell sorting *J. Theor. Biol.* **37** 43–73

Grainge C L, Lau L C K, Ward J A, Dulay V, Lahiff G, Wilson S, Holgate S, Davies D E and Howarth P H 2011 Effect of bronchoconstriction on airway remodeling in asthma *N. Engl. J. Med.* **364** 2006–15

Graner F and Glazier J A 1992 Simulation of biological cell sorting using a two-dimensional extended Potts model *Phys. Rev. Lett.* **69** 2013–6

Grashoff C *et al* 2010 Measuring mechanical tension across vinculin reveals regulation of focal adhesion dynamics *Nature* **466** 263–66

Green J B A 2008 Sophistications of cell sorting *Nat. Cell Biol.* **10** 375–7

Gregory P A *et al* 2008 The miR-200 family and miR-205 regulate epithelial to mesenchymal transition by targeting ZEB1 and SIP1 *Nat. Cell Biol.* **10** 593–601

Gregory P A, Bracken C P and Smith E *et al* 2011 An autocrine TGF-β/ZEB/miR-200 signaling network regulates establishment and maintenance of epithelial–mesenchymal transition *Mol. Biol. Cell* **22** 1686–98

Griffioen A W, Damen C A, Martinotti S, Blijham G H and Groenewegen G 1996 Endothelial intercellular adhesion molecule-1 expression is suppressed in human malignancies: the role of angiogenic factors *Cancer Res.* **56** 1111–7

Griffiths J R, Stevens A N, Iles R A, Gordon R E and Shaw D 1981 31P-NMR investigation of solid tumours in the living rat *Biosci. Rep.* **1** 319–25

Grigore A D, Jolly M K, Jia D, Farach-Carson M C and Levine H 2016 Tumor budding: the name is EMT. Partial EMT *J. Clin. Med.* **5** 51

Grosse-Wilde A, Fouquier d'Hérouël A, McIntosh E, Ertaylan G, Skupin A, Kuestner R E, del Sol A, Walters K A and Huang S 2015 Stemness of the hybrid epithelial/mesenchymal state in breast cancer and its association with poor survival *PLoS ONE* **10** 51

Gu Z, Noss E H, Hsu V W and Brenner M B 2011 Integrins traffic rapidly via circular dorsal ruffles and macropinocytosis during stimulated cell migration *J. Cell Biol.* **193** 61–70

Guck J, Schinkinger S and Lincoln B *et al* 2005 Optical deformability as an inherent cell marker for testing malignant transformation and metastatic competence *Biophys. J.* **88** 3689–98

Gumbiner B M 2005 Regulation of cadherin-mediated adhesion in morphogenesis *Nat. Rev. Mol. Cell Biol.* **6** 622–34

Gundem G, Van Loo P and Kremeyer B *et al* 2015 The evolutionary history of lethal metastatic prostate cancer *Nature* **520** 353–7

Gupta P B, Kuperwasser C, Brunet J P, Ramaswamy S, Kuo W L, Gray J W, Naber S P and Weinberg R A 2005 The melanocyte differentiation program predisposes to metastasis after neoplastic transformation *Nat. Genet.* **37** 1047–54

Haberkorn U, Strauss L G, Reisser C, Haag D, Dimitrakopou-lou A, Zigler S, Oberdorfer F, Rudat V and VanKaick G 1991 Glucoseuptake, perfusion, and cell proliferation in head and neck tumors: relation of positron emission tomography to flow cytometry *J. Nucl. Med.* **32** 1548–55

Haeger A, Krause M, Wolf K and Friedl P 2014 Cell jamming: collective invasion of mesenchymal tumor cells imposed by tissue confinement *Biochim. Biophys. Acta* **1840** 2386–95

Haeger A, Wolf K, Zegers M M and Friedl P 2015 Collective cell migration: guidance principles and hierarchies *Trends Cell Biol.* **25** 556–66

Hale C M, Shrestha A L, Khatau S B, Stewart-Hutchinson P J, Hernandez L, Stewart C L, Hodzic D and Wirtz D 2008 Dysfunctional connections between the nucleus and the actin and microtubule networks in laminopathic models *Biophys. J.* **95** 5462–75

Hales C T 2001 The honeycomb conjecture *Discrete Comput. Geometry* **25** 1–22

Hanahan D and Weinberg R A 2000 The hallmarks of cancer *Cell* **100** 57–70

Hanahan D and Weinberg R A 2011 Hallmarks of cancer: the next generation *Cell* **144** 646–74

Hardee M E, Dewhirst M W, Agarwal N and Sorg B S 2009 Novel imaging provides new insights into mechanisms of oxygen transport in tumors *Curr. Mol. Med.* **9** 435–41

Harris A K 1976 Is cell sorting caused by differences in the work of intercellular adhesion? A critique of the Steinberg hypothesis *J. Theor. Biol.* **61** 267–85

Harris A L 2002 Hypoxia—a key regulatory factor in tumour growth *Nat. Rev. Cancer* **2** 38–47

Hartwell K A, Muir B, Reinhardt F, Carpenter A E, Sgroi D C and Weinberg R A 2006 The Spemann organizer gene, Goosecoid, promotes tumor metastasis *Proc. Natl Acad. Sci. USA* **103** 18969–74

Hase K, Shatney C, Johnson D, Trollope M and Vierra M 1993 Prognostic value of tumor 'budding' in patients with colorectal cancer *Dis. Colon Rectum.* **36** 627–35

Hashim A I, Zhang X, Wojtkowiak J W, Martinez G V and Gillies R J 2011 Imaging pH and metastasis *NMR Biomed.* **24** 582–91

Hatzikirou H, Basanta D, Simon M, Schaller K and Deutsch A 2012 'Go or grow': The key to the emergence of invasion in tumour progression? *Math. Med. Biol.* **29** 49–65

Hay E D 1995 An overview of epithelio–mesenchymal transformation *Acta Anat.* **154** 8–20

Hellman A, Zlotorynski E, Scherer S W, Cheung J, Vincent J B, Smith D I, Trakhtenbrot L and Kerem B 2002 A role for common fragile site induction in amplification of human oncogenes *Cancer Cell* **1** 89–97

Henkes S, Fily Y and Marchetti M C 2011 Active jamming: self-propelled soft particles at high density *Phys. Rev.* E **84** 040301

Henkes S, van Hecke M and van Saarloos W 2010 Critical jamming of frictional grains in the generalized isostaticity picture *Europhys. Lett.* **90** 14003

Hertwig O 1893 Ueber den Werth der ersten Furchungszellen für die Organbildung des Embryos. Experimentelle Studien am Frosch-und Tritonei *Arch. Mikrosk. Anat.* **42** 662–807

Hida K, Hida Y, Amin D N, Flint A F, Panigrahy D, Morton C C and Klagsbrun M 2004 Tumor-associated endothelial cells with cytogenetic abnormalities *Cancer Res.* **64** 8249–55

Hockel S, Schlenger K, Vaupel P and Hockel M 2001 Association between host tissue vascularity and the prognostically relevant tumor vascularity in human cervical cancer *Int. J. Oncol.* **19** 827–32

Holgate S T 2008 The airway epithelium is central to the pathogenesis of asthma *Allergol. Int.* **57** 1–10

Holmes S J 1914 The behavior of the epidermis of amphibians when cultivated outside the body *J. Exp. Zool.* **17** 281–95

Holtfreter J 1939 Gewebeaffinität, ein Mittel der embryonalen Formbildung *Arch. Exp. Zellforsch.* **23** 169–209

Homer R J and Elias J A 2005 Airway remodeling in asthma: therapeutic implications of mechanisms *Physiology* **20** 28–35

Hong T, Watanabe K, Ta C H, Villarreal-Ponce A, Nie Q and Dai X 2015 An Ovol2-Zeb1 mutual inhibitory circuit governs bidirectional and multi-step transition between epithelial and mesenchymal states *PLoS Comput. Biol.* **11** 51

Horwitz A R and Parsons J T 1999 Cell migration–movin' on *Science* **286** 1102–3

Hsu P P and Sabatini D M 2008 Cancer cell metabolism: Warburg and beyond *Cell* **134** 703–7

Huang B, Jolly M K, Lu M, Tsarfaty I, Onuchic J N and Ben-Jacob E 2015 Modeling the transitions between collective and solitary migration phenotypes in cancer metastasis *Sci. Rep.* **5** 17379

Huang J, Zarnitsyna V I, Liu B, Edwards L J, Jiang N, Evavold B D and Zhu C 2010 The kinetics of two-dimensional TCR and pMHC interactions determine T-cell responsiveness *Nature* **464** 932–6

Huang R Y J, Wong M K and Tan T Z *et al* 2013 An EMT spectrum defines an anoikis-resistant and spheroidogenic intermediate mesenchymal state that is sensitive to e-cadherin restoration by a src-kinase inhibitor, saracatinib (AZD0530) *Cell Death Dis.* **4** e915

Huang S and Ingber D E 2005 Cell tension, matrix mechanics, and cancer development *Cancer Cell* **8** 175–6

Huber H L and Koessler K K 1922 The pathology of bronchial asthma *Arch. Intern. Med.* **30** 689–760

Hufnagel L, A Teleman A A, Rouault H, Cohen S M and Shraiman B I 2007 On the mechanism of wing size determination in fly development *Proc. Natl Acad. Sci. USA* **104** 3835

Hutson M S, Tokutake Y, Chang M S, Bloor J W, Venakides S, Kiehart D P and Edwards G S 2003 Forces for morphogenesis investigated with laser microsurgery and quantitative modeling *Science* **300** 145–9

Hynes R O 2002 Integrins: bidirectional, allosteric signaling machines *Cell* **110** 673–87

Ikeda A, Berthier L and Sollich P 2012 Unified study of glass and jamming rheology in soft particle systems *Phys. Rev. Lett.* **109** 018301

Imai T 1960 Growth patterns in human carcinoma. Their classification and relation to prognosis *Obstet. Gynecol.* **16** 296–308

Ince T A, Richardson A L, Bell G W, Saitoh M, Godar S, Karnoub A E, Iglehart J D and Weinberg R A 2007 Transformation of different human breast epithelial cell types leads to distinct tumor phenotypes *Cancer Cell* **12** 160–70

Ingber D E 1997 Tensegrity: the architectural basis of cellular mechanotransduction *Annu. Reb. Physiol* **59** 575–99

Itoh Y, Ito N, Nagase H and Seiki M 2008 The second dimer interface of MT1-MMP, the transmembrane domain, is essential for ProMMP-2 activation on the cell surface *J. Biol. Chem.* **283** 13053–62

Jacob K, Sollier C and Jabado N 2007 Circulating tumor cells: detection, molecular profiling and future prospects *Expert Rev. Proteomics* **4** 741–56

Jain R K 1987 Transport of molecules in the tumor interstitium: a review *Cancer Res.* **47** 3039–51

Jass J R, Barker M, Fraser L, Walsh M D, Whitehall V L J, Gabrielli B, Young J and Leggett B A 2003 APC mutation and tumour budding in colorectal cancer *J. Clin. Pathol.* **56** 69–73

Jeevan D S, Cooper J B, Braun A, Murali R A J and Jhanwar-uniyal M 2016 Molecular pathways mediating metastases to the brain via epithelial-to-mesenchymal transition: genes, proteins, and functional analysis *Anticancer Res.* **36** 523–32

Jenkins P F, Benfield G F and Smith A P 1981 Predicting recovery from acute severe asthma *Thorax* **36** 835–41

Jensen D H, Dabelsteen E, Specht L, Fiehn A M, Therkildsen M H, Jønson L, Vikesaa J, Nielsen F C and von Buchwald C 2015 Molecular profiling of tumour budding implicates TGF-β-mediated epithelial-mesenchymal transition as a therapeutic target in oral squamous cell carcinoma *J. Pathol.* **236** 505–16

Jia D, Jolly M K, Boareto M, Parsana P, Mooney S M, Pienta K J, Levine H and Ben-Jacob E 2015 OVOL guides the epithelial–hybrid–mesenchymal transition *Oncotarget* **6** 15436–48

Jiang J, Jia P, Zhao Z and Shen B 2014 Key regulators in prostate cancer identified by co-expression module analysis *BMC Genomics* **15** 1015

Jolly M K, Boareto M, Huang B, Jia D, Lu M, Ben-Jacob E, Onuchic J N and Levine H 2015 Implications of the hybrid epithelial/mesenchymal phenotype in metastasis *Front. Oncol.* **5** 1–19

Jolly M K, Huang B, Lu M, Mani S A, Levine H and Ben-Jacob E 2014 Towards elucidating the connection between epithelial–mesenchymal transitions and stemness *J. R. Soc. Interface* **11** 20140962

Jolly M K, Jia D, Boareto M, Mani S A, Pienta K J, Ben-Jacob E and Levine H 2015 Coupling the modules of EMT and stemness: a tunable 'stemness window' model *Oncotarget.* **6** 25161–74

Jolly M K, Tripathi S C, Jia D, Mooney S M, Celiktas M, Hanash S M, Mani S A, Pienta K J, Ben-Jacob E and Levine H 2016 Stability of the hybrid epithelial/mesenchymal phentoype *Oncotarget* **7** 27067–84

Jones R G and Thompson C B 2009 Tumor suppressors and cell metabolism: a recipe for cancer growth *Genes Dev.* **23** 537–48

Jones S *et al* 2008 Comparative lesion sequencing provides insights into tumor evolution *Proc. Natl Acad. Sci. USA* **105** 4283–8

Jordan N V, Johnson G L and Abell A N 2011 Tracking the intermediate stages of epithelial–mesenchymal transition in epithelial stem cells and cancer *Cell Cycle* **10** 2865–73

Jung A, Schrauder M, Oswald U, Knoll C, Sellberg P, Palmqvist R, Niedobitek G, Brabletz T and Kirchner T 2001 The invasion front of human colorectal adenocarcinomas shows co-localization of nuclear beta-catenin, cyclin D1, and p16INK4A and is a region of low proliferation *Am. J. Pathol.* **159** 1613–7

Jung Y S, Liu X W, Chirco R, Warner R B, Fridman R and Kim H R C 2012 TIMP-1 induces an EMT-like phenotypic conversion in MDCK cells independent of its MMP-inhibitory domain *PLoS ONE* **7** 51

Kabla A J 2012 Collective cell migration: leadership, invasion and segregation *J. R. Soc. Interface* **9** 3268–78

Kainulainen T, Hakkinen L, Hamidi S, Larjava K, Kallioinen M, Peltonen J, Salo T, Larjava H and Oikarinen A 1998 Laminin-5 expression is independent of the injury and the microenvironment during reepithelialization of wounds *J. Histochem. Cytochem.* **46** 353–60

Kalluri R and Weinberg R A 2009 The basics of epithelial-mesenchymal transition *J. Clin. Invest.* **119** 1420–8

Kam Y and Quaranta V 2009 Cadherin-bound β-catenin feeds into the Wnt pathway upon adherens junctions dissociation: evidence for an intersection between β-catenin pools *PLoS One* **24** e4580

Kamoun W S, Chae S S, Lacorre D A, Tyrrell J A, Mitre M, Gillissen M A, Fukumura D, Jain R K and Munn L L 2010 Simultaneous measurement of RBC velocity, flux, hematocrit and shear rate in vascular networks *Nat. Methods* **7** 655–60

Kanai Y, Oda T, Tsuda H, Ochiai A and Hirohashi S 1994 Point mutation of the E-cadherin gene in invasive lobular carcinoma of the breast *Japan J Cancer Res.* **85** 1035–9

Kasza K E, Farrell D L and Zallen J A 2014 Spatiotemporal control of epithelial remodeling by regulated myosin phosphorylation *Proc. Natl Acad. Sci. USA* **111** 11732–7

Kedrin D, Gligorijevic B, Wyckoff J, Verkhusha V V, Condeelis J, Segall J E and van Rheenen J 2008 Intravital imaging of metastatic behavior through a mammary imaging window *Nat. Methods* **5** 1019–21

Kennedy K M and Dewhirst M W 2010 Tumor metabolism of lactate: the influence and therapeutic potential for MCT and CD147 regulation *Future Oncol.* **6** 127–48

Khuon S, Liang L, Dettman R W, Sporn P H, Wysolmerski R B and Chew T L 2010 Myosin light chain kinase mediates transcellular intravasation of breast cancer cells through the underlying endothelial cells: a three-dimensional FRET study *J. Cell Sci.* **123** 431–40

Kienast Y, von Baumgarten L, Fuhrmann M, Klinkert W E, Goldbrunner R, Herms J and Winkler F 2010 Real-time imaging reveals the single steps of brain metastasis formation *Nat. Med.* **16** 116–22

Kim J H, Serra-Picamal X and Tambe D T *et al* 2013 Propulsion and navigation within the advancing monolayer sheet *Nat. Mater.* **12** 856

Kim S and Hilgenfeldt S 2015 Cell shapes and patterns as quantitative indicators of tissue stress in the plant epidermis *Soft Matter.* **11** 7270–5

Kirsch M, Strasser J, Allende R, Bello L, Zhang J and Black P M 1998 Angiostatin suppresses malignant glioma growth *in vivo Cancer Res.* **58** 4654–9

Klagsbrun M and Moses M A 1999 Molecular angiogenesis *Chem. Biol.* **6** R217–24

Klein M, Seeger P, Schuricht B, Alper S L and Schwab A 2000 Polarization of Na+/H+ and Cl −/HCO3−exchangers in migrating renal epithelial cells *J. Gen. Physiol.* **115** 599–608

Knight D A, Lane C L and Stick S M 2004 Does aberrant activation of the epithelial–mesenchymal trophic unit play a key role in asthma or is it an unimportant sideshow? *Curr. Opin. Pharmacol.* **4** 251–6

Knudson A G 1971 Mutation and cancer: statistical study of retinoblastoma *Proc. Natl Acad. Sci. USA* **68** 820–3

Koch T M, Münster S, Bonakdar N, Butler J P and Fabry B 2012 3D traction forces in cancer cell invasion *PLoS ONE* **7** e33476
Koelzer V H, Langer R, Zlobec I and Lugli A 2016 Tumor budding in colorectal cancer—ready for diagnostic practice? *Hum. Pathol.* **47** 4–19
Kohler I, Bronsert P, Timme S, Werner M, Brabletz T, Hopt U T, Schilling O, Bausch D, Keck T and Wellner U F 2015 Detailed analysis of epithelial–mesenchymal transition and tumor budding identifies predictors of long-term survival in pancreatic ductal adenocarcinoma *J. Gastroenterol. Hepatol.* **30** 78–84
Konerding M A, Malkusch W, Klapthor B, van Ackern C, Fait E, Hill S A, Parkins C, Chaplin D J, Presta M and Denekamp J 1999 Evidence for characteristic vascular patterns in solid tumours: quantitative studies using corrosion casts *Br. J. Cancer* **80** 724–32
Koukourakis M I, Giatromanolaki A, Harris A L and Sivridis E 2006 Comparison of metabolic pathways between cancer cells and stromal cells in colorectal carcinomas: a metabolic survival role for tumor-associated stroma *Cancer Res.* **66** 632–7
Kraeling B M, Wiederschain D G, Boehm T, Rehn M, Mulliken J B and Moses M A 1999 The role of matrix metalloproteinase activity in the maturation of human capillary endothelial cells *in vitro J. Cell Sci.* **112** 1599–609
Krens S F G, Moellmert S and Heisenberg C P 2011 Enveloping cell-layer differentiation at the surface of zebrafish germ-layer tissue explants *Proc. Natl Acad. Sci. USA* **108** E9
Krieg M, Arboleda-Estudillo Y, Puech P H, Kaefer J, Graner F, Mueller D J and Heisenberg C P 2008 Tensile forces govern germ-layer organization in zebrafish *Nat. Cell Biol.* **10** 429–36
Krishnan R, Klumpers D D and Park C Y 2011 Substrate stiffening promotes endothelial monolayer disruption through enhanced physical forces *Am. J. Physiol. Cell Physiol.* **300** C146–54
Krishnan R, Park C Y and Lin Y C *et al* 2009 Reinforcement versus fluidization in cytoskeletal mechanoresponsiveness *PLoS One* **4** e5486
Kroemer G and Pouyssegur J 2008 Tumor cell metabolism: cancer's Achilles' heel *Cancer Cell* **13** 472–82
Kumar A, Coleman I and Morrissey C *et al* 2016 Substantial interindividual and limited intraindividual genomic diversity among tumors from men with metastatic prostate cancer *Nat. Med.* **22** 369–78
Kumar S and Weaver V M 2009 Mechanics, malignancy, and metastasis: the force journey of a tumor cell *Cancer Metas. Rev.* **28** 113–27
Kuvaja P, Wuertz S Ø, Talvensaari-Mattila A, Brünner N, Paakko P and Turpeenniemi-Hujanen T 2007 High serum TIMP-1 correlates with poor prognosis in breast carcinoma—a validation study *Cancer Biomark.* **3** 293–300
Labernadie A, Kato T and Brugués A *et al* 2017 A mechanically active heterotypic E-cadherin/N-cadherin adhesion enables fibroblasts to drive cancer cell invasion *Nat. Cell Biol.* **19** 224–37
Ladoux B 2009 Cells guided on their journey *Nature Physics* **5** 377–8
Laemmermann T and Sixt M 2009 Mechanical modes of 'amoeboid' cell migration *Curr. Opin. Cell Biol.* **21** 636–44
Laevsky G and Knecht D A 2003 Cross-linking of actin filaments by myosin II is a major contributor to cortical integrity and cell motility in restrictive environments *J. Cell Sci.* **116** 3761–70

Lagana A, Vadnais J, Le P U, Nguyen T N, Laprade R, Nabi I R and Noel J 2000 Regulation of the formation of tumor cell pseudopodia by the Na+/H+ exchanger NHE1 *J. Cell Sci.* **113** 3649–62

Lambrecht B N and Hammad H 2012 The airway epithelium in asthma *Nat. Med.* **18** 684–92

Lammerding J, Schulze P C, Takahashi T, Kozlov S, Sullivan T, Kamm R D, Stewart C L and Lee R T 2004 Lamin A/C deficiency causes defective nuclear mechanics and mechanotransduction *J. Clin. Invest.* **113** 370–8

Laemmermann T *et al* 2008 Rapid leukocyte migration by integrin-independent flowing and squeezing *Nature* **453** 51–5

Laplante C and Nilson L A 2006 Differential expression of the adhesion molecule Echinoid drives epithelial morphogenesis in *Drosophila Development* **133** 3255–64

Laser-Azogui A, Diamant-Levi T, Israeli S, Roytman Y and Tsarfaty I 2014 Met-induced membrane blebbing leads to amoeboid cell motility and invasion *Oncogene* **33** 1788–98

Lazaar A L and Panettier R A Jr 2003 Is airway remodeling clinically relevant in asthma? *Am. J. Med.* **115** 652–59

Lee J S, Hale C M, Panorchan P, Khatau S B, George J P, Tseng Y, Stewart C L, Hodzic D and Wirtz D 2007 Nuclear lamin A/C deficiency induces defects in cell mechanics, polarization, and migration *Biophys. J.* **93** 2542–52

Lee K H, Choi E Y and Hyun M S *et al* 2008 Role of hepatocyte growth factor/c-Met signaling in regulating urokinase plasminogen activator on invasiveness in human hepatocellular carcinoma: a potential therapeutic target *Clin. Exp. Metas.* **25** 89–96

Lee S W, Jung K H, Jeong C H, Seo J H, Yoon D K, Suh J K, Kim K W and Kim W J 2011 Inhibition of endothelial cell migration through the down-regulation of MMP-9 by A-kinase anchoring protein 12 *Mol. Med. Rep.* **4** 145–9

Legant W R, Miller J S, Blakely B L, Cohen D M, Genin G M and Chen C S 2010 Measurement of mechanical tractions exerted by cells in three-dimensional matrices *Nat. Methods* **7** 969–71

Lei K, Zhang X, Ding X, Guo X, Chen M, Zhu B, Xu T, Zhuang Y, Xu R and Han M 2009 SUN1 and SUN2 play critical but partially redundant roles in anchoring nuclei in skeletal muscle cells in mice *Proc. Natl Acad. Sci. USA* **106** 10207–12

Leroy P and Mostov K E 2007 Slug is required for cell survival during partial epithelial-mesenchymal transition of HGF-induced tubulogenesis *J. Cell Sci.* **18** 1943–52

Levental K R, Yu H and Kass L *et al* 2009 Matrix crosslinking forces tumor progression by enhancing integrin signaling *Cell* **139** 891–906

Levin S M 2002 The tensegrity-truss as a model for spine mechanics: biotensegrity *J. Mech. Med. Biol.* **2** 3750

Li B and Sun S X 2015 Coherent motions in confluent cell monolayer sheets *Biophys. J.* **107** 1532–41

Li H, Gerald W L and Benezra R 2004 Utilization of bone marrow-derived endothelial cell precursors in spontaneous prostate tumors varies with tumor grade *Cancer Res.* **64** 6137–43

Li X Y, Ota I, Yana I, Sabeh F and Weiss S J 2008 Molecular dissection of the structural machinery underlying the tissue-invasive activity of membrane type-1 matrix metalloproteinase *Mol. Biol. Cell* **19** 3221–33

Li X-F, Carlin S, Urano M, Russell J, Ling C C and O'Donoghue J A 2007 Visualization of hypoxia in microscopic tumors by immunofluorescent microscopy *Cancer Res.* **67** 7646–53

Liang F, Cao W, Wang Y, Li L, Zhang G and Wang Z 2013 The prognostic value of tumor budding in invasive breast cancer *Pathol. Res. Pract.* **209** 269–75

Linder S, Wiesner C and Himmel M 2011 Degrading devices: invadosomes in proteolytic cell invasion *Annu. Rev. Cell Dev. Biol.* .**27** 185–211
Liu A J and Nagel S R 1998 Jamming is not just cool any more *Nature* **396** 21–2
Liu A J and Nagel S R 2010 The jamming transition and the marginally jammed solid *Annu. Rev. Condens. Matter Phys.* **1** 347
Liu C H, Nagel S R, Schecter D A, Coppersmith S N, Majumdar S, Narayan O and Witten T A 1995 Force fluctuations in bead packs *Science* **269** 513–5
Liu J, He X, Qi Y, Tian X, Monkley S J, Critchley D R, Corbett S A, Lowry S F, Graham A M and Li S 2011 Talin1 regulates integrin turnover to promote embryonic epithelial morphogenesis *Mol. Cell. Biol.* **31** 3366–77
Liu Z, Tan J L, Cohen D M, Yang M T, Sniadecki N J, Ruiz S A, Nelson C M and Chen C S 2010 Mechanical tugging force regulates the size of cell-cell junctions *Proc. Natl Acad. Sci. USA* **107** 9944–9
Loftus J C and Liddington R C 1997 Cell adhesion in vascular biology. New insights into integrin-ligand interaction *J. Clin. Invest.* **99** 2302–6
Lokody I 2014 Tumor-promoting tissue mechanics *Nat. Rev. Cancer* **14** 296–7
Lovisa S, LeBleu V S and Tampe B *et al* 2015 Epithelial-to-mesenchymal transition induces cell cycle arrest and parenchymal damage in renal fibrosis *Nat. Med.* **21** 998–1009
Lozzio C B and Lozzio B B 1975 Human chronic myelogenous leukemia cell-line with positive Philadelphia chromosome *Blood* **45** 321–34
Lu M, Jolly M K, Levine H, Onuchic J N and Ben-Jacob E 2013 MicroRNA-based regulation of epithelial–hybrid–mesenchymal fate determination *Proc. Natl Acad. Sci. USA* **110** 18174–9
Lu P, Weaver V M and Werb Z 2012 The extracellular matrix: a dynamic niche in cancer progression *J. Cell Biol.* **196** 395–406
Ludwig M G, Vanek M, Guerini D, Gasser J A, Jones C E, Junker U, Hofstetter H, Wolf R M and Seuwen K 2003 Proton-sensing G-protein-coupled receptors *Nature* **425** 93–8
Lugli A, Karamitopoulou E and Zlobec I 2012 Tumour budding: a promising parameter in colorectal cancer *Br. J. Cancer* **106** 1713–7
Lugli A, Karamitopoulou E and Panayiotides I *et al* 2009 CD8+ lymphocytes/tumour-budding index: an independent prognostic factor representing a 'pro-/anti-tumour' approach to tumour host interaction in colorectal cancer *Br. J. Cancer* **101** 1382–92
Lundgren K, Nordenskjoeld B and Landberg G 2009 Hypoxia, Snail and incomplete epithelial–mesenchymal transition in breast cancer *Br. J. Cancer* **101** 1769–81
Luo W R, Gao F, Li S Y and Yao K T 2012 Tumour budding and the expression of cancer stem cell marker aldehyde dehydrogenase 1 in nasopharyngeal carcinoma *Histopathology* **61** 1072–81
Ly D P, Zazzali K M and Corbett S A 2003 *De novo* expression of the integrin $\alpha 5\beta 1$ regulates $\alpha v\beta 3$-mediated adhesion and migration on fibrinogen *J. Biol. Chem.* **278** 21878–85
Maître J L, Berthoumieux H, Krens S F, Salbreux G, Juelicher F, Paluch E and Heisenberg C P 2012 Adhesion functions in cell sorting by mechanically coupling the cortices of adhering cells *Science* **338** 253–6
Maniotis A J, Folberg R, Hess A, Seftor E A, Gardner L M, Peer J, Trent J M, Meltzer P S and Hendrix M J 1999 Vascular channel formation by human melanoma cells *in vivo* and *in vitro*: vasculogenic mimicry *Am. J. Pathol.* **155** 739–52

Manning M L, Foty R A, Steinberg M S and Schoetz E M 2010 Coaction of intercellular adhesion and cortical tension specifies tissue surface tension *Proc. Natl Acad. Sci. USA* **107** 12517–22

Marshall B T, Long M, Piper J W, Yago T, McEver R P and Zhu C 2003 Direct observation of catch bonds involving cell-adhesion molecules *Nature* **423** 190–3

Martinez-Zaguilan R, Lynch R M, Martinez G M and Gillies R J 1993 Vacuolar-type H(þ)-ATPases are functionally expressed in plasma membranes of human tumor cells *Am. J. Physiol.* **265** C1015–29

Maruthamuthu V and Gardel M L 2014 Protrusive activity guides changes in cell–cell tension during epithelial cell scattering *Biophys. J.* **107** 555–63

Maruthamuthu V, Sabass B, Schwarz U S and Gardel M L 2011 Cell-ECM traction force modulates endogenous tension at cell-cell contacts *Proc. Natl Acad. Sci. USA* **108** 4708–13

Masugi Y, Yamazaki K, Hibi T, Aiura K, Kitagawa Y and Sakamoto M 2010 Solitary cell infiltration is a novel indicator of poor prognosis and epithelial-mesenchymal transition in pancreatic cancer *Hum. Pathol.* **41** 1061–8

Mattsson J, Wyss H M, Fernandez-Nieves A, Miyazaki K, Hu Z, Reichman D R and Weitz D A 2009 Soft colloids make strong glasses *Nature* **462** 83–6

Matus D Q, Lohmer L L, Kelley L C, Schindler A J, Kohrman A Q, Barkoulas M, Zhang W, Chi Q and Sherwood D R 2015 Invasive cell fate requires G1 cell-cycle arrest and histone deacetylase-mediated changes in gene expression *Dev. Cell* **35** 162–74

Max N, Harbaum L, Pollheimer M J, Lindtner R A, Kornprat P and Langner C 2016 Tumour budding with and without admixed inflammation: two different sides of the same coin? *Br. J. Cancer* **114** 368–71

Mayor R and Etienne-Manneville S 2016 The front and rear of collective cell migration *Nat. Rev. Mol. Cell Biol.* **17** 97–109

McCarthy K, Maguire T, McGreal G, McDermott E, O'Higgins N and Duffy M J 1999 High levels of tissue inhibitor of metalloproteinase-1 predict poor outcome in patients with breast cancer *Int. J. Cancer* **84** 44–8

McCarty O J, Mousa S A, Bray P F and Konstantopoulos K 2000 Immobilized platelets support human colon carcinoma cell tethering, rolling, and firm adhesion under dynamic flow conditions *Blood* **96** 1789–97

McConkey D J and Orrenius S 1996 Signal transduction pathways in apoptosis *Stem Cells* **14** 619–31

McDonald D M and Choyke P L 2003 Imaging of angiogenesis: from microscope to clinic *Nat. Med.* **9** 713–25

McLean L A, Roscoe J, Jorgensen N K, Gorin F A and Cala P M 2000 Malignant gliomas display altered pH regulation by NHE1 compared with nontransformed astrocytes *Am. J. Physiol. Cell Physiol.* **278** C676–88

Merlo L M F, Pepper J W, Reid B J and Maley C C 2006 Cancer as an evolutionary and ecological process *Nat. Rev. Cancer* **6** 923–35

Mertz A, Banerjee S, Che Y, German G K, Xu X, Hyland C, Marchetti M C, Horsley V and Dufresne E R 2012 Scaling of traction forces with the size of cohesive cell colonies *Phys. Rev. Lett.* **108** 198101

Meyer A S, Hughes-Alford S K, Kay J E, Castillo A, Wells A, Gertler F B and Lauffenburger D A 2012 2D protrusion but not motility predicts growth factor-induced cancer cell migration in 3D collagen *J. Cell Biol.* **197** 721–9

Micalizzi D S, Farabaugh S M and Ford H L 2010 Epithelial–mesenchymal transition in cancer: parallels between normal development and tumor progression *J. Mammary Gland Biol. Neoplasia* **15** 117–34

Micchelli C A and Blair S S 1999 Dorsoventral lineage restriction in wing imaginal discs requires notch *Nature* **401** 473–6

Mierke C T 2009 The role of vinculin in the regulation of the mechanical properties of cells *Cell Biochem. Biophys.* **53** 115–26

Mierke C T 2011 Cancer cells regulate biomechanical properties of human microvascular endothelial cells *J. Biol. Chem.* **286** 40025–37

Mierke C T 2013a The integrin $\alpha v\beta 3$ increases cellular stiffness and cytoskeletal remodeling dynamics to facilitate cancer cell invasion *New J. Phys.* **15** 015003

Mierke C T 2013b Physical break-down of the classical view on cancer cell invasion and metastasis *Eur. J. Cell Biol.* **92** 89–104

Mierke C T 2014 The fundamental role of mechanical properties in the progression of cancer disease and inflammation *Rep. Prog. Phys.* **77** 076602

Mierke C T, Frey B, Fellner M, Herrmann M and Fabry B 2011 Integrin $\alpha 5\beta 1$ facilitates cancer cell invasion through enhanced contractile forces *J. Cell Sci.* **124** 369–83

Mierke C T, Kollmannsberger P, Paranhos-Zitterbart D, Smith J, Fabry B and Goldmann W H 2008b Mechano-coupling and regulation of contractility by the vinculin tail domain *Biophys. J.* **94** 661–70

Mierke C T, Kollmannsberger P, Zitterbart D P, Diez G, Koch T M, Marg S, Ziegler W H, Goldmann W H and Fabry B 2010 Vinculin facilitates cell invasion into three-dimensional collagen matrices *J. Biol. Chem.* **285** 13121–30

Mierke C T, Rosel D, Fabry B and Brabek J 2008c Contractile forces in tumor cell migration *Eur. J. Cell Biol.* **87** 669–76

Mierke C T, Zitterbart D P, Kollmannsberger P, Raupach C, Schlotzer-Schrehardt U, Goecke T W, Behrens J and Fabry B 2008a Breakdown of the endothelial barrier function in tumor cell transmigration *Biophys. J.* **94** 2832–46

Mills J C, Stone N L, Erhardt J and Pittman R N 1998 Apoptotic membrane blebbing is regulated by myosin light chain phosphorylation *J Cell Biol.* **140** 627–36

Mitrovic B, Schaeffer D F, Riddell R H and Kirsch R 2012 Tumor budding in colorectal carcinoma: time to take notice *Mod. Pathol.* **25** 1315–25

Monier B, Pélissier-Monier A and Sanson B 2011 Establishment and maintenance of compartmental boundaries: role of contractile actomyosin barriers *Cell. Mol. Life Sci.* **68** 1897–910

Monthus C and Bouchaud J P 1996 Models of traps and glass phenomenology *J. Phys. A: Math. Gen.* **29** 3847

Moolgavkar S H and Knudson A G 1981 Mutation and cancer: a model for human carcinogenesis *J. Natl Cancer Inst.* **66** 1037–52

Morbini P, Inghilleri S, Campo I, Oggionni T, Zorzetto M and Luisetti M 2011 Incomplete expression of epithelial–mesenchymal transition markers in idiopathic pulmonary fibrosis *Pathol. Res. Pract.* **207** 559–67

Morita T, Nagaki T, Fukuda I and Okumura K 1992 Clastogenicity of low pH to various cultured mammalian cells *Mutat. Res.* **268** 297–305

Morodomi T, Isomoto H, Shirouzu K, Kakegawa K, Irie K and Morimatsu M 1989 An index for estimating the probability of lymph node metastasis in rectal cancers. Lymph node metastasis and the histopathology of actively invasive regions of cancer *Cancer* **63** 539–43

Mouw J K *et al* 2014 Tissue mechanics modulate microRNA-dependent PTEN expression to regulate malignant progression *Nature Med.* **20** 360–7
Munjal A, Philippe J M, Munro E and Lecuit T 2015 A self-organized biomechanical network drives shape changes during tissue morphogenesis *Nature* **524** 351–5
Nagai T and Honda H 2001 A dynamic cell model for the formation of epithelial tissues *Philos. Mag.* B **81** 699
Nagy J A, Chang S H, Dvorak A M and Dvorak H F 2009 Why are tumor blood vessels abnormal and why is it important to know? *Br. J. Cancer* **100** 865–9
Neff N T, Lowrey C, Decker C, Tovar A, Damsky C, Buck C and Horwitz A F 1982 A monoclonal antibody detaches embryonic skeletal muscle from extracellular matrices *J. Cell Biol.* **95** 654–66
Newman S A 2012 Physico-genetic determinants in the evolution of development *Science* **338** 217–9
Ng M R, Besser A, Brugge J S and Danuser G 2014 Mapping the dynamics of force transduction at cell-cell junctions of epithelial clusters *Elife* **3** 1–29
Ni R, Stuart M A C and Dijkstra M 2013 Pushing the glass transition towards random close packing using self-propelled hard spheres *Nat. Commun.* **4** 2704
Niessen C M and Gumbiner B M 2002 Cadherin-mediated cell sorting not determined by binding or adhesion specificity *J. Cell Biol.* **156** 389–99
Nieto M A 2013 Epithelial plasticity: a common theme in embryonic and cancer cells *Science* **342** 1234850
Ninomiya H, David R, Damm E W, Fagotto F, Niessen C M and Winklbauer R 2012 Cadherin-dependent differential cell adhesion in *Xenopus* causes cell sorting *in vitro* but not in the embryo *J. Cell Sci.* **125** 1877–83
Niwa Y, Yamada S, Koike M, Kanda M, Fujii T, Nakayama G, Sugimoto H, Nomoto S, Fujiwara M and Kodera Y 2014 Epithelial to mesenchymal transition correlates with tumor budding and predicts prognosis in esophageal squamous cell carcinoma *J. Surg. Oncol.* **110** 764–69
Nnetu K D, Knorr M, Kaes J and Zink M 2012 The impact of jammingon boundaries of collectively moving weak-interacting cells *New J. Phys.* **14** 115012
Nnetu K, Knorr M, Pawlizak S, Fuhs T and Kaes J 2013 Slow and anomalous dynamics of an MCF-10A epithelial cell monolayer *Soft Matter* **9** 9335–41
Notbohm J, Banerjee S, Utuje K J C, Gweon B, Jang H, Park Y, Shin J, Butler J P, Fredberg J J and Marchetti M C 2016 Cellular contraction and polarization drive collective cellular motion *Biophys. J.* **110** 2729–38
Nowell P C 1976 The clonal evolution of tumor cell populations *Science* **194** 23–8
Okado Y, Aoki M, Hamasaki M, Koga K, Sueta T, Shiratsuchi H, Oda Y, Nakagawa T and Nabeshima K 2015 Tumor budding and laminin5-γ2 in squamous cell carcinoma of the external auditory canal are associated with shorter survival *Springerplus* **4** 814
Oliver M, Kovats T, Mijailovich S, Butler J, Fredberg J and Lenormand G 2010 Remodeling of integrated contractile tissues and its dependence on strain-rate amplitude *Phys. Rev. Lett.* **105** 158102
Olsson P and Teitel S 2013 Athermal jamming versus thermalized glassiness in sheared frictionless particles *Phys. Rev.* E **88** 010301

Onder T, Gupta P B, Mani S A, Yang J, Lander E S and Weinberg R A 2008 Loss of E-cadherin promotes metastasis via multiple downstream transcriptional pathways *Cancer Res.* **68** 3645–54

Otrock Z K, Mahfouz R A R, Makarem J A and Shamseddine A I 2007 Understanding the biology of angiogenesis: review of the most important molecular mechanisms *Blood Cells Mol. Dis.* **39** 212–20

Padera T P, Stoll B R, Tooredman J B, Capen D, di Tomaso E and Jain R K 2004 Pathology: cancer cells compress intra tumor vessels *Nature* **427** 695

Paget S 1989 The distribution of secondary growths in cancer of the breast *Lancet* **1** 571–3

Palecek S P, Loftus J C, Ginsberg M H, Lauffenburger D A and Horwitz A F 1997 Integrin–ligand binding properties govern cell migration speed through cell–substratum adhesiveness *Nature* **385** 537–40

Palmqvist R, Rutegard J N, Bozoky B, Landberg G and Stenling R 2000 Human colorectal cancers with an intact p16/cyclin D1/pRb pathway have up-regulated p16 expression and decreased proliferation in small invasive tumor clusters *Am. J. Pathol.* **157** 1947–53

Paluch E K and Raz E 2013 The role and regulation of blebs in cell migration *Curr. Opin. Cell Biol.* **25** 582–90

Park J A, Fredberg J J and Drazen J M 2015a Putting the squeeze on airway epithelia *Physiology* **30** 293–303

Park J A, Kim J H and Bi D 2015b Unjamming and cell shape in the asthmatic airway epithelium *Nat. Mater.* **14** 1040–8

Park S M, Gaur A B, Lengyel E and Peter M E 2008 The miR-200 family determines the epithelial phenotype of cancer cells by targeting the E-cadherin repressors ZEB1 and ZEB2 *Genes Dev.* **22** 894–907

Patan S 2004 Vasculogenesis and angiogenesis *Cancer Treat. Res.* **117** 3–32

Paterson E L, Kazenwadel J, Bert A G, Khew-Goodall Y, Ruszkiewicz A and Goodall G J 2013 Down-regulation of the miRNA-200 family at the invasive front of colorectal cancers with degraded basement membrane indicates EMT is involved in cancer progression *Neoplasia* **15** 180–91

Pathak A and Kumar S 2013 Transforming potential and matrix stiffness co-regulate confinement sensitivity of tumor cell migration *Integr. Biol. (Camb.)* **5** 1067–75

Pattabiraman D R, Bierie B, Kober K I, Thiru P, Krall J A, Zill C, Reinhardt F, Tam W L and Weinberg R A 2016 Activation of PKA leads to mesenchymal-to-epithelial transition and loss of tumor-initiating ability *Science* **351** aad3680

Pawar S C, Demetriou M C, Nagle R B, Bowden G T and Cress A E 2007 Integrin $\alpha 6$ cleavage: a novel modification to modulate cell migration *Exp. Cell Res.* **313** 1080–9

Pawlizak S *et al* 2015 Testing the differential adhesion hypothesis across the epithelial-mesenchymal transition *New J. Phys.* **17** 083049

Peinado H, Olmeda D and Cano A 2007 Snail, Zeb and bHLH factors in tumor progression: an alliance against the epithelial phenotype? *Nat. Rev. Cancer* **7** 415–28

Perou C M, Sørlie T and Eisen M B *et al* 2000 Molecular portraits of human breast tumours *Nature* **406** 747–52

Petheram I S, Jones D A and Collins J V 1979 Patterns of recovery of airflow obstruction in severe acute asthma *Postgrad. Med. J.* **55** 877–80

Petitjean L, Reffay M, Grasland-Mongrain E, Poujade M, Ladoux B, Buguin A and Silberzan P 2010 Velocity fields in a collectively migrating epithelium *Biophys. J.* **98** 1790–800

Petrie R J, Gavara N, Chadwick R S and Yamada K M 2012 Nonpolarized signaling reveals two distinct modes of 3D cell migration *J. Cell Biol.* **197** 439–55

Phillips H M and Davis G S 1978 Liquid tissue mechanics in amphibian gastrulation: germ-layer assembly in Rana pipiens *Am. Zool.* **18** 81–93

Phillips H M and Steinberg M S 1978 Embryonic tissues as elasticoviscous liquids: I. Rapid and slow shape changes in centrifuged cell aggregates *J. Cell Sci.* **30** 1–20

Phillips H M, Steinberg M S and Lipton B H 1977 Embryonic tissues as elasticoviscous liquids: II. Direct evidence for cell slippage in centrifuged aggregates *Dev. Biol.* **59** 124–34

Pinheiro C, Longatto-Filho A, Ferreira L, Pereira S M M, Etlinger D, Moreira M A R, Jube L F, Queiroz G S, Schmitt F and Baltazar F 2008 Increasing expression of monocarboxylate transporters 1 and 4 along progression to invasive cervical carcinoma *Int. J. Gynecol. Pathol.* **27** 568–74

Poincloux R, Collin O, Lizarraga F, Romao M, Debray M, Piel M and Chavrier P 2011 Contractility of the cell rear drives invasion of breast tumor cells in 3D Matrigel *Proc. Natl Acad. Sci. USA* **108** 1943–8

Ponti A, Machacek M, Gupton S L, Waterman-Storer C M and Danuser G 2004 Two distinct actin networks drive the protrusion of migrating cells *Science* **305** 1782–6

Potter V R 1958 The biochemical approach to the cancer problem *Fed. Proc* **17** 691–7

Poujade M, Grasland-Mongrain E, Hertzog A, Jouanneau J, Chavrier P, Ladoux B, Buguin A and Silberzan P 2007 Collective migration of an epithelial monolayer in response to a model wound *Proc. Natl Acad. Sci. USA* **104** 15988–93

Pouyssegur J, Chambard J C, Franchi A, Paris S and Van Obberghen-Schilling E 1982 Growth factor activation of an amiloride-sensitive Na+/H+ exchange system in quiescent fibroblasts: coupling to ribosomal protein S6 phosphorylation *Proc. Natl Acad. Sci. USA* **79** 3935–9

Prager-Khoutorsky M, Lichtenstein A, Krishnan R, Rajendran K, Mayo A, Kam Z, Geiger B and Bershadsky A D 2011 Fibroblast polarization is a matrix-rigidity-dependent process controlled by focal adhesion mechanosensing *Nat. Cell Biol* .**13** 1457–65

Puech P H, Taubenberger A, Ulrich F, Krieg M, Muller D J and Heisenberg C P 2005 Measuring cell adhesion forces of primary gastrulating cells from zebrafish using atomic force microscopy *J. Cell Sci.* **118** 4199–206

Puliafito A, Hufnagel L, Neveu P, Streichan S, Sigal A, Fygenson D K and Shraiman B I 2012 Collective and single cell behavior in epithelial contact inhibition *Proc. Natl Acad. Sci. USA* **109** 739–44

Putney L K and Barber D L 2003 Na–H exchange-dependent increase in intracellular pH times G2/M entry and transition *J. Biol. Chem.* **278** 44645–9

Qian B-Z and Pollard J W 2010 Macrophage diversity enhances tumor progression and metastasis *Cell* **141** 39–51

Radisky D C 2011 miR-200c at the nexus of epithelial-mesenchymal transition, resistance to apoptosis, and the breast cancer stem cell phenotype *Breast Cancer Res.* **13** 110

Raghunand N, He X, van Sluis R, Mahoney B, Baggett B, Taylor C W, Paine-Murrieta G, Roe D, Bhujwalla Z M and Gillies R J 1999 Enhancement of chemotherapy by manipulation of tumour pH *Br. J. Cancer* **80** 1005–11

Rape A D, Guo W H and Wang Y L 2011 The regulation of traction force in relation to cell shape and focal adhesions *Biomaterials* **32** 2043–51

Raupach C, Paranhos-Zitterbart D, Mierke C, Metzner C, Mueller A F and Fabry B 2007 Stress fluctuations and motion of cytoskeletal-bound markers *Phys. Rev.* E **76** 011918
Reffay M, Parrini M C, Cochet-Escartin O, Ladoux B, Buguin A, Coscoy S, Amblard F, Camonis J and Silberzan P 2014 Interplay of RhoA and mechanical forces in collective cell migration driven by leader cells *Nat. Cell Biol.* **16** 217–23
Reitman Z J and Yan H 2010 Isocitrate dehydrogenase 1 and 2 mutations in cancer: alterations at a crossroads of cellular metabolism *J. Natl Cancer Inst.* **102** 932–41
Renkawitz J and Sixt M 2010 Mechanisms of force generation and force transmission during interstitial leukocyte migration *EMBO Rep.* **11** 744–50
Reshkin S J, Bellizzi A, Caldeira S, Albarani V, Malanchi I, Poignee M, Alunni-Fabbroni M, Casavola V and Tommasino M 2000 Na+/H+ exchanger-dependent intracellular alkalinization is an early event in malignant transformation and plays an essential role in the development of subsequent transformation-associated phenotypes *FASEB J.* **14** 2185–97
Rhim A D *et al* 2012 EMT and dissemination precede pancreatic tumor formation *Cell* **148** 349–61
Ribatti D 2009 Endogenous inhibitors of angiogenesis: a historical review *Leuk. Res.* **33** 638–44
Ribeiro A S and Paredes J 2015 P-cadherin linking breast cancer stem cells and invasion: a promising marker to identify an 'intermediate/metastable' EMT state *Front. Oncol* **4** 371
Rice J 2012 The rude awakening *Nature* **485** 55–7
Ricono J M, Huang M, Barnes L A, Lau S K, Weis S M, Schlaepfer D D, Hanks S K and Cheresh D A 2009 Specific cross-talk between epidermal growth factor receptor and integrin $\alpha v\beta 5$ promotes carcinoma cell invasion and metastasis *Cancer Res.* **69** 1383–91
Ridley A J 2011 Life at the leading edge *Cell* **145** 1012–22
Riedle S, Kiefel H, Gast D, Bondong S, Wolterink S, Gutwein P and Altevogt P 2009 Nuclear translocation and signalling of L1-CAM in human carcinoma cells requires ADAM10 and presenilin/gamma-secretase activity *Biochem. J.* **420** 391–402
Riveline D, Zamir E, Balaban N Q, Schwarz U S, Ishizaki T, Narumiya S, Kam Z, Geiger B and Bershadsky A D 2001 Focal contacts as mechanosensors: externally applied local mechanical force induces growth of focal contacts by an mDia1-dependent and ROCK-independent mechanism *J. Cell Biol.* **153** 1175–86
Robson E J D, Khaled W T, Abell K and Watson C J 2006 Epithelial-to-mesenchymal transition confers resistance to apoptosis in three murine mammary epithelial cell lines *Differentiation* **74** 254–64
Roche W R, Beasley R, Williams J H and Holgate S T 1989 Subepithelial fibrosis in the bronchi of asthmatics *Lancet* **333** 520–4
Rodriguez I and Basler K 1997 Control of compartmental affinity boundaries by hedgehog *Nature* **389** 614–8
Rohani N, Canty L, Luu O, Fagotto F and Winklbauer R 2011 EphrinB/EphB signaling controls embryonic germ layer separation by contact-induced cell detachment *PLoS Biol.* **9** e1000597
Rolli M, Fransvea E, Pilch J, Saven A and Felding-Habermann B 2003 Activated integrin alphavbeta3 cooperates with metalloproteinase MMP-9 in regulating migration of metastatic breast cancer cells *Proc. Natl Acad. Sci. USA* **100** 9482–7
Roos A and Boron W F 1981 Intracellular pH *Physiol. Rev.* **61** 296–434
Roesel D *et al* 2008 Up-regulation of Rho/ROCK signaling in sarcoma cells drives invasion and increased generation of protrusive forces *Mol. Cancer Res.* **6** 1410–20

Rout U K, Wang J, Paria B C and Armant D R 2004 $\alpha 5\beta 1$, $\alpha V\beta 3$ and the platelet-associated integrin $\alpha IIb\beta 3$ coordinately regulate adhesion and migration of differentiating mouse trophoblast cells *Dev. Biol.* **268** 135–51

Rowe R G and Weiss S J 2009 Navigating ECM barriers at the invasive front: the cancer cell–stroma interface *Annu. Rev. Cell Dev. Biol.* **25** 567–95

Rubashkin M G, Ou G and Weaver V M 2014 Deconstructing signaling in three dimensions *Biochemistry* **53** 2078–90

Rubio C A 2008 Arrest of cell proliferation in budding tumor cells ahead of the invading edge of colonic carcinomas a preliminary report *Anticancer Res.* **28** 2417–20

Ruddell A, Croft A, Kelly-Spratt K, Furuya M and Kemp C J 2014 Tumors induce coordinate growth of artery, vein, and lymphatic vessel triads *BMC Cancer* **14** 354

Runz S, Mierke C T, Joumaa S, Behrens J, Fabry B and Altevogt P 2008 CD24 induces localization of beta1 integrin to lipid raft domains *Biochem. Biophys. Res. Commun.* **365** 35–41

Sabeh F *et al* 2004 Tumor cell traffic through the extracellular matrix is controlled by the membrane-anchored collagenase MT1-MMP *J. Cell Biol.* **167** 769–81

Sabeh F, Shimizu-Hirota R and Weiss S J 2009 Protease-dependent versus independent cancer cell invasion programs: three-dimensional amoeboid movement revisited *J. Cell Biol.* **185** 11–9

Sadati M, Taheri Qazvini N, Krishnan R, Park C Y and Fredberg J J 2013 Collective migration and cell jamming *Differentiation* **86** 121–5

Sampson V B, David J M, Puig I, Patil P U, de Herreros A G, Thomas G V and Rajasekaran A K 2014 Wilms' tumor protein induces an epithelial–mesenchymal hybrid differentiation state in clear cell renal cell carcinoma *PLoS ONE* **9** 51

Sanitarias I, Valdés F and Álvarez A M *et al* 2002 The epithelial mesenchymal transition confers resistance to the apoptotic effects of transforming growth factor β in fetal rat hepatocytes *Mol. Cancer Res.* **1** 68–78

Sarioglu A F, Aceto N and Kojic N *et al* 2015 A microfluidic device for label-free, physical capture of circulating tumor cell clusters *Nat. Methods* **12** 685–91

Sato Y, Yoshizato T and Shiraishi Y *et al* 2013 Integrated molecular analysis of clear-cell renal cell carcinoma *Nat. Genet.* **45** 860–7

Savagner P 2010 The epithelial–mesenchymal transition (EMT) phenomenon *Ann. Oncol.* **21** vii89–92

Savagner P 2015 Epithelial–mesenchymal transitions: from cell plasticity to concept elasticity *Curr. Top. Dev. Biol.* **112** 273–300

Savagner P, Kusewitt D F, Carver E A, Magnino F, Choi C, Gridley T and Hudson L G 2005 Developmental transcription factor slug is required for effective reepithelialization by adult keratinocytes *J. Cell. Physiol.* **202** 858–66

Sawada K *et al* 2008 Loss of E-cadherin promotes ovarian cancer metastasis via α5-integrin, which is a therapeutic target *Cancer Res.* **68** 2329–39

Scarpa E, Szabó A, Bibonne A, Theveneau E, Parsons M and Mayor R 2015 Cadherin switch during EMT in neural crest cells leads to contact inhibition of locomotion via repolarization of forces *Develop. Cell* **34** 421–34

Scarpa E and Mayor R 2016 Collective cell migration in development *J. Cell Biol.* **212** 143–55

Schall P, Weitz D A and Spaepen F 2007 Structural rearrangements that govern flow in colloidal glasses *Science* **318** 1895–9

Schaller M D 2001 Paxillin: a focal adhesion-associated adaptor protein *Oncogene* **20** 6459–72

Schelter F, Halbgewachs B, Baeumler P, Neu C, Görlach A, Schroetzlmair F and Krueger A 2011 Tissue inhibitor of metalloproteinases-1-induced scattered liver metastasis is mediated by hypoxia-inducible factor-1α *Clin. Exp. Metas.* **28** 91–9

Schliekelman M J, Taguchi A and Zhu J *et al* 2015 Molecular portraits of epithelial, mesenchymal and hybrid states in lung adenocarcinoma and their relevance to survival *Cancer Res.* **75** 1789–800

Schmalhofer O, Brabletz S and Brabletz T 2009 E-cadherin, β-catenin, and ZEB1 in malignant progression of cancer *Cancer Metas. Rev.* **28** 151–66

Schmid R S, Shelton S, Stanco A, Yokota Y, Kreidberg J A and Anton E S 2004 alpha3beta1 integrin modulates neuronal migration and placement during early stages of cerebral cortical development *Development* **131** 6023–31

Schonichen A, Webb B A, Jacobson M P and Barber D L 2013 Considering protonation as a posttranslational modification regulating protein structure and function *Annu. Rev. Biophys.* **42** 289–314

Schoetz E M, Burdine R D, Juelicher F, Steinberg M S, Heisenberg C P and Foty R A 2008 Quantitative differences in tissue surface tension influence zebrafish germ layer positioning *HFSP J.* **2** 42–56

Schramm H M 2014 Should EMT of cancer cells be understood as epithelial–myeloid transition? *J Cancer* **5** 125–32

Schuijs M J, Willart M A and Vergote K *et al* 2015 Farm dust and endotoxin protect against allergy through A20 induction in lung epithelial cells *Science* **349** 1106–10

Segre P N, Prasad V, Schofield A B and Weitz D A 2001 Glasslike kinetic arrest at the colloidal–gelation transition *Phys Rev Lett.* **86** 6042–5

Seltmann K, Fritsch A, Kaes J A and Magin T M 2013 Keratins significantly contribute to cell stiffness and impact invasive behavior *PNAS* **110** 18507–12

Semenza G L 2003 Targeting HIF-1 for cancer therapy *Nat. Rev. Cancer* **3** 721–32

Semenza G L 2008 Tumor metabolism: cancer cells give and take lactate *J. Clin. Invest.* **118** 3835–37

Semenza G L 2010a HIF-1: upstream and downstream of cancer metabolism *Curr. Opin. Genet. Dev.* **20** 51–6

Semenza G L 2010b Defining the role of hypoxia-inducible factor 1 in cancer biology and therapeutics *Oncogene* **29** 625–34

Sepulveda N, Petitjean L, Cochet O, Grasland-Mongrain E, Silberzan P and Hakim V 2013 Collective cell motion in an epithelial sheet can be quantitatively described by a stochastic interacting particle model *PLoS Comput. Biol.* **9** e1002944

Serra-Picamal X, Conte V, Vincent R, Anon E, Tambe D, Bazellieres E, Butler J P, Fredberg J and Trepat X 2012 Mechanical waves during tissue expansion *Nat. Phys.* **8** 628–34

Shamir E R, Pappalardo E, Jorgens D M, Coutinho K, Tsai W T, Aziz K, Auer M, Tran P T, Bader J S and Ewald A J 2014 Twist1-induced dissemination preserves epithelial identity and requires E-cadherin *J. Cell Biol.* **204** 839–56

Silbert L E 2010 Jamming of frictional spheres and random loose packing *Soft Matter.* **6** 2918–24

Singh B, Schneider M, Knyazev P and Ullrich A 2009 UV-induced EGFR signal transactivation is dependent on proligand shedding by activated metalloproteases in skin cancer cell lines *Int. J. Cancer* **124** 531–9

Slaga T J 1983 Overview of tumor promotion in animals *Environ Health Perspect* **50** 3–14

Sollich P 1998 Rheological constitutive equation for a model of soft glassy materials *Phys Rev* E **58** 738–59

Soumya S S, Gupta A, Cugno A, Deseri L, Dayal K, Das D, Sen S and Inamdar M M 2015 Coherent motion of monolayer sheets under confinement and its pathological implications *PLoS Comput. Biol.* **11** e1004670

St Croix B *et al* 2000 Genes expressed in human tumor endothelium *Science* **289** 1197–202

Staple D B, Farhadifar R, Roeper J C, Aigouy B, Eaton S and Juelicher F 2010 Mechanics and remodelling of cell packings in epithelia *Eur. Phys. J.* E **33** 117–27

Starr D A and Han M 2003 ANChors away: an actin based mechanism of nuclear positioning *J. Cell Sci.* **116** 211–16

Starr D A, Hermann G J, Malone C J, Fixsen W, Priess J R, Horvitz H R and Han M 2001 unc-83 encodes a novel component of the nuclear envelope and is essential for proper nuclear migration *Development* **128** 5039–50

Steeg P S 2006 Tumor metastasis: mechanistic insights and clinical challenges *Nat. Med.* **12** 895–904

Steinberg M S 1962a On the mechanism of tissue reconstruction by dissociated cells: I. Population kinetics, differential adhesiveness, and the absence of directed migration *Proc. Natl Acad. Sci. USA* **48** 1577–82

Steinberg M S 1962b Mechanism of tissue reconstruction by dissociated cells: II. Time course of events *Science* **137** 762–63

Steinberg M S 1962c On the mechanism of tissue reconstruction by dissociated cells: III. Free energy relations and the reorganization of fused, heteronomic tissue fragments *Proc. Natl Acad. Sci. USA* **48** 1769–76

Steinberg M S 1963 Reconstruction of tissues by dissociated cells. Some morphogenetic tissue movements and the sorting out of embryonic cells may have a common explanation *Science* **141** 401–8

Steinberg M S 1964 The problem of adhesive selectivity in cellular interactions *Cellular Membranes in Development* (22nd Symposium of the Society for the Study of Development and Growth) Locke M (New York: Academic) pp 321–66

Steinberg M S 1970 Does differential adhesion govern self-assembly processes in histogenesis? Equilibrium configurations and the emergence of a hierarchy among populations of embryonic cells *J. Exp. Zool.* **173** 395–434

Steinberg M S 1978 Cell–cell recognition in multicellular assembly: levels of specificity *Symp. Soc. Exp. Biol.* **32** 25–49

Steinberg M S 2007 Differential adhesion in morphogenesis: a modern view *Curr. Opin. Genet. Dev.* **17** 281–86

Steinberg M S and Poole T J 1982 Liquid behavior of embryonic tissues *Cell Behavior* ed R Bellairs, A S G Curtis and G Dunn (Cambridge: Cambridge University Press) pp 583–607

Steinberg M S and Takeichi M 1994 Experimental specification of cell sorting, tissue spreading, and specific spatial patterning by quantitative differences in cadherin expression *Proc. Natl Acad. Sci. USA* **91** 206–9

Steinway S N, Zañudo J G T, Michel P J, Feith D J, Loughran T P and Albert R 2015 Combinatorial interventions inhibit TGFβ-driven epithelial-to-mesenchymal transition and support hybrid cellular phenotypes *NPJ Syst. Biol. Appl.* **1** 15014

Stewart-Hutchinson P J, Hale C M, Wirtz D and Hodzic D 2008 Structural requirements for the assembly of LINC complexes and their function in cellular mechanical stiffness *Exp. Cell Res.* **314** 1892–905

Stuewe L, Mueller M, Fabian A, Waning J, Mally S, Noel J, Schwab A and Stock C 2007 pH dependence of melanoma cell migration: protons extruded by NHE1 dominate protons of the bulk solution *J. Physiol.* **585** 351–60

Sulsky D, Childress S and Percus J K 1984 A model of cell sorting *J. Theor. Biol.* **106** 275–301

Swartz M A and Fleury M E 2007 Interstitial flow and its effects in soft tissues *Annu. Rev. Biomed. Eng.* **9** 229–56

Swetha G, Chandra V, Phadnis S and Bhonde R 2011 Glomerular parietal epithelial cells of adult murine kidney undergo EMT to generate cells with traits of renal progenitors *J. Cell. Mol. Med.* **15** 396–413

Swietach P, Patiar S, Supuran C T, Harris A L and Vaughan-Jones R D 2009 The role of carbonic anhydrase 9 in regulating extracellular and intracellular pH in three-dimensional tumor cell growths *J. Biol. Chem.* **284** 20299–310

Szabo A, Runnep, Mehes E, Twal W O, Argraves W S, Cao Y and Czirok A 2010 Collective cell motion in endothelial monolayers *Phys. Biol.* **7** 046007

Szabo B, Szöllösi G J, Gönci B, Jurányi Z, Selmeczi D and Vicsek T 2006 Phase transition in the collective migration of tissue cells: experiment and model *Phys. Rev.* E **74** 061908

Taddei M L, Giannoni E, Morandi A, Ippolito L, Ramazzotti M, Callari M, Gandellini P and Chiarugi P 2014 Mesenchymal to amoeboid transition is associated with stem-like features of melanoma cells *Cell Commun. Signal* **12** 24

Taira T, Ishii G and Nagai K *et al* 2012 Characterization of the immunophenotype of the tumor budding and its prognostic implications in squamous cell carcinoma of the lung *Lung Cancer* **76** 423–30

Takeichi M 1988 The cadherins: cell–cell adhesion molecules controlling animal morphogenesis *Development* **102** 639–55

Tam W L and Weinberg R A 2013 The epigenetics of epithelial–mesenchymal plasticity in cancer *Nat. Med.* **19** 1438–49

Tambe D T *et al* 2011 Collective cell guidance by cooperative intercellular forces *Nat. Mater.* **10** 469–75

Tambe D T, Hardin C C and Angelini T E *et al* 2011 Collective cell guidance by cooperative intercellular forces *Nat. Mater.* **10** 469–75

Tambe D, Smolensky A V, Knoll A H, Butler J P and Fredberg J J 2009 Reinforcement versus fluidization in cytoskeletal mechanoresponsiveness *PLoS ONE* **4** e5486

Tania M, Khan M A and Fu J 2014 Epithelial to mesenchymal transition inducing transcription factors and metastatic cancer *Tumor Biol.* **35** 7335–42

Tannock I F and Rotin D 1989 Acid pH in tumors and its potential for therapeutic exploitation *Cancer Res.* **49** 4373–84

Tannock I F, de Wit R and Berry W R *et al* 2004 Docetaxel plus prednisone or mitoxantrone plus prednisone for advanced prostate cancer *N. Engl. J. Med.* **351** 1502–12

Technau M and Roth S 2008 The *Drosophila* KASH domain proteins Msp-300 and Klarsicht and the SUN domain protein klaroid have no essential function during oogenesis *Fly (Austin)* **2** 82–91

Teicher B A and Fricker S P 2010 CXCL12 (SDF-1)/CXCR4 pathway in cancer *Clin. Cancer Res.* **16** 2927–31

Temin H M 1988 Evolution of cancer genes as a mutation-driven process *Cancer Res* **48** 1697–701

Theveneau E, Steventon B, Scarpa E, Garcia S, Trepat X, Streit A and Mayor R 2013 Chase-and-run between adjacent cell populations promotes directional collective migration *Nat. Cell Biol.* **15** 763–72

Thiery J P 2002 Epithelial–mesenchymal transitions in tumor progression *Nat. Rev. Cancer* **2** 442–54

Thiery J P 2003 Epithelial–mesenchymal transitions in development and pathologies *Curr. Opin. Cell Biol.* **15** 740–46

Thiery J P and Sleeman J P 2006 Complex networks orchestrate epithelial–mesenchymal transitions *Nat. Rev. Mol. Cell Biol.* **7** 131–42

Thiery J P, Acloque H, Huang R Y J and Nieto M A 2009 Epithelial–mesenchymal transitions in development and disease *Cell* **139** 871–90

Thomas R K, Baker A C and Debiasi R M *et al* 2007 High-throughput oncogene mutation profiling in human cancer *Nat. Genet.* **39** 347–51

Thomlinson R H and Gray L H 1955 The histological structure of some human lung cancers and the possible implications for radiotherapy *Br. J. Cancer* **9** 539–49

Thompson D A W 1917 *On Growth and Form* (Cambridge: Cambridge University Press) pp 88–125

Tian X J, Zhang H and Xing J 2013 Coupled reversible and irreversible bistable switches underlying TGFβ-induced epithelial to mesenchymal transition *Biophys. J.* **105** 1079–89

Tinevez J Y, Schulze U, Salbreux G, Roensch J, Joanny J F and Paluch E 2009 Role of cortical tension in bleb growth *Proc. Natl Acad. Sci. USA* **106** 18581–6

Tooley A J, Gilden J, Jacobelli J, Beemiller P, Trimble W S, Kinoshita M and Krummel M F 2008 Amoeboid T lymphocytes require the septin cytoskeleton for cortical integrity and persistent motility *Nat. Cell Biol.* **11** 17–26

Townes P L and Holtfreter J 1955 Directed movements and selective adhesion of embryonic amphibian cells *J. Exp. Zool.* **128** 53–120

Tozluoglu M, Tournier A L, Jenkins R P, Hooper S, Bates P A and Sahai E 2013 Matrix geometry determines optimal cancer cell migration strategy and modulates response to interventions *Nat. Cell Biol.* **15** 751–62

Trappe V, Prasad V, Cipelletti L, Segre P N and Weitz D A 2001 Jamming phase diagram for attractive particles *Nature* **411** 772–5

Trepat X and Fredberg J 2011 Plithotaxis and emergent dynamics in collective cellular migration *Trends Cell Biol.* **21** 638–46

Trepat X, Deng L, An S, Navajas D, Tschumperlin D, Gerthoffer W, Butler J and Fredberg J 2007 Universal physical responses to stretch in the living cell *Nature* **447** 592–5

Trepat X, Wasserman M R, Angelini T E, Millet E, Weitz D A, Butler J P and Fredberg J J 2009 Physical forces during collective cell migration *Nat. Phys.* **5** 426–30

Tschumperlin D J, Dai G and Maly I V *et al* 2004 Mechanotransduction through growth-factor shedding into the extracellular space *Nature* **429** 83–6

Tseng Y, Lee J S, Kole T P, Jiang I and Wirtz D 2004 Micro-organization and viscoelasticity of the interphase nucleus revealed by particle nanotracking *J. Cell Sci.* **117** 2159–67

Turchi L, Loubat A, Rochet N, Rossi B and Ponzio G 2000 Evidence for a direct correlation between c-Jun NH2 terminal kinase 1 activation, cyclin D2 expression, and G(1)/S phase transition in the murine hybridoma 7TD1 cells *Exp. Cell Res.* **261** 220–8

Umbreit C, Flanjak J, Weiss C, Erben P, Aderhold C, Faber A, Sternstraeter J, Hoermann K and Schultz J D 2014 Incomplete epithelial–mesenchymal transition in p16-positive squamous cell carcinoma cells correlates with β-catenin expression *Anticancer Res.* **34** 7061–70

van Aarsen L A, Leone D R, Ho S, Dolinski B M, McCoon P E and LePage D J *et al* 2008 Antibody-mediated blockade of integrin $\alpha v\beta 6$ inhibits tumor progression *in vivo* by a transforming growth factor-β-regulated mechanism *Cancer Res.* **68** 561–70

van de Vijver M *et al* 2002 A gene-expression signature as a predictor of survival in breast cancer *N. Engl. J. Med.* **347** 1999–2009

van Sluis G L, Niers T M, Esmon C T, Tigchelaar W, Richel D J, Buller H R, Van Noorden C J and Spek C A 2009 Endogenous activated protein C limits cancer cell extravasation through sphingosine-1-phosphate receptor 1-mediated vascular endothelial barrier enhancement *Blood* **114** 1968–73

van Tetering G, van Diest P, Verlaan I, van der Wall E, Kopan R and Vooijs M 2009 Metalloprotease ADAM10 is required for Notch1 site 2 cleavage *J. Biol. Chem.* **284** 31018–27

van Wyk H C, Park J, Roxburgh C, Horgan P, Foulis A and McMillan D C 2015 The role of tumour budding in predicting survival in patients with primary operable colorectal cancer: a systematic review *Cancer Treat. Rev.* **41** 151–9

Vander Heiden M G, Cantley L C and Thompson C B 2009 Understanding the Warburg effect: the metabolic requirements of cell proliferation *Science* **324** 1029–33

Vaupel P, Kallinowski F and Okunieff P 1989 Blood flow, oxygen and nutrient supply, and metabolic microenvironment of human tumors: a review *Cancer Res.* **49** 6449–65

Vedula S R K, Leong M C, Lai T L, Hersen P, Kabla A J, Lim C T and Ladoux B 2012 Emerging modes of collective cell migration induced by geometrical constraints *Proc. Natl Acad. Sci. USA* **109** 12974–9

Vega S, Morales A V, Ocaña O H, Valdés F, Fabregat I and Nieto M A 2004 Snail blocks the cell cycle and confers resistance to cell death *Genes Dev.* **18** 1131–43

Veiga S S, Elias M C Q B, Gremski W, Porcionatto M A, da Silva R, Nader H B and Brentani R R 1997 Post-translational modifications of $\alpha 5\beta 1$ integrin by glycosaminoglycan chains. The $\alpha\beta 1$ integrin is a facultative proteoglycan *J. Biol. Chem.* **272** 12529–35

Verhaak R G, Tamayo P and Yang J Y *et al* 2013 Prognostically relevant gene signatures of high-grade serous ovarian carcinoma *J. Clin. Invest.* **123** 517–25

Vesselle H, Schmidt R A, Pugsley J M, Li M, Kohlmyer S G, Vallières E and Wood D E 2000 Lung cancer proliferation correlates with [F-18] fluorodeoxyglucose uptake by positron emission tomography *Clin. Cancer Res.* **6** 3837–44

Vicente-Manzanares M, Choi C K and Horwitz A R 2009 Integrins in cell migration—the actin connection *J. Cell Sci.* **122** 199–206

Vicsek T and Zafeiris A 2012 Collective motion *Phys. Rep.* **517** 71–140

Vicsek T, Czirók A, Eshel B-J, Cohen I and Shochet O 1995 Novel type of phase transition in a system of self-driven particles *Phys Rev Lett.* **75** 1226–29

Vincent R, Bazellières E, Pérez-González C, Uroz M, Serra-Picamal X and Trepat X 2015 Active tensile modulus of an epithelial monolayer *Phys. Rev. Lett.* **115** 248103

Vitelli V and van Hecke M 2011 Soft materials: marginal matters *Nature* **480** 325–26

Voura E B, Chen N and Siu C H 2000 Platelet–endothelial cell adhesion molecule 1 (CD31) redistributes from the endothelial junction and is not required for the transendothelial migration of melanoma cells *Clin. Exp. Metas.* **18** 527–32

Waldmann R, Champigny G, Bassilana F, Heurteaux C and Lazdunski M 1997 A proton-gated cation channel involved in acid-sensing *Nature* **386** 173–7

Wang C, Huang H, Huang Z, Wang A, Chen X, Huang L, Zhou X and Liu X 2011 Tumor budding correlates with poor prognosis and epithelial–mesenchymal transition in tongue squamous cell carcinoma *J. Oral Pathol. Med* **40** 545–51

Wang C, Lisanti M P and Liaoc D J 2011 Reviewing once more the c-myc and Ras collaboration. Converging at the cyclin D1-CDK4 complex and challenging basic concepts of cancer biology *Cell Cycle* **10** 57–67

Wang G L, Jiang B H, Rue E A and Semenza G L 1995 Hypoxia-inducible factor 1 is a basic-helix-loop-helix-PAS heterodimer regulated by cellular O_2 tension *Proc. Natl Acad. Sci. USA* **92** 5510–14

Wang H, Flach H, Onizawa M, Wei L, McManus M T and Weiss A 2014 Negative regulation of Hif1α expression and TH17 differentiation by the hypoxia-regulated microRNA miR-210 *Nat. Immunol.* **15** 393–401

Wang H, Singh D and Fliegel L 1997 The Na+/H+ antiporter potentiates growth and retinoic acid-induced differentiation of P19 embryonal carcinoma cells *J. Biol. Chem.* **272** 26545–9

Warburg O 1930 *The Metabolism of Tumours* (London: Arnold Constable)

Warburg O 1956a On the origin of cancer cells *Science* **123** 309–14

Warburg O 1956b On respiratory impairment in cancer cells *Science* **124** 269–70

Warren B A, Shubik P and Feldman R 1978 Metastasis via the blood stream: the method of intravasation of tumor cells in a transplantable melanoma of the hamster *Cancer Lett.* **4** 245–51

Watanabe K, Villarreal-Ponce A, Sun P, Salmans M L, Fallahi M, Andersen B and Dai X 2014 Mammary morphogenesis and regeneration require the inhibition of EMT at terminal end buds by Ovol2 transcriptional repressor *Dev. Cell* **29** 59–74

Weaire D L and Hutzler S 1999 *The Physics of Foams* (Oxford: Oxford University Press)

Webb B A, Chimenti M, Jacobson M P and Barber D L 2011 Dysregulated pH: a perfect storm for cancer progression *Nat. Rev. Cancer* **11** 671–7

Webb D J, Donais K, Whitmore L A, Thomas S M, Turner C E, Parsons J T and Horwitz A F 2004 FAK-Src signaling through paxillin, ERK, and MLCK regulates adhesion disassembly *Nat. Cell Biol.* **6** 154–61

Weinstein I B and Joe A K 2006 Mechanisms of disease: oncogene addiction—a rationale for molecular targeting in cancer therapy *Nat. Clin. Pract. Oncol.* **3** 448–57

Weiss L 2000 Patterns of metastasis *Cancer Metastasis Rev.* **19** 281–301

Weiss P 1961 Guiding principles in cell locomotion and cell aggregation *Exp. Cell Res.* **8** 260–81

Whitcutt M J, Adler K B and Wu R 1988 A biphasic chamber system for maintaining polarity of differentiation of culture respiratory tract epithelial cells *in vitro Cell. Dev. Biol. Anim.* **24** 420–8

Wike-Hooley J L, Haveman J and Reinhold H S 1984 The relevance of tumour pH to the treatment of malignant disease *Radiother. Oncol.* **2** 343–66

Wilk G, Iwasa M, Fuller P E, Kandere-Grzybowska K and Grzybowski B A 2014 Universal area distributions in the monolayers of confluent mammalian cells *Phys. Rev. Lett.* **112** 138104

Wilson W R and Hay M P 2011 Targeting hypoxia in cancer therapy *Nat. Rev. Cancer* **11** 393–410

Wirtz D 2009 Particle-tracking microrheology of living cells: principles and applications *Annu. Rev. Biophys.* **38** 301–26

Wittchen E S, Worthylake R A, Kelly P, Casey P J, Quilliam L A and Burridge K 2005 Rap1 GTPase inhibits leukocyte transmigration by promoting endothelial barrier function *J. Biol. Chem.* **280** 11675–82

Wolf K, Mazo I, Leung H, Engelke K, von Andrian U H, Deryugina E I, Strongin A Y, Brocker E B and Friedl P 2003a Compensation mechanism in tumor cell migration: mesenchymal–amoeboid transition after blocking of pericellular proteolysis *J. Cell Biol.* **160** 267–77

Wolf K, Mueller R, Borgmann S, Broecker E B and Friedl P 2003b Amoeboid shape change and contact guidance: T-lymphocyte crawling through fibrillar collagen is independent of matrix remodeling by MMPs and other proteases *Blood* **102** 3262–9

Wolf K, Wu Y I, Liu Y, Geiger J, Tam E, Overall C, Stack M S and Friedl P 2007 Multi-step pericellular proteolysis controls the transition from individual to collective cancer cell invasion *Nat. Cell Biol.* **9** 893–904

Wood L D *et al* 2007 The genomic landscapes of human breast and colorectal cancers *Science* **318** 1108–13

Wu J *et al* 2015 Plasminogen activator inhibitor-1 inhibits angiogenic signaling by uncoupling vascular endothelial growth factor receptor-2-$\alpha v\beta 3$ integrin cross talk *Arterioscler. Thromb. Vasc. Biol.* **35** 111–20

Wyatt T P J, Harris A R, Lam M, Cheng Q, Bellis J, Dimitracopoulos A, Kabla A J, Charras G T and Baum B 2015 Emergence of homeostatic epithelial packing and stress dissipation through divisions oriented along the long cell axis *Proc. Natl. Acad. Sci. USA* **112** 5726–31

Wyss H M, Miyazaki K, Mattsson J, Hu Z, Reichman D R and Weitz D A 2007 Strain-rate frequency superposition: a rheological probe of structural relaxation in soft materials *Phys. Rev. Lett.* **98** 238303

Xiong F, Ma W and Hiscock T W *et al* 2014 Interplay of cell shape and division orientation promotes robust morphogenesis of developing epithelia *Cell* **159** 415–27

Xu J, Wang F, Van Keymeulen A, Rentel M and Bourne H R 2005 Neutrophil microtubules suppress polarity and enhance directional migration *Proc. Natl Acad. Sci. USA* **102** 6884–9

Xu Q, Mellitzer G, Robinson V and Wilkinson D G 1999 *In vivo* cell sorting in complementary segmental domains mediated by Eph receptors and ephrins *Nature* **399** 267–71

Yamada S and Nelson W J 2007 Localized zones of Rho and Rac activities drive initiation and expansion of epithelial cell–cell adhesion *J. Cell Biol.* **178** 517–27

Yamaguchi Y, Ishii G and Kojima M *et al* 2010 Histopathologic features of the tumor budding in adenocarcinoma of the lung: tumor budding as an index to predict the potential aggressiveness *J. Thorac. Oncol.* **5** 1361–8

Yang J, Mani S A, Donaher J L, Ramaswamy S, Itzykson R A, Come C, Savagner P, Gitelman I, Richardson A and Weinberg R A 2004 Twist, a master regulator of morphogenesis, plays an essential role in tumor metastasis *Cell* **117** 927–39

Yang X, Manning M L and Marchetti M C 2014 Aggregation and segregation of confined active particles *Soft Matter* **10** 6477–84

Yen K E, Bittinger M A, Su S M and Fantin V R 2010 Cancer-associated IDH mutations: biomarker and therapeutic opportunities *Oncogene* **29** 6409–17

Yokoyama S, Matsui T S and Deguchi S 2016 New wrinkling substrate assay reveals traction force fields of leader and follower cells undergoing collective migration *Biochem. Biophys. Res. Commun.* **482** 975–9

Yoshida K and Soldati T 2006 Dissection of amoeboid movement into two mechanically distinct modes *J. Cell Sci.* **119** 3833–44

Youssef J, Nurse A K, Freund L B and Morgan J R 2011 Quantification of the forces driving self-assembly of three-dimensional microtissues *Proc. Natl Acad. Sci. USA* **108** 6993–8

Zaman M H, Trapani L M, Sieminski A L, Mackellar D, Gong H, Kamm R D, Wells A, Lauffenburger D A and Matsudaira P 2006 Migration of tumor cells in 3D matrices is governed by matrix stiffness along with cell–matrix adhesion and proteolysis *Proc. Natl Acad. Sci. USA* **103** 10889–94

Zaritsky A, Welf E S, Tseng Y Y, Angeles Rabadan M, Serra-Picamal X, Trepat X and Danuser G 2015 Seeds of locally aligned motion and stress coordi- nate a collective cell migration *Biophys. J.* **109** 2492–500

Zehnder S M, Suaris M, Bellaire M M and Angelini T E 2015a Cell volume fluctuations in MDCK monolayers *Biophys. J.* **108** 247–50

Zehnder S M, Wiatt M K, Uruena J M, Dunn A C, Sawyer W G and Angelini T E 2015b Multicellular density fluctuations in epithelial monolayers *Phys. Rev.* E **92** 032729

Zhang J, Tian X J, Zhang H, Teng Y, Li R, Bai F, Elankumaran S and Xing J 2014 TGF-β–induced epithelial-to-mesenchymal transition proceeds through stepwise activation of multiple feedback loops *Sci. Signal* **7** ra91

Zhang X, Cook P C, Zindy E, Williams C J, Jowitt T A, Streuli C H, MacDonald A S and Redondo-Muñoz J 2016 Integrin $\alpha 4\beta 1$ controls G9a activity that regulates epigenetic changes and nuclear properties required for lymphocyte migration *Nucleic Acids Res.* **44** 3031–44

Zheng X, Carstens J L, Kim J, Scheible M, Kaye J, Sugimoto H, Wu C-C, LeBleu V S and Kalluri R 2015 Epithelial-to-mesenchymal transition is dispensable for metastasis but induces chemoresistance in pancreatic cancer *Nature* **527** 525–30

Zhou E H, Martinez F D and Fredberg J J 2013 Cell rheology: mush rather than machine *Nat. Mater.* **12** 184–5

Zhou E H, Trepat X, Park C Y, Lenormand G, Oliver M N, Mijailovich S M, Hardin C, Weitz D A, Butler J P and Fredberg J J 2009 Universal behavior of the osmotically compressed cell and its analogy to the colloidal glass transition *Proc. Natl Acad. Sci. USA* **106** 10632–7

Zhu C, Yago T, Lou J Z, Zarnitsyna V I and McEver R P 2008 Mechanisms for flow-enhanced cell adhesion *Ann. Biomed. Eng.* **36** 604–21

Ziegler W H, Gingras A R, Critchley D R and Emsley J 2008 Integrin connections to the cytoskeleton through talin and vinculin *Biochem. Soc. Trans.* **36** 235–9

Zijlstra A, Lewis J, Degryse B, Stuhlmann H and Quigley J P 2008 The inhibition of tumor cell intravasation and subsequent metastasis via regulation of *in vivo* tumor cell motility by the tetraspanin CD151 *Cancer Cell* **13** 221–34

Zimmermann J, Camley B A, Rappel W-J and Levine H 2016 Contact inhibition of locomotion determines cell–cell and cell–substrate forces in tissues *Proc. Natl Acad. Sci. USA* **113** 2660–5

Chapter 2

Inflammation and cancer

Summary

The link between inflammation and cancer has been proposed a long time ago and inflammation is now a major part of current cancer research. The analysis of the biophysical parameters or the determination of the prerequisites for inflammation belong to a still growing and newly emerging research field of physical tumor-immunology, which will probably set a novel future focus in the physics of cancer field. Inflammation, especially chronic inflammation, possesses pro-tumorigenic effects and it therefore increases the probability of promoting various tumor growths and the further progression of cancer. The impact of inflammation on the metastatic potential of tumors is not yet well known. It has been hypothesized that the transendothelial migration process leading to cancer metastasis is affected by inflammatory cytokines and the associated mechanical microenvironmental changes. However, the impact of acute and chronic inflammatory processes on cancer progression is still elusive, although it has been seen that the success of cancer therapy depends on the associated inflammatory processes involved in cancer disease. The mechanical properties of the tumor surrounding stroma may also change during these inflammatory processes and subsequently alter the tumorigenic potential of distinct cancer cell types or cancer cell subpopulations. However, this is not yet well known and it requires extensive future research efforts. In this chapter, the principal cellular and molecular pathways promoting tumor progression and inhibiting tumor progression are presented and discussed in the context of inflammation. Finally, the interplay between cancer and inflammation is highlighted throughout the chapter.

2.1 Inflammation: acute and chronic

Inflammation is primarily a protective response of an organism, employing host cells, blood vessels and certain proteins to fight against tissue injury and infection. This inflammatory response aims to eliminate the initial reason for the cell injury, to clear and remove necrotic cells and tissues, and to start tissue repair. Inflammation is

doi:10.1088/978-0-7503-1753-5ch2

a potentially harmful mechanism to protect against intruders, such as bacteria, fungi and viruses, and cellular as well as tissue damage. Indeed, several components of the inflammatory process are able to destroy microbes, but they may also cause further damage to uninflamed tissue surrounding the inflammatory tissue site. The inflammatory process involves white blood cells, such as leukocytes, and plasma proteins that are normally located in the blood fluid.

The goal of the inflammatory reaction is to guide these immune cells and inflammatory proteins to the sites of the tissue injury or the infection. Chemical substances secreted by the host cells initiate the onset of the inflammatory process. Among these substances are cytokines, including chemokines (figure 2.1). The inflammatory process is precisely controlled and hence self-limited. However, there are excess inflammatory reactions that lead to an inappropriate inflammatory response even when there are no foreign intruders left. This may then lead to autoimmune disease. Hence, inflammatory responses need to be regulated and controlled precisely by the immune system in order to inhibit excessive tissue damage and the involvement of surrounding normal tissue in the inflammation. What are the signs of an inflammation? The cardinal signs are heat, red staining of the tissue, swelling, pain and then loss of function. What are the distinct markers for

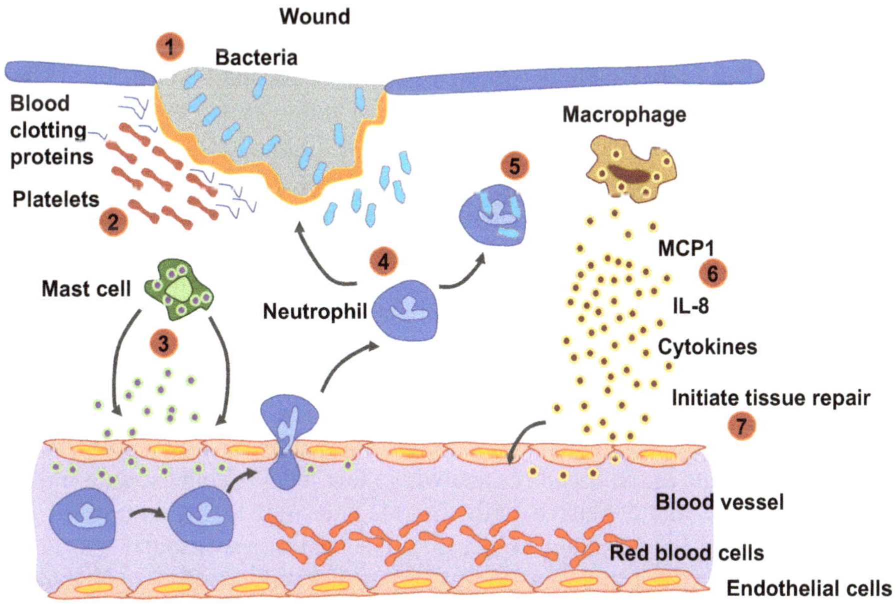

Figure 2.1. Inflammatory reaction: 1. Bacteria and other pathogens enter the tissue through a wound. 2. Platelets release blood clotting proteins at the site of the wound. 3. Mast cells secrete factors such as histamine regulating the vasodilation and vascular constriction, which cause the delivery of blood, plasma and cells to the wound. 4. Neutrophils secrete factors for killing and degrading pathogens. 5. Neutrophils and macrophages phagocytize pathogens. 6. Macrophages release hormones such as the cytokines IL-8 and MCP-1 attracting immune cells to the wound site and activate cells involved in tissue repair. 7. The inflammatory response is present until all foreign material is eliminated and the wound is closed after repair.

an acute or a chronic inflammation? First, the stimuli for acute inflammations may be infections, for example bacterial, fungal, parasitic or viral, and the toxins of microbes. Second, necrosis of tissues may be evoked by ischemia, physical or chemical injury from, for example, irradiation or environmental chemicals, and trauma. Third, the cause of inflammation can be foreign bodies, such as splinters, sutures or just dirt. Fourth, immune reactions induced by hypersensitivity reactions can evoke acute inflammation. Acute inflammation leads to vascular alterations, such as vasodilation, increased vascular permeability and elevated adhesion of leukocytes to the endothelial cell layer of blood vessels. In addition, acute inflammation induces cellular events, such as the recruitment and activation of neutrophils. In more detail, vasodilation is the reaction of blood vessels to acute inflammation and leads to alterations in the vascular diameter. Finally, these reactions decrease the entire blood pressure. The blood pressure reduction is followed by vascular leakage and edema formation, which represents the accumulation of fluid and proteins of the blood plasma in the extravascular tissue spaces (so-called interstitium). Finally, the leukocytes emigrate to extravascular tissues by undergoing sequentially the following steps, such as margination and rolling, activation of the endothelium and adhesion to the endothelial vessel lining, and transendothelial migration. Acute inflammation induces an immediate increase in vascular permeability that is facilitated by pro-inflammatory substances, such as histamine, complement factors, serotonin, bradykinin, tachykinins and nitrogen oxide species (Di Lorenzo *et al* 2009). This release of pro-inflammatory factors is simultaneously regulated with the recruitment of neutrophils to the sites of the inflammation (within the first six hours) and is followed by the next step, the recruitment of monocytes and macrophages to these inflamed tissue sites (within one day) (Walzog and Gaehtgens 2000).

The step of vasodilation alters the vessel flow in order to remove the released histamine and reduce the increased heat and redness in this tissue. Additionally, the clearance of the released histamine is important as it can also stimulate vascular smooth muscle contraction. The change in the vessel flow omits stasis, which means the slowing down of the blood vessel flow caused by hyperviscosity of the vessel fluid, the margination of circulating leukocytes and subsequently the activation of the endothelial vessel lining. This alteration follows the increased permeability of the vasculature including the formation of an early transudate (protein-poor filtrate of the blood plasma with a few cells) and an exudate (protein-rich filtrate of the blood plasma with intermediate and high amounts of cells) within extracellular tissues (Nagy *et al* 2012). The next step is vascular leakage and edema formation evoked by a change in vessel permeability caused by histamines, bradykinins and leukotrienes, which then lead to the contraction of endothelial cells and subsequently to a widening of the intercellular gaps of venules. Taken together, the endothelium lining of the vessels becomes leaky. The outpouring of the exudate (protein-rich) into the extracellular matrix of connective tissues reduces the intravascular osmotic pressure and increases the extracellular (interstitial) osmotic pressure. The last step causes the formation of edema, consisting mainly of water and ions.

Chronic inflammation has the following features. It lasts longer than acute inflammation, and other cell types such as lymphocytes, plasma cells and macrophages (monocytes) infiltrate the interstitium. Finally, this may lead to tissue destruction by inflammatory cells, which is then repaired via fibrosis and angiogenesis. The persistence of an injury or infection leads to diseases such as colitis ulcerosa and tuberculosis, which are known chronic inflammations. Prolonged exposure to a toxic agent may lead to pulmonary conditions characterized by a self-perpetuating immune reaction resulting in damage to and inflammation of tissue, for example rheumatoid arthritis, systemic lupus erythematosus and multiple sclerosis (Viatte *et al* 2013, Jacob and Stohl 2011, Leray *et al* 2010).

2.1.1 Receptors involved in leukocyte activation

Several receptors are involved in the inflammation-dependent activation of leukocytes. These are toll-like receptors (TLRs), different seven-transmembrane G-protein-coupled receptors, receptors for cytokines, such as IFN-gammaR and opsonins, together with their receptors. In addition, antibodies against specific opsonins are recognized by Fcgamma receptors (Viegas *et al* 2012). In particular, the complement system has been described as a heat-labile component of the normal blood plasma that supports the opsonization of bacteria by antibodies in order to kill them. This complements the antibacterial activity of the antibody. The complement not only acts as an effector for the antibody driven response, it can also be activated during the early onset of an infection, even in the absence of an antibody response. In more detail, the complement system consists of numerous distinct plasma proteins reacting with one another to opsonize pathogens to facilitate inflammatory responses. For example, the digestive enzyme pepsin inside cells can be secreted as an inactive precursor enzyme, the pepsinogen, which is only cleaved to pepsin in the acidic microenvironment of the stomach (termed autodigestion).

The components of the complement system, such as the precursor zymogens, are widely distributed throughout the fluids and tissues of the body. At the inflamed sites, complement factors are activated to facilitate the inflammatory progression through an enzymatic cascade. In this cascade, the cleavage of a zymogen precursor reveals an active complement enzyme. This complement enzyme cleaves its desired substrate, such as a targeted complement zymogen, which is then activated. In addition, this activated enzyme then cleaves to activate the next zymogen, and so on. Thus, the activation of a small group of complement proteins at the beginning of the inflammatory cascade is substantially amplified by each successive step of the enzymatic reaction, leading to a pronounced large complement response.

To inhibit excess activation of the complement pathway, there are regulatory mechanisms to prevemt uncontrolled complement activation. There are three distinct pathways by which the complement can be activated on the surfaces of pathogens. Although these pathways are initiated differently, they are able to generate the same effector molecules. In all of the three ways that have been discovered to date, the complement system is involved in protecting against infection. First, the complement system generates numerous activated complement proteins that are able to bind covalently to pathogens to opsonize them for the phagocytosis by phagocytes displaying complement receptors on their cell surface. Second, small fragments of distinct complement proteins can operate as chemo-attractants to recruit and accumulate phagocytes, such as macrophages, to the site of the complement activation by stimulating these phagocytes. Third, the terminal complement factors try to kill certain bacteria by generating holes (pores) in the bacterial membrane.

In summary, the complement, mainly the CR1, that recognizes breakdown products of C3 by either the classical or an alternative pathway, mediates the inflammatory process. Within the blood plasma, there are early activation proteins that are the markers of an inflammation such as the C-reactive protein (CRP), serum amyloid protein (SAP), fibronectin, fibrinogen and lectins such mannose binding lectin (MBL) (Janeway and Medzhitov 2002). They all contribute to the progression of the inflammatory cascade, which then fights against the inflammation and protects the healthy tissue and organs from that inflammation. In addition, cytokines (15–30 kD in size) are produced by stimulated lymphocytes, macrophages, the endothelium, the epithelium and the surrounding connective tissue. Chemokines (8–10 kD in size) fulfill roles as chemoattractants for specific types of leukocytes (chemotaxis process). As a secondary step, extracellular fibrillar networks are secreted by neutrophil and eosinophil granulocytes as well as mast cells in response to certain infectious pathogens and inflammatory mediators such as chemokines and cytokines (Poon *et al* 2010). Additionally, granule proteins such as the peptides of microbes and enzymes are embedded in this extracellular matrix framework. This framework provides an extracellular matrix to catch and kill microbes and induce the contact system. There are chemical mediators of inflammation that seem to be produced locally by cells at the site of primary inflammation. These cell-derived mediators are usually sequestered in intracellular granules and then they are rapidly secreted upon cellular activation by an inflammation or are synthesized *de novo* in response to a stimulus. The plasma proteins, such as complement proteins and kinins (measure blood pressure and cause dilation of blood vessels), circulate in an inactive form and are activated by an inflammation through proteolytic cleavage (Poon *et al* 2010).

2.1.2 Extravasation of inflammatory cells

Leukocyte extravasation is a step that belongs to the inflammatory cascade of acute and chronic inflammation (figure 2.2). Extravasation represents a critical step during the inflammatory response as it includes the migration of leukocytes out of the bloodstream into the targeted, inflamed tissues. At these inflamed tissue sites, leukocytes fulfill their effector function. The step of leukocyte extravasation requires the combined action of cellular adhesion receptors and chemokines to evoke the drastic morphological alterations of leukocytes and the endothelial cell vessel lining (Carman and Springer 2004). This is an active process for both cell types and leads to the extremely fast and totally efficient 'invasion' of leukocytes into the sites of inflammation, whereas the barrier function of the endothelium is not severely

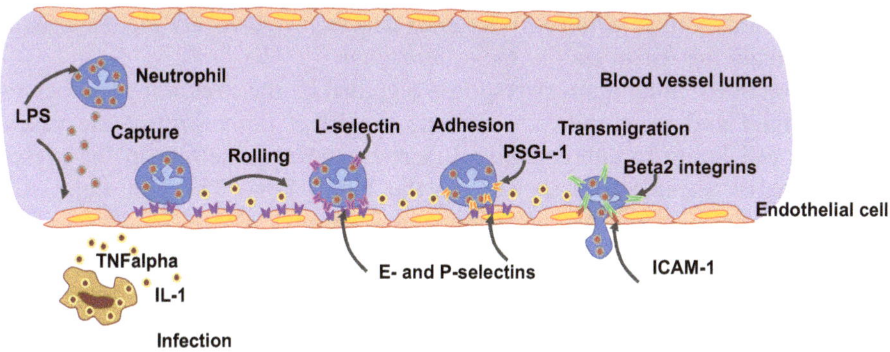

Figure 2.2. Leukocyte transendothelial migration after infection.

affected from a molecular point of view (Muller 2003). The role that the mechanical properties of this 'activated' endothelial barrier play is not yet understood. Thus, these mechanical properties of the endothelium have to be investigated in future research studies under inflammatory and non-inflammatory conditions. What is known so far? In particular, the steps of the adhesion cascade are clearly understood. Among these steps are the slow rolling motion, intraluminal crawling of leukocytes and the alternative pathways of the transcellular migration. In addition, the functional role of novel adhesion receptors is presented, as well as the spatiotemporal organization of receptors at the cell surface of membrane, together with the signaling pathways regulating the phases of the entire extravasation process.

In more detail, adhesion receptors are essential in maintaining the integrity of tissues through the regulation of many important processes, such as cellular activation, migration, growth, differentiation and death (Frenette and Wagner 1996a, 1996b). The regulation is facilitated by direct signal transduction and the alteration of intracellular signaling pathways triggered by several growth factors (Aplin *et al* 1998). It is known that cell–cell interactions are essential for the regulation of hematopoiesis (Lévesque *et al* 1999, Verfaillie 1998) and the inflammatory processes (Butcher 1991, Butcher and Picker 1996). Thus, adhesion receptors are also implicated in a wide variety of diseases, such as cardiovascular disorders and cancer involving the process of inflammation.

As the functioning of adhesion receptors is important, their coordination between adhesion receptors, the cytoskeleton and the signaling molecules is precisely regulated during the extravasation of leukocytes (a key process in the whole immune response reaction). In particular, the correct integration of outside–in and inside–out signaling in leukocytes and in the endothelial cell monolayer lining blood vessels during the extravasation step (called multi-step paradigm) is necessary for the proper function of this phenomenon (Butcher 1991, Springer 1994). The extravasation of leukocytes is not only observed during an inflammatory response, it is also found when lymphocytes recirculate to secondary lymphoid organs, but this is not addressed within this book.

The initial interactions facilitated by selectins and their ligands are the tethering and rolling of circulating leukocytes on the endothelium

The initiation of the inflammatory response is guided by the circulating leukocytes in the blood fluid and their ability to provide a contact (tethering) to the vascular endothelial cell lining to subsequently adhere to the endothelium in the presence of the dramatic shear forces exerted by the blood flow. The first steps of the extravasation process are the tethering and rolling of the leukocytes over the activated endothelial cell lining of vessel walls. The initial contact or tethering of the leukocytes is facilitated by the cell–cell interaction through selectins and their ligands, while the blood flow is still present (Alon and Ley 2008). However, it is known that selectins and their ligands interact at variable affinity to each other, which enables them to adapt rapidly to the high frequency of association and dissociation events between this leukocyte–endothelial cell connection by facilitating labile and transient tethers between the flowing leukocytes and the static endothelium (Mehta *et al* 1998, Nicholson *et al* 1998). In particular, the step of tethering decreases the velocity the leukocytes to provide their rolling over the endothelium and establish more stable interactions through integrins and their ligands. Subsequently, the adherence of leukocytes towards the endothelium is enhanced and they become tightly attached to it (Evans and Calderwood 2007).

In more detail, the P-, S- and L-selectins belong to the type 1 transmembrane glycoproteins. These selectin receptors interact in a Ca^{2+}-dependent way with fucosylated and sialylated hydrocarbon presented on the cell surface of counter-acting cells. L-selectin is mainly expressed by leukocytes and the E- and P-selectins are exposed on the cell surface of endothelial cells, when stimulated by pro-inflammatory stimuli such as tumor necrosis factor alpha (TNF-alpha). In addition, P-selectin is expressed by activated platelets (Barreiro *et al* 2004). In particular, the leukocyte selectin (L-selectin) can interact with the endothelial selectin (P- and E-selectin) ligand, the P-selectin glycoprotein ligand-1 (PSGL1) protein. However, the binding of PSGL1 to P- and E-selectin initiates the interaction of leukocytes with the endothelial lining of the blood vessel, whereas the binding of PSGL1 to L-selectin on leukocytes mediates leukocyte–leukocyte interactions, whereby the fully adhered leukocytes capture other additional non-adhered circulating leukocytes towards inflamed sites of the endothelium. In this step, it is not necessary that the non-adhered floating leukocytes express ligands for endothelial selectins (termed secondary recruitment process) (Eriksson *et al* 2001). Selectins are additionally able to bind to other glycoproteins, such as CD44 or E-selectin ligand-1 (ESL1), which are both ligands for E-selectin (Hidalgo *et al* 2007). It is suggested that each special ligand has a certain function during the capture process of neutrophils. Hence, PSGL1 is involved in the initial tethering of the leukocytes and ESL1 converts the first transient tethering behavior to a slower and more stable rolling step. In order to support secondary recruitment of leukocytes, CD44 regulates the velocity of the rolling step and interferes in the polarization step of PSGL1 and L-selectin (Hidalgo *et al* 2007). Indeed, platelets have the ability to behave as secondary recruiters for leukocytes, as they possess the capability to interact directly and simultaneously with the circulating leukocytes and the endothelial cell lining of blood vessels. Moreover, they are even able to release chemokines that were formerly immobilized on the luminal endothelial cell surface, thereby strongly supporting the adhesion process of leukocytes (von Hundelshausen *et al* 2009).

After the weak contact between leukocytes and the endothelium through selectins and their ligands, the next step is the establishment of a strong adhesion via integrins and their ligands. For example, the $\alpha 4\beta 1$- and $\alpha 4\beta 7$-integrins bind to their ligands such as the vascular cell adhesion molecule 1 (VCAM-1) and the mucosal addressin cell adhesion molecule 1(MAdCAM-1), respectively, which are both able to independently facilitate the initial tethering (Alon *et al* 1995, Berlin *et al* 1993, 1995). In line with this, the interaction between lymphocyte function-associated antigen 1 (LFA-1) and cell–cell adhesion molecule 1 (ICAM-1) supports the function of L-selectin, thereby converting the transient contact into a stable contact by reducing the velocity of the rolling step (Henderson *et al* 2001, Kadono *et al* 2002).

In general, the adhesion receptors need to be precisely localized on the surface of cells in order to fulfill their function in the correct manner during the trafficking of leukocytes (von Adrian *et al* 1995). The selectins as well as their ligands and the $\alpha 4$ integrins cluster at the tips of the microvilli of the leukocytes. For the proper tethering of the selectins, their connection to the actin cytoskeleton via connecting proteins such as α-actinin or ezrin/radixin/moesin (ERM) is necessary (Dwir *et al* 2001, Ivetic *et al* 2002, Pavalko *et al* 1995, Killock *et al* 2009).

It has been reported that selectins activate signal transduction pathways that are linked to processes such as actin cytoskeletal reorganization. These pathways are the MAPK, p56lck, Ras or Rac2 cascade (Barreiro *et al* 2004). On the other hand, PSGL1 activates different intracellular signal transduction pathways that cause an inductive effect on the activation of leukocytes, increasing the expression of molecules that are supposed to act in the following steps of the extravasation process and hence possess an effector function. Moreover, they fulfill a surprising role in the induction of tolerogenic function in dendritic cells (Urzainqui *et al* 2002, 2007, Zarbock *et al* 2007, 2008).

The prominent role of leukocyte's integrins and their ligands during activation, arrest, strong adhesion and crawling on the endothelium

In order to develop innate and acquired immunity, the leukocyte trafficking through the different tissues and organs and subsequently their interaction with other immune cells at the sites of the inflammation is necessary, and represents an important hallmark for an inflammatory response (von Andrian and Mackay 2000). Integrins have a fundamental function in cellular motility as they regulate the cell–cell and cell–extracellular matrix interactions during recirculation of leukocytes and inflammation. One feature is that integrins can alter their adhesive activity independently of their expression levels and their exposure on the cell's surface (Hynes 2002). In particular, circulating leukocytes in the bloodstream display their integrins in an inactive conformation to omit nonspecific contact with the uninflamed endothelial cell linings of blood vessels. When they enter the inflammatory focus, a rapid *in situ* activation of the integrins takes place (Campbell *et al* 1998). In this light, the spatial distribution of integrins and their ligands on

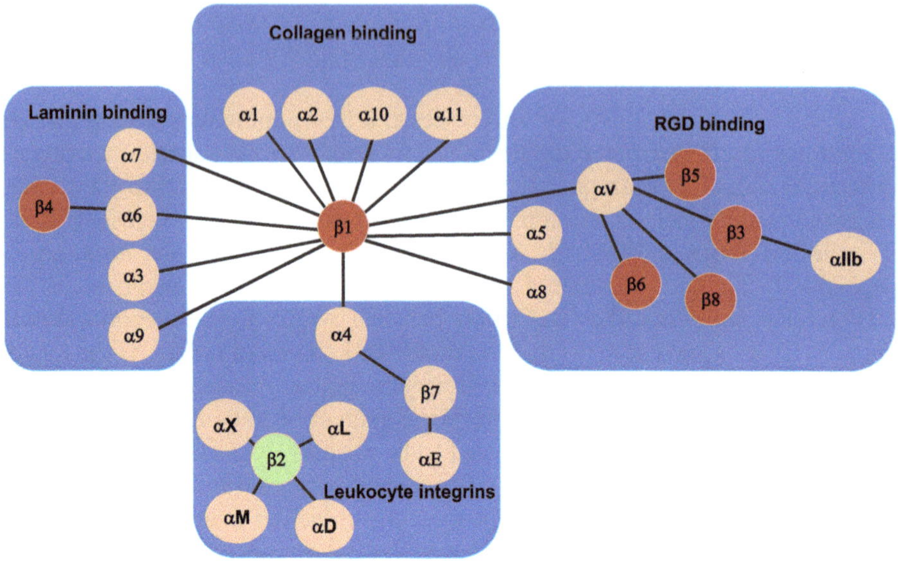

Figure 2.3. Integrin subunits and their combination to integrins.

distinct membrane regions is necessary for their full function. In order to provide the spatial distribution, a sharp regulation of the cytoskeleton is necessary, facilitating the recruitment of signaling molecules and second messengers for the activation of the cell (Vicente-Manzanares and Sánchez-Madrid 2004, Barreiro *et al* 2007).

The integrin receptors consist of α (18 different ones) and β (eight different ones) subunits building 24 heterodimeric receptors (Humphries *et al* 2006) (figure 2.3). The large number of possible integrin receptors provides dynamical adhesive properties through conformational alterations (affinity) or through changes in the spatial distribution on the cell's surface (avidity) (Carman and Springer 2003). There are at least three different conformational states of integrins hypothesized, such as a bent conformation with low affinity ('rest state'), an extended conformation (after pre-stimulation by ligand) with intermediate affinity ('partly activated state') and extended conformation with high affinity (after mechanical stimulation by the ligand, which crosslinks receptors; 'fully activated state') (Beglova *et al* 2002, Nishida *et al* 2006, Schwartz 2010). Members of the $\beta2$ integrin subfamily, such as LFA-1 (CD11a/CD18 or $\alpha L\beta2$) and the myeloid-specific integrin Mac-1 (CD11b/CD18 or $\alpha M\beta2$), and the integrins $\alpha4$ VLA-4 ($\alpha4\beta1$) and $\alpha4\beta7$, play an essential role in the adhesion of leukocytes to the endothelial cell lining of blood vessels. The majority of their ligands are transmembrane receptors of the immunoglobulin superfamily. For example, LFA-1 can bind to five intercellular adhesion molecules, such as ICAM-1 to ICAM-5, but the most important ones are ICAM-1 and ICAM-3 (Gahmberg *et al* 1990). It is known that ICAM-1 is expressed on leukocytes, dendritic cells, epithelial cells and endothelial cells. Under normal conditions, the surface the expression of ICAM-1 on these cells is low, whereas it can rapidly increase after exposure to pro-inflammatory cytokines (Dustin *et al* 1986). However,

ICAM-3 is constitutively expressed on all kinds of leukocytes (Acevedo *et al* 1993). LFA-1, can interact with other ligands such as the junctional adhesion molecule A (JAM-A), which is located on the apical region of the endothelial tight junctions. After stimulation with special pro-inflammatory factors it is partially redistributed to the apical surface of the endothelial cells (Ostermann *et al* 2002). Mac-1 binds to ICAM-1, JAM-C and the receptor for advanced glycation end-products (RAGE) (Chavakis *et al* 2003, Lamagna *et al* 2005). The integrin $\alpha4\beta1$ (VLA-4) interacts with VCAM-1 (Elices *et al* 1990), which is an adhesion receptor that is expressed *de novo* after the activation of the endothelium (Carlos and Harlan 1994) and additionally can bind to JAM-B (Cunningham *et al* 2002). In more detail, VLA-4 additionally binds to ADAM-28, thrombospondin, osteopontin, von Willebrand factor, fibronectin and the invasin bacterial protein (Mittelbrunn *et al* 2006). The $\alpha E\beta7$ integrin interacts with VCAM-1 and fibronectin, whereas it is also able to bind to MAdCAM-1, which is a receptor detected in the lymphoid tissues of the mucosa (Berlin *et al* 1993).

What role do chemokines play in the regulation of the activity of integrins?
At the onset of the tethering on the vascular endothelial cell lining, the rolling velocity of the leukocytes slows, while they are activated by the endothelial cell's luminal surface expressing immobilized chemokines and integrin ligands (Ley *et al* 2007). This leads to the arrest of leukocytes and subsequently to strong adhesion to the endothelial cell lining of vessels under normal flow characteristics (Alon *et al* 2003, Rot and von Andrian 2004). During leukocyte activation the leukocyte undergoes a morphological transformation from a rounded, circulating cell to a promigratory polarized cell with a clear cell front and rear (termed an uropod) (del Pozo *et al* 1995). This cellular polarization is helpful for the coordination of intracellular forces to facilitate the crawling step during the extravasation cascade (Geiger and Bershadsky 2002).

The chemokines can interact with the glycosaminoglycans of the apical endothelial membrane of the endothelial cell lining of blood vessels by signaling via the G-protein-coupled receptors (GPCR) located on the surface of the leukocyte's microvilli, thereby they immediately induce the outside–in signaling cascade. This cascade triggers multiple conformational alterations in the integrins (Constantin *et al* 2000, Sanchez-Madrid and del Pozo 1999, Shamri *et al* 2005). Chemokines induce complex signaling mechanisms within a short time period, which regulate the activation of integrins and support the hypothesis that there is a compartmentalized and pre-formed protein network (termed signalosomes) inside the leukocytes (Laudanna and Alon 2006). This is further supported by the presence of specific chemokines in special vascular beds that drive the selective recruitment of different leukocyte subpopulations to the side of the primary inflammation or to secondary lymphoid organs such as lymph nodes (Luster 1998). Moreover, chemokines have various effects on specific integrins, even if these are located within the same local microenvironment (Laudanna 2005).

How is the ligand-facilitated affinity of integrins altered?
The conformation of integrins is altered reversibly after chemokine-induced activation, as the conformation converts from an inactive (bent) state with low affinity to

an extended state with intermediate affinity. This process is the necessary conformation of the integrin to bind to its ligand expressed on the endothelium efficiently. In particular, integrins possess an I domain in their α subunits that changes its conformation after binding to its ligand, leading to the complete activation of the integrin and subsequently to the arrest of the leukocyte on the endothelium (Cabanas and Hogg 1993, Jun *et al* 2001, Salas *et al* 2004). Taken together, the high-affinity state of the integrin's conformation for immediate arrest of the leukocyte on the surface of the endothelial cell lining blood vessels requires the presence of bound chemokines and the ligand of the integrin (Shamri *et al* 2005, Grabovsky *et al* 2000). The $\alpha 4$ integrins represent an exception, as they contain an I-like domain within their β chains, but can directly interact with their ligands expressed on the endothelial cell's surface without the presence of a previous chemotactic trigger substance (Alon *et al* 1995).

The signaling mediated through the binding of the ligand induces the separation of the cytoplasmic regions of the subunits of the integrin, thus providing its association with the cell's cortical actin cytoskeleton. In particular, the $\alpha 4$ integrins are linked through the focal adhesion protein paxillin, whereas $\beta 2$ integrins are linked through other focal adhesion molecules, such as talin and filamin (Takala *et al* 2008). Additional integrins are recruited when a ligand is bound to the first integrin to enhance the tight adhesion of the leukocyte under blood flow stress (Dobereiner *et al* 2006). This process is termed the clustering of integrins and is triggered by the release from their tether to the cytoskeleton (regulated by protein kinase C (PKC) and calpain) to enhance the lateral mobility of integrins on the cell membrane (Stewart *et al* 1998). In different types of hematopoietic cells, the role of Rap-1 and its activator CalDAG-GEFI has been reported and also the role of kindlin-3 interacting with talin in the activation process of integrins to provide the firm adhesion of leukocytes has been shown (Mory *et al* 2008, Pasvolsky *et al* 2007). The spatial organization of the integrins is critical for their proper function. For example, the nanoclusters of LFA-1, which are not even connected to the ligand, are able to form microclusters efficiently upon ligand binding (Cairo *et al* 2006, Cambi *et al* 2006). In line with this, many studies show that the flow stress affects the integrins by reinforcing their bonds and even increasing their affinity (Marschel *et al* 2002, Zwartz *et al* 2004). The integration of chemokine-induced and external-force-induced signaling to facilitate transendothelial migration has been presented as a phenomenon called chemorheotaxis (Cinamon *et al* 2001).

How is the crawling of leukocytes altered by integrins?

The signals involved in the integrin-dependent firm adhesion of leukocytes to the endothelial cell lining of blood vessels need to be weakened to allow leukocytes to migrate to the site of the transendothelial migration initiation. In particular, the $\beta 2$ integrins are assumed to be necessary for crawling, because the blockade of $\beta 2$ integrins or their ligands evokes random migration and hence failure to find the sites for transendothelial migration at the interendothelial junctions and finally leads to defective diapedesis (Schenkel *et al* 2004). This finding is in line with *in vivo* studies investigating genetically modified mice lacking LFA-1 or Mac-1 to reveal the different underlying mechanisms for each of these $\beta 2$ integrins. The firm adhesion is triggered by LFA-1, while the crawling is facilitated by Mac-1. Both processes provide efficient cell migration (Phillipson *et al* 2006). After integrin activation through its ligand, the integrins regulate the signal transduction through different effectors to provide myosin contractility and modulate actin-remodeling GTPases as well as molecules regulating the microtubule network at both the front and the rear (uropod) of the cell. Taken together, both signals generated at the two cellular poles coordinate the motility of leukocytes (Vicente-Manzanares and Sánchez-Madrid 2004).

What is the functional role of the endothelial expressed VCAM-1 and ICAM-1 during leukocyte capture?
VCAM-1 and ICAM-1 are members of the immunoglobulin superfamily and are expressed as endothelial adhesion molecules, facilitating the binding to the integrins VLA-4 and LFA-1, respectively (Elices *et al* 1990, Marlin and Springer 1987). In more detail, ICAM-1 is slightly expressed on the non-activated endothelial cells, whereas the expression of VCAM-1 and ICAM-1 is enhanced after endothelial cell activation through pro-inflammatory cytokines such as interleukin-1 (IL-1) and tumor necrosis factor α (TNF-α) (Dustin *et al* 1986, Carlos and Harlan 1994). The binding of VCAM-1 and ICAM-1 to the actin cytoskeleton is facilitated by ezrin and moesin linking the membrane to the actin cytoskeleton to provide cortical morphogenesis and adhesion (Barreiro *et al* 2002, Heiska *et al* 1998).

Human umbilical vein endothelial cells (HUVECs) have been used to investigate the dynamics of VCAM-1 and ICAM-1 expression upon activation with TNF-α during the interaction between leukocytes and the endothelial cell lining of vessels. In particular, after the arrest of the leukocyte at the endothelial cell surface, the interaction of VCAM-1 and ICAM-1 with their ligands leads to the reorganization of the endothelial cell's cortical actin cytoskeleton and subsequently to the formation of a 3D docking structure surrounding the leukocyte. This structure includes a large accumulation of adhesion receptors and activated ezrin and moesin. Hence the docking structure should protect adhered leukocytes under normal flow conditions from detachment of the endothelial cell lining within blood vessels. This docking structure is supported by the actin cytoskeleton of the endothelial cell and contains actin-bundling proteins such as α actinin and focal adhesion proteins such as talin, paxillin and vinculin, as well as actin nucleaters such as Arp2/3. To maintain the docking structure, second messengers such as PI(4,5)P2 and the Rho/160ROCK signal transduction pathway are required (Barreiro *et al* 2002). Even in the case where only one ligand is bound to either ICAM-1 or VCAM-1, both receptors can cluster and form the endothelial docking structure. This kind of clustering occurs independently of the connection to the actin's cytoskeleton and of the assembly of ICAM-1/VCAM-1 heterodimers, because this behavior requires the inclusion of VCAM-1 and ICAM-1 in microdomains that contain many tetraspanins and hence they function as endothelial cell platforms for a special type of adhesion (Barreiro *et al* 2008). In more detail, the tetraspanins are relatively small proteins that are located in the cell membrane in a special manner, as they cross the membrane four times. In addition, their second extracellular domain interacts laterally with other neighboring integral membrane proteins (Charrin *et al* 2009). Thus, they regulate the function of the membrane by building up multiple protein domains on the cell membrane. Moreover, tetraspanins have been reported to fulfill several cellular functions such as cellular migration, homotypic and heterotypic cell–cell adhesion and the presentation of antigens (Hemler 2005, Gordon-Alonso *et al* 2006, Mittelbrunn *et al* 2002, Yáñez-Mó *et al* 1998, García-López *et al* 2005).

As microscopic analytical methods have improved significantly, the characterization of the diffusive properties of the membrane, the membrane's organization at the nanoscale and certain molecular interactions within the microdomains of the membrane have been investigated in primary human endothelial cells (Barreiro *et al* 2008). Indeed, these precise analyzes of the membrane have revealed evidence of certain endothelial adhesion platforms that differ from the so-called lipid raft structures in the cell membrane (Barreiro *et al* 2008). Using scanning electron microscopy, the membranes of endothelial cells are investigated after treatment with a specific peptide blocker of tetraspanins. It has been revealed that the nanoclustering (avidity) of VCAM-1 and ICAM-1 facilitated by the endothelial adhesion platforms is a novel molecular organization mechanism regulating the efficient adhesive ability of endothelial adhesion receptors to bind to their integrin ligands presented on the surface of leukocytes (Barreiro *et al* 2008). The functional importance of recruitment of ICAM-1 and VCAM-1 into tetraspanin microdomains on the membrane of primary human endothelial cells has been analyzed using interfering RNAs (RNAi) targeting the tetraspanins CD9 and CD151. Knockdown of CD9 and CD151 revealed that the inclusion of ICAM-1 and VCAM-1 in tetraspanin-containing domains is necessary for their proper function under a highly dynamic microenvironment such as the flow stress within blood vessels (Barreiro *et al* 2005). In a second (independent) approach, the alteration of endothelial tetraspanin microdomains by CD9-large extracellular loop (LEL)-glutathione S-transferase (GST) peptides altered the ICAM-1 and VCAM-1 functions (Barreiro *et al* 2005). In addition to VCAM-1 and ICAM-1, other adhesion receptors such as CD44, ICAM-2, JAM-A and PECAM-1 can interact with tetraspanin microdomains. In summary, tetraspanin microdomains seem to be specialized platforms that basically organize certain adhesion receptors in the special membrane regions to provide leukocyte extravasation efficiently (Barreiro *et al* 2008).

Moreover, the endothelial VCAM-1 and ICAM-1 receptors can transmit signals after binding to their ligands. The VCAM-1 receptor is supposed to play a role in the opening of interendothelial junctions to provide the extravasation of leukocytes. Thereby, VCAM-1 activates NADPH oxidase (possibly through NOX2) and produces the reactive oxygen species (ROS) that is regulated by the GTPase Rac activity, with subsequent activation of matrix metalloproteinases and reduction of VE-cadherin dependent cell–cell adhesion, which is facilitated by the Pyk-2-induced phosphorylation of β-catenin and hence supports the extravasation process (Cook-Mills 2002, van Wetering *et al* 2002, Deem and Cook-Mills 2004, van Buul *et al* 2005a, 2005b). In contrast, VCAM-1 and ICAM-1 can facilitate a rapid increase in Ca^{2+} concentration that leads to an activation of Src kinase and subsequently promotes the phosphorylation of cortactin (Etienne-Manneville *et al* 2000, Lorenzon *et al* 1998, Yang *et al* 2006a, 2006b). Additionally, ICAM-1 may activate RhoA, causing the formation of F-actin stress fibers and leading to the phosphorylation of focal adhesion kinase (FAK), paxillin and p130Cas, which are in turn involved in signaling cascades including c-Jun N-terminal kinase (JNK) and p38 pathways (Greenwood *et al* 2002, Hubbard and Rothlein 2000, Thompson *et al* 2002 Wang and Doerschuk 2002, Mierke 2013, Mierke *et al* 2017). This signaling cascade leads to increased permeability of the endothelial cell lining of blood vessels and is followed by enhanced transmigration of leukocytes through the endothelial cell lining. In addition, intercellular adhesion receptor ICAM-1 promoted the transcription of c-fos and RhoA (Thompson *et al* 2002). Finally, ICAM-1 is able to facilitate its own gene expression and additionally the expression of VCAM-1 in order to provide the transendothelial migration of leukocytes in a guided manner (Clayton *et al* 1998).

The role of integrins and their ligands during the transendothelial migration
Endothelial cell–cell junctions are pulled apart during the process of the transmigration of leukocytes to avoid injury to the endothelial monolayer or severe alterations in permeability. The permeability of the vascular beds of endothelial vessels is altered due to the specific tissue properties. In particular vascular beds of coronary, pulmonary, splanchnic and skeletal muscle consist of continuous, non-fenestrated monolayer linings of endothelial cells, which form a restrictive and strong barrier (Sukriti *et al* 2014, Aird 2007, Stevens *et al* 2008). However, organs such as the liver, kidney, and lymphatics are assembled from discontinuous and highly permeable endothelial monolayers (Mehta and Malik 2006). During the diapedesis step the leukocytes' membrane and the endothelial cells' membrane are in close contact and thereafter the membranes of the two neighboring endothelial cells interact again with each other to form cell–cell adhesions without the occurrence of any damage after the transmigration process (Muller 2003, Turowski *et al* 2008, Bamforth *et al* 1997, Carman 2009, Carman and Springer 2004, Yang *et al* 2005). When leukocytes are at a distinct location on the endothelial membrane site at which the transmigration is likely to occur, such as the preferred intercellular endothelial junctions, they generate pseudopods (termed invadosomes) between the two neighboring endothelial cells and thereby possibly explore the mechanical properties of the endothelial cell barrier. In the next step, the pseudopods are reformed into a lamella that is able to migrate into the open space between two adjacent endothelial cells on the monolayer lining of the blood vessels. During this diapedesis step, the LFA-1 molecule plays a major role as it is able to reassemble with high speed to build up a ring-shaped cluster at the contact site between the leukocyte and endothelium. At the ring-shaped cluster, LFA-1 is still able to bind to ICAM-1 (Shaw *et al* 2004) or in other cellular models to JAM-A (Woodfin *et al* 2007). After finishing the transmigration process, LFA-1 is still located in the uropod (Sandig *et al* 1997). However, other proteins, such as ICAM-2, JAM-B, JAM-C, PECAM-1 (CD31), ESAM and CD99, are involved in the transendothelial migration process. In particular, these proteins help to provide the existing endothelial cell–cell

interaction or they facilitate the leukocyte–endothelial cell interaction by additional homophilical and heterophilical interactions (Vestweber 2007, Weber *et al* 2007, Woodfin *et al* 2007, Muller 2003).

In addition to the classical diapedesis path through the endothelial cell–cell junctions (paracellular track), there is an alternative route that leukocytes use to walk through the entire individual endothelial cells (transcellular track) without altering the endothelial cell–cell junctions. The latter track may possibly involve the endothelial cells' actomyosin cytoskeleton as a guidance scaffold for the transmigration through the cytoplasm of the endothelial cell. However, the question of which track is used at different times or under different conditions is still under investigation and intense discussion. It has been suggested that the mechanical properties of the endothelial cell play a role. In addition, mechanical properties such as protrusive force generation may be of importance for the invasive transmigration behaviors of certain leukocytes. This transcellular track has been seen in the microvasculature such as the blood–brain and blood–retinal barriers as well as in the high endothelial venules of the secondary lymphoid organs (Engelhardt and Wolburg 2004, Carman and Springer 2004, Carman *et al* 2007). As the mechanism of this transcellular path during the transendothelial migration process has been investigated in more detail, the leukocytes can build up Src kinase-dependent and Wiskott–Aldrich syndrome protein (WASP) activity-dependent invasive podosomes to sense the endothelial surface. Finally, these podosomes form a transcellular pore through which the entire cell transmigrates intracellularly through the endothelial cell. In the endothelial monolayer membrane, the fusion of membranes is regulated by calcium and SNARE-containing complexes and new parts of the membrane are delivered by vesicular vacuole organelles (Carman *et al* 2007). In more detail, this refers to the translocation of ICAM-1 into caveolae after the adhesion of the leukocyte to the endothelial cell lining and the assembly of a multivesicular channel, in which ICAM-1 and caveolin-1 are located around the leukocyte's pseudopod and move directly through the entire endothelial cell. In particular, ICAM-1 and caveolin line the entire path of the leukocyte through the endothelial cell's interior and move in the direction of the basal endothelial cell's membrane (Millan *et al* 2006). In this special transmigration path, the intermediate filament vimentin is supposed to be important for the transcellular route (Nieminen *et al* 2006). The dome-shaped endothelial structures have been found to facilitate a leukocyte's transendothelial migration *in vivo* (Phillipson *et al* 2008). Taken together, the endothelial docking structures are observed to rebuild to dome-like structures that cover the leukocytes on their luminal surface of the endothelial cell lining, through which the basolateral endothelial membrane is ruptured, although the endothelial membrane is not leaky and thus can still provide the endothelial barrier function.

2.2 The dual relationship between inflammation and cancer

The inflammation of tissues is an innate immune response to alterations of tissue homeostasis. Acute and chronic inflammatory processes have the potential to crosstalk with a primary tumor at all stages of the tumor development and may

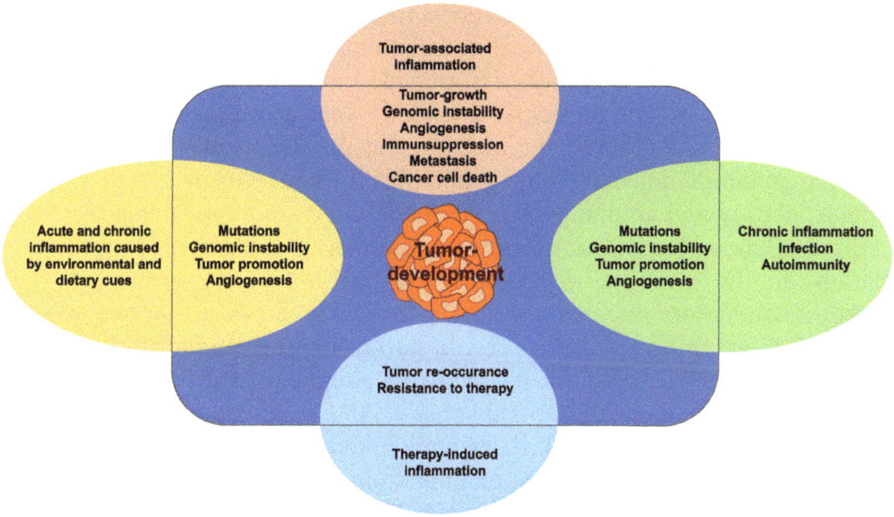

Figure 2.4. Interplay between inflammation and cancer.

have a major impact on cancer therapy (Wan *et al* 2017, Lamichhane *et al* 2017, Sun *et al* 2017). As solely less than 10% of all cancer types seem to be evoked by mutations in the germline, the cause of over 90% of cancer may be based on somatic mutations and alterations in the microenvironment surrounding primary tumors. The development of tumors is related to many factors, such as chronic infections, dietary influences, obesity, inhaled pollutants (including tobacco and asbestos), high ozone levels and autoimmune diseases, such as rheumatoid arthritis and asthma (Jemal *et al* 2010). The underlying principle of all these processes is that they are caused by a chronic inflammatory process, which is an indefinite prolonged form of a formerly protective response (acute immune response) and now even disturbs the tissue homeostasis by the uninterrupted performance of multiple rounds of the inflammatory cycle (figure 2.4) (Medzhitov 2008). In line with this, the development of cancer can be regarded in numerous cases as a deregulated formerly protective tissue repair and growth response due to tissue injury. Thus, inflammatory and neoplastic processes can co-develop into specific wounds that are no longer able to heal (Dvorak 1986). The initial connection between inflammation and tumor growth was observed by Rudolf Virchow in the nineteenth century (Virchow 1881). In his research, Virchow found that leukocyte infiltrates into the middle of primary tumors, which can be regarded as the first evidence for a hallmark of cancer described later (Hanahan and Weinberg 2011; see also chapter 1 of this book). These leukocyte infiltrates are regarded as markers of a tumor's immune surveillance and antitumor immune responses, whereas, in fact, they support tumor-suppressive and tumor-promoting effects. However, the mechanisms involved in these divergent effects are not yet well defined. Tumor infiltrating leukocytes show an altered gene expression pattern and the intratumoral abundance of various immune cell populations is manifested in their associated gene expression, which mainly supports the promotion of cancer, rather than inhibits malignant cancer progression (Danaher *et al* 2017, Lakins *et al* 2018).

There are many causes of inflammation-promoted cancer progression. However, infection is a major player in providing inflammation-mediated tumorigenesis, as approximately 20% of all cancer cases worldwide are caused by a microbial infection (Kuper *et al* 2000). At the molecular level, it is still not yet precisely known what causes the inflammation during an infection, but the signals may be different from those of an inflammation caused by no microbial or viral infection, such as noxious inhalants, obesity and autoimmunity. Although there are basic biological concepts that drive inflammation-associated cancer in various cases, there is a hypothesis that variations in the microbial ecosystems are mirrored in the variability among different inflammatory diseases and among host individuals during the manifestation of inflammation-associated cancer. In particular, common microbial elements have been characterized that are related to inflammation and the development of cancer (Grivennikov *et al* 2012, Goodwin *et al* 2011, Abdulamir *et al* 2011, Arthur *et al* 2012, Patra and Wolf 2016). Indeed, microbial organisms have been identified to facilitate the interaction between inflammation and cancer. In the following the relationship between inflammation and cancer is discussed in more detail. In addition to microbial inflammatory reactions, breast carcinomas can display high levels of lysophosphatidic acid 3 (LPA3) expression in epithelial cells or autotaxin (ATX) in stromal cells, which leads to larger tumors with nodal involvement and higher stage of the disease (Popnikolov *et al* 2012). In particular, LPA is generated from circulating lysophosphatidylcholine (LPC) facilitated by the secreted enzyme ATX. Under physiological conditions, ATX plays a role in embryonic development, tissue remodeling processes and wound healing after tissue injury.

The ATX driven LPA production induces the migration and invasion of breast cancer cells, which represents a crucial step in the malignant progression of cancer such as metastasis (Federico *et al* 2016, Gaetano *et al* 2009). Moreover, the ATX gene is one of the 40 most upregulated genes in metastatic cancers (Euer *et al* 2002). ATX expression has been found to be increased in human breast tissue compared to in breast tumors. In the next step the mechanisms that regulate the interactions of primary breast tumors with their surrounding breast tissue microenvironment need to be investigated in a mouse model. When breast tumors proliferate and grow, they produce inflammatory cytokines that induce the production of ATX in surrounding adipose tissue. In a malignant inflammatory cycle, the increase in ATX elevates the production of LPA, which in turn stimulates the production of inflammatory mediators in the adipose tissue. This in turn leads to the elevated production of ATX. This cycle can be impaired by ATX inhibition, which subsequently broke down the inflammation-facilitated interdependence between cancer cells and the surrounding adipose tissue (Benesch *et al* 2018).

However, a novel finding is that ataxia telangiectasia mutated (ATM) increased the ATX generation and secretion (Benesch *et al* 2018). Ataxia-telangiectasia (AT) is a rare autosomal recessive disorder that is evoked by mutations in the ataxia-telangiectasia gene (ATM) leading to defective repair of DNA double strand breaks. In addition to silica, also rotenone, camptothecin and H_2O_2 can activate ATX through ATM, and hence it can be supposed that ATX is part of a generalized ATM response to double-strand breaks. In respiratory epithelial cells, the DNA damage is induced by silica particles, where the end-products are reactive oxygen species (ROS), the NOD-like receptor family pyrin domain containing-3 (NLRP3) inflammasome activation, ATX and ATM (Zheng *et al* 2017). An inflammasome is a multiprotein complex that functions as a platform for the activation of inflammatory caspases leading to the maturation of pro-inflammatory cytokines such as interleukin-1β (IL-1β) and interleukin-18 (IL-18). In addition to its function as a sink for pro-inflammatory cytokines, the inflammasome induces the secretion of a myriad of leaderless proteins controlling cell proliferation and tissue repair. These DNA damage results indicate that an ATM–ATX-dependent loop propagates inflammation and the accumulation of double-strand breaks and moreover it has been revealed that even low doses of silica are effective inducers of double-strand breaks in epithelial cells that elevate the risk of lung cancer (Zheng *et al* 2017). Finally, it can be concluded that the ATM–ATX axis interconnects double-strand breaks with the silica-induced inflammation and thereby propagates these effects in epithelial cells and tissues.

2.2.1 Inflammation can cause cancer (pro-tumorigenic)

Outside–in: inflammation induces cancer promotion

It is common knowledge that inflammatory responses play decisive roles at the different tumor development stages, including tumor initiation, tumor growth, malignant tumor progression, cellular invasion and the metastasis of the cancer. Within these processes of cancer development and severe progression, inflammation regulates the immune surveillance and responses to tumor therapy. In more detail, immune cells within primary tumors facilitate extensive and dynamic crosstalk with the broad mass of cancer cells and even molecular events connecting the crosstalk between immune cells and cancer cells have been found (figure 2.5). In the following,

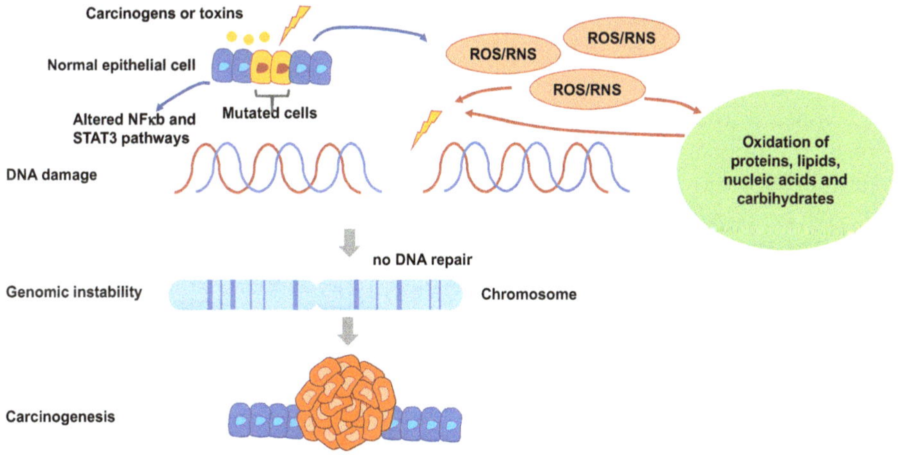

Figure 2.5. Inflammation can induce cancer. Extrinsic pathway of an inflammation-induced carcinogenesis. ROS/reactive nitrogen species (RNS) can directly oxidase proteins, lipids, nucleic acids and carbohydrates. Hence, they have an impact on the associated metabolic pathway, which may lead to further DNA damage. With chronic inflammation the function of the DNA repair system is impaired, which cause genomic instability.

this crosstalk will be discussed, as well as the principal mechanisms that connect inflammation and immunity with the development of tumors.

The process of carcinogenesis involves a controlled interplay between numerous intrinsic and extrinsic cellular signaling pathways, such as genomic instability, altered proliferation, altered stromal environment and abnormal differentiation of epithelial and mesenchymal states. A common feature of inflammation is its ability to induce all cellular and molecular mechanisms that promote tumorigenesis (Hanahan and Weinberg 2011). However, the mechanisms that regulate the neoplastic transformation of cells stimulated by inflammatory cells remain elusive. The principles behind inflammation-induced cancer will be discussed below.

Induction of proliferation and cell survival mechanisms
Excessive cancer growth leads to the loss of homeostasis in the tissue's architecture. Within a tissue the architecture is precisely controlled by soluble factors such as growth factors and their receptors, facilitating proliferation, homeostasis and cell survival. Cancer cells are able to utilize growth factor signaling using the following common strategies: they alter the process of endocytosis and the recycling of receptors, abolish negative-feedback mechanisms, reduce growth rates and empower redundant proliferative pathways acting downstream of the receptor tyrosine kinase signaling (Amit *et al* 2007, Mosesson *et al* 2008, Wilson *et al* 2012, Straussman *et al* 2012, Prahallad *et al* 2012). These growth factor receptor tyrosine kinases are known to have a wide range of incoming signals that finally lead to a few prominent signaling cascades inside the cell, such as proliferation, metabolism, survival, differentiation, cellular adhesion and motility (Casaletto and McClatchey *et al* 2012). In particular, these pathways cover the activation of signal transducers and the activators of transcription (STAT) family members such as STAT3, which are known to play a role in tumorigenesis, while they have also been linked to inflammatory processes in gastric, lung, colon, pancreatic and liver cancers (Fukuda *et al* 2011, Lesina *et al* 2011, Bollrath *et al* 2009, Grivennikov *et al* 2009, Bronte-Tinkew *et al* 2009, Gao *et al* 2007). Myeloid cells produce interleukin-6 (IL-6), which is able to activate STAT3. STAT3 expression is a crucial step during the onset tumorigenesis of colitis-associated colorectal cancer (CAC), as it enhances the proliferation of pre-malignant cells and abolishes apoptosis *in vivo* (Bollrath *et al* 2009, Grivennikov *et al* 2009, Waldner *et al* 2012). From a mechanistic point of view, STAT3 induces cell proliferation through the up-regulation of the cell cycle regulators such as cyclin D1, cyclin D2 and cyclin B, as well as the proto-oncogene MYC. Additionally, STAT3 increases cellular survival, as it increases the expression of the anti-apoptotic genes BCL2 and BCL2-like 1 (BCL2L1 encoding BCL-XL) (Bollrath *et al* 2009, Grivennikov *et al* 2009, Yu *et al* 2009).

Another regulatory function of STAT3 has been discovered. The deletion of the sphingosine kinase 2 (Sphk2) in mice has been reported to be an important part of the development of CAC, in which nuclear factor-κB (NF-κB)–IL-6–STAT3 signaling is activated by the production of sphingosine-1-phosphate (S1P). This then leads to an up-regulation of the S1P receptor 1 (S1PR1) and represents an additional amplification step for the support of the chronic inflammation and subsequently of

CAC (Liang *et al* 2013). Indeed, excessive STAT3 activation during early neoplasia is provided by immune cell signaling in the local surroundings of a primary tumor. In particular, Kras-driven pancreatic tumor models propose that STAT3 signaling plays an intrinsic role in pancreatic intraepithelial neoplasia development, even in the absence of an inflammation (Fukuda *et al* 2011). This indicates that the inflammation has a supportive effect, whereas it is not essential for tumor development. There are several cases in which epithelial STAT3 functions through the transduction of myeloid cell-derived IL-6 pro-tumorigenic signals, for example, during the transformation of pancreatic intraepithelial neoplasia to pancreatic ductal adenocarcinoma, while it plays no role during its initiation (Lesina *et al* 2011). In line with this, a low grade of inflammation mediated by obesity in the diethylnitrosamine-induced hepatocellular carcinoma (HCC) model activates tumorigenic STAT3 signaling through IL-6 and TNFα (Park *et al* 2010). A possible regulatory mechanism during tumor initiation and progression has been reported for immortalized mammary cells that show a transient activation of the oncoprotein Src. This activation of Src is triggered by an epigenetic alteration that leads to a stably transformed state, which is facilitated by activated STAT3 (Iliopoulos *et al* 2009). Indeed, Src activation leads to an NF-κB-dependent inflammatory response that is driven by increased IL-6 up-regulation and decreased expression of IL-6's negative regulator, microRNA let-7. Thus, elevated levels of IL-6 lead to the activation of STAT3 transcription, which is essential for cell transformation as well as activation of NF-κB. This behavior can be described as a positive feedback loop that is even present in the absence of the inducer (Iliopoulos *et al* 2009). In summary, an inflammatory process seems to be a supporter of oncogenic activation by increasing the NF-κB–IL-6–STAT3 signaling cascade and subsequently the proliferation of cancer cells.

What role does NF-κB signaling play in tumor development?
NF-κB activation has been reported in most tumors to be evoked by either inflammatory stimuli or oncogenic mutations leading to gene expressions that provide inflammation, proliferation and survival of the cells (Grivennikov *et al* 2010, Karin and Greten 2005, Ben-Neriah and Karin 2011). The first molecular links between inflammation and cancer were discovered in the liver (Pikarsky *et al* 2004) and the colon (Greten *et al* 2004). Additionally, this connection emphasizes the importance of the tumor microenvironment as a potential source of pro-tumorigenic and inflammatory cytokines, such as NF-κB activators and target genes.

In a mouse model of inflammation-associated HCC (Abcb4$^{-/-}$ mice, Abcb4 encodes the multidrug resistance protein 2 (MDR2)), the secretion of TNFα by surrounding tissue endothelial cells and inflammatory cells induces NF-κB expression in hepatocytes, which is needed for the progression of HCC, whereas the overall transformation of hepatocytes is not affected (Pikarsky *et al* 2004). In another model, such as the CAC, the inactivation of NF-κB signaling by an epithelial cell-specific inhibitor of NF-κB kinase-β (IKKβ) significantly decreases the occurrence of CAC, whereas the deletion of Ikbkb encoding IKKβ in myeloid cells reduces the primary tumor size (Greten *et al* 2004). In line with the first model, it has been demonstrated that TNFα secretion by leukocytes invading the lamina propria and other submucosal sites of the colon initiates CAC and its progression (Popivanova *et al* 2008). As mentioned above, NF-κB activation induces a tumorigenic autocrine signaling in the transformed cells. In particular, this autocrine loop is necessary for the oncogenic effects of Ras in keratinocytes through stimulation by IL-1α, IL-1R and myeloid differentiation primary response 88 (MYD88), which promotes in turn the activation of NF-κB (Cataisson *et al* 2012). In addition, MYD88 can regulate the colonic macrophage production of cytokines such as IL-6 and transforming growth factor-β (TGFβ), necessary for CAC development (Schiechl *et al* 2011). However, this seems to be a possible linkage between intrinsic NF-κB and extrinsic tumorigenic factors during an inflammatory process. Moreover, this hypothesis is further supported by the role of toll-like receptor 2 (TLR2) in gastric tumorigenesis through STAT3 activation (Tye *et al* 2012).

As myeloid cell-derived inflammatory cytokines provide proliferative responses during the onset of neoplasia, tissue conditions supporting the release of inflammatory cytokine are of special importance. A response to the tissue damage is the pronounced production of multiple inflammatory cytokines such as IL-22, which is crucial for the proliferation and survival of epithelial cells (Zenewicz *et al* 2008, Sonnenberg *et al* 2011). Secreted IL-22 by CD11c^{+} cells facilitate wound healing responses through the activation of STAT3 in epithelial cells of the intestine after tissue damage and colitis triggered by the dextran sodium sulfate (DSS) treatment (Pickert *et al* 2009). It has been shown that IL-22 leads to CAC development in mice lacking the IL-22 neutralizing receptor IL-22-binding protein (IL-22BP; synonymously termed IL22Rα2), indicating that IL-22BP, produced by CD11c^{+} cells downstream of inflammasome signaling, impedes CAC (Huber *et al* 2012). IL-22 secreted by HCC-infiltrating leukocytes leads to HCC through the activation of STAT3 and subsequently of downstream survival genes (Jiang *et al* 2011). Taken together, IL-22 seems to fulfill a direct tumor-promoting function, as indicated by the hepatocyte-specific expression of IL-22 in transgenic mice resulting in HCC survival and proliferation without substantially changing the overall liver inflammation (Park *et al* 2011).

One way to adopt the capabilities of cell proliferation and survival is through the inflammatory cytokine signaling and a second way is through the dedifferentiation of pre-malignant cells by the achievement of stem-cell-like properties. Therefore, a model system of constitutive WNT activation has been used, in which increased NF-κB signaling in epithelial cells increases the activation of WNT–β-catenin and hence triggers the dedifferentiation of intestinal non-stem cells, which then possesses a tumor-initiating capacity, leading to tumorigenesis in the intestine (Schwitalla *et al* 2013). A third way is the loss of the tumor suppressor adenomatous polyposis coli (APC), which can activate the GTPase Rac1 and lead to hyperproliferation of leucine-rich repeat-containing G-protein-coupled receptor 5 (LGR5)-expressing intestinal stem cells, evoked by the Rac1 regulated production of the reactive oxygen species (ROS) and the activation of NF-κB (Myant *et al* 2013). STAT3 has been reported to facilitate cancer stem cells' tumor-promoting properties in certain tissues (Marotta *et al* 2011, Ho *et al* 2012, Zhou *et al* 2007, Scheitz *et al* 2012).

Additionally, it has been demonstrated that IL-1β-dependent and IL-6-dependent inflammation may trigger the activation and migration of gastric cardiac progenitor cells, leading to the malignant transformation of Barrett's esophagus and esophageal adenocarcinoma in a mouse model (Quante *et al* 2012). Overall, the inflammatory process is able to drive preliminary neoplasia towards tumor progression by controlling the expansion of stem cells within the tissue and through acquisition of stem-cell-like properties.

What do we know about the role of inflammation in genomic destabilization?
In addition to increased proliferation and the ability to escape apoptosis, cancer cells are characterized by enhanced genomic instability followed by genomic mutagenesis. Inflammation may be a major cause of genomic instability through interference with genomic integrity. Moreover, somatic mutation caused during the inflammatory response may then function in a cell-intrinsic or cell-extrinsic manner. First, ROS and reactive nitrogen species (RNS) are secreted upon inflammatory cytokine stimulation by tissue macrophages and neutrophils, or by pre-malignant cells and lead to DNA breaks, point mutations and even more complex DNA lesions (Mantovani *et al* 2008, Campregher *et al* 2008, Mills *et al* 2003). Second, several cytokines such as TNFα, IL-1β, IL-4, IL-13 and TGFβ can induce the ectopic expression of activation-induced cytidine deaminase (AID, synonymously termed AICDA), which belongs to the DNA and RNA cytosine deaminase family and can cause the mutation of special cancer-associated genes such as TP53 (encoding p53) and MYC in order to facilitate oncogenesis (Takai *et al* 2009, Okazaki *et al* 2007, Endo *et al* 2008, Komori *et al* 2008). Inflammation can enhance the mutagenesis and indirectly elevate genomic destabilization through the downregulation of DNA repair mechanism and out-of-order checkpoints of the cell cycle, increasing random genetic alterations (Campregher *et al* 2008, Colotta *et al* 2009, Schetter *et al* 2010, Singh *et al* 2007). Therefore, mismatch repair (MMR) proteins are useful to prevent genetic instability caused by microsatellite instability. Hence MMP proteins are targeted during an inflammation, as they are repressed through several mechanisms to elevate the DNA replication errors in the genome. In particular, a multiple inflammatory substance, such as TNFα, IL-1β, prostaglandin E2 (PGE-2) and ROS, triggers the hypoxia-inducible factor 1α (HIF1α) to decrease the expression of the MMR genes MSH2 and MSH6 (Colotta *et al* 2009), whereby ROS is able to additionally decrease the MMR proteins' enzymatic activity (Schetter *et al* 2010). Inflammation can also mediate epigenetic silencing of the MMR protein MLH1 and other tumor suppressors by hypermethylation (Schetter *et al* 2010, Hahn *et al* 2008). Moreover, inflammation can inactivate the genomic surveillance evoked by p53 and the over-expression of BCL2 and MYC, leading to decreased DNA repair and hence more mutations in cancer cells (Mantovani *et al* 2008, Schetter *et al* 2010).

Taken together, it is supposed that inflammation can promote genetic instability in several ways, for example, by the ectopic expression of IL-15 inducing large granular lymphocytic leukemia through MYC and NF-κB signaling and by reducing the expression of miR-29b, which can repress the DNA methyltransferase 3B

(DNMT3B) leading to DNA hypermethylation and chromosomal instability (Mishra *et al* 2012).

How is the induction of invasion and metastasis regulated by an inflammation?
Since a loss of tissue homeostasis has not been associated with high mortality rates, around 90% of cancer mortality is due to cancer metastasis. However, a major support of metastasis is that myeloid-derived monocytes and macrophages facilitate cancer cell invasion, extravasation and finally metastatic outgrowth (Peinado *et al* 2011, Qian and Pollard 2010), although the underlying mechanisms remain unclear. In particular, IL-4-activated tumor-associated macrophages (TAMs) (Gocheva *et al* 2010, DeNardo *et al* 2009) and CC-chemokine receptor 1 $(CCR1)^+$ immature myeloid cells (Kitamura *et al* 2007) at the tumor periphery play a major role in providing local invasion of cancer cells, as they contain matrix-degrading enzymes such as MMPs, cathepsins and heparanase, which a cancer cell can utilize to disseminate and invade the surrounding microenvironment (Talmadge and Fidler 2010, Qian and Pollard 2010, Kalluri and Weinberg 2009, Murdoch *et al* 2008, Lerner *et al* 2011). In addition, the pro-inflammatory cytokine TNFα increases the vascular permeability, thereby enhancing the cancer cell transendothelial migration events (Grivennikov *et al* 2006), while the cyclooxygenase-2 (COX-2)-dependent increases prostaglandin levels and MMP levels facilitating tissue remodeling (Nguyen *et al* 2009). Taken together, the tissue microenvironment of tumors is a reservoir of multiple inflammatory substances inducing cellular transformation and tumorigenesis.

What role does acute inflammation play in mediating cancer progression?
Since long-lasting and chronic inflammation may lead to cancer, the focus is now on the role of acute inflammation in cancer progression. For example, virally encoded genes such as the oncogenes E6 and E7 of the human papilloma virus (HPV) are able to contribute to cellular transformation (Munger and Howley 2002). However, there are also numerous microbes, such as *Helicobacter pylori,* that are not able to transform cells into malignant cancer cells (Peek and Blaser 2002).

Inflammatory states caused by infection and irritation may alter tissue microenvironments, promoting genomic lesions and the initiation of a primary tumor. An effector mechanism is the production of free radicals, such as reactive oxygen intermediated (ROI), hydroxyl radical ($OH^\bullet$), superoxide ($O_2^{-\bullet}$), reactive nitrogen intermediates (RNI), nitric oxide ($NO^\bullet$) and peroxynitrite ($ONOO^-$), through the release of which the host fights against the microbial infection. Although these free radicals are anti-microbial, ROI and RNI also cause oxidative damage and nitration of DNA bases, enhancing the occurrence of DNA mutations (Hussain *et al* 2003). However, cells have mechanisms such as DNA repair, cell cycle arrest, apoptosis and senescence to control unregulated proliferation and hence accumulate mutational events in the DNA. When there is solely little DNA damage or less oncogenic activation, the cells will first try to repair the DNA in order to prevent mutations or instead they will induce programmed cell death of the mutated cell (apoptosis). In the case of excessive cell death as it occurs in infection or even in non-infectious tissue after injury, the lost cells need to be replaced by the expansion and proliferation of other neighboring cells that are normally undifferentiated precursor cells such as tissue stem cells. To fulfill this, there are at least two prerequisites. First, at least some cells must survive during the tissue injury and second, they must have the capacity to be expanded in order to maintain the cell numbers required for a properly functioning tissue. Indeed, multiple inflammatory pathways fulfill these two essential prerequisites for tissue repair (Wang *et al* 2005, Chen *et al* 2003). However, beyond the physiological role of tissue repair—playing a prominent role in host defense against infection—the inflammatory response clearly provides a strong survival and proliferative signal to initiated cells and thus may as a side-effect also facilitate the tumor promotion.

Apart from inflammation-induced biochemical mediators of cancer initiation and progression, tissue stiffness is tightly and precisely regulated under normal physiological conditions, however, it is altered during the progression of the disease. In cancer, primary solid tumors are usually stiffer than the surrounding uninvolved stromal tissue, whereas the cancer cells themselves become softer (Alibert *et al* 2017, Rianna *et al* 2017, Guck *et al* 2005, Fischer *et al* 2017, Panzetta *et al* 2017). There is increasing evidence that the stiffness of the extracellular matrix regulates cancer and stromal cell mechanical properties and function, which in turn affects cancer disease hallmarks such as angiogenesis, migration and metastasis (Chin *et al* 2016, Ringera *et al* 2017, Hope *et al* 2018). How can cancer cells and fibrosis-relevant stromal cells respond to extracellular matrix stiffness? What are the possible sensing appendages and signaling mechanisms involved? Moreover, mechanosensitive ion channels are altered in their regulation and promote the initiation of cancer, which indicates that the inflammatory microenvironments play a major role in initiation of cancer (Hope *et al* 2018). The mechanotransduction of physical cues towards cells can initiate intracellular signaling pathways in multiple cancer types and a wide range of effects have been identified that range from acute alterations such as the activation of ion channels or protein kinases to long-term alterations in the cellular phenotype, which in addition requires the induction of specific gene transcription and protein production (Doyle and Yamada 2016, Hytönen and Wehrle-Haller 2016).

Outside regulation of the inside, possibly by the microbiota

What instigates the inflammatory response? What kind of pathogen do we have? What is the cancer-causing stimulus during a microbial infection? In nearly one-fifth of all cancer cases there seems to be a direct linkage between cancer and an infectious disease. The knowledge of how oncoviruses trigger and evoke cancers is well established, however, the cancer-supporting action of microorganisms such as bacteria and parasites is less well understood, but they have been accounted for in enhanced tumor development and malignancy. The role of fungi in possibly causing cancer is not yet clear. In several types of infections, it seems to be primarily the host's inflammatory response that fights against the pathogenic infection and provides the connection between the infection caused by the pathogen and the development of cancer. In some cases, the inflammatory response even leads to the development and progression of cancer that is not directly attributable to an infectious disease; it is caused instead by the tissue damage from the chronic and uncontrolled inflammation leading to the malignant transformation of cells.

There is the most evidence for bacterial involvement in inflammation-induced cancer (for example, gastric adenocarcinoma) with respect to *H. pylori*. Specifically, the chronic gastric inflammation facilitated by *H. pylori* evokes aberrant β-catenin signaling in epithelial cells supporting malignant transformation (Franco *et al* 2005). In addition to this and other pathogens, it has been suggested that members of the commensal group of microorganisms are involved in tumor-inducing and tumor-progressive inflammation. In fact, the human body is a host to trillions of viruses, bacteria, fungi and parasites on the skin and in the oral cavity, respiratory tract, urogenital system and gastrointestinal tract (Huttenhower *et al* 2012). However, these commensal microorganisms are necessary for the human physiology under homeostatic conditions and thus they also contribute to abnormalities within inflammatory processes (Gordon 2012). Do these commensal microorganism-dependent inflammatory processes support tumor growth in a similar way to the inflammation caused by a pathogenic infection?

One study reported that commensal microorganisms induce inflammatory tumorigenesis by revealing the contribution of the TLR-signaling adaptor protein MYD88 to cancer progression in a $Apc^{Min/+}$ mouse model for spontaneous intestinal tumorigenesis and in another mouse model in which the animals were treated with repeated injections of azoxymethane (AOM) (Rakoff-Nahoum and Medzhitov 2007). These results indicate that the innate sensing of microorganisms in the intestine is essential for the appropriate regulation of inflammation and cancer development. In more detail, Erk activates as a downstream target gene myeloid differentiation primary response gene 88 (MYD88), which then drives intestinal tumorigenesis in $Apc^{Min/+}$ mice (Lee *et al* 2010). However, a role for MYD88 in intestinal tumorigenesis has been found in colitis-susceptible $IL\text{-}10^{-/-}$ mice (Uronis *et al* 2009). However, in another DSS-AOM model of inflammation-induced colorectal cancer, the $Myd88^{-/-}$ mouse model, the mice even displayed increased development of intestinal tumors, indicating that MYD88 is not essential for developing an intestinal tumor (Salcedo *et al* 2010). In contrast to the $Myd88^{-/-}$ mouse model, in the $Tlr4^{-/-}$ mouse model the animals showed no DSS-AOM-induced intestinal tumor development (Fukata *et al* 2007), while another result for the $Tlr2^{-/-}$ mouse model revealed enhanced intestinal tumorigenesis upon DSS-AOM-induction (Lowe *et al* 2010). How can we deal with these vastly contrasting results? These contradictory findings can be explained by the differences in the commensal microbiota composition in the innate immune-deficient mice. In particular, these differences in the content of the microbiota lead to diverse results ranging from promoting to impairing the tumorigenic cascade. As there are contradictory roles for TLR signaling, this may suggest that there exists a second alternative pathway of innate immune signaling, which is also dependent on MYD88, but signals through the receptors for the cytokines IL-1 and IL-18. As suggested, the $IL\text{-}18^{-/-}$ mice possess a similar susceptibility to the development of intestinal polyps compared with $Myd88^{-/-}$ mice (Salcedo *et al* 2010). Indeed, IL-18 is secreted by cells upon the activation of inflammasomes (Lin and Zhang 2017), which can be seen as an innate immune platform that consists of an upstream NOD-like receptor (NLR), the adaptor protein apoptosis-associated speck-like protein containing CARD (ASC, termed synonymously PYCARD) and caspase 1 (Elinav *et al* 2013). Due to the lack of 'cancer-protecting' IL-18 in mice, including mice deficient in caspase 1 or of the inflammasome-forming NLRs, NLRP3 and NLRP6, all these mice show a strongly enhanced susceptibility to colorectal carcinoma (Hu *et al* 2010, Zaki *et al* 2010, Chen *et al* 2011, Normand *et al* 2011). Taken together, these mice possess an altered microbial community with an altered function in the gut system (called dysbiosis), which exerts pro-inflammatory properties on the gut mucosa (Elinav *et al* 2011, Henao-Mejia *et al* 2012). Indeed, the microflora connected with inflammasome deficiency in mice has been reported to be involved in inflammation-facilitated colorectal cancer induction through the activation of the inflammatory stimuli-based, epithelial IL-6 signaling. Moreover, this activation of epithelial IL-6 signaling has also been seen in cohabitant wild-type mice (Hu *et al* 2013). These results indicate that alterations in the microbiota composition are a major factor for providing intestinal tumorigenesis upon inflammatory stimulation.

Increased tumor development has been found to be associated with an altered microbiota metagenome, which displays enhanced levels of lipopolysaccharide and flagellin. In particular, dietary emulsifier-induced alterations in the microbiome are required and sufficient to facilitate alterations in prominent proliferation and apoptosis signal transduction pathways governing the development of tumors (Viennois *et al* 2017). In summary, these findings deliver support for the concept that perturbations in the host-microbiota interplay, which leads to a low-grade inflammation of the gut, can even evoke colon carcinogenesis (Viennois *et al* 2017).

However, as suggested, intestinal dysbiosis has been connected to colorectal cancer. First, for example, cancer has been identified in the $Tbet^{-/-}$ (synonymously termed $Tbx21^{-/-}$) recombination activating gene 2 $(Rag2)^{-/-}$ mouse model for ulcerative colitis (TRUC) (Garrett *et al* 2009) and secondly, colorectal cancer has been detected even in humans exhibiting ulcerative colitis (Sobhani *et al* 2011, Marchesi *et al* 2011). In addition, there is inflammation-mediated colorectal cancer that is connected to the intestinal dysbiosis and is transmissible between mice, as the infectious part of the colorectal cancer that cannot be associated with a specific pathogenic species. What causes the cancer? It is the aberrant outgrowth and activity of the normally symbiotic (but in this special case 'dysbiosis') species that acts as a pathogenic species ('pathobionts'), which can even elevate the inflammation and induce the neoplastic outgrowth of the tissue cells.

The development of dysbiosis is supposed to be the driving force of the inflammation-associated tumor growth and hence provide a clear example of microorganism-facilitated inflammation causing a primary tumor, while in turn there are also microorganism-facilitated tumor-driven inflammatory responses. In line with this, the dysbiosis and the enhanced microbial distribution across the intestinal barrier leads to secondary and subsequently to chronic inflammation or even to neoplasia, and hence supports the development of cancer. In more detail, colorectal cancers are characterized by defective mucin production and show altered expression of the proteins regulating the maintenance of the intestinal barrier. The result of this dysregulation is increased microbial translocation to the targeted tissue location of the tumor growth and the up-regulation of IL-23 and IL-17, which supports ongoing local tumor-driven inflammation (Grivennikov *et al* 2012). One example is the model of AOM-treated colitis-susceptible IL-$10^{-/-}$ mice, in which the intestinal microbial composition is altered compared to healthy mice, and in particular commensal *Escherichia coli* is able to facilitate an invasive carcinoma that is promoted by the polyketide synthase (pks) genotoxic island. In line with these results, patients with inflammatory bowel disease (IBD) and colorectal cancer show a higher incidence of pks^{+} *E. coli* compared to healthy controls (Arthur *et al* 2012), which leads to the suggestion that only a single irregularly positioned bacterial organism can finally promote increased neoplastic growth. In addition, at least three studies propose the importance of microorganisms in supporting cancer progression. In particular, the anaerobe *Fusobacterium nucleatum* has been found in human colorectal carcinoma samples (Castellarin *et al* 2012, Kostic *et al* 2012, Strauss *et al* 2011). It has been reported that *F. nucleatum* increases colorectal neoplasia progression by supporting the recruitment of myeloid cells that are able to infiltrate the primary tumor (Kostic *et al* 2013). On the molecular level, *F. nucleatum* binds to E-cadherin, which is expressed on the cell surface of epithelial cells, and hence activates β-catenin signaling in order to promote cell proliferation (Rubinstein *et al* 2013). The enterotoxigenic *Bacteroides fragilis* (ETBF) has been shown to support inflammation-driven colon cancer (Wu *et al* 2009) by inducing spermine oxidase-dependent ROS production (Goodwin *et al* 2011). In summary, these findings propose a major role for bacterial microbiota during inflammatory processes in the induction and progression of cancer development. What roles do viral, fungal and parasitic members of the microbiota play in inflammation-induced cancer? This is not yet fully understood and much research effort is needed to reveal their role in possibly supporting cancer initiation and progression.

The liver is the ‘first pass’ organ for microbial products from the gastrointestinal tract through the portal vein and hence enhanced translocation of bacterial products to the liver represent a hallmark of chronic liver disease (Henao-Mejia *et al* 2012). Thus, the intestinal microbiota lead to the development of HCC through TLR4-mediated signaling of non-hematopoietic cells in the liver (Dapito *et al* 2012). In particular, the TLR4-induced hepatomitogen epiregulin supports carcinogenesis in the liver. In addition, obesity can facilitate alterations in the gut microbial composition, which increases deoxycholic acid (DCA) and subsequently leads to DNA damage. Increased DCA in the enterohepatic circulation enhances cytokine production by hepatic stellate cells and hence supports hepatic carcinogenesis (Yoshimoto *et al* 2013). Taken together, these results demonstrate that the inflammatory-response-driven cancer of commensal microbial elements is not locally restricted to the direct interaction sites of the primary tumor and the microorganisms.

What is the future direction of inflammatory-driven cancer research?

The idea of inflammation-caused cancer has been fully established, whereas the linking mechanisms between inflammatory processes and the various stages of tumor development are still not well understood and require further research. The molecular details coordinating the interaction between cancer cells and the inflammatory response are known, but fundamental questions have still not been answered. Why does not every chronic inflammation increase the risk of cancer? Does the location site and the tissue type of the inflammation play a role in determining whether or not the inflammatory response is able to induce cancer development?

There is indeed a difference between the organs in which the inflammation occurs and whether this site (for example, the gastrointestinal tract, liver and lung) does or does not induce alterations increasing the risk of cancer, for example, in the joints in rheumatoid arthritis and in the brain. Does this depend on the excessive interaction between the immune cells under inflammatory stimulation and cancer cells? What mechanism do commensal microbiota utilize for inflammation-induced cancer development? Does microorganism-driven inflammation regulate only early neoplasia and tumor progression, or can it also affect cancer metastasis? What is the impact of the mechanical properties of inflamed tissue? Do inflammation-altered mechanical properties lead to increased tumor development?

2.2.2 Inflammation can inhibit cancer (anti-tumorigenic)

In addition to its cancer-promoting function, inflammation can also play the opposite role and act as an anti-tumorigenic factor. This second, but no less important role of inflammation in controlling tumor development is not yet fully understood. There is evidence to suggest that inflammatory and immune systems are able to inhibit cancer progression. Currently, there are at least two natural ways to recognize cancer: the first approach is tumor immunosurveillance, through which the host can recognize and eliminate transformed cells. The second is the adaptive immune recognition of tumor-associated and specific antigens, through which the immune system can regulate and suppress the development of cancer (Smyth *et al* 2006).

Inflammation is normally suppressed in primary tumors in order to inhibit an immune response and to avoid an inflammatory response switching of the primary tumor development or impairing the malignant tumor progression. However, the underlying interaction pathways are not yet well understood. The hypothesis that inflammation inhibits cancer is not new, but not much research effort has been put into this proposal. There is still the question of whether the inflammation changes the tumor microenvironment in such a manner that the primary tumor can no longer grow into that microenvironment and is then just a subject of rejection, apoptosis and clearance. As is discussed above, an inflammation can cause cancer, and there seems to be a precisely regulated balance between tumor-promoting and tumor-antagonizing inflammatory responses and, indeed, there is a dependence of tumor-neutralizing immune responses on the optimal function of key transcription factors and tumor suppressors.

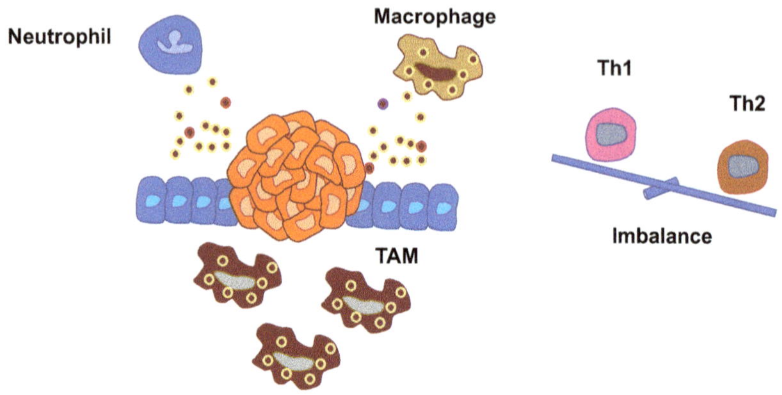

Figure 2.6. Intrinsic pathway: cancer-associated inflammation builds the tumor microenvironment. When transformed or cancerous cells are present, immune responses are induced similar to microbial intruders. Inflammasomes are activated through different mechanisms and they release interleukin-1 β (IL-1 β) and IL-18 to initiated inflammation. The alteration of pro-inflammatory cytokines will lead to abnormal polarization of T-helper cells. The imbalanced T cells and cancer-related immune cells, such as tumor-associated macrophages (TAMs), mediate the cancer progression through angiogenesis, metastasis and cancerous resistance against immune attacks.

2.2.3 Cancer induces inflammation

The pre-malignant tumors seem to be 'wound-like' (Coussens *et al* 1999), as they are similar to healing or desmoplastic tissue regarding the presence of activated platelets (figure 2.6) (Dvorak 1986, Mueller and Fusenig 2004). In particular, tumor growth is thought to be 'biphasic' (Coussens *et al* 1999). In the initial phase, the body handles early tumors as wounds. This first phase involves tumor growth facilitated by the surrounding tumor stroma as an indirect control mechanism caused by the physiological tissue repair during 'wound' healing processes. In more detail, in mouse models for skin and pancreatic carcinogenesis, matrix metalloproteases produced by bone-marrow derived cells such as mast cells can convert VEGF into an active form that is able to facilitate the angiogenic switch to promote further tumor progression (Coussens *et al* 1999, Coussens *et al* 2000, Bergers *et al* 2000). In the later tumor growth state pro-inflammatory factors such as MMPs seem to be directly regulated by the primary tumors (Coussens *et al* 1999). Indeed, this malignant transition has been observed for the regulation of inflammation by early tumors compared to late tumors, leading to spontaneous intestinal tumorigenesis in mice and humans. In particular, in early tumors COX-2 is expressed by stromal cells (Hull *et al* 1999, Sonoshita *et al* 2002), whereas in late tumors COX-2 is expressed by the dysplastic epithelium (Sheehan *et al* 1999). What is the reason for this change of the cell type expressing COX-2? One hypothesis is that there are regulatory mechanisms in tumor-associated stromal cells regulating the upper limit of their expression of tissue repair factors. Thus, these findings are supposed to be a selective pressure for cancer cells that are able to maintain these COX-2 dependent ancillary processes and are then not dependent on a wound-like stroma surrounding the primary tumor. However, the tumor-associated stroma can undergo selective pressure, as there are genetic alterations found in tumor-associated stroma (Moinfar *et al* 2000) and even the loss of p53 has been detected in tumor-associated fibroblasts (Hill *et al* 2005). Moreover, the inflammatory response may not only play a role in promoting tumor growth by increasing the angiogenesis, it may be involved in cell invasion and metastasis. In more detail, the process of angiogenesis supports the invasion and transmigration of cancer cells. Hence, matrix metalloproteases and MMP inhibitors (TIMPs) play an important role in the process of angiogenesis and help to remodel the extracellular matrix microenvironment (Egeblad and Werb 2002). In order to support cancer progression, infiltrating leukocytes may build a trail through the extracellular matrix tissue (termed countercurrent invasion theory) (Opdenakker and Van Damme 2004). The countercurrent principle of invasion has been proposed due to observations that cancer cells produce chemokines that regulate cancer progression (Opdenakker and Van Damme 1992, 1999, Balkwill and Mantovani 2001, Mareel and Leroy 2003, Coussens *et al* 2000, Van Coillie *et al* 2001). The term 'countercurrent model' for describing the regulatory role of chemokines in cancer was initially chosen because the fluxes of the primary tumor and host cells are opposite: in particular, chemoattracted leukocytes and growing blood vessels (angiogenesis) are towards the primary tumor, whereas the invasion of cancer cells is away from the primary tumor into the surrounding microenvironment and can be mediated by chemokine-induced proteolysis.

Inside–out signaling: is there an immune modulation by the cancer disease?
Cancer was formerly regarded as a cell-autonomous process. The current concept is based on genetically transformed cells that can develop into malignant neoplasms. As the tumor microenvironment has become the focus of cancer research, tumor progression seems to be regulated by biochemical and mechanical factors of the dense network of interactions between cancer cells and their surrounding tumor stroma, niche-defining cancer subcell populations and the tumor surrounding vasculature. A special feature of cancer cells is that they are able to alter the crosstalk between the tumor and its microenvironment through the regulation of the inflammatory response by soluble mediators. Cancer cells can affect the innate and the adaptive immune response through many pathways, including changing the acquired T cell response from the T-helper 1 (TH1) cell subset to the TH2 cell subset, the induction of immunosuppressive T regulatory (TReg) cells, an altered phenotype of macrophages and neutrophils to a type 2 differentiation state, as well as the stimulation of myeloid-derived suppressor cells (MDSCs) (Motz and Coukos 2013). What are the pathways for cancer cells to facilitate the antitumor immune response? Cancer cells can interact at every step of the antitumor inflammatory response by inhibiting the function of immune cells through changing their inflammatory phenotype to a regulatory (and immunosuppressive) one.

In more detail, inflammatory responses to tumor growth start with the tumor antigen that is presented by antigen-presenting cells. These immunostimulatory antigens can be of a special type that marks the onset of the primary tumor, and these antigens can additionally be overexpressed or mutated (termed neo-antigens)

(Robbins *et al* 2013). Primary tumors alter the function of antigen-presenting cells in many ways, such as through the direct tumor-facilitated regulation of the functions of myeloid cells (Gabrilovich *et al* 1996). In line with this, the tumor-derived vascular endothelial growth factor A (VEGFA) has been shown to suppress the function and maturation of dendritic cells. The VEGFA stimulation of peripheral myeloid cells leads to the expression of programmed cell death 1 ligand-1 (PDL1, also called CD274), which is an inhibitory ligand of the B7 family (Curiel *et al* 2003). Primary tumors can affect the antigen-presenting cell function through the release of TGFβ, IL-10 and IL-6 (Geissmann *et al* 1999, Steinbrink *et al* 1999, Menetrier-Caux *et al* 1998). Hence, this cytokine stimulation of primary tumors modulates the immune system to an antitumor T cell response. Indeed, this hypothesis has been confirmed in a mouse model of pancreatic ductal adenocarcinoma (PDA), where tumor-derived granulocyte–macrophage colony-stimulating factor (GM-CSF) induces the development and recruitment of granulocyte differentiation antigen 1 $(GR1)^{+}CD11b^{+}$ myeloid cells repressing antigen-specific T cells in the local microenvironment of a primary tumor (Bayne *et al* 2012). Thus, regulatory myeloid cells accumulate in the nearby tumor microenvironment and have been described and characterized as MDSCs in the peripheral blood of cancer patients. As expected, tumors can secrete CCL2, CXCL5, CXCL12 and stem cell factor (SCF, synonymously termed KIT ligand) in order to recruit MDSCs (Murdoch *et al* 2008). At their targeted site around the primary tumor, MDSCs induce their T cell-suppressive function as they secrete IL-10, TGFβ and arginase (Gabrilovich and Nagaraj 2009).

Primary tumors also actively regulate the inflammatory microenvironment through the recruitment of TReg cells. For example, Hodgkin's lymphoma cells and ovarian cancer cells release the chemokine CCL22 that binds to the CCR4 receptor and thereby recruits TReg cells to the tumor stroma (Curiel *et al* 2004). In addition, under special conditions, hypoxia ovarian cancer cells secrete CCL28 that binds to its receptor CCR10, which is expressed predominantly on the cell surface of TReg cells (Facciabene *et al* 2011). Recruited TReg cells are able to convert the local tumor microenvironment to angiogenic mechanisms in order to support oxygen supply (Facciabene *et al* 2012). Moreover, TReg cells are actively targeted towards the primary tumor, as cancer cells secrete TGFβ and adenosine (Zarek *et al* 2008).

In summary, inflammatory processes support all stages of tumor development, and hence the primary tumor is able to actively vary the immune reactions in the local tumor surrounding microenvironment.

2.2.4 Cancer inhibits inflammation

What role does COX-2 play in the interaction of cancer and inflammation?

A key hypothesis is that constitutive over-expression of cyclooxygenase-2 (COX-2) supports mammary carcinogenesis and, in turn, inhibition of COX-2 promotes breast cancer prevention and cancer therapy. The major results are as follows. There exist two primary genes encoding the cyclooxygenase, a constitutive gene (termed COX-1) and its inducible isoform (termed COX-2). In more detail, COX-1 is constitutively expressed in multiple cell types at basal levels such as epithelium, platelets or vascular smooth muscle cells. Under physiological conditions, the

COX-1 expression is constitutive and causes the sustained expression of low levels of prostaglandins maintaining homeostatis and protecting the cells. In contrast, the COX-2 gene is not transcribed and hence silent until the presence of inflammatory factors. First, COX-2 is constitutively expressed in tissue samples during the development of breast cancer and COX-2 expression is correlated with the cancer disease stage, cancer progression and finally metastasis. Second, the essential characteristics of mammary carcinogenesis, such as mutagenesis, mitogenesis, angiogenesis, decreased apoptosis, increased metastasis and immunosuppression, depend on COX-2-driven prostaglandin E2 (PGE-2) biosynthesis. Third, increased levels of COX-2 and PGE-2 expression facilitate the transcription of CYP-19 and aromatase-catalyzed estrogen biosynthesis, promoting increased mitogenesis. Fourth, extrahepatic CYP-1B1 is detected in mammary adipose tissue and converts paracrine estrogen to carcinogenic quinones supporting mutagenesis. Fifth, inhibitory substances of COX-2 decrease initial breast cancer occurrence in women and can lower the risk of cancer recurrence and mortality in women carrying breast cancer. Increases in the occurrence of breast cancer and increased breast cancer mortality are mediated by chronic inflammation of mammary adipose tissue and supported by the enhanced expression of COX-2 due to an obesity pandemic. Mammary carcinogenesis is a progressive series of highly specific cellular and molecular alterations evoked by constitutive over-expression of COX-2 and the activation of the prostaglandin cascade during the process of the primary tumor's so-called 'inflammogenesis'.

More than a hundred years ago, it was already proposed that chronic inflammation facilitates cancer development and progression by up-regulation of cellular proliferation (Virchow 1856, Virchow 1863, Balkwill and Mantovani 2001). Several models of carcinogenesis deal with inflammatory stimuli and mediators of wound healing in order to investigate the effect of inflammation on cancer development and progression (Schrieber and Rowley 1999, Coussens and Werb 2002, Philip *et al* 2004). The basic facts of the interaction between inflammation and cancer are given in table 2.1. The inducible COX-2 gene has been shown to connect cancer and inflammation in a similar manner, as it is proposed by many models of

Table 2.1. Inflammation and cancer—basic facts.

1. Chronic inflammation increases the risk of cancer.
2. Subclinical inflammation is important in increasing cancer risk.
3. Multiple types of immune and inflammatory cells are present within tumors.
4. Immune cells affect malignant cells via cytokines, chemokines, growth factors, prostaglandins, reactive oxygen and nitrogen species.
5. Inflammation affects each step of tumorigenesis.
6. In growing tumors anti- and pro-tumorigenic immune and inflammatory mechanisms coexist.
7. Signaling pathways providing pro-tumorigenic effects of inflammation are often subject to a feed-forward loop.
8. Special immune and inflammatory substances are dispensable during one stage, but absolutely critical in another stage of tumorigenesis.

carcinogenesis predicting an interaction between cancer and inflammation mediated by inflammatory stimuli and COX-2 expression (Koki *et al* 2002, Jang and Hla 2002, Harris 2002, 2007). Current knowledge indicates that COX-2 driven inflammogenesis represents a good model for breast cancer development. In addition, a general model of inflammogenesis for breast cancer is supposed to be involved in the induction of constitutive COX-2 over-expression and the up-regulation of the prostaglandin cascade. Both models will be presented in the following. There are two primary genes encoding the cyclooxygenase, a constitutive gene (COX-1) and an inducible form (COX-2) (Hla and Neilson 1992, Herschman 1994, 2002). The inducible COX-2 gene represents the switching molecule that induces the inflammatory response via COX-2 activation by an inflammatory stimulus such as alcohol, tobacco, ischemia, trauma, pressure, cytotoxins, foreign objects, bacteria, viruses, fungi and lipopolysaccharides. Moreover, COX-2 induces the biosynthesis of E-prostaglandins such as prostaglandin E2 (PGE-2), which then mediate the inflammatory response.

References and further reading

Abdulamir A S, Hafidh R R and Abu Bakar F 2011 The association of streptococcus bovis/gallolyticus with colorectal tumors: the nature and the underlying mechanisms of its etiological role *J. Exp. Clin. Cancer Res.* **30** 11

Acevedo A, del Pozo M A, Arroyo A G, Sánchez-Mateos P, González-Amaro R and Sánchez-Madrid F 1993 Distribution of ICAM-3-bearing cells in normal human tissues. Expression of a novel counter-receptor for LFA-1 in epidermal Langerhans cells *Am. J. Pathol.* **143** 774–83

Aird W C 2007 Phenotypic heterogeneity of the endothelium: I. Structure, function, and mechanisms *Circ. Res.* **100** 158–73

Alibert C, Goud B and Manneville J B 2017 Are cancer cells really softer than normal cells? *Biol. Cell* **109** 167–89

Alon R and Ley K 2008 Cells on the run: shear-regulated integrin activation in leukocyte rolling and arrest on endothelial cells *Curr. Opin. Cell Biol.* **20** 525–32

Alon R, Grabovsky V and Feigelson S 2003 Chemokine induction of integrin adhesiveness on rolling and arrested leukocytes local signaling events or global stepwise activation? *Microcirculation* **10** 297–311

Alon R, Kassner P, Carr M, Finger E, Hemler M and Springer T 1995 The integrin VLA-4 supports tethering and rolling in flow on VCAM-1 *J. Cell Biol.* **128** 1243–53

Amit I, Citri A and Shay T Y *et al* 2007 A module of negative feedback regulators defines growth factor signaling *Nat. Genet.* **39** 503–12

Aplin A E, Howe A, Alahari S K and Juliano R L 1998 Signal transduction and signal modulation by cell adhesion receptors: the role of integrins, cadherins, immunoglobulin-cell adhesion molecules, and selectins *Pharmacol. Rev.* **50** 197–263

Arthur J C, Perez-Chanona E and Mühlbauer M *et al* 2012 Intestinal inflammation targets cancer-inducing activity of the microbiota *Science* **338** 120–3

Balkwill F and Mantovani A 2001 Inflammation and cancer: back to Virchow? *Lancet* **357** 539–45

Bamforth S D, Lightman S L and Greenwood J 1997 Ultrastructural analysis of interleukin-1 beta-induced leukocyte recruitment to the rat retina *Invest. Ophthalmol. Vis. Sci.* **38** 25–35

Barreiro O, de la Fuente H, Mittelbrunn M and Sánchez-Madrid F 2007 Functional insights on the polarized redistribution of leukocyte integrins and their ligands during leukocyte migration and immune interactions *Immunol. Rev.* **218** 147–64

Barreiro O, Vicente-Manzanares M, Urzainqui A, Yáñez-Mó M and Sánchez-Madrid F 2004 Interactive protrusive structures during leukocyte adhesion and transendothelial migration *Front. Biosci.* **9** 1849–63

Barreiro O, Yáñez-Mó M, Sala-Valdes M, Gutiérrez-López M D, Ovalle S, Higginbottom A, Monk P N and Cabañas C and Sánchez-Madrid F 2005 Endothelial tetraspanin microdomains regulate leukocyte firm adhesion during extravasation *Blood* **105** 2852–61

Barreiro O, Yánez-Mó M, Serrador J M, Montoya M C, Vicente-Manzanares M, Tejedor R, Furthmayr H and Sanchez-Madrid F 2002 Dynamic interaction of VCAM-1 and ICAM-1 with moesin and ezrin in a novel endothelial docking structure for adherent leukocytes *J. Cell Biol.* **157** 1233–45

Barreiro O, Zamai M, Yánez-Mó M, Tejera E, López-Romero P, Monk P N, Gratton E, Caiolfa V R and Sánchez-Madrid F 2008 Endothelial adhesion receptors are recruited to adherent leukocytes by inclusion in preformed tetraspanin nanoplatforms *J. Cell Biol.* **183** 527–42

Bassaganya-Riera J, Viladomiu M, Pedragosa M, De Simone C and Hontecillas R 2012 Immunoregulatory mechanisms underlying prevention of colitis-associated colorectal cancer by probiotic bacteria *PLoS One* **7** e34676

Bates R C and Mercurio A M 2003 Tumor necrosis factor-α stimulates the epithelial-to-mesenchymal transition of human colonic organoids *Mol. Biol. Cell* **14** 1790–800

Bayne L J, Beatty G L, Jhala N, Clark C E, Rhim A D, Stanger B Z and Vonderheide R H 2012 Tumor-derived granulocyte-macrophage colony-stimulating factor regulates myeloid inflammation and T cell immunity in pancreatic cancer *Cancer Cell* **21** 822–35

Beglova N, Blacklow S C, Takagi J and Springer T A 2002 Cysteine-rich module structure reveals a fulcrum for integrin rearrangement upon activation *Nat. Struct. Biol.* **9** 282–7

Ben-Neriah Y and Karin M 2011 Inflammation meets cancer, with NF-κB as the matchmaker *Nat. Immunol.* **12** 715–23

Benesch M G K, Tang X, Dewald J, Dong W-F, Mackey J R, Hemmings D G, McMullen T P W and Brindley D N 2018 Tumor-induced inflammation in mammary adipose tissue stimulates a vicious cycle of autotaxin expression and breast cancer progression *FASEB J.* **29** 3990–4000

Bergers G *et al* 2000 Matrix metalloproteinase-9 triggers the angiogenic switch during carcinogenesis *Nat. Cell Biol.* **2** 737–44

Berlin C, Bargatze R, Campbell J, von Andrian U, Szabo M, Hasslen S, Nelson R D, Berg E L, Erlandsen S L and Butcher E C 1995 α4 integrins mediate lymphocyte attachment and rolling under physiologic flow *Cell* **80** 413–22

Berlin C, Berg E L, Briskin M J, Andrew D P, Kilshaw P J, Holzmann B, Weissman I L, Hamann A and Butcher E C 1993 $\alpha 4\beta 7$ integrin mediates lymphocyte binding to the mucosal vascular addressin MAdCAM-1 *Cell* **74** 185–95

Bollrath J, Phesse T J and von Burstin V A *et al* 2009 gp130-mediated Stat3 activation in enterocytes regulates cell survival and cell-cycle progression during colitis-associated tumorigenesis *Cancer Cell* **15** 91–102

Braumüller H, Wieder T, Brenner E, Assmann S, Hahn M and Alkhaled M *et al* 2013 T-helper-1-cell cytokines drive cancer into senescence *Nature* **494** 361–5

Bronte-Tinkew D M, Terebiznik M, Franco A, Ang M, Ahn D, Mimuro H, Sasakawa C, Ropeleski M J, Peek R M Jr and Jones N L 2009 Helicobacter pylori cytotoxin-associated gene A activates the signal transducer and activator of transcription 3 pathway *in vitro* and *in vivo Cancer Res.* **69** 632–9

Butcher E C 1991 Leukocyte-endothelial cell recognition: three (or more) steps to specificity and diversity *Cell* **67** 1033–6

Butcher E C and Picker L J 1996 Lymphocyte homing and homeostasis *Science* **272** 60–6

Cabanas C and Hogg N 1993 Ligand intercellular adhesion molecule 1 has a necessary role in activation of integrin lymphocyte function-associated molecule 1 *Proc. Natl Acad. Sci. USA* **90** 5838–42

Cairo C W, Mirchev R and Golan D E 2006 Cytoskeletal regulation couples LFA-1 conformational changes to receptor lateral mobility and clustering *Immunity* **25** 297–308

Cambi A, Joosten B, Koopman M, de Lange F, Beeren I, Torensma R, Fransen J A, Garcia-Parajó M, van Leeuwen F N and Figdor C G 2006 Organization of the integrin LFA-1 in nanoclusters regulates its activity *Mol. Biol. Cell* **17** 4270–81

Campbell J J, Hedrick J, Zlotnik A, Siani M A, Thompson D A and Butcher E C 1998 Chemokines and the arrest of lymphocytes rolling under flow conditions *Science* **279** 381–4

Campisi J and d'Adda di Fagagna F 2007 Cellular senescence: when bad things happen to good cells *Nat. Rev. Mol. Cell Biol.* **8** 729–40

Campregher C, Luciani M G and Gasche C 2008 Activated neutrophils induce an hMSH2-dependent G2/M checkpoint arrest and replication errors at a (CA)13-repeat in colon epithelial cells *Gut* **57** 780–7

Carlos T M and Harlan J M 1994 Leukocyte-endothelial adhesion molecules *Blood* **84** 2068–101

Carman C V 2009 Mechanisms for transcellular diapedesis: probing and pathfinding by 'invadosome-like protrusions' *J. Cell Sci.* **122** 3025–35

Carman C V and Springer T A 2003 Integrin avidity regulation: are changes in affinity and conformation underemphasized? *Curr. Opin. Cell Biol.* **15** 547–56

Carman C V and Springer T A 2004 A transmigratory cup in leukocyte diapedesis both through individual vascular endothelial cells and between them *J. Cell Biol.* **167** 377–88

Carman C V, Sage P T, Sciuto T E, de la Fuente M A, Geha R S, Ochs H D, Dvorak H F, Dvorak A M and Springer T A 2007 Transcellular diapedesis is initiated by invasive podosomes *Immunity* **26** 784–97

Casaletto J B and McClatchey A I 2012 Spatial regulation of receptor tyrosine kinases in development and cancer *Nat. Rev. Cancer* **12** 387–400

Castellarin M *et al* 2012 Fusobacterium nucleatum infection is prevalent in human colorectal carcinoma *Genome Res.* **22** 299–306

Cataisson C, Salcedo R and Hakim S *et al* 2012 IL-1R-MyD88 signaling in keratinocyte transformation and carcinogenesis *J. Exp. Med.* **209** 1689–702

Charrin S, le Naour F, Silvie O, Milhiet P E, Boucheix C and Rubinstein E 2009 Lateral organization of membrane proteins: tetraspanins spin their web *Biochem. J.* **420** 133–54

Chavakis T, Bierhaus A and Al-Fakhri N *et al* 2003 The pattern recognition receptor (RAGE) is a counterreceptor for leukocyte integrins: a novel pathway for inflammatory cell recruitment *J. Exp. Med.* **198** 1507–15

Chen G Y, Liu M, Wang F, Bertin J and Nunez G 2011 A functional role for Nlrp6 in intestinal inflammation and tumorigenesis *J. Immunol.* **186** 7187–94

Chen J, Yao Y and Gong C *et al* 2011 CCL18 from tumor-associated macrophages promotes breast cancer metastasis via PITPNM3 *Cancer Cell* **19** 541–55

Chen L W, Egan L, Li Z W, Greten F R, Kagnoff M F and Karin M 2003 The two faces of IKK and NF-kappaB inhibition: prevention of systemic inflammation but increased local injury following intestinal ischemia-reperfusion *Nat. Med.* **9** 575–81

Chien Y *et al* 2011 Control of the senescence-associated secretory phenotype by NF-κB promotes senescence and enhances chemosensitivity *Genes Dev.* **25** 2125–36

Chin L, Xia Y, Discher D E and Janmey P A 2016 Mechanotransduction in cancer *Curr. Opin. Chem. Eng.* **11** 77–84

Cinamon G, Shinder V and Alon R 2001 Shear forces promote lymphocyte migration across vascular endothelium bearing apical chemokines *Nat. Immunol.* **2** 515–22

Clayton A, Evans R A, Pettit E, Hallett M, Williams J D and Steadman R 1998 Cellular activation through the ligation of intercellular adhesion molecule-1 *J. Cell Sci.* **111** 443–53

Colotta F, Allavena P, Sica A, Garlanda C and Mantovani A 2009 Cancer-related inflammation, the seventh hallmark of cancer: links to genetic instability *Carcinogenesis* **30** 1073–81

Constantin G, Majeed M, Giagulli C, Piccio L, Kim J Y, Butcher E C and Laudanna C 2000 Chemokines trigger immediate β2 integrin affinity and mobility changes: differential regulation and roles in lymphocyte arrest under flow *Immunity* **13** 759–69

Cook-Mills J M 2002 VCAM-1 signals during lymphocyte migration: role of reactive oxygen species *Mol. Immunol.* **39** 499–508

Corthesy B, Gaskins H R and Mercenier A 2007 Cross-talk between probiotic bacteria and the host immune system *J. Nutr.* **137** 781S–90S

Coussens L M and Werb Z 2002 Inflammation and cancer *Nature* **420** 860–7

Coussens L M, Raymond W W, Bergers G, Laig-Webster M, Behrendtsen O, Werb Z, Caughey G H and Hanahan D 1999 Inflammatory mast cells up-regulate angiogenesis during squamous epithelial carcinogenesis *Genes Dev.* **13** 1382–97

Coussens L M, Tinkle C L, Hanahan D and Werb Z 2000 MMP-9 supplied by bone marrow-derived cells contributes to skin carcinogenesis *Cell* **103** 481–90

Cunningham S A, Rodríguez J M, Arrate M P, Tran T M and Brock T A 2002 JAM2 interacts with $\alpha4\beta1$. Facilitation by JAM3 *J. Biol. Chem.* **277** 27589–92

Curiel T J, Coukos G and Zou L *et al* 2004 Specific recruitment of regulatory T cells in ovarian carcinoma fosters immune privilege and predicts reduced survival *Nat. Med.* **10** 942–9

Curiel T J, Wei S and Dong H *et al* 2003 Blockade of B7-H1 improves myeloid dendritic cell-mediated antitumor immunity *Nat. Med.* **9** 562–7

Danaher P, Warren S, Dennis L, D'Amico L, White A, Disis M L, Geller M A, Odunsi K, Beechem J and Fling S P 2017 Gene expression markers of tumor infiltrating leukocytes *J. Immunother. Cancer* **5** 18

Dapito D H, Mencin A and Gwak G Y *et al* 2012 Promotion of hepatocellular carcinoma by the intestinal microbiota and TLR4 *Cancer Cell* **21** 504–16

Deem T L and Cook-Mills J M 2004 Vascular cell adhesion molecule 1 (VCAM-1) activation of endothelial cell matrix metalloproteinases: role of reactive oxygen species *Blood* **104** 2385–93

Del Pozo M A, Sánchez-Mateos P, Nieto M and Sánchez-Madrid F 1995 Chemokines regulate cellular polarization and adhesion receptor redistribution during lymphocyte interaction with endothelium and extracellular matrix. Involvement of cAMP signaling pathway *J. Cell Biol.* **131** 495–508

DeNardo D G, Barreto J B, Andreu P, Vasquez L, Tawfik D, Kolhatkar N and Coussens L M 2009 CD4+ T cells regulate pulmonary metastasis of mammary carcinomas by enhancing protumor properties of macrophages *Cancer Cell* **16** 91–102

Di Lorenzo A, Fernández-Hernando C, Cirino G, William C and Sessa W C 2009 Akt1 is critical for acute inflammation and histamine-mediated vascular leakage *Proc. Natl Acad. Sci. U. S. A.* **106** 14552–7

Dobereiner H G, Dubin-Thaler B J, Hofman J M, Xenias H S, Sims T N, Giannone G, Dustin M L, Wiggins C H and Sheetz M P 2006 Lateral membrane waves constitute a universal dynamic pattern of motile cells *Phys. Rev. Lett.* **97** 038102

Doyle A D and Yamada K M 2016 Mechanosensing via cell-matrix adhesions in 3D microenvironments *Exp. Cell Res.* **343** 60–6

Dustin M L, Rothlein R, Bhan A K, Dinarello C A and Springer T A 1986 Induction by IL 1 and interferon-gamma: tissue distribution, biochemistry, and function of a natural adherence molecule (ICAM-1) *J. Immunol.* **137** 245–54

Dvorak H F 1986 Tumors: wounds that do not heal. Similarities between tumor stroma generation and wound healing *N. Engl. J. Med.* **315** 1650–9

Dwir O, Kansas G S and Alon R 2001 Cytoplasmic anchorage of L-selectin controls leukocyte capture and rolling by increasing the mechanical stability of the selectin tether *J. Cell Biol.* **155** 145–56

Egeblad M and Werb Z 2002 New functions for the matrix metalloproteinases in cancer progression *Nat. Rev. Cancer* **2** 161–74

Elices M J, Osborn L, Takada Y, Crouse C, Luhowskyj S, Hemler M E and Lobb R R 1990 VCAM-1 on activated endothelium interacts with the leukocyte integrin VLA-4 at a site distinct from the VLA-4/fibronectin binding site *Cell* **60** 577–84

Elinav E, Henao-Mejia J and Flavell R A 2013 Integrative inflammasome activity in the regulation of intestinal mucosal immune responses *Mucosal. Immunol.* **6** 4–13

Elinav E *et al* 2011 NLRP6 inflammasome regulates colonic microbial ecology and risk for colitis *Cell* **145** 745–57

Endo Y, Marusawa H, Kou T, Nakase H, Fujii S, Fujimori T, Kinoshita K, Honjo T and Chiba T 2008 Activation-induced cytidine deaminase links between inflammation and the development of colitis-associated colorectal cancers *Gastroenterology* **135** 889–98

Engelhardt B and Wolburg H 2004 Mini-review: transendothelial migration of leukocytes: through the front door or around the side of the house? *Eur. J. Immunol.* **34** 2955–63

Eriksson E E, Xie X, Werr J, Thoren P and Lindbom L 2001 Importance of primary capture and L-selectin-dependent secondary capture in leukocyte accumulation in inflammation and atherosclerosis *in vivo J. Exp. Med.* **194** 205–18

Etienne-Manneville S, Manneville J B, Adamson P, Wilbourn B, Greenwood J and Couraud P O 2000 ICAM-1-coupled cytoskeletal rearrangements and transendothelial lymphocyte migration involve intracellular calcium signaling in brain endothelial cell lines *J. Immunol.* **165** 3375–83

Euer N, Schwirzke M, Evtimova V, Burtscher H, Jarsch M, Tarin D and Weidl U H 2002 Identification of genes associated with metastasis of mammary carcinoma in metastatic versus non-metastatic cell lines *Anticancer Res.* **22** 733–40

Evans E A and Calderwood D A 2007 Forces and bond dynamics in cell adhesion *Science* **316** 1148–53

Facciabene A, Motz G T and Coukos G 2012 T-regulatory cells: key players in tumor immune escape and angiogenesis *Cancer Res.* **72** 2162–71

Facciabene A *et al* 2011 Tumor hypoxia promotes tolerance and angiogenesis via CCL28 and Treg cells *Nature* **475** 226–30

Federico L, Jeong K J, Vellano C P and Mills G B 2016 Autotaxin, a lysophospholipase D with pleomorphic effects in oncogenesis and cancer progression *J. Lipid Res.* **57** 25–35

Fischer T, Wilharm N, Hayn A and Mierke C T 2017 Matrix and cellular mechanical properties are the driving factors for facilitating human cancer cell motility into 3D engineered matrices *Converg. Sci. Phys. Oncol.* **3** 044003

Franco A T, Israel D A, Washington M K, Krishna U, Fox J G and Rogers A B *et al* 2005 Activation of *β*-catenin by carcinogenic *Helicobacter pylori Proc. Natl Acad. Sci. USA* **102** 10646–51

Frenette P S and Wagner D D 1996a Adhesion molecules part I *N. Engl. J. Med.* **334** 1526–9

Frenette P S and Wagner D D 1996b Adhesion molecules: part II. Blood vessels and blood cells *N. Engl. J. Med.* **335** 43–5

Fukata M, Chen A and Vamadevan A S *et al* 2007 Toll-like receptor-4 promotes the development of colitis-associated colorectal tumors *Gastroenterology* **133** 1869–81

Fukuda A, Wang S C and Morris J P *et al* 2011 Stat3 and MMP7 contribute to pancreatic ductal adenocarcinoma initiation and progression *Cancer Cell* **19** 441–55

Gabrilovich D I and Nagaraj S 2009 Myeloid-derived suppressor cells as regulators of the immune system *Nat. Rev. Immunol.* **9** 162–74

Gabrilovich D I, Chen H L, Girgis K R, Cunningham H T, Meny G M, Nadaf S, Kavanaugh D and Carbone D P 1996 Production of vascular endothelial growth factor by human tumors inhibits the functional maturation of dendritic cells *Nat. Med.* **2** 1096–103

Gaetano C G, Samadi N, Tomsig J L, Macdonald T L, Lynch K R and Brindley D N 2009 Inhibition of autotaxin production or activity blocks lysophosphatidylcholine-induced migration of human breast cancer and melanoma cells *Mol. Carcinog.* **48** 801–9

Gahmberg C G, Nortamo P, Kantor C, Autero M, Kotovuori P, Hemiö L, Salcedo R and Patarroyo M 1990 The pivotal role of the Leu-CAM and ICAM molecules in human leukocyte adhesion *Cell Differ. Dev.* **32** 239–45

Gao S P *et al* 2007 Mutations in the EGFR kinase domain mediate STAT3 activation via IL-6 production in human lung adenocarcinomas *J. Clin. Invest.* **117** 3846–56

García-López M A, Barreiro O, García-Díez A, Sánchez-Madrid F and Penas P F 2005 Role of tetraspanins CD9 and CD151 in primary melanocyte motility *J. Invest. Dermatol.* **125** 1001–9

Garrett W S, Punit S, Gallini C A, Michaud M, Zhang D, Sigrist K S, Lord G M, Glickman J N and Glimcher L H 2009 Colitis-associated colorectal cancer driven by T-bet deficiency in dendritic cells *Cancer Cell* **16** 208–19

Geiger B and Bershadsky A 2002 Exploring the neighborhood: adhesion-coupled cell mechanosensors *Cell* **110** 139–42

Geissmann F, Revy P, Regnault A, Lepelletier Y, Dy M, Brousse N, Amigorena S, Hermine O and Durandy A 1999 TGF-*β* 1 prevents the noncognate maturation of human dendritic Langerhans cells *J. Immunol.* **162** 4567–75

Gocheva V, Wang H W, Gadea B B, Shree T, Hunter K E, Garfall A L, Berman T and Joyce J A 2010 IL-4 induces cathepsin protease activity in tumor-associated macrophages to promote cancer growth and invasion *Genes Dev.* **24** 241–55

Goodwin A C *et al* 2011 Polyamine catabolism contributes to enterotoxigenic bacteroides fragilis-induced colon tumorigenesis *Proc. Natl Acad. Sci. USA* **108** 15354–9

Gordon J I 2012 Honor thy gut symbionts redux *Science* **336** 1251–3

Gordon-Alonso M, Yáñez-Mó M, Barreiro O, Álvarez S, Muñoz-Fernández M A, Valenzuela-Fernández A and Sánchez-Madrid F 2006 Tetraspanins CD9 and CD81 modulate HIV-1-induced membrane fusion *J. Immunol.* **177** 5129–37

Grabovsky V, Feigelson S and Chen C *et al* 2000 Subsecond induction of α4integrin clustering by immobilized chemokines stimulates leukocyte tethering and rolling on endothelial vascular cell adhesion molecule 1 under flow conditions *J. Exp. Med.* **192** 495–506

Greenwood J, Etienne-Manneville S, Adamson P and Couraud P O 2002 Lymphocyte migration into the central nervous system: implication of ICAM-1 signalling at the blood–brain barrier *Vascul. Pharmacol.* **38** 315–32

Greten F R, Eckmann L, Greten T F, Park J M, Li Z W, Egan L J, Kagnoff M F and Karin M 2004 IKKβ links inflammation and tumorigenesis in a mouse model of colitis-associated cancer *Cell* **118** 285–96

Grivennikov S I and Karin M 2010 Inflammation and oncogenesis: a vicious connection *Curr. Opin. Genet. Dev.* **20** 65–71

Grivennikov S I, Greten F R and Karin M 2010 Immunity, inflammation, and cancer *Cell* **140** 883–99

Grivennikov S I, Kuprash D V, Liu Z G and Nedospasov S A 2006 Intracellular signals and events activated by cytokines of the tumor necrosis factor superfamily: from simple paradigms to complex mechanisms *Int. Rev. Cytol.* **252** 129–61

Grivennikov S I, Wang K, Mucida D, Stewart C A, Schnabl B and Jauch D *et al* 2012 Adenoma-linked barrier defects and microbial products drive IL-23/ IL-17-mediated tumor growth *Nature* **491** 254–8

Grivennikov S *et al* 2009 IL-6 and Stat3 are required for survival of intestinal epithelial cells and development of colitis-associated cancer *Cancer Cell* **15** 103–13

Guck J *et al* 2005 Optical deformability as an inherent cell marker for testing malignant transformation and metastatic competence *Biophys. J.* **88** 3689–98

Hahn M A, Hahn T, Lee D H, Esworthy R S, Kim B W, Riggs A D, Chu F F and Pfeifer G P 2008 Methylation of polycomb target genes in intestinal cancer is mediated by inflammation *Cancer Res.* **68** 10280–9

Hanahan D and Weinberg R A 2011 Hallmarks of cancer: the next generation *Cell* **144** 646–74

Harris R E 2002 Cyclooxygenase-2 blockade in cancer prevention and therapy: widening the scope of impact COX-2 Blockade *Cancer Prevention and Therapy* ed R E Harris (Totowa, NJ: Humana)

Harris R E 2007 Cyclooxygenase-2 (cox-2) and the inflammo-genesis of cancer *Subcell. Biochem.* **42** 93–126

Heiska L, Alfthan K, Gronholm M, Vilja P, Vaheri A and Carpen O 1998 Association of Ezrin with intercellular adhesion molecule-1 and -2 (ICAM-1 and ICAM-2). Regulation by phosphatidylinositol 4, 5-bisphosphate *J. Biol. Chem.* **273** 21893–900

Hemler M E 2005 Tetraspanin functions and associated microdomains *Nat. Rev. Mol. Cell Biol.* **6** 801–11

Henao-Mejia J, Elinav E and Jin C *et al* 2012 Inflammasome-mediated dysbiosis regulates progression of NAFLD and obesity *Nature* **482** 179–85

Henderson R B, Lim L H, Tessier P A, Gavins F N, Mathies M, Perretti M and Hogg N 2001 The use of lymphocyte function-associated antigen (LFA)-1-deficient mice to determine the role of LFA-1, Mac-1, and alpha4 integrin in the inflammatory response of neutrophils *J. Exp. Med.* **194** 219–26

Herschman H R 1994 Regulation of prostaglandin synthase-1 and prostaglandin synthase-2 *Cancer Metastasis Rev.* **13** 241–56

Herschman H R 2002 Historical aspects of COX-2 COX-2 Blockade *Cancer Prevention and Therapy* ed R E Harris (Totowa, NJ: Humana) pp 13–32

Hidalgo A, Peired A J, Wild M K, Vestweber D and Frenette P S 2007 Complete identification of E-selectin ligands on neutrophils reveals distinct functions of PSGL-1, ESL-1, and CD44 *Immunity* **26** 477–89

Hill R, Song Y, Cardiff R D and Van Dyke T 2005 Selective evolution of stromal mesenchyme with p53 loss in response to epithelial tumorigenesis *Cell* **123** 1001–11

Hla T and Neilson K 1992 Human cyclooxygenase-2 cDNA *Proc. Natl Acad. Sci. USA* **89** 7384–8

Ho P L, Lay E J, Jian W, Parra D and Chan K S 2012 Stat3 activation in urothelial stem cells leads to direct progression to invasive bladder cancer *Cancer Res.* **72** 3135–42

Hope J M, Greenlee J D and King M R 2018 Mechanosensitive ion channels: TRPV4 and P2X7 in disseminating cancer cells *Cancer J.* **24** 84–92

Hu B, Elinav E, Huber S, Booth C J, Strowig T, Jin C, Eisenbarth S C and Flavell R A 2010 Inflammation-induced tumorigenesis in the colon is regulated by caspase-1 and NLRC4 *Proc. Natl Acad. Sci. USA* **107** 21635–40

Hu B *et al* 2013 Microbiota-induced activation of epithelial IL-6 signaling links inflammasome-driven inflammation with transmissible cancer *Proc. Natl Acad. Sci. USA* **110** 9862–7

Hubbard A K and Rothlein R 2000 Intercellular adhesion molecule-1 (ICAM-1) expression and cell signaling cascades *Free Radic. Biol. Med.* **28** 1379–86

Huber S, Gagliani N and Zenewicz L A *et al* 2012 IL-22BP is regulated by the inflammasome and modulates tumorigenesis in the intestine *Nature* **491** 259–63

Hull M A, Booth J K, Tisbury A, Scott N, Bonifer C, Markham A F and Coletta P L 1999 Cyclooxygenase 2 is up-regulated and localized to macrophages in the intestine of Min mice *Br. J. Cancer* **79** 1399–405

Humphries J D, Byron A and Humphries M J 2006 Integrin ligands at a glance *J. Cell Sci.* **119** 3901–3

Hussain S P, Hofseth L J and Harris C C 2003 Radical causes of cancer *Nat. Rev. Cancer* **3** 276–85

Huttenhower C, Gevers D, Knight R, Abubucker S, Badger J H, Asif T and Chinwalla A T 2012 Structure, function and diversity of the healthy human microbiome *Nature* **486** 207–14

Hynes R O 2002 Integrins: bidirectional, allosteric signaling machines *Cell* **110** 673–87

Hytönen V P and Wehrle-Haller B 2016 Mechanosensing in cell–matrix adhesions—converting tension into chemical signals *Exp. Cell Res.* **343** 35–41

Iliopoulos D, Hirsch H A and Struhl K 2009 An epigenetic switch involving NF-κB, Lin28, Let-7 MicroRNA, and IL6 links inflammation to cell transformation *Cell* **139** 693–706

Ivetic A, Deka J, Ridley A J and Ager A 2002 The cytoplasmic tail of L-selectin interacts with members of the Ezrin–Radixin–Moesin (ERM) family of proteins: cell activation-dependent binding of Moesin but not Ezrin *J. Biol. Chem.* **277** 2321–9

Jacob N and Stohl W 2011 Cytokine disturbances in systemic lupus erythematosus *Arthritis Res. Ther.* **13** 228

Janeway C A Jr and Medzhitov R 2002 Innate immune recognition *Annu. Rev. Immunol.* **20** 197–216

Jang B C and Hla T 2002 Regulation of expression and potential carcinogenic role of cylcooxygenase-2 COX-2 blockade *Cancer Prevention and Therapy* ed R E Harris (Totowa, NJ: Humana) pp 171–84

Jemal A, Siegel R, Xu J and Ward E 2010 Cancer statistics *CA Cancer J. Clin.* **60** 277–300

Jiang R, Tan Z, Deng L, Chen Y, Xia Y, Gao Y, Wang X and Sun B 2011 Interleukin-22 promotes human hepatocellular carcinoma by activation of STAT3 *Hepatology* **54** 900–9

Jun C D, Shimaoka M, Carman C V, Takagi J and Springer T A 2001 Dimerization and the effectiveness of ICAM-1 in mediating LFA-1-dependent adhesion *Proc. Natl Acad. Sci. USA* **98** 6830–5

Kadono T, Venturi G M, Steeber D A and Tedder T F 2002 Leukocyte rolling velocities and migration are optimized by cooperative L-selectin and intercellular adhesion molecule-1 functions *J. Immunol.* **169** 4542–50

Kalluri R and Weinberg R A 2009 The basics of epithelial-mesenchymal transition *J. Clin. Invest.* **119** 1420–8

Kang T W, Yevsa T, Woller N, Hoenicke L, Wuestefeld T and Dauch D *et al* 2011 Senescence surveillance of pre-malignant hepatocytes limits liver cancer development *Nature* **479** 547–51

Karin M and Greten F R 2005 NF-κB: linking inflammation and immunity to cancer development and progression *Nat. Rev. Immunol.* **5** 749–59

Killock D J, Parsons M, Zarrouk M, Ameer-Beg S M, Ridley A J, Haskard D O, Zvelebil M and Ivetic A 2009 *In vitro* and *in vivo* characterization of molecular interactions between calmodulin, ezrin/radixin/ moesin (ERM) and L-selectin *J. Biol. Chem.* **284** 8833–45

Kim Y, Lee D, Kim D, Cho J, Yang J, Chung M, Kim K and Ha N 2008 Inhibition of proliferation in colon cancer cell lines and harmful enzyme activity of colon bacteria by *Bifidobacterium adolescentis* SPM0212 *Arch. Pharm. Res.* **31** 468–73

Kitamura T *et al* 2007 SMAD4-deficient intestinal tumors recruit CCR1+ myeloid cells that promote invasion *Nat. Genet.* **39** 467–75

Koki A T, Leahy K M, Harmon J M and Masferrer J L 2002 Cyclooxy-genase-2 and cancer COX-2 blockade *Cancer Prevention and Therapy* ed R E Harris (Totowa, NJ: Humana) pp 185–203

Komori J, Marusawa H, Machimoto T, Endo Y, Kinoshita K, Kou T, Haga H, Ikai I, Uemoto S and Chiba T 2008 Activation-induced cytidine deaminase links bile duct inflammation to human cholangiocarcinoma *Hepatology* **47** 888–96

Kostic A D *et al* 2012 Genomic analysis identifies association of *Fusobacterium* with colorectal carcinoma *Genome Res.* **22** 292–8

Kostic A D *et al* 2013 Fusobacterium nucleatum potentiates intestinal tumorigenesis and modulates the tumor-immune microenvironment *Cell Host Microbe* **14** 207–15

Kuper H, Adami H O and Trichopoulos D 2000 Infections as a major preventable cause of human cancer *J. Intern. Med.* **248** 171–83

Lakins M A, Ghorani E, Munir H, Martins C P and Shields J D 2018 Cancer-associated fibroblasts induce antigen-specific deletion of CD8+ T Cells to protect tumour cells *Nat. Commun.* **9** 948

Lamagna C, Meda P and Mandicourt G *et al* 2005 Dual interaction of JAM-C with JAM-B and α (M)β2 integrin: function in junctional complexes and leukocyte adhesion *Mol. Biol. Cell.* **16** 4992–5003

Lamichhane P *et al* 2017 IL10 Release upon PD-1 blockade sustains immunosuppression in ovarian cancer *Cancer Res.* **77** 6667–78

Laudanna C 2005 Integrin activation under flow: a local affair *Nat. Immunol.* **6** 429–30

Laudanna C and Alon R 2006 Right on the spot. Chemokine triggering of integrin-mediated arrest of rolling leukocytes *Thromb. Haemost.* **95** 5–11

Le Leu R K, Brown I L, Hu Y, Bird A R, Jackson M, Esterman A and Young G P 2005 A synbiotic combination of resistant starch and *Bifidobacterium lactis* facilitates apoptotic deletion of carcinogen-damaged cells in rat colon *J. Nutr.* **135** 996–1001

Lee S H *et al* 2010 ERK activation drives intestinal tumorigenesis in Apcmin/+ mice *Nat. Med.* **16** 665–70

Leray E, Yaouanq J, Le Page E, Coustans M, Laplaud D, Oger J and Edan G 2010 Evidence for a two-stage disability progression in multiple sclerosis *Brain* **133** 1900–13

Lerner I, Hermano E and Zcharia E *et al* 2011 Heparanase powers a chronic inflammatory circuit that promotes colitis-associated tumorigenesis in mice *J. Clin. Invest.* **121** 1709–21

Lesina M, Kurkowski M U and Ludes K *et al* 2011 Stat3/Socs3 activation by IL-6 transsignaling promotes progression of pancreatic intraepithelial neoplasia and development of pancreatic cancer *Cancer Cell* **19** 456–69

Lévesque J P, Zannettino A C, Pudney M, Niutta S, Haylock D N, Snapp K R, Kansas G S, Berndt M C and Simmons P J 1999 PSGL-1-mediated adhesion of human hematopoietic progenitors to P-selectin results in suppression of hematopoiesis *Immunity* **11** 369–78

Ley K, Laudanna C, Cybulsky M I and Nourshargh S 2007 Getting to the site of inflammation: the leukocyte adhesion cascade updated *Nat. Rev. Immunol.* **7** 678–89

Liang J, Nagahashi M and Kim E Y *et al* 2013 Sphingosine-1-phosphate links persistent STAT3 activation, chronic intestinal inflammation, and development of colitis-associated cancer *Cancer Cell* **23** 107–20

Lin C and Zhang J 2017 Inflammasomes in inflammation-induced cancer *Front. Immunol.* **8** 271

Lorenzon P, Vecile E, Nardon E, Ferrero E, Harlan J M, Tedesco F and Dobrina A 1998 Endothelial cell E- and P-selectin and vascular cell adhesion molecule-1 function as signaling receptors *J. Cell Biol.* **142** 1381–91

Lowe E L, Crother T R, Rabizadeh S, Hu B, Wang H, Chen S, Shimada K, Wong M H, Michelsen K S and Arditi M 2010 Toll-like receptor 2 signaling protects mice from tumor development in a mouse model of colitis-induced cancer *PLoS One* **5** e13027

Luster A D 1998 Chemokines—chemotactic cytokines that mediate inflammation *N. Engl. J. Med.* **338** 436–45

Mantovani A, Allavena P, Sica A and Balkwill F 2008 Cancer-related inflammation *Nature* **454** 436–44

Marchesi J R, Dutilh B E, Hall N, Peters W H, Roelofs R, Boleij A and Tjalsma H 2011 Towards the human colorectal cancer microbiome *PLoS One* **6** e20447

Mareel M and Leroy A 2003 Clinical, cellular, and molecular aspects of cancer invasion *Physiol. Rev.* **83** 337–76

Marlin S D and Springer T A 1987 Purified intercellular adhesion molecule-1 (ICAM-1) is a ligand for lymphocyte function-associated antigen 1 (LFA-1) *Cell* **51** 813–9

Maroof H, Hassan Z M, Mobarez A M and Mohamadabadi M A 2012 Lactobacillus acidophilus could modulate the immune response against breast cancer in murine model *J. Clin. Immunol.* **32** 1353–9

Marotta L L, Almendro V, Marusyk A, Shipitsin M, Schemme J and Walker S R *et al* 2011 The JAK2/STAT3 signaling pathway is required for growth of CD44+CD24– stem-cell-like breast cancer cells in human tumors *J. Clin. Invest.* **121** 2723–35

Marschel P and Schmid-Schonbein G W 2002 Control of fluid shear response in circulating leukocytes by integrins *Ann. Biomed. Eng.* **30** 333–43

McDonald B, Spicer J, Giannais B, Fallavollita L, Brodt P and Ferri L E 2009 Systemic inflammation increases cancer cell adhesion to hepatic sinusoids by neutrophil mediated mechanisms *Int. J. Cancer* **125** 1298–305

Medzhitov R 2008 Origin and physiological roles of inflammation *Nature* **454** 428–35

Mehta D and Malik A B 2006 Signaling mechanisms regulating endothelial permeability *Physiol. Rev.* **86** 279–367

Mehta P, Cummings R D and McEver R P 1998 Affinity and kinetic analysis of P-selectin binding to P-selectin glycoprotein ligand-1 *J. Biol. Chem.* **273** 32506–13

Menetrier-Caux C, Gaucherand M, Péguet-Navarro J, Plouet J, Pageaux J-F, Schmitt D and Staquet M-J 1998 Inhibition of the differentiation of dendritic cells from CD34+ progenitors by tumor cells: role of interleukin-6 and macrophage colony-stimulating factor *Blood* **92** 4778–91

Mierke C T 2013 The role of focal adhesion kinase in the regulation of cellular mechanical properties *Phys. Biol.* **10** 065005

Mierke C T, Fischer T, Puder S, Tom Kunschmann T, Soetje B and Ziegler W H 2017 Focal adhesion kinase activity is required for actomyosin contractility-based invasion of cells into dense 3D matrices *Sci. Rep.* **7** 42780

Millan J, Hewlett L, Glyn M, Toomre D, Clark P and Ridley A J 2006 Lymphocyte transcellular migration occurs through recruitment of endothelial ICAM-1 to caveola- and F-actin-rich domains *Nat. Cell Biol.* **8** 113–23

Mills K D, Ferguson D O and Alt F W 2003 The role of DNA breaks in genomic instability and tumorigenesis *Immunol. Rev.* **194** 77–95

Mishra A, Liu S, Sams G H, Curphey D P, Santhanam R and Rush L J *et al* 2012 Aberrant overexpression of IL-15 initiates large granular lymphocyte leukemia through chromosomal instability and DNA hypermethylation *Cancer Cell* **22** 645–55

Mittelbrunn M, Cabanas C and Sánchez-Madrid F 2006 Integrin $\alpha 4$ *UCSD-Nature Molecule Pages*

Mittelbrunn M, Yáñez-Mó M, Sancho D, Ursa A and Sánchez-Madrid F 2002 Cutting edge: dynamic redistribution of tetraspanin CD81 at the central zone of the immune synapse in both T lymphocytes and APC *J. Immunol.* **169** 6691–5

Moinfar F, Man Y G, Arnould L, Bratthauer G L, Ratschek M and Tavassoli F A 2000 Concurrent NF-κB activation by double-strand breaks and independent genetic alterations in the stromal and epithelial cells of mammary carcinoma: implications for tumorigenesis *Cancer Res.* **60** 2562–6

Mory A, Feigelson S W, Yarali N, Kilic S S, Bayhan G I, Gershoni-Baruch R, Etzioni A and Alon R 2008 Kindlin-3: a new gene involved in the pathogenesis of LAD-III *Blood* **112** 2591

Mosesson Y, Mills G B and Yarden Y 2008 Derailed endocytosis: an emerging feature of cancer *Nat. Rev. Cancer* **8** 835–50

Motz G T and Coukos G 2013 Deciphering and reversing tumor immune suppression *Immunity* **39** 61–73

Mueller M M and Fusenig N E 2004 Friends or foes—bipolar effects of the tumor stroma in cancer *Nat. Rev. Cancer* **4** 839–49

Muller W A 2003 Leucocyte–endothelial-cell interactions in leukocyte transmigration and the inflammatory response *Trends Immunol.* **24** 327–34

Munger K and Howley P M 2002 Human papillomavirus immortalization and transformation functions *Virus Res.* **89** 213–28

Murdoch C, Muthana M, Coffelt S B and Lewis C E 2008 The role of myeloid cells in the promotion of tumor angiogenesis *Nat. Rev. Cancer* **8** 618–31

Myant K B *et al* 2013 ROS production and NF-κB activation triggered by RAC1 facilitate WNT-driven intestinal stem cell proliferation and colorectal cancer initiation *Cell Stem Cell* **12** 761–73

Nagy J A, Dvorak A M and Dvorak H F 2012 Vascular hyperpermeability, angiogenesis and stroma generation *Cold Spring Harb. Perspect. Med.* **2** a006544

Nguyen D X, Bos P D and Massague J 2009 Metastasis: from dissemination to organ-specific colonization *Nat. Rev. Cancer* **9** 274–84

Nicholson M W, Barclay A N, Singer M S, Rosen S D and van der Merwe P A 1998 Affinity and kinetic analysis of L-selectin (CD62L) binding to glycosylation-dependent cell-adhesion molecule-1 *J. Biol. Chem.* **273** 763–70

Nickoloff B J, Ben-Neriah Y and Pikarsky E 2005 Inflammation and cancer: is the link as simple as we think? *J. Invest. Dermatol.* **124** 10–4

Nieminen M, Henttinen T, Merinen M, Marttila-Ichihara F, Eriksson J E and Jalkanen S 2006 Vimentin function in lymphocyte adhesion and transcellular migration *Nat. Cell Biol.* **8** 156–62

Nishida N, Xie C, Shimaoka M, Cheng Y, Walz T and Springer T A 2006 Activation of leukocyte $\beta 2$ integrins by conversion from bent to extended conformations *Immunity* **25** 583–94

Normand S, Delanoye-Crespin A, Bressenot A, Huot L, Grandjean T, Peyrin-Biroulet L, Lemoine Y, Hot D and Chamaillard M 2011 Nod-like receptor pyrin domain-containing protein 6 (NLRP6) controls epithelial self-renewal and colorectal carcinogenesis upon injury *Proc. Natl Acad. Sci. USA* **108** 9601–6

Okazaki I M, Kotani A and Honjo T 2007 Role of AID in tumorigenesis *Adv. Immunol.* **94** 245–73

Opdenakker G and Van Damme J 1992 Chemotactic factors, passive invasion and metastasis of cancer cells *Immunol. Today* **13** 463–4

Opdenakker G and Van Damme J 1992 Cytokines and proteases in invasive processes: molecular similarities between inflammation and cancer *Cytokine* **4** 251–8

Opdenakker G and Van Damme J 1999 Novel monocyte chemoattractants in cancer *Chemokines and Cancer* ed B Rollins (Totowa, NJ: Humana) pp 51–69

Opdenakker G and Van Damme J 2004 The counter-current principle in invasion and metastasis of cancer cells. Recent insights on the roles of chemokines *Int. J. Dev. Biol.* **48** 519–27

Orlando A, Messa C, Linsalata M, Cavallini A and Russo F 2009 Effects of Lactobacillus rhamnosus GG on proliferation and polyamine metabolism in HGC-27 human gastric and DLD-1 colonic cancer cell lines *Immunopharmacol. Immunotoxicol.* **31** 108–16

Ostermann G, Weber K S, Zernecke A, Schroder A and Weber C 2002 JAM-1 is a ligand of the beta(2) integrin LFA-1 involved in transendothelial migration of leukocytes *Nat. Immunol.* **3** 151–8

Panzetta V, Musella I, Rapa I, Volante M, Netti P A and Fusco S 2017 Mechanical phenotyping of cells and extracellular matrix as grade and stage markers of lung tumor tissues *Acta Biomater.* **57** 334–41

Park E J, Lee J H, Yu G Y, He G, Ali S R, Holzer R G, Osterreicher C H, Takahashi H and Karin M 2010 Dietary and genetic obesity promote liver inflammation and tumorigenesis by enhancing IL-6 and TNF expression *Cell* **140** 197–208

Park O *et al* 2011 *In vivo* consequences of liver-specific interleukin-22 expression in mice: implications for human liver disease progression *Hepatology* **54** 252–61

Pasvolsky R, Feigelson S W and Kilie S S *et al* 2007 A LAD-III Syndrome is associated with defective expression of the Rap-1 activator CalDAGGEFI in lymphocytes, neutrophils, and platelets *J. Exp. Med.* **204** 1571–82

Patra V and Wolf P 2016 Microbial elements as the initial triggers in the pathogenesis of polymorphic light eruption? *Exp. Dermatol.* **25** 999–1001

Pavalko F M, Walker D M, Graham L, Goheen M, Doerschuk C M and Kansas G S 1995 The cytoplasmic domain of L-selectin interacts with cytoskeletal proteins via alpha-actinin: receptor positioning in microvilli does not require interaction with alpha-actinin *J. Cell Biol.* **129** 1155–64

Peek R M Jr and Blaser M J 2002 Helicobacter pylori and gastrointestinal tract adenocarcinomas *Nat. Rev. Cancer* **2** 28–37

Peinado H, Lavotshkin S and Lyden D 2011 The secreted factors responsible for pre-metastatic niche formation: old sayings and new thoughts *Semin. Cancer Biol.* **21** 139–46

Philip M, Rowley D A and Schreiber H 2004 Inflammation as a tumor promoter in cancer induction *Semin. Cancer Biol.* **14** 433–9

Phillipson M, Heit B, Colarusso P, Liu L, Ballantyne C M and Kubes P 2006 Intraluminal crawling of neutrophils to emigration sites: a molecularly distinct process from adhesion in the recruitment cascade *J. Exp. Med.* **203** 2569–75

Phillipson M, Kaur J, Colarusso P, Ballantyne C M and Kubes P 2008 Endothelial domes encapsulate adherent neutrophils and minimize increases in vascular permeability in paracellular and transcellular emigration *PLoS One* **3** e1649

Pickert G *et al* 2009 STAT3 links IL-22 signaling in intestinal epithelial cells to mucosal wound healing *J. Exp. Med.* **206** 1465–72

Pikarsky E, Porat R M, Stein I, Abramovitch R, Amit S, Kasem S, Gutkovich-Pyest E, Urieli-Shoval S, Galun E and Ben-Neriah Y 2004 NF-κB functions as a tumor promoter in inflammation-associated cancer *Nature* **431** 461–6

Pool-Zobel B L, Neudecker C, Domizlaff I, Ji S, Schillinger U, Rumney C, Moretti M, Vilarini I, Scassellati-Sforzolini R and Rowland I 1996 Lactobacillus- and bifidobacterium-mediated antigenotoxicity in the colon of rats *Nutr. Cancer* **26** 365–80

Poon I K H, Hulett M D and Parish C R 2010 Molecular mechanisms of late apoptotic/necrotic cell clearance *Cell Death Differ.* **17** 381–97

Popivanova B K, Kitamura K, Wu Y, Kondo T, Kagaya T, Kaneko S, Oshima M, Fujii C and Mukaida N 2008 Blocking TNF-α in mice reduces colorectal carcinogenesis associated with chronic colitis *J. Clin. Invest.* **118** 560–70

Popnikolov N K, Dalwadi B H, Thomas J D, Johannes G J and Imagawa W T 2012 Association of autotaxin and lysophosphatidic acid receptor 3 with aggressiveness of human breast carcinoma *Tumour Biol.* **33** 2237–43

Prahallad A, Sun C, Huang S, Di Nicolantonio F, Salazar R, Zecchin D, Beijersbergen R L, Bardelli A and Bernards R 2012 Unresponsiveness of colon cancer to BRAF(V600E) inhibition through feedback activation of EGFR *Nature* **483** 100–3

Pribluda A, Elyada E, Wiener Z, Hamza H, Goldstein R E and Biton M *et al* 2013 A senescence-inflammatory switch from cancer-inhibitory to cancer-promoting mechanism *Cancer Cell* **24** 242–56

Qian B Z and Pollard J W 2010 Macrophage diversity enhances tumor progression and metastasis *Cell* **141** 39–51

Qian B Z, Li J, Zhang H, Kitamura T, Zhang J, Campion L R, Kaiser E A, Snyder L A and Pollard J W 2011 CCL2 recruits inflammatory monocytes to facilitate breast-tumor metastasis *Nature* **475** 222–5

Quante M, Bhagat G and Abrams J A *et al* 2012 Bile acid and inflammation activate gastric cardia stem cells in a mouse model of Barrett-like metaplasia *Cancer Cell* **21** 36–51

Rakoff-Nahoum S and Medzhitov R 2007 Regulation of spontaneous intestinal tumorigenesis through the adaptor protein MyD88 *Science* **317** 124–27

Rianna C, Kumar P and Radmacher M 2017 The role of the microenvironment in the biophysics of cancer *Semin Cell Dev Biol* **73** 107–14

Ringera P, Colob G, Faessler R and Grashoff C 2017 Sensing the mechano-chemical properties of the extracellular matrix *Matrix Biol.* **64** 6–16

Robbins P F, Lu Y C and El-Gamil M *et al* 2013 Mining exomic sequencing data to identify mutated antigens recognized by adoptively transferred tumor-reactive T cells *Nat. Med.* **19** 747–52

Rot A and von Andrian U H 2004 Chemokines in innate and adaptive host defense: basic chemokinese grammar for immune cells *Annu. Rev. Immunol.* **22** 891–928

Rubinstein M R, Wang X, Liu W, Hao Y, Cai G and Han Y W 2013 Fusobacterium nucleatum promotes colorectal carcinogenesis by modulating E-cadherin/β-catenin signaling via its FadA adhesin *Cell Host Microbe* **14** 195–206

Salas A, Shimaoka M, Kogan A N, Harwood C, von Andrian U H and Springer T A 2004 Rolling adhesion through an extended conformation of integrin αLβ2 and relation to αI and βI-like domain interaction *Immunity* **20** 393–406

Salcedo R *et al* 2010 MyD88-mediated signaling prevents development of adenocarcinomas of the colon: role of interleukin 18 *J. Exp. Med.* **207** 1625–36

Sánchez-Madrid F and del Pozo M A 1999 Leukocyte polarization in cell migration and immune interactions *EMBO J.* **18** 501–11

Sanders M E, Guarner F, Guerrant R, Holt P R, Quigley E M, Sartor R B, Sherman P M and Mayer E A 2013 An update on the use and investigation of probiotics in health and disease *Gut* **62** 787–96

Sandig M, Negrou E and Rogers K A 1997 Changes in the distribution of LFA-1, catenins, and F-actin during transendothelial migration of monocytes in culture *J. Cell Sci.* **110** 807–18

Scheitz C J, Lee T S, McDermitt D J and Tumbar T 2012 Defining a tissue stem cell-driven Runx1/Stat3 signalling axis in epithelial cancer *EMBO J.* **31** 4124–39

Schenkel A R, Mamdouh Z and Muller W A 2004 Locomotion of monocytes on endothelium is a critical step during extravasation *Nat. Immunol.* **5** 393–400

Schetter A J, Heegaard N H and Harris C C 2010 Inflammation and cancer: interweaving microRNA, free radical, cytokine and p53 pathways *Carcinogenesis* **31** 37–49

Schiechl G *et al* 2011 Tumor development in murine ulcerative colitis depends on MyD88 signaling of colonic F4/80+CD11bhighGr1low macrophages *J. Clin. Invest.* **121** 1692–708

Schrieber H and Rowley D A 1999 Inflammation and cancer *Inflammation: Basic Principles and Clinical Correlates* 3rd edn ed J I Gallin and R Snyderman (Philadelphia, PA: Williams and Wilkins) pp 1117–29

Schwartz M A 2010 Integrins and extracellular matrix in mechanotransduction *Cold Spring Harb. Perspect. Biol.* **2** a005066

Schwitalla S *et al* 2013 Intestinal tumorigenesis initiated by dedifferentiation and acquisition of stem-cell-like properties *Cell* **152** 25–38

Shamri R, Grabovsky V, Gauguet J M, Feigelson S, Manevich E, Kolanus W, Robinson M K, Staunton D E, von Andrian U H and Alon R 2005 Lymphocyte arrest requires instantaneous induction of an extended LFA-1 conformation mediated by endothelium-bound chemokines *Nat. Immunol.* **6** 497–506

Shaw S K, Ma S and Kim M B *et al* 2004 Coordinated redistribution of leukocyte LFA-1 and endothelial cell ICAM-1 accompany neutrophil transmigration *J. Exp. Med.* **200** 1571–80

Sheehan K M, Sheahan K, O'Donoghue D P, MacSweeney F, Conroy R M, Fitzgerald D J and Murray F E 1999 The relationship between cyclooxygenase-2 expression and colorectal cancer *JAMA* **282** 1254–7

Singh B, Vincent L, Berry J A, Multani A S and Lucci A 2007 Cyclooxygenase-2 expression induces genomic instability in MCF10A breast epithelial cells *J. Surg. Res.* **140** 220–6

Smyth M J, Dunn G P and Schreiber R D 2006 Cancer immunosurveillance and immunoediting: the roles of immunity in suppressing tumor development and shaping tumor immunogenicity *Adv. Immunol.* **90** 1–50

Sobhani I, Tap J, Roudot-Thoraval F, Roperch J P, Letulle S, Langella P, Corthier G, Tran Van Nhieu J and Furet J P 2011 Microbial dysbiosis in colorectal cancer (CRC) patients *PLoS One* **6** e16393

Sonnenberg G F, Fouser L A and Artis D 2011 Border patrol: regulation of immunity, inflammation and tissue homeostasis at barrier surfaces by IL-22 *Nat. Immunol.* **12** 383–90

Sonoshita M, Takaku K, Oshima M, Sugihara K and Taketo M M 2002 Cyclooxygenase-2 expression in fibroblasts and endothelial cells of intestinal polyps *Cancer Res.* **62** 6846–9

Springer T A 1994 Traffic signals for lymphocyte recirculation and leukocyte emigration: the multiple paradigm *Cell* **76** 301–14

Steinbrink K, Jonuleit H, Müller G, Schuler G, Knop J and Enk A H 1999 Interleukin-10-treated human dendritic cells induce a melanoma-antigen-specific anergy in CD8+ T cells resulting in a failure to lyse tumor cells *Blood* **93** 1634–42

Stevens T, Phan S, Frid M G, Alvarez D, Herzog E and Stenmark K R 2008 Lung vascular cell heterogeneity: endothelium, smooth muscle, and fibroblasts *Proc. Am. Thorac. Soc.* **5** 783–91

Stewart M P, McDowall A and Hogg N 1998 LFA-1-mediated adhesion is regulated by cytoskeletal restraint and by a Ca2+-dependent protease, calpain *J. Cell Biol.* **140** 699–707

Strauss J, Kaplan G G, Beck P L, Rioux K, Panaccione R, Devinney R, Lynch T and Allen-Vercoe E 2011 Invasive potential of gut mucosa-derived *Fusobacterium nucleatum* positively correlates with IBD status of the host *Inflamm. Bowel Dis.* **17** 1971–8

Straussman R *et al* 2012 Tumor micro-environment elicits innate resistance to RAF inhibitors through HGF secretion *Nature* **487** 500–4

Sukriti S, Tauseef M, Yazbeck P and Mehta D 2014 Mechanisms regulating endothelial permeability *Pulm. Circ.* **4** 535–51

Sullivan N J, Sasser A K, Axel A E, Vesuna F, Raman V, Ramirez N, Oberyszyn T M and Hall B M 2009 Interleukin-6 induces an epithelial-mesenchymal transition phenotype in human breast cancer cells *Oncogene* **28** 2940–7

Sun W *et al* 2017 A positive-feedback loop between tumour infiltrating activated Treg cells and type 2-skewed macrophages is essential for progression of laryngeal squamous cell carcinoma *Br. J. Cancer.* **117** 1631–43

Takai A, Toyoshima T and Uemura M *et al* 2009 A novel mouse model of hepatocarcinogenesis triggered by AID causing deleterious p53 mutations *Oncogene* **28** 469–78

Takala H, Nurminen E, Nurmi S M, Aatonen M, Strandin T, Takatalo M, Kiema T, Gahmberg C G, Ylaenne J and Fagerholm S C 2008 *β*2 integrin phosphorylation on Thr758 acts as a molecular switch to regulate 14-3-3 and filamin binding *Blood* **112** 1853–62

Talmadge J E and Fidler I J 2010 AACR centennial series: the biology of cancer metastasis: historical perspective *Cancer Res.* **70** 5649–69

Thompson P W, Randi A M and Ridley A J 2002 Intercellular adhesion molecule (ICAM)-1, but not ICAM-2, activates RhoA and stimulates c-fos and rhoA transcription in endothelial cells *J. Immunol.* **169** 1007–13

Turowski P *et al* 2008 Phosphorylation of vascular endothelial cadherin controls lymphocyte emigration *J. Cell Sci.* **121** 29 37

Tye H *et al* 2012 STAT3-driven up-regulation of TLR2 promotes gastric tumorigenesis independent of tumor inflammation *Cancer Cell* **22** 466–78

Uronis J M, Mühlbauer M, Herfarth H H, Rubinas T C, Jones G S and Jobin C 2009 Modulation of the intestinal microbiota alters colitis-associated colorectal cancer susceptibility *PLoS One* **4** e6026

Urzainqui A *et al* 2007 Functional role of P-selectin glycoprotein ligand 1/P-selectin interaction in the generation of tolerogenic dendritic cells *J. Immunol.* **179** 7457–65

Urzainqui A *et al* 2002 ITAM-based interaction of ERM proteins with Syk mediates signaling by the leukocyte adhesion receptor PSGL-1 *Immunity* **17** 401–12

van Buul J D, Anthony E C, Fernández-Borja M, Burridge K and Hordijk P L 2005a Proline-rich tyrosine kinase 2 (Pyk2) mediates vascular endothelial-cadherin-based cell–cell adhesion by regulating *β*-catenin tyrosine phosphorylation *J. Biol. Chem.* **280** 21129–36

van Buul J D, Fernández-Borja M, Anthony E C and Hordijk P L 2005b Expression and localization of NOX2 and NOX4 in primary human endothelial cells *Antioxid. Redox Signal.* **7** 308–17

Van Coillie E, Van Aelst I, Wuyts A, Vercauteren R, Devos R, De Wolf-Peeters C, Van Damme J and Opdenakker G 2001 Tumor angiogenesis induced by granulocyte chemotactic protein-2 as a countercurrent principle *Am. J. Pathol.* **159** 1405–14

van Wetering S, van Buul J D, Quik S, Mul F P, Anthony E C, ten Klooster J P, Collard J G and Hordijk P L 2002 Reactive oxygen species mediate Rac-induced loss of cell–cell adhesion in primary human endothelial cells *J. Cell Sci.* **115** 1837–46

van Wetering S, van Den Berk N, van Buul J D, Mul F P, Lommerse I, Mous R, ten Klooster J P, Zwaginga J J and Hordijk P L 2003 VCAM-1-mediated Rac signaling controls endothelial cell–cell contacts and leukocyte transmigration *Am. J. Physiol. Cell Physiol.* **285** C343–52

Verfaillie C M 1998 Adhesion receptors as regulators of the hematopoietic process *Blood* **92** 2609–12

Vestweber D 2007 Adhesion and signaling molecules controlling the transmigration of leukocytes through endothelium *Immunol. Rev.* **218** 178–96

Viatte S, Darren Plant D and Raychaudhuri S 2013 Genetics and epigenetics of rheumatoid arthritis *Nat. Rev. Rheumatol.* **9** 141–53

Vicente-Manzanares M and Sánchez-Madrid F 2004 Role of the cytoskeleton during leukocyte responses *Nat. Rev. Immunol.* **4** 110–22

Viegas M S, Estronca L M B B and Vieira O V 2012 Comparison of the kinetics of maturation of phagosomes containing apoptotic cells and IgG-opsonized particles *PLoS One* **7** e48391

Viennois E, Merlin D, Gewirtz A T and Chassaing B 2017 Dietary emulsifier-induced low-grade inflammation promotes colon carcinogenesis *Cancer Res.* **77** 27–40

Virchow R 1856 Reizung and Reizbarkeit *Arch. Pathol. Anat. Klin. Med.* **14** 1–63

Virchow R 1863 Aetiologie der neoplastischen Geschwulst/Pathogenie der neoplastischen Geschwulste *Die Krankhaften Geschwulste* (Berlin: August Hirschwald) pp 57–101

Virchow R 1881 An address on the value of pathological experiments *Br. Med. J* **2** 198–203

von Andrian U H and Mackay C R 2000 T-cell function and migration. Two sides of the same coin *N. Engl. J. Med.* **343** 1020–34

von Andrian U H, Hasslen S R, Nelson R D, Erlandsen S L and Butcher E C 1995 A central role for microvillous receptor presentation in leukocyte adhesion under flow *Cell* **82** 989–99

von Hundelshausen P, Koenen R R and Weber C 2009 Platelet-mediated enhancement of leukocyte adhesion *Microcirculation* **16** 84–96

Voronov E, Shouval D S, Krelin Y, Cagnano E, Benharroch D, Iwakura Y, Dinarello C A and Apte R N 2003 IL-1 is required for tumor invasiveness and angiogenesis *Proc. Natl Acad. Sci. USA* **100** 2645–50

Waldner M J, Foersch S and Neurath M F 2012 Interleukin-6—a key regulator of colorectal cancer development *Int. J. Biol. Sci.* **8** 1248–53

Walzog B and Gaehtgens P 2000 Adhesion molecules: the path to a new understanding of acute inflammation *Physiology* **15** 107–13

Wan R, Wang Z W, Li H, Peng X D, Liu G Y, Ou J M and Cheng A Q 2017 Human leukocyte antigen-G inhibits the anti-tumor effect of natural killer cells via immunoglobulin-like transcript 2 in gastric cancer *Cell Physiol. Biochem.* **44** 1828–41

Wang D, Mann J R and DuBois R N 2005 The role of prostaglandins and other eicosanoids in the gastrointestinal tract *Gastroenterology* **128** 1445–61

Wang Q and Doerschuk C M 2002 The signaling pathways induced by neutrophil-endothelial cell adhesion *Antioxid. Redox Signal.* **4** 39–47

Weber C, Fraemohs L and Dejana E 2007 The role of junctional adhesion molecules in vascular inflammation *Nat. Rev. Immunol.* **7** 467–77

Wilson T R *et al* 2012 Widespread potential for growth-factor-driven resistance to anticancer kinase inhibitors *Nature* **487** 505–9

Wolf M J, Hoos A and Bauer J *et al* 2012 Endothelial CCR2 signaling induced by colon carcinoma cells enables extravasation via the JAK2-Stat5 and p38MAPK pathway *Cancer Cell* **22** 91–105

Woodfin A, Reichel C A, Khandoga A, Corada M, Voisin M B, Scheiermann C, Haskard D O, Dejana E, Krombach F and Nourshargh S 2007 JAM-A mediates neutrophil transmigration in a stimulus-specific manner *in vivo*: evidence for sequential roles for JAM-A and PECAM-1 in neutrophil transmigration *Blood* **110** 1848–56

Woodfin A, Voisin M B and Nourshargh S 2007 PECAM-1: a multi-functional molecule in inflammation and vascular biology *Arterioscler. Thromb. Vasc. Biol.* **27** 2514–23

Wu Y, Deng J, Rychahou P G, Qiu S, Evers B M and Zhou B P 2009 Stabilization of snail by NF-κB is required for inflammation-induced cell migration and invasion *Cancer Cell* **15** 416–28

Wu S *et al* 2009 A human colonic commensal promotes colon tumorigenesis via activation of T helper type 17 T cell responses *Nat. Med.* **15** 1016–22

Wyckoff J B, Wang Y, Lin E Y, Li J F, Goswami S, Stanley E R, Segall J E, Pollard J W and Condeelis J 2007 Direct visualization of macrophage-assisted tumor cell intravasation in mammary tumors *Cancer Res.* **67** 2649–56

Xue W, Zender L, Miething C, Dickins R A, Hernando E, Krizhanovsky V, Cordon-Cardo C and Lowe S W 2007 Senescence and tumor clearance is triggered by p53 restoration in murine liver carcinomas *Nature* **445** 656–60

Yáñez-Mó M, Alfranca A, Cabanas C, Marazuela M, Tejedor R, Ursa M A, Ashman L K, de Landázuri M O and Sánchez-Madrid F 1998 Regulation of endothelial cell motility by complexes of tetraspan molecules CD81/TAPA-1 and CD151/ PETA-3 with $\alpha3\beta1$ integrin localized at endothelial lateral junctions *J. Cell Biol.* **141** 791–804

Yang L, Froio R M, Sciuto T E, Dvorak A M, Alon R and Luscinskas F W 2005 ICAM-1 regulates neutrophil adhesion and transcellular migration of TNF-α-activated vascular endothelium under flow *Blood* **106** 584–92

Yang L, Kowalski J R, Zhan X, Thomas S M and Luscinskas F W 2006a Endothelial cell cortactin phosphorylation by Src contributes to polymorphonuclear leukocyte transmigration *in vitro Circ. Res.* **98** 394–402

Yang L, Kowalski J R, Yacono P, Bajmoczi M, Shaw S K, Froio R M, Golan D E, Thomas S M and Luscinskas F W 2006b Endothelial cell cortactin coordinates intercellular adhesion molecule-1 clustering and actin cytoskeleton remodeling during polymorphonuclear leukocyte adhesion and transmigration *J. Immunol.* **177** 6440–9

Yoshimoto S, Loo T M and Atarashi K *et al* 2013 Obesity-induced gut microbial metabolite promotes liver cancer through senescence secretome *Nature* **499** 97–101

Yu H, Pardoll D and Jove R 2009 STATs in cancer inflammation and immunity: a leading role for STAT3 *Nat. Rev. Cancer* **9** 798–809

Zaki M H, Vogel P, Body-Malapel M, Lamkanfi M and Kanneganti T D 2010 IL-18 production downstream of the Nlrp3 inflammasome confers protection against colorectal tumor formation *J. Immunol.* **185** 4912–20

Zarbock A, Abram C L, Hundt M, Altman A, Lowell C A and Ley K 2008 PSGL-1 engagement by E-selectin signals through Src kinase Fgr and ITAM adapters DAP12 and FcR gamma to induce slow leukocyte rolling *J. Exp. Med.* **205** 2339–47

Zarbock A, Lowell C A and Ley K 2007 Spleen tyrosine kinase Syk is necessary for E-selectin-induced alpha(L)beta(2) integrin-mediated rolling on intercellular adhesion molecule-1 *Immunity* **26** 773–83

Zarek P E, Huang C T, Lutz E R, Kowalski J, Horton M R, Linden J, Drake C G and Powell J D 2008 A2A receptor signaling promotes peripheral tolerance by inducing T-cell energy and the generation of adaptive regulatory T cells *Blood* **111** 251–9

Zenewicz L A, Yancopoulos G D, Valenzuela D M, Murphy A J, Stevens S and Flavell R A 2008 Innate and adaptive interleukin-22 protects mice from inflammatory bowel disease *Immunity* **29** 947–57

Zheng H, Hoegberg J and Stenius U 2017 ATM-activated autotaxin (ATX) propagates inflammation and DNA damage in lung epithelial cells: a new mode of action for silica-induced DNA damage? *Carcinogenesis* **38** 1196–206

Zhou J, Wulfkuhle J, Zhang H, Gu P, Yang Y, Deng J, Margolick J B, Liotta L A, Petricoin E and Zhang Y 2007 Activation of the PTEN/mTOR/STAT3 pathway in breast cancer stem-like cells is required for viability and maintenance *Proc. Natl Acad. Sci. USA* **104** 16158–63

Zwartz G J, Chigaev A, Dwyer D C, Foutz T D, Edwards B S and Sklar L A 2004 Real-time analysis of very late antigen-4 affinity modulation by shear *J. Biol. Chem.* **279** 38277–86

Part II

The role of the mechanical properties of cancer cells in cellular invasion

During the last few decades, the mechanical properties of cells and their microenvironment have become the focus of cell biological and cancer research, by combing classical biochemical cues controlling cellular behavior with the mechanical properties of cells. The field of cell mechanics in physical oncology is vastly growing and hence the development of the experimental biophysical devices, such as active and passive measurement approaches to access mechanical properties of living cells, cell clusters and tissues, is rapidly increasing. Within part 2 of *Physics of Cancer*, the focus is, on the one hand, on how cellular mechanical properties can be probed physically and on how diverse geometries enable the assessment of different cellular properties, and on the other hand, on how the forces are generated in cells and transmitted to their surrounding microenvironment. Moreover, part 2 of this book discusses the impact of mechanical properties on cellular behavior, such as cell motility, during the progression of cancer. Several state-of-the-art biophysical methods are presented and their impact on physics-driven cancer research is highlighted. Mechanical properties such as force transmission and their generation by cells play an important and basic role in numerous biological processes, including cell motion, replication and survival. Single-molecule force spectroscopy methods, including the most common methods such as magnetic and optical tweezers, optical cell stretching as well as atomic force microscopy (AFM), have developed into powerful tools for investigating forces and possess a high impact on understanding cellular functions such as cell motility. Biological motion plays a fundamental role in many cellular processes such as the replication and segregation of DNA, which is facilitated by molecular scale forces provided by the cell's cytoskeletal architecture. Advantages and disadvantages such as limitations of the biophysical methods are discussed. Cellular motility can be eliminated or reduced through the binding of ligands to their specific receptors or through an altered folding of a polypeptide establishing a stable three-dimensional structure, which involves the formation of bonds to overcome thermal and other forces. In addition to the mechanical properties of the individual cells or cell clusters, the mechanical properties of the microenvironment play a crucial role in providing cellular homeostasis and, in addition, the ability of cells to actively sense and respond to external mechanical cues by the generation and transmission of forces such as the cytoskeletal structure formation of contractile actomyosin stress fibers. The impact of the microenvironmental mechanical properties is presented and discussed in volume 2 of this book. However, to elucidate the complex mechanical interactions of the cells and their nearby microenvironments, it is first necessary to measure the active and passive mechanical properties of the cell.

IOP Publishing

Physics of Cancer, Volume 1 (Second Edition)
Interplay between tumor biology, inflammation and cell mechanics
Claudia Tanja Mierke

Chapter 3

Cellular stiffness and deformability

Summary

Cells interact with and thereby sense their microenvironments via their cell surfaces. Hence, cellular surfaces are crucial in basic cellular processes such as signaling, communication, adhesion, migration, sensing, transport, energy generation, embryonic and tissue development, cancer metastasis, and viral as well as bacterial infections. These highly complex and multi-facetted functions of cell membrane surfaces are driven by the structural assemblies or scaffolds and their dynamic assembly. It is crucial to reveal how these cellular molecules structurally assemble and how they alter their cellular interaction with other neighboring cells and their microenvironment. Every single process of the cell's membrane surface is driven by a complicated interaction based on chemical, biological and physical cues. The typical interactions at cell membrane surfaces are of hydrophobic, hydrophilic, electrostatic, van der Waals and hydrogen bonding nature. In order to reveal insights into these interactions, it is necessary to understand how a cell surface receptor recruits its ligand, how and where the ligand binds within the huge receptor, and by which molecular mechanism the ligand switches to the functional state of its targeted receptor. Moreover, we need to understand how cellular interactions can regulate the assembly and functional state of cell surface receptors. Hence, a conceptual understanding of cellular forces goes beyond the simple description of the sensing, transduction and response of cells to mechanical cues, instead the functional dependences need to be revealed and impaired in order to prove the concept.

As single-molecule force spectroscopy has emerged as a powerful tool to investigate the forces and motions necessary for the malignant progression of cancer, this chapter focuses on this and also discusses other biophysical methods in depth, such as magnetic and optical tweezers, atomic force microscopy (AFM) and traction force microscopy. All these biophysical techniques are presented in a critical manner by highlighting their limitations and elucidating their developmental potential in the light of physics-driven cancer research. A range of force spectroscopy assays enables the analysis of inter- and intramolecular interactions of cell surfaces on the molecular scale. Based on the nanoscopic force sensor function of AFM, it is the only assay that allows us to sense and locate specific interactions of cell surfaces at a high resolution of approximately 10 nm. In addition, due to the high sensitivity of AFM-based approaches, the detection of a broad range of forces from 5 pN to 100 nN is possible, which reveals the suitability of the AFM method for the detection of forces ranging from the strength of single receptor–ligand bonds of approximately 60–80 pN to those covering the entire adhesion of cells of more than 1 nN.

3.1 How can cellular stiffness and the deformability of cells be measured?

Forces play a fundamental role in the process of cancer development. How are they facilitated on the molecular scale? When investigating the role of external physical forces on cellular behavior, a key resulting question is, how can cells sense and then integrate these physical signals into specific behavioral alterations? In addition, such behavioral alterations can even include adaption of their own cell mechanical states. As biochemical research has been focused on studying reaction kinetics, it is clearly understood that various molecules undergo conformational changes when forces are exerted on them, which subsequently leads to the exposure of additional binding sites and thereby alterations of the binding kinetics. Hence, the molecular pathways and reactions involved in mechanical force sensing or response are extensively investigated using tools and methods developed in the fields of mechanobiology and mechano-transduction.

Over the last three decades, the ability to study forces has been revolutionized by the development of biophysical techniques that allow measurement of the force generated by single molecules, ranging from cells to proteins, for example through the analysis of the displacement of marker beads. The single-molecule manipulation techniques, such as optical tweezers, magnetic tweezers, AFM, microneedle manipulation (Cluzel *et al* 1996, Kishino and Yanagida 1988), biomembrane force probing (Evans *et al* 1995) and flow induced stretching (Smith *et al* 1992, Kim *et al* 2007), ensure the measurement of forces exerted by cells, and are currently still expanding and undergo steady refinement. As the first three methods are still the most important and most commonly used biophysical methods, this chapter focuses on them and additionally highlights the optical cell stretcher and traction force microscopy.

The current state-of-the-art single-molecule manipulation techniques span six orders of magnitude in length (10^{-10}–10^{-4} m) and force (10^{-14}–10^{-8} N). These methods can manipulate cells (~100 μm), measure the RNA-polymerase advancing a single base pair (0.34 nm) along DNA (Abbondanzieri *et al* 2005) and the mechanical rupture forces of covalent bonds (of several nN) (Grandbois *et al* 1999), as well as the kinetics of nucleic-acid folding processes (~0.1 pN) (Hohng *et al* 2007). The importance of these methods comes from their individual power and the breadth that is evoked by the wide variety of applications and systems in which these methods can be used. Indeed, single cells can be probed to investigate the strength and location of receptor binding interactions (Litvinov *et al* 2002) and adhesion, or to measure traction and adhesion forces (Prass *et al* 2006). In particular, viscoelastic properties can be measured on short length scales and in small volumes (MacKintosh and Schmidt 1999), or even within living cells using a live-cell imaging approach combined with an incubation chamber (Daniels *et al* 2006). Single-molecule force and displacement measurements started with the characterization of classical motor proteins such as kinesins (Block *et al* 1990, Svoboda and Block 1994) and myosins (Simmons *et al* 1993). In particular, the stall force and processivity of these motors have been determined and led to the raising of fundamental questions concerning the connection and regulation of chemical and mechanical cycles (Tinoco and Bustamante 2002), which are now in the focus of biophysical research. The application of external forces allows the selective deregulation of the single steps in a biochemical reaction cycle that drives cellular migration through connective tissue. Detailed analyses of the force-extension relationship (for example, the elasticity of individual polymers such as nucleic acids) have been reported (Cluzel *et al* 1996, Smith *et al* 1992). The next step was to investigate unconventional nucleic-acid molecular motors, which are able to translocate or alter DNA or RNA in a special manner. At the same time, techniques were invented or further developed to mechanically disrupt molecular bonds (Tskhovrebova *et al* 1997, Kellermayer *et al* 1997, Rief *et al* 1997). The determination of rupture forces or even full force spectra enables us to measure whole bond energies, lifetimes and even entire energy landscapes (Hummer and Szabo 2005, Woodside *et al* 2006). Force spectroscopy is an established tool to measure ligand and antibody binding strength (Lim *et al* 2006) and has been used to investigate the multi-state unfolding of single proteins as well as nucleic-acid structures (Zhuang and Rief 2003). Indeed, force spectroscopy gained attention from theoretical approaches that allow us to extract detailed equilibrium thermodynamic parameters from naturally non-equilibrium pulling experiments (Hummer and Szabo 2005, Evans *et al* 2001, Jarzynski *et al* 1997).

In their ability to apply external forces and measure their displacement, single-molecule techniques exhibit the following advantages. First, the measurement of single molecules avoids the problems associated with population averaging. Second, rare or transient phenomena can now be detected, as the required resolution is available and the events can be measured often enough to ensure that this is no artifact. Third, it is possible to measure multi-state or multi-species distributions, together with static and even dynamic enzymatic heterogeneity (Lu *et al* 1998). Fourth, it is even possible to obtain kinetic rates from single-molecule measurements or by investigation of the event time distributions (Xie and Lu 1999, Hua *et al* 1997, Schnitzer and Block 1997). Fifth, the processivity of enzymes can be analyzed by intrinsically synchronizing single-molecule measurements.

3.2 Magnetic tweezers

In the last two decades, the biological and medical sciences have reached a completely new phase of measurement techniques. To date, genomics, proteomics and metabolomics have revealed numerous components, physical interactions and chemical reactions that occur in living cells under certain conditions or at a specific disease or developmental step. As all the components have been detected, they need to be integrated into networks of molecular signaling pathways, which are suggested to regulate fundamental processes. Among these processes are the orchestration of gene expression, intermediary metabolism, outside–in or inside–out signal transduction and cell-cycle control, all of which are precisely controlled in both time and space. These regulatory interacting cycles are the basis of life on a molecular scale and they are needed to reveal the pathological state of cells such as tumorigenesis in order to be able to design drugs and develop appropriate therapies.

Indeed, a wide variety of methods have been established to analyze the localization, dynamics and interactions of molecules in living cells. The uses of the green fluorescent protein together with live-cell microscopy, fluorescence recovery after photobleaching and fluorescence resonance energy transfer have all delivered a major breakthrough in the field of cancer research (Wouters *et al* 2001). In particular, these methods provide insights into molecular processes in living cells, but their limitation is that they can only passively reveal these processes in time and space.

A number of analysis methods have been invented to nanomanipulate biological systems and perform single-molecule research. AFM (Viani *et al* 1999), optical tweezers (Svoboda *et al* 1993), optical cell stretchers (Guck *et al* 2000, 2005) and magnetic tweezers (Strick *et al* 1996, Fabry *et al* 2001) are commonly used to study the biophysical behavior of individual macromolecules or living cells. In particular, molecules are usually connected to micron-sized latex and/or magnetic beads, allowing the exact movement and positioning in the nanometer-range of the molecule as well as measurement (figure 3.1). Moreover, the application of forces on these molecules or on cells that bind these molecules through their cell surface receptor can be performed in the physiologically relevant piconewton (pN) range.

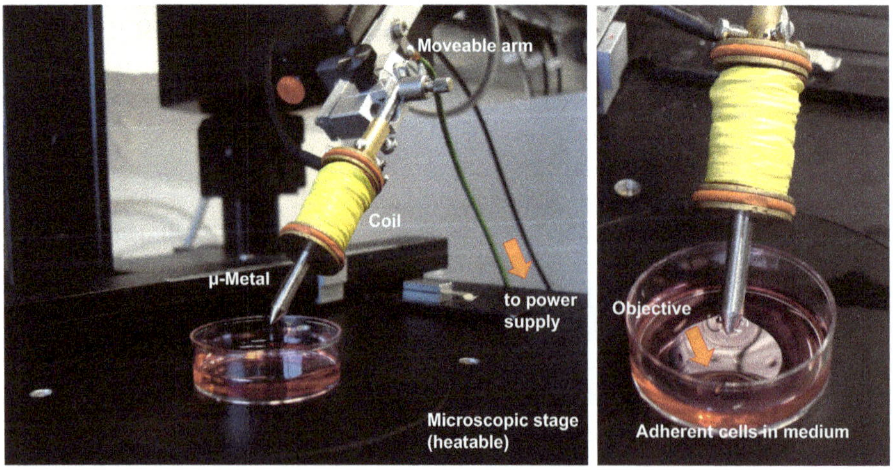

Figure 3.1. Magnetic tweezer setup in a fluorescence microscope.

In more detail, the dynamics of biomolecular systems such as the movement of a single RNA-polymerase along a DNA double helix (Davenport *et al* 2000), the movement of kinesin on microtubules in order to transport a vesicle (Howard *et al* 1989), and even the analysis of chromatin structure, can be investigated (Pope *et al* 2002). However, these techniques are still limited, as they have many problems regarding the measurement of molecular systems inside living cells or even tissues. There are special requirements for these methods to be adopted for the nanomanipulation of objects located in the cytoplasm of living cells. First, the force objects used must be small (1 μm). Second, the exerted forces should be in the relevant physiological range of a few pN or even nN. Third, it would be good to control the forces in their amplitude and direction in order to move the force probe to the targeted site.

A good improvement is the use of magnetic forces in order to meet the prerequisites for proper measurements. The magnetic tweezer technique represents a geometrically completely different approach to physically probing cells, as it is based on the interaction of small round magnetic particles (superparamagnetic beads) with an external magnetic field. A possible first report on the fundamental principle of magnetic tweezers applied to biological systems seems to be an experiment performed in 1950 by Crick and Hughes, which has been called the 'magnetic particle method' (Crick and Hughes 1950). Specifically, cultured chicken fibroblasts that had taken in ferromagnetic particles through phagocytosis were subjected to a magnetic field. Three types of particle movements were observed: twisting, dragging and prodding. Subsequently, these experiments clearly revealed that the cytoplasm of cells is not purely viscous and instead possesses elastic properties. After some time, in the 1990s, these pioneering experiments have been further refined, and magnetic tweezers were used to probe single DNA molecules to quantitatively measure force-extension curves and apply the freely jointed-chain and the worm-like-chain models to the obtained data (Smith *et al* 1992). Magnetic tweezers have now become quite common in biophysical research, in particular for single-molecule force measurements, and are widely used in various implementations (Vlaminck and Dekker 2012, Rehfeldt and Schmidt 2017).

A crucial issue for these experiments is the precise calibration of the forces acting on the superparamagnetic beads in the external inhomogeneous magnetic field. Two pioneering studies demonstrated precisely controlled forces of up to 10 nN on 4.5 μm beads and reported a mechanical response of adherent cells to the forced displacement of beads bound to specific cell surface receptors on the membranes (Bausch *et al* 1998, Bausch *et al* 1999). Subsequently, magnetic beads coated with integrin ligands such as fibronectin or RGD peptide were used to mechanically probe the connection between the extracellular matrix and the cytoskeleton. By applying forces in the range from a few pN to 1 nN, it has been shown that the focal adhesion protein vinculin plays an essential role in force transmission (Alenghat *et al* 2000). Since then, a high-force magnetic tweezer instrument has been developed by reducing the distance between the magnet and the beads to a few μm. This enables the exertion of forces up to 100 nN on 4.5 μm beads, a force that is well suited to mechanically probe living cells (Kollmannsberger and Fabry 2007, Mierke *et al* 2008a, 2008b). It has been shown that vinculin also plays a role in the force transmission of high forces and vinculin's tail is required for the transmission of these forces, whereas vinculin's head domain is not essential in mechanotransduction (Mierke *et al* 2008a and Mierke *et al* 2010). These forces are in the same range as the forces exerted by cells on their surroundings during matrix adhesion and migration and are hence suited to study the mechanical properties of such processes in detail. It is crucial in the magnetic tweezer experiment to position the magnet extremely close to the sample, which can be difficult for living cells in 3D microenvironments or for tissues due to the sterical hindrances of the microenvironment surrounding the cells.

The state-of-the-art magnetic tweezers used to analyze living cells rely mainly on microrheology (Bausch *et al* 1998, 1999). Indeed, it has been reported that large forces can be exerted on magnetic beads that are located inside living cells (Hosu *et al* 2003). In more detail, these magnetic tweezers are constructed of one or two poles and thus the manipulation of magnetic beads in different directions and even the change of one direction into another are possible. A major limitation of this method is that it is technically very difficult to establish the required speed and accuracy for the repositioning of the pole(s), which are needed for exact manipulation. Although cantilevers and optical tweezers are widely used for micromanipulating cells or biomolecules in order to determine their mechanical properties, they do not allow easy rotary motion and can even damage living cells. Hence a system of magnetic tweezers can be used that overcomes these drawbacks while retaining most of the previous dynamometric features, such as the manipulation in three dimensions at a speed of approximately 10 μm s^{-1} and rotation along the optical axis with a 10 Hz frequency. In particular, electromagnets are coupled to a microscope-based particle tracking system through a digital feedback loop. In more detail, this apparatus functions as a dynamometer that is based on either the usual calibration against the viscous drag or complete calibration using Brownian fluctuations.

There are more practical methods, consisting of multiple magnetic poles that enable us to perform 2D or 3D manipulation, while only low forces can be applied (Amblard *et al* 1996, Gosse and Croquette 2002, Huang *et al* 2002, Sacconi *et al* 2001). However, a new type of magnetic tweezer has been reported that uses micronscale magnetic poles, which can be in close neighborhood to living cells (de Vries *et al* 2005). This technique allows us to combine high forces with 2D manipulation of magnetic beads and thus molecular systems inside living cells can be investigated. In comparison to magnetic tweezers, the technique of optical tweezers has the major advantage of possible 3D manipulation in order to stimulate beads inside cells or on cell surfaces for microrheology (Caspi *et al* 2002, Laurent *et al* 2002) or molecular manipulation (Peters *et al* 1999). In contrast, the advantages of magnetic tweezers are seen rather in the assessment of intracellular properties. The restriction of optical tweezers is that they exert forces on microscopic objects, which have a higher or lower refractive index compared to the local surroundings. In particular, melanoma cells containing relatively dark intracellular objects die during optical cell stretcher experiments. Another disadvantage is that there is a myriad of objects inside a cell and thus optical tweezers cannot act selectively in the intracellular microenvironment. Moreover, the relatively high optical intensities needed for the proper measurement damage living cells (Neuman *et al* 1999).

Why is the recovery after stress application incomplete? Is this based on plastic deformations?
After the application of forces to cells, it has been observed that the recovery is incomplete (Mierke *et al* 2008a, 2008b, 2010, 2011). To distinguish between a viscous and a plastic origin for the incomplete shape recovery, a multidirectional high-force magnetic tweezer device has been developed that allows us to investigate the reversibility of the cell deformation and thus to reveal the underlying dissipative mechanism (Bonakdar *et al* 2016). In the case of a fully viscous process, it is supposed that the magnitude of the incomplete recovery process decreases when a second force of the same magnitude and duration is applied in the opposite direction to the first force. In the case of a purely plastic process, it is proposed that the magnitude of the incomplete recovery will not decrease, however, instead the magnitude will even increase with each reversion of the direction of the force application (Bonakdar *et al* 2016). In particular, it has been shown that there is an increase in the residual deformation with each alternating force cycle, which suggests a plastic and not a viscous underlying process causing the incomplete recovery. Moreover, this increase in the residual deformation amplitude versus force cycle number exhibits weak power law behavior. The cycle number seems to be roughly equivalent to the measurement time, when all cycles have similar durations. Finally, the deformation during force application displays nonlinear behavior that is non-reversible.

3.2.1 Bi-directional magnetic tweezers

A custom built magnetic tweezer combines high multiplexing, precisely bead tracking, and bi-directional force control into a flexible and stable platform for analyzing single-molecule behavior. In this setup, electromagnets are used, which provide high temporal control of force while achieving force levels similar to permanent magnets via large paramagnetic beads (Johnson *et al* 2017). The multiplexed magnetic tweezer with precision bead tracking and bi-directional force control (MMTB) possesses four electromagnetic poles: two poles are located above and two below a chamber containing paramagnetic beads that are suspended in a buffer. The μ-metal rods are wrapped with spools of several hundreds of turns of copper wire. In particular, the upper poles are located relatively close to the top surface of the chamber, with a separation distance between the poles of approximately 1 mm. The lower poles, due to spatial constraints below the chamber, are located approximately 7 mm apart and approximately 3 mm below the chamber. This position decreases the magnetic field that pulls the beads down, but it is still sufficient.

By applying a voltage potential across the coils using two identical power supplies, a magnetic field gradient is created, which acts to pull the beads in the upward (upper magnet) or downward (lower magnet) direction. In a typical experimental setup, the lower magnet is first used to pull the beads to the bottom, functionalized surface. When the current to the lower magnet is turned off, the current to the upper magnet is turned on, and the beads are pulled away from the bottom surface at a force controlled by the magnetic field. The cycle of pulling the beads toward and away from the bottom surface is known as a 'pull'. A high-speed camera is mounted on an inverted microscope and used to acquire images of the beads. A high frame rate can be employed to determine the rupture force of short-lived bonds, and then it can be switched to a lower frame rate to analyze the rupture of long-lived bonds, which additionally avoids the collection of an excess amount of data (Johnson *et al* 2017).

3.2.2 Microrheology

The physical properties such as the viscoelasticity of the cell's cytoplasm are crucial for numerous cellular processes such as the intracellular vesicle or compartment transport and cellular motility. However, the measurements of viscoelastic cellular properties are affected by the high degree of diversity among the cytoplasms of cells in a population due to different cell-cycle phases or different degrees of stimulation. Thus, large forces (in the nanonewton range) are required to determine the viscoelastic moduli of the cells. Indeed, a highly promising technique represents microrheology based on magnetic beads that are bound to the cells, and was invented roughly 40 years ago (Amblard *et al* 1996, Crick and Hughes 1950, Ziemann *et al* 1994). In former days, due to the technical difficulties in generating high local magnetic fields and magnetic field gradients, as well as the poor commercially availability of large spherical magnetic beads (10 μm in diameter), the applicability of quantitative measurements was restricted to large cells such as sea urchin eggs (Hiramoto 1969a, 1996b). However, the viscoelastic properties of smaller cells such as macrophages were measured by the torsional deformations inside of living cells that are evoked by magnetic twisting of magnetic beads (Valberg and Butler 1987, Valberg and Feldman 1987, Moeller *et al* 1997, Mijailovich *et al* 2002). This technique is termed magnetic twisting cytometry and has delivered the apparent viscosities of the cell's cytoplasm and also revealed nonlinear viscoelastic behavior. However, only a few measurements of forces exerted inside moving cells have been carried out, and hence forces inside cells are not yet known and require further effort to reveal them. The first measurement of internal forces was reported in 1995 (Guilford *et al* 1995). In this report, it was possible to use magnetic beads with a diameter of 6 μm, which were then phagocytized by macrophages. Another microrheometer measurement setup is capable of generating forces in the nanonewton range on superparamagnetic as well as ferromagnetic beads that both have diameters smaller than 5 μm. Using conventional and exact particle tracking techniques, the bead deflection was analyzed in order to determine the viscoelastic response and draw the recovery curves with a spatial resolution of 10 nm and a time resolution of t = 0.04 s. Moreover, this biophysical technique has been applied to analyze the viscoelastic moduli of fibroblasts' cell membranes (Bausch *et al* 1998). When comparing this particular method with others, which enable us to determine local measurements of viscoelasticity outside regions the cells such as the cell surface (Daily *et al* 1984, Radmacher *et al* 1996), the biophysical technique described here even allows us to perform local measurements of the inside and outside regions of single living cells.

In another report, the first local analysis of the viscoelastic moduli within the cytoplasm of macrophages was performed using viscoelastic creep experiments, in which solid and filled ferromagnetic beads with a diameter of 1.3 μm were used (Bausch *et al* 1999). In addition to this, an attempt was made to measure the displacement field generated by local forces through the single observation of the induced motion of nonmagnetic colloidal objects within cells (Bausch *et al* 1998). It has been reported that transgressing elastic fields exist in cells (Maniotis *et al* 1997), whereas there has been no evidence of long-range elastic coupling within the cells. Furthermore, by analyzing the magnetophoretic motion of the beads, to which a constant magnetic force has been applied, local active opposing forces generated by a cell's cytoplasm can be detected.

Taken together, micromanipulation and microrheology methods such as AFM, micropipettes, optical tweezers, magnetic tweezers and microneedles still have important applications in biophysics, cell biology and cancer research. Historically, the use of magnetic forces in biology has been reported, with magnetic particles inserted into protoplasts, where their movements in a magnetic field were analyzed (Heilbronn 1922). In line with this, the movement of particles in a magnetic field gradient in echinoderm eggs has been investigated (Freundlich and Seifriz 1923). In addition, Crick and Hughes analyzed the viscoelastic response of chick fibroblasts (Crick and Hughes 1950) and then both Yagi (Yagi 1961) and Hiramoto (Hiramoto 1969) further improved the magnetic particle tracking method in order to provide a reliable tool for investigating the viscoelasticity of living cells. Numerous researchers have used magnetic tweezers and even further improved and refined different selected variants of this biophysical method. Finally, the magnetic tweezer technique has become an important tool for mechanically stimulating biomolecules in soft materials such as living cells. For example, multidimensional magnetic tweezers consisting of several magnetic poles have been developed, which are used to move magnetic beads or objects in certain directions (Amblard *et al* 1996, de Vires *et al* 2005, Fisher *et al* 2005). A disadvantage of this method is that the magnitude of force that can be applied is strictly limited. An improvement has been the usage of exchangeable poles for the magnetic tweezers, in which the pole geometry can be altered depending on the desired experimental setup (Fisher *et al* 2006). In addition to the multidimensional magnetic tweezer, two-pole experimental setups can be built to apply high and even alternating forces between two opposing magnets (Sato *et al* 1983, Guilford and Gore 1992, Ziemann *et al* 1994, Barbic *et al* 2001). As a variation of this setup, Guilford and Gore (1992) used the optical tracking and current feedback approach in their two-pole magnetic trap, in which they were able to exert external forces of up to 80 nN on spherical metal beads 10 μm in diameter. Indeed, this amount of force is needed and is still sufficient for performing tissue studies, whereas it is not suitable for single-cell experiments due to the large bead size.

Another option of the setup is to use vertical magnetic tweezers and hence the vertical gradient between two magnetic poles placed above the sample object's surface. This special geometry has been used for DNA unfolding experimental approaches and adhesion experiments, however, while it is able to generate a homogenous gradient field, it is rather limited regarding the small forces, such as 0.2 nN, that can be applied to the magnetic objects (Barbic *et al* 2001, Assi *et al* 2002, Gosse and Croquette 2002, Walter *et al* 2006). It is known that the highest forces that can be exerted are generally obtained using one-pole microneedle geometric approaches. For these one-pole setups two kinds of magnetism can be implemented. A permanent magnet can be installed (Matthews *et al* 2004) or electromagnets with soft iron cores can be integrated (Bausch *et al* 1998, Alenghat *et al* 2000, Huang *et al* 2005, Overby *et al* 2005). Another fundamentally different setup is the use of homogenous magnetic fields to rotate permanently magnetized particles (magnetic twisting), as provided in magnetic twisting cytometry (Valberg and Albertini 1985, Wang *et al* 1993, Fabry *et al* 2001). A major weakness of this method is that only the small forces can be applied to cells or other samples, which are in the range of 500 pN (Mierke 2011).

A major breakthrough in the field of magnetic-based biophysical methods has been the improvement of the magnetic tweezer method by increasing the magnitude of the exerted forces by an easily reproducible setup, which can be constructed easily in all kinds of research laboratories. In more detail, this setup contains a single solenoid with a high-permeability soft iron core, called μ-metal, that is attached to a remote-controlled commercially available micromanipulator. Now, high forces of up to 100 nN can be applied to 4.5 μm superparamagnetic beads by reaching distances between the core metal and superparamagnetic bead that are in the range of 10–30 μm. A problem may be that the tweezer needle is probably not so far away from the cell carrying the attached magnetic bead and hence may come into contact with the cell, which leads additionally to a mechanical stimulation of the cell without any force application through the bead displacement in the magnetic field. In particular, the bead position has to be monitored by tracking the bead on the cell's surface, without any external force application. When the displacement of the bead follows a straight line, which is parallel to the x-axis in a certain time interval such as 10 s, the measurement can then be started by applying the external force on the superparamagnetic bead (Mierke *et al* 2008a, 2008b). During and after force application the bead displacement will be monitored and plotted over time, which is a classical creep experiment. An inherent problem of this setup for magnetic tweezers is the poor control of the force, as there is a highly nonlinear force–distance relationship. However, this problem can be solved by constantly tracking the distance between the superparamagnetic bead and the magnetic core of the electromagnet. When there is an alteration in the distance, this change can be compensated by adapting either the solenoid current or by moving the core position using the moveable micromanipulator in order to restore the original distance between the cell and the tweezer needle. Under some circumstances, both compensations revealed the best results for the measurement, as on the one hand the significant movement of the tweezer needle disturbs the medium and may induce local perturberances and, on the other hand, high currents can heat up the tweezer needle, which may increase the overall temperature of the cell setup within the culture dish. In more detail, the force is calibrated using the classical viscous drag approach (Yagi 1961). An empirical fit can be applied to describe the nonlinear relationship between the distance, the current and the exerted force using an analytical formula. However, this approach provides exact timing and force accuracy opportunities and can even be mounted on all commonly available inverted microscopes that carry a charge coupled device (CCD) camera.

There are at least two major advantages of this magnetic tweezer technique. The first is that the forces can be applied locally to certain structures of interest and targeted to certain receptors or molecules displayed on the cell's surface. The second is that the cells can be measured while cultured on a 2D surface. However, whether the cell can also be measured within 3D networks near the surface remains elusive. It seems likely that this application can be adapted to the current setups by applying higher forces towards the superparamagnetic beads and increasing the distance between the magnetic tweezers' core and the superparamagnetic bead.

3.2.3 Adhesion forces

In order to investigate adhesion forces, the detachment of receptor-bound beads can be measured by applying a step-wise increasing external force to the superpapramagnetic beads using a magnetic tweezer device. Before the measurements start, it is important to decide whether single cells, cell clusters or even cell monolayers will be analyzed, as the external applied forces propagate throughout the cell cluster or cell monolayer, indicating that the cytoskeletons of the individual cells are connected between neighboring cells (Raupach *et al* 2007). Hence the measurements of cell clusters or monolayers stimulate a broad range of cells, when one cell within this cell cluster or cell monolayer is measured, which needs to be taken into account and can be treated as a pre-stimulation. In order to omit a rather unspecific pre-stimulation, cells in vastly different regions need to be analyzed, where no stimulation through neighboring cells has occurred that possibly have been evoked by earlier measurements within a single-cell sample. For these measurements the superparamagnetic beads are coated with a ligand for a special cell surface receptor or with an antibody directed towards a certain target on the cell surface. In order to investigate the detachment of beads bound to a certain cell surface receptor (figure 3.2), the cells are seeded into culture dishes, which may or may not be surface coated with a special ligand for a distinct cell–matrix receptor. After one or two days of cell culture at 37 °C, 5% CO_2, and 95% humidity, the cells are then incubated with special ligand or antibody coated beads for 30 min under the same conditions. In more detail, the detachment of the coated superparamagnetic beads from the cells is measured during external force application ranging from 0.5 to 10 nN using magnetic tweezers. The percentage of detached beads in relation to the force required for the bead detachment is used to quantify the bead binding strength to the targeted receptor on the cell surface. The specificity of the binding and hence of the detachment forces can be determined by using an inhibitory antibody directed against the receptor or by using an excess of soluble ligand to impair the binding of bead-bound ligands, which are applied 30 min before the coated beads are added to

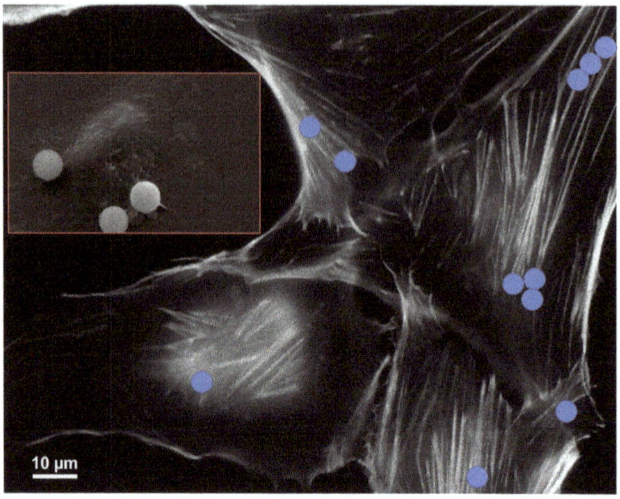

Figure 3.2. Superparamagnetic bead of diameter 4.5 μm (coated with fibronectin) bound to integrins connected to the cells' actin cytoskeleton (visibilized by Alexa Fluor 546 Phalloidin staining) on human 786-O kidney cancer cells. Inset: Scanning electron microscopic image of 4.5 μm beads bound to human MDA-MB-231 breast cancer cells.

the cells. An alternative way of determining the detachment forces will be discussed below, in the AFM section.

3.2.4 Overall cellular stiffness and fluidity

The magnetic tweezer setup is extremely well suited for creep measurements, where, for example, a staircase-like sequence of step forces ranging from 0.5 to 10 nN is applied to superparamagnetic epoxylated beads with a diameter of 4.5 μm, which are coated with the extracellular matrix protein fibronectin (100 μg ml^{-1}) in order to investigate the interaction between the integrins and their ligand fibronectin (figure 3.3) (Mierke *et al* 2008a, 2008b, McNulty *et al* 2009). For proper measurement it is important that the coated beads do not cluster after addition to the cells, otherwise they cannot be tracked accurately. The best option is to have just one bead attached to one cell in order to apply the external force locally and calculate the exact amount of force towards this cell. Thus, only those cells carrying one bead will be chosen for the measurement. Based on the bead concentration and the amount of bound protein on the bead's surface, it is possible to attach just one bead to only one cell without selecting a minority of cells for the magnetic tweezer measurement that carry only one bead.

When an external force step ΔF is applied to a cell-bound bead, the bead moves and thus a displacement $d(t)$ toward the tip of the tweezer needle can be detected. The ratio of $d(t)/\Delta F$ defines a creep response $J(t)$. The creep response $J(t)$ of the cells follows a power law with time, hence the equation is $J(t) = a(t/t_0)^b$, where both factors a and b are force-dependent and the reference time t_0 is set to 1 s. The factors a and b are then determined by a least squares fit (Mierke *et al* 2008a). The prefactor a (in units of μm nN^{-1}) characterizes the elastic properties of the cell and

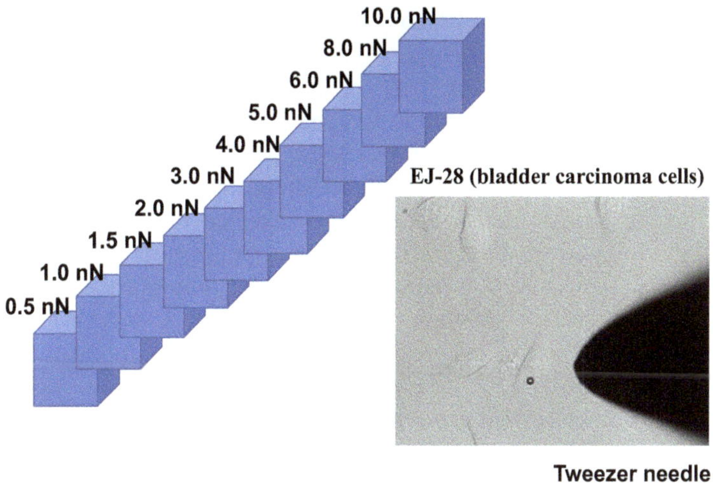

Figure 3.3. Staircase-like force pattern protocol.

corresponds to a compliance, which is indeed the reciprocal of the stiffness (Mierke *et al* 2008a). The power law exponent b reflects the dynamics of the force-bearing structures of the cell that are directly connected to the bead, such as the actomyosin cytoskeleton (Fabry *et al* 2001). In more detail, when the power law exponent b is zero, then this is indicative of a purely elastic solid. However, if b is one, then this behavior is indicative of a purely viscous fluid (Fabry *et al* 2001). As is true for quite a few soft matter materials, including cells, the power law exponent b of cells usually falls in a range between 0.1 and 0.5, whereas higher values of b are associated with a higher turnover rate of cytoskeletal structures and proteins. In addition, higher values of b are often associated with a reduced cellular stiffness, indicating that the protein turnover or the de- or reassembly of the structure affect the mechanical properties of the cell (Fabry *et al* 2001).

Magnetic tweezers have been used to investigate the functional role of vinculin in providing mechanical properties such as cellular stiffness and in influencing cytoskeletal remodeling dynamics. How does vinculin affect cytoskeletal dynamics and cellular stiffness to facilitate cellular motility? In order to investigate whether the increased 3D motility of vinculin-expressing wild-type cells is regulated through altered cellular stiffness and remodeling dynamics of the cytoskeleton, microrheology measurements are performed. In particular, using the magnetic tweezer technique, forces of up to 10 nN can be applied to FN-coated superparamagnetic beads that are bound to certain integrin receptors exposed on the cell surface. The bead displacement after the first force application or after a further step of an increased force application, normalized to the force magnitude, defines the creep response $J(t) = a(t/t_0)^b$. This creep response consists of an elastic response (cell elasticity or stiffness, $1/a$) and a frictional response (cytoskeletal fluidity, b) (Mierke *et al* 2008b, Fabry and Frederg 2003). Vinculin knockout embryonic fibroblasts (MEFvin$^{-/-}$ cells) possess a higher cytoskeletal fluidity and a lower stiffness compared with vinculin wild-type embryonic fibroblasts (MEFvin$^{fl/fl}$ cells), indicating that MEFvin$^{-/-}$ cells are more deformable. Hence this increased deformability may help them to squeeze more easily through a dense extracellular matrix compared to the stiffer MEFvin$^{fl/fl}$ cells. However, these MEFvin$^{-/-}$ cells are less invasive into 3D extracellular matrices compared to the MEFvin$^{fl/fl}$ cells. Thus, the higher cytoskeletal fluidity and increased deformability of MEFvin$^{-/-}$ cells cannot explain their decreased invasiveness (Mierke *et al* 2010).

In addition, magnetic tweezer experiments revealed that in both cell lines, the cytoskeletal fluidity and stiffness increased with elevated external forces on the cell-connected beads. However, this behavior indicates that both MEFs respond to external applied forces by showing stress stiffening and, surprisingly, increased fluidization of the actomyosin cytoskeleton, which seems to be independent of vinculin's mechano-coupling function. How does this behavior depend on the contractility of the cells? In order to answer this question, the inhibition of contractile forces after the addition of the myosin light chain kinase inhibitor ML-7 has been determined. It has been found that both cell lines displayed increased cellular fluidity, which leads to the suggestion that cellular fluidity depends mainly on cell contractility. Interestingly, after the addition of ML-7 the cellular stiffness of both MEFs decreased to similar levels, indicating that the higher initial stiffness of the MEFvin$^{fl/fl}$ cells prior to ML-7 addition is provided by higher actomyosin-mediated contractile forces, which have also been measured independently using traction microscopy (Mierke *et al* 2010, Mierke *et al* 2008). As the MEFvin$^{fl/fl}$ cells are able to generate and transmit higher contractile forces to their microenvironment compared to MEFvin$^{-/-}$ cells, this result seems to be a good explanation for the increased invasiveness into dense 3D extracellular matrices, where cells need to transmit contractile forces in order to move through the matrix network.

3.3 Optical cell stretcher

The optical stretcher (usually termed the optical cell stretcher) has become an important and useful device for studying the mechanical properties of cells that are in suspension at the measurement time point. The optical stretcher was invented by Jochen Guck and Josef A Käs. Specifically, the optical cell stretcher uses two opposing divergent laser beams directed on a cell (Guck *et al* 2001). When both beams have the exactly same intensity, there is no net force acting on the cell along the optical axis. However, in addition to a lateral optical gradient force that traps the cell on the optical axis, there is a symmetric couple of forces acting on the cell surfaces stretching the cell in the axial direction. In particular, cells are trapped in the center of the stretcher and through an increase of the laser power the forces acting on the cellular surfaces cause the axial stretching of the cells. An effective cellular compliance J can be calculated from the cell's time-dependent deformation curves. By fitting these curves with power law rheology models, such as a standard linear solid model or a Burger's model, the viscous and elastic contributions can be determined (Ekpenyong *et al* 2012).

An advantage of the optical stretcher is the possibility to combine it with microfluidic devices, which enables us to measure the compliance of cells with a high throughput. Hence it has been shown that normal, cancerous, and metastatic breast epithelial cells differ in their mechanical deformability. However, the distributions of optical deformability were so broad among one cell population that unambiguous classification of single cells was not possible (Guck *et al* 2005). A variety of physiologically adherent and non-adherent cells have been measured mechanically. Perturbed cell mechanical properties have been analyzed using biochemical drugs such as blebbistatin, a potent and selective inhibitor of the ATPase of myosin II, which hence decreases the actomyosin contractility and hence should soften the cells (Chan *et al* 2015). In contrast, the cells treated with blebbistatin became effectively stiffer, which is contrary to measurements by AFM and the dual optical trap (Schlosser *et al* 2015). In addition, it was reported that in addition to to the optical deformation of a cell, there is also a significant thermal effect on the sample, which additionally induces rapid alterations in the cell's temperature (Kießling *et al* 2013). When investigating this thermal effect, the optical stretcher setup allows us to perform a novel class of experiments termed thermo-rheology, which is intended to act far beyond purely mechanical deformation.

The important limitation for the measurement of adherent cells is now overcome, as a detailed and precise measurement protocol can be applied to these cells to obtain promising results. These cells still display differences in their cellular mechanics, even when they are rounded up, indicating that some of the cellular mechanical properties are adhesion-independent. However, there are adherent cell types, where differences in cellular mechanics are detected in the adherent state, and these differences could be confirmed using an optical cell stretcher device. Thus, a major criticism of the optical cell stretcher method has been removed for many cell types including genetically altered or even stimulated cell types, and hence it can be used for adherent cell types and non-adherent cell types. There seems to be even more potential for this method, when it is combined with a commercially available cell sorter unit, or when antibodies labeled with fluorescent molecules, which are directed to the protein of interest, during the (very short) cell stretching process are used in order to combine the deformability of cells to a certain cell surface receptor expression or protein expression intracellularly. There is hope that it will even be possible to measure whole tissue samples in a native manner without any additional and hence artificial cell culture.

3.3.1 A short introduction to the historical development of the optical stretcher

The development of the laser trap technique in biomedical research, such as cancer research, began at least five decades ago, and these laser traps have been implemented to manipulate small soft matter objects ranging in size from atoms to cells (Ashkin 1970, Chu 1991, Svoboda and Block 1994). The fundamental general principle underlying laser traps is the momentum transfer from the laser light to the sample. Due to Newton's second law, this momentum transfer exerts a force on the sample. Before the optical cell stretcher was established, these optical forces evoked by the momentum transfer were only used to trap a sample. The best-known optical trap is the one laser beam gradient trap, termed optical tweezers

(Ashkin *et al* 1986). Optical tweezers have developed as an intriguing tool for cell and molecular biological research. In particular, they have been used for trapping cells (Ashkin *et al* 1987, Ashkin and Dziedzic 1987), measuring the forces exerted by molecular motors such as myosin or kinesin (Block *et al* 1990, Shepherd *et al* 1990, Kuo and Sheetz 1993, Simmons *et al* 1993, Svoboda *et al* 1993), determining the swimming forces of sperm (Tadir *et al* 1990, Colon *et al* 1992) and studying the polymeric properties of single DNA strands (Chu 1991).

The optical cell stretcher was developed later and, in contrast to the optical tweezers technique, it is based on a double laser beam optical trap (Ashkin 1970, Constable *et al* 1993). In this setup, two opposed, but slightly divergent and identical laser beams with a Gaussian intensity profile, can trap a sample such as a living cell flowing in a liquid object by holding it in the middle of the two laser beams (figure 3.4). This trapping of a sample is stable when the total net force on the sample is zero and restoring. This special condition is provided exactly when the refractive index of the sample is higher than that of the surrounding fluid and when the beam sizes are larger than the trapped sample. However, in extended samples such as living cells, the momentum transfer primarily occurs at the cell's surface. Indeed, the total force acting on the center of gravity is zero because the geometry of the two-beam optical trap is symmetric and all the resulting surface forces cancel each other out. Nevertheless, when the sample is sufficiently elastic, as with living cells, the forces acting on the cell's surface stretch the sample along the beam axis (figure 3.4) (Guck *et al* 2000). At first glance, this optical stretching may seem to be counter-intuitive. However, there is a simple explanation. It is current knowledge that light, including laser light, carries momentum, and when a ray of light is reflected or refracted at an interface between two media with different refractive indices, the light will change the direction or velocity and subsequently its momentum is altered. Due to the law of momentum conservation, some momentum is transferred from the

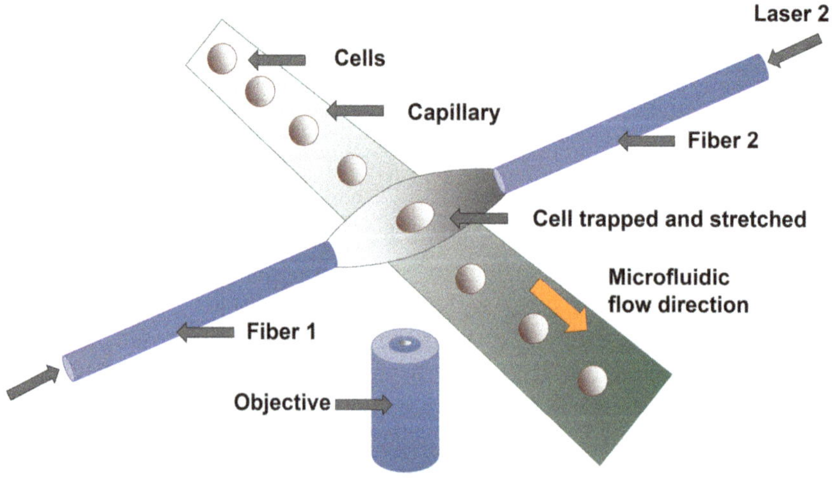

Figure 3.4. Optical cell stretcher principle.

laser light to the sample's interface and in order not to disobey Newton's second law a force is exerted on the sample's surface.

To illustrate what happens in the sample when the two laser beams are applied to it, we can consider that a ray of light passes through a cube of optically denser material. What happens? As the light enters the dielectric object, it gains momentum so that the surface gains momentum in the opposite direction. In addition, the light also loses momentum when it leaves the dielectric object: at this point, the opposite surface gains momentum in the direction of the laser light propagation. Moreover, the reflection of the laser light on both surfaces of the sample also leads to a momentum transfer on both surfaces in the direction of laser light propagation. This contribution to the surface forces is smaller than the contribution that arises from the increase of the laser light's momentum inside the cube. In summary, the two resulting surface forces at the front and back sides of the sample are in opposite directions and thus stretch the sample (Guck *et al* 2000). However, the asymmetry between the surface forces of the sample leads to a total force that acts on the center of the cube. If there is a second, identical ray of laser light, as in the optical cell stretcher approach, which passes through the cube from the opposite side, there is no total force on the cube, but the forces on the surface generated by the two rays of laser light are then additive. In contrast to asymmetric trapping geometries, where the total force is equal to the trapping force (called optical traps), the optical cell stretcher exploits surface forces to stretch the samples. High laser light power, in the range 800–1200 mW, can be used in each beam, which leads to surface forces of up to hundreds of piconewtons acting on the sample's surface. However, there is no excessive radiation damage to the cells observed by the laser beams, as these beams are not focused when using the optical cell stretcher, which reduces the light flux through the analyzed cells compared to other optical traps such as optical tweezers. In several reports, the optical cell stretcher has been used to demonstrate cellular deformability due to altered gene expression, altered cell age or altered aggressiveness of cancer cells (Guck *et al* 2005, Seltmann *et al* 2013, Mierke *et al* 2017, Fischer *et al* 2017, Kunschmann *et al* 2017). An example of a measurement setup is a 1 s trap, 2 s stretch and 2 s relaxation (figure 3.5) and hence a high throughput of cells can be achieved.

Human erythrocytes, such as red blood cells (RBCs), have served as a simple model systems and have thus been analyzed. These RBCs have some advantages for use as a model system for optical cell stretching, because they have no internal organelles, hemoglobin is homogeneously distributed within their cytoplasm and in addition they have a nearly spherical shape when they are osmotically swollen. Taken together, they model very closely an isotropic, soft, dielectric sphere without any internal structures, such that they can be used for stress profile calculations. Moreover, they are soft cells and hence their deformations due to optical stretching can be easily analyzed. Another advantage of the use of RBCs is that their elastic properties are well known (Bennett 1985, 1990, Mohandas and Evans 1994). In particular, the only elastic component of RBCs is a thin biomembrane that is composed of a phospholipid bilayer. On the inner side of the membrane is a triagonal network composed of spectrin filaments and on the outside there is a

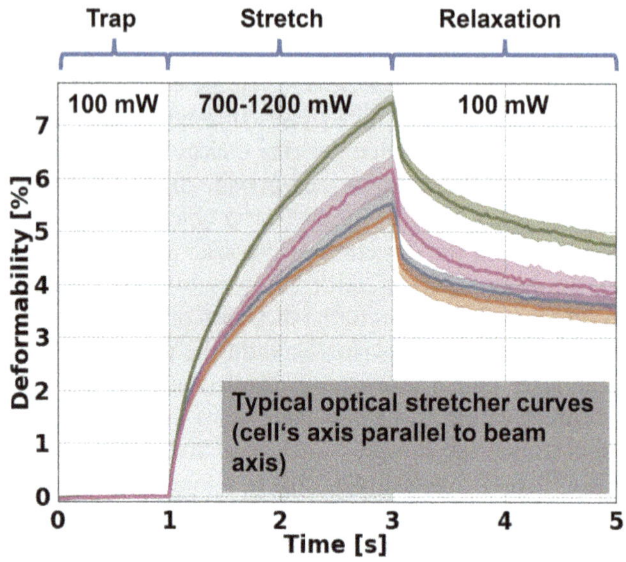

Figure 3.5. Optical cell stretcher measurement protocol.

glycocalyx network (Mohandas and Evans 1994). In more detail, the ratio between cell radius r and the thickness h of the membrane is $r/h = 100$. The bending energy is very small compared to the stretching energy and hence can be neglected. Linear membrane theory predicts the deformations of RBCs upon exertion of surface stresses using the optical cell stretcher technique.

A much better sample are the BALB 3T3 fibroblasts that, dissimilar to RBCs, have an extensive 3D network (termed the cytoskeleton) of protein filaments throughout the entire cytoplasm with elastic properties (Lodish *et al* 1995). This cytoskeleton contains semiflexible actin microfilaments, rod-like microtubules and flexible intermediate filaments that build up a complicated 3D scaffold network with the help of accessory proteins (Adelman *et al* 1968; Pollard 1984, Elson 1988, Janmey 1991). It has been shown that the classical concepts used in polymer physics fail to explain how these three types of filaments provide the mechanical stability of the cells (MacKintosh *et al* 1995), but it is known that in most cells microfilaments are a principal determinant of mechanical strength and stability (Stossel 1984, Janmey *et al* 1986, Sato *et al* 1987, Elson 1988). However, over the last decade this hypothesis has changed and there is evidence that intermediate filaments also contribute to the mechanical properties of cells (for example, deformability) in order to provide cellular functions such as motility (Seltmann *et al* 2013).

The actin cortex is a thick ($r/h = 10$) homogeneous layer located underneath the cell membrane. When the cells adhere to an extracellular matrix substrate or cell culture plastic, they form additional bundles of actin microfilaments (termed F-actin stress fibers) that then lead to the formation of focal adhesion plaques (termed focal adhesions) spanning over the entire cytoplasm of the cell. The dynamic remodeling of the F-actin cytoskeleton facilitates cellular functions such as motility and cytoplasmic cleavage, which occurs as the last step of the mitosis process (Pollard 1986, Carlier 1998, Stossel *et al* 1999).

In order to show that the actin cytoskeleton is responsible for the mechanical properties of cells, they are pronouncedly softened upon addition of actin-disrupting drugs such as cytochalasin D (Petersen *et al* 1982, Pasternak and Elson 1985) and gelsolin (Cooper *et al* 1987), indicating the importance of actin in providing the mechanical properties of cells. Frequency-dependent AFM-based microrheology revealed that fibroblasts exhibit the same viscoelastic response as homogeneous artificial actin networks *in vitro* (Mahaffy *et al* 2000). Another experimental approach using rat embryo fibroblasts showed that actin plays an important role in providing the mechanical responses upon the deformation caused by glass needles (Heidemann *et al* 1999). In particular, actin and microtubules were tagged with green fluorescent proteins and the role of these two cytoskeletal components in determining the cell's shape during mechanical deformation was visualized directly using a fluorescent microscope. Indeed, actin has been reported to be almost exclusively responsible for the cell's elastic response, whereas the microtubules displayed solely fluid-like behavior.

In nonmitotic (interphase) cells, the microtubules are located radially outward from the microtubule-organizing center in close vicinity to the cell's nucleus (Lodish *et al* 1995). Microtubules can be seen as tracks for motor proteins such as dynein and kinesin, which are able to transport vesicles through the cell's cytoplasm. Another function of microtubules is the separation of chromosomes during mitosis (Mitchison *et al* 1986, Mitchison 1992). The third type of main cytoskeletal component is intermediate filaments, which are flexible polymers that are partly specific to certain differentiated cell types such as keratinocytes (Herrmann and Aebi 1998, Janmey *et al* 1998). Vimentin is an intermediate filament that is expressed in mesenchymal cells such as fibroblasts. In more detail, vimentin fibers terminate at the cell's nuclear membrane and at hemidesmosomes (cell–matrix adhesion) or desmosomes (cell–cell adhesion) on the cell's membrane. Another type of an intermediate filament is lamin, which builds up the nuclear lamina (a polymer cortex underneath the nuclear membrane) (Aebi *et al* 1986). Intermediate filaments have been reported to be co-localized with microtubules, suggesting an interaction between these two cytoskeletal components. In the past, both microtubules and intermediate filaments have been reported to be less important in providing the elastic strength and structural response of cells upon which external stress is being applied (Petersen *et al* 1982, Pasternak and Elson 1985, Heidemann *et al* 1999, Rotsch and Radmacher 2000). It is suggested that intermediate filaments are important in large deformations that cannot be achieved with deforming stresses in the range of several pascals. In more detail, intermediate filaments are shown to be more important for cellular elasticity in adherent cells compared to suspended cells, where the initially fully extended filaments become slack (Janmey *et al* 1991, Wang and Stamenovic 2000). However, this is still under investigation and needs to be further confirmed by additional experiments.

A quantitative description of the cytoskeletal contribution to a cell's viscoelasticity is still elusive. The optical cell stretcher can measure the viscoelastic properties of the entire cell's cytoskeleton and is hence able to shed light on the regulation of cellular elasticity. The entire cytoskeleton is an intricate polymer network and a structural framework that provides the shape of a cell and its mechanical rigidity (Elson 1988, Lodish *et al* 2000). It has been shown that enormous external pressures can be endured by cells with their integrity still being preserved (Janmey *et al* 1991). The momentum carried by light and the forces it can exert on material objects such as cells is fast and even not noticeable on macroscopic length scales. The relatively small forces exerted by light are ideally suited to the deformation of microscopic small objects such as living cells and may additionally serve as a detection marker (prognostic marker) for cytoskeleton-altering diseases such as cancer (Guck *et al* 2005).

The capability of cells to withstand external deforming stresses is essential for tissue integrity and homeostasis and has led to the development of many techniques to probe cellular elasticity. In particular, AFM (Radmacher *et al* 1996), manipulation with microneedles (Felder and Elson 1990), microplate manipulation (Thoumine and Ott 1997) and cell poking (Daily *et al* 1984) are not able to detect small variations in cellular elasticity because these biophysical devices have a high spring constant compared to the elastic modulus of the sample material. However, the AFM technique has been significantly improved for the measurement of cellular elasticity by attaching micron-sized beads to the cantilever's scanning tip in order to decrease the pressure that is applied to the cell sample (Mahaffy *et al* 2000). There is a minor problem with the micropipette aspiration of cellular segments (Discher *et al* 1994) and the displacement of surface-attached microspheres (Wang *et al* 1993), as they can lead to inaccurate measurements if the cellular membrane detaches from the cytoskeleton and hence uncouples the membrane's stiffness from the overall cellular stiffness during the deformation step. In addition, a major disadvantage is that all of these techniques are very laborious and can only analyze the elasticity over a small part of a cell's surface. The whole cellular elasticity can instead be measured indirectly by determining the compression and the shear moduli of densely packed cell pellets (Elson 1988, Eichinger *et al* 1996) or by using microarray assays (Carlson *et al* 1997). Microarray assays can be used to dissect the impact of various interacting factors on cellular functions and phenotypes by simultaneously synthesizing more than 1000 unique microenvironments with automated nanoliter liquid-dispensing technology (Ranga *et al* 2014). Using this novel 3D microarray technique, the combined effects of matrix elasticity, proteolytic degradability and three distinct classes of signaling proteins on cells (such as mouse embryonic stem cells) can be observed simultaneously to unveil a comprehensive map of the interactions responsible for self-renewal. This novel approach can be used to reveal multifactorial 3D cell–matrix interactions, depending on the microenvironmental properties, such as matrix stiffness (Ranga *et al* 2014). A limitation of this technique is that these measurements only represent an average value rather than a reliable single-cell measurement and depend on adhesion forces such as cell–cell and cell–matrix adhesion. Alternatively, the optical cell stretcher is a technique that is able to solve most of the problems of the other biophysical techniques described above, but in addition it allows us to measure large numbers of individual cells by the incorporation of an automated flow chamber fabricated using modern soft lithography techniques in order to guide the detached cells through the laser beams.

Using optical cell stretching, the cytoskeletal properties of the cells can be measured, providing not only the mechanical rigidity, but also many important cellular functions, such as cellular motility and transendothelial migration behavior (Lodish *et al* 2000). The cytoskeleton composed of actin, microtubules and intermediate filaments works together with accessory proteins regulating cellular motility, ribosomal and vesicle transport, mitosis and mechano-sensing and mechano-transduction processes (Wang *et al* 1993, Wirtz and Dobbs 1990). All of these cellular functions are precisely regulated by cytoskeletal elements that are finely tuned and even synchronized with the overall cellular function. Subsequently, alterations of the cellular function during differentiation (Olins *et al* 2000) or because of a certain disease are manifested in changes to the cytoskeleton. For example, cytoskeletal alterations can evoke capillary clogs in circulatory problems (Worthen *et al* 1989) and lead to various blood diseases, such as sickle-cell anemia, hereditary spherocytosis or immune hemolytic anemia (Bosch *et al* 1994, Williamson *et al* 1985), while genetic disorders of intermediate filaments and their cytoskeletal networks can cause problems with the skin, hair, liver and colon, in addition to motor neuron diseases such as amyotrophic lateral sclerosis (Fuchs and Cleveland 1998, Kirfel *et al* 2003).

However, the most common example of the malignant transformation of cells is cancer, during which the morphology of cells changes due to alterations of the cytoskeleton. These changes can be utilized for diagnostic approaches of cancer. During the cell's malignant transformation from a fully mature and postmitotic state to a replicating, highly motile and immortal cancer cell, the cytoskeleton undergoes changes from an ordered and rigid structure to a highly irregular and compliant state. In particular, the alterations include a decreased quantity of constituent polymers and accessory proteins, as well as a restructuring of the cellular network (Ben-Zéev 1985, Cunningham *et al* 1992, Katsantonis *et al* 1994, Moustakas and Stournaras 1999, Rao and Cohen 1991). As expected, these cytoskeletal alterations serve as markers for the malignancy of cells mirrored in their increased replication and enhanced motility, both of which are inconsistent with a less dynamic cytoskeleton. In summary, these alterations in the cytoskeletal components and their concentrations as well as the cytoskeletal structure seem to be manifested in the overall mechanical properties of a cell. Hence, measuring a cell's rigidity may provide information about its malignant state and can be used as a novel marker of cellular malignancy.

There are currently only a few techniques capable of assessing cellular mechanical properties, but they all suggest a correlation between cellular rigidity and the cellular malignant stage. From a historical point of view, the state-of-the-art technique has been the micropipette aspiration method (Hochmuth 2000). Indeed, using this micropipette aspiration technique, a 50% reduction in the elasticity of malignantly transformed fibroblasts as compared to their healthy controls has been observed (Ward *et al* 1991). As an alternative method, AFM has been used for the investigation of cellular rigidity (Mahaffy *et al* 2000, Rotsch *et al* 1999).

In particular, normal human bladder epithelial cell lines and complimentary cancerous cell lines have been analyzed using AFM (Lekka *et al* 1999). They found that the rigidity of cancerous cells and their healthy counterparts differed by an order of magnitude. It has been reported that for small indentations (up to 500–1000 nm) the actin filaments are mainly responsible for providing the mechanical properties of the cells, whereas the disruption of microtubules has no pronounced effect on the mechanical properties of cells measured by AFM (Rao and Cohen 1991, Yamazaki *et al* 2005). Cells possess *in vitro* Young's modulus values in the range of 1–100 kPa (Radmacher 1997, Dulińska *et al* 2006), which includes vastly different types of cells such as vascular smooth muscle cells, fibroblasts, bladder cells, RBCs and epithelial cells. As such different cell types are being measured, the large variation in the Young's modulus is fully justified.

However, the Young's modulus varied even within human lung carcinoma cells from 13 kPa to 150 kPa (Weisenhorn *et al* 1993). Additional studies reported, for example, that the vinculin-deficient F9 mouse embryonic carcinoma cells possess a slightly lower Young's modulus (2.5 ± 1.5 kPa) compared to the wild-type cells (3.8 ± 1.1 kPa). Perhaps this was due to an altered cytoskeletal organization of actin (Goldmann and Ezzell 1996), indicating a prominent functional role for the focal adhesion protein vinculin as an essential part of the cytoskeletal matrix scaffold. Later, a comparative study of the cellular mechanical properties of malignant and non-malignant cells was performed (Lekka *et al* 1999). The Young's modulus of three human cancerous bladder cell lines was one order of magnitude lower than for healthy control cells. Additionally, these measurements were confirmed in 2005 using differently aggressive human breast cancer cell lines (Guck *et al* 2005, Park *et al* 2005). The cellular deformability was shown to be sensitive enough to monitor alterations during the malignant progression of mouse fibroblasts and human breast epithelial cells from normal to cancerous cells. The distribution of the Young's modulus of normal fibroblasts was much wider than for the malignantly transformed cells.

Taken together, cellular stiffness determined via AFM is dependent on the local point of the cantilever position and varies over the whole cellular surface, which may lead to a discrepancy in the Young's modulus, not only if measured at different parts on a single cell, but also when measured for a population of cells. In addition, the cell–cell interaction can also impact on the mechanical measurement of cells and they can behave differently to single cells. Hence, experimental errors increase alongside the large degree of cellular heterogeneity and they may also be time-dependent. Comparison between different cell types is still possible, and it is even possible to distinguish between different cancer cell types displaying different malignancy states, because of their altered mechanical properties, evoked by the cell's cytoskeleton.

Other techniques, such as magnetic bead rheology (Wang *et al* 1993), microneedle probes (Zahalak *et al* 1990), microplate manipulation (Thoumine and Ott 1997), acoustic microscopes (Kundu *et al* 2000), sorting in microfabricated sieves (Carlson *et al* 1997) and the manipulation of beads attached to cells with optical tweezers (Sleep *et al* 1999), can also be used to determine the mechanical properties of cells.

The main and most common result is that malignant cells respond to the external stresses applied by becoming either less elastic (softer and more deformable) or less viscous (less resistant to flow), depending on the measurement method used. However, metastatic cancer cells have been reported to display an even lower resistance to deformation compared to healthy cells (Raz and Geiger 1982, Ward *et al* 1991). In line with this, when using an optical stretcher it has been found that malignant cancer cells are more deformable than healthy controls (Guck *et al* 2005). Metastatic cancer cells must be capable of being highly deformable as they have to be able to squeeze through the surrounding tissue matrix as they enter the circulatory systems by transendothelial migration, after which they travel to targeted distant organs in order to build up secondary tumors (Wyckoff *et al* 2000).

Finally, these results indicate that cellular elasticity is suitable for use as a cell marker and even a diagnostic factor for determining the malignant state of cancer cells and thus their aggressive potential. However, all these techniques (excluding the optical cell stretcher) have to deal with low cell throughput measurements that are not good for statistics. In addition, the mechanical contact of the probe with the device may also lead to adhesion and active cellular response and moreover special cell preparation as well as nonphysiological probe handling may cause measurement artifacts. By contrast, an optical stretcher (Guck *et al* 2001) combined with a microfluidic chamber is able to deform individual suspended cells by optically induced surface forces at rates equivalent to the flow through rate of commercially available flow cytometers and cell sorters. Thus, the high throughput rate is suitable for the statistics and leads to reliable results. In addition, it may also be possible to combine an optical cell stretcher with a cell-sorting unit, in order to separate the stimulated and stretched cells from the non-stretched cells. Taken together, the deformability of cells measured with a microfluidic optical cell stretcher is suitable as an inherent cell marker for the malignancy of cancer cells.

3.3.2 Does optical cell stretching affect the viability of stretched cells?

The viability of the cells analyzed with an optical cell stretcher is important, as dead cells are known to possess an altered cytoskeleton. Thus, radiation damage should be avoided or reduced during optical cell stretcher measurements by selecting a wavelength of 785 nm with low absorption. In particular, the mechanical properties of BALB 3T3 cells are significantly different when they are not alive compared to their healthy counterparts. In phase contrast microscopy living cells display a characteristic bright rim around their cell edge, while dead cells show no sharp contour and appear diffuse.

However, it is not obvious that two 800 mW laser beams can be used for the deformation of cells without causing any radiation damage. It is extremely important to choose the right wavelength for the cell deformation measurements in order to avoid radiation damage. If we consider using a short wavelength, this might be desirable for optical tweezers, because it may result in higher gradients and better trapping efficiencies due to the small spot size of a focused laser beam, which is about half the wavelength of the light used before. However, the main problem with short wavelengths is that they are not appropriate for avoiding cell damage from radiation, because the absorption by the chromophores in cells is high with decreasing wavelength compared to increasing wavelength in the infrared region (Svoboda and Block 1994). It has been reported that the absorption peaks of proteins are in the ultraviolet region of the electromagnetic spectrum. Therefore, optical cell stretchers with a long wavelength of 1064 nm of a neodymium-doped yttrium aluminum garnet laser, termed Nd-YAG laser, have been implemented to improve this method and achieve better results (Ashkin *et al* 1987). At first glance, this choice of wavelength seems less than optimal because 70% of a cell's weight is water (Alberts *et al* 1994), which absorbs light more strongly with increasing wavelength. Another problem is that most cells trapped with optical tweezers do not survive light powers greater than 20–250 mW, which depends on the particular cell type and the laser wavelength (Ashkin *et al* 1987, Ashkin and Dziedzic 1987, Kuo and Sheetz 1992). Optical tweezers have been reported to induce local heating of water, which is a limiting factor in optical trapping experiments in the near infrared region. In more detail, theoretical calculations predict a temperature increase of less than 3 K per 100 mW in the wavelength range of 650–1050 nm for durations of up to 10 s (Schoenle and Hell 1998). Thus, the duration of the laser exposure to the cells is reduced to 2 s in optical cell stretcher experiments. These theoretical results have been confirmed by a 0.3 °C increase in temperature per 100 mW at powers of up to 400 mW by using Chinese hamster ovary cells (CHO cells) trapped in optical tweezers at a wavelength of 1064 nm (Liu *et al* 1996).

In addition to the effects of heating due to absorption by water, the cells can also be damaged by radiation through metabolic alterations affecting cellular viability when trapped in optical tweezers. Hence, the wavelength of the laser is carefully chosen to minimize absorption by water (Kim *et al* 2015). Further to this, microfluorometric measurements on CHO cells (Liu *et al* 1996) revealed that even up to 400 mW of 1064 nm cw laser light causes either no alterations to the DNA structure or the cellular pH. However, the right choice of laser wavelength is important, as photodamage has been reported for rotating *E. coli* assays with maxima at 870 and 930 nm and minima at 830 and 970 nm (Neuman *et al* 1999). The presence of oxygen seems to be responsible for the damage, but the origin of the damage is still elusive. The damage at 785 nm is at least as small as at 1064 nm, with the latter being the most commonly used wavelength for biological trapping experiments (Neuman *et al* 1999). In addition, sensitivity to light is linearly correlated with intensity, which cancels out multi-photon processes. For an optical cell stretcher device such damage has not yet been seen. An explanation for this is that the laser beams are not focused and the power densities are lower compared to optical tweezer devices by about two orders of magnitude for the same light power. In summary, to apply higher forces much higher light powers can be used without the risk of cellular damage.

Cell cytoskeletons in suspension

This part discusses the fact that the cells measured using the optical cell stretcher are in suspension and not adhered to a substrate. For cells in the bloodstream such as leukocytes, which are generally non-adherent, the measurement of suspended cells represents no problem when analyzing them. There is the concern, however, that adherent eukaryotic cells may dissolve their actomyosin cytoskeleton when they are in suspension and thus non-adherent. The actin cytoskeleton of BALB 3T3 fibroblasts in suspension has been analyzed through TRITC-phalloidin staining using fluorescence microscopy. The result did not confirm this hypothesis, as suspended cells possess an extensive actin network throughout the whole cytoplasm. In particular, the peripheral cortical actin is visible underneath the cell membrane. The only features of the cell's cytoskeleton not observed in suspension are stress fibers, due to the lack of focal adhesion plaques in suspended cells. However, even in the absence of stress fibers BALB 3T3 cells possess a large resistance to deformation induced by the optical cell stretcher. Stress fibers are mainly reported in cells adhered to a planar 2D substrate, whereas they are less pronounced in cells directly embedded in a connective tissue matrix. Thus, the presence of many stress fibers may even be nonphysiological. Indeed, several studies have found that the influence of the cytoskeleton on the deformability of normally adherent cells, such as bovine aortic endothelial cells (Sato *et al* 1987), chick embryo fibroblasts (Thoumine and Ott 1997) and rat embryo fibroblasts (Heidemann *et al* 1999) can be determined in suspension by using, for example, an optical cell stretcher. These results are further supported by frequency-dependent AFM microrheology experiments, comparing *in vitro* actin gels and NIH3T3 fibroblasts (Mahaffy *et al* 2000). In more detail, the viscoelastic properties of the cells are similar to those of homogeneous actin gels. Thus, the elastic strength of cells seems to be stored within the actin cortex that is still assembled in non-adherent cells. As most polymer theoretic models describe isotropic actin networks *in vitro*, they can be adapted to non-adherent cells. Moreover, it seems to be likely that these theoretic models will be compared with data obtained from living cells that only possess the actin cortex and not anisotropic structures such as F-actin stress fibers.

The optical deformability of dielectric matter has not yet been the focus of biophysical research. However, the optical stretcher serves as a nondestructive optical tool for the high-throughput and quantitative deformation of cells. Indeed, the forces exerted by light on the cells due to the momentum transferred are adequate to hold as well as move the objects and even they are high enough to deform the samples. In more detail, the momentum is predominantly transferred to the surface of the sample. The total force, such as the gradient and scattering force, that traps the sample arises from the asymmetry of the resulting surface stresses when the sample is not located at its equilibrium position and thus acts on the sample's center of gravity. These trapping forces are significantly smaller than the forces on the sample's surface because the surface forces cancel nearly completely upon integration. This phenomenon is most obvious in a two-beam trap as used for the optical cell stretcher. In the case when the dielectric object (such as a cell) is located in the center of the optical trap, the total force is zero, and the forces acting on the cell's surface can be several hundred piconewtons. The forces used for cell elasticity measurements with the optical cell stretcher are in the range of those for optical tweezers and AFMs.

At first glance, the result that the surface forces pull on the surface rather than compressing it seems surprising. This result can be explained through a ray optics (RO) approach by the increase of the light's momentum as it enters a denser medium, such as the interior of a cell, and the resulting stresses on the cell's surface. Another approach is guided by the minimization of energy. In particular, it is energetically favorable for a dielectric object such as a cell to have most of its volume in the area with the highest intensity along the laser beam axis. This leads to the result that the cell is pulled to the axis. This seems to be conceptually correct, but it would be difficult to calculate.

In more detail, the magnitude of the induced forces scales linearly with the incident laser light power and with $(n - 1)$, where $n = n_{cell}/n_{medium}$ is the relative refractive index. For example, with 1 W of power from each laser beam the forces are in the range of 200–500 pN, even when the relative refractive indices of biological materials are typically relatively low, such as $n = 1.02$–1.05.

The optical deformability of cells is a cell-type specific property and hence serves as a marker for a particular cell type, for example red blood cells (RBCs) and BALB 3T3 fibroblasts. Trapped cells show no radiation damage, even when stretched with 800 mW of light in each beam, and hence show a representative cytoskeleton in suspension.

3.3.3 Biomedical application of the optical cell stretcher

There seems to be a medical application for the optical cell stretcher technique. Using it, one cell per second can be measured to determine cellular elasticity. As the lifespan of a living cell *in vitro* is limited, measurement frequency is crucial to obtain good statistics. In particular, the optical stretcher seems to be suitable for applications in the research and diagnosis of diseases such as cancer, which are evoked by alterations of the cytoskeleton. Current cancer-detection methods are based on markers and optical inspection (Sidransky 1996). An increasing disorder of the actin cytoskeleton may serve as a marker for malignancy (Koffer *et al* 1985, Takahashi *et al* 1986). Alterations in the total amount of actin and in the relative ratio of the several actin isoforms (Wang and Goldberg 1976, Goldstein *et al* 1985, Leavitt *et al* 1986, Takahashi *et al* 1986, Taniguchi *et al* 1986), an overexpression of gelsolin in breast cancer cells (Chaponnier and Gabbiani 1989) and the lack of filamin in human malignant melanoma cells may all cause actin cytoskeletal alterations (Cunningham *et al* 1992). However, all of these alterations may evoke an altered viscoelastic response of the cells that can be detected with the optical cell stretcher. Models of actin networks (MacKintosh *et al* 1995) showed that the shear modulus increases with the actin concentration raised. In particular, a slight decrease in the actin concentration leads to a detectable decrease in the cell's elasticity. The optical cell stretcher may serve as an advanced diagnostic tool in medical laboratories. Moreover, this novel analytical tool requires minimal tissue samples, which can be obtained using simple cytobrushes on the surfaces of the lung, esophagus and stomach, as well as in the cervix or through fine-needle aspiration (Dunphy and Ramos 1997, Fajardo and DeAngelis 1997).

3.3.4 The optical deformability of mouse fibroblasts

In order to demonstrate the connection between cellular function and the optical deformability of cells, the optical deformability of BALB/3T3 and SV-T2 fibroblasts was analyzed using the microfluidic optical cell stretcher. BALB/3T3 is a fibroblast cell line established from disaggregated BALB/c mouse embryos, whereas SV-T2 cells are, through a SV40 DNA virus, oncogenic, transformed cells derived from BALB/3T3 (Aaronson and Todaro 1968). These two cell lines are good model systems for studying the malignant transformation of cells (Aaronson and Todaro 1968; Thoumine and Ott 1997). Indeed, the optical deformability of the SV-T2 cells is increased compared to that of BALB/3T3 cells.

The optical deformability is time-dependent due to the viscoelastic property of cells. Thus, the cells were stretched for one second, as this timescale exploits both elastic and viscous contributions to the cellular deformability (Wottawah *et al* 2005). This difference in deformability has been obtained after analyzing only 30 cells for each cell type. In order to analyze these cells further using proteomics, slightly more cells may be required. In the future, it may be possible that microfluidic cell stretchers reach the same sample sizes and measurement rates as fluorescence-activated cell sorter (FACS) devices. As the refractive indexes are not distinguishable between the two cells, the difference in optical deformability can only be attributed to their different mechanical properties. Moreover, the optical deformability of cells can be equated with cellular compliance.

In a theoretic model for isotropic networks of semiflexible polymers, there is a strong dependence of the shear modulus on the filament concentration (Gardel *et al* 2003, Janmey *et al* 1991, Wilhelm and Frey 2003). The relative amounts of F-actin in BALB/3T3 and SV-T2 cells using Alexa-532 Phalloidin staining, and measurements of their integrated intensity distributions, have shown that the total amount of filamentous actin in malignantly transformed cells is reduced by approximately 40%. Moreover, due to the reduced cell size of the SV-T2 cells, the concentration of F-actin is reduced by approximately 50%. The reduction in the absolute amount of F-actin is followed by a remodeling of the whole actin cytoskeleton of the malignantly transformed SV-T2 cells compared to the normal BALB/3T3 cells. In summary, the decreased quantity and the altered structures of both actin cytoskeletons provide an explanation for the enhanced optical deformability of the cancerous SV-T2 cells. In line with this critically important role for F-actin, latrunculin-treated cells (inhibited actin polymerization) soften by about one half.

3.3.5 The optical deformability of human breast carcinoma cells

The optical deformability of differently aggressive and invasive cancer cells of epithelial origin can be measured using an optical cell stretcher. It has been established that MCF and MDA-MB-231 cells are model cell lines for investigating breast cancer (Johnson *et al* 1999, Soule *et al* 1973, Tait *et al* 1990, Wang *et al* 2001). In particular, MCF-10 is a non-tumorigenic human epithelial cell line isolated from the benign breast tissue of a fibrocystic disease. Although these cells are immortal, they are normal and noncancerous mammary epithelial cells. Another breast cancer cell line (adenocarcinoma) is MCF-7, which is isolated from the pleural effusion. In particular, these cells are nonmotile and nonmetastatic epithelial cancer cells. When the phorbol ester tetradecanoylphorbol-acetate (TPA) is added to MCF-7 cells (modMCF-7 cells) these are then stimulated and display a dramatic increase (18 fold) in their invasiveness and ability to metastasize (Johnson *et al* 1999). In addition, TPA induces the secretion of MMPs in MCF-7 cells, which are able to degrade the surrounding extracellular matrix of tissues and are associated with cancer metastasis and malignant progression *in vivo* (Himelstein *et al* 1994). Using an optical cell stretcher, the cancerous MCF-7 cells showed an increased deformability compared to normal MCF-10 cells and nonmetastatic MCF-7 cells.

In addition, the optical deformability of MDA-MB-231 breast cancer cells has been analyzed in the presence and absence of all-trans-retinoic acid (modMDA-MB-231), which leads to less aggressive MDA-MB-231 cells (Wang *et al* 2001). As expected, these less agressive MDA-MB-231 cells displayed reduced optical deformability by possessing reduced metastasizing capacity. The differences in the refractive indices of all five cancer cell lines lay at around 1.365 and hence cannot explain the differences in optical deformability, which instead seem to be correlated with a reduced cytoskeletal resistance to cellular deformation in cancerous cells compared to normal cells. The reduction in structural strength is consistent with their capability to migrate through connective tissue, to intravasate into the blood and lymph vessels, to circulate through these microvascular systems and possibly extravasate to migrate and build up metastases in targeted organs. Although a major cytoskeletal element in epithelial cells is the intermediate filament keratin (Kirfel *et al* 2003), it is hypothesized that the contribution of keratins to the mechanical properties is important at strains larger than those applied in these optical cell stretcher experiments (Janmey *et al* 1991, Wang and Stamenovic 2000). Thus, the differences in optical deformability are evoked by the decrease of F-actin stress fibers during the malignant transformation by approximately 30% (Katsantonis 1994). In contrast, in another study it has been shown that the keratins play a role in providing optical cell deformability (Seltmann *et al* 2013).

Alterations in optical deformability are detectable due to subtle changes in cytoskeletal composition and structure. Indeed, this dependence confirms results from polymer physics where even slight variations in the concentration of cytoskeletal filaments lead to a nonlinearly enhanced overall elasticity (Gardel *et al* 2003, Janmey *et al* 1991, Wilhelm and Frey 2003). Optical deformability can in general serve as a unique and useful biophysical cell marker, which is able to detect even small cytoskeletal changes through the altered optical properties of the cells.

The results obtained with differently invasive and aggressive cancer cell lines can be adapted to primary cells in order to obtain diagnostic markers for the invasive potential of cancer cells. Indeed, the microfluidic optical cell stretcher permits the investigation of samples of individual cells and determines their metastatic potential (Caraway *et al* 1993, Epstein *et al* 2002, Sherman and Kurman 1996). Subsequently, pure samples of isolated cancer cells can be investigated without artificial corruption using established genomic or proteomic techniques, which cannot detect small numbers of changes. Another possible field of application for an optical cell stretcher is the isolation and separation of therapeutic stem cells (Olins *et al* 2000), which cannot be identified and sorted with a microfluidic optical cell stretcher without knowing a special set of markers able to identify unique a stem cell.

The mechanical parameter, the optical deformability of cells, is an inherent cellular marker that provides a sensitive cellular mechanical-based alternative to current proteomic techniques and thus may lead to novel promising findings regarding all cellular processes that involve the mechanical properties of the cytoskeleton.

3.4 Optical tweezers

Based on focused laser light, optical tweezers can trap and manipulate small refractile objects (Ashkin 1997) and hence are ideally suited for mechanical measurements on proteins and on supramolecular assemblies such as cells. In particular, using optical tweezers it has been revealed that the step size of the kinesin motor protein is 8 nm (Svoboda *et al* 1993). The resolution of the displacement and the range of forces are ideally suited for single-molecule measurements. Hence, optical tweezers have been successfully utilized to perform measurements of the viscoelastic properties of soft materials such as proteins or DNA on the mesoscopic length scale (MacKintosh and Schmidt 1999, Mizuno *et al* 2008, Koenderink *et al* 2006). Moreover, these mechanical experiments have been laboriously further developed in order to separate the local response from a larger-scale response by employing one- and two-particle rheology of polymer networks (Buchanan *et al* 2005, Atakhorrami *et al* 2008) and investigate even non-equilibrium mechanics of living active cytoskeletal networks (Mizuno *et al* 2007).

When making use of orthogonally polarized laser beams derived from one laser, a dual tweezer setup can be developed and hence independently trap and steer two beads. Indeed, such a dual optical tweezer setup has been utilized to analyze the fluctuations and the mechanical response of suspended cells. It has been reported that the recorded fluctuations were not thermal equilibrium fluctuations and thus largely evoked by active force generation of the cell (Mizuno *et al* 2009). A suspended cell is typically mechanically associated through integrin cell surface receptors to the two beads, which are coated with extracellular matrix ligands such as fibronectin or RDG peptide. The two-bead setup can not only be utilized to investigate cellular force fluctuations, it can be also used to apply a force to the cell and to monitor the resulting deformation. By the application of Hooke's law an effective spring constant can be determined.

The tensed actomyosin cortex has been shown to alter the measured cellular effective stiffness. When cells were incubated with blebbistatin, a myosin inhibitor, they become significantly softer and thereby the transmitted cellular force fluctuations are significantly decreased (Schlosser *et al* 2015). In contrast, it has been demonstrated that blebbistatin treatment resulted in cellular stiffening (Chan *et al* 2015). However, the difference between the two experiments is the attachment of beads coated with matrix proteins to integrin receptors and upon stable bead binding, the subsequent formation of focal adhesions for bead diameters larger than 4 μm (Galbraith *et al* 2002). When a cell is suspended between two beads, the optical tweezers represent the mechanical boundary conditions and, to some extent, fulfill the role of a substrate for adherent cells. In more detail, the stiffness of the potential well generated by the optical tweezers can be seen roughly as an equivalent for the matrix elasticity. Hence, a change in laser intensity affecting the trapping potential can simulate an altered mechanical microenvironment. Instead of providing simply a passive elastic response, the optical tweezers can also be used to exert a defined force. Moreover, the cells can be kept under a constant external load using a feedback mode. When fibroblasts are kept under a constant external tension in the range of 5–35 pN, they seem to generate slow oscillatory contractile forces with periods in the range of 1–2 min. This oscillation can be stopped when cells are treated with blebbistatin inhibiting non-muscle myosin motor proteins and hence a creep response to the external load similar to that of viscoelastic materials can be measured (Schlosser *et al* 2015).

The optical cell stretcher measures the compliance of the whole suspended cell, whereas the dual optical tweezers mechanically couple to the cell through integrin clusters and focal adhesions and transmit forces directly to the actomyosin cortex. However, more information on the cell's cortex and other cytoskeletal structures is required to fully understand cellular mechanical properties. Indeed, this can be utilized, when compared to adherent cells, simple geometry of close-to-spherical suspended cells is measured in a combined setup of the dual optical tweezers with confocal microscopy or also potentially appropriate super-resolution microscopy techniques. In particular, a 3T3 NIH fibroblast in dual confocal tweezers, which is transiently transfected with LifeAct-RFP, reveals the structure of the actomyosin cortex and also of smaller actin bundles within the cytoplasm of the suspended cell.

A strong limitation for these mechanical measurements of cells and possible also cellular aggregates is the restricted force range of up to approximately 100 pN, which can be maximally applied by using optical tweezers. A solution to this problem can be specially designed core–shell particles, which displayed an increase of maximal trap force by an order of magnitude that would allow us, together with optimization of optical parameters to apply forces up to several nN (Bormuth *et al* 2008).

3.5 Microfluidic filtration and mechanical deformability

By using microfluidic filtration, the issue of non-uniform pore size in membrane filters can be resolved, when a micromachining technique is utilized to generate an array of parallel microchannels (Kim *et al* 2015). With this technique, the deformation of entire cells can be monitored under a microscope while the cells are squeezing through multiple microchannels. Hence, microfluidics seems to be a promising, cost-effective, and high-throughput method for measuring the deformability of RBCs, with a minimum amount of blood (cells) required for the analysis (Whitesides 2006, Bransky *et al* 2007, Martin *et al* 2011, Patel *et al* 2013, Shevkoplyas *et al* 2006, Ye *et al* 2010). In more detail, the microfluidic device mimics the *in vivo* capillary blood flow system that possess internal diameters of only a few micrometers and the deformability of RBCs can be analyzed by using a funnel-shaped microconstriction system, in which the bolls sample is placed and exerted to a certain pressure (Guo *et al* 2011, 2012a, 2012b). The deformability is determined by measuring the threshold pressure necessary for the blood sample to traverse the well-defined constriction. When using this microfluidic measurement system, both individual RBCs and populational assessments of cellular deformability can be performed. In particular, microfluidic systems have been employed to determine the deformability of malaria-infected RBCs (Shelby *et al* 2003, Bow *et al* 2011) and RBCs in patients with sickle-cell disease (Wang *et al* 2011).

3.6 Real-time deformation cytometry

Cellular stiffness is a highly sensitive and suitable marker for physiological and pathological alterations in cells and moreover many applications in the field of biology and medicine are possible. A new method, called real-time deformability cytometry (RT-DC), analyzes the cell stiffness at high throughput by exposing the cells located in a microfluidic channel to a shear flow that provides the mechanical phenotyping due to the single-cell deformability (Otto *et al* 2015, Mietke *et al* 2016). The deformations of cells in the channel are not only dependent on cell stiffness, but also depend on the cell's size relative to channel size (diameter). Hence, the mutual contributions of cell size and cell stiffness to cell deformation can be disentangled by a theoretical analysis involving hydrodynamics and linear elasticity theory. When performing real-time deformability cytometry on both the model spheres of known elasticity and on biological cells, it can be shown that our analytical model does not only predict deformed shapes inside the channel but also enables us to quantify cellular mechanical properties such as deformability. Thus, fast and quantitative mechanical sampling of large cellular populations becomes feasible.

Typical methods for the measurement of cell stiffness include atomic force microscopy (AFM) indentation, magnetic twisting cytometry and optical stretching (Pullarkat *et al* 2007). The cell mechanical characterizations based on these methods do not enable us to measure throughput rates beyond one cell per minute. However, the analysis of probes from a large population size and a broad heterogeneity of cells requires high-throughput methods for their classification and analysis. Several microfluidic techniques have already started to address this need (Gossett *et al* 2012, Byun *et al* 2013, Otto *et al* 2015, Lange *et al* 2015). Among them are the deformability cytometry (DC) (Gossett *et al* 2012) and RT-DC (Otto *et al* 2015), which deform cells purely by hydrodynamic interactions and without any contact with the walls of the channels.

Although the DC analyzes cells in an extensional flow rate of thousands of cells per second, it operates in a dynamic regime of Reynolds numbers of approximately 50. Hence this transfers the calculation of associated flow fields to a challenging, time-dependent and nonlinear problem. Until now, no analytical or numerical modeling has been applied to these data, which turns the cell mechanical characterization purely into a phenomenological description.

In contrast, in RT-DC smaller rates of hundreds of cells per second are reached together with a more viscous carrier medium, which reveals a fluid flow at a low

Reynolds numbers of approximately 0.1, which is now accessible to theoretical analysis. In RT-DC measurements, the suspended cells are exposed to a shear flow through a microfluidic channel with a constant speed. During this process, the cells are deformed due to the existence of strong velocity gradients within the channel's cross section (Otto *et al* 2015). The observed deformation of initially spherical cells in the flow channel depends on cellular stiffness, however, it also seems to be influenced by the flow speed, the relative size of the cell and the internal cell architecture. In order to separate the influence of each of these parameters, the quantification of the stresses exerted on the cells through the shear flow is required to apply cell mechanical models predicting the measured deformations.

References and further reading

Aaronson S A and Todaro G J 1968 Development of 3T3-like lines from Balb-c mouse embryo cultures: transformation susceptibility to SV40 *J. Cell. Physiol.* **72** 141–8

Abbondanzieri E A, Greenleaf W J, Shaevitz J W, Landick R and Block S M 2005 Direct observation of base-pair stepping by RNA polymerase *Nature* **438** 460–5

Adelman M R, Borisy G G, Shelanski M L, Weisenberg R C and Taylor E W 1968 Cytoplasmic filaments and tubules *Fed. Proc.* **27** 1186–93

Aebi U, Cohn J, Buhle L and Gerace L 1986 The nuclear lamina is a meshwork of intermediate-type filaments *Nature* **323** 560–4

Alberts B, Bray D, Lewis J, Raff M, Roberts K and Watson J D 1994 *Molecular Biology of the Cell* (New York: Garland) pp 786—861

Alenghat F J, Fabry B, Tsai K Y, Goldmann W H and Ingber D E 2000 Analysis of cell mechanics in single vinculin-deficient cells using a magnetic tweezer *Biochem. Biophys. Res. Commun.* **277** 93–9

Allen P G and Janmey P A 1994 Gelsolin displaces phalloidin from actin filaments. A new fluorescence method shows that both Ca^{2+} and Mg^{2+} affect the rate at which gelsolin severs F-actin *J. Biol. Chem.* **269** 32916–23

Amblard F, Yurke B, Pargellis A and Leibler S 1996 A magnetic manipulator for studying local rheology and micromechanical properties of biological systems *Rev. Sci. Instr.* **67** 818–27

Ashkin A 1970 Acceleration and trapping of particles by radiation pressure *Phys. Rev. Lett.* **24** 156–9

Ashkin A 1997 Optical trapping and manipulation of neutral particles using lasers *Proc. Natl Acad. Sci.* **94** 4853–60

Ashkin A and Dziedzic J M 1973 Radiation pressure on a free liquid surface *Phys. Rev. Lett.* **30** 139–42

Ashkin A and Dziedzic J M 1987 Optical trapping and manipulation of viruses and bacteria *Science* **235** 1517–20

Ashkin A, Dziedzic J M and Yamane T 1987 Optical trapping and manipulation of single cells using infrared laser beams *Nature* **330** 769–71

Ashkin A, Dziedzic J M, Bjorkholm J E and Chu S 1986 Observation of a single-beam gradient force optical trap for dielectric particles *Opt. Lett.* **11** 288–90

Assi F, Jenks R, Yang J, Love C and Prentiss M 2002 Massively parallel adhesion and reactivity measurements using simple and inexpensive magnetic tweezers *J. Appl. Phys.* **92** 5584

Atakhorrami M, Addas K M and Schmidt C F 2008 Twin optical traps for two-particle cross-correlation measurements: eliminating cross-talk *Rev. Sci. Instrum.* **79** 043103

Barbic M, Mock J J, Gray A P and Schultz S 2001 Electromagnetic micromotor for microfluidics applications *Appl. Phys. Lett.* **79** 1399–401

Barer R and Joseph S 1954 Refractometry of living cells, part I. Basic principles *Q. J. Microsc. Sci.* **95** 399–423

Barer R and Joseph S 1955a Refractometry of living cells, part II. The immersion medium *Q. J. Microsc. Sci.* **96** 1–26

Barer R and Joseph S 1955b Refractometry of living cells, part III. Technical and optical methods *Q. J. Microsc. Sci.* **96** 423–47

Bausch A R, Möller W and Sackmann E 1999 Measurement of local viscoelasticity and forces in living cells by magnetic tweezers *Biophys. J.* **76** 573–9

Bausch A R, Ziemann F, Boulbitch A A, Jacobson K and Sackmann E 1998 Local measurements of viscoelastic parameters of adherent cell membranes by magnetic bead microrheometry *Biophys. J.* **75** 2038–49

Bennett V 1985 The membrane skeleton of human erythrocytes and its implication for more complex cells *Annu. Rev. Biochem.* **54** 273–304

Bennett V 1990 Spectrin-based membrane skeleton—a multipotential adapter between plasma membrane and cytoplasm *Physiol. Rev.* **70** 1029–60

Ben-Ze'ev A 1985 The cytoskeleton in cancer cells *Biochim. Biophys. Acta* **780** 197–212

Block S M, Goldstein L S and Schnapp B J 1990 Bead movement by single kinesin molecules studied with optical tweezers *Nature* **348** 348–52

Bormuth V, Jannasch A, Ander M, van Kats C M, Avan Blaaderen A, Howard J and Schäffer E 2008 Optical trapping of coated microspheres *Opt. Express* **16** 13831–44

Bosch F H, Werre J M, Schipper L, Roerdinkholder-Stoelwinder B, Huls T, Willekens F L, Wichers G and Halie M R 1994 Determinants of red blood cell deformability in relation to cell age *Eur. J. Haematol.* **52** 35–41

Bow H, Pivkin I V, Diez-Silva M, Goldfless S J, Dao M, Niles J C, Suresh S and Han J 2011 A microfabricated deformability-based flow cytometer with application to malaria *Lab Chip* **11** 1065–73

Bransky A, Korin N, Nemirovski Y and Dinnar U 2007 Correlation between erythrocytes deformability and size: a study using a microchannel based cell analyzer *Microvasc. Res.* **73** 7–13

Brevik I 1979 Experiments in phenomenological electrodynamics and the electromagnetic energy–momentum tensor *Phys. Rep.* **52** 133–201

Buchanan M, Atakhorrami M, Palierne J F and Schmidt C F 2005 Comparing macrorheology and one- and two-point microrheology in wormlike micelle solutions *Macromolecules* **38** 8840–4

Byun S *et al* 2013 Characterizing deformability and surface friction of cancer cells *Proc. Natl Acad. Sci. USA* **110** 7580–5

Caraway N P, Fanning C V, Wojcik E M, Staerkel G A, Benjamin R S and Ordonez N G 1993 Cytology of malignant melanoma of soft parts: fine-needle aspirates and exfoliative specimens *Diagn. Cytopathol.* **9** 632–8

Carlier M F 1998 Control of actin dynamics *Curr. Opin. Cell Biol.* **10** 45–51

Carlson R H, Gabel C V, Chan S S, Austin R H, Brody J P and Winkelman J W 1997 Self-sorting of white blood cells in a lattice *Phys. Rev. Lett.* **79** 2149–52

Caspi A, Granek R and Elbaum M 2002 Diffusion and directed motion in cellular transport *Phys. Rev.* E **66** 11916-1–12

Chan C J, Ekpenyong A E, Golfier S, Li W, Chalut K J, Otto O, Elgeti J, Guck J and Lautenschläger F 2015 Myosin II activity softens cells in suspension *Biophys. J.* **108** 1856–69

Chaponnier C and Gabbiani G 1989 Gelsolin modulation in epithelial and stromal cells of mammary carcinoma *Am. J. Pathol.* **134** 597–603

Chu S 1991 Laser manipulation of atoms and particles *Science* **253** 861–6

Cluzel P, Lebrun A, Heller C, Lavery R, Viovy J L, Chatenay D and Caron F 1996 DNA: an extensible molecule *Science* **271** 792–4

Colon J M, Sarosi P G, McGovern P G, Ashkin A and Dziedzic J M 1992 Controlled micromanipulation of human sperm in three dimensions with an infrared laser optical trap: effect on sperm velocity *Fertil. Steril.* **57** 695–8

Constable A, Kim J, Mervis J, Zarinetchi F and Prentiss M 1993 Demonstration of a fiber-optical light-force trap *Opt. Lett.* **18** 1867–9

Cooper J A, Bryan J, Schwab B, Frieden C, Loftus D J and Elson E L 1987 Microinjection of gelsolin into living cells *J. Cell Biol.* **104** 491–501

Crick F H C and Hughes A F W 1950 The physical properties of cytoplasm: a study by means of the magnetic particle method *Exp. Cell Res.* **1** 37–80

Cunningham C C, Gorlin J B, Kwiatkowski D J, Hartwig J H, Janmey P A, Byers H R and Stossel T P 1992 Actin-binding protein requirement for cortical stability and efficient locomotion *Science* **255** 325–7

Daily B, Elson E L and Zahalak G I 1984 Cell poking: determination of the elastic area compressibility modulus of the erythrocyte membrane *Biophys. J.* **45** 661–82

Daniels B R, Masi B C and Wirtz D 2006 Probing single-cell micromechanics *in vivo*: the microrheology of *C. elegans* developing embryos *Biophys. J.* **90** 4712–9

Davenport R J, Wuite G J L, Landick R and Bustamante C 2000 Single-molecule study of transcriptional pausing and arrest by *E. coli* RNA polymerase *Science* **287** 2497–500

de Vries A H B, Krenn B E, van Driel R and Kanger J S 2005 micro magnetic tweezers for nanomanipulation inside live cells *Biophys. J.* **88** 2137–44

Discher D E, Mohandas N and Evans E A 1994 Molecular maps of red cell deformation: hidden elasticity and *in situ* connectivity *Science* **266** 1032–35

Dulińska I, Targosz M, Strojny W, Lekka M, Czuba P, Balwierz W and Szymoński M 2006 Stiffness of normal and pathological erythrocytes studied by means of atomic force microscopy *J. Biochem. Biophys. Methods* **66** 1–11

Dunphy C and Ramos R 1997 Combining fine-needle aspiration and flow cytometric immunophenotyping in evaluation of nodal and extranodal sites for possible lymphoma: a retrospective review *Diagn. Cytopathol.* **16** 200–6

Eichinger L, Koeppel B, Noegel A A, Schleicher M, Schliwa M, Weijer K, Wittke W and Janmey P A 1996 Mechanical perturbation elicits a phenotypic difference between dictyostelium wild-type cells and cytoskeletal mutants *Biophys. J.* **70** 1054–60

Ekpenyong A E, Whyte G, Chalut K, Pagliara S, Lautenschläger F, Fiddler C, Paschke S, Keyser U F, Chilvers E R and Guck J 2012 Viscoelastic properties of differentiating blood cells are fate- and function-dependent *PLoS One* **7** e45237

Elson E L 1988 Cellular mechanics as an indicator of cytoskeletal structure and function *Annu. Rev. Biophys. Biophys. Chem.* **17** 397–430

Epstein J B, Zhang L and Rosin M 2002 Advances in the diagnosis of oral premalignant and malignant lesions *J. Can. Dent. Assoc.* **68** 617–21

Evans E 2001 Probing the relation between force, lifetime, and chemistry in single molecular bonds *Annu. Rev. Biophys. Biomol. Struct.* **30** 105–28

Evans E and Fung Y C 1972 Improved measurements of the erythrocyte geometry *Microvasc. Res.* **4** 335–47

Evans E, Ritchie K and Merkel R 1995 Sensitive force technique to probe molecular adhesion and structural linkages at biological interfaces *Biophys. J.* **68** 2580–7

Fabry B and Fredberg J J 2003 Remodeling of the airway smooth muscle cell: are we built of glass? *Respir. Physiol. Neurobiol.* **137** 109–24

Fabry B, Maksym G N, Butler J P, Glogauer M, Navajas D and Fredberg J J 2001 Scaling the microrheology of living cells *Phys. Rev. Lett.* **87** 148102–6

Fajardo L L and DeAngelis G A 1997 The role of stereotactic biopsy in abnormal mammograms *Surg. Oncol. Clin. North Am.* **6** 285–99

Felder S and Elson E L 1990 Mechanics of fibroblast locomotion: quantitative analysis of forces and motions at the leading lamellas of fibroblasts *J. Cell Biol.* **111** 2513–26

Fischer T, Wilharm N, Hayn A and Mierke C T 2017 Matrix and cellular mechanical properties are the driving factors for facilitating human cancer cell motility into 3D engineered matrices *Converg. Sci. Phys. Oncol.* **3** 044003

Fisher J K *et al* 2006 Thin-foil magnetic force system for high-numerical-aperture microscopy *Rev. Sci. Instrum.* **77** 023702

Fisher J K, Cummings J R and Desai K V *et al* 2005 Three-dimensional force microscope: a nanometric optical tracking and magnetic manipulation system for the biomedical sciences *Rev. Sci. Instrum.* **76** 053711

Forsyth A M, Wan J, Ristenpart W D and Stone H A 2010 The dynamic behavior of chemically 'stiffened' red blood cells in microchannel flows *Microvasc. Res.* **80** 37–43

Freundlich H and Seifriz W 1923 Ueber die Elastizitaet von Solen und Gelen *Z. Phys. Chem. Stoechiom. Verwandtschaftsl.* **104** 233

Fuchs E and Cleveland D W D.W. 1998 A structural scaffolding of intermediate filaments in health and disease *Science* **279** 514–9

Galbraith C G, Yamada K M and Sheetz M P 2002 The relationship between force and focal complex development *J. Cell Biol.* **159** 695–705

Gardel M L, Valentine M T, Crocker J C, Bausch A R and Weitz D A 2003 Microrheology of entangled F-actin solutions *Phys. Rev. Lett.* **91** 158302

Goldmann W H and Ezzell R M 1996 Viscoelasticity in wild-type and vinculin-deficient (5.51) mouse F9 embryonic carcinoma cells examined by atomic force microscopy and rheology *Exp. Cell Res.* **226** 234–7

Goldstein D, Djeu J, Latter G, Burbeck S and Leavitt J 1985 Abundant synthesis of the transformation-induced protein of neoplastic human fibroblasts, plastin, in normal lymphocytes *Cancer Res.* **45** 5643–7

Gosse C and Croquette V 2002 Magnetic tweezers: micromanipulation and force measurement at the molecular level *Biophys. J.* **82** 3314–29

Gossett D R, Tse H T, Lee S A, Ying Y, Lindgren A G, Yang O O, Rao J, Clark A T and Di Carlo D 2012 Hydrodynamic stretching of single cells for large population mechanical phenotyping *Proc. Natl Acad. Sci. USA* **109** 7630–5

Grandbois M, Beyer M, Rief M, Clausen-Schaumann H and Gaub H E 1999 How strong is a covalent bond? *Science* **283** 1727–30

Guck J, Ananthakrishnan R, Mahmood H, Moon T J, Cunningham C C and Kas J 2001 The optical stretcher: a novel laser tool to micromanipulate cells *Biophys. J.* **81** 767–84

Guck J, Ananthakrishnan R, Moon T J, Cunningham C C and Kaes J 2000 Optical deformability of soft biological dielectrics *Phys. Rev. Lett.* **84** 5451–4

Guck J *et al* 2005 Optical deformability as an inherent cell marker for testing malignant transformation and metastatic competence *Biophys. J.* **88** 3689–98

Guilford W H and Gore R W 1992 A novel remote-sensing isometric force transducer for micromechanics studies *Am. J. Physiol.* **263** C700–7

Guilford W H, Lantz R C and Gore R W 1995 Locomotive forces produced by single leukocytes *in vivo* and *in vitro Am. J. Physiol.* **268** C1308–12

Guo Q, McFaul S M and Ma H 2011 Deterministic microfluidic ratchet based on the deformation of individual cells *Phys. Rev.* E **83** 051910

Guo Q, Park S and Ma H 2012a Microfluidic micropipette aspiration for measuring the deformability of single cells *Lab Chip* **12** 2687–95

Guo Q, Reiling S J, Rohrbach P and Ma H 2012b Microfluidic biomechanical assay for red blood cells parasitized by *Plasmodium falciparum Lab Chip* **12** 1143–50

Heidemann S R, Kaech S, Buxbaum R E and Matus A 1999 Direct observations of the mechanical behaviors of the cytoskeleton in living fibroblasts *J. Cell Biol.* **145** 109–22

Heilbronn A 1922 Jahrbuch für wiss *Botanik* **61** 284

Henon S, Lenormand G, Richert A and Gallet F 1999 A new determination of the shear modulus of the human erythrocyte membrane using optical tweezers *Biophys. J.* **76** 1145–51

Herrmann H and Aebi U 1998 Structure, assembly, and dynamics of intermediate filaments *Subcell. Biochem.* **31** 319–62

Himelstein B P, Canete-Soler R, Bernhard E J, Dilks D W and Muschel R J 1994 Metalloproteinases in tumor progression: the contribution of MMP-9 *Invasion Metastasis* **14** 246–58

Hiramoto Y 1969a Mechanical properties of the protoplasm of the sea urchin egg. I. Unfertilized egg *Exp. Cell Res.* **56** 201–8

Hiramoto Y 1969b Mechanical properties of the protoplasm of the sea urchin egg. II. Fertilized egg *Exp. Cell Res.* **56** 209–18

Hochmuth R M 1993 Measuring the mechanical properties of individual human blood cells *J. Biomech. Eng.* **115** 515–9

Hochmuth R M 2000 Micropipette aspiration of living cells *J Biomech* **33** 15–22

Hohng S, Zhou R, Nahas M K, Yu J, Schulten K, Lilley D M and Ha T 2007 Fluorescence-force spectroscopy maps two-dimensional reaction landscape of the holliday junction *Science* **318** 279–83

Hosu B G, Jakab K, Banki P, Toth F I and Forgacs G 2003 Magnetic tweezers for intracellular applications *Rev. Sci. Instrum.* **74** 4158–63

Howard J, Hudspeth A J and Vale R D 1989 Movement of microtubules by single kinesin molecules *Nature* **342** 154–8

Hua W, Young E C, Fleming M L and Gelles J 1997 Coupling of kinesin steps to ATP hydrolysis *Nature* **388** 390–3

Huang H, Dong C Y, Kwon H-S, Sutin J D, Kamm R D and So P T C 2002 Three-dimensional cellular deformation analysis with a two-photon magnetic manipulator workstation *Biophys. J.* **82** 2211–23

Huang H, Sylvan J, Jonas M, Barresi R, So P T C, Campbell K P and Lee R T 2005 Cell stiffness and receptors: evidence for cytoskeletal subnetworks *Am. J. Physiol. Cell Physiol.* **288** C72–80

Hummer G and Szabo A 2005 Free energy surfaces from single-molecule force spectroscopy *Acc. Chem. Res.* **38** 504–13

Jackson J D 1975 *Classical Electrodynamics* (New York: Wiley) pp 281–2

Janmey P A 1991 Mechanical properties of cytoskeletal polymers *Curr. Opin. Cell Biol.* **3** 4–11

Janmey P A, Euteneuer V, Traub P and Schliwa M 1991 Viscoelastic properties of vimentin compared with other filamentous biopolymer networks *J. Cell Biol.* **113** 155–60

Janmey P A, Peetermans J, Zaner K S, Stossel T P and Tanaka T 1986 Structure and mobility of actin filaments as measured by quasielastic light scattering, viscometry, and electron microscopy *J. Biol. Chem.* **261** 8357–62

Janmey P A, Shah J V, Janssen K P and Schliwa M 1998 Viscoelasticity of intermediate filament networks *Subcell. Biochem.* **31** 381–97

Jarzynski C 1997 Nonequilibrium equality for free energy differences *Phys. Rev. Lett.* **78** 2690–3

Johnson K C, Clemmens E, Mahmoud H, Kirkpatrick R, Vizcarra J C and Thomas W E 2017 A multiplexed magnetic tweezer with precision particle tracking and bi-directional force control *J. Biol. Eng.* **11** 47

Johnson M D, Torri J A, Lippman M E and Dickson R B 1999 Regulation of motility and protease expression in PKC-mediated induction of MCF-7 breast cancer cell invasiveness *Exp. Cell Res.* **247** 105–13

Kaes J A, Strey H, Tang J X, Finger D, Ezzell R, Sackmann E and Janmey P A 1996 F-actin, a model polymer for semiflexible chains in dilute, semidilute, and liquid crystalline solutions *Biophys. J.* **70** 609–25

Katsantonis J, Tosca A, Koukouritaki S B, Theodoropoulos P A, Gravanis A and Stournaras C 1994 Differences in the G/total actin ratio and microfilament stability between normal and malignant human keratinocytes *Cell Biochem. Funct.* **12** 267–74

Kellermayer M S Z, Smith S B, Granzier H L and Bustamante C 1997 Folding-unfolding transitions in single titin molecules characterized with laser tweezers *Science* **276** 1112–6

Kießling T R, Stange R, Käs J A and Fritsch A W 2013 Thermorheology of living cells—impact of temperature variations on cell mechanics *New J. Phys.* **15** 045026

Kim J, Lee H and Shin S 2015 Advances in the measurement of red blood cell deformability: a brief review *J. Cell. Biotechnol.* **1** 63–79

Kim S J, Blainey P C, Schroeder C M and Xie X S 2007 Multiplexed single-molecule assay for enzymatic activity on flow-stretched DNA *Nat. Methods* **4** 397–9

Kirfel J, Magin T M and Reichelt J 2003 Keratins: a structural scaffold with emerging functions *Cell. Mol. Life Sci.* **60** 56–71

Kishino A and Yanagida T 1988 Force measurements by micromanipulation of a single actin filament by glass needles *Nature* **334** 74–6

Koenderink G H, Atakhorrami M, MacKintosh F C and Schmidt C F 2006 High-frequency stress relaxation in semiflexible polymer solutions and networks *Phys. Rev. Lett.* **96** 138307

Koffer A, Daridan M and Clarke G 1985 Regulation of the microfilament system in normal and polyoma virus transformed cultured (BHK) cells *Tissue Cell* **17** 147–59

Kollmannsberger P and Fabry B 2007 BaHigh-force magnetic tweezers with force feedback for biological applications *Rev. Sci. Instrum.* **78** 114301

Kundu T, Bereiter-Hahn J and Karl I 2000 Cell property determination from the acoustic microscope generated voltage versus frequency curves *Biophys. J.* **78** 2270–9

Kunschmann T, Puder S, Fischer T, Perez J, Wilharm N and Mierke C T 2017 Integrin-linked kinase regulates cellular mechanics facilitating the motility in 3D extracellular matrices *Biochim. Biophys. Acta Mol. Cell Res.* **1864** 580–93

Kuo S C and Sheetz M P 1992 Optical tweezers in cell biology *Trends Cell Biol.* **2** 116–8

Kuo S C and Sheetz M P 1993 Force of single kinesin molecules measured with optical tweezers *Science* **260** 232–4

Lange J R, Steinwachs J, Kolb T, Lautscham L A, Harder I, Whyte G and Fabry B 2015 Microconstriction arrays for high-throughput quantitative measurements of cell mechanical properties *Biophys. J.* **109** 26–34

Laurent V M, Henon S, Planus E, Fodil R, Balland M, Isabey D and Gallet F 2002 Assessment of mechanical properties of adherent living cells by bead micromanipulation: comparison of magnetic twisting cytometry vs optical tweezers *J. Biomech. Eng.* **124** 408–21

Leavitt J, Latter G, Lutomski L, Goldstein D and Burbeck S 1986 Tropomyosin isoform switching in tumorigenic human fibroblasts *Mol. Cell. Biol.* **6** 2721–6

Lekka M, Laidler P, Gil D, Lekki J, Stachura Z and Hrynkiewicz A Z 1999 Elasticity of normal and cancerous human bladder cells studied by scanning force microscopy *Eur. Biophys. J.* **28** 312–6

Lim C T, Zhou E H, Li A, Vedula S R K and Fu H X 2006 Experimental techniques for single cell and single molecule biomechanics *Mater. Sci. Eng.* C **26** 1278–88

Litvinov R I, Shuman H, Bennett J S and Weisel J W 2002 Binding strength and activation state of single fibrinogen–integrin pairs on living cells *Proc. Natl Acad. Sci. USA* **99** 7426–31

Liu Y, Sonek G J, Berns M W and Tromberg B J 1996 Physiological monitoring of optically trapped cells: assessing the effects of confinement by 1064-nm laser tweezers using microfluorometry *Biophys. J.* **71** 2158–67

Lodish H, Baltimore D, Berk A, Zipurski S L, Matsudaira P and Darnell J 1995 *Molecular Cell Biology* (New York: Scientific American Books) pp 1051–9

Lodish H B, Berk A, Zipursky S L, Matsudaira P, Baltimore D and Darnell J E 2000 *Molecular Cell Biology* (New York: W H Freeman and Company)

Lu H P, Xun L and Xie X S 1998 Single-molecule enzymatic dynamics *Science* **282** 1877–82

MacKintosh F C and Schmidt C F 1999 Microrheology *Curr. Opin. Colloid Interface Sci.* **4** 300–7

MacKintosh F C, Kaes J A and Janmey P A 1995 Elasticity of semiflexible biopolymer networks *Phys. Rev. Lett.* **75** 4425–8

Mahaffy R E, Shih C K, MacKintosh F C and Kaes J A 2000 Scanning probe-based frequency-dependent microrheology of polymer gels and biological cells *Phys. Rev. Lett.* **85** 880–3

Maniotis A J, Chen C S and Ingber D E 1997 Demonstration of mechanical connections between integrins, cytoskeletal filaments, and nucleoplasm that stabilize nuclear structure *Proc. Natl Acad. Sci. USA* **94** 849–54

Martin J D, Marhefka J N, Migler K B and Hudson S D 2011 Interfacial rheology through microfluidics *Adv. Mater.* **23** 426–32

Matthews B D, Overby D R, Alenghat F J, Karavitis J, Numaguchi Y, Allen P G and Ingber D E 2004 Mechanical properties of individual focal adhesions probed with a magnetic microneedle *Biochem. Biophys. Res. Commun.* **313** 758–64

Mazurkiewicz Z E and Nagorski R T 1991 *Shells of Revolution* (New York: Elsevier) pp 360–9

McNulty A L, Weinberg J B and Guilak F 2009 Inhibition of matrix metalloproteinases enhances *in vitro* repair of the meniscus *Clin. Orthop. Relat. Res.* **467** 1557–67

Mierke C T, Fischer T, Puder S, Tom Kunschmann T, Soetje B and Ziegler W H 2017 Focal adhesion kinase activity is required for actomyosin contractility-based invasion of cells into dense 3D matrices *Sci. Rep.* **7** 42780

Mierke C T, Frey B, Fellner M, Herrmann M and Fabry B 2011 Integrin $\alpha 5\beta 1$ facilitates cancer cell invasion through enhanced contractile forces *J. Cell Sci.* **124** 369–83

Mierke C T, Kollmannsberger P, Zitterbart D P, Diez G, Koch T M, Marg S, Ziegler W H, Goldmann W H and Fabry B 2010 Vinculin facilitates cell invasion into three-dimensional collagen matrices *J. Biol. Chem.* **285** 13121–30

Mierke C T, Kollmannsberger P, Zitterbart D P, Smith J, Fabry B and Goldmann W H 2008a Mechano-coupling and regulation of contractility by the vinculin tail domain *Biophys. J.* **94** 661–70

Mierke C T, Zitterbart D P, Kollmannsberger P, Raupach C, Schlötzer-Schrehardt U, Goecke T W, Behrens J and Fabry B 2008b Breakdown of the endothelial barrier function in tumor cell transmigration *Biophys. J.* **94** 2832–46

Mietke A, Otto O, Girardo S, Rosendahl P, aubenberger A, Golfier S, Ulbricht E, Aland S, Guck J and Fischer-Friedrich E 2016 Extracting cell stiffness from real-time deformability cytometry: theory and experiment *Biophys. J.* **109** 2023–203

Mijailovich S M, Kojic M, Zivkovic M, Fabry B and Fredberg J J 2002 A finite element model of cell deformation during magnetic bead twisting *J. Appl. Physiol.* **93** 1429–36

Mitchison T J 1992 Compare and contrast actin filaments and microtubules *Mol. Biol. Cell* **3** 1309–15

Mitchison T, Evans L, Schulze E and Kirschner M 1986 Sites of microtubule assembly and disassembly in the mitotic spindle *Cell* **45** 515–27

Mizuno D, Bacabac R, Tardin C, Head D and Schmidt C F 2009 High-resolution probing of cellular force transmission *Phys. Rev. Lett.* **102** 168102

Mizuno D, Head D A, MacKintosh F C and Schmidt C F 2008 Active and passive microrheology in equilibrium and nonequilibrium systems *Macromolecules* **41** 7194–202

Mizuno D, Tardin C, Schmidt C F and MacKintosh F C 2007 Nonequilibrium mechanics of active cytoskeletal networks *Science* **315** 370–3

Moeller W, Takenaka S, Rust M, Stahlhofen W and Heyder J 1997 Probing mechanical properties of living cells by magnetopneumography *J. Aerosol Med.* **10** 173–86

Mohandas N and Evans E 1994 Mechanical properties of the red cell membrane in relation to molecular structure and genetic defects *Annu. Rev. Biophys. Biomol. Struct.* **23** 787–818

Moustakas A and Stournaras C 1999 Regulation of actin organisation by TGF-beta in H-ras-transformed fibroblasts *J. Cell Sci.* **112** 1169–79

Bonakdar N, Gerum R, Kuhn M, Spoerrer M, Lippert A, Schneider W, Aifantis K E and Fabry B 2016 Mechanical plasticity of cells *Nat. Mater.* **15** 1090–4

Neuman K C, Chadd E H, Liou G F, Bergman K and Block S M 1999 Characterization of photodamage to *Escherichia coli* in optical traps *Biophys. J.* **77** 2856–63

Olins A L, Herrmann H, Lichter P and Olins D E 2000 Retinoic acid differentiation of HL-60 cells promotes cytoskeletal polarization *Exp. Cell Res.* **254** 130–42

Otto O, Rosendahl P and Mietke A *et al* 2015 Real-time deformability cytometry: on-the-fly cell mechanical phenotyping *Nat. Methods* **12** 199–202

Overby D R, Matthews B D, Alsberg E and Ingber D E 2005 Novel dynamic rheological behavior of individual focal adhesions measured within single cells using electromagnetic pulling cytometry *Acta Biomater.* **1** 295–303

Park S, Koch D, Cardenas R, Kaes J A and Shih C K 2005 Cell motility and local viscoelasticity of fibroblasts *Biophys. J.* **89** 4330–42

Pasternak C and Elson E L 1985 Lymphocyte mechanical response triggered by cross-linking surface receptors *J. Cell Biol.* **100** 860–72

Patel K V, Mohanty J G, Kanapuru B, Hesdorffer C, Ershler W B and Rifkind J M 2013 Association of the red cell distribution width with red blood cell deformability *Oxygen Transport to Tissue 34Advances in Experimental Medicine and Biology* 765 (Berlin: Springer) pp 211–6

Peters I M, van Kooyk Y, van Vliet S J, de Grooth B G, Figdor C G and Greve J 1999 3D single-particle tracking and optical trap measurements on adhesion proteins *Cytometry* **36** 189–94

Petersen N O, McConnaughey W B and Elson E L 1982 Dependence of locally measured cellular deformability on position on the cell, temperature, and cytochalasin B *Proc. Natl Acad. Sci. USA* **79** 5327–31

Pollard T D 1984 Molecular architecture of the cytoplasmic matrix *Kroc. Found. Ser.* **16** 75–86

Pollard T D 1986 Assembly and dynamics of the actin filament system in nonmuscle cells *J. Cell Biochem.* **31** 87–95

Pope L H, Bennink M L and Greve J 2002 Optical tweezers stretching of chromatin *J. Muscle Res. Cell Motil.* **23** 397–407

Prass M, Jacobson K, Mogilner A and Radmacher M 2006 Direct measurement of the lamellipodial protrusive force in a migrating cell *J. Cell Biol.* **174** 767–72

Pullarkat P A, Fernández P A and Ott A 2007 Rheological properties of the eukaryotic cell cytoskeleton *Phys. Rep.* **449** 29 53

Radmacher M 1997 Measuring the elastic properties of biological samples with the AFM *IEEE Med. Eng. Biol.* **16** 47–57

Radmacher M, Fritz M, Kacher C M, Cleveland J P and Hansma P K 1996 Measuring the viscoelastic properties of human platelets with the atomic force microscope *Biophys. J.* **70** 556–67

Ranga A, Gobaa S, Okawa Y, Mosiewicz K, Negro A and Lutolf M P 2014 3D niche microarrays for systems-level analyses of cell fate *Nat. Commun.* **5** 4324

Rao K M and Cohen H J 1991 Actin cytoskeletal network in aging and cancer *Mutat. Res.* **256** 139–48

Raz A and Geiger B 1982 Altered organization of cell-substrate contacts and membrane-associated cytoskeleton in tumor cell variants exhibiting different metastatic capabilities *Cancer Res.* **42** 5183–90

Raupach C, Zitterbart D P, Mierke C T, Metzner C, Müller F A and Fabry B 2007 Stress fluctuations and motion of cytoskeletal-bound markers *Phys. Rev.* E **76** 011918

Rehfeldt F and Schmidt C F 2017 Physical probing of cells *J. Phys. D: Appl. Phys.* **50** 463001

Rief M, Gautel M, Oesterhelt F, Fernandez J M and Gaub H 1997 Reversible unfolding of individual titin immunoglobulin domains by AFM *Science* **276** 1109–12

Roosen G 1977 A theoretical and experimental study of the stable equilibrium positions of spheres levitated by two horizontal laser beams *Opt. Commun.* **21** 189–95

Rotsch C, Jacobson K and Radmacher M 1999 Dimensional and mechanical dynamics of active and stable edges in motile fibroblasts investigated by using atomic force microscopy *Proc. Natl. Acad. Sci. USA* **96** 921–6

Rotsch C and Radmacher M 2000 Drug-induced changes of cytoskeletal structure and mechanics in fibroblasts: an atomic force microscopy study *Biophys. J.* **78** 520–35

Sacconi L, Romano G, Ballerini R, Capitanio M, De Pas M, Dunlap D, Giuntini M, Finzi L and Pavone F S 2001 Three-dimensional magneto-optic trap for micro-object manipulation *Opt. Lett.* **26** 1359–61

Sato M, Levesque M J and Nerem R M 1987 An application of the micropipette technique to the measurement of the mechanical properties of cultured bovine aortic endothelial cells *J. Biomech. Eng.* **109** 27–34

Sato M, Wong T Z and Allen R D 1983 Rheological properties of living cytoplasm: endoplasm of *Physarum plasmodium J. Cell Biol.* **97** 1089–97

Schlosser F, Rehfeldt F and Schmidt C F 2015 Force fluctuations in three-dimensional suspended fibroblasts *Philos. Trans. R. Soc.* B **370** 20140028

Schnitzer M J and Block S M 1997 Kinesin hydrolyses one ATP per 8-nm step *Nature* **388** 386–90

Schoenle A and Hell S W 1998 Heating by absorption in the focus of an objective lens *Opt. Lett.* **23** 325–7

Seltmann K, Fritsch A W, Kaes J A and Magin T M 2013 Keratins significantly contribute to cell stiffness and impact invasive behavior *Proc. Natl Acad. Sci. USA* **110** 18507–12

Shelby J P, White J, Ganesan K, Rathod P K and Chiu D T 2003 A microfluidic model for single-cell capillary obstrucstion by *Plasmodium falciparum*-infected erythrocytes *Proc. Natl Acad. Sci. USA* **100** 14618–22

Shepherd G M, Corey D P and Block S M 1990 Actin cores of hair-cell stereocilia support myosin motility *Proc. Natl Acad. Sci. USA.* **87** 8627–31

Sherman M E and Kurman R J 1996 The role of exfoliative cytology and histopathology in screening and triage *Obstet. Gynecol. Clin. North Am.* **23** 641–55

Shevkoplyas S S, Yoshida T, Gifford S C and Bitensky M W 2006 Direct measurement of the impact of impaired erythrocyte deformability on microvascular network perfusion in a microfluidic device *Lab Chip* **6** 914–20

Sidransky D 1996 Advances in cancer detection *Sci. Am.* **275** 104–9

Simmons R M, Finer J T, Warrick H M, Kralik B, Chu S and Spudich J A 1993 Force on single actin filaments in a motility assay measured with an optical trap *Adv. Exp. Med. Biol.* **332** 331–6

Sleep J, Wilson D, Simmons R and Gratzer W 1999 Elasticity of the red cell membrane and its relation to hemolytic disorders: an optical tweezers study *Biophys. J.* **77** 3085–95

Smith S B, Finzi L and Bustamante C 1992 Direct mechanical measurements of the elasticity of single DNA molecules by using magnetic beads *Science* **258** 1122–6

Soule H D, Vazguez J, Long A, Albert S and Brennan M 1973 A human cell line from a pleural effusion derived from a breast carcinoma *J. Natl. Cancer Inst.* **51** 1409–16

Stossel T P 1984 Contribution of actin to the structure of the cytoplasmic matrix *J. Cell Biol.* **99** 15s–21s

Stossel T P, Hartwig J H, Janmey P A and Kwiatkowski D J 1999 Cell crawling two decades after Abercrombie *Biochem. Soc. Symp.* **65** 267–80

Strey H, Peterson M and Sackmann E 1995 Measurements of erythrocyte membrane elasticity by flicker eigenmode decomposition *Biophys. J.* **69** 478–88

Strick T R, Allemand J F, Bensimon D, Bensimon A and Croquette V 1996 The elasticity of a single supercoiled DNA molecule *Science* **276** 1835–7

Svoboda K and Block S 1994 Force and velocity measured for single kinesin molecules *Cell* **77** 773–84

Svoboda K and Block S M 1994 Biological applications of optical forces *Annu. Rev. Biophys. Biomol. Struct.* **23** 147–285

Svoboda K, Schmidt C F, Schnapp B J and Block S M 1993 Direct observation of kinesin stepping by optical trapping interferometry *Nature* **365** 721–7

Tadir Y, Wright W H, Vafa O, Ord T, Asch R H and Berns W M 1990 Force generated by human sperm correlated to velocity and determined using a laser generated optical trap *Fertil. Steril.* **53** 944–7

Tait L, Soule H D and Russo J 1990 Ultrastructural and immunocytochemical characterization of an immortalized human breast epithelial cell line, MCF-10 *Cancer Res.* **50** 6087–94

Takahashi K, Hein V I, Junker J L, Colburn N H and Rice J M 1986 Role of cytoskeleton changes and expression of the H-ras oncogene during promotion of neoplastic transformation in mouse epidermal JB6 cells *Cancer Res.* **46** 5923–32

Taniguchi S, Kawano T, Kakunaga T and Baba T 1986 Differences in expression of a variant actin between low and high metastatic B16 melanoma *J. Biol. Chem.* **261** 6100–6

Thoumine O and Ott A 1997 Time scale dependent viscoelastic and contractile regimes in fibroblasts probed by microplate manipulation *J. Cell Sci.* **110** 2109–16

Tinoco I and Bustamante C 2002 The effect of force on thermodynamics and kinetics of single molecule reactions *Biophys. Chem.* **101–102** 513–33

Tskhovrebova L, Trinick J, Sleep J A and Simmons R M 1997 Elasticity and unfolding of single molecules of the giant muscle protein titin *Nature* **387** 308–12

Ugural A C 1999 *Stresses in Plates and Shells* (New York: McGraw-Hill) pp 339–409

Valberg P A and Albertini D F 1985 Cytoplasmic motions, rheology, and structure probed by a novel magnetic particle method *J. Cell Biol.* **101** 130–40

Valberg P A and Butler J P 1987 Magnetic particle motions within living cells. Physical theory and techniques *Biophys. J.* **52** 537–50

Valberg P A and Feldman H A 1987 Magnetic particle motions within living cells. Measurement of cytoplasmic viscosity and motile activity *Biophys. J.* **52** 551–61

van de Hulst H C 1957 *Light Scattering by Small Particles* (New York: Wiley) pp 172–6

Viani M B, Schaffer T E, Chand A, Rief M, Gaub H E and Hansma P K 1999 Small cantilevers for force spectroscopy of single molecules *J. Appl. Phys.* **86** 2258–62

Vlaminck I D and Dekker C 2012 Recent advances in magnetic tweezers *Annu. Rev. Biophys.* **41** 453–72

Walter N, Selhuber C, Kessler H and Spatz J P 2006 Cellular unbinding forces of initial adhesion processes on nanopatterned surfaces probed with magnetic tweezers *Nano Lett.* **6** 398–402

Wang E and Goldberg A R 1976 Changes in microfilament organization and surface topography upon transformation of chick embryo fibroblasts with Rous sarcoma virus *Proc. Natl Acad. Sci. USA* **73** 4065–9

Wang N and Stamenovic D 2000 Contribution of intermediate filaments to cell stiffness, stiffening, and growth *Am. J. Physiol. Cell Physiol.* **279** C188–94

Wang N, Butler J P and Ingberg D E 1993 Mechanotransduction across the cell surface and through the cytoskeleton *Science* **260** 1124–7

Wang R, Ding H, Mir M, Tangella K and Popescu G 2011 Effective 3D viscoelasticity of red blood cells measured by diffraction phase microscopy *Biomed. Opt. Express* **2** 485–90

Wang Q, Lee D, Sysounthone V, Chandraratna R A S, Christakos S, Korah R and Wieder R 2001 1,25-dihydroxyvitamin D3 and retonic acid analogues induce differentiation in breast cancer cells with function- and cell-specific additive effects *Breast Cancer Res. Treat.* **67** 157–68

Ward K A, Li W I, Zimmer S and Davis T 1991 Viscoelastic properties of transformed cells: role in tumor cell progression and metastasis formation *Biorheology* **28** 301–13

Weisenhorn A L, Khorsandi M, Kasas S, Gotzos V and Butt H J 1993 Functional imaging of early markers of disease *Nanotechnology* **4** 106–13

Whitesides G M 2006 The origins and the future of microfluidics *Nature* **442** 368–73

Wilhelm J and Frey E 2003 Elasticity of stiff polymer networks *Phys. Rev. Lett.* **91** 108103

Williamson J R, Gardner R A, Boylan C W, Carroll G L, Chang K, Marvel J S, Gonen B, Kilo C, Tran-Son-Tay R and Sutera S P 1985 Microrheologic investigation of erythrocyte deformability in diabetes mellitus *Blood* **65** 283–8

Wirtz H R and Dobbs L G 1990 Calcium mobilization and exocytosis after one mechanical stretch of lung epithelial cells *Science* **250** 1266–9

Woodside M T, Behnke-Parks W M, Larizadeh K, Travers K, Herschlag D and Block S M 2006 Nanomechanical measurements of the sequence-dependent folding landscapes of single nucleic acid hairpins *Proc. Natl Acad. Sci. USA* **103** 6190–5

Worthen G S, Schwab 3rd B, Elson E L and Downey G P 1989 Mechanics of stimulated neutrophils: cell stiffening induces retention in capillaries *Science* **245** 183–6

Wottawah F, Schinkinger S, Lincoln B, Ananthakrishnan R, Romeyke M, Guck J and Käs J 2005 Optical rheology of biological cells *Phys. Rev. Lett.* **94** 098103

Wouters F S, Vermeer P J and Bastiaens P I H 2001 Imaging biochemistry inside cells *Trends Cell Biol.* **11** 203–11

Wyckoff J B, Jones J G, Condeelis J S and Segall J E 2000 A critical step in metastasis: *in vivo* analysis of intravasation at the primary tumor *Cancer Res.* **60** 2504–11

Xie X S and Lu H P 1999 Single-molecule enzymology *J. Biol. Chem.* **274** 15967–70

Xu W, Mezencev R, Kim B, Wang L, McDonald J and Sulchek T 2012 Cell stiffness is a biomarker of the metastatic potential of ovarian cancer cells *PLoS One* **7** e46609

Yagi K 1961 The mechanical and colloidal properties of *Amoeba* protoplasm and their relations to the mechanism of amoeboid movement *Comput. Biochem. Physiol.* **3** 73–80

Yamazaki D, Kurisu S and Takenawa T 2005 Regulation of cancer cell motility through actin reorganization *Cancer Sci.* **7** 379–86

Ye T, Li H and Lam K 2010 Modeling and simulation of microfluid effects on deformation behavior of a red blood cell in a capillary *Microvasc. Res.* **80** 453–63

Zahalak G I, McConnaughey W B and Elson E L 1990 Determination of cellular mechanical properties by cell poking, with an application to leukocytes *J. Biomech. Eng.* **112** 283–94

Zhuang X W and Rief M 2003 Single-molecule folding *Curr. Opin. Struct. Biol.* **13** 88–97

Ziemann F, Raedler J and Sackmann E 1994 Local measurements of viscoelastic moduli of entangled actin networks using an oscillating magnetic bead micro-rheometer *Biophys. J.* **66** 2210–2

Chapter 4

Cell–cell and cell–matrix adhesion strength, local cell stiffness and forces

Summary

Cell–cell and cell–matrix adhesions are not only crucial to maintain tissue morphogenesis and provide homeostasis, they also activate signal transduction pathways, mediating many important cellular functions, including cell survival, gene expression, motility (single or collective), differentiation and adhesion. Gene mutations in adhesion receptors can evoke developmental disorders and cause different diseases such as cancer. Moreover, the cell–cell and cell–matrix adhesion strength influences the mechanical properties of cells, such as their cellular stiffness and contractile force generation. It is still under investigation which of the three main cytoskeletal structural components—actin microfilaments, keratin intermediate filaments or microtubules—contributes the most to the mechanical properties of cancer cells. Is it possible to dissect the contribution of each individual structural component to the overall mechanical properties? The answer to this question may be difficult to give, as the microfilaments can interact with intermediate filaments and microtubules, and this raises the question of whether one cytoskeletal structural component can replace another. However, quantitative methods to measure cell adhesion are therefore required to reveal how cells regulate cell–cell adhesion and cell–matrix during development and how aberrations in cell–cell and cell–matrix adhesion contribute to disease. In the following, this is described and how adhesion strength, cellular stiffness and forces can be measured with state-of-the-art biophysical methods is discussed. In particular, an overview of different *in vitro* techniques to study cell–cell and cell–matrix adhesion is presented such as quantitative single-cell methods based on atomic force microscopy (AFM) and a variant, AFM-based single-cell force spectroscopy (SCFS), dual micropipette aspiration (DPA), microdroplets functionalized with adhesion receptors, traction force microscopy and Förster resonance energy transfer (FRET)-based molecular

doi:10.1088/978-0-7503-1753-5ch4

tension sensors are introduced to visualize intracellular mechanical forces acting on cell–cell or cell–matrix adhesion sites.

4.1 Atomic force microscopy

AFM is basically a biophysical technique for acquiring images and other information from a wide variety of objects, such as living cells, at extremely high resolution in the nanometer range. The technique works by scanning a very sharp probe (end radius approximately 10 nm) along the object's surface. The scanning of the probe can be performed very carefully in order to keep the force between the cantilever carrying the probe and surface of the object of interest at a set, low level. In more detail, the probe is a silicon (Si) or silicon nitride (Si_3N_4) cantilever with a sharp integrated tip, and the vertical bending (termed deflection) of the cantilever due to the forces acting on the tip can be detected by a laser that is focused on the back of the cantilever. The laser is reflected by the cantilever and detected by a photodetector. The movement of the laser beam on the photodetector is a measure of the movement of the probe. This kind of setup is called an optical lever. During the measurement procedure, the probe is moved over the sample by a scanner (typically a piezoelectric element), which is able to make extremely precise movements. The extremely high resolution of the AFM is achieved through the combination of the sharp tip, the very sensitive optical lever and the very precise scanner movements, which are also combined with the precise control of probe–sample forces.

AFM is a type of very high-resolution scanning probe microscopy, with a resolution in the range of a nanometer, and it is thus more than 1000 times better than the optical diffraction limit (Hirschfeld and Huber 2010, Grazioso *et al* 2010, Lang *et al* 2004, Cappella and Dietler 1999). The precursor to AFM was the scanning tunneling microscope, developed by Gerd Binnig and Heinrich Rohrer in the early 1980s (earning them the Nobel Prize for physics). The AFM was invented by Binnig and the first experimental implementation was performed by Binnig, Quate and Gerber in 1986 (Binnig *et al* 1986). Three years later, the AFM became commercially available. It is one of the best tools for imaging, measuring and manipulating soft matter such as cells, tissues and proteins at the nano-scale. The surface of the sample is scanned by the mechanical probe (the cantilever). The highly precise scanning of the sample's surface is mediated by piezoelectric elements that facilitate tiny, accurate and precise movements on (electronic) command. In some special variations, electric potentials can be scanned using conducting cantilevers. In advanced types of AFM, the current can be passed through the tip to probe the electrical conductivity or transport of the underlying surface of the sample, but this is much more challenging, with few research groups reporting consistent data (Binnig *et al* 1986). The electron micrograph of a used AFM cantilever has an image width of approximately 100 μm and 30 μm.

An AFM can operate in one of three modes, the non-contact, contact and tapping modes, depending on the cantilever's tip motion.

1. The non-contact mode or frequency modulation AFM (dynamic mode) (termed NC-AFM, close contact AFM or FM-AFM).

2. The contact (static) mode (termed C-AFM or CMAFM).
3. The tapping (dynamic) mode (TMAFM), also termed intermittent contact (IC-AFM) or AC mode, vibrating mode or amplitude modulation AFM (AM-AFM).

The contact mode is often called a static mode. The tapping and non-contact modes are, by contrast, described as dynamic, because the cantilever is oscillated. This is done by adding an extra piezoelectric element, which oscillates up and down at somewhere between 5–400 kHz relative to the cantilever holder. There is still a little confusion over terminology, because the tapping mode was trademarked by one AFM manufacturer. Thus, this mode has many other names, such as the intermittent contact (IC-AFM), AC or vibrating mode, which are all the same. The modes of the AFM can also be named based on their detection mechanisms: the tapping mode is then called amplitude modulation AFM (AM-AFM) and the non-contact mode is called frequency modulation AFM (FM-AFM). The main difference between the tapping mode and non-contact modes (NC-AFM) is that in the former the tip of the probe of the cantilever touches the sample and moves completely away from it in each oscillation cycle. In NC-AFM, the cantilever stays close to the sample during the whole measurement and has a pronouncedly smaller oscillation amplitude. The NC-AFM is more sensitive to small oscillations of the cantilever and can be used in close contact (almost touching the sample). Compared to other AFM techniques, the NC-AFM has a true atomic resolution. The AM-AFM is typically used for the tapping mode, in which the cantilever tip taps the sample during each oscillation cycle. This has turned out to be the most stable mode for use in air and hence it is currently the most commonly used mode for most applications.

Nontopographic modes
In addition to the topographic modes that collect images, many modes have been reported to measure other properties of the sample, such as the magnetic field or the potential difference between the sample and the cantilever probe.

- Magnetic force microscopy (MFM) measures the distribution of the magnetic field within the sample.
- Kelvin probe microscopy (KPM) measures the contact potential difference across the sample.
- Force spectroscopy measures individual molecular interactions.
- The nanoindentation method measures the hardness or softness of the sample.
- Thermal modes determine thermal parameters, such as thermal conductivity on the nano-scale level.

The AFM consists of a cantilever with a sharp tip (termed the probe) at its end that is used to scan the surface of the sample. The cantilever consists of Si or Si_3N_4 and a tip radius of curvature on the order of nanometers. When the cantilever's tip is brought in close to a sample's surface, the forces between the tip and the sample deflect off the cantilever, according to Hooke's law. The forces measured in AFM include mechanical contact forces, van der Waals forces, capillary forces, chemical bonding, electrostatic forces, magnetic forces (MFM), Casimir forces and solvation forces. Together with force, additional quantities can be measured simultaneously through the use of specialized types of probes using scanning thermal microscopy, scanning joule expansion microscopy and photothermal microspectroscopy. The deflection of the cantilever carrying the probe is usually measured using a laser spot reflected from the top surface of the cantilever into an array of photodiodes. Other useful methods use optical interferometry, capacitive sensing or piezoresistive AFM cantilevers. These cantilevers are fabricated with piezoresistive elements acting as a strain gauge. For example, a Wheatstone bridge can be used to detect strain in the AFM cantilever due to deflection, but this method is not as sensitive as laser deflection or interferometry.

An AFM topographical scan of a glass surface can be performed. Thus, the features of the glass on the microscale and nano-scale can be determined and they represent the roughness of the glass surface. An image space in 3D (x, y, z) can be detected and the size is 20 μm × 20 μm × 420 nm. However, if the cantilever's tip is scanned at a constant height, there is a risk that the tip will stretch and damage the surface. In order to avoid this, a feedback mechanism is typically incorporated to adjust the tip-to-sample distance to have a constant force between the tip and the sample.

The tip or sample is mounted on a 'tripod' of three piezocrystals, with each being responsible for scanning in the x, y and z directions. In 1986, as the AFM was invented, a new piezoelectric scanner (termed the tube scanner) was developed for use in scanning tunneling microscopy (STM), which is based on the principle of quantum tunneling. Then, tube scanners were used for AFMs. In particular, a tube scanner is able to scan the sample in the x, y and z directions using a single tube piezo consisting of a single interior contact and four external contacts. A major advantage of the tube scanner is the better vibrational isolation that results, on the one hand, from the higher resonant frequency of the single-crystal construction and, on the other hand, from a low resonant frequency isolation stage. However, a disadvantage is that the x–y motion may lead to unwanted z motion, resulting in distortion (Binnig and Smith 1986).

The AFM can be operated in a number of modes. There are two major groups: static ('contact') modes and a variety of dynamic ('non-contact' or 'tapping') modes, in which the cantilever is vibrated.

What role does the cantilever's probe play?

The AFM probe is a widely distributed and well-established measuring device with a sharp tip on the free-swinging end of a cantilever, which protrudes from a holder plate. In more detail, the dimensions of the cantilever are in the range of micrometers and the radius of the tip is in the range of a few nanometers. The holder plate (termed the holder chip) often has a size of 1.6 mm by 3.4 mm, supporting the AFM probe with tweezers and hence it fits into the corresponding holder clips on the scanning head of the atomic microscope. This device is most commonly called the AFM probe, but other names include the AFM tip and

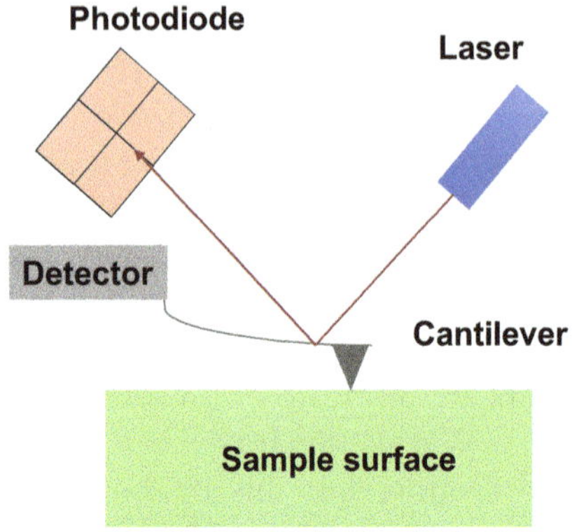

Figure 4.1. AFM principles.

cantilever (using the name of a single part as the name for the whole device) (figure 4.1). Taken together, an AFM probe is a particular type of scanning probe microscopy (SPM) device.

Most of the AFM probes used consistent of Si, although borosilicate glass and Si_3N_4 can also be used. During the AFM measurement the tip is very close to the surface of the object (sample) and then the cantilever is deflected by the interaction between the tip and the surface. Then a spatial map of the interaction between the cantilever and the sample is recorded by measuring the deflection at many points of a 2D surface. Indeed, several types of interaction can be detected. Depending on the interaction to be measured, the surface of the tip of the AFM probe needs to be modified with a coating such as extracellular matrix proteins. Gold is one possible coating that serves for covalent bonding of biological molecules and the detection of their interaction with a surface. In addition, diamond coatings can be used for increased wear resistance and magnetic coatings for detecting the magnetic properties of the investigated surface. Moreover, the surface of the cantilevers can also be altered. These coatings are used to increase the reflectance of the cantilever and subsequently to improve the deflection signal (Bryant *et al* 1988).

Non-contact mode AFM (also termed modulation AFM)

This mode is called the non-contact mode (NC-AFM), in which the tip of the cantilever does not touch the surface of the sample. In addition, this non-contact mode can also be named after the detection mode, and is then called frequency modulation AFM. In this mode, the cantilever can be oscillated at its resonant frequency, which is called the frequency modulation mode, or it can be oscillated above the resonant frequency, where the amplitude of oscillation is typically a few nanometers (less than 10 nm) down to a few picometers, which is then termed the amplitude modulation mode.

In particular, the van der Waals forces are strongest at a tip distance of 1–10 nm above the surface of the sample and they decrease when any other long-range force extends the resonance frequency of the cantilever above the surface of the sample. Moreover, this decrease in the resonant frequency is restored by using a feedback loop system in order to maintain a constant oscillation amplitude or frequency by adjusting the distance between the cantilever's tip and the sample's surface. In addition, measuring the tip-to-sample distance at each 2D data point (in the x and y dimensions) enables the scanning software to construct a topographic image of the sample's surface.

An advantage of non-contact mode AFM is that neither tip nor sample degradation effects occur, however, these effects have been observed in some cases after numerous scans with contact AFM. Therefore, non-contact AFM is preferable to contact AFM when soft samples such as living cells are measured. However, in the case of rigid samples there seems to be no difference between contact and non-contact images. If a few monolayers of an adsorbed fluid are located on the surface of a rigid sample, the images of contact and non-contact AFM are substantially different. In the contact mode the cantilever's tip will penetrate the liquid layer in order to measure the underlying surface, whereas in the non-contact mode the cantilever's tip will oscillate above the adsorbed fluid layer in order to image the fluid and the rigid surface.

The dynamic mode operation is based on frequency modulation, in which a phase-locked loop is utilized to track the cantilever's resonance frequency, whereas in the more common amplitude modulation a servo loop acts to keep the cantilever excitation to a defined amplitude.

In frequency modulation, alterations in the oscillation frequency deliver information about the interactions between the tip and the sample. The frequency can be detected with very high sensitivity and hence the frequency modulation mode allows us to use very stiff cantilevers. Another advantage is that stiff cantilevers provide stability, even very close to the surface, and hence this was the first AFM technique that revealed data with true atomic resolution under ultra-high vacuum conditions.

In amplitude modulation, alterations in the oscillation amplitude or phase deliver the feedback signal for imaging. In particular, in amplitude modulation alterations in the phase of oscillation are used to distinguish between different types of surface materials. The amplitude modulation can be performed in the non-contact as well as in the intermittent contact regime (tapping mode). In the dynamic contact mode, the cantilever is oscillated in such a manner that the distance between the cantilever's tip and the sample's surface is altered. In addition, amplitude modulation can be performed in the non-contact regime in order to image with atomic resolution by using pronouncedly stiff cantilevers and small amplitudes in an ultra-high vacuum environment.

The AFM beam deflection method

The beam deflection method is the most prominent one for cantilever deflection measurements. In more detail, laser light from a solid-state diode hits the back of the cantilever, is reflected off the back of the cantilever and is collected by a position-sensitive detector (PSD) that contains two closely spaced photodiodes. The output signals are collected with a differential amplifier. An angular displacement of the cantilever leads to an unequal light distribution between the two photodiodes, because one collects more light than the other. This produces an output signal that is based on the difference between the photodiode signals normalized by their sum, which is proportional to the cantilever deflection. In particular, it can detect cantilever deflections of less than 10 nm until the thermal noise provides the limitation. However, a long beam path of several centimeters can easily amplify alterations in the beam angle.

Other deflection measurement methods
Several other methods for measuring beam deflection have been established. They are described briefly in the following.

Piezoelectric detection. The cantilevers are made from quartz or another piezo-electric material and they are able to detect deflection directly as an electrical signal. Indeed, cantilever oscillations of up to 10 pm have been reported using this method.

Laser Doppler vibrometry. A laser Doppler vibrometer is used to obtain highly accurate deflection measurements for an oscillating cantilever, which is only performed in the non-contact mode. However, this method is expensive and hence rarely used.

The scanning tunneling microscope (STM). The STM is based on quantum tunneling. When a conducting tip is placed close to the sample's surface, a bias (voltage difference) applied between the two will enable electrons to tunnel through the vacuum between them. In particular, the resulting tunneling current is a function of tip position, the applied voltage and the local density of states of the sample. The current is detected as the tip's position scans across the sample's surface and is hence usually displayed as an image. The first atomic microscope used a complete STM system containing its own feedback mechanism to detect deflection. In particular, this method is difficult to implement and is also slow to respond to deflection alterations compared to the modern methods.

Optical interferometry. Optical interferometry is used to measure the deflection of the cantilever. Due to the nanometer scale deflections measured in AFM, the interferometer measures in the sub-fringe regime, hence, any drift even in the laser power or wavelength has a strong impact on the measurement and therefore must be omitted. For these reasons, optical interferometer measurements are performed with great care using index matching fluids between optical fiber junctions and very stable lasers. Thus, optical interferometry is rarely used as it is difficult to perform and expensive.

Piezoelectric (capacitive) detection. Metal-coated cantilevers can build a capacitor with another contact located behind the cantilever. In particular, deflection alters the distance between the contacts and can be measured as a change in capacitance.

Piezoresistive detection. This technique is similar to piezoelectric detection, but it uses piezoresistive cantilevers in order to record the detection. This is not often used as the piezoresistive detection dissipates energy from the system, affecting the quality factor Q of the resonance that characterizes how underdamped the resonator is (Roiter and Minko 2005).

Force spectroscopy. In addition to imaging, another major application of AFM is force spectroscopy, which is the direct measurement of tip–sample interaction forces as a function of the distance between the tip and sample. In this method, the AFM tip is extended towards and retracted from the sample's surface and in parallel the deflection of the cantilever is monitored as a function of piezoelectric displacement. Thus, a force–distance curve is recorded. These measurements have been used successfully to measure nano-scale contacts, atomic bonding, van der Waals forces and Casimir forces, dissolution forces in liquids and single-molecule stretching, and also rupture forces. In addition, AFM can be used to perform measurements in an aqueous microenvironment. For instance, the dispersion force due to a polymer adsorbed on the substrate can be determined. Indeed, forces of the order of a few piconewtons can be measured with a vertical distance resolution of better than 0.1 nm. Moreover, force spectroscopy can be performed with either static or dynamic modes. In the dynamic modes, information about the cantilever vibration is monitored in addition to the static deflection (Geisse 2009). The problems with this technique are the absence of the direct measurement of the tip–sample separation and the requirement for low-stiffness cantilevers, which tend to snap and stick to the sample's surface, when measuring soft material such as cells. These problems are solvable, as an AFM has been developed that directly measures the tip–sample separation. Hence, the snap-in can be reduced by measuring in liquids or by using stiffer cantilevers, whereas in the latter case a more sensitive deflection sensor is needed to monitor even small alterations in the cantilever deflection. In addition, by applying a small dither to the tip, the stiffness (in the form of a force gradient) of the bond can then be measured.

Atomic force microscopy applications. Force spectroscopy is used in biophysics to analyze the mechanical properties of living materials, such as tissues and cells.

Can even individual surface atoms be identified? Indeed, the AFM can be used to image and manipulate atoms as well as structures on many surfaces. The atom at the apex of the tip can sense individual atoms on the underlying surface by forming incipient chemical bonds with each atom. As these chemical interactions subtly alter the tip's vibration frequency, they can be detected and hence mapped. This principle has been used to distinguish between atoms of Si, tin and lead on an alloy surface by comparing these small atomic values to values obtained from large-scale density functional theory (DFT) simulations. The principle is to first measure these forces precisely for each type of atom expected to be located in the sample and then to compare these values with the forces obtained by DFT simulations. As expected, it has been found that the tip interacted most strongly with Si atoms and interacted 23% and 41% less strongly with tin and lead atoms, respectively. Thus, each different type of atom can be identified in the matrix when the cantilever's tip is moved across the surface (Gross *et al* 2009).

Contact mode

In this mode, the tip is moved (or dragged) across the surface of the sample and the contours of the surface are measured using either the deflection of the cantilever directly or, more commonly, the feedback signal, which is required to keep the cantilever at a constant position.

As the measurement of a static signal is subject to noise and drift, low-stiffness cantilevers are used to increase the deflection signal. However, in close proximity to the sample's surface the attractive forces are strong and hence lead to a 'snap-in' of the cantilever to the sample's surface. The contact mode AFM is usually performed at a depth where the overall force is repulsive, which is the case in firm contact with the solid surface below any adsorbed layers.

The tapping mode

Single polymer chains of a thickness of 0.4 nm can be recorded in a tapping mode under aqueous media with different pH values.

In ambient conditions, most samples develop a liquid meniscus layer. Thus, it is necessary to keep the probe tip close enough to the sample in order to detect short-range forces while the tip is prevented from sticking to the surface, which represents a major problem for the non-contact dynamic mode in ambient conditions. Thus, the dynamic contact mode (also termed the intermittent contact, AC or tapping mode) has been developed to solve this problem (Willemsen *et al* 2000).

In the tapping mode, the cantilever is driven to oscillate up and down close to its resonance frequency by a small piezoelectric element mounted in the AFM tip holder, similar to the non-contact mode. However, the amplitude of this oscillation is even larger than 10 nm and is typically between 100 and 200 nm. The force interactions acting on the cantilever when the tip approaches close to the surface include van der Waals forces, dipole–dipole interactions and electrostatic forces, which all cause the amplitude of this oscillation to decrease as the tip comes into close proximity with the sample. However, the height of the cantilever above the sample is precisely controlled by an electronic servo that uses the piezoelectric actuator. The servo adjusts the height to maintain the oscillation amplitude of the cantilever precisely when it is scanned over the sample. A tapping AFM image is hence gained by imaging the force of the intermittent contacts of the tip with the sample surface. This tapping method decreases the damage to the surface and the tip compared to the amount of damage caused in contact mode. Hence, the tapping mode is gentle enough even for the visualization of supported lipid bilayers or adsorbed single polymer molecules, such as 0.4 nm thick chains of synthetic polyelectrolytes under a liquid medium. With defined scanning parameters, the conformation of single molecules can remain unaltered for hours.

The first AFM

The AFM's usefulness has certain limitations, as is also the case for other biophysical devices. When determining whether analyzing a sample with an AFM is appropriate, there are various advantages and disadvantages that have to be considered.

Advantages. The AFM has several advantages over scanning electron microscopy (SEM). Compared to the electron microscope, which provides a 2D projection or a

2D image of a sample, the AFM provides a 3D surface profile. In addition, samples analyzed by AFM do not require any special treatment, such as metal/carbon coatings, which would irreversibly alter or even damage the sample, and they do not typically suffer from charging artifacts in the final image.

While an electron microscope needs an expensive vacuum environment for proper operation, most AFM modes can work perfectly well in ambient air or even a liquid environment, which allows us to investigate biological macromolecules and even living organisms. In principle, AFM can even provide a higher resolution than SEM. Moreover, the AFM has been shown to provide true atomic resolution in ultra-high vacuum and in liquid environments. Thus, high resolution AFM is comparable in resolution to STM and transmission electron microscopy. In addition, AFM can be combined with a variety of optical microscopy techniques such as fluorescent microscopy, which increases its applicability to a wide field of experiments. Indeed, combined AFM–optical instruments have been applied primarily in the biological sciences, but have also found a niche in some materials applications, such as those involving photovoltaics research (Nishida *et al* 2008).

Disadvantages. A disadvantage of AFM compared with SEM is the single scan image size. In particular, in one pass the SEM is able to image an area on the order of square millimeters with a depth of field on the order of millimeters, whereas the AFM can only image a maximum height in the order of 10–20 μm and a maximum scanning area of about 150 × 150 μm. An approach to improve the scanned area size for AFM is the use of parallel probes in a fashion similar to that of millipede data storage. In addition, the scanning speed of an AFM is a limitation. Traditionally, an AFM cannot scan images as fast as an SEM. An AFM requires several minutes for a typical scan, whereas an SEM is capable of scanning at near real time, although at relatively low quality. The relatively slow rate of scanning during AFM imaging mostly leads to thermal drift in the image, making the AFM less well suited for measuring accurate distances between topographical features on the image.

However, several fast-acting designs have been suggested to increase microscope scanning productivity, such as video AFM, which means that reasonable quality images are being obtained with video AFM at a video rate. This video rate is even faster than the average SEM. In order to eliminate image distortions induced by thermal drift, several methods have become available.

AFM images may also be affected by nonlinearity, hysteresis and creep of the piezoelectric material and cross-talk between the x, y, z axes, which seems to require software enhancement and filtering. However, filtering may flatten out real topographical features. Special AFMs utilize real-time correction software or closed-loop scanners, which practically eliminate these problems. Some AFMs also use separated orthogonal scanners (as opposed to a single tube), which additionally serve to eliminate some of the cross-talk problems. In addition, an AFM artifact arises from a tip with a high radius of curvature with respect to the special feature that is to be visualized.

Similar to any other imaging techniques, there is the possibility of image artifacts, which could be induced by an unsuitable tip, a poor operating environment or even by the sample itself. These image artifacts are usually unavoidable. However, their occurrence and effect on the results can be reduced through various methods. Artifacts resulting from a too coarse tip can be caused, for instance, by inappropriate handling or through collisions with the sample by either scanning too fast or having an unreasonably rough surface, which may cause wearing of the tip.

One major AFM artifact is steep sample topography. Due to the nature of AFM probes, they cannot normally measure steep walls or overhangs. However, specially produced cantilevers and AFMs can be used to modulate the probe sideways as well as up and down (as with dynamic contact and non-contact modes) to measure sidewalls, at the cost of more expensive cantilevers, lower lateral resolution and additional artifacts.

Piezoelectric scanners. AFM scanners are made from piezoelectric material, which expands and contracts proportionally to an applied voltage. Whether they elongate or contract depends upon the polarity of the voltage applied. The scanner is constructed by combining independently operated piezoelectrodes for x, y and z in a single tube, forming a scanner that can manipulate samples and probes with extreme precision in 3D. Moreover, independent stacks of piezos can be used instead of a tube, resulting in decoupled x, y and z movements.

Scanners are characterized by their sensitivity, which is the ratio of piezo movement to piezo voltage, for example, by how much the piezo material extends or contracts per applied volt. Due to differences in material or size, the sensitivity varies from scanner to scanner. In more detail, the sensitivity varies non-linearly with respect to the scan size. Piezo scanners exhibit more sensitivity at the end than at the beginning of a scan. In particular, this causes the forward and reverse scans to behave differently and display hysteresis between the two scan directions. However, this can be corrected by applying a non-linear voltage to the piezoelectrodes, which leads to a linear scanner movement and calibrates the scanner accordingly. One disadvantage of this approach is that it requires recalibration because the precise non-linear voltage needed for correct non-linear movement will alter as the piezo ages. However, this problem is easily solved by adding a linear sensor to the sample's or the piezo's platform, which is then able to detect the true movement of the piezo element. Thus, deviations from the ideal movement can be detected by the sensor and corrections applied to the piezo drive the signal to correct for a non-linear piezo movement. This design is known as a closed-loop AFM. In turn, nonsensored piezo AFMs are referred to as open loop AFMs.

It is common knowledge that the sensitivity of piezoelectric materials decreases exponentially with time. However, this causes most of the alterations in sensitivity to occur in the initial stages of the scanner's life. In particular, piezoelectric scanners are run for approximately 48 h before they are shipped from the factory so that they have passed the point where they may have large alterations in sensitivity. As the scanner ages, the sensitivity will alter less with time and the scanner needs to be recalibrated only in rare cases, although various manufacturer manuals recommend monthly to semi-monthly calibration of open loop AFMs (Lapshin 1998).

Receptor–ligand interactions are known to play a crucial role in biological systems and their measurement forms an important part of modern pharmaceutical development. Numerous assay formats are available that can be used to screen and quantify receptor ligands. An overview for both radioactive and non-radioactive assay technologies will be presented, with special emphasis on the latter. While radioreceptor assays are fast, easy to use and reproducible, their major disadvantage is that they are hazardous to human health, produce radioactive waste, require special laboratory conditions and are thus rather expensive on a large scale. This has promoted the development of non-radioactive assays based on optical methods like fluorescence polarization, fluorescence resonance energy transfer or surface plasmon resonance. In the light of their application in high-throughput screening environments, there has been special emphasis on so-called 'mix-and-measure' assays, which do not require separation of bound from free ligand.

The advent of recombinant production of receptors has contributed to the increased availability of specific assays and some aspects of the recombinant receptors expression will be covered briefly. Applications of receptor–ligand binding assays relate to screening and the quantification of pharmaceuticals in biological matrices. Drug discovery is a highly complex and costly process, which demands integrated efforts in several relevant aspects, involving innovation, knowledge, information, technologies and expertise. The shift from traditional to genomics- and proteomics-based drug research has fundamentally transformed the key strategies in the pharmaceutical industry addressed to the design of new chemical entities for drug candidates against a variety of biological targets. Thus, drug discovery has moved toward more rational strategies based on increased understanding of the fundamental principles of protein–ligand interactions. The combination of the knowledge of several 3D protein structures and of hundreds to thousands of small molecules has attracted attention for the application of structure- and ligand-based drug design approaches. In line with this, virtual screening technologies have substantially enhanced the impact of computational methods applied to chemistry and biology. The goal of applying such methods is to reduce large compound databases and to select a limited number of promising candidates for drug design. A novel concept for a protein–ligand docking simulator using virtual reality (VR) technologies, in particular tactile sense technology, has been designed and a prototype has been developed. Most conventional docking simulators are based on numerical differential calculations of the total energy between a protein and a ligand.

However, the basic concept of this method differs from that of conventional simulators. This particular design utilizes the force between a ligand and a protein instead of the total energy. The most characteristic function of the system is its ability to touch and sense the electrostatic potential field of a protein molecule. The surface of a protein of interest can be scanned using a globular probe, which is facilitated by the electrostatic charge and controlled through a force feedback device. The electrostatic force between the protein and the probe is calculated in real time and immediately fed back into the force feedback device. The operator can easily search interactively for positions where the probe is strongly attracted to the force field. These positions can be regarded as candidate sites where functional groups of ligands corresponding to the probe can bind to the target protein. However, certain limitations remain, for example only twenty protein atoms can be used to generate the electrostatic field. Moreover, the system can only use globular probes, excluding drug molecules or small chemical groups from being simulated.

4.1.1 Cellular stiffness

It is well established that cellular functions are essentially determined by structure. At different hierarchy levels the structural organization of cells is characterized by special mechanical properties (figure 4.2). It is obvious that the cell structure should be different for a variety of physiological processes, such as cell differentiation, growth and adhesion, and under pathogenesis, such as oxidative stress and attacks by viruses or parasites. For a long time, there have been two approaches for the investigation of cellular mechanical properties: (i) the cell mechanical properties are studied integrally, with the cell considered as a single whole object, and (ii) the mechanical properties of the cellular structural components are studied in detail, using isolated lipid bilayers, biomembranes and cytosolic proteins. However, it has become possible to probe the micro- and nanomechanical properties of cellular structures and to investigate the spatial distribution of the mechanical properties of these special cellular structures within a single-cell using the AFM technique (Hansma 2001). AFM is a novel method for high-resolution imaging of any surface, including those of living and fixed cells (Bischoff and Hein 2003, 2004, 2005). This powerful technique can also be used for the characterization of the mechanical, electrical and magnetic characteristics of samples to be studied both qualitatively and quantitatively. In particular, the AFM operation is based on the detection of repulsive and/or attractive surface forces. The interaction between the sample surface and a tip of the cantilever located very close to the sample surface corresponds to the forces between the atoms of the sample and those of the tip that scans its surface. The contrast of the image is provided by monitoring the forces of interaction between the tip and the surface. The fabricated tip is placed under a flexible cantilever responsible for the signal transduction. The interaction between

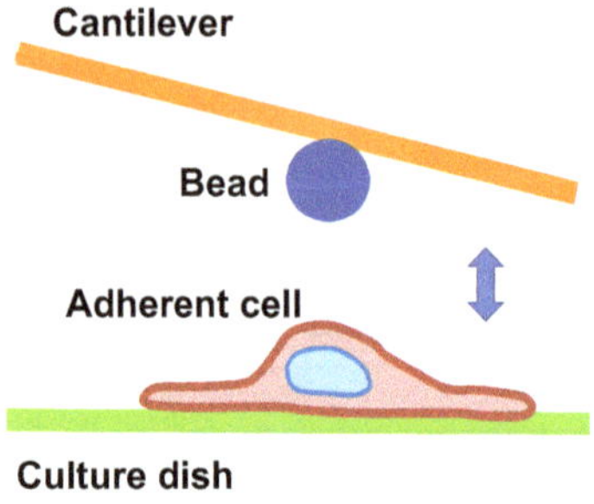

Figure 4.2. Adherent cell stiffness measurements. Using AFM cantilevers to which a 5.9 μm bead was glued, the stiffness of adherent cells can be measured.

the sample and the tip causes bending or twisting of the cantilever in a manner proportional to the interaction force. A small laser, which is focused on the cantilever, detects any bending or twisting of it. In more detail, the reflection of the laser beam is focused on a photodiode detector. Then the interaction of the sample with the tip is measured by the variation in the reflected beam's point of incidence on the photodiode. The deflection of the cantilever by interaction with features on the sample surface is monitored during scanning and is translated into a 3D image of the surface (Mozafari *et al* 2005).

An AFM can be operated in a number of diverse imaging modes, depending on the nature of the interaction between the tip and sample surface. The mechanical properties of the cell surface and the subsurface layers can be determined either by contact mode AFM techniques, such as force modulation, lateral force microscopy and force-curve analysis, or by phase imaging in tapping mode AFM, such as the intermittent or semi-contact modes.

Probing of the cell surface by AFM techniques can detect heterogeneities of the mechanical properties of the surface at the nanolevel and subsurface layers of cells. In particular, the resolution of AFM in air at the vertical direction is 0.1–0.5 nm and at the horizontal direction it is approximately 1–5 nm, depending on the rigidity of the sample. A horizontal resolution can be reached for living cells in an aqueous medium even at a range of several tens of nanometers due to the softness of the cell membrane. The thickness of cellular membranes is known to be approximately 5–10 nm. When analyzing the cell's heterogeneities using AFM techniques, it is possible to draw a picture of the cellular structure of certain specific regions within a single cell. However, the sensitivity and resolution of the AFM method depends on tip and cantilever characteristics such as radius, shape and material type (Alessandrini and Facci 2005).

The basic AFM technique for the quantitative study of the mechanical characteristics of cells and tissues is force spectroscopy (termed force-curve analysis). By recording the force value and the vertical deflection of the cantilever, the probe approaches the surface under investigation at a fixed point and thereby usually performs a force-curve analysis. Then, the force value versus the distance between the probe and the surface can be plotted for this case. The force curve contains information about long- and short-range interactions and forms the basis for estimating the sample's Young's modulus. There is a serious problem in estimating the absolute value of a cellular Young's modulus using AFM force-curve analysis, because of the question of which mechanical model is appropriate and should be chosen.

To date, the Hertz model has been used in the majority of studies devoted to the evaluation of the Young's modulus of cells. In principle, the Hertz model describes the simple case of the elastic deformation of two perfectly homogeneous smooth bodies touching each other under a specific load (Hertz 1881, Johnson 1985). The two important assumptions of the Hertz model are: (i) the indenter must have a parabolic shape and (ii) the indented sample such as a cell is assumed to be extremely thick in comparison to the indentation depth. The first assumption remains a valid one, if the spherical tip radius is much larger than the indentation depth ($h < 0.3R$) (Mahaffy *et al* 2000).

When the tip of an AFM nanoscope is approximated by a sphere with the radius R, then the force on cantilever $F(h)$ is given by equation (4.1):

$$F(h) = \frac{4\sqrt{R}}{3}E^{*}h^{3/2}, \tag{4.1}$$

where h is the depth of the indentation and E^{*} the effective modulus of a system tip sample, which is calculated from equation (4.2):

$$\frac{1}{E^{*}} = \frac{1-\nu_{\text{tip}}^{2}}{E_{\text{tip}}} + \frac{1-\nu_{\text{sample}}^{2}}{E_{\text{sample}}}, \tag{4.2}$$

in which E_{tip}, ν_{tip} and E_{sample}, ν_{sample} are the Young's modules (E) and the Poisson ratios (ν) for the materials of the tip and sample, respectively. However, if the material of the tip is considerably harder than the sample, equation (4.3) is used (Vinckier and Semenza 1998):

$$E^{*} \approx \frac{E_{\text{sample}}}{1-\nu_{\text{sample}}^{2}}. \tag{4.3}$$

The Sneddon's variation of the Hertz model (equation (4.4)) is used for the cone type tip of an AFM cantilever (Laurent *et al* 2005):

$$F(h) = \frac{2}{\pi}\tan\alpha\frac{E_{\text{sample}}}{1-\nu_{\text{sample}}^{2}}h^{2}, \tag{4.4}$$

where α is the half-opening angle of the AFM tip.

Although the indentation depth in the case of AFM probing of the cell is in the range of hundreds of nanometers, which is too high to be an appropriate depth for the Hertz model, it has been demonstrated in many studies that the Hertz model still describes the experimental results sufficiently well. The original Hertz theory did not allow for adhesion of the indenter to the material. Thus, the Hertz theory has been improved for that specific case (Johnson *et al* 1971). The Hertz model was mainly used for an estimation of the static Young's modulus of cells, whereas the dynamic Young's modulus was sometimes used for the characterization of cells' elastic properties (Mahaffy *et al* 2004). As the cell surface is heterogeneous, consisting of a network of cell membrane and submembrane structures, the cellular Young's modulus evaluation using the Hertz model assumes an error.

The second model used for investigating cell elastic properties is based on force spectroscopy data, but it is founded on the theory of elastic shells (ES) (A-Hassan *et al* 1998, Scheffer *et al* 2001, Timoshenko and Woinowsky-Krieger 1970). This theory considers cells as shells filled with liquid. In such an approach, the effective Young's modulus can be evaluated from the relationship between the effective Young's modulus, shell thickness and the bending modulus. A serious problem for such an evaluation procedure is determining the particular boundary conditions for the calculation of any constants involved in the main relation and defining the tip–sample contact radius. This has not yet been fully solved.

Among other theoretical models used in AFM-based evaluation of cells' mechanical features, the finite element model should be mentioned; it is the most popular model for analysis of elasticity in engineering (Ohashi *et al* 2002).

Hence, the elasticity parameter values calculated using various models differ from each other. For example, the mechanical properties of endothelial cells exposed to shear stress have revealed different Young's modulus values in respect of various models (Ohashi *et al* 2002). The Young's modulus values calculated using a finite element model appeared to be significantly higher: from 12.2 to 18.7 kPa with exposure to shear stress. The modulus value calculated using the Hertz model reflects the same tendency, however, different values are revealed, such as 0.87 and 1.75 kPa for control and sheared endothelial cells, respectively.

The AFM experimental approach known as force integration to equal limits (FIEL) mapping, used to produce quantitative maps of relative cell elasticity, was developed in 1998 (A-Hassan *et al* 1998). The FIEL theory assumes a simple relationship between the values of the work done by the AFM cantilever during an indentation and the elastic constants at different surface positions.

In more detail, the collection of force curves over a certain area ensures the development of the elasticity map of the cell's surface. The surface elasticity map can also be performed using either force modulation (static mode) or phase imaging (tapping mode) techniques. The characteristics of cantilever oscillation, such as the amplitude and phase shift, carry information about local elastic and friction properties of the sample in both modes. The image of the changes to the oscillation characteristics can be represented by a map of the relative mechanical properties of the cell surface.

However, indirect information regarding the elastic properties of the cell surface can be obtained by recording the lateral force map of the surface of interest. The AFM cantilever's lateral deflections, such as torsion, arise because of either alterations to the surface slope or the heterogeneity of the surface's frictional properties. The lateral force map is analyzed simultaneously with the sample's surface topography to elaborate the specificity of the elastic property map. In particular, the AFM is used to probe the elasticity of mammalian cells in order to investigate their temporal and spatial structural dynamics under physiological and pathological processes. A reliable parameter to compare different cell types under different experimental conditions, the Young's modulus, has been determined using AFM measurements.

Young's modulus and its AFM measurement in living cells

The elastic modulus value of living cells varies widely. It is evident that this reflects the real variability of the parameter and also the imperfection of AFM measurement methods, as well as the numerical estimation of cell elasticity. However, almost a decade of progress in this area allows the generalization of some methodological factors that have a significant influence on the Young's modulus value. The problems connected with cell specificity will be discussed below. Under a change of the external conditions, the elasticity of cell membranes alters much more strongly than the morphology of the cell.

The first factor is a question of AFM sample handling. Erythrocytes illustrate the progress in this acute area. The AFM method has been used to investigate both living (Nowakowski *et al* 2001, Kamruzzahan *et al* 2004) and fixed erythrocytes in either air (Gould *et al* 1990) or in a buffer (Butt *et al* 1990). Living cells are rather soft and delicate for AFM, when probing under physiological conditions. Their drying, freezing and fixing with chemical agents can improve the AFM images and AFM indentation results. However, these procedures alter the cell structure, viability and elasticity. The Young's modulus values for erythrocytes treated with 5% formalin solution are increased ten-fold (119.5 kPa) over those for viable (native) erythrocytes (16.05 kPa) (Mozhanova *et al* 2003). The transverse stiffness of cardiomyocytes is also increased by a factor of 16 after fixing with formalin (Shroff *et al* 1995). Comparing a variety of methods for preparing erythrocyte ghosts for AFM studies showed that air drying is not suitable even after fixation in glutaraldehyde (Takeuchi *et al* 1998). On the other hand, fixation may enhance the images of cell structures such as the cytoskeleton (Shroff *et al* 1995, Hofmann *et al* 1997). Moreover, the highest resolution for cells (such as 10 nm) may only be achieved in air, which assumes cell fixation before AFM probing. The high mobility of the erythrocyte shape in buffer solutions leads to smearing of the AFM image and the maximal spatial resolution of living erythrocytes becomes poor at about 200 nm (Mozhanova *et al* 2003).

The standard AFM technique for cell elasticity measurement is based on indentation of the cells firmly attached to the substrate. For reliable indentation results, a firm substrate–cell contact is required, which is a problem for nonadhered cells in solution. A good approach for the immobilization of native erythrocytes in a liquid is the attachment to a glass surface previously modified with poly-L-lysine solution. Poly-L-lysine provides accurate localization of red blood cells on the glass surface due to the electrostatic interaction between the negatively charged cell surface and the positively charged poly-L-lysine layer. However, poly-L-lysine can promote membrane rearrangement, with the formation of a specific membrane deformation pattern within the contact area (Dulinska *et al* 2006). A novel method of indentation for leukemia cells placed at special microwells has been presented (Rosenbluth *et al* 2006). This method provides the mechanical immobilization of cells, but it also has an influence on the estimated Young's modulus value. In this case, the cell deformation is described precisely by the elastic model based on Hertzian mechanics.

The second factor is related to the heterogeneity of the mechanical properties of cells. However, there are pronounced variations of the elastic modulus values at different cell regions. It has been reported that the elastic modulus value of the human umbilical vein endothelial cells is 7.22 kPa directly over the nucleus, dropping to only 2.97 kPa over the cell body in close proximity to the nucleus and 1.27 kPa on the cell body near the cell's edge (Mathur *et al* 2000). Moreover, it has been found that the cell body of bovine pulmonary artery endothelial cells is two- to three-fold softer compared to the cell periphery (Costa and Yin 1999). A corresponding study on cardiomyocytes also revealed that cells are softer at the nuclear region and become stiffer toward the periphery (Shroff *et al* 1995). Mapping of the Young's modulus across the living chicken cardiocytes indicated that the stress fibers are characterized by the presence of areas with a stiffness of 100–200 kPa embedded in softer parts of the cell, with elastic modulus values between 5 and 30 kPa (Hofmann *et al* 1997). The elasticity map images of living astrocytes such as the glial cells of nerve tissue showed that the cell membrane above the nucleus is softer (2–3 kPa) than the surroundings and that the cell membrane above ridge-like structures reflecting F-actin is stiffer (10–20 kPa) than the surroundings. In the elasticity map images of fixed astrocytes, the elasticity value of cells was found to be relatively uniform (200–700 kPa), irrespective of the inner structure of cells (Yamane *et al* 2000). The variation of the elastic modulus value within a single cell is often a subject of study in itself.

The third factor influencing elastic modulus measurement is connected with cell thickness. However, substrate contributions can be neglected if the AFM tip is never indented by more than 10% of the cell thickness (Mathur *et al* 2001). When the cell compartment under the investigation is very thin (<1000 nm), as in the case of a lamellipodium, it is necessary to implement special challenges for accurate measurement of its viscoelastic behavior. Thus, the reported AFM-based microrheology method ensures the estimation of the viscoelastic constants of thin parts of a cell (<1000 nm) and those of thick areas, applying two different models, one for well-adhered regions and another model for nonadhered regions (Mahaffy *et al* 2004).

Cellular elasticity and the functional mechanics of endothelial cells

Although the Young's modulus determined by AFM techniques must be assessed carefully as absolute values, it is also very useful as a relative parameter in special experiments. Thus, the Young's modulus can be used successfully to investigate a variety of cellular functions.

Vascular endothelial cells are an interesting system for analyzing cell mechanics and cytoskeletal dynamics. These cells are located in mechanically active micro-environments and need to withstand shear stress, blood pressure and any alteration in pressure due to breathing cycles. The earliest AFM work on living aortic endothelial cells studied the effects of shear stress on cellular organization and other factors that may have an impact on the mechanical response of cells to flow (Barbee *et al* 1994, Barbee 1995, Sato *et al* 2000, Ohashi *et al* 2002). It has been found that the local elastic parameters of aortic endothelial cells increase significantly (from 0.87 kPa to 1.75 kPa) with exposure to shear stress (Ohashi *et al* 2002). The average elastic modulus values of bovine pulmonary artery endothelial cells (BPAECs) are in a similar range of 0.2–2 kPa (Pesen and Hoh 2005). However, a difference in the mechanical properties of rabbit endothelium has been demonstrated (Miyazaki and Hayashi 1999). Moreover, cells are stiffer in the medial wall of aortic bifurcation compared to the endothelium located in the lateral wall. Using AFM and human umbilical vein endothelial cells (HUVECs), it has been shown that the cell responds globally to the localized applied force over the cell edge and the nucleus (Mathur *et al* 2000). Thus, it has been concluded that the nuclear region of the cell appears to be stiffer than the rest of the cell body, although the nucleus appears to be offset from the basal surface. In contrast, the focal adhesion movement upon the

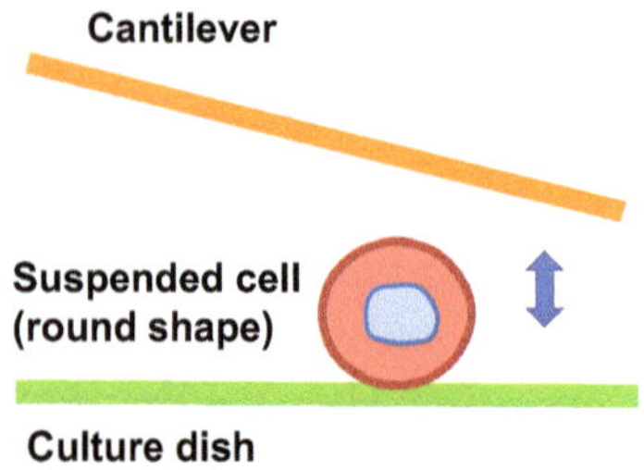

Figure 4.3. Suspended cell stiffness measurements. Using AFM flat and uncoated cantilevers the stiffness of suspended round cells can be measured by pressing the cantilever onto the cell surface.

apical cell surface perturbation suggests a connection between the nucleus and the focal adhesions through the cytoskeleton (Mathur *et al* 2000).

Besides measuring of adherent cells, it is also possible to prove the mechanical properties of non-adherent cells, in order to determine the adhesion-independent mechanical properties of cells. To determine the stiffness of suspended cells, a flat cantilever is used (figure 4.3). Indeed, the cell stiffness in suspended cells can be measured and reveals different values dependent on the cell type and the migratory capacity of the cells (Mierke *et al* 2017, Fischer *et al* 2017, Kunschmann *et al* 2017). In particular, it has been shown that kinase function deficient focal adhesion kinase ($FAK^{R454/R454}$) cells displayed higher stiffness values, whereas these cells are less invasive compared to FAK control cells (Mierke *et al* 2017). In contrast, when these cells are measured as adherent cells with a cantilever to which a 5.9 μm bead is glued and then coated with fibronectin, the $FAK^{R454/R454}$ cells are softer compared to FAK control cells. This result is in line with integrin linked kinase knock-out cells ($ILK^{-/-}$), which displayed as adherent cells a lower stiffness compared with ILK control cells. In addition, both $FAK^{R454/R454}$ and $ILK^{-/-}$ cells are less invasive compared to their respective control cells (Mierke *et al* 2017, Kunschmann *et al* 2017). However, invasive MDA-MB-231 human breast cancer cells exhibit as suspended and adherent cells less stiffness compared to less invasive MCF-7 human breast cancer cells or healthy control MCF-10A human epithelial cells using both types of AFM measurements (Fischer *et al* 2017).

4.1.2 Adhesion forces between cells

Historically, cell–cell adhesion has been studied through fluorescent microscopy and through molecular biological approaches and biochemical methods (Behrens *et al* 1989). However, quantification of cell–cell adhesion forces was not yet possible. A classical micropipette aspiration method has now been used to address quantitatively the probe's cell–cell adhesion strength or the cellular stiffness of suspended cells (figure 4.4) (Chu *et al* 2014). The microaspiration (see section 4.5 below) method has delivered quantitative datasets, but this system has been restricted to a limited number of cells and to sophisticated biophysical laboratories, and hence it is not suitable as a routine device and thus the experimental throughput has been restricted to small numbers of repeated experiments.

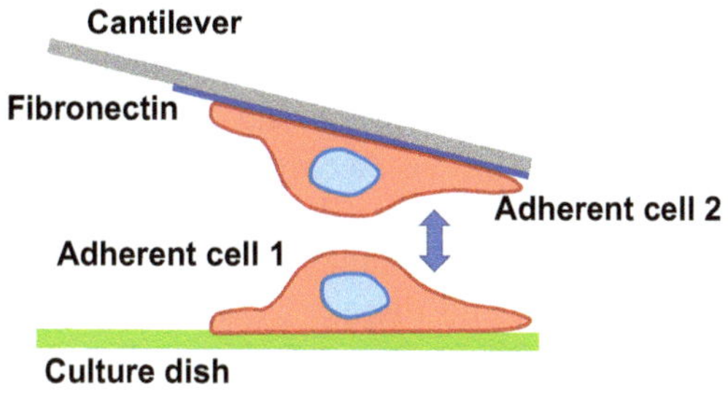

Figure 4.4. Cell–cell adherent force measurements of two adherent cells. Using AFM flat cantilevers to which a cell was attached, the cell–cell adhesion forces between two adherent cells can be measured by pressing the two cells together for a certain time. In the next step the cell–cell adhesions were ruptured by retracting the cantilever until all connections were broken.

In order to investigate cell–cell adhesions the AFM technique can be used. This AFM method was developed in 1986 (Binnig and Quate 1986). Its advantage is that structures at atomic resolution and the forces between them can be measured. A major advantage it has over STM is that insulators can also be analyzed (Binnig and Quate 1986). However, cell–cell or cell–matrix adhesion strength measurements with an AFM are restricted to the cantilever type used, as the amount of forces that can be applied to rupture the cell–cell or cell–adhesion connections depends on the material stiffness of the specific cantilever. Moreover, the distance (height) between the cell and the cantilever determines whether bond ruptures can be measured or not, as the bonds need to be fully ruptured solely by increasing the distance between the cantilever and the cell. The cell–cell adhesion technique uses a cell glued to the cantilever tip by binding to fibronectin that has been coated on the cantilever tip prior to cell adhesion to the cantilever. The cell attached to the cantilever serves as a probe for the adherent cells in the culture dish. In the next step, the cantilever is brought in close contact to an adherent cell and is pressed against it with a certain amount of force, for example 0.5 nN up to 2.5 nN, and incubated for 20 s or even much longer in order to establish cell–cell adhesions (Mierke *et al* 2017, Kunschmann *et al* 2017).

Quantitative analysis of cellular interactions with the extracellular microenvironment is necessary to reveal how cells regulate adhesion during the development and maintenance of multicellular organisms and how alterations in cell adhesion may cause diseases. Indeed, a practical guide to quantify the adhesive strength of living animal cells to various substrates using AFM-based single-cell force spectroscopy (SCFS) has been developed (Friedrichs *et al* 2013). How can we control cell states and attachment to the AFM cantilever? How can we functionalize the cantilever in order to perform reliable SCFS measurements? What role do various receptor expressions on the surfaces of two different cells types play? How can we conduct adhesion measurements in a particular cell? How can we analyze and interpret the recorded SCFS data? This section will answer all these questions and is intended to assist researchers in performing reliable AFM-based SCFS measurements.

The adhesive interactions of cells with their microenvironment trigger signal transduction pathways, which regulate important cellular processes such as cell migration, gene expression, cell survival, tissue organization and differentiation (Legate *et al* 2009, Weber *et al* 2011). Accordingly, mutations in genes encoding adhesion receptors (Fässler *et al* 1996, Sheppard 2000) or adhesion-associated components (Monkley *et al* 2000, Montanez *et al* 2008, Dowling *et al* 2008) may lead to developmental disorders and diseases. Consequently, methods that enable the characterization of cell adhesion are important for cell biological, clinical, pharmaceutical, biophysical and biomaterials research, such as tissue engineering and regeneration.

Due to the central importance of these processes, a variety of assays have been developed and established in order to characterize cell adhesion. Among these, the SCFS methods are most suitable for directly quantifying cell adhesion forces at the cellular level and on length scales discriminating between the contribution of single molecules (Benoit *et al* 2000, Helenius *et al* 2008, Taubenberger *et al* 2007). A brief overview of the most common methods applied to characterize cell adhesion is presented, which focuses mainly on AFM-based SCFS (AFM–SCFS). In particular, the technical basis of AFM–SCFS, with an emphasis on clear descriptions of experimental procedures and pitfalls, is discussed and, moreover, how AFM–SCFS data can be evaluated and analyzed is described in detail. This background information will help in understanding the principles underlying AFM–SCFS, as well as the possibilities and limitations of the method for the quantification of cellular interactions.

Biophysical methods applicable to characterize cell adhesion

Insight into the mechanical interactions between cells and their microenvironment can be obtained using different artificial cell culture substrates. Evidence has been provided that migrating cells exert compressive forces on culture substrates, for example through the finding that fibroblasts introduce wrinkles into thin Si rubber film substrates (Harris *et al* 1980). With knowledge of the substrate's stiffness and the length of the microscopic wrinkles, the forces exerted by the cells can be estimated on the order of nanonewtons (nN) (Beningo *et al* 2002). In order to reveal even local deformations, the spatial resolution of this wrinkling substrate approach was further improved by embedding micrometer-sized beads into the substrate (Oliver *et al* 1998). In line with this, other studies showed that dynamic deformations of various other substrates were the result of contractile and adhesive cellular forces (Burton and Taylor 1997, Lee *et al* 1994). Although these substrate-based methods were revealed to be important for understanding the mechanical interactions between cells and their microenvironment, they do not directly provide values for the adhesion strength of the cells and their substrates.

Methods for examining cell adhesion strength essentially focus on measuring the capability of cells to remain attached when exposed to an external detachment force. The most common adhesion assay, the plate-and-wash assay, relies on seeding cells onto substrates of interest, washing off 'non-adherent' cells with physiological buffers after a certain adhesion time interval and counting the remaining adherent cells (Klebe 1974). Plate-and-wash assays have identified key adhesion components and generated valuable insights into the mechanisms regulating adhesion (Amano *et al* 1997, Ridley and Hall 1992, Sieg *et al* 2000). However, these specific assays provide no information on adhesion strength and hence report only the initial rate of attachment of cells to the substrate, as the formation of more than ten receptor–ligand bonds is sufficient to prevent their removal from the plate (Boettiger and Wehrle-Haller 2010).

However, several semi-quantitative adhesion assays have been developed to apply controlled shear stress to adherent cells. For instance, in special flow chamber assays, shear stress is exerted on cells by a homogenous buffer flow (Kaplanski *et al* 1993). In spinning-disc assays, both controlled centrifugal forces and shear flow generated by rotation are applied to adherent cells (Garcia *et al* 1997b). Both assays are reported to provide reproducible and controllable results. Nevertheless, these techniques have limitations, as the resistance of cells to detachment by flow and centrifugal forces depends not only on the number, distribution and strength of the adhesion bonds present, but also on the spreading area and the surface topography of these cells. Thus, the adhesive strength of the cells to their substrate can only be estimated and not directly measured.

In order to quantitatively analyze the interaction forces of cells with given substrates, sensitive SCFS assays can be performed. In brief, SCFS assays allow adhesive interaction forces and binding kinetics to be measured under physiologically relevant conditions on the cellular level as well as on the single-molecule level. In micropipette-based manipulation assays a single cell, which is caught through the suction pressure at the tip of a micropipette, is brought into contact with an adhesive surface and finally retracted to measure the adhesive forces, which have been established during the adhesion time. Several micropipette-based experimental techniques that operate at both cellular and molecular levels have been developed, including the step-pressure technique (Sung *et al* 1986), the biomembrane force probe (Evans *et al* 1995) and the micropipette aspiration technique (Evans *et al* 1976). All these methods were applied to investigate surface receptor expression, membrane tether formation from single cells and single molecule or bond dynamics (Shao and Hochmuth 1996). However, the disadvantage of these micropipette-based techniques is either the low force resolution (nN range; step-pressure technique) or the low detectable forces (from approximately 10 pN to approximately 1 nN of a biomembrane force probe). In addition, optical tweezers, which trap nano- or micrometer-sized particles in the center of a laser focus, can in principle be employed to analyze cell–substrate interactions (Thoumine *et al* 2000, Choquet *et al* 1997). However, this method is restricted because of the difficulty of measuring forces higher than 100–200 pN. Among all of these assays, AFM-based SCFS is currently the most versatile method to analyze the adhesive interactions of cells with other cells, proteins and surfaces. This is because SCFS offers a large range of detectable forces, from 10 pN to approximately 100 nN and offers precise spatial (1 nm to 100 μm) and temporal (0.1 s to more than 10 min) control over the adhesion experiment and the experimental parameters (Helenius *et al* 2008).

The development and applications of SCFS

Initially, AFM-based SMFS was used to investigate the interaction of isolated receptors attached to the AFM cantilever, with ligand-decorated surfaces (Florin *et al* 1994), and also to measure the interaction of cantilever-linked ligands with receptors bound to a supporting surface (Lehenkari and Horton 1999). For these SMFS experiments the tip of the AFM is modified with a ligand (or receptor) of interest, then the tip is brought forward until the ligand (or receptor) on the tip binds to the receptor (or ligand) attached to the supporting surface. After a certain contact time the tip is withdrawn in order to rupture the receptor–ligand bond. In more detail, the rupture force is detected by the deflection of the AFM cantilever to which the tip is attached. The same principle of measurement is used to quantify local receptor–ligand interaction forces on a cellular surface (Grandbois *et al* 2000). Based on this design principle, a SCFS setup has been developed where the overall adhesion of the cantilever's tip to an immobilized and adherent cell is measured (Lehenkari and Horton 1999). As expected, this setup has certain restrictions. In order to overcome these limitations, an inverted SCFS setup has been developed (Benoit *et al* 2000). In more detail, this SCFS setup is simply inverted, as a living cell is attached to an AFM cantilever, thus converting the cell into a 'probe'. Using this inverted setup the interactions of this 'cellular probe' with a sample of interest can be quantified (Thie *et al* 1998, Li *et al* 2003, Puech *et al* 2005).

The SCFS was designed a long time ago as a tool to quantify cell adhesion (Benoit *et al* 2000, Helenius *et al* 2008, Benoit and Gaub 2000, Krieg *et al* 2008). Pioneering experiments have revealed the adhesion strength between two cells of *Dictyostelium discoideum* to single-molecule resolution (Benoit *et al* 2000). These experiments have been performed for important proteins regulating the differential, adhesive behavior of zebrafish mesendodermal progenitor cells to fibronectin, which led to an insight into germ layer formation and separation (Puech *et al* 2005). Subsequently, the adhesiveness between gastrulating zebrafish cells derived from different germ layers was quantified (Krieg *et al* 2008). In this manner, the mechanisms underlying cellular sorting during gastrulation can be analyzed and, in particular, the contribution of diverse cell adhesion strengths and cell cortex tensions in germ layer organization can be exhibited. In another series of experiments, the $\alpha2\beta1$ integrin facilitated adhesion of Chinese hamster ovary (CHO) cells to nanoscopically structured collagen type I matrices was investigated using recombinant knock-out cells and blocking substrates (Taubenberger *et al* 2007). In still another set of experiments, the role of the integrin activator tissue-type plasminogen activator (TPA) in strengthening integrin–cytoskeleton connections and enhancing $\alpha2\beta1$-integrin avidity was determined (Tulla *et al* 2008). These examples and another example (Helenius *et al* 2008) indicated the applicability of SCFS to reveal the dynamic strengthening of cell adhesive bonds. The characterization of how cells form cell adhesions during the initial binding of individual cell adhesion molecules, cluster on the cell surface and provide adhesion strengthening is essential to understand the establishment and regulation of cell adhesion (Taubenberger *et al* 2007, Cuerrier *et al* 2009, Friedrichs *et al* 2007, 2008, Selhuber-Unkel *et al* 2008). However, the spatial assembly of, for instance, extracellular matrix molecules also plays a role in cell adhesion, spreading and differentiation (Trappmann *et al* 2012). In this respect, the importance of molecular spacing has been investigated in modulating adhesion strength by measuring the force required to harvest an adhering cell from a nanopatterned surface by an adhesive cantilever (Selhuber-Unkel *et al* 2010). Taken together, these studies clearly demonstrate the versatility of SCFS in investigating different aspects of cell adhesion.

Experimental configurations of SCFS measurements
As mentioned earlier, there are two basic configurations for SCFS measurements, which are distinguishable in the relationship between the cell and the cantilever.

(1) A cell adhered to a substrate is touched by a functionalized tipless AFM cantilever (alternatively a pointed tip or a micrometer-sized bead may be used) for a defined contact time with a defined force and is then retracted from the cell (Grandbois *et al* 2000, Munoz Javier *et al* 2006).
(2) A cell adhered to an AFM cantilever is brought into contact with functionalized substrates for a defined contact time and is then retracted from the substrate. This configuration can also be used to determine the cell–cell adhesions between two cells, when the cell-carrying cantilever is placed onto an adherent cell for a certain time interval.

In more detail, configuration (1) has the advantage of testing one type of cantilever functionalization on several cells and thus decreases the influence of outlier cells. However, there are still various disadvantages to this approach. First, cellular contacts often leave behind debris, which may contaminate the functionalized cantilever tip and thus reduce the specificity of the probed interaction (Culp *et al* 1979, Cohen *et al* 2003). Consequently, for long-term cell adhesion contacts this method provides just one reliable force measurement. In the case of highly structured cell surface topographies, such as filopodia, microvilli, microridges and lipid rafts, the matching of the geometry of the cantilever's tip surface (pyramid, cone or sphere) with the local geometry of the cell's surface strongly affects the effective surface area interacting between the cantilever's tip and cell and subsequently the determined adhesive forces. Finally, adherent cells show diverse behavior in the spreading area and cellular polarization, which can have pronounced effects on the local density and identity of the cell surface receptors. However, seeding cells onto defined micropatterned substrates may reduce the variability in cell spreading and polarization, whereas the risk of cantilever contamination by debris still remains. For these reasons, this configuration may not increase the number of reliable data points, which is very important for measuring long-term adhesion contacts. Nevertheless, for short and weak contacts this approach can assist in mapping local adhesion spots on the cells (Grandbois *et al* 2000). A special application of this configuration is the harvesting of already adherent cells by addressing a cell with a cantilever, which has an even greater affinity for the cell compared to the cell's underlying adhesive substrate (Selhuber-Unkel *et al* 2010). This special format supports the investigation of cell adhesion over time points ranging from minutes to hours prior to the detachment measurement.

Configuration (2) has the advantage that the same cell can be probed at different surfaces and even at different local spots of the same surface in order to overcome the problem of contamination by cell debris left from prior adhesion events. In more detail, the cell attached to the cantilever is not allowed to spread on it, because this flat configuration of the cell would then also eliminate differences in cell area and polarization. The disadvantages of this configuration are the time-consuming and difficult additional procedure of immobilizing a living cell to the cantilever and the possibility of influencing the overall state of the cell through the adhesive contact to the functionalized AFM cantilever, which indeed occupies adhesion receptors (Friedrichs *et al* 2010). Moreover, single cells can be in different states regarding cell cycle and differentiation and thus display distinct adhesive properties that may be difficult to compare. This fact requires the control of the state of the cells as tightly as possible by ensuring a homogeneous preparation protocol for each cell type and state, and also that many measurements from numerous different cells need to be collected in order to obtain an 'average' adhesion response.

A special application of configuration (2) is the investigation of cell–cell adhesion by approaching a second cell or even a whole cell layer as well as tissue resection sample as a substrate with a cantilever carrying an attached cell. The advantages of cell–cell measurements are the perfectly prepared surfaces, regarding the orientation, functionality and natural microenvironment of the interacting molecules. However, cells usually interact through many molecules of various kinds, thus masking the signal arising from the interaction of interest through a mixture of intercellular cell–cell bonds. Moreover, variable responses from even two individual cells, rather than only one cell, contribute to enhanced variability in the collected data. Nevertheless, for some experiments the exact cellular contributions are the subject of investigation.

4.1.3 Adhesion forces between a cell and the extracellular matrix

Cell adhesion is the central mechanism that ensures the structural integrity of tissues. It is usually studied in cell culture, where the most prominent cell–matrix adhesion structures are the so-called focal adhesions or focal contacts (Gumbiner 1996). Focal contacts consist of large patches of transmembrane adhesion receptors such as the integrin family (Critchley 2000). These integrin patches within the cell membrane can have lateral sizes of several micrometers. On the extracellular side, integrins bind to ligands, such as the extracellular matrix protein fibronectin (Ruoslahti 1996). On the intracellular side, the receptors are linked to the actin cytoskeleton via a cytoplasmic plaque composed of many diverse proteins, such as talin, vinculin, paxillin and α-actinin (Zamir and Geiger 2001). This connection of the extracellular matrix to the cytoskeleton, which is often organized in the form of stress fibers, allows the transmission of forces between the cells and the extracellular matrix through focal contacts. A key player for force generation and transmission is vinculin (Mierke *et al* 2008b, Mierke 2009, Mierke *et al* 2010).

Cells use focal adhesions or focal contacts to integrate biochemical and mechanical information about their microenvironment and regulate the organization of their cytoskeleton. In more detail, cell behavior is largely affected by the mechanical properties of the extracellular matrix and extracellular mechanical signals. For instance, the stiffness of elastic substrates provides their focal contact formation, cell elasticity, cell migration and cell differentiation (Discher *et al* 2005, Engler *et al* 2004a, 2004b, 2004c, Pelham and Wang 1997, Solon *et al* 2007). When intracellular force generation is inhibited by the myosin II inhibitor blebbistatin, these effects are eliminated (Discher *et al* 2005). On the other hand, stimulating cells by an external shear force can initiate a substantial growth of focal contacts (Paul *et al* 2008, Riveline *et al* 2001). These observations demonstrate the mechanosensitive nature of focal contacts and their importance for cellular decision-making. Mechanosensing through focal contacts seems to be based on a mechanobiochemical feedback loop involving force-induced signal transduction through small GTPases from the Rho family and activation of force generation in the actin cytoskeleton (Geiger and Bershadsky 2002, Besser and Schwarz 2007). However, the molecular basis of the mechanosensing process is still elusive. It has been suggested that the integrin–fibronectin bond as well as talin, one of the integrin binding partners in the cytoplasmic plaque, undergoes conformational alterations on stretching (del Rio *et al* 2009, Friedland *et al* 2009). Thus, investigating the mechanical properties of focal contacts is a major step to fully reveal the underlying mechanism of the mechanosensing process.

Micro- and nanopatterned substrates with a controlled distribution and density of binding sites for integrin receptors have emerged as a suitable tool for determining the impact of the extracellular matrix network structure on cellular behavior. For instance, the forces that cells actively exert at focal contact sites (Balaban *et al* 2001, Tan *et al* 2003) and their dependence of cell spreading, polarization, motility, proliferation and focal contact formation on the extracellular matrix pattern (Lehnert *et al* 2004, Singhvi *et al* 1994) as well as on the nano-scale density (Maheshwari *et al* 2000, Massia and Hubbell 1991) of integrin binding sites have been analyzed. In more detail, it has been revealed that it is not only the average density of integrin binding sites underneath the cell body, but also the details of their lateral spacing at the nano-scale that seem to control focal contact assembly and cell behavior at the microscale. For an integrin binding site spacing larger or equal to 73 nm, focal contact assembly is inhibited and hence the cells do not spread. However, for a spacing larger or equal to 58 nm, focal contacts and actin stress fibers are assembled and indeed cells adopt a well-spread, pancake-like shape (Arnold *et al* 2004, Cavalcanti-Adam *et al* 2007). On surfaces with a nano-scale gradient of binding site spacing, the cells are able to sense effective differences in binding site spacing down to 1 nm over the cell diameter (Arnold *et al* 2008). These results indicate that cell adhesion is extremely sensitive to the organization of molecular binding sites at the nano-scale level.

In order to investigate the role of the nano-scale spacing of individual integrin binding sites for the detachment strength of adhered cells and the stability of focal contacts under load, nanopatterned substrates can be used. These substrates are prepared using diblock copolymer nanolithography, functionalized with the cyclic RGD peptide c[RGDfK(Ahx-Mpa)], which can bind even to the $\alpha v\beta 3$-integrin with high affinity (Kantlehner *et al* 2000, Pfaff *et al* 1994). In this way, the minimal spacing of individual $\alpha v\beta 3$-integrin molecules bound to the surface is precisely controlled. This, in turn, controls the structure and formation of focal contacts and subsequently the cell spreading. In order to quantify cell adhesion and focal contact strength, an AFM in conjunction with phase contrast and fluorescence microscopy is used, making it possible to investigate the force-induced detachment of spread cells and focal contacts. Indeed, the AFM has become a standard tool to investigate the adhesion strength of specific receptor–ligand bonds, both for single molecules and for multiple bonds in the context of single cells (Ludwig *et al* 2008, Müller *et al* 2009). However, in most studies on cell adhesion the contact time between cell and substrate is restricted to only a few minutes (Benoit *et al* 2000, Puech *et al* 2005, Taubenberger *et al* 2007). Studies on long-term adhesion strength with contact times of several hours have been carried out primarily with techniques such as the spinning disk apparatus (Garcia *et al* 1997b) and centrifugation assays (Thoumine *et al* 1996), which can only provide information about the average adhesion strength in cell ensembles. Thus, the AFM-based method for studying long-term adhesion here significantly extends the timescale and precision for probing the adhesion strength of individual cells. Moreover, it allows analysis of the stability of subcellular components such as focal contacts, which only develop after long adhesion times.

Previous studies have shown the cooperative effect of integrin spacing for cell adhesion strength during the initial cell–substrate contact (Selhuber-Unkel *et al* 2008), which means before mature focal contacts are assembled. Using this novel method, the role of integrin spacing has been investigated for the long-term stability of cell adhesion and the elasticity of the cell body. Indeed, it has been found that focal contact formation plays the key role in controlling local cell adhesion strength and cell stiffness. Interestingly, these experiments reveal that focal contacts preferentially break at their intracellular side. This suggests that the weak link between the cell and extracellular matrix is the one between the integrin receptors and the cytoskeleton rather than the cell–extracellular matrix connection. It has been further investigated whether there is a dependence of the rupture force of focal contacts on the loading rate. Applying a theoretical model for the rupture of parallel bonds under shared load, two different scaling regimes for the dependence of cluster stability on the loading rate can be identified. The quantitative information obtained from analyzing focal contact stability on substrates with different integrin binding site spacing indicates that there seems to be a maximum packing density for the proteins in the cytoplasmic adhesion plaque. Indeed, the intermolecular spacing of integrins at the nanometer scale plays an essential role for cell adhesion strength and elasticity.

RGD-functionalized nanopatterns providing binding sites for individual integrin molecules have been used to investigate the role of integrin binding site spacing for cell elasticity, cell adhesion strength and focal contact stability in mature cell adhesion. However, these long-term adhesions require very different experiments compared to initial adhesion, as the cells are spread out on the substrates and have formed stable, micrometer-sized adhesion clusters. Due to the large adhesion strength at this timescale, standard protocols that are normally used in AFM-based single-cell force microscopy experiments (Benoit *et al* 2000, Puech *et al* 2005, Taubenberger *et al* 2007, Selhuber-Unkel *et al* 2008, Wojcikiewicz *et al* 2004) have to be adapted to this special situation.

Thus, a method that is capable of investigating cell adhesion forces at adhesion timescales of 5–7 h using AFM in conjunction with optical microscopy has been developed (Selhuber-Unkel *et al* 2010). In particular, the AFM cantilever is functionalized with fibronectin so that a cytoskeletal connection between the adhesion points at the cantilever and at the surface can be established. Then the cytoskeleton can transmit forces from the dorsal side of the cell to the adhesion zones at the surface. As fibronectin also binds $\alpha v \beta 3$ integrins, similar to the RGD on the substrates, the cell–cantilever contact can diminish the cell–substrate binding strength by a depletion of the integrin contacts. However, this effect is limited because the cantilever touches the dorsal side of the cell for no longer than 10–15 min before cell detachment. However, fibronectin also binds to other integrins in the cell membrane (Hynes 2002) and hence a large excess of fibronectin-binding molecules is available in the membrane (Akiyama and Yamada 1985).

Observing the fluorescence of YFP-marked paxillin molecules in the cytoplasmic plaque during detachment of cells, it has been found that only part of the paxillin is removed. This result indicates that focal contacts predominantly break on their intracellular side, which means the link between receptors and the actomyosin cytoskeleton is breaking instead of the receptor–extracellular matrix bonds. This result is reminiscent of studies on cell migration, where an integrin releases at the rear end of the cell has been revealed (Palecek *et al* 1998, Zimmermann *et al* 1999, 2001). An intracellular breakage of focal contacts has also been observed through de-roofing cells with short ultrasonic bursts (Franz and Müller 2005). In more detail, the remaining parts of the focal contacts on the surface still contain paxillin and F-actin and the structure of the focal contact remains intact even in the rare event of actin stress fiber rupture. However, the inhomogeneous distribution of actin in a focal contact (Zaidel-Bar *et al* 2004) can be an explanation for the partial detachment of paxillin.

A well-known response of focal contacts stimulated by an external force with a tangential component is that they grow in the direction of the applied force (Paul *et al* 2008, Riveline *et al* 2001, Zaidel-Bar *et al* 2005). However, in these experiments, a lateral growth of focal contacts has never been observed. There may be several reasons for this. First, the detachment process was probably too fast for the cell to pronouncedly alter the focal adhesion area, in particular as the stress gradually built up so that the effective time where high stress was applied to the cells was less then 1 min. Second, the force was applied to the cell in the vertical direction so that the tangential component was possibly too small to induce a lateral growth of the focal contact. However, it would be interesting to further reveal the mechanisms of focal contact growth and rupture as a function of the direction and magnitude of applied load.

It is known that cells adapt their stiffness to the microenvironment, mainly through reorganization of the actin cytoskeleton (Solon *et al* 2007). Indeed, the stiffness of the cell body before each detachment step can be determined by measuring the slope of the force-extension curve. On average, it has been shown that this stiffness is increased by a factor of four due to focal contact formation and the binding site density is supposed to play a minor role in determining this parameter. However, this result seems to be plausible, as the state of the actin cytoskeleton is tightly coupled to the focal contact formation. Without the presence of focal contacts, the actin monomers form a loose network of fibers in the intracellular space, which tends to condensate close to focal contacts in order to form stress fibers (Arnold *et al* 2004, Gardel *et al* 2008).

The experimental approach was complemented by a theoretical investigation of focal contact stability. A mean field description was used to determine the rupture strength of the adhesion clusters. Although the determination of the loading rate on a focal contact was based on a number of assumptions, the experimental results indeed reproduced the two theoretically predicted scaling regimes very well and gave reasonable values for the parameters. However, the focal contact rupture force density (F_0/A_0) does not increase in direct proportion to the binding site density. As the focal contact rupture forces are measurements of the cohesion strength of the focal contact plaque rather than a measurement of the adhesion strength to the extracellular connection, this observation suggests the existence of a maximum packing density of the adhesion plaque. This maximum packing density is reached on substrates with an integrin spacing of 28 nm. This observation should be further investigated, for example by investigating the fluorescence intensity of different focal contact proteins as a function of binding site density. However, the hierarchical organization of the focal contact plaque suggests a well-defined spacing as well as a precise orientation of its proteins and hence the existence of a maximum packing density is not surprising.

Taken together, it has been shown that the intermolecular spacing of integrins at the nanometer scale plays a crucial role for cell adhesion strength, cell elasticity and the stability of focal contact and that it is possible to investigate the detachment forces of well-adhered individual cells. These results, together with the method presented, are a good basis for further experiments investigating related questions concerning the mechanical properties of focal contacts, the role and mechanism of focal contact rupture and growth, and the influence of intermolecular spacing at different timescales.

4.2 Traction forces

Cell adhesion and migration are reported to depend on the transmission and generation of actomyosin-dependent contractile forces through focal adhesions to the extracellular matrix of connective tissue. Indeed, during the last two decades, experimental and computational advances have improved the resolution and reliability of 2D and 3D traction force microscopy (figure 4.5 traction forces). It has been suggested that forces exerted by cells drive the migration and invasion of cancer cells through the extracellular matrix or through an endothelial barrier when entering or leaving blood or lymphoid vessels (Mierke *et al* 2008a, Mierke *et al* 2011a, 2011b).

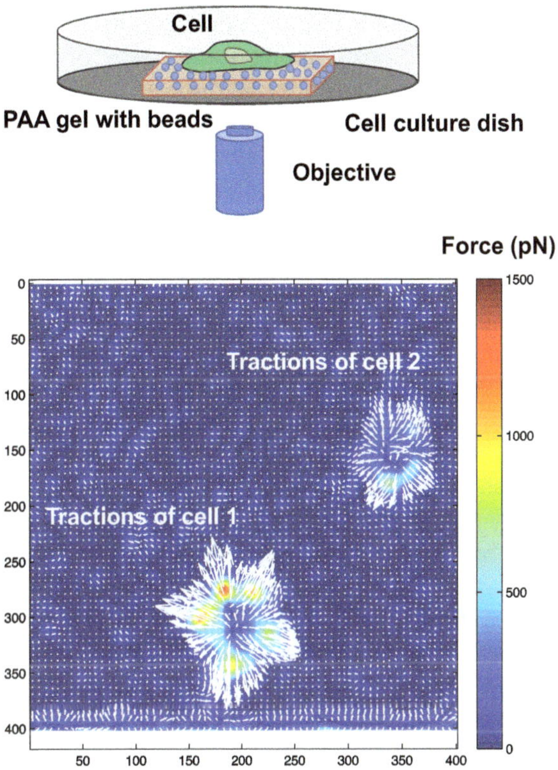

Figure 4.5. Traction force microscopy. (Top panel) Schematic drawing of the setup. In a culture dish a PAA of distinct stiffness was prepared, in which nanometer to micrometer-sized beads are embedded on the outermost (closest to the cell) part of the gel. After surface coating of the PAA gels with laminin or fibronectin, the cell can adhere to them and displace the embedded bead markers by exerting forces onto the gel. (Bottom panel) representative traction force map (force vectors) around cell 1 and cell 2 adhered to a PAA gel of known Young's modulus after 40 min of cell adhesion (as control serves an image of a relaxed PAA that is free of cells after cell detachment using a Trypsin/Cytochalasin D solution).

4.2.1 2D forces on planar substrates

First, the use of two different colored nanobeads as fiducial markers in polyacrylamide gels is discussed and how the displacement field can be computationally extracted from the fluorescence data is explained. Second, different improvements regarding standard methods for force reconstruction from the displacement field are presented, such as the boundary element method, Fourier-transform traction cytometry and traction reconstruction with point forces. Using extensive data simulation, it can be demonstrated that the spatial resolution of the boundary element method can be improved considerably by splitting the elastic field into near, intermediate and far field. Moreover, Fourier-transform traction cytometry requires considerably less computer time, whereas it can achieve a comparable resolution only when combined with Wiener filtering or appropriate regularization schemes. However, both methods seem to underestimate forces, in particular at small adhesion sites. The traction reconstruction with point forces does not suffer from this restriction, but is only applicable to stationary and well-developed adhesion sites. Third, both advantages are combined and used for the reconstruction of fibroblast traction with a spatial resolution of approximately 1 μm.

It has been hypothesized that physical force plays a crucial role as a regulator of many cellular processes, including cell adhesion and migration (Geiger and Bershadsky 2002, Discher *et al* 2005, Orr *et al* 2006, Vogel and Sheetz 2006, Schwarz 2007). In particular, the dynamics of establishing actomyosin-generated traction force and its transmission in an elastic microenvironment seem to be essential for the way cells sense and respond to the mechanical properties of their microenvironment. Thus, measuring the cellular traction forces exerted on elastic substrates upon cell–matrix adhesion is an essential tool for investigating and understanding the regulation of cell adhesion and migration in a quantitative way (Beningo and Wang 2002, Roy *et al* 2002).

The traction force microscopy method was pioneered at least 30 years ago using thin flexible Si sheets as a first wrinkling assay to provide a qualitative measure for the mechanical activity of cells (Harris *et al* 1980). In contrast, in order to perform quantitative studies, it is essential to suppress wrinkling, which has been achieved by using films under prestress (Lee *et al* 1994, Dembo *et al* 1996) and later it has also been achieved for thick films attached to a cover slide (Dembo and Wang 1999, Pelham and Wang 1999). Now the state-of-the-art method is to investigate traction forces on thick elastic substrates, which ensures the reconstruction of cellular traction forces. In more detail, the films are usually prepared from polyacrylamide (PAA) or polydimethylsiloxane (PDMS) coated with adhesive ligands such as fibronectin or collagen, where fibronectin is preferable to collagen as its coating procedure works better. RGD peptide can also be used for the coating, and it is even better suited than fibronectin. The use of PAA has the advantage that its stiffness can be tuned easily over the physiologically relevant range from 100 Pa to 100 kPa (Engler *et al* 2004a, 2004b, Yeung *et al* 2005). PDMS, which is difficult to prepare with a bulk stiffness below 10 kPa, has the advantage that it can easily be micropatterned (Balaban *et al* 2001, Goffin *et al* 2006, Cesa *et al* 2007). An alternative to flat elastic substrates is the pillar assay, which represents an array of microfabricated PDMS-pillars deforming easily under cellular traction due to their small diameter (Tan *et al* 2003, du Roure *et al* 2005, Cai *et al* 2006, Saez *et al* 2007). As each pillar represents a localized force sensor, the pillar assay ensures a simple, albeit spatially constrained, readout of the forces. For cell adhesion, which is not spatially confined, however, flat elastic substrates are the method of choice.

A setup for traction force microscopy on flat elastic substrates has to be combined with different experimental and computational techniques. In more detail, a cell adheres to a flat substrate and thereby exerts force through its sites of adhesion. In particular, the resulting deformations in the substrate are tracked by monitoring the individual movement of embedded marker beads. For PAA gels, fluorescent micro-beads are usually used, and these are embedded near the surface of the gel. In some applications they are even collected in the uppermost layer of the gels through centrifugation of the inverted PAA gels before the polymerization. For PDMS, micropatterning can be performed to create a pattern that is easily detected under phase contrast microscopy. The displacement field has to be extracted from a pair of images, one image showing the substrate as it has deformed under cell traction and one reference image showing the undeformed substrate without any cell traction acting on the gel. In general, there are two ways to perform the image processing: either one directly tracks the movement of the fiducial markers (called particle tracking velocimetry (PTV)) or one uses a cross-correlation function to obtain the local motion statistically (called particle image velocimetry (PIV)). The next step is to reconstruct the cellular traction pattern from the displacement field. For synthetic substrates such as PAA or PDMS, elastic behavior can be assumed, as they are homogeneous, isotropic and linear materials. In more detail, during cell adhesion in tissue culture the cell is rather flat and hence force in the normal direction can be neglected. However, this is a crude assumption that still needs to be proven. Then both the displacement field $u(x)$ and the traction stress field $f(x)$ are 2D in the plane of the substrate ($x = x_1, x_2$). They are related by the following integral equation (4.5):

$$u_i(x) = \int \sum_j G_{ij}(x - x') f_j(x')\, dx'. \tag{4.5}$$

Given the experimental displacement $u(x)$ and the relevant Green function $G_{ij}(x)$, one simply needs to invert the equation (4.5) to obtain the desired traction field $f(x)$. For thick films, the substrate can be approximated by an elastic half-space and hence one can use the Boussinesq Green function as in equation (4.6) (Landau and Lifshitz 1970):

$$\begin{aligned} G_{ij}(x) &= \frac{(1+\nu)}{\pi E}\left[(1+\nu)\frac{\delta_{ij}}{r} + \nu\frac{x_i x_j}{r^3}\right] \\ &= \frac{(1+\nu)}{\pi E r^3}\begin{pmatrix} (1-\nu)r^2 + \nu x^2 & \nu xy \\ \nu xy & (1-\nu)r^2 + \nu y^2 \end{pmatrix}, \end{aligned} \tag{4.6}$$

where ν and E represent the Poisson ratio and Young's modulus, respectively, and $r = |x|$. For clarity, the Green tensor has been written here in both the index notation and in full form. As is typical for a 3D elastic Green function, it scales as ~1 r^{-1} with distance, thus it seems to be long-ranged and has a singularity at the origin. Equation (4.6) is a Fredholm integral equation of the first kind with a weakly singular kernel. The long-ranged nature of the kernel implies that the direct problem corresponds to a smoothing operation. This suggests that the inverse problem may be very sensitive to small differences in the displacement field, depending on the exact nature of the experimental data. In more detail, noise in the experimental data can result from different sources, such as elastic inhomogeneities in the substrate, insufficient coupling between marker beads and polymer matrix, deficiencies in the optical setup and lack of accuracy in the tracking routines.

Three standard methods have been established to calculate the force from the displacement. Both the boundary element method (BEM) (Dembo *et al* 1996, Dembo and Wang 1999) and Fourier-transform traction cytometry (FTTC) (Butler *et al* 2002) approximate the integral on a grid (discretized methods). While BEM effectively corresponds to inverting a large system of linear equations in real space, FTTC uses the fact that the relevant system of linear equations is much smaller in Fourier space, thus facilitating inversion. Traction reconstruction with point forces (TRPF) (Balaban *et al* 2001, Cesa *et al* 2007, Schwarz *et al* 2002) uses additional experimental knowledge about the precise location of the adhesion sites, which can, for instance, be obtained by fluorescence data on proteins localizing to focal adhesions such as vinculin or paxillin (Balaban *et al* 2001, Schwarz *et al* 2002) or by reflection interference contrast microscopy (Cesa *et al* 2007). Then the integral in equation (4.6) converts into a simple sum. In the past, there has been disagreement about the advantages and disadvantages of these diverse methods. In particular, which of the two discretized methods is better in respect to resolution and reliability remains an open question. Moreover, different solutions have been proposed to deal with the issue of experimental noise, but a systematic and detailed analysis of these approaches is still missing.

Different experimental and computational advances are presented here, which together provide much higher resolution and greater reliability in traction force microscopy than was possible formerly. A new method to track the deformations of a PAA substrate has been utilized experimentally in which two differently colored nanobeads are used simultaneously as fiducial markers. In order to extract the corresponding displacement field, a new image processing method has been adopted in which particle tracking velocimetry (PTV) and particle image velocimetry (PIV) are combined. Regarding the computational reconstruction of the traction field, variants of all three standard techniques have been implemented, including BEM, FTTC and TRPF, in order to compare their performance systematically using extensive data simulation. In particular, an improved version of BEM has been adapted and compared with new variants of FTTC. Analysis of simulated data revealed the importance of filtering for FTTC and indicated that certain variants of FTTC can perform almost as well as BEM while being much more efficient in terms of computer time required. However, as a disadvantage, both discretized methods are found to be strongly biased regarding small adhesion sites. TRPF does not suffer from this limitation, but its underlying assumption of accurate knowledge of the adhesion site location restricts its applicability. Finally, it has been demonstrated with fibroblasts that the overall setup result in a spatial resolution that constitutes a 5–10 fold improvement over earlier methods (Selhuber-Unkel *et al* 2010).

Two distinct kinds of fluorescent marker beads improve the measured displacement field

The spatial resolution of the reconstructed traction field depends directly on the spatial resolution with which the displacement field is sampled. The use of a multispectral confocal spinning disk microscope enables the acquisition of images from different fluorescence channels at high resolution. In order to achieve the highest possible information content, two differently colored beads can be densely embedded in the PAA gel. Indeed, the number of features is much higher if a combination of both channels is used. As the quality of images usually differs from channel to channel, it is impossible to interlace both extracted vector fields when tracking each channel separately using correlation-based PTV.

According to the Nyquist–Shannon sampling theorem, the optimal spatial resolution that can be achieved is determined by half the sampling frequency. For one- and two-channel tracking, this should be roughly 1.5 μm and 1 μm, respectively. To check whether this limit can be reached, the capability of the different computational methods to distinguish very small traction sources is analyzed. In particular, the resolution is defined by the Rayleigh criterion, as the minimum distance between two point forces at which they can still be separated. The results of Fourier-transform traction cytometry (FTTC) after four different noise treatments, such as Gaussian filtering, Wiener filtering, zero- and second-order regularization, showed after visual inspection that very weak traction is not reproduced. The reconstructions in several cases revealed that nearby forces can no longer be separated. Indeed, all the methods provide a clear representation of the overall traction pattern and strong point forces can in any case be distinguished clearly. The differences in resolution between BEM and FTTC are mainly due to mesh geometry, which is a slight advantage for the BEM. Moreover, the choice of the filter procedure strongly influences the point-force resolution. Indeed, the new method using two colors results in a much denser and more precise sampling of the displacement field compared to the traditional approach with one color.

To determine the detection limit of the traction reconstruction quantitatively, a ten-point force traction pattern is generated and measured for a high substrate stiffness of E = 10 kPa and the reconstructed traction field is determined outside the sites of adhesion. These values represent a background in the traction pattern, which potentially masks small cellular traction forces. An increase in the noise level only leads to a slight increase in the detection limit. However, increasing the absolute level of the displacement field by increasing the original traction also leads to a strong increase in the traction background. For instance, for a maximum displacement of approximately eight pixels the detection limit is around 200 Pa, corresponding to a force of 0.2 nN μm^{-2}. Indeed, for the same displacement, but for a doubled substrate stiffness of $E = 20$ kPa, the traction background increases by a factor of two (such as 400 Pa), which corresponds to a force of 0.4 nN μm^{-2}. Thus, small forces cannot be resolved when the overall magnitude of traction is high, regardless of the noise level.

BEM and FTTC produce comparable results

In order to obtain a more general quantitative comparison between the different variants of traction force microscopy, three different scores for circular adhesions are measured with different noise levels in the displacement field. For all three scores used—deviation of traction magnitude (DTM), deviation of traction magnitude in the surrounding (DTMS) and deviation of traction angle (DTA)—it is the case that the closer its value to zero, the better the performance. In the absence of noise, filtering or regularization cannot improve the results of FTTC. Indeed, FTTC works comparably well compared to BEM with zero-order regularization, whereas BEM with first-order regularization works relatively poorly. Moreover, this result persists in the presence of noise, thus in general BEM is best together with zero-order regularization. For FTTC, it has been shown that Wiener filtering is always better than Gaussian filtering and that zero-order regularization is always better than second-order regularization. Thus, these three methods—Wiener filtering, zero-order regularization and BEM—are equally suitable. However, there are dramatic differences in the required computation times. On a standard desktop computer BEM requires several hours and in addition needs large storage resources, whereas FTTC only requires seconds. In addition, the programming effort is pronouncedly reduced for FTTC, which can be encoded in a few pages, whereas the BEM is normally spread out over several subroutines. Taken together, FTTC with Wiener filtering or zero-order regularization is reliable and the most efficient method.

The traction magnitude of small adhesion sites is underestimated with discretized methods

The deviation of traction magnitude (DTM) is always negative for both BEM and FTTC, hence the significant underestimation of traction is a persistent challenge with the discretized methods. The causes of systematic and random underestimation are manifold, but depend strongly on the sampling density of the displacement field, as suggested by the Nyquist criterion. Indeed, the analysis of adhesion sites smaller than two mesh sizes is really problematic. The difference between reconstructing traction force using BEM with zero-order regularization and FTTC with Wiener filtering on an irregular versus regular grid is clearly visible: in both cases, the overall traction pattern is nicely reproduced. In more detail, the DTM is plotted as a function of adhesion size (measured in units of mesh size) for noise levels of 0% and 10%. Below a critical size, there is an inverse relation between DTM and adhesion size, suggesting that force reconstruction is only reliable for large adhesions. For 0% noise, the critical value is approximately two mesh sizes, whereas it shifts to larger values for higher noise levels. In the presence of 10% noise, only adhesions with a size approximately four times larger than the mesh size are reconstructed in a reliable manner. For smaller adhesions, the DTM becomes strongly negative, thus the traction at these adhesions (and hence the overall strain energies) is strongly underestimated. In conclusion, it is important to start with a dense sampling of the displacement field, as this leads to small mesh sizes and thus reliable force reconstruction for small adhesions and even better estimates for the strain energy.

The traction reconstruction with point forces is precise, but depends on correct localization of focal adhesions

To improve the force reconstruction at small adhesions, traction reconstruction with point forces (TRPF) is used, which starts with the assumption of a highly localized force (Schwarz *et al* 2002). However, the DTM depends much less on the size of the adhesion, which leads to a reliable force reconstruction over a large range of adhesion sizes. The critical point in this method is that the localization of the adhesions has to be very precise. When the adhesions are not placed well, the quantitative analysis using DTM and DTA shows directional errors and strong magnitude fluctuations. Finally, this finding restricts the applicability of the method to cells with few and distinct focal adhesions.

High-resolution reconstruction for fibroblast traction measurements

How can the whole setup be used to determine fibroblast tractions? In order to obtain the required spatial resolution for the displacement field, it is proposed to image only a limited region of the cell. Thus, a region of interest has been selected, which does not have any additional nearby adhesions that might alter the traction pattern inside the region of interest. Indeed, the mesh size of approximately 500 nm results in a spatial resolution of around 1 μm, because the computational reconstruction is combined with adequate filtering. For the substrate stiffness E = 15.6 kPa and the given fibroblast strength, the traction background is approximately 500 Pa, corresponding to a force of 0.5 nN μm^{-2}. In the reconstruction forces of up to 10 kPa are found frequently, corresponding to a force of 10 nN μm^{-2}, which is 20-fold above the detection limit. For the BEM the computational mesh is an irregular grid, confined by the wedge-like region including the cell contour. For FTTC, force vectors are reconstructed on a square lattice covering the whole image. Both the BEM and FTTC methods provide very similar results. In this case (TRPF), one point is first been selected for each adhesion, whereas for very large adhesions two points are selected. The resulting forces are in visual agreement with the results from BEM and FTTC. Direct comparison between the three methods for the small region indicates the difference between them: one force vector for TRPF, an irregular pattern for BEM and a regular pattern for FTTC. It shows that the spatial resolution of BEM and FTTC is indeed around 1 μm.

Established and newly developed procedures to reconstruct cellular traction forces on flat elastic substrates have been discussed. Furthermore, the approach that guarantees an optimal spatial resolution and magnitude reconstruction for diverse experimental conditions has been indicated. As an initial step, the known experimental traction force protocols have been implemented so that they can be used for very small and differently colored fluorescent beads under a confocal microscope. The use of two kinds of distinctly colored beads densely embedded in the gel ensures the extraction of displacement fields with an average mesh size of 500 nm, setting the basic scale for the spatial resolution of the force reconstruction as 1 μm. This resolution is a major improvement (by a factor of approximately 5–10) over earlier work.

It is common knowledge that cellular adhesion sizes vary considerably, ranging from a few hundred nanometers in nascent adhesions of migrating cells to tens of microns for supermature focal adhesions in myofibroblasts. For small adhesions, the mesh size of the displacement data is larger than the adhesion feature size and thus the Nyquist frequency determining the theoretical upper restriction for the resolution of traction forces is typically too low to capture all the details of the exerted traction field. There is still an open question as to whether there is a special computational method to provide the best results in reconstructing cellular traction fields. The main methods to investigate cellular traction fields are BEM (Dembo *et al* 1996, Dembo and Wang 1999) and FTTC (Butler *et al* 2002). For both methods, different improvements have been proposed, such as analytical integration procedures and adaptive mesh generation for BEM, and different filtering and regularization schemes for both BEM and FTTC.

An intrinsic feature of both methods is that solving the inverse stress–strain problem is equivalent to multiplying each Fourier component of the displacement field by its respective wave number. Indeed, high frequency noise will thus be amplified and has to be removed while leaving as much as possible of the original signal untouched. Thus, different strategies can be used. On the one hand, it seems reasonable to impose a smoothness constraint on the displacement field and filter it before the calculation of traction stress is performed. This technique only works with the Fourier-transform methods. On the other hand, it is better to choose a good solution *a posteriori* by inspection of its properties. However, regularization constrains the resulting traction field and one can fine-tune it by compensating the above-mentioned effect of noise amplification to a desired level in the solution.

FTTC is shown to work best with Wiener filtering and zero-order regularization in simulated data, whereas the analysis of real data leads to the suggestion that the regularization is a more robust approach. As expected, BEM mostly works well with zero-order regularization. An overall comparison of both approaches leads to the conclusion that FTTC, when combined with a proper filtering procedure, is in large parts comparable with BEM regarding resolution. The initial advantages of the boundary element approach, resulting from the exact incorporation of irregular data, are lost in the presence of noise. Aliasing (minimizing the distortion artifacts) and boundary effects are not very prominent with boundary element methods, but can also limit the performance of FTTC. However, the great advantage of FTTC is the very small run time compared with the BEM.

Both discretized methods suffer from a systematic underestimation of traction forces at small adhesion sites. It would be interesting to know whether traction force reconstruction with point forces (TRPF) can measure force magnitudes properly independent of adhesion size. The concept involved, of *a priori* localized traction sources, does, in fact, serve to avoid the above bias. However, this concept is only applicable when the main assumption of this method, a reasonable localization of all traction sources, can be ensured. In general, traction measurement systems which do not permit the presumably force-facilitated, yet temporary, assembly of very small adhesion sites seem to be a slightly ill-fated choice when the whole spectrum of forces and morphologies of cellular adhesion sites is being investigated in migrating cells.

Keeping in mind that several mechanisms may contribute to force development at focal adhesions, a high spatial resolution is required for a quantitative understanding of traction forces. From the work just presented, it seems to be clear that dense displacement fields are the most important strategy to achieve this goal. However, a major concern seems to be that incorporating many marker beads in the gel contradicts the assumption of linear elasticity in the substrate, as the beads may alter the mechanical properties of the matrix. In more detail, inclusions in an elastic material are known to perturb the Young's modulus and Poisson ratio only in a small surrounding shell of the scale of the bead diameter (Jasiuk *et al* 1997). Thus, this effect only seems to be a problem when large beads such as microbeads are used.

In summary, high-resolution traction force microscopy has been reported to rely on the combination of advances in substrate preparation, image processing and computational force reconstruction. The systematic and quantitative comparison shows that, depending on the specific situation and the resources available, different approaches can be useful. For example, BEM with zero-order regularization seems to be the best choice to obtain high-resolution traction patterns for migrating cells possessing many small adhesions and with a small noise level in the displacement data. When the computer time is restricted, FTTC with zero-order regularization is an equivalent alternative, especially at higher noise levels. For stationary cells in which focal adhesions can be precisely localized, TRPF is both computationally cheap and reliable.

4.2.2 3D forces within a 3D collagen matrix scaffold

Quantitative measurements of cell-generated forces have been developed and can be performed easily when the cells are cultured on 2D substrates (Mierke *et al* 2008a, 2008b; Mierke *et al* 2011b). Indeed, a technique has been developed to quantitatively measure the 3D traction forces exerted by cells that are fully encapsulated within well-defined elastic hydrogel matrices. This approach has been used to measure tractions from a variety of cell types and diverse contexts and thus reveal patterns of force generation that are attributable to morphologically distinct regions of cells as they migrate into the surrounding extracellular matrix.

It has been known for a long time that cells are constantly probing, pushing and pulling on the surrounding extracellular matrix of connective tissues, for instance to migrate and invade through it. Therefore, the cells need to apply forces to the extracellular microenvironment in order to sense the matrix's mechanical properties. In particular, these cell-generated forces facilitate cell migration as well as tissue morphogenesis and maintain the intrinsic mechanical balance of tissues (Dembo and Wang 1999, Keller *et al* 2003). These forces not only guide mechanical and structural events, but also facilitate signal transduction pathways promoting functions ranging from cell proliferation to stem cell differentiation (Huang *et al* 1998, McBeath *et al* 2004). Hence, precise measurement of the spatial and temporal nature of these forces is needed in order to reveal when and where mechanical events play a prominent role in physiological and pathological settings.

Indeed, methods employing planar 2D elastic surfaces or arrays of flexible cantilevers have mapped with subcellular resolution and the forces that cells generate to their substrates have been determined (Balaban *et al* 2001, Butler *et al* 2002, Tan *et al* 2003, Mierke *et al* 2008a, 2008b, Mierke *et al* 2011). However, many processes are altered when cells are removed from native 3D microenvironments and are then cultured on 2D substrates. At a structural level, cells encapsulated within a 3D matrix display dramatically different morphology, cytoskeletal organization and focal adhesion structure compared to those cultured on 2D substrates (Cukiermann *et al* 2001). However, even the initial requirements for cells to adhere and spread against a 2D substrate are different from the invasive process essential for cells to extend inside a 3D matrix. In more detail, these differences suggest that dimensionality alone may significantly impact how cellular forces are generated and transduced into biochemical or structural alterations. Although the mechanical properties of 3D extracellular matrices and the cellular forces generated within a 3D microenvironment have been shown to regulate many cellular functions (Pampaloni *et al* 2007), the quantitative measurement of cellular forces within a 3D context has to be demonstrated.

In particular, the traction stresses (force per area), hereafter tractions, exerted by cells embedded within a hydrogel matrix can be determined quantitatively (Legant *et al* 2011, 2013). GFP-expressing fibroblasts were encapsulated within mechanically well-defined polyethylene glycol (PEG) hydrogels that incorporate proteolytically degradable domains in the polymer backbone and pendant adhesive ligands (Miller *et al* 2010). In more detail, the incorporation of adhesive and degradable domains enables the cells to invade, spread and adopt physiologically relevant cellular morphologies. These hydrogels had a Young's modulus of 600–1000 Pa, a range similar to those of commonly used extracellular matrices, such as reconstituted collagen or Matrigel and to *in vivo* tissues such as mammary and brain tissue (Paszek *et al* 2005, Discher *et al* 2005). As expected, cells in 3D PEG gels deformed the nearby surrounding matrix, which was visualized by tracking the displacements of 60 000–80 000 fluorescent beads in the close neighborhood of each cell. In particular, bead displacements were determined relative to a reference, stress-free, state of the gel after lysing the cell with a detergent. Deformations of 20%–30% peak principal strain have been measured in much of the cell surrounding the hydrogel. The largest strains (even up to 50%) occurred in close vicinity to long impressive extensions, which is indeed consistent with the observations of strong forces exerted by these regions on 2D substrates (Chan *et al* 2008). As the mechanics of the PEG hydrogels showed no substantial dependence on strain or frequency, a linear elasticity theory and the finite element method was used to determine the cellular tractions that would give rise to the measured bead displacements. In brief, a finite element mesh of the hydrogel is generated that surrounds the cell from confocal images. In particular, a discretized Green's function was constructed by applying unit tractions to each facet on the surface of the cell mesh and solving the finite element equations in order to calculate the induced bead displacements. As usual, the standard regularization methods for ill-posed, over-determined linear systems of equations were then used to compute the tractions exerted by the cell. The time needed to calculate a single data set is approximately 4.5 h. However, this calculation time can be reduced dramatically by using a simplified finite element mesh of the cell and hydrogel. Indeed, these lower resolution datasets still capture the fundamental character of higher-resolution measurements.

Simulated traction fields are used to validate the approach and to characterize its spatial resolution. The experimental noise for the bead displacements was determined by measuring the displacement of the beads in cell-free regions of the hydrogel before and after detergent treatment. In addition, the surface discretization noise was measured from multiple discretizations of the same cells. These datasets then were superimposed onto the displacements generated by simulated loadings prior to the traction reconstruction. In this setting, the percentage of traction recovered was proportional to the magnitude and characteristic length of the simulated loadings, defined as the average period of spatial oscillation. Indeed, for all cases, the presence of noise reduced recovery accuracy by approximately 20%–30%. Despite these limitations, the recovered tractions can still capture the essential periodic features of the most spatially complex simulated loadings with characteristic lengths of spatial variation down to only 10 μm.

In the next step, the tractions from living cells encapsulated within 3D hydrogels are measured. Indeed, it has been found that cells have exerted tractions in the range of 100–5000 Pa, with strong forces located predominantly near the tips of long extensions. For all measurements, the forces were in static equilibrium, with a typical error of approximately 1%–5% of the total force applied by the cell. Further analysis revealed that these tractions were minimally altered by possible variations in local hydrogel mechanics or by uncertainty in the recorded bead displacements. Measurements of cellular forces on 2D surfaces have generally been limited to shear loadings, although some studies have measured small forces exerted normal to the planar surface (Maskarinec *et al* 2009, Hur *et al* 2009). However, it is still elusive whether these relationships might be altered for cells inside 3D extracellular matrices. Cells encapsulated within a 3D matrix have been reported to exert predominantly shear tractions, although small normal tractions were additionally present in close proximity to the cell body. In order to investigate whether patterns of force are associated with specific regions of cells, the magnitude and angle of tractions is quantified with respect to the center of mass of the cell. In general, tractions increased as a function of distance from the center of mass. Interestingly, cells encapsulated in hydrogels with a Young's modulus of approximately 1000 Pa generated stronger tractions than those in approximately 600 Pa hydrogels, indicating that cells adapt to the mechanical properties of their microenvironment. However, the observed differences in tractions were not due to an overall increase in total cellular contractility, as measured by the net contractile moment, whereas they were most prominent in strong inward tractions near the tips of long extensions. This reveals a local and non-linear reinforcement of cellular contractility due to the substrate rigidity and suggests that such regions may be hubs for force-facilitated mechanotransduction in 3D settings. The cell bodies showed no bias in the traction angle. However, strong tractions became progressively aligned back toward the center of mass in more well spread regions of the cell, for instance, near the tips of long extensions. These patterns of force were typically reflected in multiple cell types, whereas they could be altered by cell–cell proximity or culture as a multicellular aggregate. The neighboring 3T3 cells preferentially extended away from each other, whereas proliferating multicellular tumor spheroids exerted outward normal tractions on the extracellular matrix.

In more detail, a subset of extensions was observed that displayed strong tractions several microns behind the leading tip, whereas the tractions at the tip itself were substantially lower. As such traction profiles are similar to those determined behind the leading edge of a lamellipodia for a migrating cell on a 2D substrate, it can be hypothesized that such regions represent invading or growing cellular extensions in 3D. To analyze this possibility, the tractions from time-lapse images of cells as they invaded the surrounding hydrogel have been determined. Indeed, tractions at the tips of growing extensions were pronouncedly lower compared to the strong tractions exerted by proximal regions of the same extension. However, normal forces pushing into the extracellular matrix have not been detected in these extensions, which suggests that a local inhibition of myosin-dependent contractility allows tip encouragement. Strong tractions have been detected from small extensions on the cell face opposite to the invading extensions. These stable extensions exhibited very different force distributions compared to the growing extensions and often lack the characteristic drop in force near the leading edge and hence may correspond to an anterior–posterior polarity axis build-up within the cell.

Taken together, these data suggest that cells in 3D matrices probe their surrounding extracellular matrix primarily through strong inward tractions near the tips of long slender extensions such as protrusions. Importantly, it has been demonstrated that this 3D force measurement technique is applicable to many different cell types, cell–cell interactions and even to multicellular tumor structures where both tumor growth and invasion have previously been revealed to be mechanoresponsive (Paszek *et al* 2005). Because synthetic hydrogels possess similar elastic moduli compared to *in vivo* tissues (Paszek *et al* 2005, Discher *et al* 2005) and can hence support a wide range of cellular functions (Lutolf and Hubbell 2005), this approach will enable investigations into the role of cellular forces in a variety of biological settings.

4.3 Lipid drops as stress sensors

When there are no externally applied stresses, the principle is that an interface minimizes its own surface tension, and hence a liquid drop will adapt a spherical shape (Sugimura *et al* 2016). Therefore, when a lipid drop is introduced between cells in a tissue, stresses are exerted by neighboring cells or by the surrounding extracellular matrix counteracting the surface tension of the drop and subsequently deforming it (Campàs *et al* 2014). In more detail, the drops serve as stress transducers, similarly to mechanical elements, that are incorporated into the tissue, which in turn pushes and deforms the drops within it. Moreover, confocal scanning microscopy can be utilized to analyze the three-dimensional drop shape, which enables their inference with the anisotropic stresses locally exerted on the drop. When the tissue embedded drops are small enough compared with tissue size, they can stay in a distinct tissue location for days without affecting the normal embryonic development and hence alterations in stresses over a distinct period of time can be determined.

The drops used for stress measurements can be fluorescently labeled. In particular, lipid drops are made of a bio-compatible oil, cell-sized and coated with specific ligands for cell adhesion receptors and hence both tensile and compressive stresses can be determined (Paluch 2015). The drop size should be under a certain limit of 600 μm for fluorocarbon oil in order to prevent deformation by evoked gravity and above a size of typically 10 μm in diameter, bead internalization by cells is no longer frequently observed and hence cannot impair the results.

The contour curvatures of the lipid drops can typically be determined with ±5% accuracy, which depends on the resolution of the images used for data analysis. In particular, the value of the interfacial tension of the drop can be tuned in order to widen the range of the measurable stresses from approximately 0.1–60 kPa. When the interfacial tension of the drop is decreased, the drop deforms more upon a stress application. The main limitation of the lipid drop method is the prior knowledge of the interfacial tension. Hence, the interfacial tension needs to be determined for the isolated drop before its introduction into tissues. The accuracy of the interfacial tension of drops is around 20% for the minimal interfacial tensions used (~3 mN m^{-1}). However, this experimental setup is based on the assumption that the interfacial tension remains constant and uniform when the drop, which is introduced between cells, comes into close contact with proteins, which can adsorb on the drop's surface. When the drop's surface has been saturated before the experimental start with surfactants, which leave no free space for the adsorption of proteins, such a hypothesis is still reasonable but should be validated independently. This method has been applied to cultured aggregates of tooth mesenchymal and mammary epithelial cells as well as living tissue explants, which revealed that the anisotropic stresses generated by mammary epithelial cells are driven by the myosin II activity (Campàs *et al* 2014).

However, this method can be varied by using a soft polymer bead in a microfluidic setup. The advantage is that the bead stiffness can be more precisely controlled and calibrated than the stiffness of a lipid drop. Another advantage is that it seems to be tunable in a larger range, and it is then more compressible and thus it allows us to measure the pressure or traction in the surrounding tissue. As an alternative way to determine tissue stiffness, gel sensors can be used externally instead of internally to finally yield accurate tissue-scale force determinations. For instance, by embedding *Xenopus* embryo tissue in soft agarose gels of various stiffnesses, it has been demonstrated that dorsal tissues undergoing convergent extension facilitate the upregulation of their force production when embedded in a stiffer microenvironment (Zhou *et al* 2015).

4.4 Dual micropipette aspiration (DPA)

The microaspiration techniques represents a simple and highly reproducible biophysical method to determine cellular stiffness (or inversely the deformability) by exerting a negative pressure to a cell membrane and monitoring the membrane deformability due to well-defined pressure. Microaspiration was first used to characterize the elastic properties of sea-urchin eggs for enlightening the mechanisms of cell division (Mitchinson and Swann 1954) and second to investigate the mechanical properties of red blood cells (Rand and Burton 1964). Thereafter, microaspiration has been utilized to reveal the mechanical properties of a broad variety of cell types (Discher *et al* 1994, Schmid-Schönlein *et al* 1981, Evans and Kuhan 1984, Sato *et al* 1987). Moreover, these microaspiration techniques have been combined with 3D imaging in order to analyze the stiffness of substrate-adhered cells and the alterations in cell shape or structural morphology (Byfield *et al* 2004, 2006a, 2006b).

In more detail, microaspiration is implemented by pulling a glass capillary to generate a micropipette with a very small tip (2–50 μm diameter depending on the size of a cell or a tissue resection probe), which is connected to a pneumatic pressure transducer and brought into close contact with a cell by using a microscope (Oh *et al* 2012). When the tip of the pipette reaches the cell surface, a negative pressure is applied to the pipette by using a pneumatic pressure transducer that exerts a well-defined pressure on the cell membrane. In response to this exerted pressure, the cell membrane is aspirated into the pipette and hence the progressive membrane deformation or 'membrane projection' into the pipette can be determined as a function of time. The underlying basic principle of this approach is that the degree of membrane deformation due to a defined mechanical force is a function of the membrane stiffness. In particular, the stiffer the membrane is, the slower the rate of membrane deformation and the shorter the steady-state aspiration length of the cell. In more detail, the technique requires isolated cells, but cells can be analyzed in suspension as well as adhered to a substrate. Apart from cells, large organelles and liposomes can also be investigated.

In more detail, the analysis can be performed by comparing maximal membrane deformations that are evoked through the application of a certain pressure for different cell populations or specific experimental conditions. A 'stiffness coefficient' can be determined by plotting the aspirated length of cell membrane deformation over the exerted pressure. In addition, the data can be analyzed to determine the Young's modulus of the cells (E), which represents the most common parameter to characterize the stiffness of a broad variety of materials. The cell membranes of eukaryotic cells can be regarded as a bi-component system where the membrane lipid bilayer covers and hence envelops the submembrane cytoskeleton, termed the cell cortex, and hence the cytoskeleton provides the mechanical scaffold of the cell membrane that mainly contributes to deformability of the cellular envelope. Thus, this microaspiration approach allows us to determine the mechanical properties of the submembrane cytoskeleton.

In terms of alternative approaches to AFM, microaspiration provides a useful tool and represents an inexpensive alternative to AFM, which is still commonly used to determine cell stiffness (Okajima 2011, Wang *et al* 1993). Unlike AFM which provides the local membrane rigidity, the microaspiration device reveals a global measure of the cellular deformability.

Micropipette-based assays represent an alternative method to quantitate cell–cell adhesion forces. In the so-called dual micropipette assay (DPA), two micropipettes are attached to a pressure control system and each holds a single cell under relatively gentle aspiration (Chu *et al* 2004, Daoudi *et al* 2004, Sung *et al* 1986). In more detail, the cells are then brought in close contact for a defined contact time to initiate the formation of cellular adhesion. Subsequently, on one micropipette the high aspiration is regulated to be sufficiently strong in order to stably hold the cell pair, whereas in the other micropipette the suction pressure is steadily increased in discrete steps. After each increase in a distinct pressure step, the pipettes are moved further apart in order to separate the two cells and to rupture the intercellular connections (bonds). This step of the measurement protocol is repeated until the aspiration level in the second micropipette is strong enough to evoke the full separation of the two cells. The cell–cell separation force can be determined from the aspiration pressure that is needed to rupture the cell–cell interaction during the retraction of the micropipette. When investigating two cells that are not yet interacting, the cell–cell contact time can be precisely adjusted and even the influence of the contact time on the adhesion strength can be analyzed. As an alternative approach, pre-formed cell–cell doublets can be separated using DPA, which eliminates the need to wait for a defined contact time until the cell separation experiment is started, but in this approach the contact history of the cell–cell interaction is rather undefined. When comparing DPA to AFM, the force resolution of the step-pressure DPA assay (approximately in the nanonewton range) is pronouncedly lower than AFM-based SCFS (approximately in the piconewton range) and hence DPA cannot provide the resolution of individual molecular interactions between cells. Nevertheless, DPA allow us to accurately quantify the overall adhesion forces between two cells and is not restricted at the upper force range. When single-molecule force resolution is needed, the so-called biomembrane force probe (BFP) can be employed, which is a special variant of the dual micropipette assay that utilizes a tensed erythrocyte as a force sensor (Evans *et al* 1995).

The BFP represents an ultrasensitive force measuring device and has yielded unique knowledge of the molecular mechanisms underlying the single receptor–ligand interactions, however, it has been less frequently utilized to investigate developmental processes that occur primarily on the cell and tissue scale.

DPA has been used in quantitating cadherin-facilitated adhesion in tissue cultures (Chu *et al* 2004, 2006, Martinez-Rico *et al* 2005), but is also applicable for studies aiming to reveal the developmentally required cellular adhesion. The role of tissue cohesion for the cell migration behavior of facial branchiomotor neurons (FBMNs) has been investigated in zebrafish embryonic hindbrain using the DPA assay (Stockinger *et al* 2011). Microaspiration measurements on isolated cells showed that the cohesion is higher among FBMNs compared to the cohesion among surrounding neuroepithelial cells. However, alterations in neuroepithelial cell cohesion by interfering with the function of cadherin-2 severely impacts the FBMN migration *in vivo*, indicating that neuroepithelial cell cohesion provides the FBMN migration, probably by hindering the FBMN movement. Together with high-resolution optical imaging, many insights are obtained from *in vitro* adhesion force measurements thus helping to reveal the mechanisms controlling *in vivo* cell migration in developing embryos.

An advantage of DPA is that the accumulation of fluorescent fusion proteins associated with cell adhesion in the cell–cell interaction zone can be easily monitored by fluorescent microscopy and related to the separation force measurements (Chu *et al* 2004, Maitre *et al* 2012). Moreover, the horizontal arrangement of the probed two associated cells also enables us to analyze cell shape alterations during contact and after cell separation, which reveals unique insights into the cortex tension at the cell–cell interface (Maitre *et al* 2012). In particular, membranes at dissolved cell–cell contacts have been shown to exhibit a rapid decrease in curvature, indicating decreased cortex tension and subsequently reduced contractile forces in these areas. Moreover, reduced cortex tension at cell–cell contacts supports the maximization of the contact area between spherical cells and represents the driving factor for the enlargement of cell–cell contacts, while cadherin adhesion receptors acts to mechanically couple the cells (Maitre and Heisenberg 2011).

4.5 Förster resonance energy transfer (FRET)-based molecular tension sensors

Cell surface expressed cadherins facilitate cell adhesion between neighboring cells. They are additionally involved in the regulation of the mechanotransduction through their cytoplasmic domain and drive the interactions with catenin proteins and associated cytoskeletal regulatory as well as linker proteins. The E-cadherin/catenin complex has been shown to serve as a mechanical stress sensor and to regulate the actomyosin contractility (Borghi *et al* 2012, Ladoux *et al* 2010, le Duc *et al* 2010, Taguchi *et al* 2011, Yonemura *et al* 2010). The visualization of mechanical forces seems to be an intriguing method to analyze mechanical forces exerted at the cadherin cell–cell contacts through their transformation into optical signals using a FRET-based molecular tension sensor module (TSMod) (Borghi *et al* 2012). In more detail, the TSMod is based on a monomeric teal fluorescence protein (mTFP)/monomeric enhanced yellow fluorescence protein (mEYFP) FRET pair, connected via a 40 amino acid elastic (GPGGA)8 domain that is derived from the spider silk protein flagelli-form and acts as an entropic nanospring (Becker *et al* 2003, Grashoff *et al* 2010, Wang *et al* 2011). In particular, the tension sensor module was integrated between the transmembrane and catenin-binding domain of the cytoplasmic domain of E-cadherin (Borghi *et al* 2012). The FRET approach is highly sensitive to the distance between the two fluorophores and enhanced tension hence causes a reduced FRET efficiency, as the two fluorophores are too far from each other. This method was established using Madin–Darby canine kidney (MDCK) cells by showing that the FRET efficiency of EcadTSMod at cell–cell contact sites was pronouncedly reduced compared to soluble cytoplasmic TSMod or to a control EcadTSMod deletion construct, which lacks the cytoplasmic domain of E-cadherin. Hence, these results indicate that the EcadTSMod is under tension when incorporated into the cell–cell contacts, and subsequently that this tension specifically depends on the catenin-binding domain of E-cadherin. It has been revealed that the E-cadherin cytoplasmic domain is under 1–2 pN of constitutive load from the cytoskeleton, which is very similar to the results obtained from measurements of vinculin in cell–matrix sites (Grashoff *et al* 2010). However, as E-cadherin is under actomyosin-dependent tension even in regions of the cell membrane containing no cell–cell adhesion sites, it can be suggested that the cadherin/catenin complex acts as a constitutive membrane anchor to the cortical actomyosin cytoskeleton similarly to the focal adhesion protein vinculin in cell–matrix adhesions (Borghi *et al* 2012). Moreover, it has been speculated that membrane attachment to the cortex by the cadherin–catenin complex may be highly important for single wound closure in *Drosophila* and also in cell membrane blebbing during zebrafish epiboly (Abreu-Blanco *et al* 2011, Schepis *et al* 2012).

It has been shown that FRET tensions sensors can also be used to analyze developmental processes involving cell–cell adhesion. In particular, the alteration of N-cadherin-facilitated cell–cell adhesion and tension is required for collective migration of *Xenopus* CNCC (Kuriyama *et al* 2014), whereas the recruitment of vinculin to cell–cell adhesion complexes drives the mechanotransduction processes

(Ishiyama *et al* 2013, Miyake *et al* 2006, Sumida *et al* 2011, Taguchi *et al* 2011, Yonemura *et al* 2010). In order to reveal indirectly the N-cadherin dependent tension at cell–cell contacts, vinculin has been utilized as a tension sensor (Grashoff *et al* 2010). Both the inhibition of N-cadherin expression and the actomyosin contractility evoked a release of junctional tension. However, inhibition of the lysophosphatidic acid (LPA) signaling regulating the internalization of N-cadherin, resulted in increased accumulation and clustering of N-cadherin at cell–cell contacts and enhanced tension through the focal adhesion protein and mechano-coupling protein vinculin (Kashef and Franz 2015). The increased N-cadherin-facilitated cell–cell adhesion may explain why CNCC are unable to disperse after depletion of LPA signaling, which subsequently leads to impaired CNCC migration *in vivo*. In turn, inhibition of N-cadherin expression, its homotypic binding activity or endocytosis rescues the CNCC migration in cells that are deficient of LPA signaling (Kuriyama *et al* 2014). Thus, the adaptions of cell–cell adhesion strength and membrane tension are crucial for CNCC migration *in vivo*. When the vinculin tension sensor is utilized, it needs to be taken into account that vinculin localizes to both cell–cell contact sites and focal adhesions, which are the cell–matrix sites at the cell–substrate interface. Thus, in FRET experiments determining junctional tension, it is necessary to place the confocal imaging plane across the cell–cell contacts at the apical site of the cell, which omits the detection of vinculin-facilitated tension at basal focal adhesion sites. Apart from N-cadherin, *Xenopus* CNCC also express the classical cadherin-11, which induces the migration of CNCC (Becker *et al* 2013, Borchers *et al* 2001, Kashef *et al* 2009). The introduction of the TSMod into the cytoplasmic domain of cadherin-11 may provide a novel tool to analyze the possible role of cadherin-11 in facilitating cellular tension during CNCC migration. Taken together, their usage as a sensor may help to directly detect the regions within the cell membrane through which force is transduced, possible through alteration of the extracellular matrix microenvironment.

References and further reading

A-Hassan E, Heinz W F, Antonik M D, D'Costa N P, Nageswaran S, Schoenenberger C-A and Hoh J N 1998 Relative microelastic mapping of living cells by atomic force microscopy *Biophys. J.* **74** 1564–78

Abreu-Blanco M T, Verboon J M and Parkhurst S M 2011 Cell wound repair in *Drosophila* occurs through three distinct phases of membrane and cytoskeletal remodeling *J. Cell Biol.* **193** 455–64

Akiyama S K and Yamada K M 1985 The interaction of plasma fibronectin with fibroblastic cells in suspension *J. Biol. Chem.* **260** 4492–500

Alessandrini A and Facci P 2005 AFM: a versatile tool in biophysics *Meas. Sci. Technol.* **16** R65–92

Amano M, Chihara K, Kimura K, Fukata Y, Nakamura N, Matsuura Y and Kaibuchi K 1997 *Science* **275** 1308–11

Arnold M, Cavalcanti-Adam E A, Glass R, Bluemmel J, Eck W, Kantlehner M, Kessler H and Spatz J P 2004 Activation of integrin function by nanopatterned adhesive interfaces *Chem. Phys. Chem.* **5** 383–8

Arnold M *et al* 2008 Induction of cell polarization and migration by a gradient of nanoscale variations in adhesive ligand spacing *Nano Lett.* **8** 2063–9

Balaban N Q *et al* 2001 Force and focal adhesion assembly: a close relationship studied using elastic micropatterned substrates *Nat. Cell Biol.* **3** 466–72

Barbee K A 1995 Changes in surface topography in endothelial monolayers with time at confluence: influence on subcellular shear stress distribution due to flow *Biochem. Cell Biol.* **73** 501–5

Barbee K A, Davies P F and Lal R 1994 Shear stress-induced reorganization of the surface topography of living endothelial cells imaged by atomic force miroscopy *Circ. Res.* **74** 163–71

Becker N, Oroudjev E, Mutz S, Cleveland J P, Hansma P K, Hayashi C Y, Makarov D E and Hansma H G 2003 Molecular nanosprings in spider capture-silk threads *Nat. Mater.* **2** 278–83

Becker S F, Mayor R and Kashef J 2013 Cadherin-11 mediates contact inhibition of locomotion during *Xenopus* neural crest cell migration *PLoS One* **8** e85717

Behrens J, Mareel M M, Van Roy F M and Birchmeier W 1989 Dissecting tumor cell invasion: epithelial cells acquire invasive properties after the loss of uvomorulin-mediated cell–cell adhesion *J. Cell Biol.* **108** 2435–47

Bell G I 1987 Models for the specific adhesion of cells to cells *Science* **200** 618–27

Beningo K A and Wang Y-L 2002 Flexible substrata for the detection of cellular traction forces *Trends Cell Biol.* **12** 79–84

Beningo K A, Lo C-M and Wang Y-L 2002 Flexible polyacrylamide substrata for the analysis of mechanical interactions at cell-substratum adhesions *Methods Cell Biol.* **69** 325–39

Benoit M and Gaub H 2000 Measuring cell adhesion forces with the atomic force microscope at the molecular level *Cells Tissues Organs* **172** 174–89

Benoit M, Gabriel D, Gerisch G and Gaub H E 2000 Discrete interactions in cell adhesion measured by single-molecule force spectroscopy *Nat. Cell Biol.* **2** 313–7

Berdyyeva T K, Woodworth C D and Sokolov I 2005 Human epithelial cells increase their rigidity with ageing *in vitro*: direct measurements *Phys. Med. Biol.* **50** 81–92

Besser A and Schwarz U S 2007 Coupling biochemistry and mechanics in cell adhesion: a model for inhomogeneous stress fiber contraction *New J. Phys.* **9** 425

Binnig G and Quate C F 1986 Atomic force microscope *Phys. Rev. Lett.* **56** 930–4

Binnig G and Smith D P E 1986 Single-tube three-dimensional scanner for scanning tunneling microscopy *Rev. Sci. Instrum.* **57** 1688

Binnig G, Quate C F and Gerber C 1986 Atomic force microscope *Phys. Rev. Lett.* **56** 930–3

Bischoff G and Hein H-J (ed) 2003 *Micro- and Nanostructures of Biological Systems*vol 1 (Aachen: Shaker)

Bischoff G and Hein H-J (ed) 2004 *Micro- and Nanostructures of Biological Systems*vol 2 (Aachen: Shaker)

Bischoff G and Hein H-J (ed) 2005 *Micro- and Nanostructures of Biological Systems*vol 3 (Aachen: Shaker)

Boettiger D and Wehrle-Haller B 2010 Integrin and glycocalyx mediated contributions to cell adhesion identified by single cell force spectroscopy *J. Phys. Condens. Matter* **22** 194101

Borchers A, David R and Wedlich D 2001 *Xenopus* cadherin-11 restrains cranial neural crest migration and influences neural crest specification *Development* **128** 3049–60

Borghi N, Sorokina M, Shcherbakova O G, Weis W I, Pruitt B L, Nelson W J and Dunn A R 2012 E-cadherin is under constitutive actomyosin-generated tension that is increased at cell–cell contacts upon externally applied stretch *Proc. Natl Acad. Sci. USA* **109** 12568–73

Braet F, de Zanger R, Seynaeve C, Baekeland M and Wisse E 2001 A comparative atomic force microscopy study on living skin fibroblasts and liver endothelial cells *J. Electron Microsc.* **50** 283–90

Bryant P J, Miller R G and Yang R 1988 Scanning tunneling and atomic force microscopy combined *Appl. Phys. Lett.* **52** 2233–5

Burton K and Taylor D L 1997 Traction forces of cytokinesis measured with optically modified elastic substrata *Nature* **385** 450–4

Bushell G R, Cahill C, Clarke F M, Gibson C T, Myhra S and Watson G S 1999 Imaging and force–distance analysis of human fibroblasts *in vitro* by atomic force microscopy *Cytometry* **36** 254–64

Butler J P, Tolic-Norrelykke I M, Fabry B and Fredberg J J 2002 Traction fields, moments, and strain energy that cells exert on their surroundings *Am. J. Physiol. Cell Physiol.* **282** C595–605

Butt H-J, Wolff E K, Gould S A C, Northern B D, Peterson C M and Hansma P K 1990 Imaging cells with the atomic force microscope *J. Struct. Biol.* **105** 54–61

Byfield F, Aranda-Aspinoza H, Romanenko V G, Rothblat G H and Levitan I 2004 Cholesterol depletion increases membrane stiffness of aortic endothelial cells *Biophys. J.* **87** 3336–43

Byfield F J, Hoffman B D, Romanenko V G, Fang Y, Crocker J C and Levitan I 2006a Evidence for the role of cell stiffness in modulation of volume-regulated anion channels *Acta. Physiol.* **187** 285–94

Byfield F J, Tikku S, Rothblat G H, Gooch K J and Levitan I 2006b OxLDL increases endothelial stiffness, force generation, and network formation *J. Lipid Res.* **47** 715–23

Cai Y *et al* 2006 Nonmuscle myosin IIa-dependent force inhibits cell spreading and drives F-actin flow *Biophys. J.* **91** 3907–20

Campàs O, Mammoto T, Hasso S, Sperling R A, O'Connell D, Bischof A G, Maas R, Weitz D A, Mahadevan L and Ingber D E 2014 Quantifying cell-generated mechanical forces within living embryonic tissues *Nat. Methods* **11** 183–9

Cappella B and Dietler G 1999 Force-distance curves by atomic force microscopy *Surf. Sci. Rep.* **34** 1–104

Cavalcanti-Adam E A, Volberg T, Micoulet A, Kessler H, Geiger B and Spatz J P 2007 Cell spreading and focal adhesion dynamics are regulated by spacing of integrin ligands *Biophys. J.* **92** 2964–74

Cesa C, Kirchgessner N, Mayer D, Schwarz U S, Hoffmann B and Merkel R 2007 Micropatterned silicone elastomer substrates for high resolution analysis of cellular force patterns *Rev. Sci. Instrum.* **78** 034301

Chan C E and Odde D J 2008 Traction dynamics of filopodia on compliant substrates *Science* **322** 1687–91

Choquet D, Felsenfeld D P and Sheetz M P 1997 Extracellular matrix rigidity causes strengthening of integrin-cytoskeleton linkages *Cell* **88** 39–48

Chu Y-S, Thomas W A, Eder O, Pincet F, Perez E, Thiery J P and Dufour S 2014 Force measurements in E-cadherin-mediated cell doublets reveal rapid adhesion strengthened by actin cytoskeleton remodeling through Rac and Cdc42 *J. Cell Biol.* **167** 1183–94

Chu Y S, Eder O, Thomas W A, Simcha I, Pincet F, Ben-Ze'ev A, Perez E, Thiery J P and Dufour S 2006 Prototypical type I E-cadherin and type II cadherin-7 mediate very distinct adhesiveness through their extracellular domains *J. Biol. Chem.* **281** 2901–10

Chu Y S, Thomas W A, Eder O, Pincet F, Perez E, Thiery J P and Dufour S 2004 Force measurements in E-cadherin-mediated cell doublets reveal rapid adhesion strengthened by actin cytoskeleton remodeling through Rac and Cdc42 *J. Cell Biol.* **167** 1183–94

Chung K-H and Ki D-E 2007 Wear characteristics of diamond-coated atomic force microscope probe *Ultramicroscopy* **108** 1–10

Cohen M, Klein E, Geiger B and Addadi L 2003 Organization and adhesive properties of the hyaluronan pericellular coat of chondrocytes and epithelial cells *Biophys. J.* **85** 1996–2005

Collinsworth A M, Zhang S, Kraus W E and Truskey G A 2002 Apparent elastic modulus and hysteresis of skeletal muscle cells throughout differentiation *Am. J. Physiol. Cell Physiol.* **283** C1219–27

Costa K D and Yin F C P 1999 Analysis of indentation: implications for measuring mechanical properties with atomic force microscopy *J. Biomech. Eng.* **121** 462–71

Critchley D R 2000 Focal adhesions—the cytoskeletal connection *Curr. Opin. Cell Biol.* **12** 133–9

Cuerrier C M, Benoit M, Guillemette G, Gobeil F and Grandbois M 2009 Real-time monitoring of angiotensin II-induced contractile response and cytoskeleton remodeling in individual cells by atomic force microscopy *Pflug. Arch.* **457** 1361–72

Cukierman E, Pankov R, Stevens D R and Yamada K M 2001 Taking cell–matrix adhesions to the third dimension *Science* **294** 1708–12

Culp L A, Murray B A and Rollins J B 1979 Fibronectin and proteoglycans as determinants of cell-substratum adhesion *Supramol. Struct.* **11** 401–27

Daoudi M, Lavergne E, Garin A, Tarantino N, Debre P, Pincet F, Combadiere C and Deterre P 2004 Enhanced adhesive capacities of the naturally occurring Ile249-Met280 variant of the chemokine receptor CX3CR1 *J. Biol. Chem.* **279** 19649–57

del Rio A, Perez-Jimenez R, Liu R, Roca-Cusachs P, Fernandez J M and Sheetz M P 2009 Stretching single talin rod molecules activates vinculin binding *Science* **323** 638–41

Dembo M and Wang Y-L 1999 Stresses at the cell-to-substrate interface during locomotion of fibroblasts *Biophys. J.* **76** 2307–16

Dembo M, Oliver T, Ishihara A and Jacobson K 1996 Imaging the traction stresses exerted by locomoting cells with the elastic substratum method *Biophys. J.* **70** 2008–22

Discher D E, Janmey P and Wang Y-L 2005 Tissue cells feel and respond to the stiffness of their substrate *Science* **310** 1139–43

Discher D E, Mohandas N and Evans E A 1994 Molecular maps of red cell deformation: hidden elasticity and *in situ* connectivity *Science* **266** 1032–5

Domke J, Dannohl S, Parak W J, Muller O, Aicher W K and Radmacher M 2000 Substrate dependent differences in morphology and elasticity of living osteoblasts investigated by atomic force microscopy *Colloids Surf.* B **19** 367–79

Dowling J J, Gibbs E, Russell M, Goldman D, Minarcik J, Golden J A and Feldman E L 2008 Kindlin-2 is an essential component of intercalated discs and is required for vertebrate cardiac structure and function *Circ. Res.* **102** 423–31

du Roure O, Saez A, Buguin A, Austin R H, Chavrier P, Silberzan P and Ladoux B 2005 Force mapping in epithelial cell migration *Proc. Natl Acad. Sci. USA* **102** 2390–5

Dulinska I, Targosz M, Strojny W, Lekka M, Czuba P, Balwierz W and Szymonski M 2006 Stiffness of normal and pathological erythrocytes studied by means of atomic force microscopy *J. Biochem. Biophys. Methods* **66** 1–11

Engler A J, Griffin M A, Sen S, Boennemann C G, Sweeney H L and Discher D E 2004a Myotubes differentiate optimally on substrates with tissue-like stiffness: pathological implications for soft or stiff microenvironments *J. Cell Biol.* **166** 877–87

Engler A J, Richert L, Wong J Y, Picart C and Discher D E 2004b Surface probe measurements of the elasticity of sectioned tissue, thin gels and polyelectrolyte multilayer films: correlations between substrate stiffness and cell adhesion *Surf. Sci.* **570** 142–54

Engler A, Bacakova L, Newman C, Hategan A, Griffin M and Discher D 2004c Substrate compliance versus ligand density in cell on gel response *Biophys. J.* **86** 617–28

Erdmann T and Schwarz U S 2004 Adhesion clusters under shared linear loading: a stochastic analysis *Europhys. Lett.* **66** 603–9

Erdmann T and Schwarz U S 2004 Stability of adhesion clusters under constant force *Phys. Rev. Lett.* **92** 108102

Evans E A, Waugh R and Melnik L 1976 Elastic area compressibility modulus of red cell membrane *Biophys. J.* **16** 585–95

Evans E and Ritchie K 1997 Dynamic strength of molecular adhesion bonds *Biophys. J.* **72** 1541–55

Evans E and Kuhan B 1984 Passive material behavior of granulocytes based on large deformation and recovery after deformation tests *Blood* **64** 1028–35

Evans E, Ritchie K and Merkel R 1995 Sensitive force technique to probe molecular adhesion and structural linkages at biological interfaces *Biophys. J.* **68** 2580–7

Fässler R, Georges-Labouesse E and Hirsch E 1996 Genetic analyses of integrin function in mice *Curr. Opin. Cell Biol.* **8** 641–6

Fischer T, Wilharm N, Hayn A and Mierke C T 2017 Matrix and cellular mechanical properties are the driving factors for facilitating human cancer cell motility into 3D engineered matrices *Converg. Sci. Phys. Oncol.* **3** 044003

Florin E L, Moy V T and Gaub H E 1994 Adhesion forces between individual ligand–receptor pairs *Science* **264** 415–7

Franz C M and Müller D J 2005 Analyzing focal adhesion structure by atomic force microscopy *J. Cell Sci.* **118** 5315–23

Friedland J C, Lee M H and Boettiger D 2009 Mechanically activated integrin switch controls $\alpha 5\beta 1$ function *Science* **323** 642–4

Friedrichs J, Helenius J and Müller D J 2010 Stimulated single-cell force spectroscopy to quantify cell adhesion receptor crosstalk *Proteomics* **10** 1455–62

Friedrichs J, Legate K R, Schubert R, Bharadwaj M, Werner C, Müller D J and Benoit M 2013 A practical guide to quantify cell adhesion using single-cell force spectroscopy *Methods* **60** 169–78

Friedrichs J, Manninen A, Müller D J and Helenius J 2008 Galectin-3 regulates integrin $\alpha 2\beta 1$-mediated adhesion to collagen-I and -IV *J. Biol. Chem.* **283** 32264–72

Friedrichs J, Torkko J M, Helenius J, Teräväinen T P, Füllekrug J, Müller D J, Simons K and Manninen A 2007 Contributions of galectin-3 and -9 to epithelial cell adhesion analyzed by single cell force spectroscopy *J. Biol. Chem.* **282** 29375–83

Fritz M, Radmacher M and Gaub H E 1993 *In vitro* activation of human platelets triggered and probed by atomic force microscopy *Exp. Cell Res.* **205** 187–90

Fritz M, Radmacher M and Gaub H E 1994 Granula motion and membrane spreading during activation of human platelets imaged by atomic force microscopy *Biophys. J.* **66** 1328–34

García A J, Ducheyne P and Boettiger D 1997a Quantification of cell adhesion using a spinning disc device and application to surface-reactive materials *Biomaterials* **18** 1091–8

Garcia C R S, Takeuschi M, Yoshioka K and Miyamoto H 1997b Imaging *Plasmodium falciparum*-infected ghosts and parasite by atomic force microscopy *J. Struct. Biol.* **119** 92–8 Q1

Gardel M L, Sabass B, Ji L, Danuser G, Schwarz U S and Waterman C M 2008 Traction stress in focal adhesions correlates biphasically with actin retrograde flow speed *J. Cell Biol.* **183** 999–1005

Geiger B and Bershadsky A 2002 Exploring the neighborhood: adhesion-coupled cell mechanosensors *Cell* **110** 139–42

Geisse N A 2009 AFM and combined optical techniques *Mater. Today* **12** 40–5

Gladilin E, Micoulet A, Hosseini B, Rohr K, Spatz J and Eils R 2007 3D finite element analysis of uniaxial cell stretching: from image to insight *Phys. Biol.* **4** 104–13

Goffin J M, Pittet P, Csucs G, Lussi J, Meister J J and Hinz B 2006 Focal adhesion size controls tension-dependent recruitment of α-smooth muscle actin to stress fibers *J. Cell Biol.* **172** 259–68

Gould S A C, Drake B and Prater C B *et al* 1990 From atoms to integrated circuit chips, blood cells, and bacteria with the atomic force microscope *J. Vac. Sci. Technol.* **A8** 369–73

Grandbois M, Dettmann W, Benoit M and Gaub H E J 2000 Affinity imaging of red blood cells using an atomic force microscope *Histochem. Cytochem.* **48** 719–24

Grashoff C *et al* 2010 Measuring mechanical tension across vinculin reveals regulation of focal adhesion dynamics *Nature* **466** 263–6

Grazioso F, Patton B R and Smith J M 2010 A high stability beam-scanning confocal optical microscope for low temperature operation *Rev. Sci. Instrum.* **81** 093705-4

Gross L, Mohn F, Moll N, Liljeroth P and Meyer G 2009 The chemical structure of a molecule resolved by atomic force microscopy *Science* **325** 1110–4

Gumbiner B M 1996 Cell adhesion: the molecular basis of tissue architecture and morphogenesis *Cell* **84** 345–57

Haga H, Nagayama M, Kawabata K, Ito E, Ushiki T and Sambongi T 2000 Time-lapse viscoelastic imaging of living fibroblasts using force modulation mode in AFM *J. Electron Microsc.* **49** 473–81

Hansma H G 2001 Surface biology of DNA by atomic force microscopy *Annu. Rev. Phys. Chem.* **52** 71–92

Harris A K, Wild P and Stopak D 1980 Silicone rubber substrata: a new wrinkle in the study of cell locomotion *Science* **208** 177–9

Helenius J, Heisenberg C-P, Gaub H E and Muller D J 2008 Single-cell force spectroscopy *J. Cell Sci.* **121** 1785–91

Hertz H 1881 Ueber den Kontakt elastischer Koerper *J. Reine Angew. Math.* **92** 156

Hirschfeld V and Huebner C G 2010 A sensitive and versatile laser scanning confocal optical microscope for single-molecule fluorescence at 77 K *Rev. Sci. Instrum.* **81** 113705-1–7

Hofmann U G, Rotsch C, Parak W J and Radmacher M 1997 Investigating the cytoskeleton of chicken cardiocytes with the atomic force microscope *J. Struct. Biol.* **119** 84–91

Huang S, Chen C S and Ingber D E 1998 Control of cyclin D1, p27(Kip1) and cell cycle progression in human capillary endothelial cells by cell shape and cytoskeletal tension *Mol. Biol. Cell* **9** 3179–93

Hur S S, Zhao Y, Li Y S, Botvinick E and Chien S 2009 Live cells exert 3-dimensional traction forces on their substrata *Cell Mol. Bioeng.* **2** 425–36

Hynes R O 2002 Integrins: bidirectional, allosteric signaling machines *Cell* **110** 673–87

Ishiyama N, Tanaka N and Abe K *et al* 2013 An autoinhibited structure of α-catenin and its implications for vinculin recruitment to adherens junctions *J. Biol. Chem.* **288** 15913–25

Jasiuk I, Sheng P and Tsuchida E 1997 A spherical inclusion in an elastic half-space under shear *J. Appl. Mech.* **64** 471–9

Johnson K L 1985 *Contact Mechanics* (Cambridge: Cambridge University Press)

Johnson K L, Kendall K and Roberts A D 1971 Surface energy and the contact of elastic solids *Proc. R. Soc. Lond.* A **324** 301–21

Kamruzzahan A S, Kienberger F, Stron C M, Berg J, Huss R, Ebner A, Zhu R, Rankl C, Gruber H J and Hinterdorfer P 2004 Imaging morphological details and pathological differences of red blood cells using tapping-mode AFM *Biol. Chem.* **385** 955–60

Kantlehner M, Schaffner P, Finsinger D, Meyer J, Jonczyk A, Diefenbach B, Nies B, Hölzemann G, Goodman S L and Kessler H 2000 Surface coating with cyclic RGD peptides stimulates osteoblast adhesion and proliferation as well as bone formation *ChemBioChem* **1** 107–14

Kaplanski G, Farnarier C, Tissot O, Pierres A, Benoliel A M, Alessi M C, Kaplanski S and Bongrand P 1993 Granulocyte-endothelium initial adhesion. Analysis of transient binding events mediated by E-selectin in a laminar shear flow *Biophys. J.* **64** 1922–33

Kashef J and Franz C M 2015 Quantitative methods for analyzing cell–cell adhesion in development *Dev. Biol.* **401** 165–74

Kashef J, Kohler A, Kuriyama S, Alfandari D, Mayor R and Wedlich D 2009 Cadherin-11 regulates protrusive activity in *Xenopus* cranial neural crest cellsupstream of Trio and the small GTPases *Genes Dev.* **23** 1393–8

Keller R, Davidson L A and Shook D R 2003 How we are shaped: the biomechanics of gastrulation *Differentiation* **71** 171–205

Klebe R J 1974 Isolation of a collagen-dependent cell attachment factor *Nature* **250** 248–51

Koo L Y, Irvine D J, Mayes A M, Lauffenburger D A and Griffith L G 2002 Co-regulation of cell adhesion by nanoscale RGD organization and mechanical stimulus *J. Cell Sci.* **115** 1423–33

Krieg M, Arboleda-Estudillo Y, Puech P-H, Kaefer J, Graner F, Müller D J and Heisenberg C-P 2008 Tensile forces govern germ-layer organization in zebrafish *Nat. Cell Biol.* **10** 429–36

Kunschmann T, Puder S, Fischer T, Perez J, Wilharm N and Mierke C T 2017 Integrin-linked kinase regulates cellular mechanics facilitating the motility in 3D extracellular matrices *Biochim. Biophys. Acta* **1864** 580–93

Kuriyama S, Theveneau E, Benedetto A, Parsons M, Tanaka M, Charras G, Kabla A and Mayor R 2014 *In vivo* collective cell migration requires an LPAR2-dependent increase in tissue fluidity *J. Cell Biol.* **206** 113–27

Ladoux B, Anon E, Lambert M, Rabodzey A, Hersen P, Buguin A, Silberzan P and Mege R M 2010 Strength dependence of cadherin-mediated adhesions *Biophys. J.* **98** 534–42

Landau L D and Lifshitz E M 1970 Theory of elasticity *Course of Theoretical Physics* vol 7, 2nd edn (Oxford: Pergamon)

Lang K M, Hite D A, Simmonds R W, McDermott R, Pappas D P and Martinis J M 2004 Conducting atomic force microscopy for nanoscale tunnel barrier characterization *Rev. Sci. Instrum.* **75** 2726–31

Lapshin R V 1998 Automatic lateral calibration of tunneling microscope scanners *Rev. Sci. Instrum.* **69** 3268–76

Laurent V M, Kasas S, Yersin A, Schaeffer T E, Catsicas S, Dietler G, Verkhovsky A B and Meister J-J 2005 Gradient of rigidity in the lamellipodia of migrating cells revealed by atomic force microscopy *Biophys. J.* **89** 667–75 le Duc Q, Shi Q, Blonk I, Sonnenberg A, Wang N, Leckband D and de Rooij J 2010 Vinculin potentiates E-cadherin mechanosensing and is recruited to actin-anchored sites within adherens junctions in a myosin II-dependent manner *J. Cell Biol.* **189** 1107–15

Lee J, Leonard M, Oliver T, Ishihara A and Jacobson K 1994 Traction forces generated by locomoting keratocytes *J. Cell Biol.* **127** 1957–64

Legant W R, Choi C K, Miller J S, Shao L, Gao L, Betzig E and Chen C S 2013 Multidimensional traction force microscopy reveals out-of-plane rotational moments about focal adhesions *Proc. Natl Acad. Sci. USA* **110** 881–6

Legant W R, Miller J S, Blakely B L, Cohen D M, Genin G M and Chen C S 2011 Measurement of mechanical tractions exerted by cells within three-dimensional matrices *Nat. Methods* **7** 969–71

Legate K R, Wickstroem S A and Faessler R 2009 Genetic and cell biological analysis of integrin outside–in signaling *Genes Dev.* **23** 397–418

Lehenkari P P and Horton M A 1999 Single integrin molecule adhesion forces in intact cells measured by atomic force microscopy *Biochem. Biophys. Res. Commun.* **259** 645–50

Lehnert D, Wehrle-Haller B, David C, Weiland U, Ballestrem C, Imhof B A and Bastmeyer M 2004 Cell behavior on micropatterned substrata: limits of extracellular matrix geometry for spreading and adhesion *J. Cell Sci.* **117** 41–52

Lekka M, Laidler P, Gil D, Lekki J, Stachura Z and Hrynkiewicz A Z 1999 Elasticity of normal and cancerous human bladder cells studied by scanning force microscopy *Eur. Biophys. J.* **28** 312–6

Lekka M, Laidler P, Ignacak J, Labedz M, Lekki J, Struszczyk H, Stachura Z and Hrynkiewicz A Z 2001 The effect of chitosan on stiffness and glycolytic activity of human bladder cells *Biochim. Biophys. Acta* **1540** 127–36

Li F, Redick S D, Erickson H P and Moy V T 2003 Force measurements of the $\alpha5\beta1$ integrinfibronectin interaction *Biophys. J.* **84** 1252–62

Liang X, Mao G and Simon Ng K Y 2004 Probing small unilamellar EggPC vesicles on mica surface by atomic force microscopy *Colloids Surf. B: Biointerfaces* **34** 41–51

Lieber S C, Aubry N, Pain J, Diaz G, Kim S-J and Vatner S F 2004 Aging increases stiffness of cardiac myocytes measured by atomic force microscopy nanoindentation *Am. J. Physiol. Heart Circ. Physiol.* **287** H645–51

Ludwig T, Kirmse R, Poole K and Schwarz U S 2008 Probing cellular microenvironments and tissue remodeling by atomic force microscopy *Pflug. Arch.* **456** 29–49

Lussi J W, Falconnet D, Hubbell J A, Textor M and Csucs G 2006 Pattern stability under cell culture conditions—a comparative study of patterning methods based on PLL-g-PEG background passivation *Biomaterials* **27** 2534–41

Lutolf M P and Hubbell J A 2005 Synthetic biomaterials as instructive extracellular microenvironments for morphogenesis in tissue engineering *Nat. Biotechnol.* **23** 47–55

Mahaffy R E, Park S, Gerde E, Kaes J A and Shih S K 2004 Quantitative analysis of the viscoelastic properties of thin regions of fibroblasts using atomic force microscopy *Biophys. J.* **86** 1777–93

Mahaffy R E, Shih C K, MacKintosh F C and Kaes J A 2000 Scanning probe-based frequency-dependent microrheology of polymer gels and biological cells *Phys. Rev. Lett.* **85** 880–3

Maheshwari G, Brown G, Lauffenburger D A, Wells A and Griffith L G 2000 Cell adhesion and motility depend on nanoscale RGD clustering *J. Cell Sci.* **113** 1677–86

Maitre J L, Berthoumieux H, Krens S F, Salbreux G, Julicher F, Paluch E and Heisenberg C P 2012 Adhesion functions in cell sorting by mechanically coupling the cortices of adhering cells *Science* **338** 253–6

Maitre J L and Heisenberg C P 2011 The role of adhesion energy in controlling cell–cell contacts *Curr. Opin. Cell Biol.* **23** 508–14

Martinez-Rico C, Pincet F, Perez E, Thiery J P, Shimizu K, Takai Y and Dufour S 2005 Separation force measurements reveal different types of modulation of E-cadherin-based adhesion by nectin-1 and -3 *J. Biol. Chem.* **280** 4753–60

Maskarinec S A, Franck C, Tirrell D A and Ravichandran G 2009 Quantifying cellular traction forces in three dimensions *Proc. Natl Acad. Sci. USA* **106** 22108–13

Massia S P and Hubbell J A 1991 An RGD spacing of 440 nm is sufficient for integrin $\alpha V\beta 3$-mediated fibroblast spreading and 140 nm for focal contact and stress fiber formation *J. Cell Biol.* **114** 1089–100

Mathur A B, Collinsworth A M, Reichert W M, Kraus W E and Truskey G A 2001 Endothelial, cardiac muscle and skeletal muscle exhibit different viscous and elastic properties as determined by atomic force microscopy *J. Biomech.* **34** 1545–53

Mathur A B, Truskey G A and Reichert W M 2000 Atomic force and total internal reflection fluorescence microscopy for the study of force transmission in endothelial cells *Biophys. J.* **78** 1725–35

McBeath R, Pirone D M, Nelson C M, Bhadriraju K and Chen C S 2004 Cell shape, cytoskeletal tension, and RhoA regulate stem cell lineage commitment *Dev. Cell* **6** 483–95

Merkel R, Nassoy P, Leung A, Ritchie K and Evans E 1999 Energy landscapes of receptor-ligand bonds explored with dynamic force spectroscopy *Nature* **397** 50–3

Micoulet A, Spatz J P and Ott A 2005 Mechanical response analysis and power generation by single-cell stretching *ChemPhysChem* **6** 663–70

Mierke C T 2009 The role of vinculin in the regulation of the mechanical properties of cells *Cell Biochem Biophys.* **53** 115–26

Mierke C T 2011 Cancer cells regulate biomechanical properties of human microvascular endothelial cells *J. Biol. Chem.* **286** 40025–37

Mierke C T, Bretz N and Altevogt P 2011b Contractile forces contribute to increased glycosylphosphatidylinositol-anchored receptor CD24-facilitated cancer cell invasion *J. Biol. Chem.* **286** 34858–71

Mierke C T, Fischer T, Puder S, Tom Kunschmann T, Soetje B and Ziegler W H 2017 Focal adhesion kinase activity is required for actomyosin contractility-based invasion of cells into dense 3D matrices *Sci. Rep.* **7** 42780

Mierke C T, Frey B, Fellner M, Herrmann M and Fabry B 2011a Integrin $\alpha 5\beta 1$ facilitates cancer cell invasion through enhanced contractile forces *J. Cell Sci.* **124** 369–83

Mierke C T, Kollmannsberger P, Paranhos-Zitterbart D, Smith J, Fabry B and Goldmann W H 2008b Mechano-coupling and regulation of contractility by the vinculin tail domain *Biophys. J.* **94** 661–70

Mierke C T, Kollmannsberger P, Zitterbart D P, Diez G, Koch T M, Marg S, Ziegler W H, Goldmann W H and Fabry B 2010 Vinculin facilitates cell invasion into three-dimensional collagen matrices *J. Biol. Chem.* **285** 13121–30

Mierke C T, Zitterbart D P, Kollmannsberger P, Raupach C, Schlotzer-Schrehardt U, Goecke T W, Behrens J and Fabry B 2008a Breakdown of the endothelial barrier function in tumor cell transmigration *Biophys. J.* **94** 2832–46

Miller J S, Shen C J, Legant W R, Baranski J D, Blakely B L and Chen C S 2010 Bioactive hydrogels made from step-growth derived PEG-peptide macromers *Biomaterials* **31** 3736–43
Mitchinson J M and Swann M M 1954 The mechanical properties of the cell surface: I. The cell elastimeter *J. Exp. Biol.* **31** 443–60
Miyake Y, Inoue N, Nishimura K, Kinoshita N, Hosoya H and Yonemura S 2006 Actomyosin tension is required for correct recruitment of adherens junction components and zonula occludens formation *Exp. Cell Res.* **312** 1637–50
Miyazaki H and Hayashi K 1999 Atomic force microscopic measurement of the mechanical properties of intact endothelial cells in fresh arteries *Med. Biol. Eng. Comput.* **37** 530–6
Monkley S J, Zhou X H, Kinston S J, Giblett S M, Hemmings L, Priddle H, Brown J E, Pritchard C A, Critchley D R and Fässler R 2000 Disruption of the talin gene arrests mouse development at the gastrulation stage *Dev. Dyn.* **219** 560–74
Montanez E, Ussar S, Schifferer M, Boesl M, Zent R, Moser M and Fässler R 2008 Kindlin-2 controls bidirectional signaling of integrins *Genes Dev.* **22** 1325–30
Mozafari M R, Reed C J, Rostron C and Hasirci V 2005 A review of scanning probe microscopy investigations of liposome–DNA complexes *J. Liposome Res.* **15** 93–107
Mozhanova A A, Nurgazizov N I and Bukharaev A A 2003 Local elastic properties of biological materials studied by SFM *Proc. Nizhni Novgorod* pp 266–7
Müller D J, Helenius J, Alsteens D and Dufrêne Y F 2009 Force probing surfaces of living cells to molecular resolution *Nat. Chem. Biol.* **5** 383–90
Muñoz Javier A *et al* 2006 *Small* **2** 394–400
Murakoshi M, Yoshida N, Iida K, Kumano S, Kobayashi T and Wada H 2006 Local mechanical properties of mouse outer hair cells: atomic force microscopic study *Auris Nasus Larynx* **33** 149–57
Nagayama M, Haga H and Kawabata K 2001 Drastic change of local stiffness distribution correlating to cell migration in living fibroblasts *Cell Motil. Cytoskel.* **50** 173–9
Nishida S, Kobayashi D, Sakurada T, Nakazawa T, Hoshi Y and Kawakatsu H 2008 Photothermal excitation and laser Doppler velocimetry of higher cantilever vibration modes for dynamic atomic force microscopy in liquid *Rev. Sci. Instrum.* **79** 123703
Nowakowski R, Luckham P and Winlove P 2001 Imaging erythrocytes under physiological conditions by atomic force microscopy *Biochim. Biophys. Acta* **1514** 170–6
Oh M-J, Kuhr F, Byfield F and Levitan I 2012 Micropipette aspiration of substrate-attached cells to estimate cell stiffness *J. Vis. Exp.* **67** 3886
Ohashi T, Ishii Y, Ishikawa Y, Matsumoto T and Sato M 2002 Experimental and numerical analyses of local mechanical properties measured by atomic force microscopy for sheared endothelial cells *Biomed. Mater. Eng.* **12** 319–27
Okajima T 2011 Atomic force microscopy for the examination of single cell rheology *Methods Mol. Biol.* **736** 303–29
Oliver T, Jacobson K and Dembo M 1998 Design and use of substrata to measure traction forces exerted by cultured cells *Methods Enzymol.* **298** 497–521
Orr A W, Helmke B P, Blackman B R and Schwartz M A 2006 Mechanisms of mechanotransduction *Dev. Cell* **10** 11–20
Palecek S P, Huttenlocher A, Horwitz A F and Lauffenburger D A 1998 Physical and biochemical regulation of integrin release during rear detachment of migrating cells *J. Cell Sci.* **111** 929–40
Paluch E K (ed) 2015 Biophysical methods in cell biology *Methods Cell Biol.* 125 1–488

Pampaloni F, Reynaud E G and Stelzer E H 2007 The third dimension bridges the gap between cell culture and live tissue *Nat. Rev. Mol. Cell Biol.* **8** 839–45
Pasche S, Textor M, Meagher L, Spencer N D and Griesser H J 2005 Relationship between interfacial forces measured by colloid-probe atomic force microscopy and protein resistance of poly(ethylene glycol)-grafted poly(L-lysine) adlayers on niobia surfaces *Langmuir* **21** 6508–20
Paszek M J *et al* 2005 Tensional homeostasis and the malignant phenotype *Cancer Cell* **8** 241–54
Paul R, Heil P, Spatz J P and Schwarz U S 2008 Propagation of mechanical stress through the actin cytoskeleton toward focal adhesions: model and experiment *Biophys. J.* **94** 1470–82
Pelham R J and Wang Y-L 1999 High resolution detection of mechanical forces exerted by locomoting fibroblasts on the substrate *Mol. Biol. Cell* **10** 935–45
Pelham R J Jr and Wang Y 1997 Cell locomotion and focal adhesions are regulated by substrate flexibility *Proc. Natl Acad. Sci. USA* **94** 13661–5
Pesen D and Hoh J H 2005 Micromechanical architecture of the endothelial cell cortex *Biophys. J.* **88** 670–9
Pfaff M, Tangemann K, Müller B, Gurrath M, Müller G, Kessler H, Timpl R and Engel J 1994 Selective recognition of cyclic RGD peptides of NMR defined conformation by αIIbβ3, αVβ3, and α5β1 integrins *J. Biol. Chem.* **269** 20233–8
Puech P-H, Taubenberger A, Ulrich F, Krieg U M, Mueller D J and Heisenberg C-P 2005 Measuring cell adhesion forces of primary gastrulating cells from zebra fish using atomic force microscopy *J. Cell Sci.* **118** 4199–206
Radmacher M, Fritz M, Kacher C M, Cleveland J P and Hansma P K 1996 Measuring the viscoelastic properties of human platelets with the atomic force microscope *Biophys. J.* **70** 556–67
Rand R P and Burton A C 1964 Mechanical properties of the red cell membrane. I. Membrane stiffness and intracellular pressure *Biophys. J.* **4** 115–35
Raupach C, Paranhos-Zitterbart D, Mierke C, Metzner C, Mueller A F and Fabry B 2007 Stress fluctuations and motion of cytoskeletal-bound markers *Phys. Rev.* E **76** 011918
Ridley A J and Hall A 1992 The small GTP-binding protein rho regulates the assembly of focal adhesions and actin stress fibers in response to growth factors *Cell* **70** 389–99
Riveline D, Zamir E, Balaban N Q, Schwarz U S, Ishizaki T, Narumiya S, Kam Z, Geiger B and Bershadsky A D 2001 Focal contacts as mechanosensors: externally applied local mechanical force induces growth of focal contacts by an mDia1-dependent and ROCK-independent mechanism *J. Cell Biol.* **153** 1175–86
Roiter Y and Minko S 2005 AFM single molecule experiments at the solid–liquid interface: *in situ* conformation of adsorbed flexible polyelectrolyte chains *J. Am. Chem. Soc.* **127** 15688–9
Rosenbluth M J, Lam W A and Fletcher D A 2006 Force microscopy of nonadherent cells: a comparison of leukemia cell deformability *Biophys. J.* **90** 2994–3003
Rotsch C and Radmacher R 2000 Drug-induced changes of cytoskeletal structure and mechanics in fibroblasts: an atomic force microscopy study *Biophys. J.* **78** 520–35
Rotsch C, Braet F, Wisse E and Radmacher M 1997 AFM imaging and elasticity measurements on living rat liver macrophages *Cell Biol. Int.* **21** 685–96
Rotsch C, Jacobson K and Radmacher R 1999 Dimensional and mechanical dynamics of active and stable edges in motile fibroblasts investigated by using atomic force microscopy *Proc. Natl Acad. Sci. USA* **96** 921–6
Roy P, Rajfur Z, Pomorski P and Jacobson K 2002 Microscope-based techniques to study cell adhesion and migration *Nat. Cell Biol.* **4** E91–6

Ruoslahti E 1996 RGD and other recognition sequences for integrins *Annu. Rev. Cell Dev. Biol.* **12** 697–715

Saez A, Ghibaudo M, Buguin A, Silberzan P and Ladoux B 2007 Rigidity-driven growth and migration of epithelial cells on microstructured anisotropic substrates *Proc. Natl Acad. Sci. USA* **104** 8281–6

Sasaki S, Morimoto M, Haga H, Kawabata K, Ito E, Ushiki T, Abe K and Sambongi T 1998 Elastic properties of living fibroblasts as imaged using force modulation mode in atomic force microscopy *Arch. Histol. Cytol.* **61** 57–63

Sato H, Kataoka N, Kajiya F, Katano M, Takigawa T and Masuda T 2004 Kinetic study on the elastic change of vascular endothelial cells on collagen matrices by atomic force microscopy *Colloids Surf.* B **34** 141–6

Sato H, Nagayama K, Kataoka N, Sasaki M and Hane K 2000 Local mechanical properties measured by atomic force microscopy for cultured bovine endothelial cells exposed to shear stress *J. Biomech.* **33** 127–35

Sato M, Levesque M J and Nerem R M 1987 Micropipette aspiration of cultured bovine aortic endothelial cells exposed to shear stress *Arteriosclerosis* **7** 276–86

Scheffer L, Bitler A, Ben-Jacob E and Korenstein R 2001 Atomic force pulling: probing the local elasticity of the cell membrane *Eur. Biophys. J.* **30** 83–90

Schepis A, Sepich D and Nelson W J 2012 αE-catenin regulates cell–cell adhesion and membrane blebbing during zebrafish epiboly *Development* **139** 537–46

Schmid-Schönbein G W, Sung K L, Tözeren H, Skalak R and Chien S 1981 Passive mechanical properties of human leukocytes *Biophys. J.* **36** 243–56

Schrot S, Weidenfeller C, Schäffer T E, Robenek H and Galla H J 2005 Influence of hydrocortisone on the mechanical properties of the cerebral endothelium *in vitro Biophys. J.* **89** 3904–10

Schwarz U S 2007 Soft matters in cell adhesion: rigidity sensing on soft elastic substrates *Soft Matter* **3** 263–6

Schwarz U S, Balaban N Q, Riveline D, Bershadsky A, Geiger G and Safran S A 2002 Calculation of forces at focal adhesions from elastic substrate data: the effect of localized force and the need for regularization *Biophys. J.* **83** 1380–94

Selhuber-Unkel C, Erdmann T, López-García M, Kessler H, Schwarz U S and Spatz J P 2010 Cell adhesion strength is controlled by intermolecular spacing of adhesion receptors *Biophys. J.* **98** 543–51

Selhuber-Unkel C, López-García M, Kessler H and Spatz J P 2008 Cooperativity in adhesion cluster formation during initial cell adhesion *Biophys. J.* **95** 5424–31

Shao J Y and Hochmuth R M 1996 Micropipette suction for measuring piconewton forces of adhesion and tether formation from neutrophil membranes *Biophys. J.* **71** 2892–901

Sheppard D 2000 *In vivo* functions of integrins: lessons from null mutations in mice *Matrix Biol.* **19** 203–9

Shroff S G, Saner D R and Lal R 1995 Dynamic micromechanical properties of cultured rat atrial myocytes measured by atomic force microscopy *Am. J. Physiol. Cell Physiol.* **269** C286–92

Sieg D J, Hauck C R, Ilic D, Klingbeil C K, Schaefer E, Damsky C H and Schlaepfer D D 2000 FAK integrates growth-factor and integrin signals to promote cell migration *Nat. Cell Biol.* **2** 249–56

Simon A, Cohen-Bouhacina T, Porte V C, Aime J P, Amedee J, Bareille R and Baquey C 2003 Characterization of dynamic cellular adhesion of osteoblasts using atomic force microscopy *Cytometry* A **54** 36–47

Singhvi R, Kumar A, Lopez G P, Stephanopoulos G N, Wang D I, Whitesides G M and Ingber D E 1994 Engineering cell shape and function *Science* **264** 696–8

Solon J *et al* 2007 Fibroblast adaptation and stiffness matching to soft elastic substrates *Biophys. J.* **93** 4453–61

Spatz J P, Moessmer S, Hartmann C, Moeller M, Herzog T, Krieger M, Boyen H-G, Ziemann P and Kabius B 2000 Ordered deposition of inorganic clusters from micellar block copolymer films *Langmuir* **16** 407–15

Staedler B, Falconnet D, Pfeiffer I, Hook F and Voros J 2004 Micropatterning of DNA-tagged vesicles *Langmuir* **20** 11348–54

Stockinger P, Maitre J L and Heisenberg C P 2011 Defective neuroepithelial cell cohesion affects tangential branchiomotor neuron migration in the zebrafish neural tube *Development* **138** 4673–83

Sugawara M, Ishida Y and Wada H 2002 Local mechanical properties of guinea pig outer hair cells measured by atomic force microscopy *Hear. Res.* **174** 222–9

Sugawara M, Ishida Y and Wada H 2004 Mechanical properties of sensory and supporting cells in the organ of Corti of the guinea pig cochlea—study by atomic force microscopy *Hear. Res.* **192** 57–64

Sugimura K, Lenne P-F and Graner F 2016 Measuring forces and stresses in situ in living tissues *Development* **143** 186–96

Sumida G M, Tomita T M, Shih W and Yamada S 2011 Myosin II activity dependent and independent vinculin recruitment to the sites of E-cadherin-mediated cell–cell adhesion *BMC Cell Biol.* **12** 48

Sung K L, Sung L A, Crimmins M, Burakoff S J and Chien S 1986 Determination of junction avidity of cytolytic T cell and target cell *Science* **234** 1405–8

Taguchi K, Ishiuchi T and Takeichi M 2011 Mechanosensitive EPLIN-dependent remodeling of adherens junctions regulates epithelial reshaping *J. Cell Biol.* **194** 643–56

Takai E, Costa K D, Shaheen A, Hung C T and Guo X E 2005 Osteoblast elastic modulus measured by atomic force microscopy is substrate dependent *Ann. Biomed. Eng.* **33** 963–71

Takeuchi M, Miyamoto H, Sako Y, Komizu H and Kuzumi A 1998 Structure of the erythrocyte membrane skeleton as observed by atomic force microscopy *Biophys. J.* **74** 2171–83

Tan J L, Tien J, Pirone D M, Gray D S, Bhadriraju K and Chen C S 2003 Cells lying on a bed of microneedles: an approach to isolate mechanical force *Proc. Natl Acad. Sci. USA* **100** 1484–9

Taubenberger A, Cisneros D A, Friedrichs J, Puech P H, Muller D J and Franz C M 2007 Revealing early steps of $\alpha 2\beta 1$ integrin-mediated adhesion to collagen type I by using single-cell force spectroscopy *Mol. Biol. Cell* **18** 1634–44

Thie M, Röspel R, Dettmann W, Benoit M, Ludwig M, Gaub H E and Denker H W 1998 Interactions between trophoblast and uterine epithelium: monitoring of adhesive forces *Hum. Reprod.* **13** 3211–9

Thoumine O, Kocian P, Kottelat A and Meister J J 2000 Short-term binding of fibroblasts to fibronectin: optical tweezers experiments and probabilistic analysis *Eur. Biophys. J.* **29** 398–408

Thoumine O, Ott A and Louvard D 1996 Critical centrifugal forces induce adhesion rupture or structural reorganization in cultured cells *Cell Motil. Cytoskel.* **33** 276–87

Timoshenko S P and Woinowsky-Krieger S 1970 *Theory of Plates and Shells* (New York: McGraw-Hill)

Tolomeo J F, Steele C R and Holley M C 1996 Mechanical properties of the lateral cortex of mammalian auditory outer hair cells *Biophys. J.* **71** 421–9

Trappmann B, Gautrot J E and Connelly J T *et al* 2012 Extracellular-matrix tethering regulates stem-cell fate *Nat. Mater.* **11** 642–9

Tulla M, Helenius J, Jokinen J, Taubenberger A, Müller D J and Heino J 2008 TPA primes $\alpha 2\beta 1$ integrins for cell adhesion *FEBS Lett.* **582** 3520–4

Vinckier A and Semenza G 1998 Measuring elasticity of biological materials by atomic force microscopy *FEBS Lett.* **430** 12–6

Vogel V and Sheetz M 2006 Local force and geometry sensing regulate cell functions *Nat. Rev. Mol. Cell Biol.* **7** 265–75

Wada H, Usukura H, Katori Y, Kakehata S, Ikeda K and Kobayashi T 2003 Relationship between the local stiffness of the outer hair cell along the cell axis and its ultrastructure observed by atomic force microscopy *Hear. Res.* **177** 61–70

Wang N, Butler J P and Ingber D E 1993 Mechanotransduction across the cell surface and through the cytoskeleton *Science* **260** 1124–7

Wang Y, Meng F and Sachs F 2011 Genetically encoded force sensors for measuring mechanical forces in proteins *Commun. Integr. Biol.* **4** 385–90

Weber G F, Bjerke M A and DeSimone D W 2011 Integrins and cadherins join forces to form adhesive networks *J. Cell Sci.* **124** 1183–93

Willemsen O H, Snel M M E, Cambi A, Greve J, De Grooth B G and Carl G 2000 Figdor biomolecular interactions measured by atomic force microscopy *Biophys. J.* **79** 3267–81

Wojcikiewicz E P, Zhang X and Moy V T 2004 Force and compliance measurements on living cells using atomic force microscopy (AFM) *Biol. Proceed. Online* **6** 1–9

Wu H W, Kuhn T and Moy V N 1998 Mechanical properties of L929 fibro-blasts measured by atomic force microscopy: effects of anticytoskeletal drugs and membrane crosslinking *Scanning* **20** 389–97

Xiong J P, Stehle T, Diefenbach B, Zhang R, Dunker R, Scott D L, Joachimiak A, Goodman S L and Arnaout M A 2001 Crystal structure of the extracellular segment of integrin $\alpha V\beta 3$ *Science* **294** 339–45

Xu X and Raman A 2007 Comparative dynamics of magnetically, acoustically and Brownian motion driven microcantilevers in liquids *J. Appl. Phys.* **102** 034303

Yamane Y, Shiga H, Haga H, Kawabata K, Abe K and Ito E 2000 Quantitative analyses of topography and elasticity of living and fixed astrocytes *J. Electron Microsc.* **49** 463–71

Yeung T, Georges P C, Flanagan L A, Marg B, Ortiz M, Funaki M, Zahir N, Ming W, Weaver V and Janmey V A 2005 Effects of substrate stiffness on cell morphology, cytoskeletal structure, and adhesion *Cell Motil. Cytoskelet.* **60** 24–34

Yonemura S, Wada Y, Watanabe T, Nagafuchi A and Shibata M 2010 α-catenin as a tension transducer that induces adherens junction development *Nat. Cell Biol.* **12** 533–42

Yoshikawa Y, Yasuike T, Yagi A and Yamada T 1999 Transverse elasticity of myofibrils of rabbit skeletal muscle studied by atomic force microscopy *Biochem. Biophys. Res. Commun.* **256** 13–9

Zaidel-Bar R, Cohen M, Addadi L and Geiger B 2004 Hierarchical assembly of cell–matrix adhesion complexes *Biochem. Soc. Trans.* **32** 416–20

Zaidel-Bar R, Kam Z and Geiger B 2005 Polarized downregulation of the paxillin-p130CAS-Rac1 pathway induced by shear flow *J. Cell Sci.* **118** 3997–4007

Zamir E and Geiger B 2001 Molecular complexity and dynamics of cell–matrix adhesions *J. Cell Sci.* **114** 3583–90

Zamir E, Katz B Z, Aota S, Yamada K M, Geiger B and Kam Z 1999 Molecular diversity of cell–matrix adhesions *J. Cell Sci.* **112** 1655–69

Zhang S, Kraus W E and Truskey G A 2004 Stretch-induced nitric oxide modulates mechanical properties of skeletal muscle cells *Am. J. Physiol. Cell Physiol.* **287** C292–9

Zhou J, Pal S, Maiti S and Davidson L A 2015 Force production and mechanical accommodation during convergent extension *Development* **142** 692–701

Zimmermann H, Richter E, Reichle C, Westphal I, Geggier P, Rehn U, Rogaschewski S, Bleiss W and Fuhr G R 2001 Mammalian cell traces—morphology, molecular composition, artificial guidance and biotechnological relevance as a new type of 'bionanotube' *Appl. Phys.* **73** 11–26

Zimmermann H, Hagedorn R, Richter E and Fuhr G 1999 Topography of cell traces studied by atomic force microscopy *Eur. Biophys. J.* **28** 516–25

IOP Publishing

Physics of Cancer, Volume 1 (Second Edition)
Interplay between tumor biology, inflammation and cell mechanics
Claudia Tanja Mierke

Chapter 5

Cell surface tension, the mobility of cell surface receptors and their location in specific regions

Summary

The shape fluctuations of the cell membrane that occur in all cells are continuously detected and hence are proposed to determine the function of the membrane. Although it has been shown how membrane fluctuations are affected by cellular activity in adherent cells, the spatial regulation of these fluctuations and the associated alterations in the mechanical properties of the membrane are still not clearly understood. The cell surface tension seems to play a crucial a role on the selection of an aggressive and invasive cancer cell subpopulation that is able to disseminate from the primary tumor in the microenvironment and propagate through all the steps of the metastatic cascade for the malignant tumor progression. Similar to cells rounding up during mitosis, cancer cells enhance their surface tension through an increase of the contractile actomyosin cortex based on elevated internal hydrostatic pressure. In the scenario of a liquid cellular interior, the surface tension is purely related to the local curvature and the hydrostatic pressure difference by Laplace's law. Moreover, the classical differential adhesion hypothesis fails in describing the experimental results and it is refined by including surface tension. The diffusion on the cell surface is not purely Brownian motion, as the mobility of cell surface receptors and their structural organization in special compartments (regions) within the cell surface membrane is constrained. These confinements restrict the lateral motility of cell surface receptors and are hence termed lipid rafts. The impact of these confinements alters the entire motility of cancer cells. Moreover, the location of cell surface receptors within the membrane and their anchoring to the cell's actomyosin cytoskeleton affects cellular mechanical properties such as membrane fluctuations. Although the actin polymerization or myosin II activity individually increases fluctuations, the cortex in unperturbed cells stretches out the cell membrane and thereby dampens membrane fluctuations. When these

doi:10.1088/978-0-7503-1753-5ch5

fluctuations are fitted with common models, the dampening of these fluctuations is provided by the spatial confinement of the cell's cortex. However, decreased fluctuations during mitosis or upon ATP-depletion/stabilization of actomyosin cortex are associated with elevated surface tension. Fluctuation maps of the cell membrane and also local temporal autocorrelation functions both reveal ATP-dependent transient short-range (less than 2 μm) heterogeneities. However, it is still elusive what role these fluctuation heterogeneities play during the malignant progression of cancer.

5.1 Surface tension

During animal morphogenesis, large-scale cellular movements occur, which lead to the rearrangement, mutual spreading and compartmentalization of cell populations in specific structural arrangements. The morphogenetic cell rearrangements, including cell sorting and mutual tissue spreading, have been compared with the behavior of other soft matter, such as immiscible liquids, which turned out to be very similar. Based on this strong similarity, it has been suggested that entire tissues act as liquids and are characterized by a specific surface tension, which is a collective, macroscopic property of groups of mobile, cohering (connected) cells (figure 5.1). How can tissues generate surface tension? Several different theories have been formulated to explain how mesoscopic cell properties such as cell–cell adhesion and the contractility of cell interfaces may regulate surface tension of tissues. It has been proposed that cell–cell adhesion and cellular contractility contribute to tissue surface tension, but there is not yet a model that describes and predicts the dependence of tissue surface tension on these two mesoscopic parameters. However, there is an approach that shows that the ratio of adhesion to cortical tension regulates the amount of tissue surface

Figure 5.1. Cell surface receptor and cell surface tension. The scanning electron microscopic images of 786-O human kidney cancer cells shows the membrane receptors and proteins as dot-like structures on the curved and hence tensioned membrane surface.

tension. This approach seems to be based on a minimal model that relies on the regulatory feedback between mechanical energy and geometry and predicts the shapes of aggregate surface cells. In addition, this model has been experimentally verified. In particular, this model indicates that there is a crossover from adhesion-driven to cortical-tension-driven behavior as a function of the ratio between these two influencing factors. Moreover, the increased surface tension is correlated with decreased fluctuations on ATP-depletion/stabilization of cortex (Biswas *et al* 2017). In particular, ATP-dependent cellular activities increase the nanoscale z-fluctuations of the cell membrane, whereas the membrane is stretched out laterally, which in turn alters the surface tension.

Cell membrane deformations can affect multiple cellular functions such as motility, cell division, vesicle trafficking and mechano-response (Zimmerberg and Kozlov 2006, McMahon and Gallop 2005). Fluctuation spectra of deformations can capture parameters that are responsible for the deformability of the cell membrane. Moreover, fluctuations are based on the viscoelastic properties of the membrane and they require thermal energy and ATP-driven processes. The temporal range of fluctuations is broad, as slow (10 s) actin waves can drive large wavelength fluctuations (100 nm–10 mm) at cell's edges and the basal membrane (Giannone *et al* 2004, Doebereiner *et al* 2006, Chen *et al* 2009). In addition, they are accompanied by fluctuations with smaller amplitudes (5–50 nm) that are usually prominently observed at the basal membrane (Pierres *et al* 2008, Monzel *et al* 2015, Cardoso Dos Santos *et al* 2016) and are mainly based on thermal alterations. Although larger amplitudes of these fluctuations are required in cell spreading and motility (Giannone *et al* 2004, Doebereiner *et al* 2006, Dubin-Thaler *et al* 2004), thermal fluctuations with amplitudes less than 10 nm can affect the spatial organization of molecules on membranes, which causes the assembly of nano-domains or the formation of invaginations (Pezeshkian *et al* 2017, Fenz *et al* 2017).

In addition to cell edges, fluctuations within the other regions in the basal membrane of nucleated adherent cells have been well investigated (Pierres *et al* 2008, Monzel *et al* 2015, Cardoso Dos Santos *et al* 2016, Tarantola *et al* 2011, Lo *et al* 1993). There exists a broad variety of imaging techniques assessing the fluctuations of the basal membrane, which turns out to be an ideal platform to analyze the mechanical properties and functional implications of these shorter fluctuations. In particular, the dynamics of the basal membrane can be determined due to the height of the fluctuations (distance from the substrate in the z-direction) (Monzel and Sengupta 2016). It has been found that the amplitudes can be tens of nanometers and these fluctuations are altered by ATP-depletion and perturbations of the cytoskeleton (Pierres *et al* 2008, Monzel *et al* 2015, Cardoso Dos Santos *et al* 2016, Tarantola *et al* 2011, Lo *et al* 1993). A weakness of these studies is that they have either explored the power spectral density of fluctuations within cell monolayers with no spatial resolution (Tarantola *et al* 2011, Lo *et al* 1993) or have been focused on solely temporal (Monzel *et al* 2015) or spatial (Cardoso Dos Santos *et al* 2016) aspects in single cells, whereas they rarely combine temporal and spatial effects for a broad range of time- or length-scales (Chen *et al* 2009, Pierres *et al* 2008). Indeed, the combination of both seems to be highly recommended, as has been demonstrated in red blood cells for the discrimination of nonthermal from thermal fluctuations (Park *et al* 2010, Rodríguez-García *et al* 2015). In order to understand the spatial regulation of lipid/protein organization by fluctuations, their spatial heterogeneities cannot be simply ignored.

At the basal membrane of nucleated adherent cells, ATP-dependent processes increase the amplitude of fluctuations and the distributions are then non-Gaussian (Monzel *et al* 2015). However, how ATP-dependent cellular processes alter the fluctuation distribution and the heterogeneity of fluctuations is yet to be clearly revealed and remains to be established. Moreover, the impact of the actomyosin network on the spatio-temporal regulation of the fluctuations also remains unclear, as two reports presented opposite effects of ATP such as an increase (Pierres *et al* 2008) or a decrease of the fluctuations (Tarantola *et al* 2011). Unlike the spectrin network of red blood cells, actin cytoskeletal structures within the cytoplasm of a single nucleated adherent cell builds heterogeneous patterns (the cytoskeletal architecture), which alter the height profile (termed undulations) of the cell membrane (Bereiter-Hahn *et al* 1979). How does the actomyosin cortex, including the actin polymerization and myosin II motor activity, regulate the distribution of the membrane fluctuations and their heterogeneities? In order to reveal the effect of the cortex on fluctuations, the spatial distribution of the membrane fluctuations need to be varied through changes of the actomyosin cortex and its mechanical properties, which then need to be analyzed simultaneously. Using total internal reflection fluorescence (TIRF), these issues have been investigated in HeLa, Chinese hamster ovary cells CHO (epithelial), and C2C12 (myoblast) cells (Biswas *et al* 2017). The noninvasive imaging technique, termed interference reflection microscopy (IRM) (Abercrombie and Dunn 1975, Curtis 1964, Godwin *et al* 1989, Raedler and Sackmann 1993, Limozin and Sengupta 2009), can be adapted to measure the spatio-temporal parameters of membrane z-fluctuations at high z-resolution, whereas the xy-resolution is limited by diffraction. In particular, the depth of the fluctuation analysis is limited to 100 nm in the z-direction. Using pharmacological drug-based perturbations of the ATP-driven processes and the actomyosin cortex, their impact on the fluctuations can be investigated and subsequently the temporal spectra can be related to existing theoretical models.

The role of tissue surface tension needs to be revealed, as tissue surface tension is assumed to act as a driving force for many biological processes. Knowing its cellular origins can help to manipulate the tissue organization using distinct drug treatments. Indeed, it is well established that many tissues act similarly to liquids over long timescales. Cell tracking *in vivo* and *in vitro* has revealed large-scale flows, exchange of nearest neighbors in a cellular aggregate and rounding-up as well as fusion of cell aggregates (Schoetz 2008). Surface tension is a macroscopic rheological property and thus can be determined by a tissue surface tensiometer (TST) (Schoetz 2008, Foty *et al* 1994, 1996, Forgacs *et al* 1998, Schoetz *et al* 2008, Mgharbel *et al* 2009, Norotte *et al* 2008, Davis *et al* 1997) or by the classical biophysical micropipette aspiration technique (Guevorkian *et al* 2010). In particular, the surface tension may be the driving factor for tissue self-organization in embryogenesis (Davis *et al* 1997, Armstrong 1989, Holtfreter 1944, Steinberg 1996) and cancer (Foty *et al* 1998, Foty and Steinberg 1997). In line with this, the special manner of cell sorting within tissues and tissue spreading can be indeed explained in terms of tissue surface tensions, which can be altered between different cell types (Schoetz 2008, Foty *et al* 1986, Forgacs *et al* 1998, Schoetz *et al* 2008, Davis *et al* 1997, Duguay *et al* 2003, Borghi and Nelson 2009).

Over the last three decades, there have been only two prominent opposing theories describing the mesoscopic origin of tissue surface tension. One theory, the differential adhesion hypothesis (DAH), predicts that, similar to ordinary fluids, tissue surface tension is proportional to the intensity of the adhesive energy between the constituent cells. The constituent cells are treated as simple point objects. The DAH explains the motility of cells during tissue morphogenesis using thermodynamic principles. The theory treats tissues as liquids consisting of migrating cells with different amounts of surface tension. In order to minimize their interfacial free energy, the cells assemble spontaneously into aggregates, leading finally to tissue reorganization. According to the DAH theory, during tissue morphogenesis, the migrating cells displayed a very similar behavior to a mixture of different liquids. The theory was originally primarily developed to understand cell-sorting behavior in the embryos of vertebrates, but it can also be applied to other morphogenetic phenomena, for example those occurring during the malignant progression of cancer, such as metastasis, epithelial-to-mesenchymal transition and wound-healing processes. In particular, the DAH relies on quantitative differences in the strength of cell surface adhesion.

The DAH has proven successful in numerous studies with different cell lines (Foty *et al* 1984, 1986, Forgacs *et al* 1998, Schoetz *et al* 2008, Duguay *et al* 2003), malignant tissues (Foty *et al* 1998, Foty and Steinberg 1997) and embryonic tissues (Schoetz 2008, Schoetz *et al* 2008, Davis *et al* 1997, Borghi and Nelson 2009) and hence is widely accepted in the field of biophysical research (Steinberg 1996, Lecuit and Lenne 2007). In addition, it has been experimentally verified that there is a linear relationship between the expression of adhesion molecules on the cell surface and the levels of tissue surface tension (Foty and Steinberg 2005).

However, experimental data obtained with AFM (Krieg *et al* 2008) and TST (Manning *et al* 2010) postulate a dependence of the surface tension on activity of the actomyosin cytoskeleton in the cell, supporting an alternative theory to DAH in which cortical tension in individual cells is supposed to be the important factor for providing the surface tension of tissues. This second hypothesis on the origin of tissue surface tension is termed the differential interfacial tension hypothesis (DITH), and it was formulated by Harris (1976), Brodland (2003) and Graner (1993). In particular, the DITH relates tissue surface tension to the tension along individual cellular interfaces. The DITH theories are appealing, as they recognize, in contrast to the DAH theory, that individual cells cannot be treated as point objects. Thus, a cell's mechanical energy alterations due to its shape and cortical tension obviously play a role in this energy balance.

It has been reported that interfacial tensions arise from a balance of adhesion, cortical tension and cortical elasticity (Lecuit and Lenne 2007, Krieg *et al* 2008, Farhadifar *et al* 2007, Paluch and Heisenberg 2009). However, the exact nature of this interplay is still elusive. A model has been developed that specifies a relationship between surface tension and the ratio of adhesion to cortical tension; this needs to be experimentally verified. With the DITH model, it can be explained why the 'simple' DAH is still so successful, although it is now replaced by other models. In addition, the DITH model predicts regimes where the DAH breaks down and shows that alterations in tissue surface tension are accompanied by alterations in the shapes of surface cells as a function of the ratio between adhesion and cortical tension. One of two main implications is that when $\gamma/\beta < 2$, surface cells do not stretch out and the DAH is essentially correct. Indeed, surface cells form fewer focal adhesions compared to interior cells and this effect is the primary contribution to the surface tension (similar to fluids). Although cells alter their shapes, the surface tension varies nearly linearly with the effective adhesion. Thus, adhesion in the DAH must correspond to the net energetic contribution of contacting surfaces, which depends on the free energy of cadherin binding complexes and local alterations to the cortical tension caused by these bonds. Taken together, the balance between the cortical tension based on the actomyosin cytoskeleton and cell adhesion seems to be crucial for determining surface cell morphologies.

The second implication is that the analogy with fluids is no longer applicable when surface cells stretch to form additional nearest-neighbor contacts. In contrast to fluids, if the surface cells (outermost cells) stretch to have the same net contact perimeter compared to cells in the bulk, there is no adhesive contribution to the surface tension and the DAH must fail. However, this observation does not depend on a theoretical model. In particular, it has been shown that in high-adhesion zebrafish aggregates, surface cells possess a precisely defined, reproducible cross-sectional area that is significantly larger than the cross-sectional area of bulk cells. In more detail, this areal fraction is consistent with the assumption that surface cells provide the same surface area of contact when they are located in the bulk. Thus, the surface tension of these aggregates seems not to be strongly dependent on the cell adhesion.

In particular, the observation that stretched surface cells exist leads to the consideration of cadherin diffusion and its impact on surface tension. In another model, the adhesive energy density is considered to be constant along contacting interfaces and thus in stretched cells cadherin molecules need to diffuse to the much larger contact interface to maintain the same cadherin density. However, in compact surface cells cadherins are not recruited additionally to the contact interface by inducing their migration. This may be a good first approximation because there are no excess cadherins to bind with on the surface of the bulk cells and hence compact cells can obtain no additional energy from such a receptor migration. Indeed, heterogeneities in cadherin density could lead to unique minimum energy cell shapes. However, these interactions have not yet been investigated and hence are still elusive.

The theoretical model

A minimal mechanical model has been developed that is based on two experimental systems: (i) zebrafish embryonic tissues and (ii) a P-cadherin-transfected L-cell line (LP2). A confocal section of a zebrafish aggregate shows that cells in the bulk are roughly polyhedral with sharp corners, display an aspect ratio of unity and no obvious polarization. The rate of cell divisions in zebrafish aggregates is low and cells within a single tissue type have approximately the same size. The model hence enforces a constant volume for individual cells of $V = 1$ and all cell volumes are normalized by the average volume of a single cell (Manning *et al* 2010).

As the goal is to understand the collective behavior of cell populations, one can focus on the coarse-grained mechanical properties of individual cells, such as cortical tension and adhesion. As in other cell models (Simons and Ikonen 1997, Bouchaud and Georges 1990, Ghosh 1991), the energy is associated with cell–cell contacts, where the energy is proportional to the surface area of contact between cells: $W_{ad} = (\Gamma/2)P_C$, where P_C is the surface area (perimeter in 2D) in contact with other cells and Γ is the free energy per unit surface area for cadherin (adhesive) bonds.

Moreover, the response of single cells to low-frequency pressures and forces may be characterized by a cortical tension (Simons 1997, Ghosh 1991, Slattery 1995): $W_{cort} = \beta_{PT}$, where P_T is the total surface area of a cell and β is the aggregate surface tension. Indeed, feedbacks between the adhesion molecule and cytoskeletal dynamics are abundant, which indicates that the cortical tension along contacting interfaces (β_C) may be different than that along non-contacting interfaces (β_{NC}). As the cortical-tension energy and adhesive energy both scale linearly with the surface area, feedback can be adjusted in a simple manner. In more detail, the term $\beta = \beta_{NC}$ is the cortical tension of a cell in the absence of any cell–cell adhesions, which is the quantity that is measured in single-cell pipette aspiration experiments (Slattery 1995). The effective adhesion γ represents the total energetic contribution of contacting surfaces. In particular, this is defined as the difference between the free energy of the adhesive bonds per unit area (Γ) and local alterations of the cortical tension near an interface $2(\beta_C - \beta_{NC})$. Moreover, the coarse-grained mechanical energy for each cell in an aggregate is given by equation (5.1):

$$W_{cell} = (\beta - \gamma/2)P_C + \beta P_{NC}|V{=}1, \tag{5.1}$$

where $P_{NC} = P_T - P_C$ is the surface area of the non-contacting interface. Furthermore, $\beta - \gamma/2$ is half the interfacial tension of cell–cell adhesion and β is the interfacial tension of cell–culture medium interfaces. It has been found that cells in the interior of aggregates exhibit polyhedral shapes with sharp corners. This means that when $\gamma/\beta < 2$, the cortical elasticity only contributes less to the energy on the relevant long timescales over which the surface tension is a meaningful quantity (Manning *et al* 2010). Thus equation (5.1) omits elastic terms and is valid when $\gamma/\beta < 2$.

Previous methods for approximating tissue surface tension assumed that individual cells do not alter their individual shapes. When the shapes are unaltered, the surface tension is the difference between the interfacial tensions of cell–cell and cell–culture medium interfaces (Simson *et al* 1998, Ghosh 1991). However, confocal images of the equatorial plane of spherical zebrafish aggregates indicate that the surface cell shape depends on the tissue surface tension. Moreover, there seems to be an interplay between the mechanical energy and geometry, which cannot be ignored.

Thus, the strategy will be to find cellular shapes and configurations that locally minimize the mechanical energy and then calculate the surface tension of those configurations.

When an aggregate of cells is compressed, the surface area of the aggregate increases and the cells, which were formerly located in the bulk, are then exposed to the surface. Thus, in an exact analogy with fluids, the response of the tissue to alterations in surface area is the difference in energy ΔW between a cell in the bulk and a cell on the surface multiplied by the number of cells per unit area (or length in 2D) at the surface. The number of cells per unit area is equal to unity divided by the projected area A_{proj} (or projected length in 2D) of a single cell onto the surface of the aggregate:

$$\sigma = (W_{surf} - W_{bulk})/A_{proj}, \tag{5.2}$$

where σ is the surface tension.

This quantity is difficult to calculate for large aggregates, as the total energy depends on the precise geometry of each individual cell, which is in force balance with other cells under the constant volume constraint. However, this can be solved for an ordered 2D system.

Active processes enable the cells to exchange neighbors and explore many possible configurations, indicating that they are not confined to the global minimum energy configuration. In contrast, the collection of cells explores a large number of local minima and these configurations are disordered, such as the typical configurations of a fluid or a jammed granular material. Numerical simulations have revealed that the surface tension for 2D disordered aggregates is related in a simple manner to ordered structures.

Indeed, two minimal structures generated by this procedure have been found (Feder *et al* 1996). For small values of γ/β, surface cells are rounded, as they minimize their total perimeter at the expense of cell–cell contacts, whereas for large values of γ/β the cells are flat, as they tend to maximize neighbor cell–cell adhesions.

Because the surface tension is a change in energy divided by a projected length, both of these quantities are calculated for individual cells in each numerical simulation. Indeed, the projected length, which corresponds to the macroscopic perimeter of the entire aggregate, is the only quantity altered by the disorder in the limit of small γ/β. The total area cannot change, as the number of cells within a fixed area is constant. The macroscopic shape of a 2D hexagonal ordered crystal has hexagonal symmetry, whereas the disordered structure is spherically symmetric, suggesting that the macroscopic perimeter is altered similar to the ratio of the perimeter of a hexagon to a circle of the same area: $\sigma_{disorder} = 1.05\sigma_{order}$.

Indeed, there is excellent agreement at small values of γ/β and systematic deviations at large values of γ/β can be explained easily. The geometry serves as a major constraint on the macroscopic surface tension when $\gamma = 2\beta$. In more detail, as γ approaches 2β, the tension along cell–cell interfaces approach zero and the force balance requires that even the cell–culture medium interface is flat. Under these conditions, the macroscopic surface tension is identical to the cortical tension at $\gamma = 2\beta$.

LP2 and zebrafish cell-shape alterations

This prediction of surface cell-shape alterations can be investigated experimentally in LP2 cells by applying actin-depolymerizing drugs such as cytochalasin D and latrunculin A to the cell aggregates and determining their surface tension using a TST and imaging the aggregates with a scanning electron microscope (SEM). The control aggregate has a relatively high surface tension of 3.16 erg cm^{-2} (work per area). In particular, the cells on the surface of this aggregate are so flat that they cannot be distinguished from each other and the model predicts that the ratio of adhesion to cortical tension is high (Manning *et al* 2010). As expected, cell aggregates treated with actin-depolymerizing drugs showed reduced cortical tension and cell–cell adhesion as the actin anchor of cadherin bonds is weakened. Thus, the macroscopic surface tension is pronouncedly lower. In addition, it is important to note that the effect of actin-depolymerizing drugs on tissue surface tension is reversible and that rounded surface cells are still viable cells.

Confocal images of zebrafish surface cells indicate that this shape alteration is more substantial than going from round to flat. Although this model suggests that structures with $n_{\text{stretch}} = 1$ are the minimum energy states, aggregates with high surface tension often contain surface cells that stretch over multiple bulk cells. In more detail, these stretched surface cells, however, do not express epithelial markers on their cell surface, they instead express the same tissue-type-specific markers as bulk cells (Jovin and Vaz 1989). In addition, they are also indistinguishable from bulk cells in their behavior, as stretched surface cells become intermixed with bulk cells during aggregate fusion (Jovin and Vaz 1989) and the surface cells continue to diffuse in and out of the surface layer. Based on these observations, the model will be improved in order to account for these stretched cells and thus investigate the phenomenon quantitatively in control experiments (Manning *et al* 2010).

Modeling of stretched cell shapes

When the adhesion is larger than the cortical tension, the mechanical model equation (5.1) predicts that cells will begin to spread out because the line tension is negative, $dW/dP = -\gamma/2 + \beta < 0$, thus cells will continue to increase their perimeter without limit. However, this perimeter increase cannot continue until infinity, as other mechanical forces may cause the cells to abolish their expansion (Manning *et al* 2010).

What additional restoring force explains the experimental findings? Adhesion molecule regulation generates a plausible restoring force. In more detail, the amount of energy a single cell can obtain by increasing its perimeter seems to be restricted by the number of cadherins on its cell surface, which is in turn regulated. Although the exact form of regulation is still elusive, its energetic contribution can be expanded as a Taylor series in the difference between the actual adhesive energy and the adhesive energy of a cell with the preferred (optimal) number of cadherins. Keeping to second-order terms, this adds a term αP^2_{C} to equation (5.1). Another possible restoring force is the elasticity, which is generated by a denser cortical network underneath the cell–cell interfaces. In line with this, it would also add a term proportional to P^2_{C} (Sheets *et al* 1997, Edidin 1997). The analysis does not depend on the mechanistic origin of these forces, but requires that the first higher-order term is of the form $P^2{}_C$:

$$W_{\text{cell}} = (\beta - \gamma/2)P_C + \alpha P_C^2 + \beta P_{NC}|V{=}1 \tag{5.3}$$

The lowest energy states corresponding to this energy are functionally complicated, because the $P^2{}_C$ introduces a length-scale that is not necessarily the same as the length-scale introduced by the incompressibility constraint. For instance, in 2D the smallest perimeter structure is hexagonal, whereas the preferred perimeter introduced by the second-order term does not need to be a hexagon. However, in agreement with experimental observations, a simple assumption is that both length-scales are the same and therefore α increases linearly from zero with γ/β for $\gamma/\beta > 2$ (Manning *et al* 2010). In the next step, the surface cells stretch to reach the same contact area as cells in the bulk and thus the 'covering ratio' for ordered 2D packings can be calculated. Each bulk cell contacts a surface cell over a length $1/6L$ and hence a surface cell covers approximately three bulk cells, as the upper surface is exposed towards the surrounding medium. The calculation for the 3D microenvironment is similar and the projected area of surface cells is 3.7 times greater than for bulk cells.

Zebrafish tissue surface cells

The covering ratio for the stretched surface cells has been determined in disordered 3D zebrafish ectoderm aggregates. After analysis of hand-segmented confocal slices in order to determine an estimate of the projected areas of each cell, the first result is that stretched surface cells possess a preferred size that is reproducible from aggregate to aggregate and there is a pronounced difference between surface cells and bulk cells. Indeed, the projected areas for the surface cells are much larger than for the bulk cells, as the ratio between the projected areas is 3.7 ± 0.4, which is in line with the theoretical prediction of 3.7 (Manning *et al* 2010). However, this stretching effect is not due to an increase in the volume of surface cells, as using $V_{\text{cell}} \sim A_{\text{proj}}\, h$ (where h is the distance between the cell's top and bottom), it has been found that bulk cells span on average eight to nine z-direction slices and surface cells span only over three slices, and hence the A_{proj} for surface cells is about three times larger than for bulk cells.

What are the theoretical predictions for the surface tension in this case? In particular, a specific value for α is used to adapt the contact length for bulk and surface cells to equal the values and calculate the surface tension of ordered 2D aggregates for a wide range of values of γ/β and n_{stretch}. As expected, the experimentally observed surface tension and surface cell shape will correspond to the minimum solution at each value of γ/β. In more detail, for $\gamma/\beta < 2$ the compact surface cell shapes are optimal, whereas for $\gamma/\beta > 2$ the surface cells stretch across three interior bulk cells. The surface tension exhibits a crossover at $\gamma/\beta \sim 2$ from adhesion-dominated behavior (according to DAH) to a dependence on cortical tension and other mechanical effects.

Can the stretched surface cell states be explained by a minimal model? In a regime where the adhesion is stronger than the cortical tension, stretched surface cells have been shown to be the minimum energy structures as long as there is still a restoring force regulating the areas of cells in contact. However, there seem to be at least two plausible mechanisms for such a restoring force, such as (i) the adhesion molecule regulation and (ii) the elasticity of the cortical network. But these assumptions have not been directly confirmed experimentally. In addition, when calculating the surface tension, a particular value for the magnitude of this restoring force (α) is chosen, which is based on the assumption that the contact area preferred by the restoring force is the same area as that for bulk cells with sharp corners. When this assumption is relaxed, the surface tension will still exhibit a crossover (γ/β approximately 2), however, the exact nature of the crossover will then be altered. In a future experiment, both the existence and the magnitude of the restoring force can be investigated by using laser ablation in order to destroy individual cell–cell interfaces in low- and high-adhesion aggregates and analyzing the structural relaxation. It has been proposed that the anisotropy of the network response and the magnitude of the structural relaxation can be used to extract the relative magnitudes of cortical elasticity compared to interfacial tension (Farhadifar *et al* 2007).

Using a model that includes only adhesion and cortical tension (see equation (5.1)), stretched surface cells are not the minimal energy structures. Despite this, as γ/β approaches two, the energies of stretched states become closer to that for the unstretched states and active processes would then allow surface cells to explore these so-called metastable configurations. However, these metastable stretched states are supposed to have a wide range of projected areas, which are weighted towards smaller area ratios, as these possess lower energy. The fact that the zebrafish surface cells have a specific, reproducible area fraction, which is significantly larger than unity, indicates that these structures are not metastable, whereas they still display a preferred contact area with other neighboring cells as predicted by the model (see equation (5.3)).

In order to fully interpret the available experimental data with this model, the magnitude of the adhesive tension should be compared to that of the cortical tension in individual cells. However, the net effect of adhesive contacts on interfacial tension (which is denoted γ) depends on both the free energy of adhesive molecule bonds and the alterations to the cortical tension along the contacting interface. Thus, it is difficult to determine how alterations in the expression levels or the activity of cadherins, actin or myosin affect γ. Surface tension has been reported to increase linearly with the numbers of surface cadherins (Foty and Steinberg 2005), which is in agreement with the DAH and this novel model, when γ/β increases linearly with the number of cadherins. However, because the interaction between actin and cadherin-facilitated adhesion is a highly dynamic process, which is regulated by α-catenin (Drees *et al* 2005), it seems possible that elevated cadherin expression significantly increases cortical tension and γ/β remains unknown. In line with this, when the surface tension is decreased by cytoskeletal drugs, both cortical tension and adhesive energy are reduced, as the cadherin bonds are stabilized by the cortical network (Imamura *et al* 1999, McClay *et al* 1981, Chu *et al* 2004). At least one attempt to dissect the connection of cortical tension and adhesion of individual cells has been reported, using AFM measurements (Krieg *et al* 2008). Although in principle AFM is a suitable technique for those measurements, the thermal drift remains a technical challenge that still prevents the required long timescale measurements. Indeed, the results are on timescales of seconds and thus are too short to be relevant for the interpretation of tissue surface tension, which unfortunately becomes valid only for long timescales on the order of tens of minutes (Krieg *et al* 2008). At these short timescales (of seconds), the actin cytoskeleton cannot remodel (Adams *et al* 1996, Angres *et al* 1996) and thus the AFM probes the cytoskeleton elasticity and not exclusively the cortical tension. In addition, cadherin bonds strengthen pronouncedly over time after initiation, thus the adhesion dynamics over long timescales are substantially different from those on short timescales (Borghi and Nelson 2009, Imamura *et al* 1991, McClay *et al* 1981). Taken together, the latter seems to be a major obstacle for measuring cortical tension.

To solve this problem, a set of experiments able to evaluate the cortical-tension β and the effective adhesion γ is presented. One possible approach for measuring the cortical tension is the micropipette aspiration technique (Evans and Yeung 1989, Chu *et al* 2004). However, in the near future it seems it will be possible to reduce AFM drift to perform single-cell AFM experiments on the relevant long timescales, hence enabling the measurement of repulsive force generated by the pure cortical tension. The determination of the effective adhesion γ is more difficult. A semi-quantitative approach for investigating the effect of γ/β would be the use of cell lines that are engineered to express a controlled number of fluorescently labeled adhesion molecules, actin, myosin and actin-associated proteins as performed previously for Madin–Darby canine kidney cells (Yamada and Nelson 2007). Tissue surface tension and surface cell geometries can then be calculated as a function of the ratio of the densities of these molecules and used to investigate the predicted crossover in surface cell shapes and energy contributions at $\gamma/\beta = 2$ in this novel model. Another point is that laser ablation experiments can be used to estimate the interfacial tension along cell–cell interfaces (Farhadifar *et al* 2007). Moreover, it would be interesting to adapt the shape–energy, which is functionally given by equation (5.1), to a Monte Carlo or cellular Potts model (Graner and Glazier 1992, Mombach *et al* 1995) approach with activated dynamics and compare simulated cell sorting based on this interaction potential with the experimentally available data. However, these approaches will need to be confirmed by future studies and for the present model it is suggested that surface cell shapes can be used to estimate the ratio between adhesion and cortical tension in an experimental aggregate.

Disordered cellular structures have been detected in many problems in physics and biology. The fact that an analytic expression for their surface energy can be calculated is unexpected, but highly useful. For instance, equation (5.1) with $\gamma = \beta$ describes a dry foam and thus this method can be used to calculate the surface energies of finite 2D foams as a function of the cluster size. An extension of this work also predicts how cells in the bulk alter their shape from spheres, when there is no adhesion, to polyhedra with sharp corners, when the adhesion is high.

Thus, a minimal model has been developed that relates tissue surface tension to the mechanical properties of the individual cells, such as cortical tension, cell–cell adhesion and incompressibility. In addition, this model predicts how a crossover from the DAH to significant cortical-tension dependence may occur, which is an important consideration when designing and developing novel drugs to modulate the mechanical behavior of cells. Both the DAH and the DITH have been developed in order to explain cell-sorting experiments *in vitro*. Integrating both surface-tension-based hypotheses into a single framework, the model then predicts not only that cells sort out according to the surface tension of their aggregates, but also that this surface tension exhibits a crossover from a regime where intercellular adhesion is dominant, to another regime, where cortical tension is dominant.

Osmotic stress evokes one of the most fundamental challenges to living cells. In particular, the largely inextensible cell membrane of eukaryotic cells can easily rupture under in-plane tension, searching for sophisticated strategies to respond immediately to osmotic stress. How do epithelial cells react and adapt mechanically to the exposure to hypotonic and hypertonic solutions in the context of a confluent monolayer? In order to answer this question, site-specific indentation experiments in conjunction with tether pulling on individual cells have been performed with an AFM to reveal the spatio-temporal alterations in membrane tension and surface area. It has been observed that cells compensate for an increase in lateral tension due to hypo-osmotic stress by sacrificing excess of membrane area, which is stored in protrusions and invaginations such as microvilli and caveolae. At mild hypotonic conditions lateral tension increases and is hence partly compensated for by surface regulation, for example, the cell sacrifices some of its membrane reservoirs. A loss of membrane–actin anchors occurs upon exposure to stronger hypotonic solutions, giving rise to a drop in lateral tension. A tension release recovers on longer timescales by an increasing endocytosis, which efficiently removes excess membrane from the apical side to restore the initial pre-stress. Hypertonic solutions lead to the shrinkage of cells and a collapse of the apical membrane onto the actomyosin cortex. In addition, exposure to distilled water leads to stiffening of cells due to removal of excess surface area and tension increase, which is caused by elevated osmotic pressure across the cell membrane.

The plasma membrane of cells is a highly dynamic and strongly regulated 2D liquid crystal. Many cellular processes, such as endo- and exocytosis (Dai *et al* 1997, Apodaca 2002), cell migration (Sheetz and Dai 1996), cell spreading (Gauthier *et al* 2011, Raucher and Sheetz 2000) and mitosis (Raucher and Sheetz 1999, Chugh *et al* 2017), are regulated by an intrinsic feature of the cell membrane and in particular by the membrane tension. The cell membrane tension encompasses the in-plane tension of the lipid bilayer and the membrane–cytoskeleton adhesion, which is actively regulated by the contractile actomyosin cortex (figure 5.2) (Dai and Sheetz 1999). The interplay between the cell membrane and its actin cortex enables cells to withstand mechanical alterations from the microenvironment. Due to the fact that cell membranes are thin and fragile structures, sophisticated feedback mechanisms based on tension homeostasis are required to maintain an intact cellular shell. For instance, osmotic stress is a physiologically relevant mechanical stimulus, as animal cells have to withstand substantial fluctuations in the osmolarity of external fluids, which leads to a considerable pressure difference between the cytosol and the microenvironment. Osmotic pressure forces the cell to adapt quickly in order to

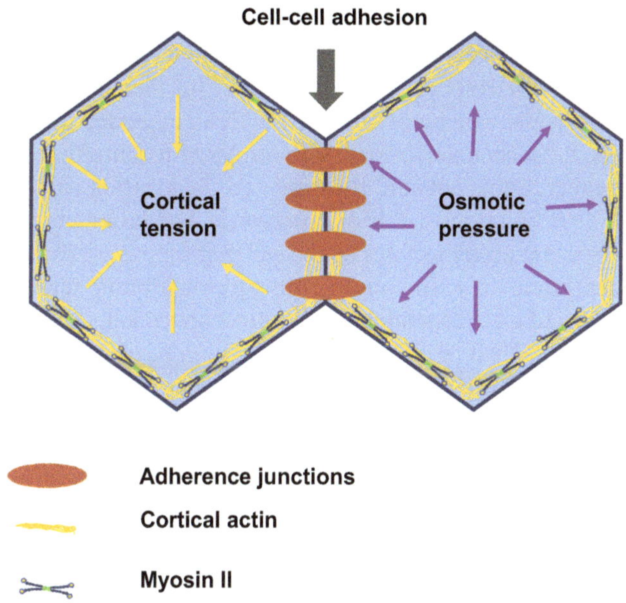

Figure 5.2. Cell surface tension and cortical actin.

avoid damage to the largely inextensible cell membrane. When cells are exposed to a hypo-osmotic solution, they increase their volume due to the influx of water and hence they extend their projected surface area. Animal cells have been reported to increase their surface area by a factor of three and ten times their volume, depending on the particular cell type (Groulx *et al* 2006). As membranes cannot bear large strains (3–4%), they require regulatory processes to maintain the overall cell membrane tension below the lysis tension. More specifically, tension-driven surface area regulation is necessary to accommodate alterations in tension. In particular, high tension can be buffered through an excess of membrane area, whereas a decrease in membrane area is triggered when the tension decreases (Morris and Homann 2001). In order to provide sufficient membrane area, the cells store excess membrane in reservoirs such as microvilli and caveolae, which can vastly unfold and hence buffer membrane tension. Indeed, it has been shown that caveolae can serve as membrane reservoirs to compensate for an increase in membrane tension, which has been induced by swelling (Kozera *et al* 2009).

However, a comprehensive picture of membrane homeostasis is still missing, as it is difficult to measure surface area and tension at the same time in order to analyze the regulative relationship between the two parameters. To solve this, a unique combination of site-specific indentation followed by tether pulling has been combined to acquire the tension and excess area of the apical side of epithelial cells, which can be spatio-temporally resolved. Indeed, polar epithelial cells such as intestine- or kidney-derived epithelial cells are suitable for investigating the impact of microenvironmental physicochemical stimuli on alterations in membrane tension, as they frequently face alterations in the chemical potential. In particular,

Madin–Darby canine kidney cells II (MDCK II) are grown to confluence and challenged by different osmotic solutions to determine the mechanical response of the cell membrane with respect to tension-buffering membrane reservoirs. Thus, indentation experiments, which are analyzed with an extended liquid droplet model for adherent cells, have been performed with an AFM to simultaneously assess local alterations in the membrane tension, the time and the location to monitor the excess membrane area as a function of osmotic pressure. The latter task is performed by measuring the apparent area modulus of the cell membrane mirroring the amount of stored excess membrane area. In conjunction with membrane-tether-pulling experiments, which represent an independent mechanical approach, it has been shown how epithelial cells adjust their surface area in response to tension alterations. In addition, morphological alterations due to cell swelling and shrinking have been observed, showing distinct alterations in tension-buffering membrane reservoirs such as microvilli. Moreover, it has been found that MDCK II cells use their microvilli to generate excess membrane that is readily consumed to accommodate increasing tension. The lysis of the cell membrane is prevented through loosening the membrane–actin connections and sacrificing membrane reservoirs. However, exposure of the cells to distilled water exhausts all existing reservoirs and consequently leads to cell lysis and death. The long-term observation of tension and surface area reveals that membrane tension largely recovers at the expense of membrane area taken up through an increased endocytosis rate.

5.2 The mobility of surface receptors

It has been revealed from fluorescence recovery after photobleaching (FRAP) experiments that the mobility of most cell surface receptors is much smaller than expected for a free diffusion of proteins within a fluid lipid bilayer. In more detail, single-particle tracking experiments have been performed to uncover the complexity of the local constraints to free diffusion. Indeed, evidence has been obtained for several different processes, such as domain-limited diffusion, temporary confinement and anomalous diffusion. The type of motion of a given cell surface receptor will pronouncedly influence the rate of any functional process, which requires movement in the plane of the membrane. For instance, anomalous diffusion greatly reduces the distance traveled by a receptor on a timescale of minutes.

The lateral movement of several membrane proteins is essential to their appropriate function. This may include the movement towards a specific site on the cell membrane as in receptor-facilitated endocytosis or the formation of transient as well as long-lived interactions between different cell surface receptors. Restrictions on receptor movement, confining functionally related proteins to close proximity, are necessary for their function and regulation (Mierke *et al* 2011a, 2011b). An understanding of the mobility of membrane proteins is essential for revealing the mechanisms and especially the kinetics of several membrane-associated functions.

The lateral mobility of a wide variety of membrane proteins has been determined by the method of FRAP (figure 5.3) (Jovin and Vas 1989, Peters 1991). Typically, a small area (1–2 μm diameter) of fluorescent-labeled receptors is photobleached using

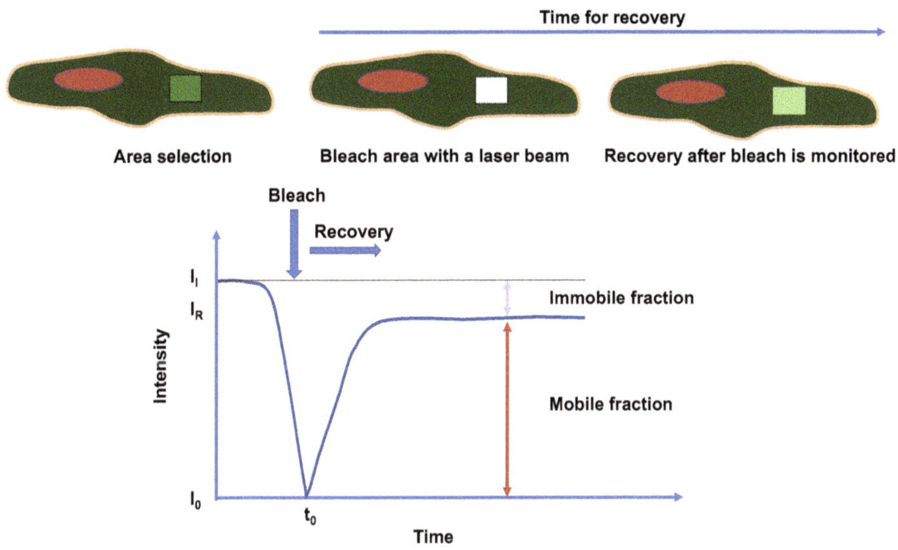

Figure 5.3. Experimental setup of a typical FRAP experiment.

a focused laser beam. The fluorescence in the bleached area is recorded and recovers in a certain time interval due to diffusion of unbleached molecules into this area. The recovery of the fluorescence is normally incomplete when measurements are performed on the cell membrane of living cells. These experiments are usually interpreted using two components. One fraction is immobile on the timescale of the experiment and another is mobile, and this is characterized by a diffusion coefficient smaller than that for the unconfined diffusion.

It has become possible to record the movements of individual cell surface receptors using the technique of single-particle tracking (SPT) (Cherry 1992, Sheets *et al* 1995, Saxton and Jacobson 1997). SPT involves the attachment of a small particle (typically 11–40 nm in diameter) to the targeted protein. In more detail, two types of particles have been used: fluorescent particles, which are imaged by low-light-level fluorescence microscopy; and gold particles, which can be imaged using differential interference contrast microscopy. The movement of proteins in the cell membrane of living cells can be observed by tracking the individual particle positions through a sequence of images. When the particles are well separated compared with the resolution of the optical microscope, the positions of the particles can be determined with high precision, hence the spatial resolution of SPT is 10–20 nm (compared with ~1 μm in a FRAP experiment).

SPT measurements have been performed with a number of receptors on different cell types. In particular, these experiments have revealed a considerable complexity in the motion of individual molecules. The methods for analyzing SPT data mostly depend on comparing movements over different timescales (Anderson *et al* 1992, Saxton 1993, 1994, 1995, 1996, Kusumi *et al* 1993, Simson *et al* 1995): for random diffusion, $\langle r^2 \rangle / t$ is independent of time, where $\langle r^2 \rangle$ is the mean-square displacement measured over a time interval t. In particular, an increase in $\langle r^2 \rangle / t$ with time is indicative of directed motion, whereas a decrease corresponds to some form of constrained diffusion. Individual tracks have to be carefully analyzed. Using Monte Carlo simulations, it has been shown that random movements produce apparently non-random behavior with high probability (Saxton 1993). Various statistical tests, however, have revealed that non-random movements of membrane proteins occur commonly. The statistical analysis can be applied to individual tracks when there is a sufficiently large number of data points (Qian *et al* 1991), which ensures classification into receptor subpopulations exhibiting random diffusion, constrained diffusion, directed motion or immobility. As an alternative method, different populations on the same cell type may be inferred from analysis of the experimental probability distribution of particle displacements (Anderson *et al* 1992, Wilson *et al* 1996, Schutz *et al* 1997).

Constrained diffusion: domain-limited diffusion

The simplest conceptual model for constrained diffusion seems to be the domain model (figure 5.4). According to this model, barriers to free diffusion split the cell membrane into domains. In more detail, receptors undergo random diffusion within a certain domain, with a diffusion coefficient similar to that of free diffusion in a lipid bilayer. In particular, long-range diffusion occurs much more slowly and depends on the rate at which receptors can hop or switch between special domains. The first evidence for the existence of these domains was obtained from a FRAP experiment in which the 'immobile fraction' was revealed to increase with increasing size of the bleached spot (Yechiel and Edidin 1987). In line with this, SPT experiments with gold-labeled transferrin receptors on normal rat kidney epithelial cells (NRK cells) revealed trajectories, which visually give a strong impression of the receptors hopping between different domains of a few hundred nanometers' diameter (Sako and Kusumi 1994). Barriers that can build the walls of domains

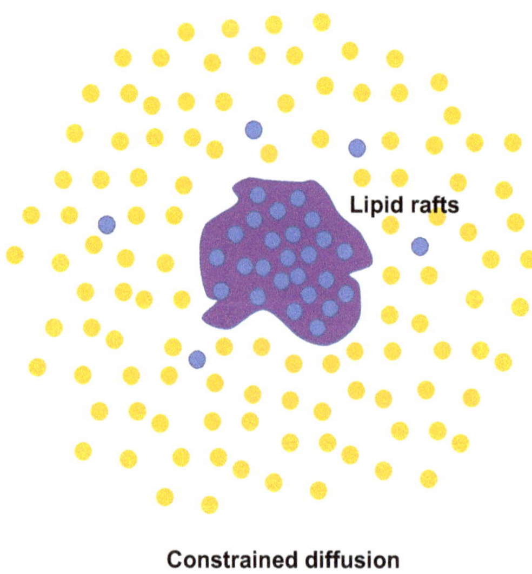

Figure 5.4. Constrained diffusion of molecules on the cell membrane.

have been detected by experiments in which gold-labeled receptors are dragged across the cell surface by laser tweezers (Edidin *et al* 1991, Sako and Kusumi 1995, Sako *et al* 1998). In addition, SPT and laser tweezer experiments with E-cadherin indicate that these molecules may be either corralled by the cytoskeleton or tethered to it (Sako *et al* 1998).

Constrained diffusion: temporary confinement
The trajectories of both the neural cell adhesion molecule (NCAM) and the T-lymphocyte differentiation marker Thy-1 have been analyzed using a temporary confinement model (Simson *et al* 1995, 1998). In this model, molecules perform free random diffusion, which is interspersed with periods of confinement within regions of about 300 nm diameter. These regions seem to be similar to the domains proposed by Sako and Kusumi (1994), but they are also explainable using a variety of other mechanisms (Simson *et al* 1998). Thus, the temporary confinement domains may consist of clusters of integral membrane proteins, between which the mobile protein is entangled. In the case of the GPI-anchored receptor Thy-1, it has been suggested (Sheets *et al* 1997) that transient confinement zones may be glycolipid-rich regions, and these may correspond to the detergent-insoluble membrane fractions observed in biochemical experiments (Simons and Ikonen 1997). In addition, the evidence for lipid microdomains in cell membranes has previously been presented in another study (Edidin 1997).

Constrained diffusion: anomalous diffusion
Alternatively, the constrained diffusion may be interpreted using an anomalous diffusion model. Studies of transport in disordered systems have revealed many examples where anomalous diffusion occurs (Bouchaud and Georges 1990). In particular, anomalous diffusion in cell membranes may result from obstacles and traps, such as binding sites with a broad distribution of binding energies and escape times (Saxton 1996).

SPT measurements with fluorescent low-density lipoprotein (LDL) bound to LDL receptors or via Fab to IgE receptors have been interpreted and confirmed using an anomalous diffusion model (Ghosh 1991, Slattery 1995, Feder *et al* 1996). Additional data for LDL receptors are indeed consistent with anomalous diffusion (Anderson *et al* 1992). However, these experiments are difficult to perform as LDL can bind non-specifically and this binding has to be eliminated (Goldstein and Brown 1977). These findings suggest that specifically bound LDL may bind weakly to other cellular components, such as the extracellular matrix, which then interferes with the interpretation of the SPT data and hence complicates it pronouncedly.

Detailed studies of the mobility of MHC class I molecules have been performed on HeLa cells (Smith *et al* 1998). In more detail, SPT studies were performed using R-phycoerythrin coupled to Fab derived from a monoclonal antibody to MHC class I (Smith *et al* 1998). However, this probe is the smallest (11×8 nm) used for SPT experiments on cells and hence the non-specific binding is negligible, so that the risk of a perturbing effect of the particle is minimized. Another advantage of this probe is that it is monovalent, as it is purified as a 1:1 complex of Fab and R-phycoerythrin. This circumvents complications caused by cross-linking, which may occur with multivalent probes.

A disadvantage of R-phycoerythrin is that it photobleaches rather readily, thus limiting the number of images that can be obtained in an SPT experiment (Sako *et al* 1998). In experiments with HeLa cells, the time interval between images has been varied from 4 to 60 s, hence data were obtained for time intervals from 4 s to 20 min. The displacements r over a given time interval t were fitted to the probability distribution:

$$P(r)dr = [r/2Dt][\exp(-r^2/4Dt)]dr. \tag{5.4}$$

In more detail, equation (5.4) assumes random diffusion and yields the diffusion coefficient D. Determination of D over a range of times provides a prediction for different types of motion (Saxton 1996). For normal diffusion, D is independent of the time, whereas for anomalous diffusion D (strictly $\langle r^2 \rangle/4t$) decreases over all times. For instance, the domain-hopping model predicts anomalous diffusion whilst molecules are confined within a domain, followed by a crossover to normal diffusion as long-range motion becomes limited by the rate of hopping between domains. However, when a population of molecules performs directed motion, the histogram develops a second peak at longer times (Wilson *et al* 1996).

The SPT experiments with MHC class I on HeLa cells revealed strong evidence for anomalous diffusion. In particular, plots of $\log D$ versus $\log t$ for data obtained from two experiments on different timescales show that the negative slope of these plots demonstrates that diffusion is anomalous over all times covered by the experiments. However, as is usual for single-cell experiments, there is indeed some cell-to-cell variability in the parameters, while the negative slope was consistently observed. Fitting the data as follows (Feder *et al* 1996)

$$D = D_0 t^{\alpha - 1} \tag{5.5}$$

gave a value for α of about 0.5 ($\alpha = 1$ for normal diffusion).

However, the domain-hopping and anomalous diffusion models are not necessarily incompatible. The results obtained for MHC class I on HeLa cells can be explained by domains when the distribution of escape times from the domains is sufficiently broad. In addition, a range of models involving obstacles and binding sites can also predict these results (Saxton 1996). It has been proposed several times that diffusion confinements in membranes have to arise from binding to cytoskeletal components, immobile transmembrane proteins or the extracellular matrix (Sheets *et al* 1995). These constraints may be caused by binding or obstruction. However, it is plausible that there is considerable variability amongst receptors and cell types in their individual hindrance factors, which constrain mobility, and it is unlikely that a single model would be universally applicable.

The relationship between FRAP and SPT

The results of SPT experiments raise questions about the interpretation of FRAP experiments. It has been proposed that anomalous diffusion might occur in cell membranes as a consequence of long-time tails in the jump rate of diffusing molecules (Nagle 1992). Moreover, the effect of long tail kinetics has been analyzed on FRAP measurements and indeed this showed that the diffusion coefficient as well as the immobile fraction determined by conventional means depends on the length-scale and timescale of the individual experiment. In addition, FRAP data for IgE receptors have been analyzed on rat basophilic leukemia cells using both the conventional model of random diffusion with an immobile fraction and a model in which all receptors simply undergo anomalous diffusion (Feder *et al* 1996). They found that the two models fitted the experimental data equally well. In addition, they also performed simulations hypothesizing that FRAP experiments are not suitable to distinguish between the two models.

A further issue is whether there is quantitative agreement between FRAP and SPT measurements. There is some evidence that there may be agreement, but, due to a number of problems, it is difficult to perform a valid comparison. The simplest experimental system consists of lipid diffusion in model lipid membranes. Indeed, a fair agreement has been revealed between values of D measured by SPT and FRAP for lipids labeled with a single fluorophore (Schmidt *et al* 1996), but another study observed a two to four times lower D for SPT of lipids labeled with gold particles (Lee *et al* 1991). The lower D seems to be related to the multivalency of the particles.

An advantage of fluorescence SPT is that comparison with FRAP is feasible under exactly identical experimental conditions. In addition, experiments with fluorescent low density lipid (LDL) particles attached via IgE to IgE receptors are performed (Feder *et al* 1996). In SPT experiments, 27% of the cell surface receptors have been classified as immobile. In particular, the mobile receptors mainly exhibited anomalous diffusion with mean values of α = 0.64 and D measured over 1 s of 0.96×10^{-10} cm^2 s^{-1}. In comparison, FRAP experiments using the same probe lead to $\alpha = 0.15$ and D (1 s) = 1.4×10^{-10} cm^2 s^{-1}, when the data are fitted to the anomalous diffusion model. A complication of these experiments was that the IgE receptors seemed to be much more mobile when FRAP experiments were performed with fluorescein-labeled IgE serving as the probe for unlabeled IgE receptors. Moreover, this suggests that the attachment of the LDL particle pronouncedly perturbs the receptor mobility. However, it will be important to determine whether this problem is restricted to LDL or occurs even with other particles used for SPT.

The new insights obtained regarding the movement of cell surface receptors have profound implications for membrane function. In particular, a variety of functional processes require receptors to form associations or move to specific sites, such as coated pits. Theoretical analyses of such processes have generally assumed normal diffusion with diffusion coefficients derived from the mobile fraction observed in FRAP experiments. When, however, receptors undergo anomalous diffusion, then processes occur over longer distances and timescales may be dramatically slowed down. It has been worked out that the receptors are often coupled through G proteins in order to overcome slow diffusion in the membrane (Peters 1988). The existence of anomalous diffusion supports this case strongly. Moreover, it has been proposed that altered mobility of integrins can impair cell adhesion by reducing the ability of the receptors to associated to clusters (Yauch *et al* 1977). On the other hand, restrictions on diffusion, by whatever mechanism, seem to be advantageous. Indeed, receptors, which are delivered to the cell membrane through vesicle transport may not move far from the fusion site. This could provide an explanation for the finding that functionally related receptors remain in close proximity to each other.

Additional experiments need to be performed to investigate whether SPT and FRAP data can be incorporated within the same theoretical model. SPT experiments indicate that the random diffusion plus the immobile fraction model is inappropriate for evaluating FRAP experiments. However, doubts about possible effects of the particle in SPT experiments have not yet been completely dispelled. One advantage will be to dispense with particles by tracking fluorescent-labeled antibodies. In line with this, it has become possible to image and track single fluorophores, although it has been shown so far only in a model system (Schmidt *et al* 1996). In addition, IgG bound to MHC class I has been imaged on HeLa cells. In more detail, the IgG was labeled at a ratio of 10 fluorophores per IgG without loss of specific binding, indicating that tracking experiments for cells with fluorescent-labeled antibodies are indeed feasible.

5.3 Specific membrane regions as a location for surface receptors

Nearly one third of the human genome encodes proteins that are membrane-associated, such as transmembrane, lipid-anchored or peripheral membrane proteins. The protein diffusion (described by the diffusion constant D) not only provides the distribution of proteins across the membrane surface, it also determines their ability to interact. In particular, the forward reaction kinetics of two generic membrane proteins is ascertained by the sum of their individual diffusion constants. Thus, the mobility of proteins within the cell membrane determines their interaction capabilities, for example in signal transduction events. The interaction efficiency is further determined by the local protein concentration, which determines the collision probability. Thus, two slowly diffusing proteins interact inefficiently unless their local protein concentration is very high. To draw a complete picture of membrane proteins in a complex cellular process such as receptor signaling, it is not sufficient to know the binding affinities of ligands and kinases, but only by quantifying protein dynamics and concentrations can the interactions and functions be fully revealed.

The kinetic properties of a membrane protein and the types of motion it can undergo are critically influenced by the lateral organization of the cell membrane. The fluid mosaic model of Singer and Nicolson proposed that the cell membrane is laterally homogenous and hence membrane-associated proteins can diffuse freely using Brownian motion in the plane of the lipid bilayer (Singer and Nicolson 1972) and hence they are randomly distributed on the cell surface (figure 5.5). In this random diffusion model, the diffusion constant D is only dependent on the hydrodynamic radius of the protein, which is assumed to be cylindrical, and hence the viscosity of the membrane can be described by the Saffman–Delbrueck model (Saffman and Delbrueck 1975). In practical terms, Brownian motion means that the distance a molecule diffuses, measured as the mean-square displacement (MSD) in time t in a 2D membrane, relates linearly to the diffusion coefficient D: $\mathrm{MSD} = 4Dt$ (Singer and Nicolson 1972).

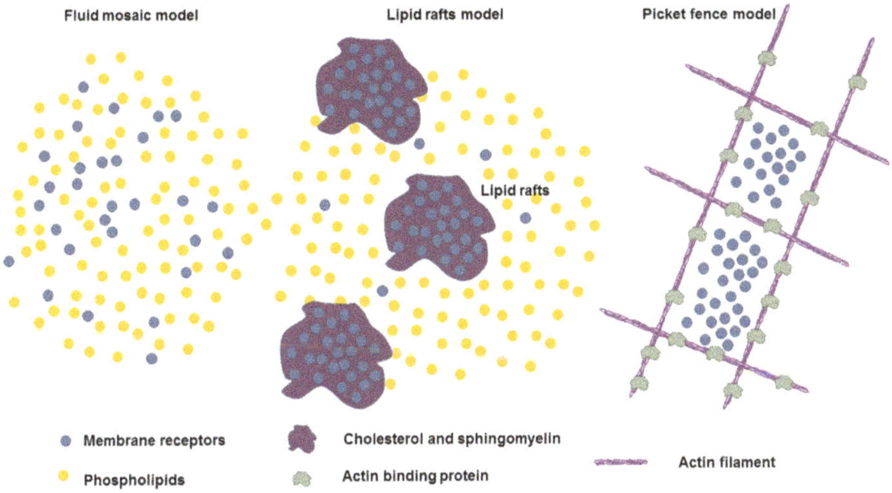

Figure 5.5. Three models for confined diffusion of membrane receptors on the cell membrane.

However, cell membranes such as the plasma membrane are complex lipid–protein composites in which protein diffusion becomes anomalous (Weiss *et al* 2003), with diffusion coefficients an order of magnitude lower than those observed in artificial membrane systems (Bacia *et al* 2004). This retardation is possibly due to protein–protein interactions (Douglass and Vale 2005), active transport, cytoskeleton interactions and confinement (Kusumi *et al* 2005), as well as membrane lipid microdomains (Simons and Ikonen 1997). In particular, lipid rafts are defined as small, heterogeneous, highly dynamic, sterol- and sphingolipid-enriched domains (Pike 2006) that are hypothesized to facilitate protein–protein interactions through their ability to selectively accumulate proteins and control the diffusion via their internal viscosity and/or boundary properties.

In heterogeneous cell membranes, the diffusion coefficient D can be divided into local and global values, as diffusion over a short distance and time period such as within a raft is clearly different from that over longer distances and timescales such as when covering many rafts. Thus, by simply changing the observation area and recording time intervals, one can obtain different values. Indeed, this phenomenon has been observed and hence it has been concluded that there seem to be protein-rich membrane domains $\sim 1\ \mu m$ in size in the cell membrane of endothelial cells (Yechiel and Edidin 1987). Hence by quantifying the local and global diffusion behavior of proteins and lipids, it is possible to build more detailed physical models of cell membranes and understand the underlying principles of their organization. Indeed, there seems to be a connection between membrane organization and protein diffusion in different scenarios that are neither comprehensive nor mutually exclusive.

The accuracy in measuring protein dynamics in living cells has been a major limitation. However, fluorescence microscopy is the only suitable approach to characterize protein dynamics in living cells and tremendous progress has been made in hardware, automated acquisition and analysis algorithms. In principle, there are three different techniques that can be used for the measurement and analysis of molecular diffusion in cell membranes. These are FRAP, SPT and fluorescence correlation spectroscopy (FCS). The goal is to explain how these have deepened the knowledge of membrane rafts and the global organization of the cell membrane.

Brownian protein diffusion and membrane viscosity

FRAP has been used to measure diffusion coefficients in living cells. In more depth, an area of the sample is first quickly photobleached by intense laser exposure. As mobile fluorophores from outside the bleached region diffuse into the dark area, the fluorescence intensity is observed to recover. Indeed, the speed of this recovery is related to the fluorophore mobility and thus the diffusion coefficient can be determined. The fraction of the fluorescence that is recovered after long timescales is termed the mobile fraction (Mf) and this gives information on the percentage of molecules that can freely diffuse. By using the volume of the laser focal spot for bleaching, more localized diffusion parameters can be derived that are distinct from the global measurements, in which the laser spot is used to bleach a larger area such as a disk or rectangle. In more detail, the size and geometry of the bleached area has to be taken into account when interpreting and comparing FRAP results.

FRAP studies were employed to analyze the hypothesized effects of lipid rafts on diffusion (Kenworthy *et al* 2004). If rafts are small immobile domains of high-viscosity, raft-favoring proteins with a stable raft association would have low mobile fractions and diffusion coefficients and hence different raft markers would have similar diffusion properties. Proteins with a high affinity for membrane rafts, including the dual-palmitoylated transmembrane protein linker for activation of T cells (LAT) and farnesylated/palmitoylated H-Ras, and proteins that target raft constituents, such as cholera toxin subunit B (CTxB), which binds and clusters ganglioside GM1 (Brown 2006), were determined, however, to have pronouncedly diverse diffusion coefficients. It was thus concluded that the type of membrane anchor, and not purely the raft association phenomenon, determines the diffusion coefficient of proteins in the cell membrane (Kenworthy *et al* 2004).

Interestingly, in the apical membranes of MDCK epithelial cells at room temperature, only the raft-associated proteins LAT and glycosylphosphatidylinositol (GPI)-anchored proteins showed complete fluorescence recovery in bleached regions, whereas 30% of non-raft proteins were immobile (Meder *et al* 2006). However, the latter formerly immobile proteins became mobile at 37 °C. This temperature dependence was explained as the formation of a discontinuous liquid-ordered phase at lower temperatures, which segregates and confines membrane proteins. Lipid raft proteins can diffuse freely in either phase, whereas non-raft proteins are excluded from ordered-phase domains, which hence act as obstacles for their own free diffusion. Here, FRAP was sensitive enough to detect differences between proteins that could dynamically continue diffusion in raft domains, but only because the raft coverage was close to the percolation threshold. However, there is currently no direct evidence that lipid phase separation does indeed occur in native *in vivo* cell membranes.

Due to spatial resolution restrictions, FRAP studies cannot detect the impact of small raft domains on protein diffusion, while they are able to confirm the general relationship between membrane viscosity and protein dynamics. As expected, diffusion rates for both raft and non-raft markers slows down with decreasing temperature, as the membranes become less fluid (Kenworthy *et al* 2004). Cholesterol has been reported to reduce the diffusion rate of Ras proteins due to increased membrane viscosity, but so does cholesterol depletion by using methyl-β-cyclodextrin (mβCD) (Goodwin *et al* 2005). It is known that the removal of cholesterol, particularly with mβCD, not only alters membrane viscosity, but it can also hinder diffusion inside the membrane (Shvartsman *et al* 2006), possibly due to induction of gel-like regions (Nishimura *et al* 2006) and reorganization of the actin cytoskeleton (Kwik *et al* 2003).

The spatial limitation has been overcome by simultaneously recording FRAP and anisotropy recovery after photobleaching (ARAP) of GPI-anchored proteins (Goswami *et al* 2008). Alterations in fluorescence anisotropy are an indication of the fluorescence resonance energy transfer (FRET) between identical fluorophores (homoFRET). While monomers diffuse freely, nanoclusters of only a few GPI-anchored proteins are essentially immobile and thus are non-randomly distributed across the membrane. Together with the concentration-independent, temperature-dependent and cholesterol-dependent nature of the conversion between monomers and clusters, this study indicates that the formation of clusters is actively regulated by cortical actin and myosin contractility (Goswami *et al* 2008). However, the extent to which protein multimerization correlates with lipid domains can only be resolved when both processes are recorded simultaneously but independently from each other.

The ability of FRAP to detect the global behavior of macromolecular complexes was demonstrated for the epidermal growth factor receptor (EGFR) (Lajoie *et al* 2007). By systematically measuring the diffusion rate and the mobile fraction, it was discovered that the EGFR association with a galectin lattice overrides the immobilization of the receptor in caveolae and hence the tumor suppressor function of caveolin-1. Subsequently, the regulation of EGFR diffusion rates determines its ligand responsiveness, which seems to be of high importance. EGFR signaling was suppressed when the receptor was sequestered into immobile caveolae, whereas EGFR signaling was enhanced, relative to unconfined diffusion, when the receptor's diffusion rate was retarded by the galectin lattice. Taken together, the data thus indicate that a receptor's mobility, as regulated by the membrane properties, has a strong influence on its receptor–ligand interactions and function.

The major restriction of FRAP studies is that subpopulations that diffuse differently cannot easily be distinguished, as FRAP is essentially an ensemble or bulk measurement. In an elegant theoretical paper, it was shown that the existence of high-viscosity islands such as rafts affects the long-range mobility of both raft- and non-raft-partitioning proteins (Nicolau *et al* 2006). Moreover, the theoretical model predicted that all proteins, with or without raft affinity, diffuse pronouncedly more slowly when rafts are present in the cell membrane. It is interesting to note that protein diffusion in cell membranes is several times slower than in artificial bilayer constructs composed solely of a few lipid species. For instance, the lipid dye diI-C18 has diffusion coefficients of ~1.4 $\mu m^2 s^{-1}$ in the cell membranes of human embryonic kidney (HEK) cells (Bacia *et al* 2004), whereas the same dye has a diffusion coefficient of ~6.3 $\mu m^2 s^{-1}$ in dioleoylphosphatidylcholine (DOPC)-containing artificial bilayers (Kahya *et al* 2003). In addition to raft domains, the diffusion within the membrane of cells seems to be confined by the underlying membrane cytoskeleton and the extracellular matrix, which suggest a more complex diffusion behavior, with at least two different subpopulations. However, these systems must be tackled using microscopy techniques with single-molecule sensitivity.

Anomalous sub-diffusion: single-molecule techniques

SPT can be used to characterize the anomalous sub-diffusion and the relative diffusion rates in different regions of the cell membrane. In SPT (Schmidt *et al* 1996), individual molecules are labeled, imaged and tracked, and the trajectories are analyzed (Saxton and Jacobson 1997, Sergé *et al* 2008). When only a few, precisely separated molecules are fluorescent, their location can be pinpointed with great accuracy by simply calculating the center of the observed Gaussian point-spread function (Ober *et al* 2004). The localization in consecutive image frames is then connected to form trajectories and additionally the diffusion properties can be determined from the mean-square displacement (MSD).

However, the first publication that showed that raft proteins are anchored into ~30 nm domains, which diffuse as an entity across the cell surface, employed tracking of membrane proteins with latex beads and a laser trap (Pralle *et al* 2000). Due to the fact that the diffusion coefficient of membrane constituents is related to their hydrodynamic radius, large tags such as antibody-coated gold beads may pronouncedly alter diffusion characteristics, particularly if protein cross-linking cannot be omitted. Most studies have been tailored to track only single proteins (Suzuki *et al* 2005) or have exploited the improved signal to noise ratio (S/N) of total internal reflection fluorescence (TIRF) microscopy to image fluorescent fusion constructs or small molecule dyes directly.

The main disadvantages of SPT are that there is an inherent trade-off between spatial and temporal resolution and that statistical analysis is somehow limited unless a large number of molecules are tracked in each experiment. Thus, to map the entire cell surface, localization and connecting molecules into trajectories have to be achieved for a high density of molecules. There has been no standardized tracking algorithm that can simultaneously account for high particle density, heterogeneous particle motion, particle interactions (merging and splitting) and temporary disappearance, such as from blinking. However, progress has been made in addressing this (Sergé *et al* 2008, Jaqaman *et al* 2008), but the interpretation of trajectories will continue to rely on simulations (Wieser *et al* 2008) until robust algorithms have been established.

The advantages of SPT are that it enables molecular diffusion coefficients to be calculated and different diffusion modes of the same molecule to be identified. Monte Carlo simulations have identified how modes of movement differ from Brownian motion when molecules are actively transported or confined in immobile as well as in diffusing domains (Wieser *et al* 2008).

Many examples that demonstrate the power of SPT have addressed the influence of post-translational lipid modifications and raft affinity (Douglass and Vale 2005), protein–protein interactions (Suzuki *et al* 2007a, 2007b) and the actin cytoskeleton (Andrews *et al* 2008) on protein diffusion. Similar to FRAP experiments, SPT was used to assess the contribution of lipid modifications to protein dynamics on the surface of T cells. Palmitoylation of LAT or dual acylation of the Src-family kinase Lck did not alter their diffusion pronouncedly, but the trapping of these signaling proteins in cholesterol-independent clusters of the co-receptor CD2 did slow them by approximately twofold (Douglass and Vale 2005). Such differential modes of movement can only be detected by dual-color SPT, and not with FRAP. In the case of the T cell plasma membrane, protein–protein interactions have to be factored into the interpretation of modes of motion, which seems to be the case for other cell types where stimulation of surface receptors results in multi-molecular protein complexes. Dual-color tracking of the GPI-anchored receptor CD59 in the outer leaflet and signaling proteins anchored to the inner leaflet also revealed how ligand clustering of the receptor induces temporary immobilizations (named STALLs (stimulation-induced temporary arrest of lateral diffusion)) that serve as short-lived platforms for signal transduction activities (Suzuki *et al* 2007a, 2007b).

Another recent example of SPT revealing the mechanism of receptor dynamics is the tracking of individual Fcε RI receptors using non-bleachable quantum dots (Andrews *et al* 2008). Different modes of motion, such as immobile, free, directed and confined, were all detected for the same receptor. Indeed, the confined motion was attributed to the presence of the actin cytoskeleton, which was found to dynamically restrict location of the receptor into micron-sized domains.

Transient confinement zones (TCZs), detected by SPT, define areas where an observed molecule stays much longer than expected from the average diffusion coefficient and are thought to bear some resemblance to lipid rafts. TCZs are typically 200–300 nm in diameter (Simson *et al* 1998), preferentially trap GPI-anchored proteins and glycosphingolipids (Sheets *et al* 1997) and are cholesterol-dependent (Dietrich *et al* 2002). Whether the viscosity differential inside and outside the raft is a sufficient mechanical diffusion barrier is still under investigation (Dietrich *et al* 2001), particularly since TCZs appear to be temperature-independent (Dietrich *et al* 2002). Fluorescent fusion constructs of only the membrane anchor regions of H-Ras (raft), K-Ras (non-raft) and Lck (raft) have been tracked. These three constructs showed similar diffusion coefficients, assuming Brownian motion, however, two populations of diffusing molecules were observed (Lommerse *et al* 2006). The major population displayed similar diffusion times to small molecule membrane dyes (0.6–1.6 μm^2 s^{-1}). However, a second population (16% for Lck and 27% for K-Ras) was confined to domains of roughly 200 nm. In this instance, cholesterol depletion did not affect confinement. For H-Ras, the mobile fraction was 73%, with a diffusion coefficient of 0.53 μm^2 s^{-1} (SPT) or 0.48 μm^2 s^{-1} (FRAP) (Lommerse *et al* 2004). Interestingly, the 200 nm confinement of the slow-diffusing fraction was only observed for active GTP-bound and 'raft-associated' Ras, but not the inactive GDP-bound form (observed by both mutagenesis and insulin activation of H-Ras), indicating that confinement is not purely controlled by lipid–lipid interactions (Lommerse *et al* 2006, Prior *et al* 2001). A similar confinement was shown for a G-protein coupled odorant receptor with ~50% of the receptor being confined to larger domains (300–550 nm), ~30% to small domains (180–200 nm) and the remaining 20% being either immobile or freely diffusing (Jacquier *et al* 2006). For a different G-protein-coupled receptor, the μ-opioid one, it was found that 90% of the receptors diffuse within a domain that diffuses itself by a motion, which has been termed 'walking confined diffusion' (Daumas *et al* 2003). Taken together, these studies suggest that confinement areas are unique to each of the analyzed receptors, in turn suggesting that specific receptor interactions define these zones rather than generic 'raft domains'.

As with the observation area size in FRAP, the sampling frequency in SPT experiments can impact on the values of the diffusion coefficients and data interpretation. By developing SPT with ultra-high temporal resolution (25 μs instead of the typical video rate of 30 ms), it was possible to detect a different type of diffusion, 'hop-diffusion', in which the entire membrane is compartmentalized into 30–250 nm compartments (Kusumi *et al* 2005, Fujiwara *et al* 2002, Murase *et al* 2004). In particular, proteins and lipids are slowed down when they hop from one compartment to another, whereas within each compartment the diffusion is not pronouncedly lower compared to that observed for free diffusion. Thus, these membrane compartments are distinctly different to the TCZs formed by lipid microdomains (Kusumi *et al* 2004) and are considered to be the result of an underlying membrane cytoskeleton to which the transmembrane proteins are anchored, creating a 'picket fence' within the membrane (Morone *et al* 2006). However, it should be kept in mind that in order to distinguish between Brownian motion and hop-diffusion a time resolution of tens of microseconds is required, without such a time resolution the hop-diffusion is simply interpreted as free diffusion with a slower average diffusion coefficient.

Multi-parameter imaging: combining diffusion with concentration

A key finding for raft characteristics through theoretical calculation is that global diffusion retardation caused by high-viscosity islands increases the rate of protein–protein interactions (Nicolau *et al* 2006). Moreover, the existence of raft domains with high viscosity can elevate the local concentration of proteins with a low average density but high raft affinity due to their specific membrane anchors. The maximal collision rate, which is a measure of signaling efficiency, is achieved when 10% of the membrane is covered by small (approximately 6 nm diameter), mobile domains and $_{\mathrm{Draft}}/D_{\text{non-raft}} = 0.5$. Indeed, values of $_{\mathrm{Draft}}/D_{\text{non-raft}} = 0.3$–$0.5$ have been reported previously in cells (Pralle *et al* 2000, Dietrich *et al* 2002). Raft domains of that nature also increase the collision rate of proteins modestly, even if they have no affinity for the raft domain. It has been demonstrated that this may have important biological consequences for different types of membrane domains, such as caveolae (large and immobile) and GPI-protein-enriched domains (small and mobile). However, it has been predicted that experiments require a time resolution of 5–10 ms in order to detect these small raft domains by SPT, which is at the technical limit of most set-ups.

An alternative single-molecule technique is FCS, which was first demonstrated in 1974 as a method for analyzing molecular kinetics (Elson and Magde 1974). In its most basic form, the femto-liter illumination volume of a laser scanning confocal microscope can be held stationary. The intensity fluctuations as a low concentration of fluorescent molecules diffuse in and out of this spot are then recorded using high-speed detectors (Schwille *et al* 1999). From this time series of intensity measurements, an autocorrelation analysis allows molecular diffusion rates to be obtained with high statistical accuracy and over a wide range of timescales. In more detail, modern detectors achieve a range from microseconds up to seconds (Digman *et al* 2005). In particular, this method uses low laser powers and low probe concentrations, making it applicable even to living cell measurements, with minimized perturbation to the cell membrane. Several related techniques have been developed recently (and continue to be developed), such as two-color cross-correlation spectroscopy (FCCS) (Bacia *et al* 2006), image correlation spectroscopy (ICS) (Hebert *et al* 2005) and raster-scanned FCS (raster ICS or RICS) (Digman *et al* 2005). Indeed, these novel methods have the potential to map the spatial distribution of diffusion coefficients at selected membrane regions, as demonstrated for the mobility of paxillin at focal adhesions (Digman *et al* 2008).

An advantage of FCS is that it not only provides extract diffusion coefficients and identifies anomalous diffusion, it can also measure the number of molecules in the observed spot. For instance, fluorescence fluctuation analysis has been used to determine the degree of clustering of the EGF receptor (Saffarian *et al* 2007) and the stoichiometry of protein complexes (Chen and Muller 2007). Thus, FCS has further developed into a multi-parameter imaging mode that has been used to describe the dynamics and oligomerization of GPI-anchored receptors (Malengo *et al* 2008), adhesion receptors and tetraspanins (Barreiro *et al* 2008).

In order to reveal the relative contributions of membrane lipid microdomains and the actin cytoskeleton, variable spot-size FCS and a defined 'FCS diffusion law' have been used, extracting structural information from below the diffraction limit of optical microscopy (Lenne *et al* 2006). Microdomains and cytoskeletal confinement impart different deviations to the relationship between the diameter of the observation spot and the transit time that a molecule requires to diffuse through that spot (Wawrezinieck *et al* 2005). On a plot of diffusion time against area, dynamic partitioning results in a positive intercept with the *y*-axis, whereas a meshwork produces a negative intercept (unhindered diffusion passes through the origin) (Lenne *et al* 2006). Indeed, this was shown theoretically and confirmed by the observation of Thy-1, which is raft-associated, has a positive intercept, and transferrin receptors, and which is a transmembrane protein that is hindered by the cytoskeleton and has a negative intercept with the *y*-axis. In particular, this approach has been used to demonstrate that the protein kinase Akt dynamically partitions into microdomains in cells via its pleckstrin homology (PH) domain (Lasserre *et al* 2008).

FCS has been combined with total internal reflection (TIR-FCS) excitation (Starr and Thompson 2001, Thompson and Steele 2007). Such a setup enables measurement of the local probe concentration and the local translational mobility within the membranes, as well as kinetic rate constants, which describe the association and dissociation of ligands with membrane receptors. The improved accuracy of TIR-

FCS over FCS has been demonstrated by imaging a farnesylated variant of green fluorescent protein (GFP) (Ohsugi *et al* 2006). While SPT under TIR illumination cannot easily track cytosolic proteins due to their 3D motion, TIR-FCS was able to quantify the proportion of the GFP construct bound to the membrane. Further technical improvements to FCS measurements specific to membranes have been accomplished (Ries and Schwille 2008).

An exciting prospect is the integration of time-resolved single photon counting (TCSPC) techniques with unrestricted data acquisition to store spatial, temporal, spectral and intensity information (Koberling *et al* 2008). In particular, this enables the correlation of dependences between various fluorescence parameters. For instance, FCS autocorrelation curves can be constructed for fluorescent events of a certain fluorescence lifetime. This not only helps to remove background noise and autofluorescence, thus providing more accurate diffusion measurements, it also enables a single experiment to give information about the kinetics of, for instance, monomers and dimers, local concentrations of proteins and their different states, their relative stoichiometry and affinities. However, can this information be combined with mathematical models to reveal a more detailed understanding of how proteins function within the cell membrane?

Taken together, the technical advances in fluorescence microscopy hardware and analysis have led to the quantification of the dynamics of membrane proteins in living cells. FRAP, FCS and SPT can go beyond traditional fluorescence imaging and have turned microscopes into molecular measurement devices that can operate on many spatial and temporal length-scales. Synergistically, all these approaches have developed a picture where the diffusion characteristics of membrane proteins display complex and subtle behavior. The influence of membrane association type on protein diffusion, diffusion restrictions imparted by the membrane cytoskeleton and anchored-protein fences, protein oligomerization, clustering and protein–protein interactions, and dynamic partitioning into small, heterogeneous and mobile lipid microdomains have been measured with remarkable accuracy, given the resolution limits of optical microscopy. The arrival of super-resolution microscopy approaches, such as stimulated emission depletion (STED), photoactivated localization microscopy (PALM) and stochastic optical reconstruction microscopy (STORM) have further improved the accuracy of dynamic measurements, as demonstrated by PALM being combined with SPT (Manley *et al* 2008) and STED with FCS (Eggeling *et al* 2009). The reduced observation volume of STED detected differences between the FCS autocorrelation curve of phosphoethanolamine and that of sphingomyelin, whose diffusion is transiently trapped in cholesterol-enriched complexes (Eggeling *et al* 2009).

However, the interdependence of membrane organization and protein diffusion is not yet well understood. Moreover, current models do not adequately take into account the active transport systems, such as directed flow, membrane budding and recycling. In order to investigate these parameters, multi-dimensional microscopy techniques have to be performed, as they have the ability to extract more precise information about molecular diffusion rates and other membrane parameters. To date, there has only been a limited understanding of how local variations in membrane fluidity affect protein concentration, oligomerization and diffusion. However, these questions can be addressed by integrating FCS or SPT with the imaging of the membrane lipid order using microenvironmentally sensitive probes such as Laurdan (Gaus *et al* 2006) or di-4-ANEPPDHQ (Jin *et al* 2006). Finally, quantitative multi-parameter microscopy can further reveal how global cellular manipulations and treatments act on the localized and molecular organization of membrane domains and compartments, which in turn influence protein dynamics, interactions and functions.

5.4 Role of the cortex confinement on membrane diffusion

The concept of the classical Singer–Nicolson fluid mosaic model for biological membranes (Singer and Nicolson 1972) is still a milestone in cell membrane research. However, other methods for investigating membrane dynamics have revealed more insights that even contradict this model. One example is the assumption that the proteins and lipids within the membrane undergo Brownian diffusion, which turned out to be inaccurate (Fujiwara *et al* 2002, Kusumi *et al* 2005, Wawrezinieck *et al* 2005, Eggeling *et al* 2009, Jacobson *et al* 1995, Lingwood and Simons 2010, Wieser *et al* 2007). Moreover, using different methods it has been revealed that the lateral motion of molecules within the membrane is restricted by various mechanisms (figure 5.6).

The different confinements are evoked by different membrane-organizing principles: first, interactions of the molecules with transient self-assemblies of specific lipids and lipid-anchored proteins in so-called 'lipid rafts' confines the diffusion (Lingwood and Simons 2010, Dietrich *et al* 2002, Jacobson *et al* 2007). Second, direct or indirect interactions of the molecules with actin-cytoskeleton based

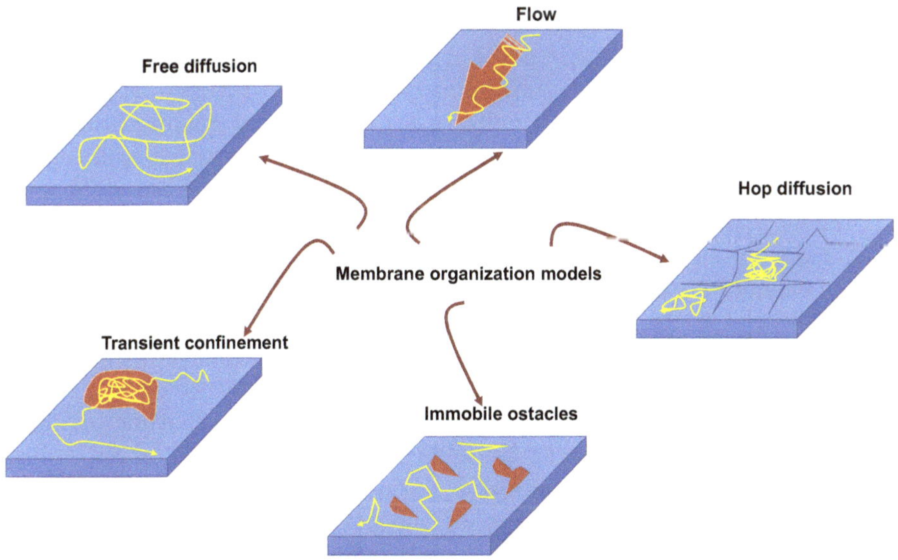

Figure 5.6. Different membrane organization models and the impact on diffusion of membrane constituents.

barriers or anchors such as cytoskeleton-anchored proteins restrict their diffusion (Wawrezinieck *et al* 2005, Eggeling *et al* 2009, Jacobson *et al* 1995, Edidin *et al* 1991, Gowrishankar *et al* 2012, Kwik *et al* 2003, Andrews *et al* 2008, Jaqaman *et al* 2011).

Third, the curvature of the membrane affects the diffusion of the molecules (Hatzakis *et al* 2009, Roux *et al* 2005). The second point has been addressed by different methods and it has indeed been shown that various membrane proteins are confined by the actin cytoskeleton (Gowrishankar *et al* 2012, Andrews *et al* 2008, Jaqaman *et al* 2011). In line with this, based on single-cell particle (SPT) experiments the phospholipid diffusion within the plasma membrane may even be constrained, which may possibly be based on the cortical actin cytoskeleton (Fujiwara *et al* 2002, Clausen and Lagerholm 2013). Based on this finding the 'picket-fence' model has been revealed (Kusumi *et al* 2005), which hypothesizes that direct anchoring of transmembrane proteins (called the pickets) to cortical cytoskeletal filaments (called the fences) underneath the cell membrane builds confinements such as barriers. These indirectly hinder the diffusion of nearby membrane proteins and lipids. These specific barriers evoke the compartmentalization of the cell membrane. In each compartment the molecules can solely diffuse freely, whereas the crossing from one compartment to another is restricted and hence creates a compartmentalized or so-called 'hop'-diffusion. The premise of compartmentalization of membrane proteins and lipids seem to be highly interesting as it is very attractive to associate the individual localized proteins to specific signaling cues (Kusumi *et al* 2005). In particular, the diffusion of integral membrane proteins within distinct compartments has been found to increase the interaction probability of less abundant proteins, which in turn induces certain cellular signaling pathways (Andrews *et al* 2008, Jaqaman *et al* 2011).

The picket-fence model for compartmentalization of phospholipids in the cell membrane contains several obstacles, which decrease its broad acceptance (Wieser *et al* 2007, Abbott 2005, Clausen and Lagerholm 2011). However, in principle, the compartmentalized phospholipid diffusion has only been revealed in SPT experiments employing gold particles (Fujiwara *et al* 2002) and quantum dots (QDs) (Clausen and Lagerholm 2013) in order to access the sub-millisecond temporal resolution regime needed for the detection of compartmentalized diffusion. A weakness of these probes is that they are very artifact-prone based on their large size and on the difficulty in validating the probe specificity to target molecules. Hence, it cannot be determined precisely whether or not these probes alter the native target molecule in its mobility (diffusion) by steric hindrance or trigger the oligomerization of the target molecules (causing a confinement) (Fujiwara *et al* 2002, Clausen and Lagerholm 2011). Indeed, the validity of SPT reports on compartmentalized diffusion has been questioned by showing that the irregularity of cell membrane topography can cause an artificial compartmentalized diffusion using this technique (Adler *et al* 2010).

In order to resolve this problem of the actin-cytoskeleton-modulated lipid compartmentalized diffusion, a stimulated emission depletion fluorescence correlation spectroscopy (STED-FCS) (Eggeling *et al* 2009, Kastrup *et al* 2005, Mueller *et al* 2011) has been utilized to probe the diffusion of a phospholipid analog, which is labeled with a small and potentially less invasive organic dye, in the cell membrane of living cells. STED-FCS provides a systematic probing of molecular diffusion for observation spot sizes that range from a diffraction-limited 240 nm down to below 40 nm, which is in the range of the size of the postulated actin-cytoskeleton-facilitated compartments (Fujiwara *et al* 2002, Kusumi *et al* 2005, Wawrezinieck *et al* 2005). A phospholipid di-palmitoyl-phosphoethanolamine (DPPE) labeled with the fluorescent dye Atto647N at its head group has been used for these experiments. However, the label has been found to alter some of the lipids characteristics, such as preference of DPPE changes from forming more ordered to more disordered membrane microenvironments (Mueller *et al* 2011, Honigmann *et al* 2014, Sezgin *et al* 2012). As an advantage over other dyes, the Atto647N dye modification has been shown using STED-FCS not to interfere with the diffusion dynamics and lipid–lipid interactions in living cells (Eggeling *et al* 2009, Mueller *et al* 2011). In line with this, previous STED-FCS experiments using living PtK2 and HeLa cells have been shown that diffusion of the DPPE analog is mainly free and hardly impaired or reduced by transient interactions with other membrane molecules such as slow-moving or cytoskeleton-connected proteins or by the curvature of the membrane (Eggeling *et al* 2009, Mueller *et al* 2011, Honigmann *et al* 2014). Based on these results this lipid analog seems to be a very good candidate for analyzing the effect on phospholipid diffusion constricted by the cortical cytoskeleton-dependent membrane partition or other hindrances.

The experimental studies have been performed by using two adherent cell types such as NRK fibroblasts and Ink4a/Arf ($^{-/-}$) mouse embryo fibroblasts (termed IA32 MEFs) (Andrade *et al* 2015, Wu *et al* 2012). These cell types have been selected based on the results of previous work using SPT in combination with either gold particles or QDs as lipid labels that proposed compartmentalized phospholipid diffusion in these cell types (Fujiwara *et al* 2002, Clausen and Lagerholm 2011). Moreover, both cell types contain very distinct actin-rich regions either in-between a distinct stress-fiber network supporting strong adhesion sites to the cover glass surface (NRK), or in the form of prominent exertion of lamellipodia in front of a large, thin lamella (IA32) (Wu *et al* 2012). Thus, STED-FCS provides the possibility to detect compartmentalized lipid diffusion and another advantage is that even minimally invasive phospholipid probes can be used, which are spatially constrained within compartments of the cell membrane.

Compartmentalized diffusion of lipids in cells and simulation

How can the compartmentalized diffusion be resolved? To confirm the applicability of STED-FCS for detecting compartmentalized diffusion, simulated diffusion within a heterogeneous lattice with a characteristic average compartment length L has been performed *in silico*. In more detail, molecules diffuse freely within compartments with a diffusion coefficient D_{free}, while the transposition of compartment boundaries is possible with a distinct 'hopping probability', termed P_{hop}. In particular, simulated trajectories are transformed into intensity traces, which are auto-correlated to generate *in silico* FCS curves for different observation spots. After fitting these curves using standard STED-FCS analysis, the average apparent diffusion coefficient D_{app} of the molecules can be obtained as a function of the observation spot diameter *d* (Mueller *et al* 2011). Indeed, the resulting dependence of the apparent diffusion coefficient D_{app} on the diameter *d* of the observation spot shows that molecular diffusion is spatial constrained (Wawrezinieck *et al* 2005, Mueller *et al* 2011). Within these simulations, free diffusion (P_{hop} = 1) is characterized by D_{app}, which is independent of *d*, whereas compartmentalized diffusion (P_{hop} < 1) is indicated by a decrease of D_{app} with increasing *d*. However, measurements at small observation spots primarily reveal the free diffusion within the compartments ($D_{app} = D_{free}$), whereas measurements at large observation spots (in the range of compartment sizes) showed reduced apparent diffusion coefficient $D_{app} < D_{free}$, as the compartments represent barriers for molecular transits through the observation spot. In particular for $P_{hop} \ll 1$, which represents a strong confinement, the compartmentalized diffusion can be separated easily from free diffusion using STED-FCS. Moreover, experimental STED-FCS measurements using NRK fibroblasts and IA32 MEFs revealed that both cell types, in which the diffusion has been measured at thin peripheral areas of the cells (within 2–5 μm from the cell edge) displayed a clear compartmentalized diffusion in both apical and basal membranes, which is indicated by a significant decrease of $D_{app}(d)$ at large observation diameters.

What is the effect of actin and Arp2/3 on compartmentalized diffusion?
In order to reveal the underlying mechanism of the lipid compartmentalized diffusion, the actin and the actin nucleating and branching protein Arp2/3 has been inhibited by adding pharmacological inhibitors to NRK and IA32 cells using STED-FCS. In particular, the actin-cytoskeleton modulation has been evoked by treating cells with either latrunculin B (LatB) or CK-666 (Nolen *et al* 2009, Fritzsche *et al* 2013). LatB treatment impairs the polymerization of all F-actin networks, whereas CK-666 inhibits solely the actin nucleator Arp2/3 (Goley and Welch 2006) and hence decreases the cytoskeleton branching. The diffusion of the DPPE analog after LatB addition was faster at $d = 240$ nm in both cell types, but still weak compartmentalized diffusion was determined for measurements at d = 240 nm and $d \approx 40$ nm. The diffusion of DPPE in both cell types treated with CK-666 was pronouncedly faster at $d = 240$ nm and compartmentalized diffusion was no longer detectable.

What are the effects of myosin and cholesterol on compartmentalized diffusion?
In contrast, blebbistatin inhibition of myosin II did not alter the DPPE diffusion in either NRK nor IA32 cells, indicating that compartmentalized diffusion is independent of myosin-based contractility. To reveal the effect of membrane cholesterol on the underlying mechanism of compartmentalized diffusion, cholesterol has been decreased by adding the cholesterol oxidase (Coase). Similar to myosin II inhibition, cholesterol depletion using COase did not affect the DPPE diffusion indicating that compartmentalization of DPPE is independent of cholesterol-dependent interactions. Finally, the compartmentalized diffusion of DPPE seem to be dependent on the cortical actin cytoskeleton and is particularly dependent on Arp2/3-based

cortical actin networks. In order to reveal whether the observed results are applicable to parts closer to the cell body, STED-FCS measurements are performed.

Dissimilar to IA32 cells, NRK cells are relatively large cell bodies, where STED-FCS specifically measures the basal plasma membrane. These measurements revealed that a faster and slightly less compartmentalized diffusion in deeper regions, as compared to the measurements near the cell edge, which still the combines effects of diffusion in the basal and apical plasma membranes. This result may be explained by a less branched F-actin network beneath the cell body compared to the cell edge or by a less branched F-actin network in the basal plasma membrane, compared to the apical membrane (Xu *et al* 2012).

As compartmentalized diffusion of DPPE has been detected in IA32 and NRK cells, the STED-FCS measurements in PtK2 cells revealed free diffusion (Eggeling *et al* 2009, Mueller *et al* 2011). After CK-666 treatment, even in PtK2 cells diffusion of DPPE was significantly faster than in untreated control cells, indicating that the lipid diffusion is compartmentalized by Arp2/3-dependent cortical actin networks. This result leads to the suggestion that either the compartment size L in unperturbed PtK2 is smaller than the accessible spatial sampling range of STED-FCS or that the compartment strength P_{hop} is weaker than that of NRK cells and IA32 MEFs. Another support for the role of the Arp2/3 complex in lipid compartmentalization has been found by analyzing Arp2/3-depleted MEFs. To confirm the effects of Arp2/3 depletion, the experiments are performed using IA32 MEFs depleted of p34Arc and Arp2, which are two essential subunits of the Arp2/3 complex (IA32 2xKD) (Wu *et al* 2012). In more detail, these cells display no lamellipodia and contain, as demonstrated by electron microscopy, a much sparser actin network in the lamella, which possesses only frequent filopodia at the cell's edge (Wu *et al* 2012). STED-FCS measurements in deeper cell regions revealed that these cells display free diffusion. However, this result is in contrast to the pivotal role of Arp2/3 in impairing the lipid diffusion in the cell membrane. For small observation areas, the DPPE diffusion was slower in IA32 2xKD MEFs than in either WT or CK-666 treated IA32 MEFs. Hence it can be hypothesized that the reason for this discrepancy is due to enhanced membrane curvature as a consequence of less sub-membranous cortical actin mechanical support, which is also seen in the phenotypic observation that the IA32 2xKD MEFs possess a defective cell volume response to osmotic stress alterations (Wu *et al* 2013). However, no suitable biophysical methods are available to examine the relation between impaired diffusion and an enhanced membrane curvature obstacle in live cells. When the Saffman–Delbrueck model (Saffman and Delbrueck 1975) has been applied to these observations, the viscosity of the cell membrane within the confined regions seem to be identical in both cell types.

The actin cytoskeleton has been shown to be not a single structure but rather composed of distinctly diverse substructures (Fritzsche *et al* 2013). Most important for diffusion of receptors or lipids in the cell membrane seems to be the cortical actin cytoskeleton, which is the actin network directly below the cell membrane, and which directly couples to the cell membrane through a variety of actin binding proteins and cytoplasmic domains of certain membrane proteins (Fritzsche *et al* 2014). However, a direct fluorescence imaging of the cortical cytoskeleton is currently still not possible in cells under physiological conditions, as it is smaller than the limited axial resolution of the available microscopes. However, these results indicated a prominent role of the cortical actin in constraining the phospholipid diffusion and thereby evoking the compartmentalized diffusion in the cell membrane.

Dissimilar to other cytoskeletal structures, the composition of the cell cortex is poorly known (Salbreux *et al* 2012). In particular, a specific role in providing compartmentalized diffusion fulfills the Arp2/3 complex as a component of the cortical actin networks (Wu *et al* 2013, Bovellan *et al* 2014, Yang *et al* 2014). Moreover, the Arp2/3 is highly crucial for the generation of dense, highly branched F-actin networks at the leading edge of lamellipodia and/or at adhesion sites (Wu *et al* 2012, Romer *et al* 2006), and orchestrates multiple functions performed by the actin cytoskeleton (Goley and Welch 2006). The Arp2/3 inhibitor CK-666 and Arp2/3-depleted mammalian cells have shown that in addition to the role of Arp2/3 in providing branched actin networks, the Arp2/3-branched actin network contributes to various cellular processes including matrix sensing, cytoplasmic streaming, spindle positioning and cell–cell junction regulation (Wu *et al* 2012, Goley and Welch 2006, Wu *et al* 2013, Yang *et al* 2014, Chaigne *et al* 2013, Yi *et al* 2011, Zhou *et al* 2013). In addition, it has been proposed that membrane curvature is responsible for the reported hop-diffusion (Alder *et al* 2010), although this has never been demonstrated to affect STED-FCS experiments. However, the effect of the cell curvature on hop-diffusion cannot be completely ruled out.

Taken together, cortical actin-assisted compartmentalization seems to be the most accurate model for the general constraints that limit lipid diffusion in the absence of specific molecular interactions. However, the determination of the precise structural and molecular mechanisms by which the cytoskeleton causes the lipid compartmentalization is still elusive (possibly provided by steric effects of the transmembrane proteins connected to cortical actin filaments (Fujiwara *et al* 2002), possibly associated with alterations in membrane order (Honigmann *et al* 2014) and the mechanisms involving fundamental signaling processes). In conclusion, models for membrane organization that ignore the symbiosis between the cell membrane and the cytoskeleton are far too oversimplified (Gowrishankar *et al* 2012).

References and further reading

Abbott A 2005 Cell biology: hopping fences *Nature* **433** 680–83

Abercrombie M and Dunn G A 1975 Adhesions of fibroblasts to substratum during contact inhibition observed by interference reflection microscopy *Exp. Cell Res.* **92** 57–62

Adams C L, Nelson J W and Smith S J 1996 Quantitative analysis of cadherin–catenin–actin reorganization during development of cell–cell adhesion *J. Cell Biol.* **135** 1899–911

Adler J, Andrew I S, Novak P, Korchev Y E and Parmryd I 2010 Plasma membrane topography and interpretation of single-particle tracks *Nat. Methods* **7** 170–1

Anderson C M, Georgiou G N, Morrison I E G, Stevenson G V and Cherry R J 1992 *J. Cell Sci.* **101** 415–25

Andrade D M, Clausen M P, Keller J, Mueller V, Wu C, Bear J E, Hell S W, Lagerholm B C and Eggeling C 2015 Cortical actin networks induce spatio-temporal confinement of phospholipids in the plasma membrane—a minimally invasive investigation by STED-FCS *Sci. Rep.* **5** 11454

Andrews N L, Lidke K A, Pfeiffer J R, Burns A R, Wilson B S, Oliver J M and Lidke D S 2008 Actin restricts Fcε RI diffusion and facilitates antigen-induced receptor immobilization *Nat. Cell Biol.* **10** 955–63

Angres B, Barth A and Nelson W J 1996 Mechanism for transition from initial to stable cell–cell adhesion: kinetic analysis of E-cadherin-mediated adhesion using a quantitative adhesion assay *J. Cell Biol.* **134** 549–57

Apodaca G 2002 Modulation of membrane traffic by mechanical stimuli *Am. J. Physiol. Renal. Physiol.* **282** 179–90

Armstrong P B 1989 Cell sorting out: the self-assembly of tissues in vitro *Crit. Rev. Biochem. Mol. Biol.* **24** 119–49

Bacia K, Kim S A and Schwille P 2006 Fluorescence cross-correlation spectroscopy in living cells *Nat. Methods* **3** 83–9

Bacia K, Scherfeld D, Kahya N and Schwille P 2004 Fluorescence correlation spectroscopy relates rafts in model and native membranes *Biophys. J.* **87** 1034–43

Barreiro O, Zamai M, Yanez-Mo M, Tejera E, Lopez-Romero P, Monk P N, Gratton E, Caiolfa V R and Sanchez-Madrid F 2008 Endothelial adhesion receptors are recruited to adherent leukocytes by inclusion in preformed tetraspanin nanoplatforms *J. Cell Biol.* **183** 527–42

Bereiter-Hahn J, Fox C H and Thorell B 1979 Quantitative reflection contrast microscopy of living cells *J. Cell Biol.* **82** 767–79

Biswas A, Alex A and Sinha B 2017 Mapping cell membrane fluctuations reveals their active regulation and transient heterogeneities *Biophys. J.* **113** 1768–81

Borghi N and Nelson J W 2009 Intercellular adhesion in morphogenesis: molecular and biophysical considerations *Curr. Top. Dev. Biol.* **89** 1–32

Bouchaud J-P and Georges A 1990 Anomalous diffusion in disordered media: statistical mechanisms, models and physical applications *Phys. Rep.* **195** 127–293

Bovellan M, Romeo Y and Biro M *et al* 2014 Cellular control of cortical actin nucleation *Curr. Biol.* **24** 1628–35

Brodland G W 2003 New information from aggregate compression tests and its implications for theories of cell sorting *Biorheology* **40** 273–7

Brown D A 2006 Lipid rafts, detergent-resistant membranes and raft targeting signals *Physiology* **21** 430–9

Cardoso Dos Santos M, Déturche R, Vézy C and Jaffiol R 2016 Topography of cells revealed by variable-angle total internal reflection fluorescence microscopy *Biophys. J.* **111** 1316–27

Chaigne A, Campillo C and Gov N S *et al* 2013 A soft cortex is essential for asymmetric spindle positioning in mouse oocytes *Nat. Cell Biol.* **15** 958–66

Chen Y and Muller J D 2007 Determining the stoichiometry of protein heterocomplexes in living cells with fluorescence fluctuation spectroscopy *Proc. Natl Acad. Sci.* **104** 3147–52

Chen C H, Tsai F C, Wang C C and Lee C H 2009 Three-dimensional characterization of active membrane waves on living cells *Phys. Rev. Lett.* **103** 238101

Cherry R J 1992 Keeping track of cell surface receptor *Trends Cell Biol.* **2** 242–4

Chu Y-S, Thomas W A, Eder O, Pincet F, Perez E, Thiery J P and Dufour S 2004 Force measurements in E-cadherin-mediated cell doublets reveal rapid adhesion strengthened by actin cytoskeleton remodeling through Rac and Cdc42 *J. Cell Biol.* **167** 1183–94

Chugh P, Clark A G, Smith M B, Cassani D A D, Dierkes K, Ragab A, Roux P P, Charras G, Salbreux G and Paluch E K 2017 Actin cortex architecture regulates cell surface tension *Nat. Cell Biol.* **19** 689–97

Clausen M and Lagerholm C 2011 The probe rules in single particle tracking current *Protein Pept. Sci.* **12** 699–713

Clausen M P and Lagerholm B C 2013 Visualization of plasma membrane compartmentalization by high-speed quantum dot tracking *Nano Lett.* **13** 2332–7

Curtis A S G 1964 The mechanism of adhesion of cells to glass. A study by interference reflection microscopy *J. Cell Biol.* **20** 199–215

Dai J and Sheetz M P 1999 Membrane tether formation from blebbing cells *Biophys. J.* **77** 3363–70

Dai J, Ting-Beall H P and Sheetz M P 1997 The secretion-coupled endocytosis correlates with membrane tension changes in RBL 2H3 cells *J. Gen. Physiol.* **110** 1–10

Daumas F, Destainville N, Millot C, Lopez A, Dean D and Salome L 2003 Confined diffusion without fences of a G-protein-coupled receptor as revealed by single particle tracking *Biophys. J.* **84** 356–66

Davis G S, Phillips H M and Steinberg M S 1997 Germ-layer surface tensions and 'tissue affinities' in Rana pipiens gastrulae: quantitative measurements *Dev. Biol.* **192** 630–44

Dietrich C, Bagatolli L A, Volovyk Z N, Thompson N L, Levi M, Jacobson K and Gratton E 2001 Lipid rafts reconstituted in model membranes *Biophys. J.* **80** 1417–28

Dietrich C, Yang B, Fujiwara T, Kusumi A and Jacobson K 2002 Relationship of lipid rafts to transient confinement zones detected by single particle tracking *Biophys. J.* **82** 274–84

Digman M A, Brown C M, Horwitz A R, Mantulin W W and Gratton E 2008 Paxillin dynamics measured during adhesion assembly and disassembly by correlation spectroscopy *Biophys J.* **94** 2819–31

Digman M A, Brown C M, Sengupta P, Wiseman P W, Horwitz A R and Gratton E 2005 Measuring fast dynamics in solutions and cells with a laser scanning microscope *Biophys. J.* **89** 1317–27

Doebereiner H-G, Dubin-Thaler B J, Hofman J M, Xenias H S, Sims T N, Giannone G, Dustin M L, Wiggins C H and Sheetz M P 2006 Lateral membrane waves constitute a universal dynamic pattern of motile cells *Phys. Rev. Lett.* **97** 038102

Douglass A D and Vale R D 2005 Single-molecule microscopy reveals plasma membrane microdomains created by protein–protein networks that exclude or trap signaling molecules in T cells *Cell* **121** 937–50

Drees F, Pokutta S, Yamada S, Nelson W J and Weis W I 2005 α-catenin is a molecular switch that binds E-cadherin-β-catenin and regulates actin-filament assembly *Cell* **123** 903–15

Dubin-Thaler B J, Giannone G, Döbereiner H G and Sheetz M P 2004 Nanometer analysis of cell spreading on matrix-coated surfaces reveals two distinct cell states and STEPs *Biophys. J.* **86** 1794–806

Duguay D, Foty R A and Steinberg M S 2003 Cadherin-mediated cell adhesion and tissue segregation: qualitative and quantitative determinants *Dev. Biol.* **253** 309–23

Edidin M 1997 Lipid microdomains in cell surface membranes *Curr. Opin. Cell Biol.* **7** 528–32

Edidin M, Kuo S C and Sheetz M P 1991 Lateral movements of membrane glycoproteins restricted by dynamic cytoplasmic barriers *Science* **254** 1379–82

Eggeling C, Ringemann C and Medda R *et al* 2009 Direct observation of the nanoscale dynamics of membrane lipids in a living cell *Nature* **457** 1159–63

Elson E L and Magde D 1974 Fluorescence correlation spectroscopy: conceptual basis and theory *Biopolymers* **13** 1–27

Evans E and Yeung A 1989 Apparent viscosity and cortical tension of blood granulocytes determined by micropipet aspiration *Biophys. J.* **56** 151–60

Farhadifar R, Roeper J C, Aigouy B, Eaton S and Juelicher F 2007 The influence of cell mechanics, cell–cellinteractions, and proliferation on epithelial packing *Curr. Biol.* **17** 2095–104

Feder T J, Brust-Mascher I, Slattery J P, Baird B and Webb W W 1996 Constrained diffusion or immobile fraction on cell surfaces: a new interpretation *Biophys. J.* **70** 2767–73

Fenz S F, Bihr T, Schmidt D, Merkel R, Udo S, Sengupta K and Smith A-S 2017 Membrane fluctuations mediate lateral interaction between cadherin bonds *Nat. Phys.* **13** 906–13

Forgacs G, Foty R A, Shafrir Y and Steinberg M S 1998 Viscoelastic properties of living embryonic tissues: a quantitative study *Biophys. J.* **74** 2227–34

Foty R A and Steinberg M S 2005 The differential adhesion hypothesis: a direct evaluation *Dev. Biol.* **278** 255–63

Foty R A, Corbett S A, Schwarzbauer J E and Steinberg M S 1998 Effects of dexamethasone on cadherin-mediated cohesion of human fibrosarcoma HT-1080 cells *Cancer Res.* **58** 3586–9

Foty R A, Forgacs G, Pfleger C M and Steinberg M S 1994 Liquid properties of embryonic tissues: measurement of interfacial tensions *Phys. Rev. Lett.* **72** 2298–301

Foty R A, Pfleger C M, Forgacs G and Steinberg M S 1996 Surface tensions of embryonic tissues predict their mutual envelopment behavior *Development* **122** 1611–20

Foty R A and Steinberg M S 1997 Measurement of tumor cell cohesion and suppression of invasion by E- or P-cadherin *Cancer Res.* **57** 5033–6

Fritzsche M, Lewalle A, Duke T, Kruse K and Charras G 2013 Analysis of turnover dynamics of the submembranous actin cortex *Mol. Biol. Cell* **24** 757–67

Fritzsche M, Thorogate R and Charras G 2014 Quantitative analysis of ezrin turnover dynamics in the actin cortex *Biophys. J.* **106** 343–53

Fujiwara T, Ritchie K, Murakoshi H, Jacobson K and Kusumi A 2002 Phospholipids undergo hop diffusion in compartmentalized cell membrane *J. Cell Biol.* **157** 1071–82

Gaus K, Zech T and Harder T 2006 Visualizing membrane microdomains by Laurdan 2-photon microscopy *Mol. Membr. Biol.* **23** 41–8

Gauthier N C, Fardin M A, Roca-Cusachs P and Sheetz M P 2011 Temporary increase in plasma membrane tension coordinates the activation of exocytosis and contraction during cell spreading *Proc. Natl Acad. Sci. USA* **108** 14467–72

Ghosh R N 1991 Mobility and clustering of individual low-density lipoprotein receptor molecules on the surface of human skin fibroblasts *PhD Thesis* Cornell University

Giannone G, Dubin-Thaler B J, Döbereiner H G, Kieffer N, Bresnick A R and Sheetz M P 2004 Periodic lamellipodial contractions correlate with rearward actin waves *Cell* **116** 431–43

Godwin S L, Fletcher M and Burchard R P 1989 Interference reflection microscopic study of sites of association between gliding bacteria and glass substrata *J. Bacteriol.* **171** 4589–94

Goldstein J L and Brown M S 1977 *Annu. Rev. Biochem.* **46** 897–930

Goley E D and Welch M D 2006 The ARP2/3 complex: an actin nucleator comes of age *Nat. Rev. Mol. Cell Biol.* **7** 713–26

Goodwin J S, Drake K R, Remmert C L and Kenworthy A K 2005 Ras diffusion is sensitive to plasma membrane viscosity *Biophys. J.* **89** 1398–410

Goswami D, Gowrishankar K, Bilgrami S, Ghosh S, Raghupathy R, Chadda R, Vishwakarma R, Rao M and Mayor S 2008 Nanoclusters of GPI- anchored proteins are formed by cortical actin-driven activity *Cell* **135** 1085–97

Gowrishankar K, Ghosh S, Saha S, Rumamol C, Mayor S and Rao M 2012 Active remodeling of cortical actin regulates spatiotemporal organization of cell surface molecules *Cell* **149** 1353–67

Graner F 1993 Can surface adhesion drive cell-rearrangement? Part I: biological cell-sorting *J. Theor. Biol.* **164** 455–76

Graner F and Glazier J A 1992 Simulation of biological cell sorting using a two-dimensional extended Potts model *Phys. Rev. Lett.* **69** 2013–6

Groulx N, Boudreault F, Orlov S and Grygorczyk R 2006 Membrane reserves and hypotonic cell swelling *J. Membr. Biol.* **214** 43–56

Guevorkian K, Colbert M-J, Durth M, Dufour S and Brochard-Wyart F 2010 Aspiration of biological viscoelastic drops *Phys. Rev. Lett.* **104** 218101

Harris A K 1976 Is cell sorting caused by differences in the work of intercellular adhesion? A critique of the Steinberg hypothesis *J. Theor. Biol.* **61** 267–85

Hatzakis N S, Bhatia V K, Larsen J, Madsen K L, Bolinger P Y, Kunding A H, Castillo J, Gether U, Hedegård P and Stamou D 2009 How curved membranes recruit amphipathic helices and protein anchoring motifs *Nat. Chem. Biol.* **5** 835–41

Hebert B, Costantino S and Wiseman P W 2005 Spatiotemporal image correlation spectroscopy (STICS) theory, verification, and application to protein velocity mapping in living CHO cells *Biophys. J.* **88** 3601–14

Holtfreter J 1944 A study of the mechanics of gastrulation *J. Exp. Zool.* **95** 171–212

Honigmann A, Sadeghi S, Keller J, Hell S W, Eggeling C and Vink R 2014 A lipid bound actin meshwork organizes liquid phase separation in model membranes *eLife* **3** e01671

Honigmann A, Mueller V, Ta H, Schoenle A, Sezgin E, Hell S W and Eggeling C 2014 Scanning STED-FCS reveals spatiotemporal heterogeneity of lipid interaction in the plasma membrane of living cells *Nat. Commun.* **5** 5412

Imamura Y, Itoh M, Maeno Y, Tsukita S and Nagafuchi A 1999 Functional domains of α-catenin required for the strong state of cadherin-based cell adhesion *J. Cell Biol.* **144** 1311–22

Jacobson K, Mouritsen O G and Anderson R G 2007 Lipid rafts: at a crossroad between cell biology and physics *Nat. Cell Biol.* **9** 7–14

Jacobson K, Sheets E D and Simson R 1995 Revisiting the fluid mosaic model of membranes *Science* **268** 1441–2

Jacquier V, Prummer M, Segura J-M, Pick H and Vogel H 2006 Visualizing odorant receptor trafficking in living cells down to the single-molecule level *Proc. Natl Acad. Sci.* **103** 14325–30

Jaqaman K, Loerke D, Mettlen M, Kuwata H, Grinstein S, Schmid S L and Danuser G 2008 Robust single-particle tracking in live-cell time-lapse sequences *Nat. Methods* **5** 695–702

Jaqaman K, Kuwata H, Touret N, Collins R, Trimble W S, Danuser G and Grinstein S 2011 Cytoskeletal control of CD36 diffusion promotes its receptor and signaling function *Cell* **146** 593–606

Jin L, Millard A C, Wuskell J P, Dong X, Wu D, Clark H A and Loew L M 2006 Characterization and application of a new optical probe for membrane lipid domains *Biophys. J.* **90** 2563–75

Jovin T and Vaz W L C 1989 Rotational and translational diffusion in membranes measured by fluorescence and phosphorescence methods *Methods Enzymol.* **172** 471–513

Kahya N, Scherfeld D, Bacia K, Poolman B and Schwille P 2003 Probing lipid mobility of raft-exhibiting model membranes by fluorescence correlation spectroscopy *J. Biol. Chem.* **278** 28109–15

Kastrup L, Blom H, Eggeling C and Hell S W 2005 Fluorescence fluctuation spectroscopy in subdiffraction focal volumes *Phys. Rev. Lett.* **94** 178104

Kenworthy A K, Nichols B J, Remmert C L, Hendrix G M, Kumar M, Zimmerberg J and Lippincott-Schwartz J 2004 Dynamics of putative raft-associated proteins at the cell surface *J. Cell Biol.* **165** 735–46

Koberling F, Kramer B, Tannert S, Ruttinger S, Ortmann U, Patting M, Wahl M, Ewers B, Kapusta P and Erdmann R 2008 Recent advances in time-correlated single-photon counting *Proc. SPIE* **6862** 09

Kozera L, White S and Calaghan S 2009 Caveolae act as membrane reserves which limit mechanosensitive I(Cl, swell) channel activation during swelling in the rat ventricular myocyte *PLoS One* **4** e8312

Krieg M, Arboleda-Estudillo Y, Puech P H, Kaefer J, Graner F, Mueller D J and Heisenberg C P 2008 Tensile forces govern germ-layer organization in zebrafish *Nat. Cell Biol.* **10** 429–36

Kusumi A, Koyama-Honda I and Suzuki K 2004 Molecular dynamics and interactions for creation of stimulation-induced stabilized rafts from small unstable steady-state rafts *Traffic* **5** 213–30

Kusumi A, Nakada C, Ritchie K, Murase K, Suzuki K, Murakoshi H, Kasai R S, Kondo J and Fujiwara T 2005 Paradigm shift of the plasma membrane concept from the two-dimensional continuum fluid to the partitioned fluid: high-speed single-molecule tracking of membrane molecules *Ann. Rev. Biophys. Biomol. Struct.* **34** 351–78

Kusumi A, Sako Y and Yamamoto M 1993 Confined lateral diffusion of membrane receptors as studied by single particle tracking (nanovid microscopy). Effects of calcium-induced differentiation in cultured epithelial cells *Biophys. J.* **65** 2021–40

Kwik J, Boyle S, Fooksman D, Margolis L, Sheetz M P and Edidin M 2003 Membrane cholesterol, lateral mobility, and the phosphatidylinositol 4,5-bisphosphate-dependent organization of cell actin *Proc. Natl Acad. Sci.* **100** 13964–9

Lajoie P, Partridge E A, Guay G, Goetz J G, Pawling J, Lagana A, Joshi B, Dennis J W and Nabi I R 2007 Plasma membrane domain organization regulates EGFR signaling in tumor cells *J. Cell Biol.* **179** 341–56

Lasserre R, Guo X-J and Conchonaud F *et al* 2008 Raft nanodomains contribute to Akt/PKB plasma membrane recruitment and activation *Nat. Chem. Biol.* **4** 538–47

Lecuit T and Lenne P F 2007 Cell surface mechanics and the control of cell shape, tissue patterns and morphogenesis *Nat. Rev. Mol. Cell Biol.* **8** 633–44

Lee G M, Ishihara A and Jacobson K A 1991 Direct observation of brownian motion of lipids in a membrane *Proc. Natl Acad. Sci. USA* **88** 6274–8

Lenne P-F, Wawrezinieck L, Conchonaud F, Wurtz O, Boned A, Guo X- J, Rigneault H, He H-T and Marguet D 2006 Dynamic molecular confinement in the plasma membrane by micro-domains and the cytoskeleton meshwork *EMBO J.* **25** 3245–56

Limozin L and Sengupta K 2009 Quantitative reflection interference contrast microscopy (RICM) in soft matter and cell adhesion *ChemPhysChem* **10** 2752–68

Lingwood D and Simons K 2010 Lipid rafts as a membrane-organizing principle *Science* **327** 46–50

Lo C M, Keese C R and Giaever I 1993 Monitoring motion of confluent cells in tissue culture *Exp. Cell Res.* **204** 102–9

Lommerse P H M, Blab G A, Cognet L, Harms G S, Snaar-Jagalska B E, Spaink H P and Schmidt T 2004 Single-molecule imaging of the H-Ras membrane-anchor reveals domains in the cytoplasmic leaflet of the cell membrane *Biophys. J.* **86** 609–16

Lommerse P H M, Snaar-Jagalska B E, Spaink H P and Schmidt T 2005 Single-molecule diffusion measurements of H-Ras at the plasma membrane of live cells reveal microdomain localization upon activation *J. Cell Sci.* **118** 1799–809

Lommerse P H M, Vastenhoud K, Pirinen N J, Magee A I, Spaink H P and Schmidt T 2006 Single-molecule diffusion reveals similar mobility for the Lck, H-Ras and K-Ras membrane anchors *Biophys. J.* **91** 1090–7

Malengo G, Andolfo A, Sidenius N, Gratton E, Zamai M and Caiolfa V R 2008 Fluorescence correlation spectroscopy and photon counting histogram on membrane proteins: functional dynamics of the glycosylphosphatidylinositol-anchored urokinase plasminogen activator receptor *J. Biomed. Opt.* **13** 031215

Manley S, Gillette J M, Patterson G H, Shroff H, Hess H F, Betzig E and Lippincott-Schwartz J 2008 High-density mapping of single-molecule trajectories with photoactivated localization microscopy *Nat. Methods* **5** 155–7

Manning M L, Foty R A, Steinberg M S and Schoetz E-M 2010 Coaction of intercellular adhesion and cortical tension specifies tissue surface tension *Proc. Natl Acad. Sci. USA* **107** 12517–22

McClay D R, Wessel G M and Marchase R B 1981 Intercellular recognition: quantitation of initial binding events *Proc. Natl Acad. Sci. USA* **78** 4975–9

McMahon H T and Gallop J L 2005 Membrane curvature and mechanisms of dynamic cell membrane remodelling *Nature* **438** 590–6

Meder D, Moreno M J, Verkade P, Vaz W L C and Simons K 2006 Phase coexistence and connectivity in the apical membrane of polarized epithelial cells *Proc. Natl Acad. Sci.* **103** 329–34

Mgharbel A, Delanoë-Ayari H and Rieu J-P 2009 Measuring accurately liquid and tissue surface tension with a compression plate tensiometer *HFSP J.* **3** 213–21

Mierke C T, Frey B, Fellner M, Herrmann M and Fabry B 2011a Integrin α5β1 facilitates cancer cell invasion through enhanced contractile forces *J. Cell Sci.* **124** 369–83

Mierke C T, Bretz N and Altevogt P 2011b Contractile forces contribute to increased glycosylphosphatidylinositol-anchored receptor CD24-facilitated cancer cell invasion *J. Biol. Chem.* **286** 34858–71

Mombach J C M, Glazier J A, Raphael R C and Zajac M 1995 Quantitative comparison between differential adhesion models and cell sorting in the presence and absence of fluctuations *Phys. Rev. Lett.* **75** 2244–7

Monzel C and Sengupta K 2016 Measuring shape fluctuations in biological membranes *J. Phys. D: Appl. Phys.* **49** 243002

Monzel C, Schmidt D, Kleusch C, Kirchenbüchler D, Seifert U, Smith A S, Sengupta K and Merkel R 2015 Measuring fast stochastic displacements of bio-membranes with dynamic optical displacement spectroscopy *Nat. Commun.* **6** 8162

Morone N, Fujiwara T, Murase K, Kasai R S, Ike H, Yuasa S, Usukura J and Kusumi A 2006 Three-dimensional reconstruction of the membrane skeleton at the plasma membrane interface by electron tomography *J. Cell Biol.* **174** 851–62

Morris C E and Homann U 2001 Cell surface area regulation and membrane tension *J. Membr. Biol.* **179** 79–102

Mueller V, Ringemann C, Honigmann A, Schwarzmann G, Medda R, Leutenegger M, Polyakova S, Belov V N, Hell S W and Eggeling C 2011 STED nanoscopy reveals molecular details of cholesterol- and cytoskeleton-modulated lipid interactions in living cells *Biophys. J.* **101** 1651–60

Murase K, Fujiwara T, Umemura Y, Suzuki K, Lino R, Yamashita H, Saito M, Murakoshi H, Ritchie K and Kusumi A 2004 Ultrafine membrane compartments for molecular diffusion as revealed by single molecule techniques *Biophys. J.* **86** 4075–93

Nagle J F 1992 Long tail kinetics in biophysics? *Biophys. J.* **63** 366–70

Nicolau Jr D V, Burrage K, Parton R G and Hancock J F 2006 Identifying optimal lipid raft characteristics required to promote nanoscale protein–protein interactions on the plasma membrane *Mol. Cell Biol.* **26** 313–23

Nishimura S Y, Vrljic M, Klein L O, McConnell H M and Moerner W E 2006 Cholesterol depletion induces solid-like regions in the plasma membrane *Biophys. J.* **90** 927–38

Nolen B J, Tomasevic N, Russell A, Pierce D W, Jia Z, McCormick C D, Hartman J, Sakowicz R and Pollard T D 2009 Characterization of two classes of small molecule inhibitors of Arp2/3 complex *Nature* **460** 1031–4

Norotte C, Marga F, Neagu A, Kosztin I and Forgacs G 2008 Experimental evaluation of apparent tissue surface tension based on the exact solution of the Laplace equation *Europhys. Lett.* **81**–46003

Ober R J, Ram S and Ward E S 2004 Localization accuracy in single-molecule microscopy *Biophys. J.* **86** 1185–200

Ohsugi Y, Saito K, Tamura M and Kinjo M 2006 Lateral mobility of membrane-binding proteins in living cells measured by total internal reflection fluorescence correlation spectroscopy *Biophys. J.* **91** 3456–64

Paluch E and Heisenberg C-P 2009 Biology and physics of cell shape changes in development *Curr Biol.* **19** R790–9

Park Y, Best C A, Auth T, Gov N S, Safran S A, Popescu G, Suresh S and Feld M S 2010 Metabolic remodeling of the human red blood cell membrane *Proc. Natl Acad. Sci. USA* **107** 1289–94

Peters R 1988 Lateral mobility of proteins and lipids in the red cell membrane and the activation of adenylate cyclase by β-adrenergic receptors *FEBS Lett.* **234** 1–7

Peters R 1991 *New Techniques of Optical Microscopy and Microspectroscopy* ed R J Cherry (Basingstoke: Macmillan) pp 199–228

Pezeshkian W, Gao H, Arumugam S, Becken U, Bassereau P, Florent J C, Ipsen J H, Johannes L and Shillcock J C 2017 Mechanism of Shiga toxin clustering on membranes *ACS Nano* **11** 314–24

Pierres A, Benoliel A M, Touchard D and Bongrand P 2008 How cells tiptoe on adhesive surfaces before sticking *Biophys. J.* **94** 4114–22

Pike L J 2006 Rafts defined: a report on the Keystone symposium on lipid rafts and cell function *J. Lipid Res.* **47** 1597–8

Pralle A, Keller P, Florin E-L, Simons K and Horber J K H 2000 Sphingolipid–cholesterol rafts diffuse as small entities in the plasma membrane of mammalian cells *J. Cell Biol.* **148** 997–1008

Prior I A, Harding A, Yan J, Sluimer J, Parton R G and Hancock J F 2001 GTP-dependent segregation of H-ras from lipid rafts is required for biological activity *Nat. Cell Biol.* **3** 368–75

Qian H, Sheetz M P and Elson E 1991 Single particle tracking. Analysis of diffusion and flow in two-dimensional systems *Biophys. J.* **60** 910–21

Raedler J O and Sackmann E 1993 Imaging optical thicknesses and separation distances of phospholipid vesicles at solid surfaces *J. Phys. II* **3** 727–48

Raucher D and Sheetz M P 1999 Membrane expansion increases endocytosis rate during mitosis *J. Cell Biol.* **144** 497–506

Raucher D and Sheetz M P 2000 Cell spreading and lamellipodial extension rate is regulated by membrane tension *J. Cell Biol.* **148** 127–36

Ries J and Schwille P 2008 New concepts for fluorescence correlation spectroscopy on membranes *Phys. Chem. Chem. Phys.* **10** 3487–97

Rodríguez-García R, López-Montero I, Mell M, Egea G, Gov N S and Monroy F 2015 Direct cytoskeleton forces cause membrane softening in red blood cells *Biophys. J.* **108** 2794–806

Romer L H, Birukov K G and Garcia J G N 2006 Focal adhesions: paradigm for a signaling nexus *Circ. Res.* **98** 606–16

Roux A, Cuvelier D, Nassoy P, Prost J, Bassereau P and Goud B 2005 Role of curvature and phase transition in lipid sorting and fission of membrane tubules *EMBO J.* **24** 1537–45

Saffarian S, Li Y, Elson E L and Pike L J 2007 Oligomerization of the EGF receptor investigated by live cell fluorescence intensity distribution analysis *Biophys. J.* **93** 1021–31

Saffman P G and Delbrueck M 1975 Brownian motion in biological membranes *Proc. Natl Acad. Sci.* **72** 3111–3

Sako Y and Kusumi A 1994 Compartmentalized structure of the plasma membrane for receptor movements as revealed by a nanometer-level motion analysis *J. Cell Biol.* **125** 1251–64

Sako Y and Kusumi A 1995 Barriers for lateral diffusion of transferrin receptor in the plasma membrane as characterized by receptor dragging by laser tweezers: fence versus tether *J. Cell Biol.* **129** 1559–74

Sako Y, Nagafuchi A, Tsukita S, Takeichi M and Kusumi A 1998 Cytoplasmic regulation of the movement of E-cadherin on the free cell surface as studied by optical tweezers and single particle tracking: corralling and tethering by the membrane skeleton *J. Cell Biol.* **140** 1227–40

Salbreux G, Charras G and Paluch E 2012 Actin cortex mechanics and cellular morphogenesis *Trends Cell Biol.* **22** 536–45

Saxton M J 1993 Lateral diffusion in an archipelago. Single-particle diffusion *Biophys. J.* **64** 1766–80

Saxton M J 1994 Anomalous diffusion due to obstacles: a Monte Carlo study *Biophys. J.* **66** 394–401

Saxton M J 1996 Anomalous diffusion due to obstacles: a Monte Carlo study *Biophys. J.* **70** 1250–62

Saxton M J and Jacobson K 1997 Single-particle tracking: applications to membrane dynamics *Ann. Rev. Biophys. Biomol. Struct.* **26** 373–99

Saxton M J 1995 Single-particle tracking: effects of corrals *Biophys. J.* **69** 389–98

Schmidt T, Schutz G J, Baumgartner W, Gruber H J and Schindler H 1996 Imaging of single molecule diffusion *Proc. Natl Acad. Sci.* **93** 2926–9

Schoetz E M, Burdine R D, Juelicher F, Steinberg M S, Heisenberg C P and Foty R A 2008 Quantitative differences in tissue surface tension influence zebra fish germlayer positioning *HFSP J.* **2** 1–56

Schoetz E-M 2008 *Dynamics and Mechanics of Zebra Fish Embryonic Tissues—A Study of the Physical Properties of Zebra Fish Germlayer Cells and Tissues and Cell Dynamics During Early Embryogenesis* (Saarbruecken: Mueller)

Schutz G J, Schindler H and Schmidt T 1997 *Biophys. J.* **73** 1073–80

Schwille P, Korlach J and Webb W W 1999 Fluorescence correlation spectroscopy with single-molecule sensitivity on cell and model membranes *Cytometry* **36** 176–82

Sergé A, Bertaux N, Rigneault H and Marguet D 1997 Dynamic multiple-target tracing to probe spatiotemporal cartography of cell membranes *Nat. Methods* **5** 687–94

Sergé A, Bertaux N, Rigneault H and Marguet D 2008 Dynamic multiple-target tracing to probe spatiotemporal cartography of cell membranes *Nat. Methods* **5** 687–94

Sezgin E, Levental I and Grzybek M *et al* 2012 Partitioning, diffusion, and ligand binding of raft lipid analogs in model and cellular plasma membranes *Biochim. Biophys. Acta* **1818** 1777–84

Sheets E D, Lee G M, Simson R and Jacobson K 1997 Transient confinement of a glycosyl-phosphatidylinositol-anchored protein in the plasma membrane *Biochemistry* **36** 12449–58

Sheets E D, Simson R and Jacobson K 1995 New insights into membrane dynamics from the analysis of cell surface interactions by physical methods *Curr. Opin. Cell Biol.* **7** 707–14

Sheetz M P and Dai J 1996 Modulation of membrane dynamics and cell motility by membrane tension *Trends Cell. Biol.* **6** 85–9

Shvartsman D E, Gutman O, Tietz A and Henis Y I 2006 Cyclodextrins but not compactin inhibit the lateral diffusion of membrane proteins independent of cholesterol *Traffic* **7** 917–26

Simons K and Ikonen E 1997 Functional rafts in cell membranes *Nature* **387** 569–72

Simson R, Sheets E D and Jacobson K 1995 Detection of temporary lateral confinement of membrane proteins using single-particle tracking analysis *Biophys. J.* **69** 989–93

Simson R, Yang B, Moore S, Doherty P, Walsh F and Jacobson K 1998 Structural mosaicism on the submicron scale in the plasma membrane *Biophys. J.* **74** 297–308

Singer S J and Nicolson G L 1972 The fluid mosaic model of the structure of cell membranes *Science* **175** 720–31

Slattery J P 1995 Lateral mobility of FceRI on rat basophilic leukemia cells as measured by single particle tracking using a novel bright fluorescent probe *PhD Thesis* Cornell University

Smith P R, Wilson K M, Morrison I E G, Cherry R J and Fernandez N 1998 *MHC Biochemistry and Genetics* ed N Fernandez and G Butcher (Oxford: Oxford University Press) pp 133–51

Starr T E and Thompson N L 2001 Total internal reflection with fluorescence correlation spectroscopy: combined surface reaction and solution diffusion *Biophys. J.* **80** 1575—84

Steinberg M S 1996 Adhesion in development: an historical overview *Dev. Biol.* **180** 377–88

Suzuki K G N, Fujiwara T K, Edidin M and Kusumi A 2007a Dynamic recruitment of phospholipase Cγ at transiently immobilized GPI-anchored receptor clusters induces IP3-Ca2+ signaling: single-molecule tracking study 2 *J. Cell Biol.* **177** 731–42

Suzuki K G N, Fujiwara T K, Sanematsu F, Iino R, Edidin M and Kusumi A 2007b GPI-anchored receptor clusters transiently recruit Lyn and Gαfor temporary cluster immobilization and Lyn activation: single-molecule tracking study 1 *J. Cell Biol.* **177** 717–30

Suzuki K, Ritchie K, Kajikawa E, Fujiwara T and Kusumi A 2005 Rapid hop diffusion of a G-protein-coupled receptor in the plasma membrane as revealed by single-molecule techniques *Biophys. J.* **88** 3659–80

Tarantola M, Sunnick E, Schneider D, Marel A K, Kunze A and Janshoff A 2011 Dynamic changes of acoustic load and complex impedance as reporters for the cytotoxicity of small molecule inhibitors *Chem. Res. Toxicol.* **24** 1494–506

Thompson N L and Steele B L 2007 Total internal reflection with fluorescence correlation spectroscopy *Nat. Protoc.* **2** 878–90

Wawrezinieck L, Rigneault H, Marguet D and Lenne P-F 2005 Fluorescence correlation spectroscopy diffusion laws to probe the submicron cell membrane organization *Biophys. J.* **89** 4029–42

Weiss M, Hashimoto H and Nilsson T 2003 Anomalous protein diffusion in living cells as seen by fluorescence correlation spectroscopy *Biophys. J.* **84** 4043–52

Wieser S, Axmann M and Schutz G J 2008 Versatile analysis of single- molecule tracking data by comprehensive testing against Monte Carlo simulations *Biophys. J.* **95** 5988–6001

Wieser S and Schutz G J 2008 Tracking single molecules in the live cell plasma membrane-do's and don't's *Methods* **46** 131–40

Wieser S, Moertelmaier M, Fuertbauer E, Stockinger H and Schutz G 2007 (Un)confined diffusion of CD59 in the plasma membrane determined by high-resolution single molecule microscopy *Biophys. J.* **92** 3719–28

Wilson K W, Morrison I E G, Smith P R, Fernandez N and Cherry R J 1996 Single particle tracking of cell-surface HLA-DR molecules using R-phycoerythrin labeled monoclonal antibodies and fluorescence digital imaging *J. Cell Sci.* **109** 2101–9

Wu C, Asokan S B, Berginski M E, Haynes E M, Sharpless N E, Griffith J D, Gomez S M and Bear J E 2012 Arp2/3 is critical for lamellipodia and response to extracellular matrix cues but is dispensable for chemotaxis *Cell* **148** 973–87

Wu C, Haynes E M, Asokan S B, Simon J M, Sharpless N E, Baldwin A S, Davis I J, Johnson G L and Bear J E 2013 Loss of Arp2/3 induces an NF-κB-dependent, nonautonomous effect on chemotactic signaling *J. Cell Biol.* **203** 907–16

Xu K, Babcock H P and Zhuang X 2012 Dual-objective STORM reveals three-dimensional filament organization in the actin cytoskeleton *Nat. Methods* **9** 185–8

Yamada S and Nelson W J 2007 Localized zones of Rho and Rac activities drive initiation and expansion of epithelial cell–cell adhesion *J. Cell Biol.* **178** 517–27

Yamada S, Pokutta S, Drees F, Weis W I and Nelson W J 2005 Deconstructing the cadherin–catenin–actin complex *Cell* **123** 889–901

Yang W, Thein S, Lim C-Y, Ericksen R E, Sugii S, Xu F, Robinson R C, Kim J B and Han W 2014 Arp2/3 complex regulates adipogenesis by controlling cortical actin remodelling *Biochem. J.* **464** 179–92

Yauch R L, Felsenfeld D P, Kraeft S K, Chen L B, Sheetz M P and Hemler M E 1997 Mutational evidence for control of cell adhesion through integrin diffusion/ clustering, independent of ligand binding *J. Exp. Med.* **186** 1347–55

Yechiel E and Edidin M 1987 Micrometer-scale domains in fibroblast plasma membranes *J. Cell Biol.* **105** 755–60

Yi K, Unruh J R, Deng M, Slaughter B D, Rubinstein B and Li R 2011 Dynamic maintenance of asymmetric meiotic spindle position through Arp2/3-complex-driven cytoplasmic streaming in mouse oocytes *Nat. Cell Biol.* **13** 1252–8

Zhou K, Muroyama A, Underwood J, Leylek R, Ray S, Soderling S H and Lechler T 2013 Actin-related protein2/3 complex regulates tight junctions and terminal differentiation to promote epidermal barrier formation *Proc. Natl Acad. Sci.* **110** E3820–9

Zimmerberg J and Kozlov M M 2006 How proteins produce cellular membrane curvature *Nat. Rev. Mol. Cell Biol.* **7** 9–19

IOP Publishing

Physics of Cancer, Volume 1 (Second Edition)
Interplay between tumor biology, inflammation and cell mechanics

Claudia Tanja Mierke

Chapter 6

Cytoskeletal remodeling dynamics

Summary

The identification of how cells in 3D microenvironments or natural tissues sense and respond to biophysical stimuli is a longstanding challenge in the physics of cancer and tissue engineering. Cytoskeletal remodeling dynamics are involved in many cellular processes, such as cell motility, transendothelial migration and cell–matrix or cell–cell adhesion receptor recycling. In particular, the connection between the cell's cytoskeleton and the surrounding microenvironment is crucial for the proper function of the cells, including the mechanical properties regulating cellular motility. In this chapter, a strategy is provided to analyze morphological and mechanical responses of contractile cells, such as fibroblasts or cancer cells, in a 3D microenvironment. The cells are able to react to mechanical stimulation such as cell stretch through specific, cell-wide mechanisms such as a staged retraction, cell stiffening and reinforcement. Cellular retraction responses upon mechanical cell stretch occurred for all orientations of stress fibers and cellular protrusions relatively independent of the stretch direction, whereas the reinforcement responses, including the extension of cellular processes and enhanced stress fiber formation, occurred predominantly in the direction of the stretch in order to withstand deformation or stretching. An unreported role of F-actin clumps has been seen, in which clumps act as possible F-actin reservoirs for retraction and reinforcement responses during stretch. These responses upon stretch seem to be related to a model of cellular sensitivity to local physical cues. Moreover, these observations suggest mechanisms for the global actin cytoskeletal remodeling in nonmuscle cells and hence provide insights into cellular responses acting in pathologies such as fibrosis, hypertension or cancer including metastasis.

6.1 Cytoskeletal remodeling dynamics within unperturbed cells

The dissipation and fluctuations of Brownian degrees of freedom can be applied to living matter such as cells under nonequilibrium conditions. The dissipation in the

doi:10.1088/978-0-7503-1753-5ch6

cell's actomyosin cytoskeleton, which acts on a microscopic scale as cytoskeletal and polymerization motors, and the fluctuations in membrane proturberances are supposed to regulate their migration through dense 3D extracellular matrices. The exploration of the underlying mechanisms for cell migration will provide a tool for hindering or promoting the motility of cells in human diseases such as cancer and, subsequently, lead to a paradigm for proposing enhanced motility of distinct cancer cell types.

In particular, the simplest modules of living systems such as Brownian motors are unlikely to have emerged by chance (random event) or energy minimization, as they are too complex (Vale and Milligan 2000). Many protein-based molecular motors in the cells may represent Brownian motors including the Brownian ratchet as a simplest candidate for which Brownian motion is essential. Cells are evolutionarily designed Brownian molecular motors in complex fluids that migrate through connective tissues by altering several parameters in order to be optimally adapted to microenvironmental conditions. These molecular motors are responsible for cellular shape alterations and hence cellular polarization, which represents a symmetry-breaking event.

The migration of cells is basically the interaction between a macroscopic cell body and multiple thermal baths. Dissipation within the cell's actin cytoskeleton caused by cytoskeletal rearrangements drives cellular migration on flat substrates (Bursac *et al* 2005, Raupach *et al* 2007, Mierke *et al* 2008). In more detail, cells harvest chemical or thermal energy to perform persistent motion. A migrating cell seems to change its shape from a symmetric to an asymmetric and polarized shape which may be provided by forces such as thermodynamic forces. There exists an analogy between Brownian ratchets and mechanical motion of proteins in living matter and possibly cells. Multiple ratchet-like processes in proteins seem to be the driving factors for cellular functions such as cell motility (Lui *et al* 2014). Moreover, cellular perturberances are observed in migrating cells and they fluctuate during migration on 2D and within 3D environments, whereas they are absent in less motile cells. However, the role of these cytoskeletal rearrangements and membrane perturberances in regulating cellular motility in distinct 3D environments is not yet well understood. In addition, the cell's microenvironment for cellular motility, such as structural or mechanical effects, hinders or restricts cellular motility and hence plays a major role for cellular functions involving motility and the malignant progression of cancer such as metastasis.

6.2 Cytoskeletal remodeling dynamics upon mechanical stretching

How cells respond to changes in mechanical forces is crucial for the maintenance of homeostasis, during the development of tissues (Geiger *et al* 2001, Vogel and Sheetz 2006, Taber 2009) and under pathological conditions such as cardiomyopathy (Rohr 2009) and asthma (Fredberg 2002, Waters *et al* 2002, Fredberg 2004, Shore and Fredberg 2005, Lenormand and Fredberg 2006, An and Fredberg 2007, Fabry and Fredberg 2007, Nguyen and Fredberg 2008, Discher *et al* 2009). The mechanical responses are particularly required in cells that fulfill a mechanical function, such as myofibroblasts, which are a unique group of smooth muscle-like fibroblasts acting in oncogenesis, inflammation, wound contraction and fibrosis. The transition of fibroblasts into the myofibroblastic phenotype occurs due to microenvironmental changes caused by factors including mechanical stress, transforming growth factor β (TGF-β), and extra domain-A (EDA) of fibronectin, which is a splice variant of fibronectin (Gabbiani 2003, Wight and Potter-Perigo 2011).

In particular, in living cells cultured on two-dimensional (2D) substrata, it has been shown that a range of mechanical and structural responses of cells are evoked upon mechanical perturbation. These mechanically induced cellular responses include cytoskeletal fluidization due to a single stretch, which is followed by a release of the stretch (Trepat *et al* 2007, Krishnan *et al* 2009), and by cytoskeletal reinforcement in response to the single sustained stretch (Trepat *et al* 2007) or localized stressing (Krishnan *et al* 2009). These cytoskeletal alterations and the associated reorientation of cellular tractions can be seen over time-scales that are much shorter than those associated with reorientation of a cell body on a 2D substrate (Krishnan *et al* 2012).

A major question is how cells cultured on 2D substrata can mirror the behaviors of the same cells in a natural tissue environment *in vivo*. Hence, in order to investigate the effect of a 3D microenvironment, 3D engineered tissue constructs have been developed as model systems, in which cellular biophysical properties can be analyzed. Indeed, cytoskeletal fluidization (depolymerization) and reinforcement (polymerization) induced after rapidly applied, sustained stretch have been revealed in cells cultured within 3D engineered tissue scaffolds (Nekouzadeh *et al* 2008). However, the time courses and mechanisms by which these responses on stretch are evoked have not yet been revealed for living cells in a 3D microenvironment. To elucidate the mechanisms, a system has been established to stretch and then hold engineered tissue constructs over the objective of a scanning confocal fluorescence microscope. Using this system, the cytoskeletal dynamics can be monitored and quantified in a fraction of cells within these 3D engineered tissue matrices, in which the cells can be identified due to expression of the m-cherry LifeAct, as they are transfected with m-cherry LifeAct, a fluorescent F-actin reporter, prior to the experiment (Lee *et al* 2012).

Indeed, myofibroblasts can react to stretch via specific, cell-wide mechanisms involving retraction and reinforcement. In particular, retraction responses have been seen for all orientations of stress fibers and cellular protrusions have been exerted in the stretch direction, while reinforcement responses such as the extension of cellular processes and stress fiber formation been observed predominantly in the direction of the stretch. Moreover, a previously unreported role of F-actin clumps has been detected, with these clumps acting as F-actin reservoirs for the retraction and reinforcement phases during the cell stretch. What are the underlying mechanical mechanisms of these responses of cellular sensitivity to local physical cues?

Prior to the mechanical stretch exertion to most cells, they displayed a diffuse cytoskeleton composed of randomly arranged actin filaments and aggregations of actin 'reservoirs' and only a few cells showed visible stress fibers. Moreover, cells displayed staged responses to mechanical stretch. The retraction responses of the cells employ the depolymerization of actin stress fibers, the retraction of filopodia-like cellular protrusions and the creation and increase of F-actin reservoirs. Cellular protrusions in all directions can be retracted. Stress fibers in all directions can depolymerize. The F-actin reservoirs, newly formed or grown, are often located near the retracting cellular protrusions or depolymerizing stress fibers.

The reinforcement responses require the polymerization and thickening of actin stress fibers, extension of cellular protrusions, and a reduction in the sizes and numbers of F-actin reservoirs, which are often near the extension site of cellular protrusions. Actin reservoirs seems to feed the growth of stress fibers. In particular, the extension of cellular protrusions and growth of stress fibers has been predominately observed in the direction of applied stretch.

Cells in 3D engineered tissue matrices display reactions to stretch that can be highly diverse, but follow a unifying set of underlying principles, with the local physical cues transducing cell-wide alterations. Cellular responses encompass retraction and then, in multiple cases, reinforcement leading to fiber alignment towards the direction of a stretch over the course of several minutes. The retraction responses have been found for stress fibers and cellular protrusions, which are oriented in any direction, whereas the reinforcement have been observed predominantly in the direction of the exerted stretch on the 3D engineered tissue matrix scaffolds. However, no retraction responses have been found after 5% stretch that leads to the hypothesis that a critical strain threshold is required for a retraction response and also that a weak stretch event leads to a stabilization of stress fibers and cellular protrusions. The dynamic nature of cellular responses upon mechanical stretch can be divided into three general classes of behavior. First, monotonic reinforcement causing the extension of stress fibers and the exertion of cellular protrusions into the direction of the 3D engineered tissue scaffold stretch. Second, monotonic retraction employs the depolymerization of stress fibers and the retraction of cellular processes. Third, retraction, which is followed by reinforcement, has been found frequently, whereas the opposite behavior has never been observed.

6.3 Dynamic cell-level responses derive from local physical cues

A mathematical model applying three governing principles to stress fiber behavior is able to predict a broad range of experimental observations. The first two principles are still derived from the observations of cells cultured on 2D substrates. The first principle is that the stress fibers collapse when the stretch of the cells is outside a preferred range: both tensile and compressive strains can cause the depolymerization of stress fibers through rupture or buckling, respectively. The second principle is that stress near an adhesion site between the cell and the extracellular matrix guides the reinforcement. For the 3D engineered tissue matrices, the extracellular matrix stiffness is also increased with stress, so it is not possible to distinguish between the effects of the initial stress and second event stiffening in this experiment. The third principle is derived from the detection of clusters of F-actin within cells, which provide a reliable mechanism by which cells can reinforce rapidly. These F-actin clusters are included indirectly in the mathematical model, as a conservation law for the quantity of actin in stress fibers is omitted.

These principles help to explain the observations in several ways. Retraction responses occur for stress fibers in any orientation, as a combination of two factors. First, the Poisson contraction predicts that a tensile strain evokes a transverse compressive strain of nearly equal magnitude and, second, as a distribution in the prestretch of stress fibers emerges over a cell's mechanical loading history (Small *et al* 1998), the stress fibers near the critical point for depolymerization in tension or compression can point in all orientations. A region exists at ±45° to the loading direction in which strains due to the 3D engineered tissue scaffold loading are small, but an increased association of stress fibers in these orientations that withstand the mechanical stretch has not yet been found. Although this effect was not evident in the experiments, it cannot be ruled out, as the sample size of cells containing clear features of stress fibers oriented at ±45° to the loading direction was very small.

The time course of reinforcement responses after stretch are in line with a law of growth rates increasing with poststretch stress in the extracellular matrix near a cell–matrix adhesion site. The model predicted that even three different types of temporal response are possible, regardless of the cellular shape. Based on the model, the selection of a specific type of response is based on the probability that stress fibers stay intact in the direction of stretch, which is predicted by the degree of variation in the prestretch of cells.

This was found to be true for the two extremes: cells exhibiting no natural axis and cells with a highly defined natural axis. In both cases, the variation in prestretch in stress fibers is altered between the three detected types of temporal response. Indeed, monotonic reinforcement has been predicted for cells with little variation in prestretch, with few depolymerizing stress fibers and many adhesion sites supporting the stress fiber driven cell growth. The monotonic retraction is predicted for some cases of high variation of prestretch when even a sufficient fraction of stress fibers is simply ruptured due to tension in order to decrease the stress level at the focal adhesion sites below a stress level in which stress fiber growth can occur. The retraction is followed by reinforcement, as it has been predicted for an intermediate variation, whereas the reinforcement after retraction never occurred using this model. This is can be explained by a separation of the time-scales for retraction and reinforcement.

The model additionally predicted the polarization of cells into the direction of 3D engineered tissue scaffold stretch. The reinforcement is observed predominantly in the direction of stretch in this model, where stresses at the adhesion sites emerge then to sufficiently high levels. Stress fibers, which are aligned in the stretch direction and withstand the 3D engineered tissue scaffold stretch, play a key role for reinforcement and polarization responses, as seen in the sectors near $\theta = 0°$. Moreover, these sectors become denser as they are filled with more stress fibers, whereas the 3D engineered tissue scaffold is kept isometric unless the 3D engineered tissue scaffold stretch evokes its depolymerization.

Cells in a 3D engineered tissue scaffold exhibit specific responses to mechanical stimulation that are predicted through a few general principles that can be implemented in a simple model. Other additional factors, which have not yet been modeled, are certainly critical for cellular responses. The active contraction of cells is in certain cases initiated in response to stress, and the fibrosity can either increase or decrease as cells pull back against their adhesion sites. The magnitude of active cellular contraction measured at the level of a 3D engineered tissue scaffold is rather independent of the magnitude of mechanical stretch in these scaffolds, which is in line with previous observations (Wakatsuki *et al* 2000). It has been pointed out that the model aimed to identify possible driving forces for the major findings. In one case a cell is retracted into a ball-like shape, then elongated into a spindle-shaped cell. However, the latter case cannot be modeled using the introduced model, and hence it is mentioned to highlight that the scope of this model is somewhat restricted to a specific set of phenomena that seem to cover the vast majority of observations. However, the governing principles reported can predict the major trends and might even broadly affect the development and remodeling of entire tissues.

6.4 Cytoskeletal dynamics in 3D differ from those observed in 2D

Many of the cellular behaviors which have been detected in 3D engineered tissue constructs possess direct analogs in observations from cells cultured on 2D substrates. Indeed, the poststretch retraction responses have been found to be consistent with the Fredberg model (Trepat *et al* 2007) of cytoskeletal fluidization, which has been induced exclusively through mechanical stimulation in 2D culture.

Consistent with these findings of the Fredberg model, the time scale for stress fiber depolymerization was also too short to be identified by the 3D approaches. The dependence of the cellular behavior on the distribution of prestained stress fibers, as predicted by the Fredberg model, was also consistent with the Kaunas models for cells loaded cyclically on a 2D substrate (Kaunas *et al* 2005, 2006, Hsu *et al* 2010, Kaunas and Hu 2009). Although results are not available for the allowable stretch range of stress fibers in cells within 3D engineered tissue scaffolds, by applying the range that has been reported by the Kaunas group to the models allows us to replicate the experiment results qualitatively. Moreover, the realignment of cells and the increased protrusion of filopodia in the direction of sustained stretch is also consistent with this work, as well as many other findings and models predicting the polarization of endothelial cells adherent to flexible 2D substrates, on which very low frequency stretch has been exerted (Zemel *et al* 2007, 2010a, 2010b, De and Safran 2008, De *et al* 2010). The concept of stress-driven alteration of actin stress fiber density is supported by the 2D Deshpande model (Deshpande *et al* 2006), and the concept of the alignment of stress fibers due to elastic interactions with the extracellular matrix is in line with other results (De *et al* 2010, Friedrich *et al* 2011).

However, several differences still exist. In particular, contrary to results of cells in 2D, the retraction and reinforcement responses were both detected during isometric stretches, with the retraction responses pointing in all directions. In 2D, the F-actin depolymerization has been found frequently in unloading of the actin cytoskeleton following a stretch step (Chen *et al* 2010). The F-actin depolymerization was detected in 3D following both sustained stretch and also in compressed stress fibers, which is dissimilar to the reinforcement and fluidization effects observed upon stretch and compression, respectively, in 2D (Wu and Feng 2015).

6.5 Nano-scale particle tracking

Nano-scale single-particle tracking (NSPT) is an effective method to analyze the dynamics of molecules in living cells. NSPT exists in two dimensions as well as three, with 3D-NSPT being more sophisticated than 2D-NSPT. 2D-NSPT has been used in numerous biological and medical research fields, for example in the analysis of membrane dynamics (Fujiwara *et al* 2002, Murase *et al* 2004, Kusumi *et al* 2005, Raupach *et al* 2007, Mierke 2013), to reveal the virus infection mechanism (Lakadamyali *et al* 2003, Joo *et al* 2011, Liu *et al* 2011a, 2012) and to discover the intracellular or intercellular transport dynamics of molecules (He *et al* 2010, Liu *et al* 2011b, Wang *et al* 2012). For 2D-NSPT, the particle center of mass can be determined with a precision of several nanometers by combining it with a 2D Gaussian fitting method (Kural *et al* 2005, Liang *et al* 2007). The main limitation of the 2D-NSPT method is that the particle can only be tracked within a thin focal plane, otherwise the tracking mechanism will fail. The nano-scale tracking of beads inside cells is of great interest (figure 6.1) as it is important for determining the cytoskeletal remodeling dynamics within cells that migrate through a dense 3D extracellular matrix microenvironment. Thus, 3D-NSPT tracking that captures the full spatiotemporal behavior of the beads embedded on the living cells is becoming more and more important for investigating the cellular microenvironment. Several microscopy techniques have been developed for 3D localization in the SPT field (Katayama *et al* 2009, Wells *et al* 2010, Dupont and Lamb 2011, Juette *et al* 2013,

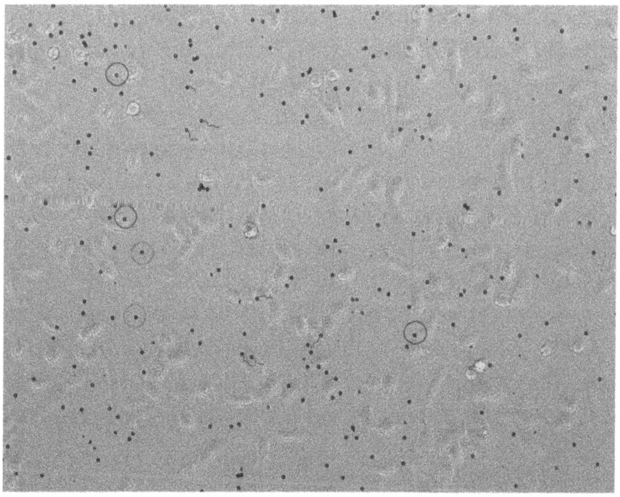

Figure 6.1. Nano-scale particle tracking. Adherent cancer cells (red circle, examples for cell-bound beads, green are beads connected to the substrate) are incubated with 4.5 μm beads coated with fibronectin for 15 min and the bead displacement (blue line) is measured for several minutes.

van den Broek 2013), resulting in many single-particle applications (Ram *et al* 2008, Jones *et al* 2011, Planchon *et al* 2011, Han *et al* 2012, Quirin *et al* 2012). There are basically two main types of techniques for tracking individual particles in 3D: (1) 3D tracking by postanalyzing the recorded image data and (2) real-time 3D tracking via feedback approaches (Dupont and Lamb 2011). The first approach is useful for tracking multiple particles simultaneously. Most versions require modifications to a standard microscope and use of a 2D Gaussian fitting algorithm to achieve super-resolution localization precision, including astigmatic imaging (Holtzer *et al* 2007, Huang *et al* 2008), multifocal plane (Ram *et al* 2008), biplane detection (Juette *et al* 2008) and a double-helix point spread function (Pavani *et al* 2009, Thompson *et al* 2010). However, with the development of sensitive cameras and bright fluorescent tags, a high number of 2D images at different z-positions can be recorded, which seems to be the most straightforward method for 3D single-particle imaging and tracking in living cells.

Although several corresponding algorithms have been developed for locating the particle center from a series of 3D image z-stacks (Kubitscheck *et al* 1996, Patwardhan 2003, Ruthardt *et al* 2011), the frequently used centroid and Gaussian nonlinear least-squares fitting methods are suitable for 3D-SPT. In more detail, the high-speed centroid method has very low localization precision and high edge sensitivity. Although the iterative Gaussian fitting methods are considered to be the most accurate algorithm, they are not suitable for 3D-SPT because the large data volumes require too much time to analyze. Additionally, the initial values should be set for fitting parameters before calculation, which is a very tedious process. Thus, it has been considered whether an efficient algorithm could be developed for 3D-SPT with high computation speed and high precision.

There is a localization algorithm based on a radial-symmetry method that can solve the challenges of 3D-SPT mentioned above. Although the whole-pixel and sub-pixel localizations using the radial-symmetry method in 2D have been described in previous reports (Loy and Zelinsky 2003, Parthasarathy 2012), such a method for 3D and sub-pixel localization has not been reported so far. The localization accuracy and computation speed of the method is primarily assessed based on two applications from theoretical and practical perspectives: (1) localization of simulated 3D particle images by wide-field and confocal microscopy and (2) 3D-SPT of a quantum dot (QD)-labeled influenza virus in living host cells. However, this novel method determines the 3D particle center with low edge sensitivity and sub-pixel precision, which is similar to the Gaussian nonlinear least-squares fitting method. Without any iterative or numerical fitting steps, the computation speed of this method is about two orders of magnitude higher than that of the Gaussian fitting method and thus similar to that of the centroid method. These features make this algorithm a very competitive method for 3D-SPT applications.

The 3D single-particle localization method

The 3D point spread function (PSF) of the Born–Wolf model (Born and Wolf 1980) is introduced to simulate the 3D charge-coupled device (CCD) images of individual particles. In particular, the axial intensity distribution of a 3D-PSF has the shape of a 'butterfly' in typical wide-field microscopy. Moreover, the intensity of the 'wings' is dramatically reduced for confocal microscopy, whereas the 'body' is an approximate ellipsoid, which can be characterized by the ratio of axial resolution (R_z) to lateral resolution (R_{xy}):

$$\frac{R_z}{R_{xy}} = \frac{3.28n}{N_A}, \tag{6.1}$$

where n represents the refractive index of the objective medium and the particular N_A is the numerical aperture of the objective (Pawley 2006). Based on this approximation and by scaling of the 3D images in the axial direction according to the ratio, images with the intensity distribution of an approximate sphere are obtained. This 3D radial-symmetry localization method is based on the fact that the gradient line at each pixel in a 3D sphere would intersect theoretically with the true center of the 3D sphere, which has a distance of 0 to all gradient lines. Considering any optical noise and deformation error, the center is estimated using the point which has the least total distance to all gradient lines. Moreover, a displacement-weighted method is implemented to reduce the influence of the intensity of the wings, leading to a more accurate approach regarding the high-intensity areas.

Accuracy for wide-field microscopy

The performance of this method is evaluated for wide-field microscopy using 1000 3D CCD images, which are simulated with a signal-to-noise ratio (S/N) of 20 and with the true centers distributed randomly between ±0.5 pixels in the x-, y- and z -directions. By tracking 1000 images with a centered $7 \times 7 \times 11$ pixel region (pixel size: $100 \times 100 \times 150$ nm), the 3D radial-symmetry method leads to a mean lateral error of 0.011 pixels and an axial error of 0.044 pixels, much smaller than that of the centroid method (lateral error: 0.077 pixels, axial error: 0.163 pixels) and in a similar range to that of the Gaussian fitting method (lateral error: 0.011 pixels, axial error: 0.045 pixels). The 3D scatter plots of the errors in 3D further illustrate that the error distribution of this method is similar to that of the Gaussian fitting method and more concentrated compared to that of the centroid method. Meanwhile, the accuracy of the 3D radial-symmetry method has been investigated without scaling the 3D image in the axis direction by using simulated 3D CCD images with a S/N of 20. The accuracy of the 3D radial-symmetry method (lateral error: 0.018 pixels, axial error: 0.089 pixels) is lower than that of the Gaussian fitting method, indicating that the scaling process is very important for improving the precision of the 3D radial-symmetry method.

The intensity distribution of the scaled image of a 3D particle shows that the intensity in the axial direction is not only the shape of a sphere, but it also contains halos (the wings of the 3D-PSF image). In order to eliminate the influence of the wings of the 3D particle image on the localization accuracy, the centers have been determined with different lateral sizes. The total error plots show that the accuracy decreases with the shrinking of the lateral size. Moreover, the accuracy is similar to that of the Gaussian fitting method for large lateral size and even better for small size, indicating that this approach can reduce the surrounding interference.

The accuracy of this method is further determined using simulated 3D images with S/N of 3–100. The error plots indicate that the lateral accuracy is slightly lower than that of the Gaussian fitting method for low S/N, whereas it is higher for high S/N and the axial accuracy has the same trend. The total errors for the three localization methods have been revealed. The total error is 7.06 nm for the radial-symmetry method, 7.14 nm for the Gaussian fitting method and 28.44 nm for the centroid method at a S/N of 20. The total error plots indicate that the accuracy of the 3D radial-symmetry method is much better than that of the centroid method and seems to be as good as that of the Gaussian fitting method over almost the entire range and even better for high S/N. Moreover, the average computation time is 0.39 ms for the radial-symmetry method, 0.16 ms for centroid method and 39.12 ms for the Gaussian fitting method, which is about 100 times longer than that of the radial-symmetry method. Indeed, this is a key attribute of this method for processing a mass of 3D images in the SPT field.

The accuracy of confocal microscopy

Taking the z-stacks by confocal microscopy is a widely used tool for 3D particle imaging and tracking. Thus, the accuracy of different localization methods is obtained by using simulated 3D CCD images for confocal microscopy. Hence, the 3D radial-symmetry method (lateral error: 0.009 pixels, axial error: 0.036 pixels, total error: 5.58 nm) is still as accurate as the Gaussian fitting method (lateral error: 0.008 pixels, axial error: 0.038 pixels, total error: 5.59 nm) and much better compared to the centroid method (lateral error: 0.064 pixels, axial error: 0.097 pixels, total error: 18.59 nm) for 1000 CCD images with a S/N of 20. As shown above, a series of simulated images with a wide range of S/N values has been examined. Thus, it has been found that the accuracy of this method is slightly lower than that of the Gaussian fitting method for low S/N, but even higher in the axial direction for high S/N. Indeed, these results demonstrate that the localization accuracy for three methods is improved by using the 3D confocal CCD images and thus the 3D radial-symmetry method is applicable for confocal microscopy. With the decrease of axial size, the accuracy decreases similarly for the three localization methods. Taken together, this method is as accurate as the Gaussian fitting method.

The accuracy estimated by actual experiments

In order to investigate locally the ability of cells to react mechanically under particular distortions or imperfections, the algorithm using 3D CCD images from actual experiments has been refined. In particular, the positions of multiple fluorescent beads have been detected in 3D images using different methods. However, the true centers could not be obtained in actual experiments, but the positions located by the radial-symmetry method are consistent with those obtained with the Gaussian fitting method. In addition, the 3D-SPT technique has been utilized in order to investigate the behavior of a QDs-labeled influenza virus in live Madin–Darby canine kidney cells and analyze the trajectory of a single virus precisely. It has been observed that the trajectory of radial-symmetry-based tracking is very similar to that of the Gaussian fitting method. It has to be pointed out that the 3D radial-symmetry method took only 0.76 s to calculate the particle positions of the trajectory (273 frames), whereas 47.77 s were needed for the Gaussian fitting method. The relationship of mean square displacement (MSD) with time interval is highly important for analyzing the motion mode and calculating the relative motional parameters, such as the diffusion coefficient and fitting velocity (Saxton and Jacobson 1997, Liu *et al* 2012). Indeed, it has been found that the MSD–time plot for radial-symmetry-based tracking overlaps very significantly with that of the Gaussian fitting method. In addition, the results obtained further indicate that the radial-symmetry-based method is as accurate as the Gaussian fitting method in the actual experiments, but it has a much higher computation speed.

Although z-stacks microscopy has been widely used for 3D single-particle tracking, the relatively rapid algorithms for high-resolution single-particle localization are only rarely provided. A noniterative localization algorithm has been developed that first utilizes the scaling of 3D image in the axial direction and then focuses on evaluating the radial-symmetry center of the scaled image to achieve the desired single-particle localization. In the first step, the deformation of a 3D image of single particles is accomplished by scaling the image in the axial direction based on the ratio of axial resolution to lateral resolution. By evaluating the accuracy using simulated 3D particle images, it can be confirmed that the scaling step is essential for improving the accuracy of this localization method. In practice, the ratio of the axial resolution to the lateral resolution may differ from microscope to microscope. However, the localization method can be adjusted slightly with different microscopes for higher precision. Localizing the particle position in the second step is based on the particle center that has the least total distance to all gradient lines at each pixel in the scaled 3D image. In addition, a displacement-weighted method is applied to improve the localization precision pronouncedly, which can reduce the influence of the edge intensity. Finally, the combination of these two steps achieves 3D single-particle localization with a sub-pixel localization accuracy and sub-millisecond computation time.

Compared with the frequently used centroid and Gaussian fitting methods (Kubitscheck *et al* 1996, Patwardhan 2003, Ruthardt *et al* 2011), this approach seems to be more precise than the centroid method in its localization precision and it also has the same precision as the 3D Gaussian fitting method, although the computation speed is two orders of magnitude higher compared to that for the Gaussian fitting method. High computation speed and precise localization ability are the main contributions of the 3D radial-symmetry method. Meanwhile, although the radial-symmetry method has been used in whole-pixel and sub-pixel localization in 2D (Loy and Zelinsky 2003, Parthasarathy 2012), the present 3D radial-symmetry method is a significant expansion of the radial-symmetry-based algorithm in 3D and is expected to be exploited in some new algorithms. Moreover, by tracking a QD-labeled influenza virus in living cells, it was further confirmed that the 3D radial-symmetry method is suitable for rapidly localizing the position of single viruses with a high precision in the 3D-SPT field. As a main advantage, this approach can considerably reduce the time and cost for processing the large volume of data of 3D images, which makes it particularly suited to 3D high-precision single-particle tracking, 3D single-molecule imaging and new microscopy techniques.

Single-particle tracking in 3D in a living cell microenvironment holds the promise of revealing important new biological insights. However, conventional microscopy-based imaging techniques are not well suited for fast 3D tracking of single particles in cells. An imaging modality multifocal plane microscope (MUM) has been developed to image fast intracellular dynamics in 3D in living cells. Thus, an algorithm is introduced, the MUM localization algorithm (MUMLA), in order to determine the 3D position of a point source that is imaged using MUM. MUMLA has been validated through simulated and experimental data and indeed showed that the 3D position of QDs can be determined over a wide spatial range. Moreover, it has been shown that MUMLA provides the best possible accuracy for determining the 3D position. This analysis demonstrates that MUM overcomes the poor depth discrimination of the conventional microscope and thus paves the way for high-accuracy tracking of nanoparticles in a living cell microenvironment. Using MUM and MUMLA the full 3D trajectories of QD-labeled antibody molecules undergoing endocytosis in living cells, from the cell membrane to the sorting endosome deep inside the cell, have been reported.

Fluorescence microscopy of living cells represents a major tool for analyzing intracellular trafficking events. Using state-of-the-art microscopy techniques, only a single focal plane can be imaged at a particular time. Membrane protein dynamics can be investigated in one focal plane and the significant advances can be regarded as an advantage of fluorescence microscopy (Zenisek *et al* 2000, Kwik *et al* 2003). However, cells are 3D objects and intracellular trafficking pathways are typically not constrained to one focal plane. If the dynamics are not restricted to one focal plane, the currently available technology is inadequate for the detailed investigation of fast intracellular dynamics (Ehrlich *et al* 2004, Ober *et al* 2004, Oheim 2004, Hua *et al* 2006, Rutter and Hill 2006). For instance, significant advances have been made in the investigation of events that precede endocytosis at the cell membrane (Conner and Schmid 2003, Maxfield and McGraw 2004, Pelkmans and Zervial 2005).

However, the dynamic events post-endocytosis typically cannot be imaged, as they occur outside the focal plane that is set to image only the cell membrane. Classical approaches based on altering the focal plane are often ineffective in such situations, as the focusing devices are relatively slow in comparison to many of the intracellular dynamics (Schütz *et al* 2001, Thomann *et al* 2002, Arhel *et al* 2006). In addition, the focal plane may frequently fail to be in the right place at the right time, thus missing important aspects of the dynamic events.

However, modern microscopy techniques have generated significant interest in investigating the intracellular trafficking pathways at the single-molecule level (Sako *et al* 2000, Ober *et al* 2004). In particular, single-molecule experiments overcome the averaging effects and thus provide information that is not accessible using conventional bulk studies. However, the 3D tracking of single molecules poses several challenges. In addition to whether or not images of the single molecule can be captured while it undergoes potentially highly complex 3D dynamics (Moerner

2007), the question arises of whether the 3D location of the single molecule can be determined and how accurately this can be performed.

Several imaging techniques have been proposed to determine the *z*-position of a single molecule/particle. Approaches (Speidel *et al* 2003, Toprak *et al* 2007) that use out-of-focus rings of the 3D-PSF to infer the *z*-position are not capable of tracking QDs (Toprak *et al* 2007) and pose several challenges, especially for live-cell imaging applications, since the out-of-focus rings can be detected only when the particle is at a certain depth. Moreover, a large number of photons must be collected so that the out-of-focus rings can be detected above the background, which severely compromises the temporal resolution. Similar problems are also encountered with the approach that infers the *z*-position from out-of-focus images acquired in a conventional fluorescence microscope (Aguet *et al* 2005). Moreover, this approach is applicable only at certain depths and hence is problematic, when the point source is close to the plane of focus. The technique based on encoding the 3D position using a cylindrical lens (Kao and Verkman 1994, Holtzer *et al* 2007, Huang *et al* 2008) is somewhat limited in its spatial range to 1 μm in the *z*-direction (Holtzer *et al* 2007). Moreover, this technique uses epi-illumination and thus poses the same problems as conventional epifluorescence microscopy in tracking events that fall outside even one focal plane. The approach based on *z*-stack imaging in order to determine the 3D position of a point source (Schütz *et al* 2001, Watanabe and Higuchi 2007) has limitations in terms of the acquisition speed and the accuracy achievable for the location estimates, and thus poses problems for investigating fast and highly complex 3D dynamics. It should be pointed out that the above-mentioned techniques have not been able to image the cellular microenvironment with which the point sources interact. This is particularly important for gaining useful biological information, such as identification of the final destination of the single molecules of interest. Confocal/two-photon particle tracking approaches, which scan the sample in 3D, can only track one or very few particles within the cell and require therefore high photon emission rates of the beads (Levi *et al* 2005).

One of the key requirements for 3D tracking of single molecules within a cellular microenvironment is that the targeted molecule must be continuously tracked for extended periods of time at high spatial and temporal precision. However, conventional fluorophores such as organic dyes and fluorescent proteins typically have a limited fluorescent on-time (typically 1–10 s), after which they are irreversibly photobleached, thus severely limiting the duration over which the tagged molecule can be tracked. However, the use of QDs, which are extremely bright and photostable fluorescent labels when compared to conventional fluorophores, enables continuous tracking of single molecules for extended periods of time (several minutes or even hours). There have been several reports on single QD tracking within a cellular microenvironment, for example on the cell membrane (Dahan *et al* 2003, Crane and Verkman 2008) or inside the cells (Nan *et al* 2005, Courty *et al* 2006, Cui *et al* 2007). All of these reports have focused on QD tracking in 2D. However, the 3D tracking of QDs in cells has been problematic due to the above-mentioned challenges that relate to imaging fast 3D dynamics with conventional microscopy-based techniques.

Rapid progress has been made recently in developing localization-based super-resolution imaging techniques (Lidke *et al* 2005, Betzig *et al* 2006, Hess *et al* 2006, Rust *et al* 2006, Egner *et al* 2007). In particular, these techniques typically use photoactivated fluorescent labels and exploit the fact that the location of a point source can be determined with a very high (nanometer) level of accuracy (Thompson *et al* 2002, Ober *et al* 2004). This is in conjunction with the working assumption that during photoactivation, sparsely distributed (thus spatially well separated) labels are turned on, enabling the retrieval of nano-scale positional and distance information for the point sources well below Rayleigh's resolution limit.

Originally demonstrated in 2D fixed-cell samples, these techniques have also been extended to 3D imaging of noncellular/fixed-cell samples (Folling *et al* 2007, Huang *et al* 2008, Juette *et al* 2008) and more recently have even been used to track single molecules in 2D in living cells (Hess *et al* 2007, Manley *et al* 2008, Shroff *et al* 2008). However, single molecules were tracked only for a short period of time because of the use of conventional fluorophores, which are susceptible to rapid photobleaching. Moreover, live-cell imaging was carried out using conventional microscopy-based imaging approaches, with their 3D tracking problems in terms of imaging events that fall outside the plane of focus. Thus, these techniques do not support the long-term, continuous (time-lapse) 3D imaging of fluorophores, which limits their applicability to 3D tracking in living cells.

An imaging modality, multifocal plane microscopy (MUM) has been developed to ensure 3D subcellular tracking within a living cell microenvironment (Prabhat *et al* 2004, 2007). In MUM, the sample is simultaneously imaged at distinct focal planes. This is achieved by placing detectors at specific distances in the microscope's emission-light path. The sample can be concurrently illuminated in the epifluorescence mode and also in the total internal reflection fluorescence (TIRF) mode. In MUM, the temporal resolution is determined by the frame rate of the camera that images the corresponding focal plane, which does not produce a realistic limitation, given current camera technology. Indeed, MUM has been performed to investigate the exocytic pathway of immunoglobulin G molecules from the sorting endosome to exocytosis on the cell membrane (Prabhat *et al* 2007), as facilitated by the receptor FcRn interactions with IgG (Ghetie and Ward 2000). Prior results addressed the problem of providing qualitative data, for instance, the imaging of the dynamic events at different focal planes within a cell. A major obstacle to high-accuracy 3D location estimation is the poor depth discrimination of a conventional microscope. However, the z-position, which is the position of the point source along the optical axis, is difficult to determine, and this is particularly the case when the point source is closely located to focus. Aside from this, the accuracy with which the 3D location of the point source can be determined is of fundamental importance. The latter is particularly relevant in live-cell imaging applications, where the S/N is typically very poor.

In the following, a methodology for determining the 3D coordinates of single fluorescent point source images is presented using MUM in living cells. The specifics of MUM acquisition exploit the fact that for each point in time more than one image of the point source is available, each at a different focal level. By appropriately exploiting this data structure it can be demonstrated that estimates are obtained and significantly more accurate compared to classical approaches, particularly in the case when the point source is near the focus in one of the focal planes. Moreover, simulations and experimental data have revealed that the proposed MUM localization algorithm (MUMLA) is applicable over a wide spatial range (approximately 2.5 μm depth) and produces estimates whose standard deviations are very close to the theoretically best possible level. This analysis shows that MUM overcomes the poor depth discrimination of the conventional microscope and thus paves the way for high-accuracy tracking of nanoparticles in a living cell microenvironment.

It should be pointed out that MUM supports multicolor imaging. This has enabled us to image QDs in 3D and also to image, at the same time, the cellular microenvironment with which the QD-labeled molecules interact. The latter was realized by labeling the cellular structures with spectrally distinct fluorescent fusion proteins. As is shown here, this has enabled us to track the fate of QD-labeled antibody molecules from endocytosis at the cell membrane to its delivery into the sorting endosome inside the cell in another focal plane.

The impressive developments in optical microscopy have made this technology increasingly compatible with other biological studies. Fluorescence microscopy in particular has contributed to investigation of the dynamic behaviors of living specimens and it can resolve objects with nanometer precision and resolution due to super-resolution imaging. Additionally, single-particle tracking provides information on the dynamics of individual proteins at the nanometer scale both *in vitro* and in cells. The development of fluorescent probes has complemented advances in microscopy technologies. The QD, a semiconductor fluorescent nanoparticle, is particularly suitable for single-particle tracking and super-resolution imaging.

Fluorescence microscopy has become standard for investigating the dynamic behavior of biological phenomena such as the expression, movement and localization of proteins and other molecules (Ellinger 1940, Lichtman and Conchello 2005, Drummen 2012, Miyawaki 2013, Peter *et al* 2014). However, optical diffraction limits the spatial resolution to several 100 nanometers, thus no information is obtained on many details for these phenomena (Abbe 1873). Two technologies have overcome this limitation and permit the observation of even smaller nano-scale dynamics: single-particle tracking (Ritchie and Kusumi 2003, Saxton 2009, Chenouard *et al* 2014) and super-resolution microscopy (Schermelleh *et al* 2010, Galbraith and Galbraith 2011, Leung and Chou 2011). Single-particle tracking ensures the observation of the precise position of single fluorescent probes conjugated to separate target proteins over a 2D plane. In addition, super-resolution microscopy provides highly resolved optical images beyond the aforementioned spatial resolution.

In order to conduct the above imaging techniques, it is often required to label the target protein with a fluorescent probe. Fluorescent proteins are most popular for this purpose because of their simple and easy labeling procedure in living cells (Shimomura and Johnson 1692, Tsien 1998, Nifosí *et al* 2007). In more detail, organic dyes are also commonly used because of their wide application (Wombacher and Cornish, 2011, Wysocki and Lavis 2011, Terai and Nagano 2013). Inorganic nanoparticles are forming another group of probes that gains attention and is made of semiconductors, metals and silicon (Ruedas-Rama *et al* 2012, Chinnathambi *et al* 2014, Cupaioli *et al* 2014). Although usually larger than fluorescent proteins and organic dyes, inorganic nanoparticles generally have stronger and more stable fluorescence profiles, which makes them suitable and even applicable not only to basic research, but also to clinical studies (Byers and Hitchman 2010, Choi and Frangioni 2010, Saadeh *et al* 2014, Wang and Wang 2014). In more detail, these same properties make them well suited for single-particle tracking methods (Chang *et al* 2008, Saxton 2008, Barroso 2011, Bruchez 2011, Clausen and Lagerholm 2011, Ruthardt *et al* 2011, Pierobon and Cappello 2012, Kairdolf *et al* 2013, Petryayeva *et al* 2013).

This section is focused on advanced microscopy using QDs, perhaps the most studied inorganic nanoparticles for biological applications (Pilla *et al* 2012). Single-particle tracking using QDs has reached 3D (x, y, z) (Genovesio *et al* 2006, Holtzer *et al* 2007, Watanabe and Higuchi 2007, Watanabe *et al* 2007, Ram *et al* 2008, 2012, Wells *et al* 2008, 2010, Yajima *et al* 2008) and more recently has even reached four dimensions (x,y,z,θ) (Ohmachi *et al* 2012, Watanabe *et al* 2013). For all their benefits, QDs have major drawbacks, such as high blinking (Nirmal *et al* 1996, van Sark *et al* 2001, Schlegel *et al* 2002, Hohng and Ha 2004, Ko *et al* 2011) and a spectral blue-shift during observation (Nirmal *et al* 1996, van Sark *et al* 2002, Hoyer *et al* 2011), which complicate the continuous tracking of the single particle and emerge due to photo-oxidation while under high-power illumination. These limitations have stimulated research into new super-resolution microscopy methods (Lidke *et al* 2005, Dertinger *et al* 2009, Watanabe *et al* 2010, Chien *et al* 2011, Hoyer *et al* 2011, Deng *et al* 2014).

QDs as fluorescent labeling probes

A QD is a semiconductor nanocrystal with electronic characteristics that depend on its size and shape (Rossetti *et al* 1980, Ekimov and Onushchenko 1981). Due to their unique characteristics and ease of synthesis, QDs have not only been applied to biomedical research, but also to engineering- and industry-related fields, such as transistors, solar cells, LEDs and diode lasers (Pilla *et al* 2012). QDs used in biological studies have a core–shell structure; the most famous is the cadmium selenide (CdSe) core and zinc sulfide (ZnS) shell (Dabbousi *et al* 1997, Bruchez *et al* 1998, Chan and Nie 1998, Pilla *et al* 2012). As a major advantage, this structure results in QDs having narrow emission spectra, but wide absorption spectra. There are two important criteria when applying QDs to biological studies: solubility and conjugating capability (Li *et al* 2010). Highly fluorescent QDs are usually synthesized in organic solvents in coordination with compounds such as tri-n-octylphosphine oxide (TOPO) or alkylamine. These compounds coat the QD, in order to make it too hydrophobic to be dissolved in water. Thus, further surface coating or exchange with hydrophilic compounds is needed for use in biological assays. Furthermore, upon becoming water soluble, the surface of the QD must have reactive groups such as amino and carboxyl chains in order for the QD to conjugate with the target biological sample. The surface coating not only contributes to the water solubilization, but also to the stabilization of the fluorescence of the QD in water because the photophysical properties are affected by the surface coating (Kuno *et al* 1997, Kloepfer *et al* 2005). Some surface coating methods even suppress the blinking that is a major drawback of QDs (Hohng and Ha 2004, Fomenko and Nesbitt 2008, Mandal and Tamai 2011, Zhang *et al* 2013).

There are two main ways to prepare water-soluble QDs (Erathodiyil and Ying 2011, Zhang and Clapp 2011). The first is to encapsulate a hydrophobic QD with an amphiphilic polymer or phospholipid (Dubertret *et al* 2002, Gao *et al* 2005, Li *et al* 2010, Tomczak *et al* 2013). The second is a ligand-exchange method in which the capping hydrophobic ligands are exchanged with hydrophilic ligands (Gerion *et al* 2001, Guo *et al* 2003, Pinaud *et al* 2004, Kim *et al* 2005, Nann 2005, Jiang *et al* 2006, Dubois *et al* 2007). While the water-solubilized QD obtained by the first method is more stable and suitable for commercialization, its size increases to about 20–40 nm, which increases the risk of steric hindrance against the function of the target protein (Li *et al* 2010). The ligand-exchange method is inferior in stability, but it has a simpler synthesis process and produces a smaller QD. The thin coating layer is another advantage of this method, as it reduces the risk of steric effects that could compromise the function of the protein upon conjugation with the QD.

Many coating reagents exist for the ligand-exchange method, including mercaptocarboxylic acid (Jiang *et al* 2006), carbon disulfide (Dubois *et al* 2007), thiosilanol (Gerion *et al* 2001), dendrimer (Guo *et al* 2003), peptide (Pinaud *et al* 2004), phosphine oxide (Kim *et al* 2005), and polyethylenimine (Nann 2005). The coating reagents can also functionalize QDs for specific purposes. Examples of this include β-cyclodextrin for ion-sensing (Palaniappan *et al* 2004), cyclodextrin for redox-active substrates (Palaniappan *et al* 2006) and cyclodextrin thiol for pH sensing (Cao *et al* 2006). Usually glutathione can be used as the coating compound because of its relatively easy preparation, which only requires the mixing of hydrophobic QDs with an aqueous glutathione solution (Jin *et al* 2008, Tiwari *et al* 2009). In particular, glutathione-coated QDs have two reactive groups (amino and carboxyl) that ensure easy conjugation with the target protein and show no cytotoxity (Tiwari *et al* 2009).

Fluorescence microscopy for nano-scale measurements

The microscopy introduced requires a regular wide-field fluorescence microscope and no complicated optical principles or special devices. However, because nano-scale measurements require a high S/N, a highly photon-sensitive camera, such as an electron multiplying charge-coupled device (EMCCD), is necessary. More recently, complementary metal–oxide–semiconductor (sCMOS) cameras have become available as alternatives (Huang *et al* 2011, Long *et al* 2012, Ma *et al* 2013). The vibration and/or stage drift of the microscope should of course be considered, as these can lead to artifacts in the measurement by obscuring the behavior and the structure of the target. Consequently, the microscope should be set on a vibration-isolation table and built with minimum height and maximum rigidity to reduce any vibration. As thermal expansion of the metals composing the microscope causes drifts in the stage and focus position, microscopes made of metals with lower thermal expansion such as invar (a nickel–iron alloy) are generally preferred. In addition, the drifts can be further suppressed by setting the microscopic system in a room with constant temperature and humidity.

One strategy for reducing vibrations is discussed in the following. Specifically, transition images of a silica bead with 1 μm diameter absorbed on a cover slip surface were acquired with excess illumination so that the camera gain could be set to zero. The frame rate was 2.0 ms, the images were acquired for 1.0 s, and the precise position (x, y) of the bead was calculated by image analysis. Moreover, in the usual initial setup, the position of the bead was kept stable within 0.7 nm in the x-axis and 0.4 nm in the y-axis. When a screw to fix the CCD camera was loosened, the vibration increased to 0.8 nm in both axes. Normally, a mono-objective revolver is used, but when instead a commercially available size-position revolver was used, the vibration was enhanced in the y-axis to 2 nm. Thus, rigid construction of the microscope is paramount for nano-scale measurements and observations.

Single-particle tracking using QDs

Single-particle tracking is well adapted for investigation of motor and membrane proteins, because resolving nano-scale movements is necessary for understanding the protein function (Ritchie and Kusumi 2003, Park *et al* 2007, Toprak and Selvin 2007, Saxton 2009). Although the resolution of conventional fluorescence microscopes is constrained by the diffraction limit, the 2D position of a single particle can be determined by calculating the weight center of the image of the fluorescent spot. The fluorescence emitted from a fluorescent probe forms a PSF that can be fitted with a Gaussian distribution where I_0 and (x_0, y_0) are the fluorescence intensity and the position of the fluorescing center, respectively. Moreover, σ is the radial standard deviation of the Gaussian function and C is the background fluorescence. Indeed, this analysis can be used to measure the center position of the image (Kubitscheck *et al* 2000, Cheezum *et al* 2001, Thompson *et al* 2002, Small and Stahlheber 2014). Although there are other common methods for determining the center, including cross-correlation, sum-absolute difference and the simple centroid, Gaussian fitting has the highest robustness at low S/N, which is why it is commonly used (Thompson *et al* 2002). In this case, the actual fitting computation is performed using the Levenberg–Marquardt method (Levenberg 1944).

The calculation precision from Gaussian fitting depends strongly on the photon number that the detection device receives from the emission of the fluorescent probe and can be as small as a few nanometers (Deschout *et al* 2014, Small and Stahlheber 2014). The method described above is called fluorescence imaging with one-nanometer accuracy (FIONA) and has become a standard method (Yildiz *et al* 2003, Yildiz and Selvin 2005, Park *et al* 2007, Hoffman *et al* 2011). However, the number of photons emitted by single organic dyes and fluorescent protein molecules before photobleaching, about 110,000 (Kubitscheck *et al* 2000), is too low for the observation of protein movement over a long time. As the cause of photobleaching is thought to be oxygen collisions with the dye molecule in its excited state, it can be reduced by the addition of oxygen scavengers (Sambongi *et al* 1999, Adachi *et al* 2000). Thus, the photon number from a single dye molecule can be increased to

1.4 million photons before photobleaching (Yildiz and Selvin 2005). Meanwhile, QDs show little photobleaching and strong fluorescence even in the absence of scavengers (Bruchez *et al* 1998). Although nonfluorescent nanoparticles such as gold nanoparticles are becoming increasingly popular for precise and long-term tracking using absorption (Kusumi *et al* 2005, Lasne *et al* 2006) or scattering (Nishikawa *et al* 2010), the QD is still preferred in biological studies because of its wider color spectrum.

The relationship between the tracking precision and the average number of photons emitted from a QD has been investigated. The tracking precision was defined as the standard deviation of 100 data points obtained with a QD immobilized on a glass surface. However, the experimental accuracy was a little lower compared to the theoretical expectation because of high blinking, thus it was still 2 nm when the photon number from a QD was 15,000 per exposure. In order to demonstrate the potential of single-particle tracking as a biological tool, the movement of kinesin, a microtubule-mediated motor protein was analyzed. In more detail, the motor domain of the kinesin was fused with biotin career protein (BCCP) and conjugated with a QD via biotin–avidin affinity. The QD-labeled kinesins were then bound to microtubules adsorbed onto a cover slip. Upon adding 1 mM ATP, the QD was seen to move unidirectionally along the microtubule without detaching, which is indeed consistent with kinesin using ATP to move. The unidirectional movement of kinesin was composed of successive 8 nm steps. Taken together, FIONA using QDs ensures a quantitative measurement for nano-scale tracking of proteins at the single molecular level.

3D single-particle tracking using QDs

The original FIONA only measured movement on a spatial plane, but it has since been expanded to three spatial dimensions. For this purpose, a 3D image under a microscope is obtained by scanning the objective lens along the focal axis with an actuator (Watanabe and Higuchi 2007, Wells *et al* 2008). This scanning, however, reduces the temporal resolution of the tracking. In order to solve this problem, 3D tracking methods without the objective scanning have been developed (Genovesio *et al* 2006, Holtzer *et al* 2007, Watanabe *et al* 2007, Ram *et al* 2008, 2012, Wells *et al* 2010, Jia *et al* 2014). Multifocal plane microscopy uses the difference of distinct optical pathways to estimate the z-position by simultaneously obtaining the fluorescence intensities of several focal images (Toprak *et al* 2007, Watanabe *et al* 2007, Dalgarno *et al* 2010, Juette and Bewersdorf 2010, Ram *et al* 2012). Similarly, 3D tracking using a photon-limited double-helix response system with a spatial light modulator, which has two twisting lobes along the optical axis of the image, results in a single fluorescent probe appearing as two fluorescent spots from which the z-position can be determined (Pavani *et al* 2009, Lew *et al* 2010).

One of the simplest 3D tracking methods intentionally generates astigmatism (Kao and Verkman 1994, Holtzer *et al* 2007, Izeddin *et al* 2012). Here, a pair of convex and concave cylindrical lenses is inserted into the optical pathway before the detection device (Watanabe *et al* 2013). In more detail, these lenses generate different optical path lengths along the x- and y-axes, resulting in a measurable relationship between the z-position of the particle and the ellipticity of the PSF. In order to calculate the ellipticity in addition to the 2D position, the approximation formula below is used, where σ_x and σ_y are the radial standard deviations of the Gaussian function along the x- and y-axes, respectively. The ellipticity is defined as the ratio of the full width at half maximum (FWHM) of the 2D Gaussian in the x- and y-axes due to the different focal lengths. Alteration of the distance between the convex and concave cylindrical lenses permits astigmatism for optimal tracking resolution. When the detection device receives 15 000 photons from a fluorescent probe, 3D tracking with precision of 2 nm in the x and y-axes and 5 nm along the z-axis is reached. However, a reliable range is limited to a field view of between −800 and 800 nm. This drawback is common to several 3D tracking methods. A new 3D tracking method based on Airy beams, however, may overcome this. Here, a diffraction-free self-bending PSF is applied to a two-channeled detection system (Jia *et al* 2014) and the z-position is translated to the distance difference of the two x-positions of the two channels. In particular, this method elongates the dynamic range of 3D tracking to 3 μm. Regardless of the 3D tracking method, the key is to extract z information from the xy projection.

The 4D single particle using polarized QDs

As significant as acquiring the third spatial dimension is, 3D single-particle tracking ignores any rotational movement performed by the protein. In order to acquire the orientation, fluorescence anisotropy can be used, as the fluorescence emissions are of unequal intensities along the P and S polar axes (P- and S-polarization), which are defined by the polarizing beam-splitter, as described below (Werver 1953, Albrecht 1961, Harms *et al* 1999). In more detail, anisotropy is defined as $(I_p - I_s)/(I_p + I_s)$, where I_p and I_s are the intensities in P- and S-polarization, respectively (Harms *et al* 1999). Indeed, anisotropy measurements have successfully tracked the rotatory dynamics of single protein molecules *in vitro* (Sase *et al* 1997, Forkey *et al* 2003) and in cells (Mizuno *et al* 2011). The fluorescence anisotropy of a QD depends on the aspect ratio of its shape (Peng *et al* 2000, Hu *et al* 2001, Deka *et al* 2009). Taking advantage of this property, a highly polarized rod-shaped QD (Qrod) can be synthesized by elongating the CdS shell along one-axis of the CdSe core (Peng *et al* 2000, Hu *et al* 2001). The anisotropy alterations in Qrod fluorescence can be described as a sine function and the angular position by the arcsine function. The tracking precision of the orientation was about 1–2° when the photon number from a Qrod was 15 000. By utilizing this anisotropy technique, a fourth dimension, the angular (θ) component, can be added to the orthogonal three coordinate axes described by single-particle tracking.

In the 4D tracking system, a polarizing beam-splitter is set before the cylindrical lens pair in the 3D tracking optics to divide the fluorescent image into S- and P-polar channels (Watanabe *et al* 2013). For 3D tracking, the P- and S-polarized images are summed before the 3D position is calculated. However, a small gap is generated if the two channels are not completely overlapped, leading to an asymmetrical relationship between the respective FWHM values of the x- and y-axes. The 3D position can be determined by fitting the merged PSF with a 2D Gaussian function, as mentioned above, and the orientation can be determined by the ratio of the intensities of the S- and P-polarized images. Thus, x, y, z and θ are obtained simultaneously with an acquired image. However, when the number of photons from a single Qrod is about 10 000 and the z-position is near zero, the calculated precision for the X, Y, Z and θ-positions is at most 5, 7, 9 nm and 1°, respectively.

4D tracking is used to investigate the movement of a membrane protein conjugated with a Qrod via antibody affinity (Watanabe *et al* 2013). Isolated Qrods moving on the membrane are identified under a fluorescence microscope. The different intensities in the P- and S-polarized images indicate that the Qrod is inclined against the optical axis. One circular and one elliptical spot indicate that the two Qrods are at distinct z-positions. In particular, one Qrod shows a half-moon like motion in the x- and y-axes, accompanied by highly fluctuating movements along the z-axis and fast rotational motion before endocytosis. However, this observation suggests that this protein's lateral diffusion is constrained by the membrane undercoat, but that it could rotate freely along the cell membrane. In the cytoplasm, a membrane protein seems to be moving along tracks such as microtubules, in 3D and slowly rotating helically.

Another 4D tracking method was developed to obtain x, y, θ and ϕ coordinates, the last of which provides information on the out-of-plane tilt angle (Ohmachi *et al* 2012). In this method, single Qrods are imaged as four crowded fluorescent spots by dividing the beam path using a beam-splitter and two Wollaston prisms. Otherwise, the orientation of the individual fluorescent probe can be directly estimated using the dipole emission patterns of a defocused image (Bartko and Dickson 1999a, 1999b, Fourkas 2001, Böhmer and Enderlein 2003, Lieb *et al* 2004), an approach that was successfully applied to 4D tracking of a motor protein (Toprak *et al* 2006). The combination of the Wollaston prism method with defocusing could achieve comprehensive tracking of all rotatory and translational movements of a particular molecule in a living cell.

Quantum dots

Super-resolution microscopy describes the resolution of two objects closer than the diffraction limit of light (Schermelleh *et al* 2010, Galbraith and Galbraith 2011, Leung and Chou 2011). It can be classified into two main categories. The first is based on the photo-transition of a fluorescent probe between its radiative and nonradiative states in order to confine the fluorescence emission into a sub-diffraction-limit-sized volume. This approach is known as RESOLFT (REversible Saturable OpticaL Fluorescence Transitions) and was initially proposed and demonstrated by STED (STimulated Emission Depletion), which exploits the stimulated emission phenomenon of a fluorescent dye (Hell and Wichmann 1994, Klar and Hell 1999). RESOLFT can also be realized by other photoreactions, including those from a ground-state transition phenomenon (ground-state depletion (GSD)) (Hell and Kroug 1995, Bretschneider *et al* 2007), the saturation of fluorescence excitation (SAX: SAturated eXcitation) (Fujita *et al* 2007) and reversibly photoswitchable fluorescent proteins (Hofmann *et al* 2005). RESOLFT can also be combined with structured illumination microscopy (SIM) (Heintzmann and Cremer 1999, Gustafsson 2000) in order to provide a wide-field imaging capability with super-resolution (Heintzmann, 2003, Gustafsson 2005).

The second category is based on the separate detection of individual single fluorescent probes in the time or spectra domain and can be further decomposed into different concepts. One, known as spectral precision distance microscopy (SPDM), localizes individual probes precisely over the many frames of sequentially obtained images (Bornfleth *et al* 1998, Lemmer *et al* 2008). Stochastic optical reconstruction microscopy (STORM) (Rust *et al* 2006) and fluorescence photoactivation localization microscopy (FPALM) (Betzig *et al* 2006) are both SPDM-based techniques that utilize repeated activation–deactivation cycles of photoswitchable fluorophores, such that the fluorescence spots on an obtained image are completely discrete.

Another method from the second category is blinking-based super-resolution (BBS). In more detail, BBS relies on the randomness and non-Gaussian properties of blinking, which means stochastic processing can be performed to localize individual fluorescent probes. The first report of BBS used independent component analysis, which is a computational method that decomposes a multivariate signal into independent non-Gaussian signals (Lidke *et al* 2005). Other BBS-based techniques use temporal high-order cumulant (super-resolution optical fluctuation imaging (SOFI)) (Dertinger *et al* 2009), temporal high-order variance (variance imaging for super-resolution (VISion)) (Watanabe *et al* 2010), spatial covariance (spatial covariance reconstructive (SCORE)) (Deng *et al* 2014) or Bayesian statistics (Cox *et al* 2011). A great advantage of SPDM and BBS is that they only need a relatively simple fluorescent microscope and no complicated optics.

QDs are the most compatible with BBS owing to their strong blinking phenomenon. Supposing that there are two adjoining QDs independently and randomly fluctuating, the moment that one QD emits and the other does not is a stochastic event (Dertinger *et al* 2009). To decrease the required number of images, a highly fluctuating QD has been developed in which the switching frequency between the on and off states has been increased greatly by optimizing the shell thickness to promote more interaction between the CdSe core and oxygen atoms in water. Although the quantum yield of this QD is less than that of standard QDs, it still has sufficient intensity and stability when exposed to high-power illumination and no long off state has been detected. Hence, we can easily obtain a super-resolved image by only labeling the target protein and calculating the fluctuation of the blinking-enhanced QDs. In this case, the spatial resolution has improved from 267 to 154 nm using SOFI and only 100 images (Watanabe *et al* 2010).

However, conventional optical microscopy can quantitatively acquire 3D position and orientation information at the nano-scale from the shape of the PSF and the polarization characteristics of QDs and Qrods. The amount of spatial information can be increased through analyzing the stochastic fluctuations of the fluorescence. Thus, the fluorescence of a probe attached to a molecule can provide information about the molecular phenomena or state. Increasing the intensity, stability and blinking of QDs and its derivatives will improve the acquisition.

However, super-resolution microscopy and single-particle tracking have made it possible to resolve and follow two objects closer together than the diffraction limit of light. The result is quantitative information on the dynamics of biological phenomena at the nano-scale level. Even more details for the dynamics can be acquired with the above technologies by using QDs and their derivatives as probes for labeling the molecules of interest. Moreover, the PSF and the polarization characteristics of the QDs can be used to gain comprehensive information on both the position and orientation of the targeted molecule. As this information can be extracted from the stochastic properties of the fluorescence, increasing the intensity, stability and blinking of QDs should provide even more quantitative details about the dynamics.

6.6 FRAP

After the beginning of the genomic era, a major challenge in the field of cell biology has been to understand the dynamics of proteins during signal transduction events regulating cellular responses and functions. In more detail, it is now possible to perform live-cell assays of protein diffusion utilizing the advances in protein labeling and fluorescence microscopy. Two classic techniques, fluorescence recovery after photobleaching (FRAP) and fluorescence correlation spectroscopy (FCS), are well-known and convenient tools for analyzing protein diffusion (Reits and Neefjes 2001, Medina and Schwille 2002). Noninvasive fluorescent tagging with green fluorescent protein (GFP) or variants has increased the use of these techniques.

In FRAP, fluorescent proteins in a small region of interest in a cell expressing them are permanently photobleached using a high-powered laser beam and the subsequent movement of surrounding nonphotobleached fluorescent proteins into the photobleached area is then recorded (Reits and Neefjes 2001). The rate of fluorescence recovery provides information about how quickly the fluorescent proteins can move into the bleached region. This 'mobility' is dependent on the rate of free diffusion and the active transport of the fluorescent protein. Mobility is also influenced by interactions with other proteins, which could even confine the movement of fluorescent molecules that would otherwise be freely mobile.

In FCS, the fluorescence intensity in the tiny confocal volume is measured with high repetition rates in order to detect fluctuations in the fluorescence intensity, including the contribution of Brownian motion (Medina and Schwille 2002). Through time correlation analysis of the fluorescence fluctuations, the diffusion coefficient, molecular concentration and molecular interactions can be estimated.

Although FRAP is adequate for measuring the mobile and immobile fractions of fluorescent molecules in a living cell, the measurability of diffusion rates is lower than for FCS. Moreover, the general photobleaching time required for FRAP (less than 10 ms) is sufficiently long that protein diffusion can start to occur during the photobleaching process itself, hence preventing an accurate calculation of the diffusion coefficient.

However, these problems can be overcome by using a photoswitchable fluorescent protein (PSFP) instead of a conventional fluorescent protein. The advantage of the usage of PSFPs is that they can change from dark to bright or undergo conversion of their absorption/emission wavelength when they are stimulated by light of a specific wavelength (Lukyanov *et al* 2005). Indeed, the rapid (less than 1 ms) photoswitching property of PSFPs encourages their use for protein mobility analyses. In particular, PSFPs have been used to investigate the diffusion time of a protein by using fluorescence decay measurements fitted to a single exponential curve (Theis-Febvre *et al* 2005) and to estimate the diffusion coefficient of a special protein from the alteration in the Gaussian profile of the fluorescence intensity (Ando *et al* 2002). Although these analyses are useful for relative comparison, they are not comparable to the values obtained by FCS, as the model equations to determine the diffusion coefficient are not based on the theoretical background. In more detail, the development of a method using rigorous analyses based on the FRAP theoretical model enables us to determine a wider range of diffusion coefficients, which cover the range of FCS to conventional FRAP (Matsuda 2008).

Photobleaching-based imaging
Although excitation irradiation of a sample is indispensable for fluorescence imaging, strong irradiation causes photobleaching, through which the chromophore becomes irreversibly nonfluorescent through a chemical structure alteration. In order to avoid photobleaching, excitation light power and time is reduced for general fluorescence imaging. However, for photobleaching-based imaging, bleaching is actively used. In more detail, once a region of fluorescence loss is obtained, the molecular movement is investigated by imaging the alterations in the nonfluorescent pattern.

In particular, FRAP visualizes the dynamics of fluorescent or fluorescent-labeled molecules. The method consists of photobleaching a fluorescent chromophore by irradiation of a limited region with strong excitation followed by imaging the fluorescence recovery caused by outflux of bleached molecules from that region and influx of fluorescent molecules from outside (Carrero *et al* 2003, Dundr and Misteli 2003). In general, confocal scanning-laser microscopy (CSLM), which can easily irradiate a specific region in the cell, is used for FRAP. In the past, switching the irradiation light intensity between imaging excitation and photobleaching required a significant time lag. However, recent CSLM techniques can simultaneously generate photobleaching and excitation light in order to ensure that it is possible to collect images immediately after photobleaching. By comparing the shapes of the fluorescence decay curves, qualitative evaluations can be performed to demonstrate that fast-moving molecules recover more quickly. It is even possible to perform quantitative evaluations to obtain the physiological constants associated with molecular dynamics by mathematical modeling of fluorescence recovery data (Carrero *et al* 2003, Dundr and Misteli 2003).

When the fluorescence-labeled molecules do not interact with other molecules or components in the cell, molecular dynamics are regulated by simple diffusion caused by Brownian motion. To determine the diffusion coefficients in simple diffusion, model equations based on Axelrod's model equation for 2D diffusion are generally used (Axelrod *et al* 1976). In this method, one must know the spatial distribution of photobleaching as a function of the distance from the bleach center, because the model equation for time recovery of fluorescence is expanded from the model equation for the initial bleaching profile. Thus, an image in which molecular movement is fixed (the sample is fixed in formaldehyde) is collected after photobleaching. That image is analyzed by fitting of the model equation for the initial bleaching profile and parameters relating to the bleaching radius and bleach constant are obtained (Axelrod *et al* 1976). The diffusion coefficient is calculated by model fitting using the equation for fluorescence recovery, which contains these parameters (Axelrod *et al* 1976). However, this method assumes the ideal situation in which the concentration of the fluorescence molecule is homogenous, the initial bleaching profile is adapted to the known model and diffusion is isotropic. A recently developed method treats each cellular component in confocal images, such as the cytoplasm, nucleus, plasma membrane or nuclear membrane independently from a 3D cell model. In this model, a different position-dependent porosity is assigned to each component. Then the diffusion coefficient is calculated using the lattice Boltzmann method, which simulates molecular dynamics in the cell model as movements of virtual particles (Khuen *et al* 2011). The development of analytical methods that can treat anisotropic diffusion will facilitate analyses based on the microenvironment in real cells. If molecular interactions are included in the system, fluorescence recovery becomes slower compared to simple diffusion because of the restricted motion of the labeled molecules.

When the counterpart of the interaction is immobile or has an extremely slow movement two situations may apply: (1) molecular diffusion is overwhelmingly faster than binding dissociation and the dynamics of the labeled molecule are regulated by binding dissociation (Sprague and McNally 2005); (2) binding dissociation is much faster and the labeled molecules repeat binding and diffusion in the bleached area as they leave. Thus, different model equations specific to each case are required for precise analysis. There is indeed an equation for simple binding dissociation in the former case and an equation including diffusion for the latter case (Sprague and McNally 2005).

Diffusion analysis using time correlations such as FCS do not provide information about the immobile fraction in standard analysis. However, FRAP can extract information about the immobile fraction as the fraction that does not recover with time (Carrero *et al* 2003, Dundr and Misteli 2003). FRAP analysis is additionally applicable for diffusion rates $<1\ \mu m^2\ s^{-1}$, which are difficult to analyze even using FCS. However, FRAP cannot easily analyze diffusion rates faster than $10\ \mu m^2\ s^{-1}$ because of the diffusion during photobleaching (Matsuda 2008). However, this can be overcome by analysis using photoactivatable fluorescent proteins (PAFPs).

Fluorescence loss in photobleaching (FLIP) and inverse FRAP (iFRAP)

Fluorescence loss in photobleaching (FLIP) and inverse FRAP (iFRAP) are other conventional photobleaching imaging techniques. For FLIP, instead of observing fluorescence recovery in the bleached area, the areas for measurement are set outside of the bleached area and measurement is performed with continuous photobleaching (Dundr and Misteli 2003). Generally, the bleached area (~1/4 to 1/2 of the whole cell area) is larger than that for FRAP. Differences in molecular dynamics can be observed as the difference in the starting time of the fluorescence reductions (for instance, faster molecules bleach earlier) (Dundr and Misteli 2003). There is no general method to determine the diffusion coefficient or dissociation constant for FLIP because fluorescence decay is related to the area of bleaching and the distance between the bleached and observed areas. However, it is useful for relative dynamic comparisons and is applicable for the fast diffusion, which is difficult to observe by FRAP as the initiation of the fluorescence decrease can be prolonged by setting the observation area farther from the bleaching area (Dundr and Misteli 2003). iFRAP is a method for which only a limited area remains unbleached and then the dynamics of the molecules in that area are traced (Dundr and Misteli 2003). This method is indeed effective for the analysis of molecules with small dissociation constants because molecular fluorescence is directly measured. However, it is also not suitable for the measurement of fast-diffusing molecules, as the photobleaching requires more time. In addition, there is the risk that strong irradiation over a larger area of the cell could be phototoxic and induce diffusion that may not appear under normal conditions (Dundr and Misteli 2003). As a similar image is obtained by the method based on PAFPs, which is even available for measurement of fast diffusion, so iFRAP will likely be replaced by this novel method.

Photoactivation-based imaging

Live-cell imaging technology based on fluorescent proteins has been developed and it has become common to use PSFPs, in which the fluorescence properties can be altered by photostimulation. These fluorescent proteins make it possible to alter the fluorescence properties of molecules in a limited region of the living body by photostimulation at a specific time (Patterson *et al* 2010, Miyawaki 2011). Photostimulation of PAFPs can increase fluorescence intensity from the dark state, allowing for the collection of inverted images from photobleaching techniques such as FRAP. Only activated molecules appear on the nonfluorescent background, providing high-contrast imaging. Using this advantage, PAFPs are used to measure protein dynamics in the living body through region-specific highlighting and tracing (Miyawaki 2011).

PSFPs

After the development of the photoactivatable protein (PA-GFP) from *Aequorea victoria* GFP (Patterson and Lippincott-Schwartz 2002), many fluorescent proteins that can be altered by photostimulation have been developed by genetic modification or screening from new species and hence used for live-cell imaging (Patterson *et al* 2010, Miyawaki 2011). In the initial development, PSFPs posed problems such as the formation of oligomers or the need to use low temperatures for cell culturing, however, recent PSFPs have improved and are useful for imaging (Patterson *et al* 2010). PSFPs are classified into PAFPs, which can alter from nonfluorescent to fluorescent proteins, and photoconvertible fluorescent proteins, which change their fluorescence wavelength with photostimulation. Both can be categorized further as reversible or irreversible (Patterson *et al* 2010). Although a reversible photoconvertible fluorescent protein constructed by fusion of the yellow fluorescent protein EYFP and the reversible photoactivatable red fluorescent protein rsTagRFP have been reported (Subach *et al* 2010), no single fluorescent protein has been reported thus far. The rsTagRFP is the first monomeric red fluorescent protein with reversibly photoswitchable absorbance spectra. In particular, the switching is performed by irradiation of rsTagRFP with blue (440 nm) and yellow (567 nm) light, which turns the protein fluorescence ON and OFF, respectively. Due to the photoswitchable absorbance, rsTagRFP can be utilized as an acceptor in photochromic Förster resonance energy transfer (pcFRET). Moreover, the yellow fluorescent proteins YPet and mVenus have been revealed to be highly suitable pcFRET donors for the rsTagRFP acceptor in its fusion constructs (Pletnev *et al* 2012). These photoswitchable proteins are used as genetically encoded fluorescence tags to visualize dynamics at the cellular, organelle and protein levels by highlighting and tracing (Miyawaki 2011). Moreover, these proteins also find applications in super-resolution imaging (Patterson *et al* 2010, Miyawaki 2011).

FDAP

As with imaging using photobleaching, PSFPs are available for time-lapse imaging of molecular migration and for determination of diffusion coefficients. Because reverse images are taken, diffusion coefficients can be determined by PSFPs through using procedures similar to that of FRAP (Matsuda 2008). Contrary to FRAP, which analyzes fluorescence recovery after photobleaching, analysis by PSFPs measures the fluorescence decay after photostimulation caused by outflux of photoactivated molecules. Thus, the method is known as fluorescence decay after photoactivation (FDAP) (Matsuda 2008). The fluorescence decay curve of FDAP and the fluorescence recovery of FRAP have a symmetric relationship to the time axis. However, a model function for FRAP cannot be used directly. For normalization, it is necessary to know the fluorescence intensity where all PAFPs are photoactivated. Thus, the photostimulation proceeds until the fluorescence intensity becomes saturated and the measurement of fluorescence intensity is included in the method for FRAP (Matsuda 2008). The measurements and analysis are even based on FRAP, however, there is a difference in the detectable range for diffusion coefficients. Fluorescence recovery between photobleaching and first acquisition significantly affects analysis for fast-moving molecules over 10 $\mu m^2 s^{-1}$ on FRAP. At the first image for fluorescence recovery, if the profile of the bleached molecule is significantly greater than that obtained by fixed cells, it breaks the basic assumption of the model equation for fluorescence recovery. Hence, the results of the analysis become inaccurate. On the other hand, the irradiation time for FDAP is much shorter than that for FRAP because photoswitching of PSFPs occurs with much less energy (approximately 10 times lower) compared to photobleaching. It is possible to analyze diffusion near 100 $\mu m^2 s^{-1}$ in solution by FDAP, when the stimulation is performed for 0.25 ms at a single spot and images are acquired with a high scanning speed of 4000 Hz (Matsuda 2008). In addition, it can be applied for measuring slow diffusion and binding dissociation (Plachta 2011).

FDAP does not require special equipment or software, hence, it is possible to measure a wide range of diffusion coefficients from FRAP's specialized slow diffusion to FCS's specialized fast diffusion with a general confocal microscope setup for cell observation. FDAP also contributes to research on protein dynamics as a convenient tool to obtain fluorescence decay curves that can easily be applied for relative comparison (Miyawaki 2011).

Imaging techniques using photobleaching and photoactivation of fluorescent proteins are a versatile technology for analyzing molecular dynamics. Data obtained from these methods comprise recovery or decay curves of fluorescence intensity that are easy to understand and compare. However, these curves include information on the simple diffusion of molecules and on molecular interactions. Thus, it is possible to observe interactions as incomplete recovery (or decay) due to the immobile fraction associated with intracellular structures or a significant inconsistency between the diffusion coefficient determined by curve fitting and the values estimated from molecular weight by using the Stokes–Einstein equation. However, such differences from simple diffusion can be regarded as a particular feature of the individual molecule that defines the molecular function in living cells. Although the conventional model equation was based on 2D diffusions, the real diffusion of intracellular molecules is definitely 3D and sometimes anisotropic. Thus, the development of a model equation considering these realities will allow deeper understanding of molecular dynamics in a living system.

References and further reading

Abbe E 1873 Contributions to the theory of the microscope and the microscopic perception *Arch. Mikr. Anat.* **9** 413–68

Adachi K, Yasuda R, Noji H, Itoh H, Harada Y, Yoshida M and Kinosita K Jr 2000 Stepping rotation of F1-ATPase visualized through angle-resolved single-fluorophore imaging *Proc. Natl Acad. Sci. USA* **97** 7243–7

Aguet F, Ville D V D and Unser M 2005 A maximum likelihood formalism for sub-resolution axial localization of fluorescent nanoparticles *Opt. Exp.* **13** 10503–22

Albrecht A 1961 Polarizations and assignments of transitions: the method of photoselection *J. Mol. Spectrosc.* **6** 84–108

An S S and Fredberg J J 2007 Biophysical basis for airway hyperresponsiveness *Can. J. Physiol. Pharmacol.* **85** 700–14

Ando R, Hama H, Yamamoto-Hino M, Mizuno H and Miyawaki A 2002 An optical marker based on the UV-induced green-to-red photoconversion of a fluorescent protein *Proc. Natl Acad. Sci. USA* **99** 12651–56

Arhel N, Genovesio A, Kim K A, Miko S, Perret E, Olivo-Marin J C, Shorte S and Charneau P 2006 Quantitative four-dimensional tracking of cytoplasmic and nuclear HIV-1 complexes *Nat. Methods* **3** 817–24

Axelrod D, Koppel D E, Schlessinger J, Elson E and Webb W W 1976 Mobility measurement by analysis of fluorescence photo-bleaching recovery kinetics *Biophys. J.* **16** 1055–69

Barroso M M 2011 Quantum dots in cell biology *J. Histochem. Cytochem.* **59** 237–51

Bartko A P and Dickson R M 1999a Imaging three-dimensional single molecule orientations *J. Phys. Chem.* B **103** 11237–41

Bartko A P and Dickson R M 1999b Three-dimensional orientations of polymer-bound single molecules *J. Phys. Chem.* B **103** 3053–56

Betzig E, Patterson G H, Sougrat R, Lindwasser O W, Olenych S, Bonifacino J S, Davidson M W, Lippincott-Schwartz J and Hess H F 2006 Imaging intracellular fluorescent proteins at nanometer resolution *Science* **313** 1642–45

Boehmer M and Enderlein J 2003 Orientation imaging of single molecules by wide-field epifluorescence microscopy *J. Opt. Soc. Am.* B **20** 554–9

Born M and Wolf E 1980 *Principles of Optics* (Oxford: Pergamon)

Bornfleth H, Satzler K, Elis R and Cremer C 1998 High-precision distance measurements and volume-conserving segmentation of objects near and below the resolution limit in three-dimensional confocal fluorescence microscopy *J. Microsc.* **189** 118–36

Bretschneider S, Eggeling S and Hell S W 2007 Breaking the diffraction barrier in fluorescence microscopy by optical shelving *Phys. Rev. Lett.* **98** 218103

Bruchez M Jr, Moronne M, Gin P, Weiss S and Alivisatos A P 1998 Semiconductor nanocrystals as fluorescent biological labels *Science* **281** 2013–16

Bruchez M P 2011 Quantum dots find their stride in single molecule tracking *Curr. Opin. Chem. Biol.* **15** 775–80

Bursac P, Lenormand G, Fabry B, Oliver M, Weitz D A, Viasnoff V, Butler J P and Fredberg J J 2005 Cytoskeletal remodelling and slow dynamics in the living cell *Nat. Mater.* **4** 557–61

Byers R J and Hitchman E R 2010 Quantum dots brighten biological imaging *Prog. Histochem. Cytochem.* **45** 201–37

Cao H, Chen B, Squier T C and Mayer M U 2006 CrAsH: a biarsenical multi-use affinity probe with low non-specific fluorescence *Chem. Commun.* **24** 2601–03

Carrero G, McDonald D, Crawford E, de Vries G and Hendzel M J 2003 Using FRAP and mathematical modeling to determine the *in vivo* kinetics of nuclear proteins *Methods* **29** 14–28

Chan W C W and Nie S 1998 Quantum dot bioconjugates for ultrasensitive nonisotopic detection *Science* **281** 2016–18

Chang Y P, Pinaud F, Antelman J and Weiss S 2008 Tracking bio-molecules in live cells using quantum dots *J. Biophotonics* **1** 287–98

Cheezum M K, Walker W F and Guilford W H 2001 Quantitative comparison of algorithms for tracking single fluorescent particles *Biophys. J.* **281** 2378–88

Chen C, Krishnan R, Zhou E, Ramachandran A, Tambe D, Rajendran K, Adam R M, Deng L and Fredberg J J 2010 Fluidization and resolidification of the human bladder smooth muscle cell in response to transient stretch *PloS ONE* **5** e12035

Chenouard N *et al* 2014 Objective comparison of particle tracking methods *Nat. Methods* **11** 281–9

Chien F C, Kuo C W and Chen P 2011 Localization imaging using blinking quantum dots *Analyst* **136** 1608–13

Chinnathambi S, Chen S, Ganesan S and Hanagata N 2014 Silicon quantum dots for biological applications *Adv. Healthc. Mater.* **3** 10–29

Choi H S and Frangioni J V 2010 Nanoparticles for biomedical imaging: fundamentals of clinical translation *Mol. Imaging* **9** 291–310

Clausen M P and Lagerholm B C 2011 The probe rules in single particle tracking *Curr. Protein Pept. Sci.* **12** 699–713

Conner S D and Schmid S L 2003 Regulated portals of entry into the cell *Nature* **422** 37–44

Courty S, Luccardini C, Bellaiche Y, Cappello G and Dahan M 2006 Tracking individual kinesin motors in living cells using single quantum-dot imaging *Nano Lett.* **6** 1491–95

Cox S, Rosten E, Monypenny J, Jovanovic-Talisman T, Burnette D T, Lippincott-Schwartz J, Jones G E and Heintzmann R 2011 Bayesian localization microscopy reveals nanoscale podosome dynamics *Nat. Methods* **9** 195–200

Crane J and Verkman A 2008 Long-range nonanomalous diffusion of quantum dot-labeled aquaporin-1 water channels in the cell plasma membrane *Biophys. J.* **94** 702–13

Cui B, Wu C, Chen L, Ramirez A, Bearer E L, Li W, Mobley W C and Chu S 2007 One at a time, live tracking of NGF axonal transport using quantum dots *Proc. Natl Acad. Sci. USA* **104** 13666–71

Cupaioli F A, Zucca F A, Boraschi D and Zecca L 2014 Engineered nanoparticles. How brain friendly is this new guest? *Prog. Neurobiol.* **119-120** 20–38

Dabbousi B O, Rodriguez-Viej O J and Bawendi M G 1997 (CdSe)ZnS core-shell Qdots: synthesis and characterization of a size series of highly luminescent nanocrystallites *J. Phys. Chem.* B **101** 9463–75

Dahan M, Levin S, Luccardini C, Rostaing P, Riveau B and Triller A 2003 Diffusion dynamics of glycine receptors revealed by single-quantum dot tracking *Science* **302** 442–5

Dalgarno P A, Dalgarno H I, Putoud A, Lambert R, Paterson L, Logan D C, Towers D P, Warburton R J and Greenaway A H 2010 Multiplane imaging and three dimensional nanoscale particle tracking in biological microscopy *Opt. Express.* **18** 877–84

De R and Safran S A 2008 Dynamical theory of active cellular response to external stress *Phys. Rev.* E **78** 031923

De R, Zemel A and Safran S A 2010 Theoretical concepts and models of cellular mechanosensing *Methods Cell Biol.* **98** 143–75

Deka S, Quarta A and Lupo M G *et al* 2009 CdSe/CdS/ZnS double shell nanorods with high photoluminescence efficiency and their exploitation as biolabeling probes *J. Am. Chem. Soc.* **131** 2948–58

Deng Y, Sun M, Lin P H, Ma J and Shaevitz J W 2014 Spatial covariance reconstructive (SCORE) super-resolution fluorescence microscopy *PLoS ONE* **9** e94807

Dertinger T, Colyer R, Iyer G, Weiss S and Enderlein J 2009 Fast, background-free, 3D super-resolution optical fluctuation imaging (SOFI) *Proc. Natl Acad. Sci. USA* **106** 22287–92

Deschout H, Cella Zanacchi F, Mlodzianoski M, Diaspro A and Bewersdorf J 2014 Precisely and accurately localizing single emitters in fluorescence microscopy *Nat. Methods* **11** 253–66

Deshpande V S, McMeeking R M and Evans A G 2006 A bio-chemo-mechanical model for cell contractility *Proc. Natl. Acad. Sci. USA* **103** 14015–20

Discher D, Dong C, Fredberg J J, Guilak F, Ingber D, Janmey P, Kamm R D, Schmid-Schönbein G W and Weinbaum S 2009 Biomechanics: cell research and applications for the next decade *Annals Biomed. Eng.* **37** 847–59

Drummen G P 2012 Fluorescent probes and fluorescence (microscopy) techniques-illuminating biological and biomedical research *Molecules* **17** 14067–90

Dubertret B, Skourides P, Norris D J, Noireaux V, Brivanlou A H and Libchaber A 2002 *In vivo* imaging of quantum dots encapsulated in phospholipid micelles *Science* **98** 1759–62

Dubois F, Mahler B, Dubertret B, Doris E and Mioskowski C 2007 A versatile strategy for quantum dot ligand exchange *J. Am. Chem. Soc.* **129** 482–3

Dundr M and Misteli T 2003 Measuring dynamics of nuclear proteins by photobleaching *Curr. Protoc. Cell Biol.* **13** 1–18

Dupont A and Lamb D C 2011 Nanoscale three-dimensional single particle tracking *Nanoscale* **3** 4532–41

Egner A *et al* 2007 Fluorescence nanoscopy in whole cells by asynchronous localization of photoswitching emitters *Biophys. J.* **93** 3285–90
Ehrlich M, Boll W, Van Oijen A, Haniharan R, Chandran K, Nibert M L and Kirchhausen T 2004 Endocytosis by random initiation and stabilization of clathrin-coated pits *Cell* **118** 591–605
Ekimov A I and Onushchenko A A 1981 Quantum size effect in three-dimensional microscopic semiconductor crystals *JETP Lett.* **34** 345–9
Ellinger P 1940 Fluorescence microscopy in biology *Biol. Rev.* **15** 323–47
Erathodiyil N and Ying J Y 2011 Functionalization of inorganic nanoparticles for bioimaging applications *Acc. Chem. Res.* **44** 925–35
Fabry B and Fredberg J J 2007 Mechanotransduction, asthma, and airway smooth muscle *Drug Discov. Today Disease Models* **4** 131–7
Folling J, Belov V, Kunetsky R, Medda R, Schonle A, Egner A, Eggeling C, Bossi M and Hell S W 2007 Photochromic rhodamines provide nanoscopy with optical sectioning *Angew. Chem.* **46** 6266–70
Fomenko V and Nesbitt D J 2008 Solution control of radiative and nonradiative lifetimes: a novel contribution to quantum dot blinking suppression *Nano Lett.* **8** 287–93
Forkey J N, Quinlan M E, Shaw M A, Corrie J E and Goldman Y E 2003 Three-dimensional structural dynamics of myosin V by single-molecule fluorescence polarization *Nature* **422** 399–404
Fourkas J T 2001 Rapid determination of the three-dimensional orientation of single molecules *Opt. Lett.* **26** 211–3
Fredberg J J 2002 Airway narrowing in asthma: does speed kill? *Am. J. Physiol. Lung Cell Mol. Physiol.* **283** L1179–L1180
Fredberg J J 2004 Bronchospasm and its biophysical basis in airway smooth muscle *Respir. Res.* **5** 2
Friedrich B M, Buxboim A, Discher D E and Safran S A 2011 Striated acto-myosin fibers can reorganize and register in response to elastic interactions with the matrix *Biophys. J.* **100** 2706–15
Fujita K, Kobayashi M, Kawano S, Yamanaka M and Kawata S 2007 High-resolution confocal microscopy by saturated excitation of fluorescence *Phys. Rev. Lett.* **99** 228105
Fujiwara T, Ritchie K, Murakoshi H, Jacobson K and Kusumi A 2002 Phospholipids undergo hop diffusion in compartmentalized cell membrane *J. Cell Biol.* **157** 1071–81
Gabbiani G 2003 The myofibroblast in wound healing and fibrocontractive diseases *J. Pathol.* **200** 500–3
Galbraith C G and Galbraith J A 2011 Super-resolution microscopy at a glance *J. Cell Sci.* **124** 1607–11
Gao X, Yang L, Petros J A, Marshall F F, Simons J W and Nie S 2005 *In vivo* molecular and cellular imaging with quantum dots *Curr. Opin. Biotechnol.* **16** 63–72
Geiger B, Bershadsky A, Pankov R and Yamada K M 2001 Transmembrane crosstalk between the extracellular matrix–cytoskeleton crosstalk *Nat. Rev. Mol. Cell Biol.* **2** 793–805
Genovesio A, Liedl T, Emiliani V, Parak W J, Coppey-Moisan M and Olivo-Marin J C 2006 Multiple particle tracking in 3-D+t microscopy: method and application to the tracking of endocytosed quantum dots *IEEE Trans. Image Process.* **15** 1062–70
Gerion D, Pinaud F, Williams S, Parak W, Zanchet D, Weiss S and Alivisatos A P 2001 Synthesis and properties of biocompatible water-soluble silica-coated CdSe/ZnS semiconductor quantum dots *J. Phys. Chem.* B **105** 8861–71

Ghetie V and Ward E S 2000 Multiple roles for the major histocompatibility complex class I-related receptor FcRn *Annu. Rev. Immunol.* **18** 739–66

Guo W, Li J J, Wang Y A and Peng X 2003 Conjugation chemistry and bioapplications of semiconductor box nanocrystals prepared via dendrimer bridging *Chem. Mater.* **15** 3125–33

Gustafsson M G L 2000 Surpassing the lateral resolution limit by a factor of two using structured illumination microscopy *J. Microsc.* **198** 82–7

Gustafsson M G L 2005 Nonlinear structured-illumination microscopy: wide-field fluorescence imaging with theoretically unlimited resolution *Proc. Natl Acad. Sci. USA* **102** 13081–6

Harms G S, Sonnleitner M, Schütz G J, Gruber H J and Schmidt T 1999 Single-molecule anisotropy imaging *Biophys. J.* **77** 2864–70

Han J J, Kiss C, Bradbury A R M and Werner J H 2012 Time-resolved, confocal single molecule tracking of individual organic dyes and fluorescent proteins in three dimensions *ACS Nano* **6** 8922–32

He K *et al* 2010 Intercellular transportation of quantum dots mediated by membrane nanotubes *ACS Nano* **4** 3015–22

Heintzmann R 2003 Saturated patterned excitation microscopy with two-dimensional excitation patterns *Micron* **34** 283–91

Heintzmann R and Cremer C 1999 Laterally modulated excitation microscopy: improvement of resolution by using a diffraction grating *Proc. SPIE* **3568** 185

Hell S W and Kroug M 1995 Ground-state-depletion fluorescence microscopy: a concept for breaking the diffraction resolution limit *Appl. Phys.* B **60** 495–7

Hell S W and Wichmann J 1994 Breaking the diffraction resolution limit by stimulated emission: stimulated-emission-depletion fluorescence microscopy *Opt. Lett.* **19** 780–2

Hess S T, Girirajan P K and Mason M D 2006 Ultra-high-resolution imaging by fluorescence photoactivation localization microscopy *Biophys. J.* **91** 4258–72

Hess S T, Gould T J, Gudheti M V, Maas S A, Mills K D and Zimmerberg J 2007 Dynamic clustered distribution of hemagglutinin resolved at 40 nm in living cell membranes discriminates between raft theories *Proc. Natl Acad. Sci. USA* **104** 17370–5

Hoffman M T, Sheung J and Selvin P R 2011 Fluorescence imaging with one nanometer accuracy: in vitro and *in vivo* studies of molecular motors *Methods Mol. Biol.* **778** 33–56

Hofmann M, Eggeling C, Jakobs S and Hell S W 2005 Breaking the diffraction barrier in fluorescence microscopy at low light intensities by using reversibly photoswitchable proteins *Proc. Natl Acad. Sci. USA* **102** 17565–9

Hohng S and Ha T 2004 Near-complete suppression of quantum dot blinking in ambient conditions *J. Am. Chem. Soc.* **126** 1324–5

Holtzer L, Meckel T and Schmidt T 2007 Nanometric three-dimensional tracking of individual quantum dots in cells *Appl. Phys. Lett.* **90** 053902

Hoyer P, Staudt T, Engelhardt J and Hell S W 2011 Quantum dot bluing and blinking enables fluorescence nanoscopy *Nano Lett.* **11** 245–50

Hsu H J, Lee C F, Locke A, Vanderzyl S Q and Kaunas R 2010 Stretch-induced stress fiber remodeling and the activations of JNK and ERK depend on mechanical strain rate, but not FAK *PLoS One* **5** e12470

Hu J, Li L S, Yang W, Manna L, Wang L W and Alivisatos A P 2001 Linearly polarized emission from colloidal semiconductor quantum rods *Science* **292** 2060–3

Hua W, Sheff D, Toomre D and Mellman I 2006 Vectorial insertion of apical and basolateral membrane proteins in polarized epithelial cells revealed by quantitative 3D live-cell imaging *J. Cell Biol.* **172** 1035–44

Huang B, Wang W, Bates M and Zhuang X 2008 Three-dimensional super-resolution imaging by stochastic optical reconstruction microscopy *Science* **319** 810–3

Huang Z L, Zhu H, Long F, Ma H, Qin L, Liu Y, Ding J, Zhang Z, Luo Q and Zeng S 2011 Localization-based super-resolution microscopy with an sCMOS camera *Opt. Express.* **19** 19156–68

Izeddin I, El Beheiry M, Andilla J, Ciepielewski D, Darzacq X and Dahan M 2012 PSF shaping using adaptive optics for three-dimensional single-molecule super-resolution imaging and tracking *Opt. Express.* **20** 4957–67

Jia S, Vaughan V C and Zhuang Z 2014 Isotropic three-dimensional super-resolution imaging with a self-bending point spread function *Nat. Photonics* **8** 302–6

Jiang W, Mardyani S, Fischer H and Chan W C W 2006 Design and characterization of lysine cross-linked mercapto-acid biocompatible quantum dots *Chem. Mater.* **18** 872–8

Jin T, Fujii F, Komai Y, Seki J, Seiyama A and Yoshioka Y 2008 Preparation and characterization of highly fluorescent, glutathione-coated near infrared quantum dots for *in vivo* fluorescence imaging *Int. J. Mol. Sci.* **9** 2044–61

Jones S A, Shim S H, He J and Zhuang X 2011 Fast, three-dimensional super-resolution imaging of live cells *Nat. Methods* **8** 499–508

Joo K I, Fang Y, Liu Y, Xiao L, Gu Z, Tai A, Lee C L, Tang Y and Wang P 2011 Enhanced real-time monitoring of adeno-associated virus trafficking by virus-quantum dot conjugates *ACS Nano* **5** 3523–35

Juette M F and Bewersdorf J 2010 Three-dimensional tracking of single fluorescent particles with submilli second temporal resolution *Nano Lett.* **10** 4657–63

Juette M F, Gould T J, Lessard M D, Mlodzianoski M J, Nagpure B, Bennett B T, Hess S T and Bewersdorf J 2008 Three-dimensional sub-100 nm resolution fluorescence microscopy of thick samples *Nat. Methods* **5** 527–9

Juette M F, Rivera-Molina F E, Toomre D K and Bewersdorf J 2013 Adaptive optics enables three-dimensional single particle tracking at the sub-millisecond scale *Appl. Phys. Lett.* **102** 173702

Kairdolf B A, Smith A M, Stokes T H, Wang M D, Young A N and Nie S 2013 Semiconductor quantum dots for bioimaging and biodiagnostic applications *Annu. Rev. Anal. Chem.* **6** 143–62

Kao H P and Verkman A S 1994 Tracking of single fluorescent particles in three dimensions: use of cylindrical optics to encode particle position *Biophys. J.* **67** 1291–300

Kaunas R and Hsu H J 2009 A kinematic model of stretch-induced stress fiber turnover and reorientation *J. Theor. Biol.* **257** 320–30

Kaunas R, Nguyen P, Usami S and Chien S 2005 Cooperative effects of Rho and mechanical stretch on stress fiber organization *Proc. Natl Acad. Sci. USA* **102** 15895–900

Kaunas R, Usami S and Chien S 2006 Regulation of stretch-induced JNK activation by stress fiber orientation *Cell Signal* **18** 1924–31

Khuen T, Ihalainen T O, Hyvaeluoma J, Dross N, Willman S F, Langowski J, Vihinen-Ranta M and Timonen J 2011 Protein diffusion in mammalian cell cytoplasm *PLoS One* **6** e22962

Kim S W, Kim S, Tracy J B, Jasanoff A and Bawendi M G 2005 Phosphine oxide polymer for water-soluble nanoparticles *J. Am. Chem. Soc.* **127** 4556–7

Klar T A and Hell S W 1999 Subdiffraction resolution in far-field fluorescence microscopy *Opt. Lett.* **24** 954–6

Kloepfer J A, Bradforth S E and Nadeau J L 2005 Photophysical properties of biologically compatible CdSe quantum dot structures *J. Phys. Chem.* B **109** 9996–10003

Ko H C, Yuan C T and Tang J 2011 Probing and controlling fluorescence blinking of single semiconductor nanoparticles *Nano Rev.* **2** 5895

Krishnan R, Park C Y and Lin Y C *et al* 2009 Reinforcement versus fluidization in cytoskeletal mechanoresponsiveness *PLoS One* **4** e5486

Krishnan R, Peruski Canovic E, Iordan A L, Rajendran K, Manomohan G, Pirentis A P, Smith M L, Butler J P, Fredberg J J and Stamenović D 2012 Fluidization, resolidification and reorientation of the endothelial cell in response to slow tidal stretches *Am. J. Physiol. Cell Physiol.* **303** C368–C375

Kubitscheck U, Kückmann O, Kues T and Peters R 2000 Imaging and tracking of single GFP molecules in solution *Biophys. J.* **78** 2170–9

Kubitscheck U, Wedekind P, Zeidler O, Grote M and Peters R 1996 Single nuclear pores visualized by confocal microscopy and image processing *Biophys. J.* **70** 2067–77

Kuno M, Lee J K, Dabbousi B O, Mikulec F V and Bawendi M G 1997 The band edge luminescence of surface modified CdSe nanocrystallites: probing the luminescing state *J. Chem. Phys.* **106** 9869

Kural C, Kim H, Syed S, Goshima G, Gelfand V I and Selvin P R 2005 Kinesin and dynein move a peroxisome *in vivo*: a tug-of-war or coordinated movement? *Science* **308** 1469–72

Kusumi A, Ike H, Nakada C, Murase K and Fujiwara T 2005 Single-molecule tracking of membrane molecules: plasma membrane compartmentalization and dynamic assembly of raft-philic signaling molecules *Semin. Immunol.* **17** 3–21

Kusumi A, Nakada C, Ritchie K, Murase K, Suzuki K and Murakoshi H *et al* 2005 Paradigm shift of the plasma membrane concept from the two-dimensional continuum fluid to the partitioned fluid: high-speed single-molecule tracking of membrane molecules *Annu. Rev. Biophys. Biomol. Struct.* **34** 351–78

Kwik J, Boyle S, Fooksman D, Margolis L, Sheetz M P and Edidin M 2003 Membrane cholesterol, lateral mobility, and the phosphatidylinositol 4,5-bisphosphate-dependent organization of cell actin *Proc. Natl Acad. Sci. USA* **100** 13964–9

Lakadamyali M, Rust M J, Babcock H P and Zhuang X W 2003 Visualizing infection of individual influenza viruses *Proc. Natl Acad. Sci. USA* **100** 9280–5

Lasne D, Blab G A, Berciaud S, Heine M, Groc L, Choquet D, Cognet L and Lounis B 2006 Single nanoparticle photothermal tracking (SNaPT) of 5-nm gold beads in live cells *Biophys. J.* **91** 4598–604

Lee S-L, Nekouzadeh A and Butler B *et al* 2012 Physically-induced cytoskeleton remodeling of cells in three-dimensional culture *PLoS One* **7** e45512

Lemmer P, Gunkel M, Baddeley D, Kaufmann R, Urich A, Weiland Y, Reymann J, Mueller P, Hausmann M and Cremer C 2008 SPDM: light microscopy with single-molecule resolution at the nanoscale *Appl. Phys.* B **93** 1–12

Lenormand G and Fredberg J J 2006 Deformability, dynamics, and remodeling of cytoskeleton of the adherent living cell *Biorheology* **43** 1–30

Leung B O and Chou K C 2011 Review of super-resolution fluorescence microscopy for biology *Appl. Spectrosc.* **65** 967–80

Levenberg K 1944 A method for the solution of certain non-linear problems in least squares *Q. Appl. Math.* **2** 164–8

Levi V, Ruan Q and Gratton E 2005 3D particle tracking in a two-photon microscope: application to the study of molecular dynamics in cells *Biophys. J.* **88** 2919–28

Levi V, Ruan Q Q and Gratton E 2005 3-D Particle tracking in a two-photon microscope: application to the study of molecular dynamics in cells *Biophys. J.* **88** 2919–28

Lew M D, Thompson M A, Badieirostami M and Moerner W E 2010 *In vivo* three-dimensional superresolution fluorescence tracking using a double-helix point spread function *Proc. Soc. Photo Opt. Instrum. Eng.* **7571** 75710Z

Li J, Wu D, Miao Z and Zhang Y 2010 Preparation of quantum dot bioconjugates and their applications in bio-imaging *Curr. Pharm. Biotechnol.* **11** 662–71

Liang Z Y, Xu N, Guan Y H, Xu M, He Q H, Han Q D, Zhang Y Y and Zhao X S 2007 The transport of alpha(1A)-adrenergic receptor with 33-nm step size in live cells *Biochem. Biophys. Res. Commun.* **353** 231–7

Lichtman J W and Conchello J A 2005 Fluorescence microscopy *Nat. Methods* **2** 910–9

Lidke K A, Rieger B, Jovin T M and Heintzmann R 2005 Superresolution by localization of quantum dots using blinking statistics *Opt. Exp.* **13** 7052–62

Lieb M A, Zavislan J M and Novotny L 2004 Single-molecule orientations determined by direct emission pattern imaging *J. Opt. Soc. Am.* B **21** 1210–5

Liu H, Liu Y, Liu S, Pang D W and Xiao G 2011a Clathrin-mediated endocytosis in living host cells visualized through quantum dot labeling of infectious hematopoietic necrosis virus *J. Virol.* **85** 6252–62

Liu S L, Zhang Z L, Sun E Z, Peng J, Xie M, Tian Z Q, Lin Y and Pang D W 2011b Visualizing the endocytic and exocytic processes of wheat germ agglutinin by quantum dot-based single-particle tracking *Biomaterials* **32** 7616–24

Liu S L, Zhang Z L, Tian Z Q, Zhao H S, Liu H, Sun E Z, Xiao G F, Zhang W, Wang H Z and Pang D W 2012 Effectively and efficiently dissecting the infection of influenza virus by quantum-dot-based single-particle tracking *ACS Nano* **6** 141–50

Long F, Zeng S and Huang Z L 2012 Localization-based super-resolution microscopy with an sCMOS camera part II: experimental methodology for comparing sCMOS with EMCCD cameras *Opt. Express.* **20** 17741–59

Loy G and Zelinsky A 2003 Fast radial symmetry for detecting points of interest *IEEE Trans Pat. Rec. Mach. Int.* **25** 959–73

Lui T *et al* 2014 Direct measurement of the mechanical work during translocation by the ribosome *Elife* **3** e03406

Lukyanov K A, Chudakov D M, Lukyanov S and Verkhusha V V 2005 Photoactivatable fluorescent proteins *Nat. Rev. Mol. Cell Biol.* **6** 885–91

Ma H, Kawai H, Toda E, Zeng S and Huang Z L 2013 Localization-based super-resolution microscopy with an sCMOS camera part III: camera embedded data processing significantly reduces the challenges of massive data handling *Opt. Lett.* **38** 1769–71

Mandal A and Tamai N 2011 Suppressed blinking behavior of thioglycolic acid capped CdTe quantum dot by amine functionalization *Appl. Phys. Lett.* **99** 263111

Manley S, Gillette J M, Patterson G H, Shroff H, Hess H F, Betzig E and Lippincott-Schwartz J 2008 High-density mapping of single-molecule trajectories with photoactivated localization microscopy *Nat. Methods* **5** 155–7

Matsuda T, Miyawaki A and Nagai T 2008 Direct measurement of protein dynamics inside cells using a rationally designed photoconvertible protein *Nat. Methods* **5** 339–45

Maxfield F R and McGraw T E 2004 Endocytic recycling *Nat. Rev. Mol. Cell Biol.* **5** 121–32

Medina M A and Schwille P 2002 Fluorescence correlation spectroscopy for the detection and study of single molecules *Biol. Bioessays* **24** 758–64

Mierke C T 2013 Phagocytized beads reduce the $\alpha 5\beta 1$ integrin facilitated invasiveness of cancer cells by regulating cellular stiffness *Cell Biochem Biophys.* **66** 599–622

Mierke C T, Paranhos Zitterbart D, Kollmannsberger P, Raupach C, Schlötzer-Schrehardt U, Goecke T W, Behrens J and Fabry B 2008 Breakdown of the endothelial barrier function in tumor cell transmigration *Biophys. J.* **94** 2832–46

Miyawaki A 2011 Proteins on the move: insights gained from fluorescent protein technologies *Nat. Rev. Mol. Cell Biol.* **12** 656–68

Miyawaki A 2013 Fluorescence imaging in the last two decades *Microscopy (Oxf.)* **62** 63–8

Mizuno H, Higashida C, Yuan Y, Ishizaki T, Narumiya S and Watanabe N 2011 Rotational movement of the formin mDia1 along the double helical strand of an actin filament *Science* **331** 80–3

Moerner W E 2007 New directions in single-molecule imaging and analysis *Proc. Natl Acad. Sci. USA* **104** 12596–602

Murase K, Fujiwara T, Umemura Y, Suzuki K, Iino R, Yamashita H, Saito M, Murakoshi H, Ritchie K and Kusumi A 2004 Ultrafine membrane compartments for molecular diffusion as revealed by single molecule techniques *Biophys. J.* **86** 4075–93

Nan X, Sims P A, Chen P and Xie X S 2005 Observation of individual microtubule motor steps in living cells with endocytosed quantum dots *J. Phys. Chem.* B **109** 24220–4

Nann T 2005 Phase-transfer of CdSe@ZnS quantum dots using amphiphilic hyperbranched polyethylenimine *Chem. Commun.* **7** 1735–6

Nekouzadeh A, Pryse K M, Elson E L and Genin G M 2008 Stretch-activated force shedding, force recovery, and cytoskeletal remodeling in contractile fibroblasts *J. Biomech.* **41** 2964–71

Nguyen T T and Fredberg J J 2008 Strange dynamics of a dynamic cytoskeleton *Proc. Am. Thorac. Soc.* **5** 58–61

Nifosí R, Amat P and Tozzini V 2007 Variation of spectral, structural, and vibrational properties within the intrinsically fluorescent proteins family: a density functional study *J. Comput. Chem.* **28** 2366–77

Nirmal M, Dabbousi B O and Brus L E 1996 Fluorescence intermittency in single cadmium selenide nanocrystals *Nature* **383** 802–4

Nishikawa S, Arimoto I, Ikezaki K, Sugawa M, Ueno H and Komori T *et al* 2010 Switch between large hand-over-hand and small inchworm-like steps in myosin VI *Cell* **42** 879–88

Ober R J, Martinez C, Lai X, Zhou J and Ward E S 2004 Exocytosis of IgG as mediated by the receptor, FcRn: an analysis at the single molecule level *Proc. Natl Acad. Sci. USA* **101** 11076–81

Ober R J, Ram S and Ward E S 2004 Localization accuracy in single molecule microscopy *Biophys. J.* **86** 1185–200

Oheim M 2004 A deeper look into single-secretory vesicle dynamics *Biophys. J.* **87** 1403–5

Ohmachi M, Komori Y, Iwane A H, Fujii F, Jin T and Yanagida T 2012 Fluorescence microscopy for simultaneous observation of 3D orientation and movement and its application to quantum rod-tagged myosin V *Proc. Natl Acad. Sci. USA* **109** 5294–8

Palaniappan K, Hackney S A and Liu J 2004 Supramolecular control of complexation-induced fluorescence change of water-soluble, β-cyclodextrin-modified CdS quantum dots *Chem. Commun.* **23** 2704–5

Palaniappan K, Xue C, Arumugam G, Hackney S A and Liu J 2006 Water-soluble, cyclodextrin-modified CdSe-CdS core-shell structured quantum dots *Chem. Mater.* **18** 1275–80

Park H, Toprak E and Selvin P R 2007 Single-molecule fluorescence to study molecular motors *Q. Rev. Biophys.* **40** 87–111

Parthasarathy R 2012 Rapid, accurate particle tracking by calculation of radial symmetry centers *Nat. Methods.* **9** 724–6

Patterson G H and Lippincott-Schwartz J 2002 A photoactivatable GFP for selective photolabeling of proteins and cells *Science* **297** 1873–7

Patterson G, Davidson M, Manley S and Lippincott-Schwartz J 2010 Superresolution imaging using single-molecule localization *Annu. Rev. Phys. Chem.* **61** 345–67

Patwardhan A 2003 Subpixel position measurement using 1D, 2D and 3D centroid algorithms with emphasis on applications in confocal microscopy *J. Microsc.* **186** 246–57

Pavani S R, Thompson M A, Biteen J S, Lord S J, Liu N and Twieg R J *et al* 2009 Three-dimensional, single-molecule fluorescence imaging beyond the diffraction limit by using a double-helix point spread function *Proc. Natl Acad. Sci. USA* **106** 2995–9

Pawley J 2006 *Handbook of Biological Confocal Microscopy* (New York: Springer)

Pelkmans L and Zerial M 2005 Kinase-regulated quantal assemblies and kiss-and-run recycling of caveolae *Nature* **436** 128–33

Peng X *et al* 2000 Shape control of CdSe nanocrystals *Nature* **404** 59–61

Peter S, Harter K and Schleifenbaum F 2014 Fluorescence microscopy *Methods Mol. Biol.* **1062** 429–52

Petryayeva E, Algar W R and Medintz I L 2013 Quantum dots in bioanalysis: a review of applications across various platforms for fluorescence spectroscopy and imaging *Appl. Spectrosc.* **67** 215–52

Pierobon P and Cappello G 2012 Quantum dots to tail single bio-molecules inside living cells *Adv. Drug Deliv. Rev.* **64** 167–78

Pilla V, Munin E, Dantas N O, Silva A C A and Andrade A A 2012 Photothermal spectroscopic characterization in CdSe/ZnS and CdSe/CdS quantum dots: a review and new applications *Quantum Dots—A Variety of New Applications* ed A Al-Ahmadi (London: InTech) pp 3–22

Pinaud F, King D, Moore H P and Weiss S 2004 Bioactivation and cell targeting of semiconductor CdSe/ZnS nanocrystals with phytochelatin-related peptides *J. Am. Chem. Soc.* **126** 6115–23

Plachta N, Bollenbach T, Pease S, Fraser S E and Pantazis P 2011 Oct4 kinetics predict cell lineage patterning in the early mammalian embryo *Nat. Cell Biol.* **13** 117–23

Planchon T A, Gao L, Milkie D E, Davidson M W, Galbraith J A, Galbraith C G and Betzig E 2011 Rapid three-dimensional isotropic imaging of living cells using Bessel beam plane illumination *Nat. Methods* **8** 417–23

Pletnev S, Subach F V, Dauter Z, Wlodawer A and Verkhusha V V 2012 A structural basis for reversible photoswitching of absorbance spectra in red fluorescent protein rsTagRFP *J. Mol. Biol.* **417** 144–51

Prabhat P, Gan Z, Chao J, Ram S, Vaccaro C, Gibbons S, Ober R J and Ward E S 2007 Elucidation of intracellular pathways leading to exocytosis of the Fc receptor, FcRn, using multifocal plane microscopy *Proc. Natl Acad. Sci. USA* **104** 5889–94

Prabhat P, Ram S, Ward E S and Ober R J 2004 Simultaneous imaging of different focal planes in fluorescence microscopy for the study of cellular dynamics in three dimensions *IEEE Trans. Nanobiosci.* **3** 237–42

Quirin S, Pavani S R and Piestun R 2012 Optimal 3D single-molecule localization for super-resolution microscopy with aberrations and engineered point spread functions *Proc. Natl. Acad. Sci. USA* **109** 675–9

Ram S, Kim D, Ober R J and Ward E S 2012 3D single molecule tracking with multifocal plane microscopy reveals rapid intercellular transferrin transport at epithelial cell barriers *Biophys. J.* **103** 1594–603

Ram S, Prabhat P, Chao J, Ward E S and Ober R J 2008 High accuracy 3D quantum dot tracking with multifocal plane microscopy for the study of fast intracellular dynamics in live cells *Biophys. J.* **95** 6025–43

Raupach C, Paranhos-Zitterbart D, Mierke C, Metzner C, Müller A F and Fabry B 2007 Stress fluctuations and motion of cytoskeletal-bound markers *Phys. Rev.* E **76** 011918

Reits E A and Neefjes J J 2001 From fixed to FRAP: measuring protein mobility and activity in living cells *Nat. Cell Biol.* **3** E145–E147

Ritchie K and Kusumi A 2003 Single-particle tracking image microscopy *Meth. Enzymol.* **360** 618–34

Rohr S 2009 Myofibroblasts in diseased hearts: new players in cardiac arrhythmias? *Heart Rhythm.* **6** 848–56

Rossetti R, Nakahara S and Brus L E 1980 Quantum size effects in the redox potentials, resonance Raman spectra, and electronic spectra of CdS crystallites in aqueous solution *J. Chem. Phys.* **79** 1086–8

Ruedas-Rama M J, Walters J D, Orte A and Hall E A 2012 Fluorescent nanoparticles for intracellular sensing: a review *Anal. Chim. Acta* **751** 1–23

Rust M J, Bate M and Zhuang X 2006 Sub-diffraction-limit imaging by stochastic optical reconstruction microscopy (STORM) *Nat. Methods* **3** 793–6

Ruthardt N, Lamb D C and Braeuchle C 2011 Single-particle tracking as a quantitative microscopy-based approach to unravel cell entry mechanisms of viruses and pharmaceutical nanoparticles *Mol. Ther.* **19** 1199–211

Rutter G A and Hill E V 2006 Insulin vesicle release: walk, kiss, pause … then run *Physiology (Bethesda)* **21** 189–96

Saadeh Y, Leung T, Vyas A, Chaturvedi L S, Perumal O and Vyas D 2014 Applications of nanomedicine in breast cancer detection, imaging, and therapy *J. Nanosci. Nanotechnol.* **14** 913–23

Sako Y, Minoghchi S and Yanagida T 2000 Single-molecule imaging of EGFR signaling on the surface of living cells *Nat. Cell Biol.* **2** 168–72

Sambongi Y, Iko Y, Tanabe M, Omote H, Iwamoto-Kihara A and Ueda I *et al* 1999 Mechanical rotation of the c subunit oligomer in ATP synthase (F0F1): direct observation *Science* **286** 1722–4

Sase I, Miyata H, Ishiwata S and Kinosita K Jr 1997 Axial rotation of sliding actin filaments revealed by single-fluorophore imaging *Proc. Natl Acad. Sci. USA* **94** 5646–50

Saxton M J 2008 Single-particle tracking: connecting the dots *Nat. Methods* **5** 671–2

Saxton M J 2009 *Single Particle Tracking Fundamental Concepts of Biophysics* ed T Jue (New York: Humana) pp 147–69

Saxton M J and Jacobson K 1997 Single-particle tracking: applications to membrane dynamics *Annu. Rev. Biophys. Biomol. Struct.* **26** 373–99

Schermelleh L, Heintzmann R and Leonhardt H 2010 A guide to super-resolution fluorescence microscopy *J. Cell Biol.* **190** 165–75

Schlegel G, Bohnenberger J, Potapova I and Mews A 2002 Fluorescence decay time of single semiconductor nanocrystals *Phys. Rev. Lett.* **88** 137401

Schütz G J, Axman M and Schindler H 2001 Imaging single molecules in three dimensions *Single Mol.* **2** 69–74

Shimomura O and Johnson F H 1692 Extraction, purification and properties of aequorin, a bioluminescent protein from the luminous hydromedusan *Aequorea J. Cell. Comp. Physiol.* **59** 223–39

Shore S A and Fredberg J J 2005 Obesity, smooth muscle, and airway hyperre-sponsiveness *J. Allergy Clin. Immunol.* **115** 925–7

Shroff H, Galbraith C G, Galbraith J A and Betzig E 2008 Live-cell photoactivated localization microscopy of nanoscale adhesion dynamics *Nat. Methods* **5** 417–23

Small A and Stahlheber S 2014 Fluorophore localization algorithms for super-resolution microscopy *Nat. Methods* **11** 267–79

Small J V, Rottner K, Kaverina I and Anderson K I 1998 Assembling an actin cytoskeleton for cell attachment and movement *Biochim. Biophys. Acta* **1404** 271–81

Speidel M, Jonas A and Florin E L 2003 Three-dimensional tracking of fluorescent nanoparticles with subnanometer precision by use of off-focus imaging *Opt. Lett.* **28** 69–71

Sprague B L and McNally J G 2005 FRAP analysis of binding: proper and fitting *Trends Cell Biol.* **15** 84–91

Subach F V, Zhang L, Gadella T W, Gurskaya N G, Lukyanov K A and Verkhusha V V 2010 Red fluorescent protein with reversibly photoswitchable absorbance for photochromic FRET *Chem. Biol.* **17** 745–55

Taber L A 2009 Towards a unified theory for morphomechanics *Philos. Trans.* A **367** 3555–83

Terai T and Nagano T 2013 Small-molecule fluorophores and fluorescent probes for bioimaging *Pflugers Arch.* **465** 347–59

Theis-Febvre N, Martel V, Laudet B, Souchier C, Grunwald D, Cochet C and Filhol O 2005 Highlighting protein kinase CK2 movement in living cells *Mol. Cell Biol.* **274** 15–22

Thomann D, Rines D R, Sorger P K and Danuser G 2002 Automatic fluorescent tag detection in 3D with super-resolution: application to the analysis of chromosome movement *J. Microsc.* **208** 49–64

Thompson M A, Casolari J M, Badieirostami M, Brown P O and Moerner W E 2010 Three-dimensional tracking of single mRNA particles in Saccharomyces cerevisiae using a double-helix point spread function *Proc. Natl. Acad. Sci. USA* **107** 17864–71

Thompson R E, Larson D R and Webb W W 2002 Precise nanometer localization analysis for individual fluorescent probes *Biophys. J.* **82** 2775–83

Tiwari D K, Tanaka S, Inouye Y, Yoshizawa K, Watanabe T M and Jin T 2009 Synthesis and characterization of anti-HER2 antibody conjugated CdSe/CdZnS quantum dots for fluorescence imaging of breast cancer cells *Sensors* **9** 9332–64

Tomczak N, Liu R and Vancso J G 2013 Polymer-coated quantum dots *Nanoscale* **5** 12018–32

Toprak E and Selvin P R 2007 New fluorescent tools for watching nanometer-scale conformational changes of single molecules *Annu. Rev. Biophys. Biomol. Struct.* **36** 349–69

Toprak E, Balci H, Blehm B H and Selvin P R 2007 Three-dimensional particle tracking via bifocal imaging *Nano Lett.* **7** 2043–5

Toprak E, Enderlein J, Syed S, McKinney S A, Petschek R G and Ha T *et al* 2006 Defocused orientation and position imaging (DOPI) of myosin V *Proc. Natl Acad. Sci. USA* **103** 6495–9

Trepat X, Deng L, An S S, Navajas D, Tschumperlin D J, Gerthoffer W T, Butler J P and Fredberg J J 2007 Universal physical responses to stretch in the living cell *Nature* **447** 592–5

Tsien R Y 1998 The green fluorescent protein *Annu. Rev. Biochem.* **67** 509–44

Vale R D and Milligan R A 2000 The way things move: looking under the hood of molecular motor proteins *Science* **288** 88–95

Van den Broek B, Ashkroft B, Oosterkamp T H and van Noort J 2013 Parallel nanometric 3D tracking of intracellular gold nanorods using multifocal two-photon microscopy *Nano Lett.* **13** 980–6

van Sark W G J H M, Frederix P L T, Bol A A, Gerritsen H C and Meijerink A 2002 Blueing, bleaching, and blinking of single CdSe/ZnS quantum dots *Chem. Phys. Chem.* **3** 871–9

van Sark W G J H M, Frederix P L T M and den Heuvel D J V *et al* 2001 Letter photooxidation and photobleaching of single CdSe/ZnS Quantum dots probed by room-temperature time-resolved spectroscopy *Phys. Chem.* B **105** 8281–4

Vogel V and Sheetz M 2006 Local force and geometry sensing regulate cell functions *Nat. Rev. Mol. Cell Biol.* **7** 265–75

Wakatsuki T, Kolodney M S, Zahalak G I and Elson E L 2000 Cell mechanics studied by a reconstituted model tissue *Biophys. J.* **79** 2353–68

Wang E C and Wang A Z 2014 Nanoparticles and their applications in cell and molecular biology *Integr. Biol.* **6** 9–26

Wang Z G, Liu S L, Tian Z Q, Zhang Z L, Tang H W and Pang D W 2012 Myosin-driven intercellular transportation of wheat germ agglutinin mediated by membrane nanotubes between human lung cancer cells *ACS Nano* **6** 10033–41

Watanabe T M and Higuchi H 2007 Stepwise movements in vesicle transport of HER2 by motor proteins in living cells *Biophys. J.* **92** 4109–20

Watanabe T M, Fujii F, Jin T, Umemoto E, Miyasaka M and Fujita H *et al* 2013 Four-dimensional spatial nanometry of single particles in living cells using polarized quantum rods *Biophys. J.* **105** 555–64

Watanabe T M, Fukui S, Jin T, Fujii F and Yanagida T 2010 Real-time nanoscopy by using blinking enhanced quantum dots *Biophys. J.* **99** L50–L52

Watanabe T M, Sato T, Gonda K and Higuchi H 2007 Three-dimensional nanometry of vesicle transport in living cells using dual-focus imaging optics *Biochem. Biophys. Res. Commun.* **359** 1–7

Waters C M, Sporn P H, Liu M and Fredberg J J 2002 Cellular biomechanics in the lung *Am. J. Physiol. Lung Cell. Mol. Physiol.* **283** L503–L509

Wells N P, Lessard G A and Werner J H 2008 Confocal, three-dimensional tracking of individual quantum dots in high-background environments *Anal. Chem.* **80** 9830–4

Wells N, Guillaume P, Lessard A, Goodwin P M, Phipps M E, Cutler P C, Lidke D S, Wilson B S and Werner J H 2010 Time-resolved three-dimensional molecular tracking in live cells *Nano Lett.* **10** 4732–7

Werver G 1953 Rotational Brownian motion and polarization of the fluorescence of solutions *Adv. Protein Chem.* **8** 415–59

Wight T N and Potter-Perigo S 2011 The extracellular matrix: an active or passive player in fibrosis? *Am. J. Physiol. Gastrointestinal Liver Physiol.* **301** G950–G955

Wombacher R and Cornish V W 2011 Chemical tags: applications in live cell fluorescence imaging *J. Biophotonics* **4** 391–402

Wu T and Feng J J 2015 A biomechanical model for fluidization of cells under dynamic strain *Biophys. J.* **108** 43–52

Wysocki L M and Lavis L D 2011 Advances in the chemistry of small molecule fluorescent probes *Curr. Opin. Chem. Biol.* **15** 752–9

Yajima J, Mizutani K and Nishizaka T 2008 A torque component present in mitotic kinesin Eg5 revealed by three-dimensional tracking *Nat. Struct. Mol. Biol.* **15** 1119–21

Yildiz A and Selvin P R 2005 Fluorescence imaging with one nanometer accuracy: application to molecular motors *Acc. Chem. Res.* **38** 574–82

Yildiz A, Forkey J N, McKinney S A, Ha T, Goldman Y E and Selvin P R 2003 Myosin V walks hand-over-hand: single fluorophore imaging with 1.5 nm localization *Science* **300** 2061–5

Zemel A, Rehfeldt F, Brown A E, Discher D E and Safran S A 2010a Cell shape, spreading symmetry and the polarization of stress-fibers in cells *J. Phys. Condens. Matter* **22** 194110

Zemel A, Rehfeldt F, Brown A E, Discher D E and Safran S A 2010b Optimal matrix rigidity for stress fiber polarization in stem cells *Nat. Phys.* **6** 468–73

Zemel A and Safran S A 2007 Active self-polarization of contractile cells in asymmetrically shaped domains *Phys. Rev.* E **76** 021905

Zenisek D, Steyer J A and Almers W 2000 Transport, capture and exocytosis of single synaptic vesicles at active zones *Nature* **406** 849–54

Zhang A, Dong C, Liu H and Ren J 2013 Blinking behavior of CdSe/CdS quantum dots controlled by alkylthiols as surface trap modifiers *J. Phys. Chem.* C **117** 24592–600

Zhang Y and Clapp A 2011 Overview of stabilizing ligands for biocompatible quantum dot nanocrystals *Sensors* **11** 11036–55

Part III

The role of actin filaments and intermediate filaments during cell invasion

The process of cell migration is characterized by a polarized cell and plays a role in many physiological and pathological processes such as malignant progression of cancer. The polarization of the cells is provided by the exertion of a protrusive cell front and the cell's retracting trailing edge at its rear end. Mammalian cells possess three cytoskeletal systems such as the actin cytoskeleton, the intermediate filament network, and cell division facilitating microtubules, which all contribute to the regulation of the cellular migration and invasion. The first two are described and discussed in part 3 and the latter is presented in volume 2. The actin cytoskeleton is highly dynamically as it spatially and temporally regulates the protrusion formation, focal adhesions, cellular contraction and the retraction from the cell's leading edge to its rear end. Moreover, intermediate filaments reorganize during cell migration in order to regulate the focal adhesion dynamics, cellular contractility and the nuclear stiffness. Part 3 recapitulates selected current state-of-the-art knowledge on the assembly of the actin and intermediate filamental cytoskeletal systems, includes many recent advances in understanding the mechanical properties of these filaments, how they are kept stable on larger length scales, and represents their regulation of the migratory properties of cells. Cell physiological and pathological processes are based on the regulation and coordination of both mechanical and dynamical properties of the actin or intermediate filament cytoskeleton in order to perform cellular migration and invasion.

IOP Publishing

Physics of Cancer, Volume 1 (Second Edition)
Interplay between tumor biology, inflammation and cell mechanics
Claudia Tanja Mierke

Chapter 7

Role of the actin cytoskeleton during matrix invasion

Summary

Chapter 7 introduces the structure and function of actin as the most abundant protein in cells and tissues from a biophysical point of view. The interaction of actin with actin-binding proteins such as actin regulatory proteins determines the role of actin and the actin filament network in distinct cellular structures such as lamellipodia, filopodia, invadopodia, the cortex and cell–cell adhesive interactions. The interaction of actin and myosin provides the cellular contractility, which is an important feature for the regulation of cancer cell migration and invasion. Moreover, the actomyosin interaction impacts the mechanical phenotype of the cells. In addition to the actomyosin interaction, other actin crosslinking molecules regulate cellular motility through the formation of distinct protrusive structures. In order to reveal the dynamic restructuring of F-actin in living cells, some live-cell imaging approaches are presented and compared to the established F-actin staining approaches for fixed and hence dead cells.

7.1 The actin cell cytoskeleton

The mechanical cues regulate a large variety of cellular functions and processes (Bao and Suresh 2003, Fletcher and Mullins 2010, Janmey and Miller 2011). The ability of a cell to protect itself against deformation, to transport intracellular vesicles from the membrane to cellular compartments (and the inverse) and to alter its shape during cellular migration and invasion is driven by the cytoskeleton, which is an interconnected network of filamentous polymers and regulatory proteins. Both internal and external physical forces impact the mechanical balance of the cytoskeleton by acting on it in order to cause local mechanical property alterations and responsive cellular behavior to mechanical cues. Cells can sense and react to these mechanical alterations by utilizing structural cellular key players such as the extracellular matrix,

doi:10.1088/978-0-7503-1753-5ch7

cytoskeleton and membrane. These key players are able to respond to the mechanical perturbations and hence the characterization of their roles in mechano-transduction and mechano-sensitivity has become the focus of physical biology and oncology (Hoffman *et al* 2011). In particular, the cytoskeleton is the major structural component of the cells providing the mechanical phenotype of the entire cell.

The cytoskeleton consists of three different major filaments, the actin filaments (synonymously termed microfilaments), the intermediate filaments and the micro-tubules. The most prominent component of the cytoskeleton is actin, as it reacts dynamically to mechanical deformation through remodeling of the network structure within a short time period (Brugués *et al* 2010). The actin cytoskeleton needs to act together with the flexible cell membrane as an overall supportive structural network in order to withstand cellular deformations caused by the environment, such as the extracellular matrix or cells and facilitate transmission of extracellular forces through the entire cell (Guolla *et al* 2012). In more detail, the deformation of the membrane causes chemical rearrangements through signal transduction pathways involving protein activation and intracellular signaling events (Dai and Sheetz 1999, DeMali and Burridge 2003, Farsad and De Camilli 2003, Grimm *et al* 2003, Janmey and Weitz 2004, Yang *et al* 2009). The cell membrane is directly connected to the so-called actin cortex and hence this membrane–cortex structure has been demonstrated to protect and guide the cellular mechanical properties (Charras *et al* 2006, Diz-Muñoz *et al* 2010). Additionally, the cortex fulfills a key function in regulating the cell shape during physiological functions of the cells, such as mitosis during cell division, and migration, such as wound healing processes during tissue injury (Kunda *et al* 2008, Diz-Muñoz *et al* 2010). The mechanical properties of the membrane and the cortex are closely coupled by cellular constituents and hence pronouncedly affect one another and subsequently determine how cells can respond to external forces.

The actin protein is a highly conserved and the most abundant protein in nearly all eukaryotic cells. Moreover, the actin protein is known to be involved in most protein–protein interactions than any other protein. Actin is able to polymerize from monomeric globular actin (G-actin) to filamentous (F-actin) structures, which is regulated by nucleotide hydrolysis, ions and a larger number of actin-binding proteins. All these properties reveal that actin acts as a central regulator of several cellular functions such cellular motility, maintenance of the cell's shape and polarity, and transcription activity. Moreover, the interaction of F-actin with myosin II builds the basis for the contraction of muscles. As actin plays a major cellular role, the disruption of the actin cytoskeleton or usage of the cytoskeletal scaffold as a network for invasion by pathogens such as bacteria appear often. In the following, the main two forms of actin are presented and some of the interactions regulating polymerization, filament branching and disassembly of actin are discussed.

The architecture of the actin filaments (microfilaments) is presented in detail in the following sections of this chapter. The polymerization process of microfilaments is described and how a powerstroke is exerted by interaction with filamentous actin (F-actin) and myosin is discussed. In addition, actin-interacting proteins and their function are explored in terms of biochemical and mechanical regulation.

7.2 The actin monomer

Three main actin isoforms are found in vertebrates such alpha-actin of skeletal, cardiac and smooth muscle cells, beta-actin and gamma-actin in non-muscle and muscle cells. The alpha-actin isoform is slightly different in these three cell tissue types. In general actin isoforms can be distinguished only by the difference in a few amino acids normally located at the N-terminus (5′ end) (Herman 1993) (figure 7.1). In addition, actin is post-translationally modified in different ways, such as methylation of the amino acid His73 of skeletal muscle alpha-actin and acetylation and cleaving of N-terminal methionine and cysteine residues.

After the crystal structure of G-actin complexed with DNAse I (Kabsch *et al* 1990), over 80 structures have been reported, in which G-actin without any conformational change is mainly in a complex with actin-binding proteins (ABPs) and small molecules in order to prevent actin polymerization. In more detail, actin is composed of 375 amino acids and folds in two major domains named the alpha and beta domains or outer and inner domains, respectively. Between the two major domains a hinge region exits and leads to the formation of two clefts, the upper binds the nucleotide in association with a divalent cation such as Mg^{2+} and the lower cleft is hydrophobic, representing the major binding site for many ABPs and providing longitudinal contacts between actin subunits in the actin filament (Dominguez 2004, Oda *et al* 2009, Fujii *et al* 2010). G-actin is not detected as an effective ATPase, however, F-actin in crystal structures is detected with bound ATP.

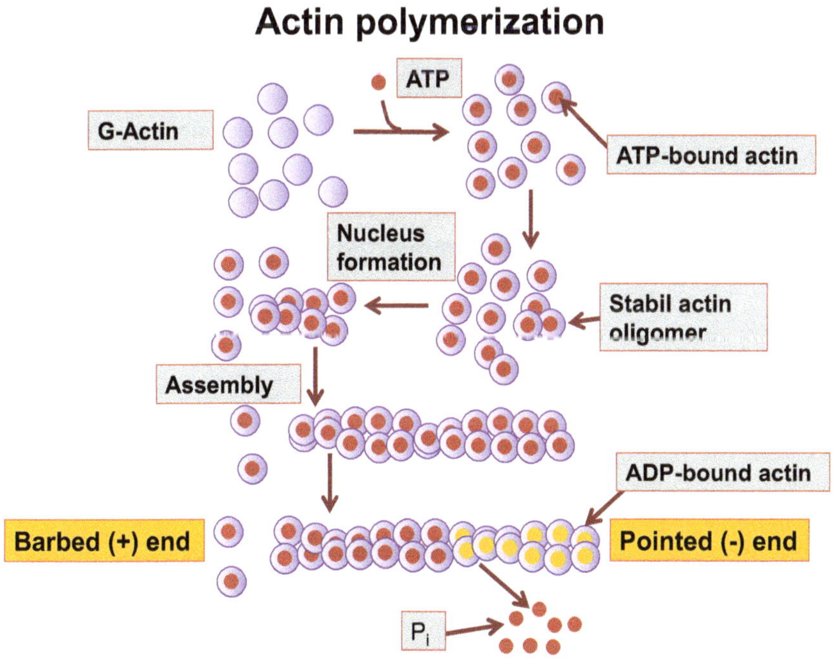

Figure 7.1. G-actin polymerization to actin filaments.

When F-actin binds ADP, it has a decreased stability due to conformational changes. As the nucleotide is located directly at the interface between the two major alpha and beta domains of actin, the hydrolysis of the nucleotide with the release of the phosphate leads to a weakening of the connection between the two major actin domains, which rotate then more freely.

7.3 The actin filaments and polymerization

The transition between G-actin and F-actin is facilitated by the hydrolysis of the ATP nucleotide, as the ATP bound state has a higher stability than the ADP bound state (figure 7.1). In *in vitro* experiments, G-actin monomers carrying ATP bind to the fast-growing barbed end (also called the + end) of the actin filament (microfilament). The hydrolysis of ATP occurs in the actin filament and bound actin monomers carrying ADP dissociate faster from the pointed end (also called the – end). This mechanism creates a steady-state of actin polymerization/depolymerization and is called the treadmilling of actin filaments (Wegner and Isenberg 1983) (figure 7.2).

The structure of G-actin and the actin subunit bound in F-actin are different. To date, it has not been possible to crystallize actin filaments and no G-actin structures can serve as a model for the conformation of an actin monomer within the actin filament. The structural details of actin filaments have been provided by x-ray fiber diagrams of orientated F-actin gels and scanning electron microscopy.

The latter method revealed that negatively stained actin fibers consist of two actin chains that are turned around each other to assemble a right-handed, two-chained helix (Hanson and Lowy 1963). The symmetry of the actin filament is a single left-handed genetic helix with approximately thirteen molecules repeating every six turns.

The individual actin monomers are globular proteins and consists of 375 amino acids (43 kD). The globular (G) actin monomer possesses tight binding sites that

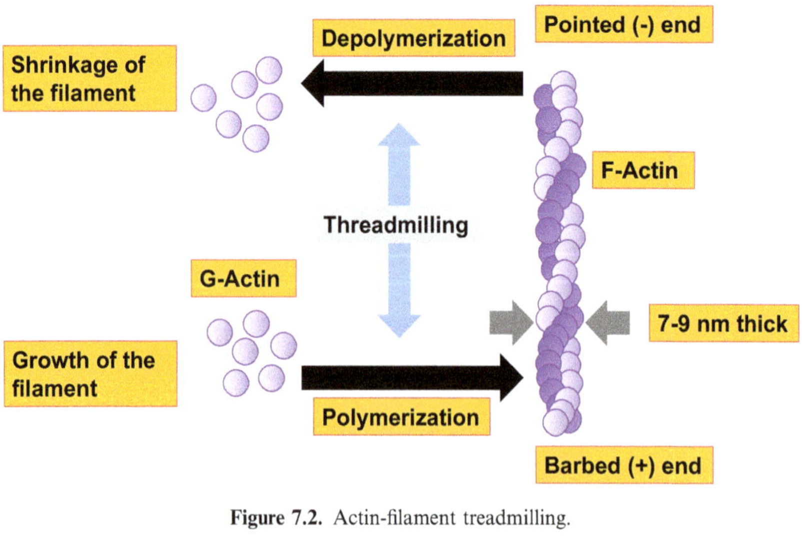

Figure 7.2. Actin-filament treadmilling.

facilitate head-to-tail interactions with two other actin monomers and hence actin monomers polymerize in order to assemble actin filaments consisting of filamentous (F) actin. Within the filaments each monomer is rotated by 166° and therefore they appear as a double-stranded helix. As all actin monomers are oriented in the same direction, the actin filaments exhibit a specific polarity and both ends of the filaments are different. One end is termed the plus or barbed end and the other one is called the minus or pointed end. The polarity of the actin filaments is necessary for both their assembly and in the establishment of a unique direction of the myosin movement on the actin filaments.

The ionic strength of actin solution is critical for the polymerization of actin filaments in *in vitro* assays. In particular, in solutions with low ionic strength, the actin filaments start to depolymerize to monomers. When the ionic strength is increased to physiological levels, actin polymerizes spontaneously to actin filaments. In more detail, the initial step in the polymerization of actin is the aggregation of three actin monomers to a small group, which is termed the nucleation process. The next step is the elongation phase, in which the actin filaments can grow by the reversible addition of monomers to the plus and the minus ends. However, the plus end elongates ten times faster than the minus end, which causes imbalanced growth of the two actin filament ends. For the filament elongation, actin monomers bind ATP, which is hydrolyzed to ADP after filament assembly. ATP is not essential for the polymerization of actin, however, actin monomers with bound ATP polymerize more easily than actin monomers with bound ADP. Hence the binding of ATP and its hydrolysis are important for facilitating the assembly of actin filaments and the remodeling of their dynamics. The polymerization of actin is reversible and, indeed, the actin filaments depolymerize by the dissociation of actin subunits, which enables the cells to rapidly decrease the amount of actin filaments under distinct conditions. Subsequently, an obvious equilibrium state is established between actin monomers and filaments that is regulated by the concentration of free actin monomers. In more detail, the integration rate of actin monomers into filaments is highly dependent on the concentration, as there exists a critical actin-monomer concentration, where the rate of the actin-monomer polymerization into filaments equals the rate of actin-filament dissociation. Thus, at this specific critical concentration, the actin monomers and filaments are in obvious equilibrium.

Two distinct types of actin structures are assembled by individual actin filaments that are termed actin bundles and actin networks and fulfill specific role within a cell (figure 7.3). Within actin bundles, the actin filaments are orientated in parallel arrays that are closely packed and crosslinked, whereas in networks the actin filaments are arranged in orthogonal arrays building three-dimensional meshes—these scaffolds are loosely crosslinked and possess properties of semisolid gels. The network formation of these actin-filament structures is regulated by a distinct set of actin-binding proteins that facilitate the connection of actin filaments in specific regular patterns. All actin-binding proteins facilitating the crosslinking of actin filaments possess at least two domains for the binding actin, which enables them to interact with actin filaments and thereby to finally crosslink them. In particular, the kind of association between the actin filaments depends on the size and structure of the

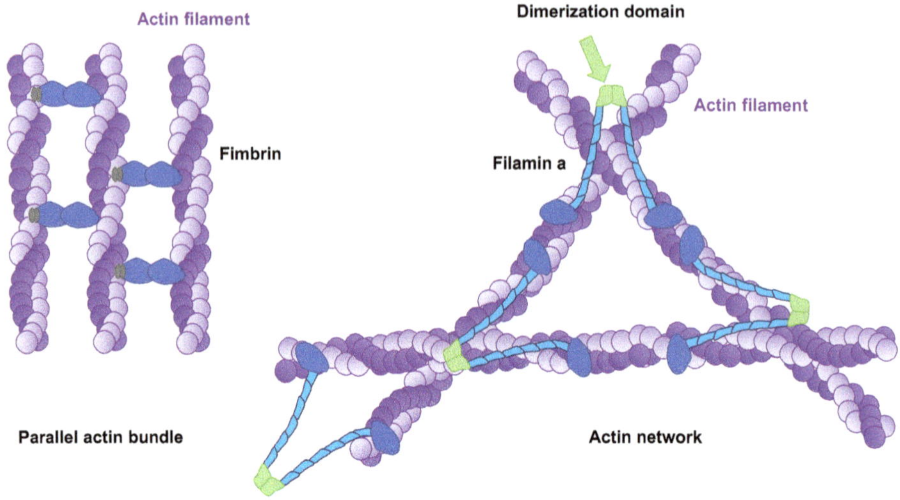

Figure 7.3. Actin bundles ain actin networks such as parallel actin bundles (left) or scaffold-like actin networks (right).

actin-crosslinking molecules (figure 7.3). The actin crosslinking proteins connecting actin filaments to bundles are termed actin-bundling proteins. They are usually small rigid proteins that hold actin filaments closely aligned to each other. In contrast, the actin-crosslinking proteins that link actin filaments to bulky networks are usually large flexible proteins, which are able to crosslink perpendicular actin filaments. In more detail, these actin-crosslinking proteins are composed of modular proteins consisting of structural units. Their actin-binding domains are usually very similar in structure, as these domains are interrupted by specific spacer sequences of varied length and flexibility. The alterations within the spacer sequences prepare the different actin-binding proteins with distinct crosslinking properties.

Dependent on the different actin-bundling proteins, two structurally and functionally distinct types of actin bundles are assembled. The first type of actin bundle consists of closely spaced actin filaments, which are aligned parallel and build the protrusions of the cell membrane such as microvilli. Within these bundles, all the actin filaments possess the same polarity, as their barbed (or plus) ends are located toward the cell membrane. Fimbrin is one example of this type of bundling protein, and is involved in the formation of these structures, and was initially isolated from intestinal microvilli and then also detected in surface perturberances of a broad variety of cell types. In more detail, fimbrin is composed of two adjacent actin-binding domains and has a molecular weight of 68 kD. Fimbrin binds as a monomer to actin filaments and thereby facilitates the close connection between two parallel filaments. An example for the second type of actin-bundling protein is alpha-actinin, which generates more loosely connected networks with spaces for the embedding of additional regulatory molecules. Hence these networks are able to generate contractions such as the actin bundles of the contractile ring structure during the division of cell in at the end of the mitosis phase. The looser structure within these so-called contractile bundles reflects the nature of the actin crosslinking protein alpha-actinin. Contrarily to fimbrin, alpha-actinin, which is 102 kD protein, binds to actin filaments solely as a dimer and each alpha-actinin within the dimeric structure has a single actin-binding site. Thus, the actin filaments that are connected by the alpha-actinin dimer are far apart from those actin filaments connected by fimbrin, as the distance between adjacent crosslinked actin filaments is 40 nm instead of 14 nm, respectively. Moreover, the increased distance between the actin filaments enables motor proteins such as myosin to interact with actin filaments within these rather loosely connected bundles and subsequently, myosin acts on the actin filaments to contract them.

In networks the actin filaments are connected by large actin-binding proteins (ABPs) such as filamin. Filamin (also called ABP-280) also binds actin as a dimer that consists of the 280 kD subunits of the filamin monomers. In more detail, the actin-binding domains and the dimerization domains are positioned at opposite ends of each dimer subunit. The filamin dimer forms a flexible V-shaped molecule containing actin-binding domains at the two ends of the dimer and subsequently, filamin connects orthogonal actin filaments through crosslinks building a loose three-dimensional network. These actin-filament networks are located underneath the cell membrane and thereby provide a structural support for the surface of the cell and its overall shape.

7.4 Actin structures: protrusions and cell cell junctions

The cells utilize their cytoskeletons to fulfill several functions that are needed under normal physiological conditions and also observed under pathological conditions such as movement, polarization, cell division and maintenance of the entire organization of multicellular tissues. In particular, the actin monomer is highly conserved and represents an essential basic structural element of the cytoskeleton assembling cable-like structures and long bars, which underlie continuous remodeling through over 100 diverse actin-binding proteins. Cells require an ongoing initiation of new actin filaments and their subsequent organization into membrane perturberances, cytoskeletal networks or stress fibers represents a key feature for the development of specialized cellular structures such as stress fibers (termed elastic contractile bundles), lamellipodia (termed sheet-like protrusions), filopodia (termed spike-like protrusions), microvilli (termed finger-like surface protrusions) and invadopodia (termed small invasive cell feet). The cytoskeleton is necessary for fulfillment of normal cellular function, whereas under pathological conditions such as cancer, the cytoskeleton of cancer cells can be altered and hence lead to dramatic changes in cell proliferation and growth rates, cellular stiffness, cellular movement and invasiveness into healthy tissue surrounding a tumor—all of which represent hallmarks of cancer (Hanahan and Weinberg 2011). What impact does actin-filament bundling have on cellular structures and functions? How are alterations in the protein activity or expression levels of actin-bundling proteins involved in the initiation of neoplasms, primary tumor formation and the malignant progression of cancer?

The actin protein is among of the most abundant proteins in eukaryotic cells and is a key player in supporting the compartmentalization within the cytoplasm and the entire motility of the cell. Moreover, actin can interact with myosin to transport cargo such as molecules and small vesicles through the cell's cytoplasm. Actin filaments are generated by the assembly of actin filaments to higher structural arrangements such as superstructures, facilitated by actin-filament-bundling proteins. Among these actin-bundling proteins are two major groups. The first group of actin-bundling proteins, such as fascin and alpha-actinin, builds parallel bundles of actin, whereas the second group of actin-bundling proteins, such as filamin, forms looser orthogonally arranged structural networks (Bartles 2000). Actin crosslinking proteins have at least two actin-binding sites and as they are mostly dimerized, the position of the actin-binding sites regulates the actin-filament arrangement and even the type of crosslinked structure assembled. In more detail, actin filaments are polarized, as they possess a fast-growing (plus and barbed end) and a slow-growing end (minus and pointed end). The maintenance of the polarity is driven by the hydrolysis of ATP (Pollard and Cooper 2009). Actin-bundling proteins can orientate in a specific manner to the actin filaments in order to regulate through this binding the distinct formation of bundles such as the creation of uniform or mixed polarized actin bundles (figure 7.3). In particular, actin-bundling proteins are usually modular and consist of repeated actin-filament-binding domains such as the calponin-homology domains (CH domain), gelsolin domains and spectrin domains.

Actin-bundling proteins fulfill an important role in cancer metastasis

In order to metastasize, malignant and aggressive cancer cells utilize actin bundles to extrude protrusions for the breakdown the basement membrane of the primary tumor and their invasion into the surrounding tumor microenvironment. After transmigration in the vasculatory system or lymphatic system, these cancer cells find their targeted new niche and form a new, secondary tumor that remains dormant for months or years and can finally start to grow aggressively upon an unknown signaling event (Hanahan and Weinberg 2011). During metastasis, cancer cells need to control their motility and adapt their adhesiveness to the requirements of their microenvironment, similar to embryonic cells during morphogenesis (Hanahan and Weinberg 2011, Roussos *et al* 2011a). In more detail, the actin cytoskeleton fulfills the role of a scaffold for signal transduction processes, provides coupling to the extracellular matrix microenvironment and functions a mechano-sensor. Does the actin bundling promote or impair the process of cancer metastasis? This question remains unanswered. However, it seems to be the case that distinct cancer cells adapt their actin bundling activity to facilitate the regulation of signal transduction processes, cellular growth, adhesion and cellular mechanical properties. Hence, the actin-bundling amount seems to be crucial for cell survival during the different phases of malignant cancer progression and subsequently metastatic spread. The mechanical stiffness usually seems to be positively correlated with the invasion and the metastatic potential (Mierke *et al* 2008, Narumiya *et al* 2009, Mierke *et al* 2011, Mierke 2013), whereas many exceptions still remain (Guck *et al* 2005, Swaminathan *et al* 2011).

The actin cytoskeleton keeps the compartmentalization within the cytoplasm and hence is a major player in providing the cell polarity. The polarity is required for normal tissue homeostasis. When the polarity is dramatically altered, the cell–cell adherens junctions may be ruptured and an epithelial to mesenchymal transition (EMT) seems to be likely to occur (Royer and Lu 2011). Moreover, physiological processes such as cell divisions are themselves polarized within tissues. If the polarity is impaired, the tissue integrity may break down which may then cause overgrowth, an aberrant invasive behavior and subsequently the initiation of neoplasms and solid tumors (Royer and Lu 2011). Actin bundling is a major contributor to the polarity of epithelial cells, as this process provides the integrity of cell–cell adherence junctions, tight junctions and protrusive structures such as microvilli and also the polarized membrane trafficking of vesicles or molecules. Finally, the knowledge of the underlying mechanisms converting the actin bundling to a metastasis promoting tool seems to be essential for impairing malignant progression of cancer.

Actin-bundle structures in normal 'healthy' and cancer cells

The processes of invasion and metastasis are facilitated by a deregulated cellular motility of malignant cancer cells (Bravo-Cordero *et al* 2014, Martin *et al* 2014). The alterations in cellular motility are based on certain ABPs that determine the assembly, function and remodeling dynamics of the actin cytoskeleton (Weaver 2008, Albiges-Rizo *et al* 2009, Gau *et al* 2015, Madsen *et al* 2015). What is the molecular basis of these cytoskeletal alterations and hence of the malignant transformation of specific proteins such as the ABPs? The roles of distinct ABPs in this process have been reported (Albiges-Rizo *et al* 2009, Madsen *et al* 2015, Gau *et al* 2015, Shishkin *et al* 2017) and some have even been selected as tumor markers (Wang *et al* 2007, Shishkin *et al* 2011, Huang *et al* 2014). The integrity of several cellular structures is based on actin bundles and the distinct actin-bundling proteins that seem to fulfill a role in the initiation and malignant progression of cancer.

Cortex

Directly below the cell membrane is a thin layer consisting of an actin-filament network and crosslinking proteins, which is called the cortex. The cortex provides a strong attachment of the actin cytoskeleton to the cell membrane that is maintained by a balanced contractility between actin and myosin, facilitating the protrusive, mesenchymal motility of the cells. In contrast, a weaker attachment of the actin cytoskeleton to the cortex without the involvement of the force of actin- and myosin-based contractility supports the exertion of membrane blebs, which represent detachment of the cell membrane form the underlying cortex, and subsequently cause blebbing migration type of the cells (Friedl and Wolf 2010). Changes in the cortical stiffness affects how the cells migrate through and into diverse microenvironments. Hence, the cortex represents a structural scaffold for the organization of transmembrane receptors and glycoproteins into crosslinked networks for providing effective signal transduction processes and ensuring the mechanical coupling between external mechanical stresses and intracellular signals.

The two non-muscle myosins IIa and IIb are the main actin-based contractile myosin motors crosslinking the actin filaments of the cell cortex and thereby regulating the cell stiffness. In particular, phosphorylation of the myosin II light chain activates the contractile activity of myosin II (Narumiya *et al* 2009). In line with this, increased levels of the myosin kinase Rho-associated protein kinase 1 (ROCK1) correlate positively with poor survival rates in breast cancer patients (Lane *et al* 2008) and the malignant progression of bladder cancer, such as poor tumor differentiation, muscle invasion by cancer cells and lymph node metastasis (Kamai *et al* 2003). A clinical study reported a positive correlation between the myosin light chain kinase facilitating the activation of myosin II with the recurrence of cancer such as cancer metastasis in non-small cell lung cancer (Minamiya *et al* 2005). There are still specific reagents for the detection of the active myosin II or myosin II isoforms required to perform defined clinical investigations (Vicente-Manzanares *et al* 2009). In addition, primary tumors affect the contractile properties of neighboring stromal cells such as fibroblasts (Wyckoff *et al* 2006, Gaggioli *et al* 2007, Sanz-Moreno *et al* 2008). In turn, enhanced contractility of stromal cells causes increased matrix stiffness and thereby additionally promotes tumor growth and malignant progression (Samuel *et al* 2011). Moreover, the matrix stiffness induces enhanced integrin adhesion, signaling and the activation of survival and growth factor signals such as the activation of focal adhesion kinase (FAK), all of which further prompt tumor progression (Frame *et al* 2010).

As myosin II builds parallel contractile actin bundles, the crosslinking protein filamin represents a long, hinged actin bundler facilitating mechanical strength and signal transduction processes in actin scaffolds near membranes (Popowicz *et al* 2006), similarly alpha-actinin cross-links actin filaments into largely spaced actin bundles that enable myosin II interaction with these actin filament bundles (figure 7.4). Indeed, filamins are mechano-sensory proteins, that regulate transcription, membrane trafficking, the function of ion channels, cell adhesion and receptor-facilitated signal transduction processes (Popowicz *et al* 2006). In more detail, filamin is located within a complex with the androgen receptor and the $\beta 1$

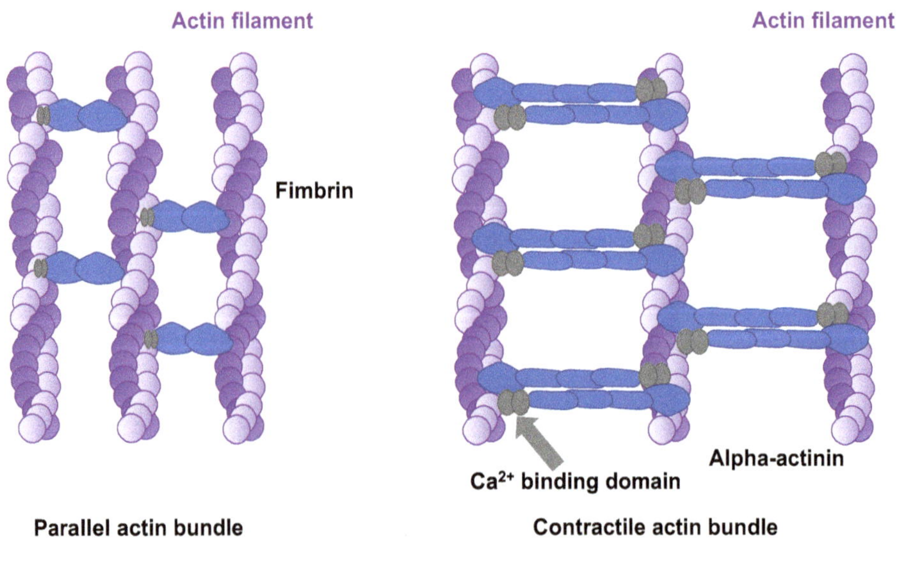

Figure 7.4. Parallel actin bundle and contractile actin bundle.

integrin subunit and regulates cellular motility based on downstream androgen signal transduction processes driving the invasion in prostate cancer (Loy *et al* 2003, Castoria *et al* 2011). Filamins can build a complex with a pro-prion protein PrP in pancreatic cancer to provide a growth advantage for cancer cells (Sy *et al* 2010) and the malignant progression of melanoma (Li *et al* 2010a). In addition, filamin alters the hepatocyte growth factor receptor (HGFR or synonymously the proto-oncogene Met) signal transduction processes that are critical for many epithelial cancers to undergo malignant transformation and hence to metastasize (Zhou *et al* 2011). Parts of the nuclear skeleton may be built by filamins, where they interact on the one hand with DNA repair complexes such as breast cancer type 1 susceptibility protein (BRCA1) (Velkova *et al* 2010) and on the other hand with cell cycle progression proteins such as cyclin D1 (CCND1) (Zhong *et al* 2010). Moreover, a secreted form of filamin has been detected in blood samples of patients with advanced metastatic breast cancer and astrocytomas supporting the hypothesis that filamin can serve as a tumor maker (Alper *et al* 2009).

Spectrins represent another class of actin crosslinking molecules within the cell cortex and have also been associated with cancer. In particular, in colorectal and pancreatic cancers, the β2-spectrin associates with transcriptional activators such as SMAD3 and SMAD4 of the transforming growth factor β (TGF-β) and thereby regulates their activity and the signal transduction pathway. TGF-β signaling performs normally the function as a tumor suppressor of colorectal cancer by suppressing growth and inducing apoptosis, whereas an impairment of β2-spectrin causes a dysregulation of TGF-β, as the Wnt signaling is activated and hence tumorigenesis is supported (Thenappan *et al* 2009, Jiang *et al* 2010). Additionally, embryonic spectrin is decreased in certain cancers and leading to a deregulation of cyclin D1 and enhanced cell cycle progression (Kitisin *et al* 2007).

Cell–cell adherence junctions

The epithelial cell layers are strongly connected by adherence junctions that contain transmembrane cadherin cell surface receptors facilitating the direct interaction of neighboring cells and intracellularly coupling of the two actin cytoskeletons (Etienne-Manneville 2011). In more detail, adherens junctions couple the cell cortex to the actin-filament bundles, which are kept at their individual location by actin-bundling proteins such as alpha-actinin and myosin II (Etienne-Manneville 2011). When these epithelial cells become cancerous and hence malignant, the adherens junctions are disassembled and enable beta-catenin to translocate from the membrane location to the nucleus, where it activates transcriptional alterations leading subsequently to EMT through the canonical Wnt signal transduction pathway (Heuberger and Birchmeier 2010). In addition, the breakdown of intercellular junctions enables cancer cells to break out of the primary solid tumor and migrate into the surrounding tissue microenvironment.

Alpha-actinin-1 and alpha-actinin-4 are localized to cell–cell contacts (Gonzalez *et al* 2001) to facilitate the regulation of the actin bundling and maintain epithelial integrity. Moreover, alpha-actinin-4 binds tight junctions and thereby recruits the tight junction proteins such and subsequently promotes the formation of tight

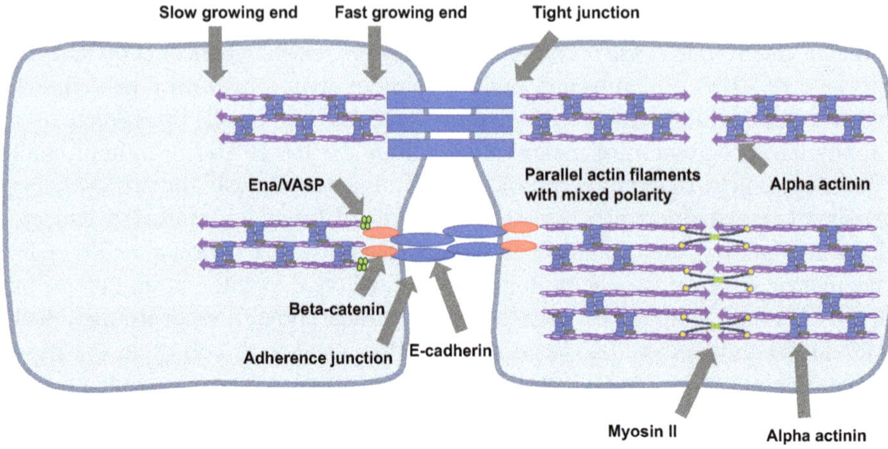

Figure 7.5. Actin network at intercellular cell–cell junctions.

junctions (Nakatsuji *et al* 2008). The tight junctions are located more apical to adherens junctions (figure 7.5) and provide the maintenance of the impermeability within epithelial tissues. A loss of alpha-actinin-4 causes a disintegrity of tight junctions and hence is correlated positively with the malignant progression of cancer such as invasion of cancer cells and metastases formation in targeted organs (Nakatsuji *et al* 2008). Increased levels of alpha-actinin-4 are associated with poor outcome or advanced disease state (Honda *et al* 1998, Patrie *et al* 2002, Honda *et al* 2004, Menez *et al* 2004, Honda *et al* 2005, Weins *et al* 2007, Yamamoto *et al* 2007, Kikuchi *et al* 2008, Welsch *et al* 2009, Yamamoto *et al* 2009, Yamada *et al* 2010) and hence its role in tight junction assembly remains not yet fully understood. However, the function of alpha-actinin-4 in the cell's leading edge protrusion may even support the migration of cells and subsequently promote cancer metastasis (Honda *et al* 1998).

7.4.1 Filopodium

The filopodia are membrane perturberances that represent thin, long, actin-based protrusions, which facilitated cell migration and promote the invasion of cancer cells (Mattila and Lappalainen 2008, Nürnberg *et al* 2011) (figure 7.6). The actin-bundling protein fascin generates parallel actin filaments in filopodia and is usually expressed in cells of mesenchymal and neural origin rather than epithelial origin (Adams 2004a, 2004b, Hashimoto *et al* 2011). However, the expression of fascin is found to be increased in epithelial cancers and is supposed to be involved in the invasion of cancer cells and cancer metastasis (Machesky and Li 2010). Fascin-facilitated actin bundling causes a strengthening of actin filaments and thereby prolongs the lifetime of filopodia acting as invasive protrusions (Li *et al* 2010b). In line with this, increased expression of fascin has been detected at the invasive front of primary solid tumors and additionally the *in vitro* decrease of fascin levels leads to impaired cellular motility and invasiveness (Hashimoto *et al* 2005, Hashimoto *et al* 2007, Li *et al* 2010b, Schoumacher *et al* 2010). Formins such as mammalian Diaphanous-related (mDia) proteins also belong to the group of filopodial proteins

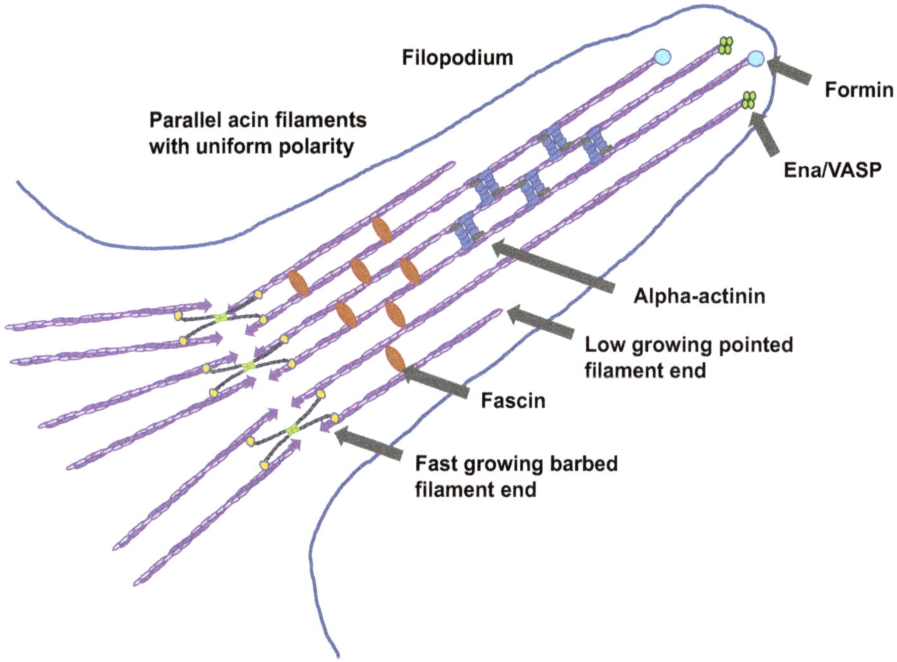

Figure 7.6. Filopodium.

and exhibit actin-nucleating as well as actin-bundling activity. In particular, the actin-binding FH2 domains of mDia1, mDia2 and mDia3 can dimerize, nucleate at both sites and perform actin bundling (Machaidze *et al* 2010). However, the role of the mDia proteins is still not well understood, similarly to the role of the other 12 mammalian formins during the initiation or progression of cancer (Nürnberg *et al* 2011). In general, the mDia formin family of Rho-effector proteins create linear actin filaments (F-actin) and regulate the modulation of microtubule dynamics in order to induce and provide the maintenance of cellular polarity (Chesarone *et al* 2010). The GTP-bound Rho can activate mDia family members through the impairment of interaction between the DID and DAD autoregulatory domains of the formins, which subsequently exposes the FH2 domain for remodeling interaction with actin and for providing the microtubule dynamics (Lash *et al* 2016).

The Ena/VASP proteins such as Mena, VASP and Evl in mammals belong to a family of proteins inducing the polymerization of actin and subsequently the formation of actin bundles. These proteins are located within filopodial tips, lamellipodia and cadherin-based cell–cell adherence junctions (Scott *et al* 2006, Breitsprecher *et al* 2008, Breitsprecher *et al* 2011) as well as in focal adhesions (Pula and Krause 2008). A splice variant of Mena, which is termed MenaINV, has been reported to be upregulated in breast and colorectal cancers, supporting the malignant progression of the tumor (Di Modugno *et al* 2004). In line with this, Mena deficiency reduces cell invasion, impairs cancer metastasis, slows down the progression of the tumor in polyoma middle-T transgenic mouse models and impairs normal breast development (Roussos *et al* 2010, 2011a, 2011b, 2011c).

7.4.2 Lamellipodium

Cellular processes such as cell migration and invasion as well as intercellular communication require actin-containing perturberances on the cell surface such as lamellipodia and filopodia. Lamellipodia represent networks of actin filaments, which are created by filament branching through the Arp2/3 complex (figure 7.7). The impairment of actin branching seems to diminish the formation and maintenance of lamellipodial actin network structures. In contrast, the regulation of nucleation or elongation of Arp2/3-independent filament subpopulations within the lamellipodial network, such as by formins or Ena/VASP family members, as well as

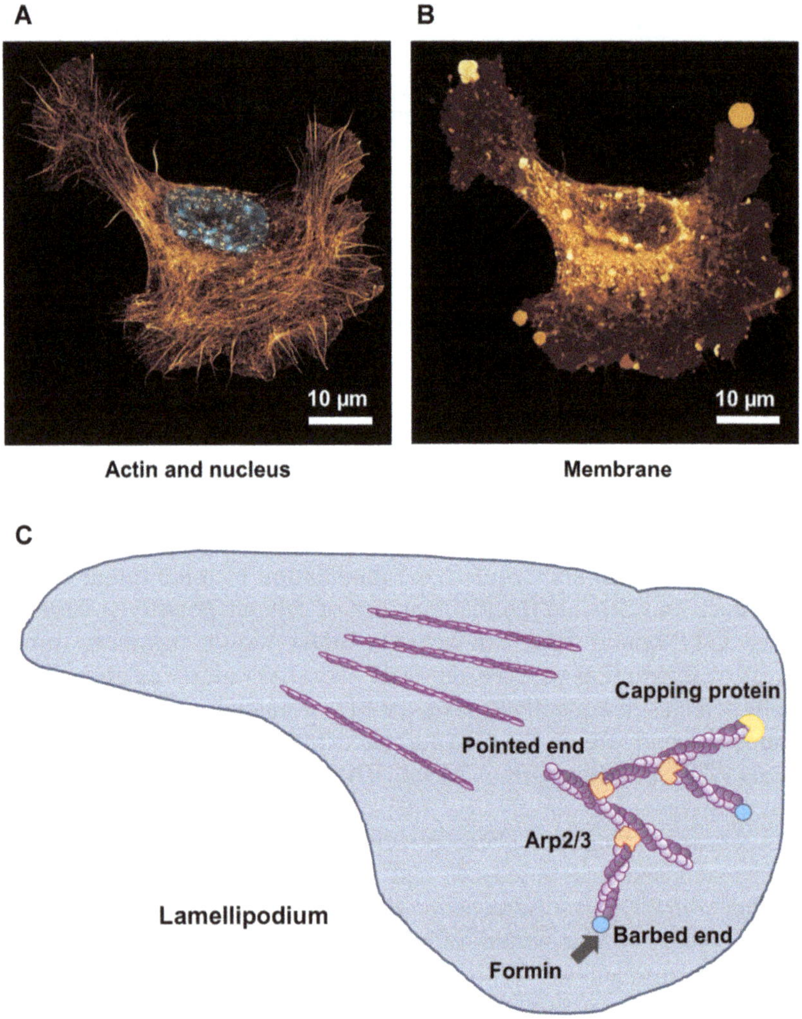

Figure 7.7. Lamellipodium. (a) A fluorescence microscopic image of a fibroblast stained with Alexa Fluor546 and Hoechst 33342 (nucleus). (b) The same cell stained with the membrane fluorescent dye DID. (c) Simple schematic drawing of the lamellipodial structure.

their contribution to the effectiveness of protrusion formation still needs more investigation and remains unclear. Several distinct formin fragments and VASP have been investigated regarding their effects on site-specific, lamellipodial versus cytosolic actin assembly and the consequences on protrusion formation. Unexpectedly, it has turned out that the expression of formin variants but not VASP diminished lamellipodial protrusion formation in B16-F1 cells to various extents (Dimchev *et al* 2017). The actin network polymerization rates display a similar trend. However, the degree of inhibition of both parameters (lamellipodial protrusion formation and actin polymerization rates) depend on the extent of cytosolic but not lamellipodial actin assembly. In line with this, an increased assembly of cytosolic actin impairs the actin monomers from being rapidly translocated and efficiently incorporated into lamellipodial protrusive structures (Dimchev *et al* 2017). Instead of exclusive regulation by actin polymerases, which function at the tips of actin filaments, the efficiency of the lamellipodial protrusion formation is precisely regulated by the delicate balance between the actin assembly in lamellipodial and cytosolic structures (Dimchev *et al* 2017).

The actin polymerization can additionally generate forces through the simple stochastic insertion of actin monomers to the barbed ends of actin filaments in actin bundles or actin networks, such as the protrusive tips of lamellipodia and filopodia (Small *et al* 2002). Essential regulators of the assembly and disassembly of actin filaments in the cells' leading edge protrusions such as lamellipodia have been revealed. Among these regulators are the branching activity possessing Arp2/3 complex, the Scar/WAVE complex regulating the Arp2/3 complex, and the Rac subfamily GTPases regulating the activation of the Scar/Wave complex and hence controlling the entire lamellipodia formation (Campellone and Welch 2010, Ridley 2011, Steffen *et al* 2014). The actin remodeling within the lamellipodium has been largely analyzed (Watanabe and Mitchison 2002, Iwasa and Mullins 2007, Lai *et al* 2008), whereas the precisely balanced homeostasis of actin assembly and disassembly of the entire cell, such as the more proximal and cytoplasmic regions and the impact on the assembly/disassembly of actin in protrusion, has not yet received similar attention. The proximal, cytoplasmic actin structures should receive more attention, as it has indeed been found that specific subcellular actin domains are able to affect the actin assembly and disassembly by one another. Moreover, competition has even been reported between different actin structures that are assembled either by the Arp2/3 complex or the formins, which represent the two main actin assembly factors (Rotty *et al* 2015, Suarez *et al* 2015).

In 2D environments, the lamellipodia are extensively described as flat networks of actin filaments, which are connected to their substratum through focal adhesions, but can migrate through lifting the membrane outermost region upwards and backwards, which is termed membrane ruffling. In 3D or in suspension, the lamellipodia protruding from the apical cell surfaces or from the surface of rounded cells are also termed membrane ruffles. The generation of actin filaments within lamellipodia, such as the branching of new actin filaments facilitated by the Arp2/3 complex, occurs at the interface between the cell membrane and the actin network (Wang 1985, Iwasa and Mullins 2007, Lai *et al* 2008). At these distinct sites various actin assembly factors such as the Arp2/3 activator WAVE and associated WAVE complex subunits (Innocenti *et al* 2004, Steffen *et al* 2004), formins such as FMNL2 and 3 (Block *et al* 2012, Kage *et al* 2017) and Ena/VASP family members (Rottner *et al* 1999, Svitkina *et al* 2003) can be detected at increased expression levels. The Arp2/3 complex and its ongoing activation seems to be an essential component of the initiation and the maintenance of lamellipodia (Suraneni *et al* 2012, Wu *et al* 2012, Koestler *et al* 2013). However, the function of each regulator for the actin-filament nucleation or elongation is not yet clearly understood. Similarly to the dendritic nucleation model of a branched actin network assembly and disassembly (Mullins *et al* 1998, Pollard and Borisy 2003), actin filaments are assembled and continue to elongate until they are capped. However, there is not much known about the relative filament numbers that grow and push in the lamellipodium compared to capped and/or depolymerizing filaments therein. Even the relevance of the ratio of actively growing actin filaments to non-productive actin filaments in lamellipodia (or in the entire cell) and their contribution to protrusion efficiency is not yet clear. Many questions remain to be answered, such as the following: what is the extent of cooperation between individual actin assembly factors such as the Arp2/3 complex, Ena/VASP or the formin family polymerases within lamellipodia?

It has been reported that formin family members seem to function both upstream, as mDia1 and mDia2 generate template filaments for Arp2/3-dependent branching (Yang *et al* 2007, Isogai *et al* 2015), and downstream or in parallel to the Arp2/3 complex driven actin-filament branching (Block *et al* 2012, Krause and Gautreau 2014, Kage *et al* 2017). Specific formins or Ena/VASP family members are each supposed to fulfill a distinct function, however, it still needs to be revealed what kind of specific activities contribute to lamellipodia formation based on their branching and/or filament capping functions. Ultimately, the efficiency of protrusion formation seems to depend on the availability of actin monomers. A limitation of actin monomers seems to be unlikely as the actin concentrations (Koestler *et al* 2009) and the actin diffusion rates (McGrath *et al* 1998, Lai *et al* 2008) *in vivo* are relatively high.

How can a spatio-temporal control of actin monomers at the actin polymerization sites be achieved? Control of the actin polymerization may be achieved by a biased forward translocation (Zicha *et al* 2003) and cell edge accumulation of actin (Vitriol *et al* 2013), similar to a sink of actin monomers for possibly intra-lamellipodial recycling processes with short traveling distances for optimized repolymerization of monomers (Cramer 1999, Vitriol *et al* 2015). However, the relevance of each individual mechanism needs to be figured out for providing effective protrusion exertion under different conditions and in distinct cell types. Although not mutually exclusive, the respective relevance of each mechanism is still not yet clear. Indeed, a straightforward, gain-of-function approach in migrating B16-F1 melanoma cells and in fibroblasts has been utilized to investigate the effects of formin variant or VASP overexpression on lamellipodium formation, such as the speed of the protrusion formation and the actin network polymerization rates and the Arp2/3 complex association to these structures. However, only very weak correlation between the amount of lamellipodial actin filaments and the efficiency of protrusion formation has been observed at each time point, whereas instead the amount of cytosolic F-actin assembly triggers the effective lamellipodium protrusion exertion (Dimchev *et al* 2017).

7.4.3 Microvilli

Microvilli are finger-like perturberances of the cell membrane and dramatically enhance the overall surface area of cells, which enables the cells to perform increased absorption and secretion of molecules. The intestinal brush-border microvilli consist of a parallel actin-bundle core that is formed by about 40 single actin filaments of uniform polarity, which are crosslinked by at least three different actin-bundling proteins such as T-plastin (synonymously termed T-fimbrin), villin and small espin (Bartles *et al* 1998, Loomis *et al* 2003). In particular, microvilli are composed of the cortical components spectrin and myosin II in their terminal web (termed the actin network) at their projection base (figure 7.8) (Brown and McKnight 2010). Moreover, microvilli are connected to the apical surface by the brush-border myosin I (McConnell and Tyska 2007). In more detail, T-plastin is a monomeric protein that is highly expressed in the small intestine and can crosslink F-actin into straight bundles (Delanote *et al* 2005, Brown and McKnight 2010). L-plastin (synonymously termed L-fimbrin) is usually only present in haematopoietic cells, however, it has been reported to be, in contrast to other studies, expressed in epithelial carcinomas and non-epithelial mesenchymal tumors (Delanote *et al* 2005). Moreover, the L-plastin expression seems to correlate positively with the stage and severity of colorectal cancers and therefore has been proposed to be a candidate for a prognostic indicator (Foran *et al* 2006, Yuan *et al* 2010). Villin facilitates the

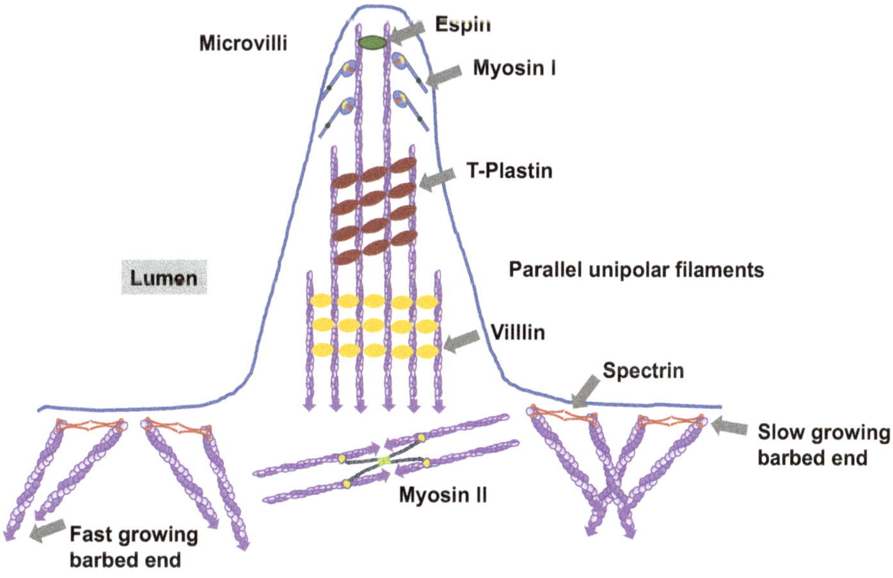

Figure 7.8. Microvilli composed of actin-bundling proteins. In differentiated intestinal epithelial cells membrane extrusions such as microvilli are formed to the luminal side of the intestine.

bundling, nucleation (including the new filament initiation), capping and severing of actin filaments in a Ca^{2+}-dependent manner (Friederich *et al* 1990). In addition, villin expression is increased in adenocarcinomas of the intestinal tract projecting brush-border microvilli (Grone *et al* 1986, Moll *et al* 1987, Suh *et al* 2005). The small espin provides the elongation of microvilli from the barbed end of the actin bundle, however, it has not been associated with cancer. Malignant cancer cells display increased numbers of microvilli exhibiting irregular morphology that can be correlated with metastatic status (Ren *et al* 1990, Ren 1991). However, still more studies are needed to reveal the functional dependence between microvilli exterion and cancer progression.

7.4.4 Invadopodium

A special kind of membrane protrusions are invadopodia that are solely detected in invasive cancer cells (Weaver 2006). They possess a mixture of bundled and branched actin (Schoumacher *et al* 2010) and can remodel the extracellular matrix environment (figure 7.9). Another kind of protrusion, called podosomes, are structurally and functionally highly similar to invadopodia, but their appearance is restricted to hematopoietic cells, endothelial cells and Src-transformed fibroblasts (Murphy and Courtneidge 2011). Both invadopodia and podosomes consist of a number of actin-bundling proteins such as fascin (Li *et al* 2010a, Schoumacher *et al* 2010), alpha-actinin, formins and Ena/VASP proteins (figure 7.9) (Murphy and Courtneidge 2011). In particular, the actin bundles extrude as protrusions into

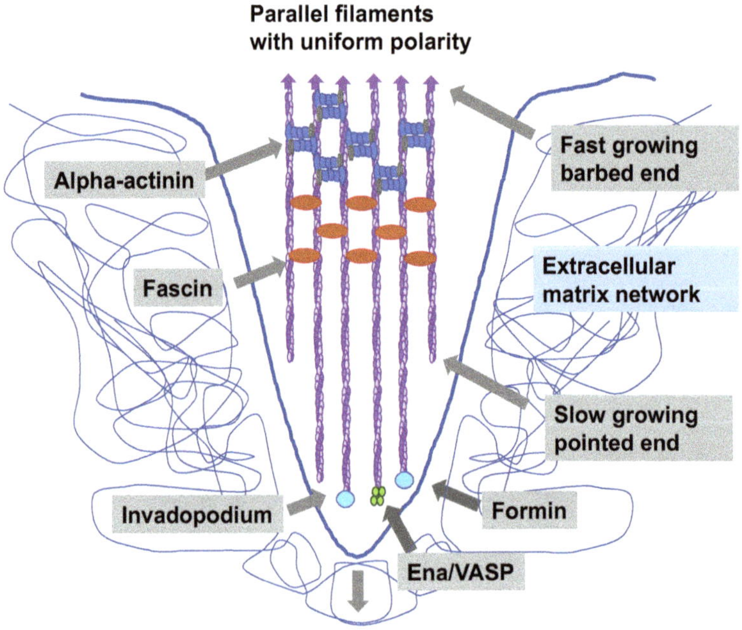

Figure 7.9. Invadopodium.

extracellular matrix and possibly release endocytic cargo such as matrix metalloproteases into the surrounding stroma (Murphy and Courtneidge 2011).

7.5 Does so-called 'cortical actin' and an actin cortex exist?

The nanoscale spacing between the cell membrane and the underlying cortical actin cytoskeleton is essential in facilitating the cellular morphology, cell mechanical properties and cellular function. However, the determination of the exact distance has been a key challenge in the field of cell biology. The current methods for determining the exact nanoscale spacing are limited in themselves as they require a complex design based on fixed samples or rely on diffraction-limited fluorescence imaging techniques whose spatial resolution is not sufficient to quantify the distances of the nanoscale length. Using a novel dual-color super-resolution STED (stimulated-emission-depletion) microscopy approach, the current limits can be overcome and hence accurate measurements of the density distribution of the cortical actin cytoskeleton can be performed, and the distance between the actin cortex and the cell membrane in living Jurkat T cells can be analyzed. An asymmetric cortical actin density distribution with a mean width of 230 (+105/−125) nm has been found (Clausen *et al* 2017). Indeed, spatial distances can be measured between the maximum density peaks of the actin cortex and the cell membrane, and they are clearly bi-modally distributed with mean values of 50 ± 15 nm and 120 ± 40 nm (Clausen *et al* 2017). Finally, the finite width of the cortex seems to be rather less than 50 nm, as the regions the cortical actin are close to 10–20 nm to the cell membrane (Clausen *et al* 2017).

Cells sense external stimuli such as mechanical property alterations of their microenvironment and convert these signals into biological responses. These responses are based on cellular processes that depend on the interactions of the cortical actin network and the cell membrane (figure 7.10). In particular, external signals are transmitted directly via signaling through distinct receptors such as the T-cell receptor expressed on the cell membrane of T cells and/or even indirectly through a mechano-transduction process, in which the physical coupling of the interface between the cell membrane and the actin cortex with the entire cell cytoskeleton provides the signaling. Further downstream these signals are translated into distinct responses such as alterations of the cell shape changes, induction of spreading, cell differentiation and cell migration (Fehon *et al* 2011, Kusumi *et al* 2012, Lewalle *et al* 2014, Fritzsche *et al* 2015). For efficient signal transduction the state of the cell membrane, the cortical actin cytoskeleton and the physical coupling of both of outermost cellular layers play a crucial role. A critical determinant for altered cell morphology, mechanics, and function is the size of the interface, which is regulated by the spatial distance between the two layers and the actin cortex density, as it supports the maintenance of the cell membrane and cortex homeostasis (Heinemann *et al* 2013, Fritzsche *et al* 2014, Saka *et al* 2014).

How dense is the cortical actin cytoskeleton? The interface between the membrane and the cortex is determined by the precise regulation through the dynamic and filamentous network scaffold, the cell's cytoskeleton. In particular, the cortex

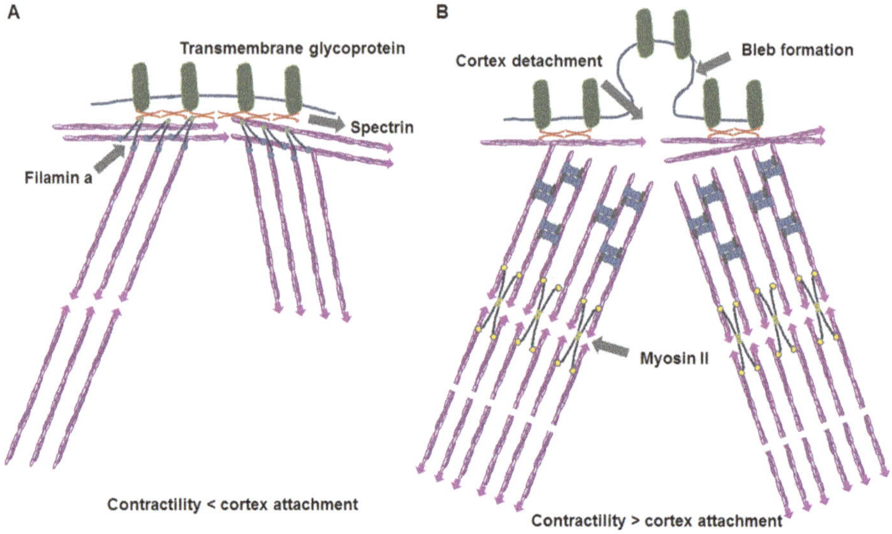

Figure 7.10. Actin-bundling proteins at the cell cortex. Cells possess a cell cortex which regulates their capacity to withstand physical forces. Thereby the cortex alters signal transduction proteins and acts as a scaffold. (a) The contractile forces are lower than the cortex attachment force and hence the membrane is still attached to the cortex. (b) Membrane blebs are formed when the attachment force between the cortex and the cell membrane is lower than the contractile forces of the cell. The cortex and the membrane ruptures and a bleb is exerted.

can be regarded as a cortical actin skeleton that is located underneath the cell membrane and thereby regulates the cellular morphology, cell mechanical properties and function, which is critically based on the dynamical remodeling of the cortex. This complex system has been proposed to be approximately 100–500 nm thick and that encompasses the polydispersed actin filaments, which continuously turn over with a constant growth of the actin filaments at their barbed (plus) ends and a defined shrinkage at their pointed (minus) ends (Clark *et al* 2013, Fritzsche *et al* 2016). Indeed, the actin kinetics characterize and subsequently determine the cortex network structure, whereas the dynamical remodeling of the filamentous actin structures provide the cell mechanical properties (Colin-York *et al* 2016). The derived cortical actin density distribution analysis at the cell membrane has revealed two different density profiles for weakly and strongly active regimes of actin polymerization with the maximum actin density closely located underneath the membrane periphery (Joanny *et al* 2013). However, this theoretical approach of the actin density analysis needs to be proven by experimental approaches. A function of cortical actin may be to insulate the cytoplasmic signaling domains of membrane–cortex spanning receptors from the binding of signaling-associated molecules, which prevents the downstream cell signal transduction processes (Dustin and Davis 2014).

What role do the architecture and the dynamical remodeling of the membrane play? Various membrane-dependent processes are regulated by the membrane–cortex interface. An example is the diffusion–reaction dynamics of membrane–cortex ligands, in particular the binding of diffusing proteins and lipids and/or receptors to their extracellular agonists, that have been shown to regulate the initial signal detection, intracellular signaling and the signal conversion into biological responses (Klann *et al* 2009, Klammt and Lillemeier 2012, Heinemann *et al* 2013). Molecules within the membrane can generally not freely diffuse, as they are commonly transiently trapped within specific confinements or compartments with different diffusion coefficients when they interact with immobile and slowly moving constituents. In particular, actin regulates the lateral diffusion by increasing the assembly of lipid and multiprotein assemblies and thereby shifting the free diffusive behavior of these molecules into a trapped, compartmentalized or so-called hop diffusion (Clausen and Lagerholm 2013, Heinemann *et al* 2013, Andrade *et al* 2015, Vicidomini *et al* 2015, Fujiwara *et al* 2016). The indirect signal transduction, such as extracellular mechanical stress via the membrane–cortex interface seems to be similarly affected by these constrictive processes of the membrane. All these processes are supposed to be influenced by two main characteristic features of the actin cytoskeleton such as the distance between the membrane and actin cortex and the density of the entire actin cortex.

What is the distance between the cell membrane and the cortex? The nanoscale spacing of the membrane–cortex interface affects the molecular activity and interactions between the membrane and the actin cortex. In more detail, the cell membrane is indeed physically tethered to the nearby underlying cortical actin network, which provides by the proteins of the ERM family (ezrin, moesin, radixin) (figure 7.11) (Charras *et al* 2006). Moreover, the kinetics of these membrane–cortex

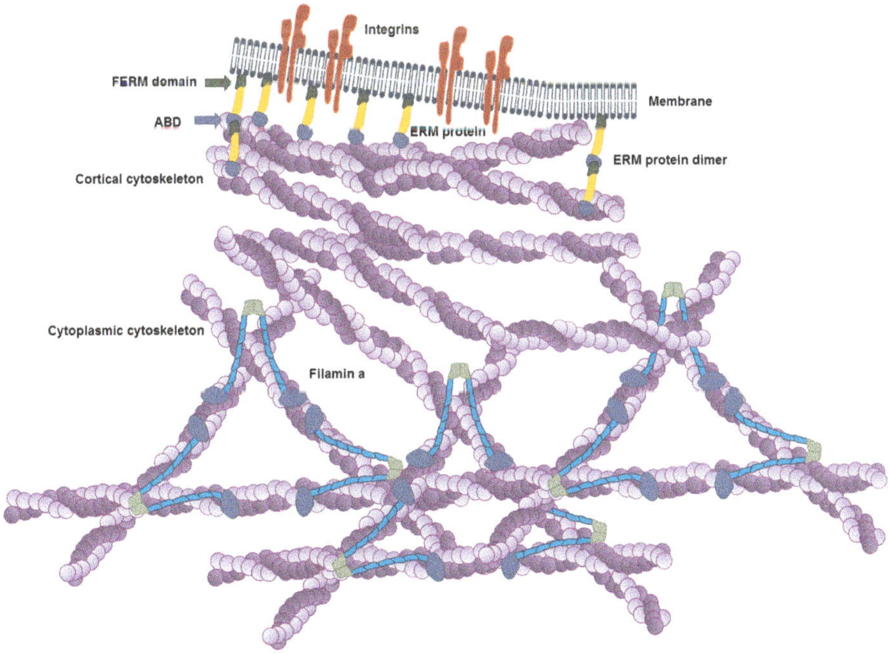

Figure 7.11. Membrane–cortex interface. The membrane lipid bilayer is connected to the actin filaments through ERM protein monomers or dimers. Cortical actin is located underneath the cell membrane and extends into the cytoplasm, where it is connected to the cytoplasmic cytoskeleton.

linkers, such as their association or dissociation rates and their ability to build oligomers, changes the binding of the cell membrane to the actin cortex (Fritzsche *et al* 2014). In turn, the local concentration of individual linkers and number of oligomers per linkage defines the precise distance between membrane and cortex, and subsequently the adhesion force. A local force of 16 pN per connection for the membrane–cortex adhesion has been proposed (Alert *et al* 2015). The hydrodynamical properties of the membrane are affected by low spatial membrane–cortex distances or a less permeable actin cortex, which in turn alters the entire architecture of the membrane (Alert *et al* 2015). In particular, enhanced membrane–cortex adhesion are observed by overexpression of constitutively active ezrin mutants such as ezrinT567A and finally leads to decreased spacing of the membrane–cortex interface and thereby creates high membrane tension and induces a spherical morphology in epithelial cancer cells such as HeLa (Liu *et al* 2012, Fritzsche *et al* 2014). Similarly, overexpression of constitutively inactive ezrin mutants such as ezrinT567D reduces the adhesion energy and hence enhances the spacing of the membrane–cortex interface leading to weakly confined cell boundaries (Liu *et al* 2012, Fritzsche *et al* 2014). In line with this, the mechano-transduction processes evoked by extracellular mechanical forces can be varied by altering the nanoscale spacing between the membrane and the actin cortex (Charras *et al* 2006, Rouven *et al* 2015).

How can the actin cortex density and membrane–cortex distance be measured? The characterization of various processes at the interface of cell membrane and the actin cortex have been studied at a coarse-grained level due to the limits of microscopical observation techniques. In particular, the structural characterization of actin networks has been a major problem (Eggeling *et al* 2009, Fritzsche *et al* 2016). The high actin density with small mesh-sizes or pore-sizes has fundamentally limited the analysis of cortical structures (Fritzsche *et al* 2013, Fritzsche *et al* 2016). The typical mesh-sizes of 10–15 nm cannot be accessed using conventional optical microscopy due the limited spatial resolution and hence scanning electron microscopy or super-resolution STORM-based optical microscopy need to be employed to visualize cortical networks (Xu *et al* 2013). However, these measurements have not revealed insights into the network structure such as the cortical actin density or the actin-filament lengths, as the high actin density impairs the precise identification of the actin-filament ends (Fritzsche *et al* 2016). The prediction of the spatial distance between the membrane and the cortex based on the crystal structure of full-length ERM proteins such as ezrin and moesin has so far failed. However, the crystallization of the dormant and active FERM domain of ezrin and moesin propose spacings in the rage of 10–50 nm for monomeric linkers (Pearson *et al* 2000, Edwards and Keep 2001, Phang *et al* 2016). Based on electron tomography reconstructions of fixed PtK2 cell cross-section samples lead to the prediction of a distance between the membrane and the cortex of less than 10 nm (Fujiwara *et al* 2016). Based on electron microscopy observations, it has been proposed that the density of the cortex is rather uniform in contact with the membrane, whereas the uniformity is abolished at the cortex–cytoplasm interface (Hanakam *et al* 1996). A weakness of these methods is that the cell samples are fixed and thereby contain well-known artifacts such as cell shrinkage and altered architecture of the cell membrane and actin networks (Wollweber *et al* 1981, Stanly *et al* 2016). Thus, the actin cortex density and membrane–cortex distance measurements need to be performed with living cells. However, conventional confocal microscopy has determined the actin cortex thickness to be approximately 190 nm by using a dual-color labeling of the actin cortex and the implementation of a theoretical description for the cortex geometry to overcome the limited resolution of their microscope (Clark *et al* 2013). For their measurements they assumed a uniform actin density and a direct connection between the cortex and the cell membrane. What would the results be, if the assumptions are omitted, as we do not know whether there is a direct connection between the membrane and the cortex or whether there exists a nanoscale spacing between them? Hence the diffraction-limited nature of the confocal experiments excludes them from further experiments.

Using dual-color super-resolution STED microscopy, it has been shown that the fluorescently labeled cell membrane and the actin cortex can be accurately localized (Clausen *et al* 2017). Due to the enhanced spatial accuracy of STED microscopy, the density distribution of the cortical actin cytoskeleton can be analyzed directly underneath the cell membrane and hence the nanoscale spacing between the two layers can be determined in living Jurkat T lymphocytes. Indeed, a bimodal distribution of the nanoscale distances between the cortex and the membrane has been found and therefore cells can dynamically alter their own membrane architecture.

7.6 The different stress-fiber types

There exist stress fibers in cells that contain bundles of parallel actin filaments displaying a mixed polarity along their length (Cramer *et al* 1997) (figure 7.12). These actin bundles interact with myosin II motors that are crosslinked by alternating zones of alpha-actinin and Ena/ VASP proteins and they are anchored at their ends towards the membrane by focal adhesion protein assembly (figure 7.12). Moreover, stress fibers connect the cytoskeleton to the extracellular matrix through focal adhesion sites, at which the integrin cell–matrix receptors span the cell membrane and build clusters such as large macromolecular hubs for signaling

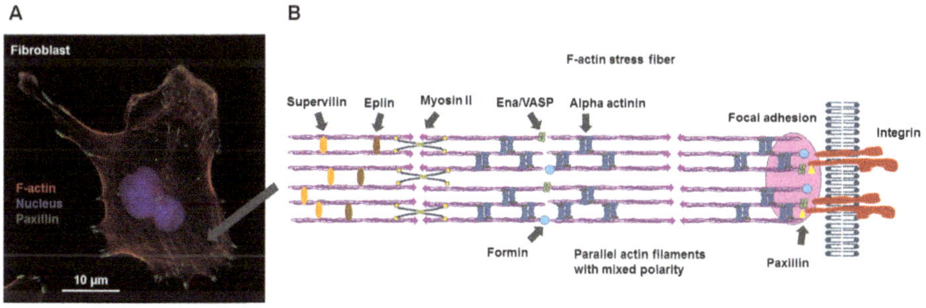

Figure 7.12. F-actin stress fibers.

and cytoskeletal proteins (Wolfenson *et al* 2009). Indeed, focal adhesions are mechano-sensitive assemblies that promote the signal transduction based on the interaction of a cell with its surrounding stroma and thereby provide cell survival and growth signals. Alpha-actinin-1 is distributed along the stress fibers in a periodic manner, binds to several focal adhesion proteins (Edlund *et al* 2001) and hence combines the actin cytoskeleton with the cell membrane. In general, alpha-actinin-4 is located at the leading edges of migrating cells within lamellipodia. Hence alpha-actinins such as alpha-actinin-4 play a role in multiple tumors such as breast (Guvakova *et al* 2002), ovarian (Yamamoto *et al* 2009), pancreas (Kikuchi *et al* 2008) and lung cancers (Menez *et al* 2004).

The epithelial protein lost in neoplasm (Eplin) also represents a component of stress fibers and possesses an actin-filament side-binding and bundling activity, which is antagonistic to the Arp2/3 complex branching activity (Maul *et al* 2003) and is decreased during the progression of cancer. Eplin contains two actin-filament-binding sites that are separated by a central LIM domain. As an immunoglobulin-repeat-containing protein palladin interacts with actin filaments and thus builds a scaffold for other stress-fiber-associated proteins such as VASP, alpha-actinin, EPS8 and the ERM (ezrin, moesin and radixin) proteins. Phalladin can be phosphorylated by the AKT1 kinase, induce actin bundling and impair *in vitro* the invasion of breast cancer cells (Chin and Toker 2010), whereas another study reported the opposite behavior, as loss of palladin inhibits the invasion of breast cancer cells (Goicoechea *et al* 2009). However, *in vivo* studies are needed to analyze whether palladin functions in metastasis. A component of focal adhesions, supervillin consist of three villin-related actin-binding sites (Wulfkuhle *et al* 1999) and possesses four predicted nuclear localization signals indicating that it may shuttle between the cytosplasm and the nucleus. Additionally, in prostate cancer, it seems to be bound to the androgen receptor and hence involved in the regulation of cell growth and androgen-dependent signal transduction processes (Sampson *et al* 2001).

There exist distinct types of actin stress fibers that represent major contractile structures within cultured eukaryotic cells (Tojkander *et al* 2012). These actomyosin bundles are pronouncedly established in fibroblasts, endothelial cells, smooth muscle and distinct cancer cell lines. In non-motile cells, the stress fibers are rather thick and stable compared to highly motile cells, in which the stress fibers are less pronounced, deceased in numbers, thinner and can more easily undergo a dynamical remodeling (Pellegrin and Mellor 2007). The polar and helical actin-filament structures grow rapidly at their barbed end and slowly at their pointed end, exhibiting a ten-fold difference (Pollard and Cooper 2009). In more detail, stress fibers are built by bundles of 10–30 actin filaments, which are connected by crosslinking proteins such as alpha-actinin by exhibiting a usually bipolar arrangement. These contractile actomyosin bundles are connected to focal adhesions through which the entire actin cytoskeleton is coupled to the extracellular matrix (Cramer *et al* 1997, Pellegrin and Mellor 2007, Naumanen *et al* 2008). Stress fibers largely vary in morphology and interaction with focal adhesion proteins. Hence, the stress fibers are grouped into at least four different categories such as dorsal stress fibers, ventral stress fibers, transverse arcs and the perinuclear actin cap (Heath 1983, Small *et al* 1998, Khatau *et al* 2009).

The dorsal stress fibers connected at their distal ends to focal adhesions. Moreover, these actin-filament bundles usually contain no myosin II (Tojkander *et al* 2011) and subsequently cannot contract. The precise organization of actin filaments in dorsal stress fibers is still elusive. However, they display a structurally related so-called 'graded polarity' of their bundles, as the distal ends consist of unipolar actin filaments with rapidly growing barbed ends facing the cell periphery, whereas the more proximal parts of the actin bundle consist of actin filaments with mixed polarity (Cramer *et al* 1997, Pellegrin and Mellor 2007). Although dorsal stress fibers cannot contract, they seem to be a platform for the assembly of other stress fibers types and provide a linkage to focal adhesions (Hotulainen and Lappalainen 2006, Tojkander *et al* 2011) (figure 7.13).

Transverse arcs consist of curved actin-filament bundles that show a periodic alpha-actinin–myosin pattern, which is a characteristic feature of contractile actomyosin bundles. As arcs do not directly bind to focal adhesions, they transfer contractile force to the surrounding microenvironment via connections with dorsal stress fibers. A main property of transverse arcs in migrating cells is that they can flow from the cell's leading edge toward the cell's interior center (Heath 1983, Small

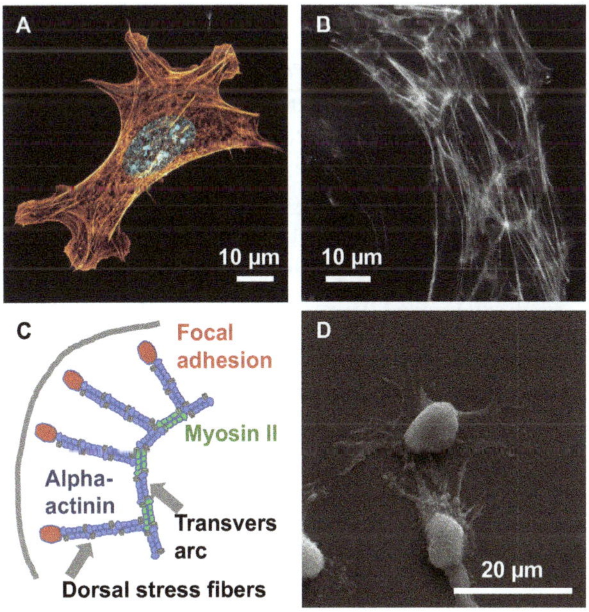

Figure 7.13. Different types of stress fibers in cells. (a) Lamellipodia in mouse embryonic fibroblasts stained with Alexa Fluor546 Phalloidin (actin, red) and Hoechst 33342 (DNA, blue). (b) Primary human dermal microvascular endothelial cells stained with Alexa Fluor546 Phalloidin (actin, white). (c) Scanning electron microscopic images of human MDA-MB-231 breast cancer cells cultured on a fibronectin coated glass substrate. (d) Schematic presentation of the stress-fiber network in lamellipodia of mesenchymal cells. Dorsal stress fibers are connected to focal adhesions at their distal end, whereas transverse arcs are curved stress fibers that flow towards the cell's center and they are anchored via adhesion indirectly through dorsal fibers.

et al 1998, Hotulainen and Lappalainen 2006), which is known as retrograde flow that is supposed to be facilitated by the continuous contraction of the arcs (Zhang *et al* 2003).

Ventral stress fibers are contractile actomyosin bundles, which are connected to focal adhesions at both ends and hence these ventral fibers build the major contractile machinery in many interphase cells (Small *et al* 1998). In particular, ventral stress fibers are usually located at the posterior parts of the cell, where cycles of contraction facilitate the rear constriction and finally cell migration and invasion (Chen 1981, Mitchison and Cramer 1996). Moreover, the perinuclear actin cap has been identified as an actin structure based on fibers that are placed over the cell nucleus and are connected to the nucleoskeleton. Hence, a major function of the perinuclear actin cap is the regulation of the nuclear shape in migrating cells. Additionally, the perinuclear actomyosin fibers may function as mechanotransducers to transmit forces from the cellular microenvironment to the cell's nucleus (Khatau *et al* 2009). Indeed, distinct stress-fiber-like structures are connected to the nuclear membrane through special membrane proteins (Luxton *et al* 2010) and thereby stabilize the position of the nucleus precisely (Nagayama *et al* 2011). Similar to the connections between canonical stress fibers and the extracellular matrix via focal adhesions, a subset of stress fibers seems to be mechanically linked to nuclear membrane proteins in order to facilitate nuclear movement during the overall cellular migration.

What are lamella within cells? Are they real?

Lamellipodium and lamella are two distinct actin-filament networks that have both been shown to support the migration of cells. The structure of the lamellipodium is based on a branched network of actin filaments that is regulated by the small Rho-GTPase Rac1, however, the origin and organization of the actin filaments of the lamella is not yet precisely known (Vallotton and Small 2009, Ydenberg *et al* 2011). In particular, the lamella of migrating cells can be synonymous with the transverse arc network, as both structures (the lamella and the transverse arcs) contain condensed actin bundles and are able to undergo retrograde flow towards the cell's center (Ponti *et al* 2004, Hotulainen and Lappalainen 2006). Moreover, both lamella and transverse arcs possess similar protein compositions such as tropomyosins, which are hardly present within the lamellipodium, but are detectable at high levels in the lamella (DesMarais *et al* 2002) as well as in transverse arcs (Tojkander *et al* 2011). Another feature supporting this view is that both arcs and the lamella are created through the condensation of lamellipodial actin filaments into arc-shaped actin bundles, which are positioned parallel to the cell's edge (Hotulainen and Lappalainen 2006, Burnette *et al* 2011, Tojkander *et al* 2011).

7.7 Actin–myosin interaction during cell migration

In muscle and cytoskeletal motor processes, the formation of the tight and hence strong interactions between actin filaments and myosins are necessary for the generation and transmission of forces.

7.7.1 Introduction to the superfamily of myosins

The myosins are motor proteins that act on actin filaments in order to generate and transmit forces. Thereby myosins require energy, which they generate by the hydrolysis of ATP. Many functions of the myosins involve cellular contractility such as cell signaling, endocytosis, vesicle trafficking and the localization of proteins or RNAs (Krendel and Mooseker 2005, Woolner and Bement 2009, Hartman and Spudich 2012). All myosins possess certain common structural and functional features such as an actin-binding head domain, which carries the myosin ATPase activity. The actin-dependent ATPase activity is employed by myosin motors for the generation of force and the translocation of a cargo along the actin filaments. In particular, myosins consist of a neck domain, which can bind light chains and a tail domain, which has a variable sequence and length, and therefore it determines the intracellular localization and cargo binding specificity of the individual myosin.

The first biochemical studies investigating myosin were on skeletal muscle myosin, termed myosin II (Szent-Gyorgyi 2004). After the identification of non-muscle myosins, the myosin family has been divided into two major groups—class I (small, mostly monomeric motors) and class II myosins (dimeric myosins similar to skeletal muscle myosin, including both muscle and non-muscle myosins). The class II myosins are seen as conventional myosins. The identification of several other non-muscle myosins markedly changed the myosin diversity (Berg *et al* 2001). The myosin superfamily of motor proteins can be subdivided into various classes. The classification is at first glance based on phylogenetic analysis of the myosin motor domain (Hodge and Cope 2000, Berg *et al* 2001, Richards and Cavalier-Smith 2005, Foth *et al* 2006, Odronitz and Kollmar 2007). Twelve classes, myosin I, II, III, V, VI, VII, IX, X, XV, XVI, XVIII and XIX have been found in vertebrates including humans. Apart from the myosin II class, myosin classes are termed unconventional myosins and fulfill their function either as monomers or dimers of heavy chains that are interacting with calmodulin-like light chains through binding via the IQ motifs in their neck region. Thereby, the light chains stabilize the neck region and perform conformational alterations in the motor domain to move along actin filaments. Indeed, the velocity of myosins and their step size are positively correlated with the neck region length (termed the lever arm) (Tyska and Warshaw 2002).

The stress fibers require tension and myosin-II-facilitated contractility, as myosin II inhibition induces the disassembly of stress fibers (Chrzanowska-Wodnicka and Burridge 1996, Bershadsky *et al* 2006). Multiple focal adhesion associated proteins are involved in the regulation of stress fibers. In particular, zyxin has been revealed to function in force-dependent actin polymerization (Hirata *et al* 2008), the mechano-sensing of stress fibers (Colombelli *et al* 2009) and repairing of stress fibers (Smith *et al* 2010).

How is the contraction of stress fibers regulated?

The myosin light chain (MLC) regulates the contractility of stress fibers (Somlyo and Somlyo 2000). The reversible phosphorylation of MLC on the Thr18 and Ser19 phosphorylation sites enhances the assembly of non-muscle myosin II filaments and increases the actin-activated ATPase activity of the myosin motor domain (Vicente-Manzanares *et al* 2009). The MLC-phosphorylation-facilitated contractility of stress fibers is regulated by at least two distinct pathways, a Ca^{2+}/calmodulin-dependent pathway and a Rho-dependent pathway (Katoh *et al* 2001a, Katoh *et al* 2001b). The Ca^{2+}/calmodulin pathway functions similarly in smooth muscle and non-muscle cells, and activates of the myosin light chain kinase (MLCK), which subsequently phosphorylates MLC. The Rho–ROCK pathway regulates the actomyosin activity either through the direct phosphorylation of MLC or through the inhibition of the phosphorylation of the myosin light chain phosphatase (MLCP), which causes its inactivation (Amano *et al* 1996, Kimura *et al* 1996, Totsukawa *et al* 2000). Finally, both phosphorylation pathways create distinct contractile responses. In particular, the Ca^{2+}/calmodulin pathway causes a local and rapid response, whereas the activation of the Rho pathway leads to more persistent response. The septin SEPT2, which is a member of a conserved family of filamentous GTPases, additionally regulates contractile structures through binding and subsequently activation of myosin II. Under physiological conditions, septins are closely associated with actin stress fibers within interphase cells, whereas they regulate the contractile ring formation in mitotic cells during cell division.

The impairment of the interaction between SEPT2 and myosin II in interphase cells leads to a loss of stress fibers, the septin-facilitated regulation of myosin II activity seems to be necessary for the proper assembly and maintenance of stress fibers (Kinoshita *et al* 2002, Joo *et al* 2007). The dephosphorylation of MLC and hence the disassembly of stress fibers is facilitated by MLCP that is connected to actin stress fibers through the interactions of its regulatory subunit MYPT1 with the myosin phosphatase Rho-interacting protein (MPRIP) (Mulder *et al* 2003, Surks *et al* 2005). The activity of MLCP can be altered by the interaction of the tumor suppressor liver kinase B1 (LKB1 or synonymously named STK11) with NUAK family kinases, which shows that phosphorylation and dephosphorylation of MLC are connected with various signaling pathways (Zagorska *et al* 2010, Vallenius *et al* 2011).

What is the role of myosin in tensional development of stress fibers and focal adhesions?

In addition to the effect of the blockage of Rho-driven myosin activity that inhibited ventral stress fibers and focal adhesion assembly in quiescent cells (Chrzanowska-Wodnicka and Burridge 1996), further support for a mechanical tension driven assembly is based on the following observations (Burridge and Guilluy 2016). Shear stress at levels equivalent to shear stress levels in arteries, has been shown to facilitate the formation of stress fibers in endothelial cells lining blood vessels (Franke *et al* 1984). More direct evidence for the growth stimulation of focal adhesions has been provided by the application of force directly to individual cells with a simple glass rod, with the growth of adhesion seen using interference reflection microscopy (IRM) optics by employing fluorescently labeled focal adhesion proteins (Riveline *et al* 2001). Moreover, the adhesion growth can be inhibited by reducing RhoA activity, which is rescued by expression of the constitutively active formin, mDia1, which is usually activated by RhoA. Thus, focal adhesion growth seems to require tension and actin polymerization, each of which is facilitated by activated RhoA. In line with this, the growth of focal adhesions mirrors the tension applied to them, which has been demonstrated by using 2D traction force microscopy to reveal a positive correlation between the size of focal adhesions and the forces which have been transmitted across them (Balaban *et al* 2001, Schwarz *et al* 2002).

When the mechanical force induces the growth and maturation of focal adhesions, it may be concluded that new components are recruited to focal adhesions as a response to force application. Indeed, individual components of focal adhesions such as vinculin have been analyzed (Galbraith *et al* 2002). Moreover, a so-called focal adhesion proteome has been identified, in which the composition of distinct proteins depends on the presence of active myosin (Kuo *et al* 2011, Schiller *et al* 2011). The hypothesis that tension increases the growth, composition and thereby maturation of focal adhesions has been further confirmed by biophysical experiments using beads coated with integrin ligands in order to apply force towards the cells. Indeed, a reinforcement of the adhesions connected to beads has been detected (Wang *et al* 1993, Choquet *et al* 1997, Galbraith *et al* 2002, Matthews *et al* 2006, Guilluy *et al* 2011). In particular, focal adhesion proteins are recruited in greater numbers to tensioned bead adhesion sites (Galbraith *et al* 2002), which is also consistent with the force promoting the focal adhesion assembly.

What is the reason for the stiffening or reinforcement occurring upon external tension application via integrins? The RhoA pathway is among the signaling pathways that are activated in response to tension application via integrins (Matthews *et al* 2006, Guilluy *et al* 2011, Zhao *et al* 2007). The enhanced actin polymerization is not exclusively regulated through the RhoA-activated formin mDia1 (Watanabe *et al* 1997), but it contributes to the stability of actin filaments, as the severing activity of cofilin is impaired through the ROCK1 dependent LIMK1 phosphorylation of cofilin, which inhibits its function (figure 7.14). Moreover, the activation of myosin II by ROCK (Amano *et al* 1996, Kimura *et al* 1996) causes a positive feedback loop and thereby increases the cellular tension. Similarly, activated myosin II that has assembled into bipolar filaments represents a multivalent crosslinker and F-actin-bundling protein (Burridge and Chrzanowska-Wodnicka 1996, Choi *et al* 2008). Moreover, F-actin exhibits enhanced affinity for myosin II when it is under tension (Uyeda *et al* 2011). In line with this, also in response to tension, F-actin decreases the binding to cofilin and it can hence withstand more pronouncedly its own severing by cofilin (Hayakawa *et al* 2011).

How might force on a focal adhesion complex increase the recruitment of new components?

In order to reveal how the force acts on the focal adhesion complex, it is necessary to identify tension-sensitive cryptic protein binding sites. These tension-sensitive cryptic protein binding sites were first identified for fibronectin (Zhong *et al* 1998). Later, these tension-sensitive cryptic binding sites were demonstrated for proteins within focal adhesions, such as talin. In particular, the application of tension to single talin molecules exposed previously hidden binding sites for vinculin (del Rio *et al* 2009).

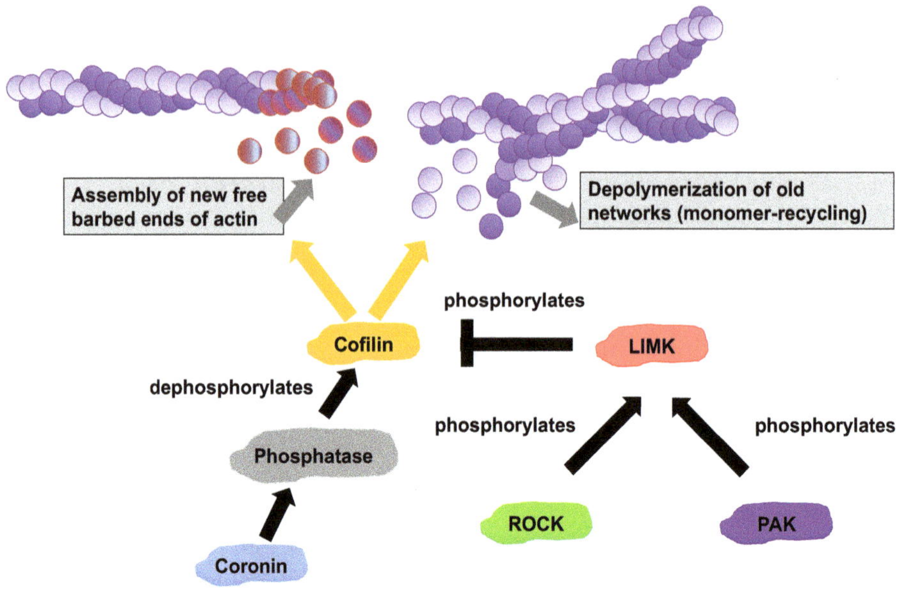

Figure 7.14. Regulation of LIMK1.

Similar to talin, other proteins may expose sites upon application of tension, facilitating new interactions (Tamada *et al* 2004). An example is p130cas, which upon mechanical stretching exposes multiple tyrosines serving as phosphorylation sites for Src (Sawada *et al* 2006). In addition, it has been found that stress fibers reinforce in response to strain applied to cells using cyclic stretch or shear stress (Yoshigi *et al* 2005). In more detail, zyxin can shuttle from focal adhesions to zones of tension within the stress fibers and initiate the local recruitment of α-actinin and vasodilator-stimulated phosphoprotein (VASP) to stress fibers, which then causes a thickening and reinforcement of stress fibers (Yoshigi *et al* 2005, Colombelli *et al* 2009). However, zyxin-dependent stress-fiber reinforcement can be found in cells exposed to external mechanical stress and it instead represents a repair mechanism limiting stress-fiber elongation and breakage at sites of excessive strain (Smith *et al* 2010). The targeting of zyxin to stress fibers depends on MAPK activation, but the mechanism involved is not yet fully clear. However, several models propose tension-dependent conformational alterations in stress-fiber-associated proteins that cause the exposure of new docking sites for zyxin (Smith *et al* 2010).

However, how the bonds between different proteins within the focal adhesion complex are facilitated or strengthened upon exposure to mechanical tension is not yet understood. A passive or an active process may drive the interaction of the tightly locally restricted focal adhesion proteins, such as the exertion of a new previously hidden binding site or a reaction towards the mechanical force causing deformation of the complex. However, so-called 'catch' bonds (Evans and Calderwood 2007), that strengthen in response to tensional force, have been revealed for distinct proteins such as for the fibronectin-binding integrin $\alpha5\beta1$ (Friedland *et al* 2009, Kong *et al* 2009). It seems to be possible to identify more catch bonds that function in focal adhesions. In line with this, it is supposed that mechanical tension signaling can modify the protein interactions by turning slip bonds into catch bonds, which finally increase the stabilization through increased tension. Conversely, the change of catch bonds into slip bonds is assumed to disassemble adhesions at the cell's rear during cell migration.

What proteins are associated with stress fibers?
The two main components of stress fibers are actin and myosin, and these create a functionally precise, regulated actomyosin structure contracting the stress fibers and hence the entire cell. Several ABPs and focal adhesion molecules are able to localize to stress fibers, and hence provide their assembly and stability. As the interactions of ABPs with stress fibers are known to be normally highly dynamic, using fluorescence recovery after photobleaching (FRAP) experiments (Schmidt and Nichols 2004, Hotulainen and Lappalainen 2006, Endlich *et al* 2009, Tojkander *et al* 2011), it is supposed that stress fibers are highly dynamic structures and the proteins associated with them undergo constant dissociation and association.

Different ABPs can be classified based on biochemical functions and their distinct localizations along the stress fibers. An example is α-actinin, an actin crosslinking protein, which shows a punctate localization pattern on the stress fibers, which is complementary to the myosin II localization within the stress fibers (Lazarides and Burridge 1975). In many non-muscle cells, the expression of the two isoforms, α-actinin-1 and -4, can be detected. In cultured cells, α-actinin-1 is found enriched in stress fibers, whereas the α-actinin-4 is more pronouncedly localized to the lamellipodia (Honda *et al* 1998). In addition to its actin-bundling activity, α-actinin can associate with kinases and signaling proteins, such as the PDZ-LIM-containing proteins, and thereby acts as a regulator for cytoskeleton-targeted signaling (Vallenius *et al* 2000, Vallenius and Makela 2002). Various actin-filament cross-linking proteins such as fascin, filamin and palladin localize to stress fibers, whereas their distinct functions within these actomyosin bundles are not yet well understood (Wang *et al* 1975, Adams 1995, Dixon *et al* 2008). In addition to their crosslinking activity, many of these proteins fulfill additional roles in regulating cytoskeletal dynamics, such as palladin which can interact with the ABPs' profilin and vaso-dilator-stimulated phosphoprotein (VASP) and may thereby operate as a scaffolding protein inducing actin dynamics in stress fibers (Boukhelifa *et al* 2004, Boukhelifa *et al* 2006).

The calponin family represents another group of multifunctional proteins that associate with stress fibers in a punctate pattern similar to α-actinin (Strasser *et al* 1993). The most studied isoform is calponin-1 (synonymously termed calponin h1), which can be detected in smooth muscle cells and regulates contractility (Winder *et al* 1998). The other calponin isoforms such as calponin-2 and -3 (synonymously termed calponin h2 and h3) are located in several non-muscle and muscle cells (Draeger *et al* 1991, Hossain *et al* 2003). In addition to the regulation of muscle contraction, calponins are proposed to crosslink and hence stabilize actin-based structures in order to facilitate cell migration (Leinweber *et al* 1999b, Danninger and Gimona 2000). In addition, calponins associate with several kinases such as extracellular-signal-regulated kinase 1/2 (ERK1/2) and protein kinase C (PKC) and seem to act as scaffolding proteins regulating cytoskeletal dynamics (Leinweber *et al* 1999a, Patil *et al* 2004).

In addition to myosin II, other proteins such as tropomyosins (TPMs) and caldesmon (CaD, synonymously known as CALD1) can manage to bind to stress fibers in a complementary pattern to α-actinin. TRMs belong to a large family of actin-binding proteins, whereas caldesmon functions in the regulation of stress-fiber contraction and reorganization (Weber and Groeschel-Stewart 1974, Lazarides 1975, Yamashiro-Matsumura and Matsumura 1988, Castellino *et al* 1995). In particular, non-muscle cells possess distinct myosin II isoforms that have specific binding partners and distinct localization patterns and among which myosin IIA and IIB are the most well-studied (Vicente-Manzanares *et al* 2009). Myosin II assembles bipolar bundles that are essential for the assembly and maintenance of stress fibers (Bao *et al* 2005, Hotulainen and Lappalainen 2006). TPMs can regulate the recruitment of myosin II to contractile stress fibers (Gunning 2008, Tojkander *et al* 2011). Within muscle cells, TPMs act together with troponin and thereby control the amount of contraction by sterical inhibition of the actin–myosin interaction in a Ca^{2+}-dependent mode (McKillop and Geeves 1993).

Additionally, TPMs regulate actin dynamics through the impairment of filament depolymerization at pointed ends and actin-filament disassembly that is driven by actin depolymerization factor (ADF or cofilin) (Broschat 1990, Ono and Ono 2002). The loss or inactivation of specific TPMs is associated with abnormal actin stress-fiber structures leading finally to increased filament disassembly and reduced myosin recruitment (Gupton *et al* 2005, Tojkander *et al* 2011). TPMs cooperate with CaD to stabilize actin filaments (Ishikawa *et al* 1989a, Ishikawa *et al* 1989b), which are both able to bind actin with its C-terminal region and myosin with its N-terminal region and hence induces the crosslinking of myosin bundles to actin filaments (Marston *et al* 1992, Katayama *et al* 1995). In particular, the interaction of CaD with actin is facilitated by phosphorylation, and during the cell cycle the CaD phosphorylation causes a disassembly of actin stress fibers (Kordowska *et al* 2006). Based on similar localization patterns of CaDs and TPMs on stress fibers and their closely linked expression levels, the functions of the two protein families within stress fibers are closely related (Yamashiro-Matsumura and Matsumura 1988, Kashiwada *et al* 1997). However, stress fibers also contain multiple other proteins acting in stress-fiber assembly, contractility or repair.

7.7.2 Myosin motors and their diverse functions

Myosin isoforms are variable in their mechanochemical properties, such as the duty ratio, which means the fraction of the ATPase cycle that is still in the strong actin-bound state, and processivity, which is the ability of myosin to move continuously (and hence persistently) along an actin filament without diffusing randomly (O'Connell *et al* 2007). In fast contractile events, the low duty ratio myosins may fulfill their function and thereby act as dynamic, short-lived tethers connecting the actin filaments and myosin cargo, whereas high duty ratio myosins have the ability to adapt faster in order to generate persistent tension. Moreover, processive myosins are optimal for the transport of organelles and macromolecules over long distances. The different mechanochemical properties and various cargo binding domains enable non-muscle myosins to fulfill many processes involved in the malignant progression of cancer, which includes several consecutive steps, such as the cell migration and invasion through the extracellular matrix, regulation of the localization of proteins and organelles, alterations of the cell shape and induction of different cell signaling pathways.

The non-muscle myosins are found in various actin-containing cellular structures such as microvilli, stereocilia, filopodia (containing linear, unipolar bundles of actin filaments), stress fibers (consisting of linear actin filaments of mixed polarity that are similar to muscle sarcomeres in their periodic organization and are connected at their ends to cell–matrix adhesions), adhesion belts (displaying parallel bundles of actin filaments along the apical portion of epithelial cells and connecting to cell–cell junctions), and even other actin assemble types (figure 7.15). The polymerization of actin polymerization in cells is driven by the activity of actin nucleators inducing the formation of short nuclei consisting of 2–3 actin monomers that start the initiation of the assembly of actin filaments. The kind of nucleator determines the structure of the actin network, such as the Arp2/3 complex inducing the formation of branched actin networks or formins nucleating parallel bundles and other nucleators that are associated with distinct subsets of ABPs or actin crosslinking proteins (Campellone and Welch 2010, Arjonen *et al* 2011, Yang and Svitkina 2011, Michael and Yap 2013, Vindin and Gunning 2013).

Distinct classes of myosin favor binding to specific actin structures. Examples are yeast and metazoan myosins II and yeast myosin V, which usually interact with parallel actin bundles nucleated by formins, which contain the ABP tropomyosin, whereas class I myosins are not even able to interact with tropomyosin-containing actin assemblies and hence preferably connect to Arp2/3-nucleated branched actin networks (Ostap 2008, Pruyne 2008, Stark *et al* 2010, Wang and Coluccio 2010, Tojkander *et al* 2011, Pollard and Lord 2014). In mammalian cells, tropomyosin can be found in stress fibers, in contractile rings during cell division and in apical adhesion belts. All these actin structures become highly contractile after myosin II binding. In contrast, class I myosins prefer binding to actin structures that are free of tropomyosin, such as branched actin networks enveloping clathrin-coated

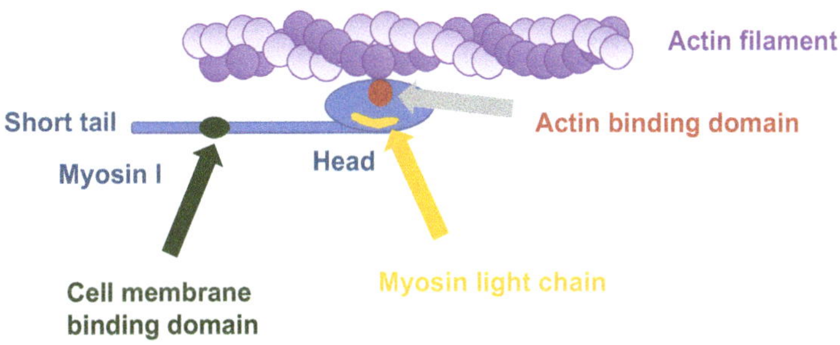

Figure 7.15. Myosin II and actin filament.

vesicles, actin filaments of nascent (immature) adherence junctions and fimbrin-crosslinked actin bundles in microvilli (figure 7.15). Another preferential interaction of a specific actin structure with a distinct myosin class has been observed in filopodia consisting of parallel bundles of actin filaments crosslinked by fascin and associated with increased levels of Myo10. The mechanism regulating the Myo10 interaction with filopodia is still elusive (Kerber and Cheney 2011). The specificity of myosin toward distinct cytoskeletal structures serves as an additional control mechanism for the activity of myosins. Examples are the alterations of the tropomyosin or fascin expression detected in cancer cells (Wang and Coluccio 2010, Arjonen *et al* 2011) that may cause alterations in myosin-facilitated cellular tension and subsequently on cellular motility.

The actomyosin system generates forces through the ATP-facilitated cyclical interaction of myosin with actin (figure 7.16). This cycle includes the binding of ATP to the strongly actin-bound myosin motor domain (MD) and thereby induces a drastic reduction in actin affinity, which causes the dissociation of the myosin from the actin filament. Based on the hydrolysis of ATP, the rebinding of the myosin to the actin filament initiates several consecutive events leading finally to the dissociation of hydrolysis products such as phosphate (Pi) and ADP, which thereby re-establishes the strong binding of the myosin-head domain to actin filaments and induces a force-generating step (termed the powerstroke) (Geeves and Holmes 2005). The powerstroke process involves the swing of the myosin's arm (through a conformational change) followed by the translocation of the distal part of the myosin molecule (and subsequently the entire thick filament in muscle) relative to the actin filament. Finally, at the end step of the force-generating pathway, the ATP-free and strongly bound actomyosin complex (termed the rigor state) is a key intermediate of the mechanochemical cycle. However, this rigor state possesses a very short lifetime and has a low abundance at physiological ATP concentrations.

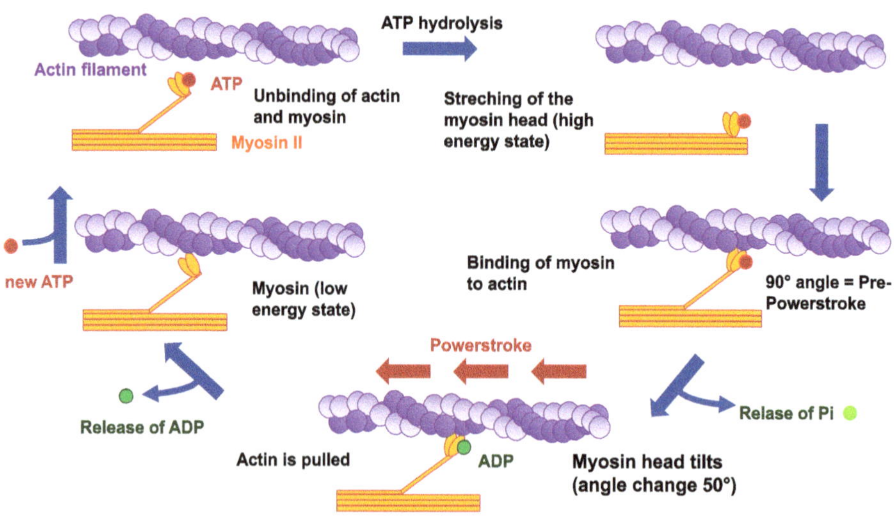

Figure 7.16. Actin and myosin cyclical interaction.

The formation of the strongly bound actomyosin complex seems to be a major contributor to the free energy necessary to perform the work during the powerstroke event (Karatzaferi *et al* 2004).

7.8 The effect of actin-bunding proteins on cell migration and invasion

These actin-bundling proteins are molecules that crosslink actin filaments into bundles. It is known that the actin cytoskeletal architecture regulates multiple cellular processes related to cancer invasion. An alteration of the relative levels of two profilin paralogs affects the architecture of the actin cytoskeleton and hence influences the physiological behavior of cancer cells. Functional divergence has been found between profilin-1, which induces membrane protrusion, motility and invasion, and profilin-2, which represses all the processes promoted by profilin-1. In order to provide these opposing phenotypes, the two profilins need to be specific to actin cytoskeletal remodeling processes. Profilin-2 performs these suppressive effects via the selective interaction with the actin polymerization regulator EVL that supports the profilin-2-driven actin cytoskeletal remodeling and subsequently reduces cell migration and invasiveness. Analyses of breast tumors indicate that lower EVL expression is associated with high invasiveness and hence poor prognosis, which reveals EVL as a potential candidate for a biomarker predicting the patient outcome.

Cellular motility needs to be precisely regulated of multicellular linearly organized processes that include the dynamical reorganization of the actin cytoskeleton. This reorganization is regulated by ABPs facilitating the nucleation, branching, elongation, bundling, severing and capping of actin filaments (Pollard and Borisy 2003, DesMarais *et al* 2005, Insall and Machesky 2009). In more detail, profilins play a key role in the regulation of actin polymerization, facilitating the conversion of ADP-actin to ATP-actin and thereby interacting with poly-L-proline domains (PPP[A/P]PPLP termed the PLP domain). These PLP domains have been detected in various actin nucleation promotion factors, actin nucleators and elongation factors acting on the barbed end of actin filaments. These factors comprise WASP/WAVE/SCAR, formins, and Ena/VASP proteins (Mena, VASP and EVL) (Reinhard *et al* 1995, Gertler *et al* 1996, Mahoney *et al* 1997, Lambrechts *et al* 2000, Ferron *et al* 2007, Jockusch *et al* 2007). The orchestrated regulation of these actin polymerization factors causes specific alterations in the architecture of the entire actin cytoskeleton. These cytoskeletal alterations control cellular processes influencing cell motility and thereby are able to either induce or impair invasive migration. Four isoforms of profilin are characterized. Since profilin-1 is ubiquitously expressed, other isoforms of profilin-1 display a more selective expression in distinct tissues. Whether profilin-2/3/4 isoforms possess distinct functional activities regarding actin polymerization or simply protect against the loss of one isoform by genetic or epigenetic alterations is not yet well understood.

Using a siRNA high-throughput screening approach, the suppression of PFN1, a ubiquitously expressed profilin isoform, has been found to inhibit cell migration in

MCF-10A mammary epithelial cells, whereas the downregulation of PFN2 increased the MCF-10A cell migration (Simpson *et al* 2008). Profilin-2 has been initially identified as a neuronal-specific isoform (Honore *et al* 1993, Witke *et al* 1998), however, it is expressed in wide variety of tissues such as breast epithelium. Despite the structural similarities in the PLP binding sites of profilin-1 and profilin-2 (Lambrechts *et al* 1997, Witke 2014, Kursula *et al* 2008), these two profilins are variable in surface charge distributions at these PLP sites (Nodelman *et al* 1999). Indeed, differences in ligand binding preferences are found between the two profilins using *in vitro* binding assays or mass spectrometry analysis of profilin-1 and profilin-2 binding proteins (Miki *et al* 1998, Witke *et al* 1998, Nodelman *et al* 1999, Lambrechts *et al* 2000, Veniere *et al* 2009). Whether these binding differences indeed alter actin-based cellular processes has to be figured out mechanistically. As expected it has been shown that the alteration of profilin-1 or profilin-2 levels causes pronouncedly different effects on the organization of the actin cytoskeleton, which then affects cell migration and invasiveness. Profilin-2 regulates the protrusive activity and migratory behavior of normal and cancer cells through supporting the EVL-based polymerization of long actin filaments, which forms contractile bundles. Moreover, the decrease of profilin-2 or EVL pronouncedly increases the invasion *in vitro* and *in vivo*, and thus the expression profiles of these actin regulators are associated with the tumor grade and invasiveness of distinct human tumors.

The assembly of stress fibers

The assembly of stress fibers is controlled in time and space through signaling pathways that function upstream of ABPs. In this process, the small GTPases RhoA, Rac1 and Cdc42 play central regulatory roles in providing actin dynamics in many eukaryotic organisms (Heasman and Ridley 2008). Among these GTPases, at least RhoA directly induces the stress-fiber assembly through its effectors, the Rho-associated protein kinase (ROCK) and the formin mDia1 (mammalian Dia1 or synonymously termed DIAPH1 or DRF1) (Watanabe *et al* 1997, Leung *et al* 1996). In particular, mDia1 regulates the polymerization of long parallel actin filaments and play a prominent role in the formation of dorsal stress fibers (Tominaga *et al* 2000, Hotulainen and Lappalainen 2006), whereas ROCK impairs ADF- or cofilin-driven disassembly of actin filaments through the activation of the LIM domain kinase 1 (LIMK1) (Maekawa *et al* 1999). In addition to the direct effects on the actin cytoskeleton, the RhoA signal transduction determines the transcription of several genes encoding cytoskeletal proteins through interaction with the myocardin-related transcription factor (MAL or synonymously termed MKL1 or MRTF-A) and serum response factor (SRF) pathway, and subsequently guides the composition of the actin cytoskeleton (Hill *et al* 1995, Miralles *et al* 2003).

The small GTPases Rac1 and Cdc42 both coordinate stress-fiber assembly indirectly. In more detail, Rac1 promotes the formation of lamellipodia and membrane ruffles through activation of the actin-nucleating complex Arp2/3, whereas Cdc42 regulates the formation of filopodia by inducing the polymerization of actin through the formin mDia2 (also known as DIAPH3 and DRF3) (Pollard 2007). Arp2/3-nucleated filaments and mDia2-nucleated actin filaments fulfill both their function as building blocks for contractile stress fibers in human osteosarcoma U2OS cells. In particular, Arp2/3 initiates the formation of α-actinin-crosslinked actin filaments, which then assemble endwise with mDia2-facilitated and tropomyosin-decorated actin filaments to finally generate transverse arcs near the cell's leading edge (Hotulainen and Lappalainen 2006, Tojkander *et al* 2011). Actin filaments, which are generated in filopodial protrusions, are shown to be recyclable for the construction of stress-fiber structures (Anderson *et al* 2008, Nemethova *et al* 2008). Both formins, mDia1 and mDia2, are also activated by the small RhoA-related GTPase RhoF (also known as ARHF and Rif), and hence seem to contribute to stress-fiber formation in distinct cell types (Fan *et al* 2010, Tojkander *et al* 2011).

Live-cell imaging microscopy has shed more light on the assembly mechanisms of different types of stress fibers. Dorsal stress fibers connected to focal adhesions at their distal ends are created through actin polymerization at focal adhesions. During the forward movement of a cell, new focal adhesions appear and the elongation of dorsal stress fibers starts from these adhesion sites (Hotulainen and Lappalainen 2006). Moreover, the polymerization of dorsal stress fibers is tightly connected to the assembly of a contractile actin stress-fiber network. During the formation of a protrusion, the cell membrane at the leading edge performs ongoing constant cycles of membrane extension and retraction. During the retraction phase, precursors of transverse arcs translocate into the lamellipodium through the process of actin and myosin bundle condensation (Burnette *et al* 2011). More precisely, these structures are generated through, on the one hand, the endwise assembly of Arp2/3-nucleated and α-actinin-crosslinked actin-filament bundles and, on the other hand, the endwise assembly of formin-nucleated actin bundles with embedded tropomyosin and myosin (Hotulainen and Lappalainen 2006, Tojkander *et al* 2011). Moreover, these nascent arcs move toward the cell's center and collide there with more immobile focal adhesions (Shemesh *et al* 2009). The coupling of actin filaments to adhesion sites may reduce the velocity of the arc precursors (Hu *et al* 2007, Gardel *et al* 2008, Burnette *et al* 2011), which then condense into mature arcs at the lamellipodium–lamella interface (Shemesh *et al* 2009). Early in the assembly process of the stress fibers, the connections between newly formed arcs and elongating dorsal stress fibers can be observed, and even several arcs can associate with a single dorsal stress fiber and flow towards the cell's center together with the still elongating dorsal stress fiber (Tojkander *et al* 2011). Dissimilar to the assembly of dorsal stress fibers and arcs, ventral stress fibers are generated from the pre-existing dorsal stress fiber and the arc network (Hotulainen and Lappalainen 2006). In particular, ventral stress fibers are formed by the fusion of two dorsal stress fibers that are attached to focal adhesions (Small *et al* 1998). A schematic model describing these mechanisms that control stress-fiber assembly in cells as U2OS is presented in figure 7.17.

7.9 The actin-binding proteins

The actin treadmilling contributes to dynamical filament assembly and disassembly, but other processes such as bound actin-binding proteins (ABPs) additionally account for the filament dynamics found in living cells. Thus, many ABPs regulate

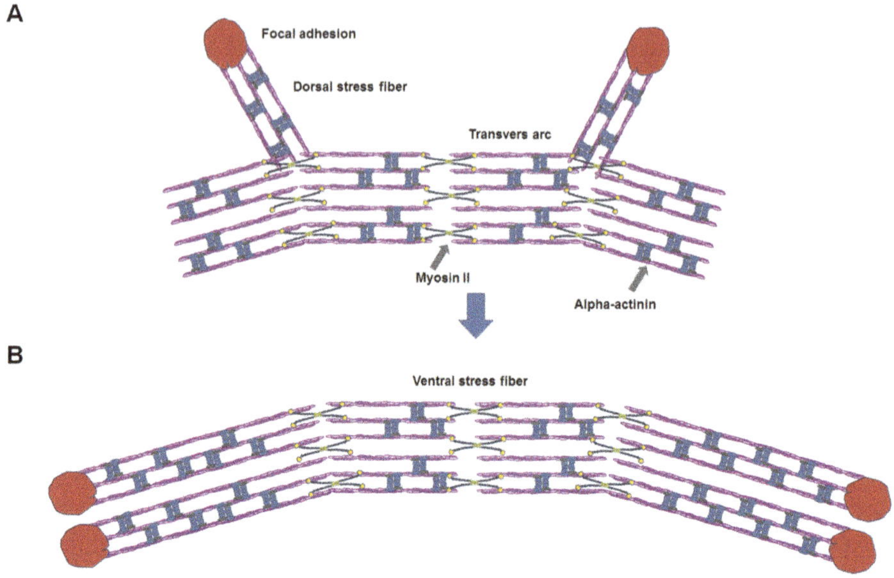

Figure 7.17. Theoretical model for the assembly of dorsal stress fibers. (a) Mature transvers arcs become connected with focal adhesions to build a stress-fiber network. The transverse arcs display a network of α-actinin crosslinks and myosin contractility. (b) Retrograde flow leads to actin scaffold localization towards the cell's center. A ventral fiber is formed from two dorsal fibers with focal adhesion attachments and a central transverse arc.

actin cytoskeletal dynamics (Pollard and Borisy 2003), but they have diverse other functions such as actin-monomer sequestration, filament capping at barbed (+) and pointed (−) ends, filament severing and finally also the crosslinking of filaments.

Although G-actin-binding proteins have several functions, are found in many structural families and bind to different actin-binding sites, the ABPs bind preferentially in the target-binding cleft of actin by inserting an amphiphilic helix (Dominguez 2004). Hence, the overall position of these helices is conserved, but the directionality of the polypeptides varies.

7.9.1 DNAse I

Actin and DNAse I build *in vivo* and *in vitro* a complex with high-affinity, which was the first crystal structure of actin to have been crystalized (Rohr and Mannherz 1978, Mannherz *et al* 1980). The binding of monomeric G-actin to the DNase I impairs their DNA-degrading activity and hence possibly inhibits the induction of programmed cell death such as apoptosis.

7.9.2 Gelsolin

Gelsolin is an actin-severing and actin-capping protein and belongs to a family with other proteins such as adseverin, villin, capG, advillin and supervillin (Silacci *et al* 2004). In more detail, gelsolin has six homologous domains (called G1 to G6), is affected by Ca^{2+}, phosphoinositides and regulated through phosphorylation at tyrosine residues. When Ca^{2+} is absent, gelsolin is inactive and stabilized by interactions masking the actin-binding sites and has hence undergone a conformational change. Two major binding sites for actin are in gelsolin's G1 and G4 domains. Both bind similarly to actin by inserting a helix into the target actin-binding cleft. For the filament severing function of Ca^{2+} activated gelsolin, a three-step model has been proposed (Burtnick *et al* 1997). The first step is that Ca^{2+} binding induces a conformational change partly exposing actin-binding sites within gelsolin. The second step involves a transitional complex in which gelsolin binds to the actin filament. In the third step, gelsolin fully exposes its actin-binding sites leading to engagement of actin protomers on two different actin-filament strands by G1 and G4 and severing of the actin filament through a coordinated pincer-like movement.

In addition to gelsolin's localization in the cytoplasm, a slightly longer gelsolin isoform can circulate in blood plasma. There gelsolin is part of a homeostatic mechanism called the actin-scavenger system (Lee and Galbraith 1992). Under certain conditions such as cell death or tissue injury, the actin protein is released into the blood. In order to remove the actin from the blood plasma the actin-scavenger system depolymerizes filamentous actin. This mechanism involves several steps. In the first step gelsolin severs actin filaments and in the second step gelsolin builds a high-affinity complex for monomeric actin together with vitamin D binding protein (DBP). In the third step this complex is largely cleared from the plasma by transport to the liver.

7.9.3 Profilin

Profilin works together with thymosin-β4 to maintain a large fraction, approximately 50%, of the cellular actin in the unpolymerized pool at a concentration of approximately 20–100 μM. The concentration is even 200-fold to 1000-fold higher than the critical concentration needed for barbed (+) end polymerization. In the cells there are commonly only low amounts of free ATP-actin present (0.1–1 μM) and actin bound to thymosin-β4-actin cannot polymerize, instead actin bound to profilin at a typical intracellular concentration of 5–40 μM is the main source for actin monomers to polymerize to filamentous actin (Pollard and Borisy 2003). Profilin has several functions and hence it plays a major role in actin-filament polymerization (dos Remedios *et al* 2003). First, profilin can induce the exchange of ADP for ATP on actin, which increases the pool of ATP-actin monomers available for actin polymerization. Second, profilin impairs the nucleation of filaments and hence provides an explanation for why filament nucleators are necessary. Third, actin monomers bound to profilin are not able to bind to the pointed (−) ends of actin filaments, however, they account for the growth of actin-filament barbed (+) ends at rates similar to unbound actin monomers. Fourth, as profilin possesses a higher affinity towards actin, it binds faster and more efficiently than thymosin-β4 to actin and, fifth, in addition to actin-binding, profilin binds simultaneously to prolin-rich sequences even with higher affinity by forming a ternary complex than to separate binding to one (Ferron *et al* 2007). The prolin-rich sequences are highly found among many cytoskeletal proteins, which is supposed to regulate the binding of profilin–actin complexes to actin filaments. Nonetheless, the elongation rate of actin filaments is enhanced through binding of profilin–actin complexes to the prolin-rich FH1 domain of formin (Kovar *et al* 2006). As with the majority of ABPs, profilin binds to the target-binding cleft of actin and localizes to its back, which enables profilin to bind its C-terminal to a prolin-rich profilin-binding sequence to WH2-related sequences such as the GAB domain of the Ena/VASP protein.

7.9.4 The actin-depolymerizing factor (ADF)/cofilin

Classification of the superfamily

Among the actin-depolymerizing factor (ADF)/cofilin superfamily, the protein cofilin-1 (Cfl-1) has gained much attention (Wang *et al* 2007, Huang *et al* 2014, Chang *et al* 2015) together with other members of the family (Estornes *et al* 2007, Bockhorn *et al* 2013). The actin-depolymerizing factors (ADF)/cofilin play a role in recycling actin monomers through actin-filament depolymerization. In some processes, such as membrane ruffling or cytokinesis, the fast turnover of the cell's cytoskeleton is favorable (Bernstein and Bamburg 2010). Cfl-1 is present in non-muscle cells, whereas Cfl-2 is found in muscle cells. Cfl-2 seem to have at least two variants that are generated by alternative splicing of a single gene CFL2 (Thirion *et al* 2001) such as Cfl-2b (present in skeletal muscle and heart) and Cfl-2a (found in various tissues). In addition, Dstn encoded by the DSTN gene has been found in various tissues. As various proteins with actin-depolymerizing activity have been found in vertebrates (Bamburg *et al* 1980, Thorstensson *et al* 1985), the actin-depolymerizing proteins are initially characterized and grouped by molecular weight (MW) into approximately 19 kDa (Bamburg *et al* 1980) or approximately 93 kDa (Thorstensson *et al* 1985). Proteins of the 19 kDa group are termed cofilins, as they form cofilaments with actin (Nishida *et al* 1984). Another similar protein in this group is destrin (Dstn or, as an old alternative name, ADF) that destroys F-actin (Vartiainen *et al* 2002). An actin-depolymerizing protein of the 93 kDa is gelsolin (Thorstensson *et al* 1985, Li *et al* 2012). The alternative name ADF is also used for gelsolin. The traditional group of cofilins in vertebrates is formed by the three closely related actin-depolymerizing proteins Cfl-1, Cfl-2, and Dstn (ADF). This family is now called ADF/cofilin family (Moon and Drubin 1995, Ono *et al* 1999, Vartiainen *et al* 2002). In addition, a special actin-binding domain (or module) of about 150 amino acids has been identified in each of the members of the ADF/cofilin family (Lappalainen *et al* 1998). These modules build specific three-dimensional structures with six-stranded mixed β-sheets and hence these modules are termed the actin-depolymerizing factor homology domains or ADF-H domains (Lappalainen *et al* 1998). Among the group are 39 proteins that have been found by searching for sequences similar to ADF/cofilins within databases and they are divided into three structurally distinct classes: the highly conserved ADF/cofilins, the twinfilins and the developmentally regulated brain proteins (drebrins)/actin-binding proteins 1 (Abp1s) (Lappalainen *et al* 1998). The first group, the ADF/cofilin superfamily, encompases the traditional cofilins such as Cfl-1, Cfl-2, Dstn, depactin and actophorin, and the members have significant structural similarity, as they possess a specific ADF-H domain.

The ADF/cofilins bind F-actin and sever the actin filaments. First, the severing of the actin filaments leads to depolymerization of actin filaments. Second, the severing of the actin filaments can cause *de novo* actin polymerization directly or indirectly through the production of free barbed ends (Bravo-Cordero *et al* 2013). ADF/cofilins can bind F-actin and G-actin in a 1:1 ratio (Nishida *et al* 1984, Lappalainen *et al* 1998), as the traditional ADF/cofilins contain two distinct actin-binding sites. One site is the G/F-site in the C-terminus and the other site is the F-site in the N-terminus. Whereas the F-site facilitates the binding of F-actin, the G/F-site is essential for the binding to both the G-actin and the F-actin (Shukla *et al* 2015). In particular, ADF/cofilins bind preferentially to ADP-forms of G- or F-actin and utilize energy from the hydrolysis of ATP in polymerization of actin (Carlier *et al* 1997). In addition to cofilins binding to actin, they can additionally bind to phosphatidylinositol 4,5-bisphosphate (PIP2) (Yonezawa *et al* 1991) and to the serine/threonine-protein kinase LIMK1 (Ivanovska *et al* 2013). Moreover, ADF/cofilins of vertebrates possess a nuclear localization site.

The function of cofilins

Cofilins of the ADF/cofilin superfamily modulate actin dynamics through the induction of actin depolymerization or polymerization through the severing of actin filaments. The effect of cofilins on actin-filament assembly or disassembly relies on the concentration of active cofilins, the relative G-actin concentration and specific protein factors. In low concentrations of ADF/cofilins, they sever the actin filaments and induce filament depolymerization. In contrast, a high concentration of ADF/cofilins is assumed to facilitate the actin nucleation and polymerization (Andrianantoandro and Pollard 2006). Indeed, cofilins promote actin polymerization by the production of free barbed ends and the supply of actin monomers. However, Cfl-1 is known to regulate actin nucleation and filament branching in synergy or even in competition with the Arp2/3 complex. In particular, the Arp2/3 protein complex is a seven-subunit complex of actin-related proteins, which promote the binding to actin, and initiate nucleation and assembly of actin branches (Dos Remedios *et al* 2003, Winder *et al* 2005). Indeed, actin branching is a key event in the generation of lamellipodia, which are required for cellular motility. Cfl-1 and Arp2/3 are reported to act in synergy in producing free barbed ends for actin polymerization (Tania *et al* 2013). In parallel, Cfl-1 reduces the affinity of the Arp2/3 complex to actin filaments inducing the dissociation of 'old' actin branches (Chan *et al* 2009). Both Cfl-1 and Cfl-2 can regulate the assembly of actomyosin complexes through inhibition of the tropomyosin and myosin II binding of actin filaments (Nishida *et al* 1984, Nakashima *et al* 2005). *In vivo* cofilins are shown to reorganize the actin cytoskeleton due to external stresses and different cell stimuli (Yahara *et al* 1996). The overexpression of cofilins causes the assembly of stress fibers. These contractile actin bundles are found in non-muscle cells fulfilling a key role in cellular contractility necessary for proper cell adhesion and migration (including the associated protrusion formation such as the assembly of lamellipodia and filopodia) and morphogenesis (Tojkander *et al* 2012). Hence, cofilins are seen as molecular modulators of development processes. Indeed, Cfl-1 and destrin are needed for ureteric bud branching morphogenesis (Kuure *et al* 2010) and, moreover, Cfl-1 is required for dynamic alterations in the cytoskeleton needed for axon engagement, which is essential for myelination of Schwann cells (Sparrow *et al* 2012). There exists evidence for the involvement of Cfl-1 (and also the Arp2/3 complex) in the regulation of axonal growth cones (Dumpich *et al* 2015). Cofilin partly regulates the cell proliferation upon response to mechanical stresses. In mammalian epithelial cells, cofilin impairs through diminishment of the cytoskeleton the remodeling activity of yes-associated protein 1 (YAP1) and translin-associated zinc finger protein 1 (TAZ1), which are both promotors of the Hippo signaling pathway and hence of organ growth and cell proliferation (Aragona *et al* 2013). In addition, Cfl-1 plays a role in development (Ohashi 2015). In summary, ADF/cofilins are key players in the regulation of actin dynamics and have a dual effect on actin filaments and may hence contribute to cellular contractility through two mechanisms: first, the local actin depolymerization and, second, the assembly of stress fibers. Taken together cofilins are key players in morphogenesis and development.

Additionally, cofilins fulfill functions in cells that are not related to actin dynamics regulation. For example, cofilins transport actin molecules possessing no nuclear localization signals to the nucleus (Nebl *et al* 1996). The activity of ADF/cofilin superfamily members can be regulated by different mechanisms such as pH, phosphatidylinositols, protein kinases and phosphatases, as well as several other proteins. ADF/cofilin superfamily member activities even depend on the cellular redox status as the F-actin-binding and depolymerizing activity of cofilins is pH-dependent. *In vitro*, in an F-actin-containing protein assembly/disassembly model system, at pH < 7.3 the concentration of monomeric actin (G-actin) is less than 1 μM, even with an excessive amount of cofilin (Yonezawa *et al* 1985). In contrast, at pH > 7.3 the concentration of G-actin increases proportionally to the concentration of added cofilin, until the total depolymerization of F-actin is achieved. In conclusion, cofilin is able to reversibly control actin polymerization and depolymerization dependent on the pH. Additionally, the pH has been shown to regulate the activity of cofilin activity *in vivo* (Bernstein *et al* 2000). pH sensitivity is not common to all ADF/cofilins in all species, as mouse Cfl-1 (dissimilar to human Cfl-1) as well as mouse Cfl-2 are not pH-independent (Vartiainen *et al* 2002).

Membrane phosphoinositides such as PIP2 can affect the regulation of ADF/cofilin activity. Cofilins directly bind to phosphatidylinositols and their PIP2-binding area overlaps with the actin-binding site, which causes the inhibition of actin binding upon PIP2 binding of cofilins (Yonezawa *et al* 1991). Alterations in PIP2 density of the cellular membrane affect the balance between membrane-bound and free, active ADF/cofilins (Zhao *et al* 2010).

In particular, the phosphorylation of Cfl-1 on a serine residue (Ser3) abolishes its binding to F- and G-actin (Moriyama *et al* 1996). Similarly results have been reported for phosphorylation of Dstn (Agnew *et al* 1995). Only the dephosphorylated (and hence active) cofilin fulfills the functions associated with the binding of actin and the protein translocations to the nucleus and mitochondrion. However, phosphorylated cofilin is necessary to activate phospholipase D1 (Bernstein and Bamburg 2010). The regulation of cofilins is facilitated by its phosphorylation or dephosphorylation through signaling pathways involving kinases and phosphatases due to extracellular signals and alterations in the microenvironment (Wang *et al* 2007, Tomasella *et al* 2014, Chang *et al* 2015). In mammals, Cfl-1 can be phosphorylated and hence inactivated by LIM-kinases such as LIMK1 and LIMK2 as well as testicular protein kinases such as TESK1 and TESK2. In turn, cofilin is dephosphorylated and hence activated by slingshot protein phosphatases (SSH1, SSH2, SSH3), protein phosphatases 1 and 2A (PP1, PP2A) and chronophin (CIN) (Aragona *et al* 2013). These events of cofilin phosphorylation and dephosphorylation are important for the alteration of actin dynamics, affecting cell motility and morphogenesis (Abe *et al* 1996, Obinata *et al* 1997). Thus, kinases and phosphatases of cofilins are critical in development. The overexpression of LIMK1 or the inactivation of SSH1 causes abnormal accumulation of F-actin and thereby incorrect cytogenesis during mitosis (Davila *et al* 2007). As LIMK1 inactivates cofilin, it decreases the formation of lamellipodium and reduces cell migration (Zebda *et al* 2000). Indeed, the addition LIMK1 inhibitor to Jurkat T cells impairs the stromal cell-derived factor (SDF) 1α-induced chemotaxis of T cells (Nishita *et al* 2002). Thus, the LIMK1-catalyzed phosphorylation of cofilin seems to be required for chemotactic response of T lymphocytes. In contrast, a non-phosphorylatable mutant cofilin can still generate protrusions and control the direction of cell migration (Ghosh *et al* 2004). What role does the phosphorylation or dephosphorylation of cofilin play? Further experiments are required to confirm the positive role of LIMK1 in migration of chemokine-stimulated Jurkat T cells. Indeed, the cell migration can be suppressed by LIMK1 knockdown, whereas a knockdown of SSH1 induces the formation of lamellipodia around the periphery of the cell upon cell stimulation (Nishita *et al* 2005). These results demonstrate that LIMK is essential for the generation of multiple lamellipodia in the initial stages of the cell response and SSH1 impairs the lamellipodial protrusion exertion for directional cell migration (Mizuno 2013). Although LIMK acts as a positive regulator of cell migration, the underlying mechanisms are still not well known.

In addition to the regulation through kinases and phosphatases, the interaction of ADF/cofilins with actin may also be directly or indirectly regulated by a large number of proteins. An example for such an interaction is the binding of cofilin to cortactin, inactivating the function of cofilin and is typically found in podosomes and invadopodia, which are actin-based protrusions that display a high dynamical remodeling and are produced by invasive cancer cells, vascular peripheral blood cells and macrophages (Beaty and Condeelis 2014, Zalli *et al* 2016). Actin-interacting protein 1 (AIP1) and cyclase-associated protein 1 (CAP1) induce the disassembly of cofilin-bound actin filaments (Zhou *et al* 2014, Nomura *et al* 2016). Coronin regulates the recruitment of cofilin to actin-filament sides and thus enhances the severing of actin filaments (Mikati *et al* 2015). The Rho GTPases are key regulators of actin dynamics such as the formation of stress fibers and additionally are associated with the regulation of ADF/cofilins through LIMK. In particular, RhoA activates the Rho-associated coiled-coil forming kinase (ROCK) that in turn phosphorylates and thereby activates LIMK. Indeed, RhoA stabilizes the stress fibers and thereby prevents depolymerization of actin filaments through the phosphorylation of cofilin and the Rho–ROCK–LIMK–cofilin pathway, altering the actin assembly in many cell types due to microenvironmental stimuli (Ohashi 2015). Epidermal growth factor (EGF) can alter the LIMK pathway or phospholipase C-facilitated hydrolysis of PIP2 and subsequently release of cofilin from membrane sequestering (Mouneimne *et al* 2004). These mechanisms, involving the activation of cofilin and the generation of free barbed ends for lamellipodial extension due to EGF stimulation, have been revealed mainly for migrating malignant cancer cells (Chan *et al* 2000, Mouneimne *et al* 2004). In contrast, an enhanced cofilin-dependent severing activity upon EGF stimulation cannot in every case be correlated with the level of dephosphorylated cofilin (Song *et al* 2006), which suggests a more complex and not yet revealed regulatory mechanism.

The cellular redox state seems to be important for the regulation of ADF/cofilins. The regulation of the redox state is provided by oxidative post-translational modifications of Cys residues such as S-glutathionylation (Fratelli *et al* 2002), disulfide bonds (Klemke *et al* 2008) and S-nitrosylation (Zhang *et al* 2015). These redox-related alterations affect the activity of cofilin and cofilin associated signaling pathways. Cofilin has been identified as a target for oxidation under oxidative stress in T cells. In more detail, the oxidation of cofilin causes the formation of intramolecular disulfide bonds and the dephosphorylation at Ser3. However, dephosphorylated oxidized cofilin can bind to F-actin, but it cannot fulfill its actin-depolymerizing function and hence the amount of F-actin increases (Klemke *et al* 2008). Instead, the oxidized cofilin translocates actin to the mitochondria, where a cytochrome-c release is triggered by the opening of the permeability transition pore. Finally, the mitochondria are damaged, which induces apoptosis (Klamt *et al* 2009).

Thus, the cellular microenvironment (such as pH, phosphoinositides and proteins including enzymes) alter pronouncedly the function of cofilin. Other members of the ADF/cofilin superfamily display only some features of regulation. In particular, twinfilins can induce filament severing, which is also pH-dependent. In contrast to ADF/cofilins, TWF-1 severs actin filaments *in vitro* at pH under 6.0 (Moseley *et al* 2006). Twinfilins are able to bind PIP2 similar to ADF/cofilins and in turn this interaction decreases their actin-binding, actin-filament severing and actin-monomer sequestering activities (Vartiainen *et al* 2003, Moseley *et al* 2006, Nevalainen *et al* 2011). Both TWF-1 and TWF-2 manage to bind to a capping protein (CP), which abolishes directly the severing activity of TWF-1 (Moseley *et al* 2006). The small GTPases Ras-related C3 botulinum toxin substrate 1 (Rac1) and cell division control protein 42 homolog (Cdc42) can both facilitate the localization of TWF-1 toward membrane ruffles and cell–cell adhesion contacts, whereas they cannot alter the localization of TWF-2 (Vartiainen *et al* 2003). The cyclin-dependent kinase 5 (Cdk5) facilitates drebrin phosphorylation that drives the cytoskeletal reorganization involved in neuronal migration. Moreover, drebrin E is phosphorylated on Ser142, and drebrin A on Ser142 or Ser342 (Tanabe *et al* 2014). The location of drebrin to the distal part of axonal filopodia and the branching in drebrin overexpressing neurons are both negatively regulated by myosin II (Ketschek *et al* 2016). ADF/cofilins and GMF-family proteins are known to be regulated by phosphorylation. In particular, GMF-G is phosphorylated at Tyr104 by Abelson tyrosine-protein kinase 1 and induces the dissociation of GMF-G from Arp2/3 leading to actin disassembly and subsequently smooth muscle contraction (Wang *et al* 2014). The subfamily of the Rho-GTPase Rac is associated with the regulation of coactosin activity. As a response to Rac induced signaling, coactosin is translocated to lamellipodia and filopodia, increasing the polymerization of actin and enhancing neural crest cell migration (Hou *et al* 2013).

The ADF/cofilin superfamily proteins perform multiple roles in cells, that range from physiological roles in proliferation and migration of mammalian cells to pathological roles during diseases such as cancer, whereas they promote the malignant phenotype of cancer cells and hence they are highly important for molecular and physical oncology. Taken together, there is no clear model of how cofilin functions in malignant cancer progression. Cfl-1 has been found in different cancer cell lines and cancer tissues such as adenocarcinomas (Yan *et al* 2007, Zhang and Tong 2010, Shishkin *et al* 2011, Wei *et al* 2012, Neely *et al* 2016), osteosarcoma (Zhang *et al* 2014), lymphoid tissue neoplasms (Jiang *et al* 2011), astrocytoma (Yan *et al* 2012), glioma (Du *et al* 2015) and neuroblastoma (Patil *et al* 2015). Finally, it can be concluded that Cfl-1 may be a common component in various cancer phenotypes.

The overexpression of Cfl-1 is correlated positively with cancer cell proliferation, invasion and metastasis (Wang *et al* 2007, Zhang and Tong 2010, Nagai *et al* 2011, Yang *et al* 2013). Moreover, the expression of the dephosphorylated and hence active form of cofilin is enhanced in aggressive and invasive cancer cells (Nagai *et al* 2011). In contrast, there are still a few opposing reports. A first example is the overexpression of Cfl-1 that repressed growth and invasion of non-small-cell lung cancer (Tsai *et al* 2015) and a second example is the phosphorylation of cofilin that has been increased in bladder cancer compared to normal bladder tissue resections or biopsies (Chung *et al* 2013). However, the molecular mechanisms of Cfl-1 during expression of the malignant phenotype of cancer cells are not yet fully understood. In cancer cells, the actin dynamics and cell motility are induced by alterations within the tumor microenvironment. EGF, transforming growth factor-α (TGF-α), stromal cell-derived factor 1 (SDF1) and heregulin have all been found to be associated with cancer cell migration and correlate positively with the malignant progression of certain tumor types (Wang *et al* 2007). EGF-induced dephosphorylation and activation of Cfl-1 enhances the F-actin-severing activity of cofilin and creates free barbed ends, which are essential in lamellipodial extension and chemotaxis toward EGF, promoting cancer invasiveness and finally metastasis (Mouneimne *et al* 2004, Wang *et al* 2006). Thus, an increased amount of dephosphorylated Cfl-1 seems to be an indicator for the malignant phenotype of cancer cells (Nagai *et al* 2011, Ito *et al* 2015, Dubois *et al* 2016). Overexpression of a non-phosphorylatable cofilin mutant (cofilin-S3A) in astrocytoma cells induced a highly invasive phenotype compared to cells expressing wild-type cofilin (Nagai *et al* 2011).

The nuclear translocation of dephosphorylated Cfl-1 supports the malignant phenotype of cells. Moreover, dephosphorylated Cfl-1 facilitates the transport of G-actin to the nucleus. In turn, nuclear actin is associated with chromatin remodeling, transcription, RNA processing, intranuclear transport and nuclear export, and provides the maintenance of the nuclear skeleton (Migocka-Patrzalek *et al* 2015). In line with this, gene expression alterations during the malignant progression of cancer may be evoked by Cfl-1 through the translocation of actin. Another mechanism for providing the malignant phenotype of cancer cells is related to Cfl-1 dephosphorylation and nuclear translocation (Samstag *et al* 1994, Samstag *et al* 1996). In nontransformed T lymphocytes, cofilin participates in a costimulatory pathway inducing T-cell proliferation in order to produce IL-2. Upon ligand attachment to accessory receptors such as CD2, cofilin is dephosphorylated and translocates in the nucleus. In malignant T lymphoma cells, dephosphorylation and nuclear translocation of cofilin can even be seen spontaneously through the constitutive activation of serine protein phosphatase. Moreover, these events cause the proliferation of T cells and inhibits apoptosis (Samstag *et al* 1994, Samstag *et al* 1996).

Cofilin activation/inactivation is challenged by alterations in the balance of kinases, phosphatases and other cofilin upstream regulatory proteins. These alterations evoke the early steps of cancer cell invasiveness and metastasis (Zebda *et al* 2000). The SSH1, which is a well-studied cofilin phosphatase, has been demonstrated to be increased in a variety of invasive cancer cells. Moreover, the overexpression of slingshot-1L (SSH1L) in pancreatic cancer showed that it contributes to cancer cell migration (Wang *et al* 2015). Indeed, this phosphatase is activated by F-actin which is formed during the formation of lamellipodia in malignant cancer cells (Kurita *et al* 2008). The phosphorylation and hence inhibition of SSH1L by protein kinase D (PDK) reduces the invasion of cancer cells (Peterburs *et al* 2009). However, the role of the kinase LIMK1 in cancer invasion and metastasis is still under strong debate, as is its role in cellular motility, as LIMK1 can either cause a decrease (Zebda *et al* 2000, Wang *et al* 2006) or an increase (Yoshioka *et al* 2003, Nishita *et al* 2005, McConnell *et al* 2011) in cellular invasiveness and metastasis. In metastatic rat mammary adenocarcinoma cells, expression of the kinase domain of LIMK1 causes nearly total phosphorylation of cofilin, which abolishes the appearance of barbed ends and lamellipodia protrusion upon EGF stimulation (Zebda *et al* 2000). The overexpression of LIMK1 decreased EGF-induced membrane protrusion and migration of rat mammary carcinoma cells (Wang *et al* 2006). In contrast, the increased LIMK1 activity induces human breast cancer progression (McConnell *et al* 2011). Moreover, it has been revealed that the amount and activity of endogenous LIMK1 was elevated in invasive breast and prostate cancer cell lines in comparison with less invasive cancer cells (Yoshioka *et al* 2003). The knockdown of LIMK1 has been found to repress the chemokine-induced assembly of the lamellipodium and impair the migration of Jurkat T cells (Nishita *et al* 2005). These data on the positive role of LIMK1 in cancer cell migration seem at first glance to contradict the mechanism of tumor progression that is correlated with Cfl-1 dephosphorylation. Hence, it has been suggested that LIMK1 regulates tumor progression via other mechanisms that are independent of cofilin. However, LIMK1 clearly determines the metastatic phenotype of cancer cells through regulation of cofilin activity and controversial effects of LIMK1 expression on migration and metastasis of cancer cells require additional experimental verification or explanations. As the LIMK1 expression separately cannot determine the motility and invasion phenotypes of carcinoma cells, the collective activity and output (the barbed end production) of the LIMK1/cofilin pathways need to be determined (Wang *et al* 2006). However, the contradictory results may be due to the different cell types investigated in these studies.

It is known that oncoproteins and tumor suppressor proteins regulate the invasive and metastatic potential of primary tumors through cofilin-regulating pathways. Among the most prominent oncoproteins is the tyrosine-protein kinase transforming protein of Rous sarcoma virus (v-Src) that is able to inhibit the function of the Rho–ROCK-LIM kinase pathway leading to the dephosporylation of Cfl-1 and subsequently to elevated levels of active Cfl-1 (Pawlak and Helfman 2002). In contrast to the oncogene Src, the tumor suppressor protein phosphoinositide phosphatase and tensin homolog (PTEN) can inactivate cofilin in cancer cells, whereas the loss of PTEN and the activation of phosphoinositide 3-kinase (PI3K) promote a differential activation of the cofilin regulators such as LIMK1 and the cofilin phosphatase SSH1L and hence cofilin dephosphorylation induces the formation of tubuin-based microtentacles increasing the metastatic risk (Vitolo *et al* 2013). Moreover, the tumor suppressor Ras has been found to interact with the domain-containing protein 1 (RASSF1A) inhibiting tumor growth by stimulation of the cofilin/PP2A-driven dephosphorylation (Dubois *et al* 2016). Hence, the role of Cfl-1 as a major participant of various signaling pathways in invasive and malignant cancer cells needs still further investigation. However, contributions of Cfl-1 to the malignant phenotype in breast cancer have been revealed (Maimaiti *et al* 2017). In conclusion Cfl-1 may represent an interesting candidate for designing new early diagnostic tools or treatment for cancer.

7.9.5 Arp2/3

Arp2/3 proteins act as nucleation sites for actin-filament assembly and, similarly to cofilin or profilin, the Arp2/3 proteins induce a rapid turnover of actin filaments and hence the remodeling of the actin cytoskeleton necessary for cellular motility and the cell's shape. The Arp2/3 complex is composed of seven subunits, named Arp2, p16, p20, p21, p34, p40 and Arp3, and it plays an important role in the regulation of the cell's actin cytoskeleton. Two components, Arp2 and Arp3, are actin-related proteins and are similar in their structure to actin monomers. Arp2 and Arp3 serve as nucleators by providing a new nucleation site for new actin filaments on existing actin filaments inducing filament branching. In particular, the nucleation core activity of Arp2/3 is activated by members of the Wiskott–Aldrich syndrome family, such as WASP, N-WASP, WAVE and WASH proteins. The Arp2/3 complex is found in specific intracellular regions such as the leading edge of migrating cells such as lamellipodia.

7.9.6 G-actin-binding RPEL domain

Actin's role in the nucleus has been proposed, but its precise role is not yet well known. Moreover, it has been supposed that actin regulates the myocardin transcriptional coactivator MAL (Miralles and Visa 2006, Olson and Nordheim 2010), which moves in and out of the nucleus to affect the activity of serum response factor (SRF) as a transcription factor. In the N-terminus of MAL is a tandem repeat of three G-actin-binding RPEL domains with the consensus motif RPxxxEL. The activity of MAL is impaired by binding of actin to RPEL domains, which inhibits MAL accumulation in the nucleus and the interaction of MAL with SRF. This system is regulated by RhoA-induced alterations of G-actin to F-actin ratios. In more detail, increased cytoplasmic G-actin levels build the inactive actin–MAL complex via two possible interactions with the RPEL domains of MAL. Thus, decreased G-actin levels due to increased actin polymerization releases MAL from its interaction with actin and leads to MAL translocation into the nucleus (Miralles and Visa 2006).

7.9.7 The effect of small molecules on actin

Several macrolides isolated from marine organisms can interact with actin and thus impair actin polymerization (Yeung and Paterson 2002). Some of these substances have strong antitumor activities and may be of interest for anticancer drugs. Several drugs serve as a useful tool in investigating actin structures, dynamics and polymerization.

The fungal toxin cytochalasin D is a substance that inhibits by its binding to the barbed (+) end of actin filaments the polymerization or depolymerization of actin monomers. This behavior leads to cell shape alterations and possibly inhibition or reduction of cellular motility, such as cell division during mitosis, demonstrating that actin polymerization is required in cell division. Like most proteins, cytochalasin D associates to the target-binding cleft of actin.

Phalloidin binds to actin filaments and stabilizes them to impair depolymerization. Thus, fluorescently labeled phalloidin is used to visualize actin filaments in fixed cells using fluorescence microscopy.

Other examples are the latrunculin A and B substances derived from sponges and nudibranchs (Yarmola *et al* 2000). They both act in contrast to other marine macrolides, as latrunculin A and B bind in the nucleotide-binding cleft of actin and hence inhibit actin polymerization.

7.9.8 Effect of pathogens on actin assembly

Human pathogens such as *Salmonella*, *Shigella* and *Listeria* are able to disrupt or hijack the cell's cytoskeleton during infection. This represents a passive pathogen invasion mechanism. However, pathogens such as *Toxoplasma gondii* use their own actin cytoskeleton in order to actively invade the host cell (Gouin *et al* 2005). In more detail, *T. gondii* uses toxofilin, which is a G-actin-binding protein released during host cell invasion. In particular, toxofilin builds a high-affinity complex with an antiparallel actin dimer by inserting a helix into the target-binding pocket of one actin monomer.

Other pathogenic bacteria produce toxins such as ADP-ribosylate actin leading to cytoskeletal disorganization and finally death (Barth and Stiles 2008, Lang *et al*

2010). The *Clostridium botulinum* C2 toxin and *Clostridium perfringens* iota-toxin alters actin at Arg177 and impair its function.

7.10 Actin-binding domains and their roles in *de novo* actin polymerization

ABPs regulate the *de novo* polymerization of actin filaments involving actin-filament nucleation and elongation factors (Pollard 2007, Campellone and Welch 2010, Dominguez 2010) such as the filament nucleator Arp2/3 and the nucleation-promoting factors such as formins, Spire, Cobl, VopL/VopF, TARP and Lmod. The polymerization time and location are controlled, which has an influence on the actin networks within the cell. Filament nucleators (except of formins) interact with actin through the WASP-homology 2 (WH2) domain. Formins interact through the formin homology 2 (FH2) domain with actin to facilitate actin nucleation and actin-filament elongation at the barbed (+) end (Goode and Eck 2007).

7.10.1 The β-thymosin/WH2 domain

The WH2 domain of actin-binding proteins is abundant, but functionally diverse (Paunola *et al* 2002, Dominguez 2007) and is located in proteins regulating the sequestration of actin monomers and cytoskeleton scaffolding (Dominguez 2010). A similar architecture has been observed in Spire, Cobl, and VopL/VopF, in which tandem WH2 domains bind three to four actin subunits to assemble an actin-filament nucleating complex. The WH2 domain consists of 17–27 amino acids and its N-terminal region builds a helix facilitating the binding to the binding cleft of actin. Downstream of this helix in the WH2 domain is an extended region with the conserved motif LKKT(V), which is also located in the linker region between the domains G1 and G2 of gelsolin.

Thymosin-β4 consists of 43 amino acids, belongs to the β-thymosin family and has a structural relationship with the WH2 domain (Paunola *et al* 2002, Dominguez 2007). Similarly to the WH2 domain, the N-terminal portion of the β-thymosin fold contains a helix which interacts with the targeted binding cleft of actin and an LKKT motif linker is located downstream. In contrast to the WH2 domain, the C-terminal region consists of a helix binding on top actin subdomains 2 and 4. The combination of N- and C-terminal helices inhibit the actin-monomer addition to the barbed (+) and pointed (−) ends and reveals thymosin-β4 as a potent actin-monomer-sequestering protein. However, thymosin-β4–actin complexes do not regulate actin-filament nucleation or elongation. Thymosin-β4 seems to be an actin buffer and hence binds actin in competitive equilibrium to profilin, which possesses higher affinity for actin monomers and is a reservoir of actin monomers for actin-filament polymerization (Pollard and Borisy 2003).

7.10.2 The FH2 domain

Formins provide cell motility and morphogenesis as well as cytokinesis by facilitating the assembly of unbranched actin networks such as filopodia, actin stress fibers and even large actin cables (Goode and Eck 2007). The diaphanous-related formins contain a profilin-binding prolin-rich region, termed the FH1 domain. The recruitment of profilin–actin by the FH1 domain increases the formin-facilitated actin-filament elongation (Kovar *et al* 2006). The FH2 domain is downstream of the prolin-rich region, and regulates actin nucleation and actin-filament elongation. In addition, the structure of the FH2 dimer has two rod-shaped domains, which are flexibly linked in a head-to-tail fashion (Xu *et al* 2004). Formin consists of a helix that interacts with the target-binding cleft of actin.

7.10.3 The WH2 domain and filament elongation

Formins promote actin nucleation and filament elongation, in contrast, the WH2-based nucleators seem not act in filament elongation. One of the WH2-containing proteins, the Ena/VASP (Drees and Gertler 2008), seems to be involved in elongation, but not in actin nucleation. Similar to formins, the Ena/VASP proteins contain a central prolin-rich region and WH2-related sequences, termed G-actin-binding (GAB) and F-actin-binding (FAB) domains, respectively. Similar to the stimulation of formin-driven filament elongation produced by profilin (Kovar *et al* 2006), it is proposed that profilin–actin stimulates Ena/VASP-driven filament elongation (Dominguez 2010). The ternary complex of profilin–actin and a fragment of VASP containing a prolin-rich segment and the GAB domain leads to the hypothesis that these complexes can be delivered directly to the barbed (+) end of elongating actin filaments.

7.11 Microscopic visualization of F-actin in fixed cells and tissue samples

The onset of actin cytoskeleton visualization involved immunofluorescence techniques, which are based on antibody staining against fixed actin structures (Lazarides and Weber 1974). In order to overcome the artifacts produced by the fixation of the cells and to perform measurements over time in living cells, the problems of antibody-based labeling techniques, such as nonspecific binding, actin scavenging, high background fluorescence or variations in the epitope among different species, need to be overcome. Hence additional nonantibody-based actin-binding components with a high specificity for F-actin have been developed. Until now, the most commonly used marker for labeling endogenous actin filaments in fixed samples, such as cells, spheroids and tissues, has remained phalloidin, which represents a fluorescently labeled derivative of the phallotoxin that is able to bind to F-actin with high affinity (Wulf *et al* 1979; Vandekerckhove *et al* 1985). A disadvantage of phalloidin is that it is restricted to the labeling of fixed cells, as it only possesses low permeability for cellular membranes and even stabilizes the actin filaments *in vivo* and *in vitro*, which excludes it from usage in living cells (Wehland *et al* 1977, Coluccio and Tilney 1984, Visegrády *et al* 2004). The actin structures visualized by fluorescent phalloidin need to be carefully assessed as the necessary fixation and permeabilization procedures may alter the results of the labeling. Moreover, due to the permeabilization and fixation protocol, the formation of artifacts has been observed, such as the artifact caused by methanol fixation, which destroys the native F-actin confirmation. Another disadvantage is that fluorescent phalloidin has been shown not to label all endogenous F-actin in native cells, as the binding of cofilin to F-actin alters the F-actin structure such that phalloidin no longer interferes with F-actin (McGough *et al* 1997, Bunnell *et al* 2011). Another major disadvantage is that at least seven actin subunits are required for the binding of phalloidin to F-actin (Kristó *et al* 2016), which impairs the labeling of short F-actin polymers.

Hence, the ectopic expression of epitope-tagged actin such as HA-, Myc- or Flag-tagged actin in combination with immunofluorescence staining has been commonly utilized to analyze the actin cytoskeleton in cells (Copeland and Treisman 2002, Miralles *et al* 2003). However, a weakness is that the expression levels of tagged actin variants need to be controlled, as their overexpression may even lead to subtle alterations in the levels of G-actin and subsequently impact the physiological actin dynamics or even initiate the polymerization of actin (Mounier *et al* 1997, Ballestrem *et al* 1998). Moreover, this approach is also not well suited for live-cell imaging approaches, as additional immunofluorescence staining needs to be performed or actin needs to be labeled with directly stainable tags such as SNAP tags (Keppler *et al* 2004, Lukinavicius *et al* 2013) or tetracysteine tags (Griffin *et al* 1998). When actin is labeled with a tetracysteine tag it can be directly visualized by using biarsenical dyes such as FlAsH and ReAsH (Martin *et al* 2005). As the tetracysteine tag has a small size of approximately 2 kDa, it may not exhibit a bulky structure that interferes with the functions of actin. However, a disadvantage is that the fission yeast formin Cdc12p, which is essential for the contractile ring formation, interferes with tetracysteine-tagged fission yeast actin, indicating that there may even be structural requirements for formins in the facilitation of actin-filament elongation (Chen *et al* 2012). Moreover, chemical toxicity evoked by the labeling, such as accumulation of FlAsH in active mitochondria (Langhorst *et al* 2006), may impact the cell's viability and is hence only used for short-term live-cell imaging.

7.12 Microscopic visualization of F-actin for live-cell-imaging during migration and invasion of cells

The main problem in the past was to observe actin monomers and actin filaments in living cells, for example, when the cells are migrating and invading in 3D extracellular matrices such as collagen gels.

7.12.1 GFP–actin derivatives for live-cell imaging

GFP derivative tags are extensively employed for live-cell analysis. The GFP–actin can be used for fluorescence recovery after photobleaching (FRAP) approaches to investigate actin remodeling dynamics and turnover within a distinct cellular F-actin structure such as the actin cortex (Clark *et al* 2013) or cellular protrusions such as lamellipodia (Koestler *et al* 2009, Melak *et al* 2017). The relatively large size of approximately 28 kDa of GFP-like proteins (Sliogeryte *et al* 2016) may cause problems such as the incorporation of GFP-tagged actin monomers into actin filaments. Hence, the GFP–actin visualization may only partly reflect the F-actin network in cells, as in yeast GFP–actin cannot incorporate efficiently in a distinct subpopulation of actin filaments (Doyle and Botstein 1996). Moreover, specific actin nucleators such as members of the formin family are known to expel GFP–actin monomers from actin nucleation and elongation, which seems to be based on steric hindrances during the formation of actin nucleations or fast and efficient barbed end actin-monomer incorporation (Wu and Pollard 2005, Vavylonis *et al* 2006, Carvalho *et al* 2009). In conclusion, similarly to any genetically encoded protein tagging, the expression of GFP–actin requires the control of the expression levels in the cell system. The GFP–actin fusion protein has been widely used to visualize the functions of actin in living cells or organisms, such as in mammalian cells (Ballestrem *et al* 1998), yeast (Doyle and Botstein 1996), *Dictyostelium discoideum* (Neujahr *et al* 1997, Westphal *et al* 1997), *Drosophila* (Verkhusha *et al* 1999) and mice (Gurniak and Witke 2007).

7.12.2 Actin-binding proteins for live-cell imaging

In 2008, a novel highly promising marker for the visualization of actin stress fibers during live-cell imaging was presented. For many fundamental biological processes such as cellular motility, proliferation, development, force transmission and generation, the observation of actin is essential and previous approaches had failed. The solution to the this problem was the development of LifeAct. LifeAct is a short peptide of 17 amino acids, which bind to filamentous actin in individual eukaryotic cells as well as in tissues. In particular, LifeAct shows no interference with actin remodeling dynamics in *in vitro* cell culture models and in *in vivo* animal models (Riedl *et al* 2008). Another advantage of LifeAct is that cells, even when they are barely transfectable can now be stained for actin using this chemically altered peptide.

In previous approaches for imaging actin dynamics, discussed above, it was attempted to inject fluorescently labeled actin, small amounts of fluorescently labeled phalloidin, which is known to be toxic for the cells and affects cellular mechanics by stabilizing actin stress fibers and thus stiffening the cells. A commonly performed alternative live-cell imaging of actin dynamics is the transfection of cells with GFP–actin fusion proteins, which involves overexpression of the physiological levels of actin within the cells under investigation. In addition, the GFP-part of the fusion protein can either be attached to the N- or C-terminus of actin, however, it can affect the function of actin (Snapp 2010). In order to reduce these effects, nontagged actin will additionally be transfected (Yamada *et al* 2005). In addition, GFP has been fused to actin-binding domains such as moesin in *Drosophila melanogaster*, ABP 120 or LimE in *Dictyostelium discoideum* (Edwards *et al* 1997, Bretschneider *et al* 2004, Lenart *et al* 2005). A major restriction for the usage is that they can only be transfected into cells, are large in size and impair the function of natural endogenous ligands for actin binding to these sites, which obvious affect their function.

The only fusion protein suitable to bind to actin fibers in the budding yeast *Saccharomyces cerevisiae* was Abp140-GFP (Asakura *et al* 1998, Yang and Pon

2002). The short peptide Abp140 has been found solely in *S. cerevisiae* and close relatives, between which it is highly conserved. The first 17 amino acids are sufficient to bind to actin and hence this peptide was named LifeAct. As the LifeAct peptide is relatively short, it serves as a C-terminal GFP fusion protein suitable as an *in vivo* marker for actin (figure 7.18). In order to access the biochemical properties, LifeAct has been fused to FITC at its N-terminus. LifeAct has affinities to G-actin and F-actin (280 ± 100 nM to G-actin and 2.3 ± 0.9 μM to F-actin) (Riedl *et al* 2008). The next focus was on answering the following question: how is the sequestering of actin by factors such as latrunculin A and profilin affected by LifeAct binding? The simple answer is that the actin sequestering is not affected by the binding of LifeAct, because it does not interfere with the binding of G-actin to F-actin fibers and uses another nonoverlapping binding site on the actin protein. Moreover, the polymerization (including nucleation and elongation phases) and depolymerization rates of actin filaments are not impaired up to 55 μM LifeAct (Cooper *et al* 1983). In addition, LifeAct shows, up to a concentration of 55 μM, no competition with myosin II and α-actinin binding leaving these interactions unperturbed. It has been reported that LifeAct is able to build a nascent helix in water and can be further stabilized by alcohol addition to form a typical alpha helix at residues two to ten similar to the physiological G-actin-interacting protein thymosin-β4 (Riedl *et al* 2008). However, the stability and function *in vivo* are tested by using this LifeAct in living cells such as fibroblasts or epithelial cells without any cytotoxic or function affecting side effects. Indeed, it specifically stained actin and recovered more rapidly after photobleaching than actin–GFP transfected cells (Riedl *et al* 2008). Moreover,

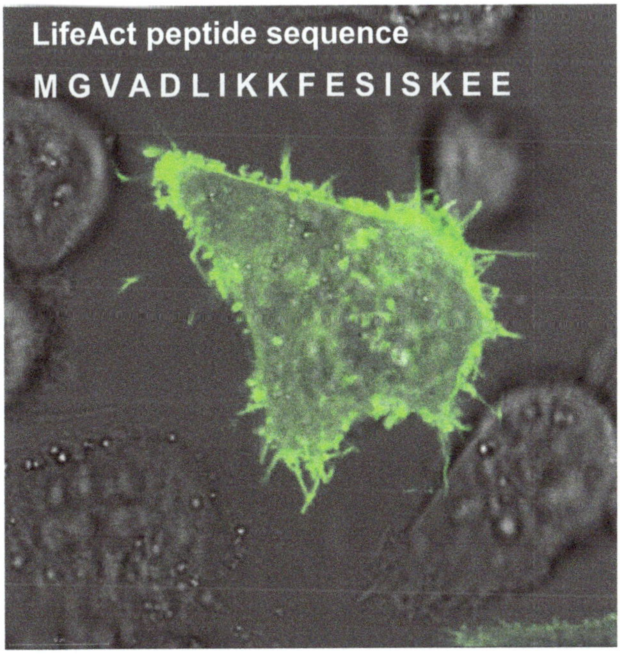

Figure 7.18. LifeAct sequence and a human MDA-MB-231 breast cancer cells stained with LifeAct.

it has been described that LifeAct–GFP can be used to visualize actin dynamics, such as in dynamics lamellipododial actin at the cell periphery and during cell division, as the typical circular actin belts are observed in epithelial cells. In addition, it has been analyzed whether LifeaAct–GFP alters actin dynamics, such as the retrograde flow of actin in lamellipodial structures, the chemotactic migration towards the highest concentrations of a chemical gradient of certain chemokine or the polarization of neuronal cells. The speed (4 μm min^{-1}) of the retrograde actin flow in lamellipodia of LifeAct–GFP expressing fibroblasts was not impaired. However, when analyzing actin–GFP expressing fibroblasts the speed, of the retrograde flow was pronouncedly reduced (Riedl *et al* 2008). The expression of LifeAct–GFP had no effect on the speed of chemotactic migration of dendrocytes in a 3D collagen matrix, whereas the expression of actin–GFP reduced the speed drastically (Riedl *et al* 2008). The neuronal polarization was also not significantly altered by the expression of LifeAct–GFP, however, when actin–GFP was expressed, the polarization was reduced. Finally, all these control experiments showed that the expression of LifeAct–GFP is preferable to the expression of the classical actin–GFP. LifeAct–GFP is also suited for paraformaldehyde-fixed cells such as MDCK and tissue sections such as mouse skeletal muscle. The LifeAct peptide represents a nontoxic and universal alternative to the phalloidin labeling of F-actin stress fibers. Additionally, the LifeAct peptide seems unlikely to compete with endogenous proteins in higher eukaryotes, as no homologous sequences have been found in these organisms. In addition to transfection, the LifeAct peptide can be brought into living cells using a scrape loading approach, in which the living cells are injured by a sharp scissor or scraper and take up by the LifeAct peptide by diffusion through the gaps (wounded membrane areas) into the still living cells.

Visualization of F-actin structures and dynamics using peptides with actin-binding domains

Several probes can be use to analyze the dynamical remodeling of the actin cytoskeleton in living cells. Except for SiR–actin, these substances possess actin-binding domains (ABDs), which are taken from yeast or human proteins and then fused to GFP derivatives. A well-known substance is the so-called LifeAct (discussed above) which is composed of a short 17 amino acid peptide derived from the yeast Abp140 protein (Riedl *et al* 2008). LifeAct has been successfully used to investigate, using live-cell imaging, cellular motility in various mammalian cells (Duleh and Welch 2012, Fiolka *et al* 2012, Rullo *et al* 2012, Grikscheit *et al* 2015, Suarez *et al* 2015). However, there are some experimental and technical drawbacks to using LifeAct as a marker for actin structures in living cells. A first disadvantage of LifeAct–GFP variants is the high amount of background fluorescence based on the high affinity of LifeAct–GFP variants toward G-actin (Riedl *et al* 2008).

A second disadvantage is that, although LifeAct has been generated as a universal marker for actin (Riedl *et al* 2008), distinct actin structures such as the specialized class of actin-based filopodia in mesenchymal cells of the developing limb bud of chick embryos (Sanders *et al* 2013) and stress-induced cofilin-bound F-actin structures in mouse striatal neuron-derived STHdh cells (Munsie *et al* 2009) are both not stained. Hence it has been hypothesized that an extensive interaction of actin filaments with actin-binding proteins such as cofilin induces conformational alterations of F-actin which in turn impair the interaction with LifeAct at its actin-binding site. A third disadvantage is that increased expression levels of LifeAct can themselves alter the formation of actin filaments. Indeed, an increased expression of LifeAct in germline cells of *Drosophila* facilitates extreme actin remodeling and finally fertility defects, whereas weak expression in somatic cells or germline cells has no effect (Spracklen *et al* 2014). A fourth disadvantage is that LifeAct alters the assembly of actin during endocytosis and cytokinesis in fission yeast (Courtemanche *et al* 2016). However, even though the underlying mechanisms are not yet clearly understood, both studies revealed that the LifeAct-facilitated effects on actin dynamics depend on the LifeAct concentration, indicating that the use of LifeAct needs more precisely defined optimization processes. In conclusion, a major advantage of LifeAct is the low toxicity when the expression level is optimal. In line with this, LifeAct has been used for the generation of transgenic animals such as mice or zebrafish (Riedl *et al* 2010, Schachtner *et al* 2012, Phang *et al* 2013, Mizoguchi *et al* 2016).

In addition to the LifeAct marker, the tandem calponin-homology domains (CH1 and CH2) of utrophin (UtrCH) can be employed as a marker for stable and dynamic F-actin investigations in living and fixed cells (Burkel *et al* 2007). UtrCH is composed of the first 261 amino acids of human utrophin, which is an actin-binding and dystrophin-related protein (Winder *et al* 1995). The two CH domains of Utr261 have been suggested to bind to adjacent actin subunits (Lin *et al* 2011). Based on the feature that actin is highly conserved, UtrCH can be utilized as a marker for actin filaments in a broad range of organisms and cell types such as mammalian cells, amphibic or echinodermal oocytes and embryos without interference with either actin organization or actin dynamics (Burkel *et al* 2007, Holubcová *et al* 2013). However, disadvantages have been reported, such as severe defects toward actin dynamics when UtrCH is expressed at higher levels, such as the breakdown of cortical actin and alterations in actin-bundle morphology, as well as female sterility during *Drosophila* oogenesis (Spracklen *et al* 2014).

Another tool for live-cell actin staining is F-tractin, which is a peptide of 43 amino acids of the rat actin-binding inositol 1,4,5-trisphosphate 3-kinase A (representing residues 10-52) (Schell *et al* 2001, Belin *et al* 2014). F-tractin is less well-studied compared to LifeAct. As F-tractin is larger in size than LifeAct, it may more easily interfere with other regulatory or accessory F-actin-binding proteins. In particular, F-tractin expression changes the actin cytoskeletal organization and morphology in a mesoderm-derived cell line of *Xenopus* (Belin *et al* 2014). In contrast to LifeAct and UtrCH, F-tractin has been seen not to disturb the actin rearrangement during *Drosophila* follicle development (Spracklen *et al* 2014), which turns it into a suitable marker for F-actin dynamical studies in *Drosophila* nurse cells. As distinct actin markers are excluded from specific actin-filament structures, the choice of the actin marker may affect the outcome of the cytoskeletal architecture and dynamics. A comparison of the distribution of GFP–actin, F-tractin, LifeAct and utrophin in different living cells (Belin *et al* 2014) revealed that actin-binding probes such as

F-tractin, LifeAct and utrophin preferably stain insufficiently in filopodia-rich regions, whereas GFP–actin has been shown to be incorporated well into filopodia. In contrast, GFP–actin is found to be excluded from stress fibers and lamellar filament structures, which demonstrates that GFP–actin is unable to interact with other endogenous actin-binding proteins. Finally, among all the tested fluorescent derivatives of actin and ABDs, F-tractin turned out to be the most accurate in reproducing F-actin structures similar to phalloidin labeling of fixed cells in a wide variety of cells (Belin *et al* 2014).

Another cell-permeable and chemically synthesized probe termed SiR–actin has been used for live-cell imaging (D'Este *et al* 2015). SiR–actin is structurally derived from Jasplakinolide, which is a toxin from the marine sponge *Jaspis johnstoni* that binds with high affinity to F-actin (Bubb *et al* 1994). An advantage of SiR–actin is that the cell transfection or protein overexpression systems are not utilized, as it can just be added to the cell culture system. A disadvantage may be that SiR–actin in live cells can stabilize F-actin or even induce the polymerization of actin, which is based on the structural similarities to Jasplakinolide. Due to this, the interpretation of results of observed F-actin structures may then be rather difficult, as these actin structures may be assembled by the actin marker itself. Hence more investigations need to be performed to reveal the full advantages and possible disadvantages of SiR–actin in comparison to commonly used live-cell imaging actin markers.

References and further reading

Abbott A 2005 Cell biology: hopping fences *Nature* **433** 680–3

Abe H, Obinata T, Minamide L and Bamburg J R 1996 *Xenopus laevis* actin-depolymerizing factor/cofilin: a phosphorylation-regulated protein essential for development *J. Cell Biol.* **132** 871–85

Adams J C 1995 Formation of stable microspikes containing actin and the 55 kDa actin bundling protein, fascin, is a consequence of cell adhesion to thrombospondin-1: implications for the anti-adhesive activities of thrombospondin-1 *J. Cell Sci.* **108** 1977–90

Adams J C 2004a Fascin protrusions in cell interactions *Trends Cardiovasc. Med.* **14** 221–6

Adams J C 2004b Roles of fascin in cell adhesion and motility *Curr. Opin. Cell Biol.* **16** 590–6

Adler J, Andrew I S, Novak P, Korchev Y E and Parmryd I 2010 Plasma membrane topography and interpretation of single-particle tracks *Nat. Methods* **7** 170–1

Agnew B J, Minamide L S and Bamburg J R 1995 Reactivation of phosphorylated actin depolymerizing factor and identification of the regulatory site *J. Biol. Chem.* **270** 17582–7

Albiges-Rizo C, Destaing O, Fourcade B, Planus E and Block M R 2009 Actin machinery and mechanosensitivity in invadopodia, podosomes and focal adhesions *J. Cell Sci.* **122** 3037–49

Alert R, Casademunt J, Brugues J and Sens P 2015 Model for probing membrane–cortex adhesion by micropipette aspiration and uctuation spectroscopy *Biophys. J.* **108** 1878–86

Alper O, Stetler-Stevenson W G and Harris L N *et al* 2009 Novel anti-filamin-A antibody detects a secreted variant of filamin-A in plasma from patients with breast carcinoma and high-grade astrocytoma *Cancer Sci.* **100** 1748–56

Amano M, Ito M, Kimura K, Fukata Y, Chihara K, Nakano T, Matsuura Y and Kaibuchi K 1996 Phosphorylation and activation of myosin by Rho-associated kinase (Rho-kinase) *J Biol Chem.* **271** 20246–9

Anderson T W, Vaughan A N and Cramer L P 2008 Retrograde flow and myosin II activity within the leading cell edge deliver F-actin to the lamella to seed the formation of graded polarity actomyosin II filament bundles in migrating fibroblasts *Mol. Biol. Cell* **19** 5006–18

Andrade D M, Clausen M P, Keller J, Mueller V, Wu C, Bear J E, Hell S W, Lagerholm B C and Eggeling C 2015 Cortical actin networks induce spatio-temporal confinement of phospholipids in the plasma membrane—a minimally invasive investigation by STED-FCS *Sci. Rep.* **5** 11454

Andrews N L, Lidke K A, Pfeiffer J R, Burns A R, Wilson B S, Oliver J M and Lidke D S 2008 Actin restricts FcεRI diffusion and facilitates antigen-induced receptor immobilization *Nat. Cell Biol.* **10** 955–10963

Andrianantoandro E and Pollard T D 2006 Mechanism of actin filament turnover by severing and nucleation at different concentrations of ADF/cofilin *Mol. Cell* **24** 13–23

Aragona M, Panciera T, Manfrin A, Giulitti S, Michielin F, Elvassore N, Dupont S and Piccolo S 2013 A mechanical checkpoint controls multicellular growth through YAP/TAZ regulation by actin-processing factors *Cell* **154** 1047–59

Arjonen A, Kaukonen R and Ivaska J 2011 Filopodia and adhesion in cancer cell motility *Cell Adhes. Migr.* **5** 421–30

Asakura T *et al* 1998 Isolation and characterization of a novel actin filament-binding protein from *Saccharomyces cerevisiae Oncogen* **16** 121–30

Balaban N Q, Schwarz U S and Riveline D *et al* 2001 Force and focal adhesion assembly: a close relationship studied using elastic micropatterned substrates *Nat. Cell Biol.* **3** 466–72

Ballestrem C, Wehrle-Haller B and Imhof B A 1998 Actin dynamics in living mammalian cells *J. Cell Sci.* **111** 1649–58

Bamburg J R, Harris H E and Weeds A G 1980 Partial purification and characterization of an actin depolymerizing factor from brain *FEBS Lett.* **121** 178–82

Bao G and Suresh S 2003 Cell and molecular mechanics of biological materials *Nat. Mater.* **2** 715–25

Bao J, Jana S S and Adelstein R S 2005 Vertebrate nonmuscle myosin II isoforms rescue small interfering RNA-induced defects in COS-7 cell cytokinesis *J. Biol. Chem.* **280** 19594–9

Barth H and Stiles B G 2008 Binary actin-ADP-ribosylating toxins and their use as molecular Trojan horses for drug delivery into eukaryotic cells *Curr. Med. Chem.* **15** 459–69

Bartles J R, Zheng L, Li A, Wierda A and Chen B 1998 Small espin: a third actin-bundling protein and potential forked protein ortholog in brush border microvilli *J. Cell Biol.* **143** 107–19

Bartles J R 2000 Parallel actin bundles and their multiple actin-bundling proteins *Curr. Opin. Cell Biol.* **12** 72–8

Beaty B T and Condeelis J 2014 Digging a little deeper: the stages of invadopodium formation and maturation *Eur. J. Cell Biol.* **93** 438–44

Belin B J, Goins L M and Mullins R D 2014 Comparative analysis of tools for live cell imaging of actin network architecture *Bioarchitecture* **4** 189–202

Berg J S, Powell B C and Cheney R E 2001 A millennial myosin census *Mol. Biol. Cell* **12** 780–94

Bernstein B W and Bamburg J R 2010 ADF/cofilin: a functional node in cell biology *Trends Cell Biol.* **20** 187–95

Bernstein B W, Painter W B, Chen H, Minamide L S, Abe H and Bamburg J R 2000 Intracellular pH modulation of ADF/cofilin proteins *Cell Motil. Cytoskelet* **47** 319–36

Bershadsky A D, Ballestrem C, Carramusa L, Zilberman Y, Gilquin B, Khochbin S, Alexandrova A Y, Verkhovsky A B, Shemesh T and Kozlov M M 2006 Assembly and mechanosensory function of focal adhesions: experiments and models *Eur. J. Cell Biol.* **85** 165–73

Block J *et al* 2012 FMNL2 drives actin-based protrusion and migration downstream of Cdc42 *Curr. Biol.* **22** 1005–12

Bockhorn J, Yee K and Chang Y F *et al* 2013 MicroRNA-30c targets cytoskeleton genes involved in breast cancer cell invasion *Breast Cancer Res. Treat.* **137** 373–82

Boukhelifa M, Moza M and Johansson T *et al* 2006 The proline-rich protein palladin is a binding partner for profilin *FEBS J.* **273** 26–33

Boukhelifa M, Parast M M, Bear J E, Gertler F B and Otey C A 2004 Palladin is a novel binding partner for Ena/VAP family members *Cell Motil. Cytoskelet.* **58** 17–29

Bovellan M, Romeo Y and Biro M *et al* 2014 Cellular control of cortical actin nucleation *Curr. Biol.* **24** 1628–35

Bravo-Cordero J J, Hodgson L and Condeelis J S 2014 Spatial regulation of tumor cell protrusions by RhoC *Cell Adhes. Migr.* **8** 263–7

Bravo-Cordero J J, Magalhaes M A, Eddy R J, Hodgson L and Condeelis J 2013 Functions of cofilin in cell locomotion and invasion *Nat. Rev. Mol. Cell Biol.* **14** 405–15

Breitsprecher D, Kiesewetter A K, Linkner J, Urbanke C, Resch G P, Small J V and Faix J 2008 Clustering of VASP actively drives processive, WH2 domain-mediated actin filament elongation *EMBO J.* **27** 2943–54

Breitsprecher D, Kiesewetter A K, Linkner J, Vinzenz M, Stradal T E, Small J V, Curth U, Dickinson R B and Faix J 2011 Molecular mechanism of Ena/VASP-mediated actin-filament elongation *EMBO J.* **30** 456–67

Bretschneider T, Diez S, Anderson K, Heuser J, Clarke M, Mueller-Taubenberger A, Koehler J and Gerisch G 2004 Dynamic actin patterns and Arp2/3 assembly at the substrate-attached surface of motile cells *Curr. Biol.* **14** 1–10

Broschat K O 1990 Tropomyosin prevents depolymerization of actin filaments from the pointed end *J. Biol. Chem.* **265** 21323–9

Brown J W and McKnight C J 2010 Molecular model of the microvillar cytoskeleton and organization of the brush border *PLoS One* **5** e9406

Brugués J, Maugis B, Casademunt J, Nassoy P, Amblard F and Sens P 2010 Dynamical organization of the cytoskeletal cortex probed by micropipette aspiration *Proc. Natl Acad. Sci. USA* **107** 15415–20

Bubb M R, Senderowicz A M, Sausville E A, Duncan K L and Korn E D 1994 Jasplakinolide, a cytotoxic natural product, induces actin polymerization and competitively inhibits the binding of phalloidin to F-actin *J. Biol. Chem.* **269** 14869–71

Bunnell T M, Burbach B J, Shimizu Y and Ervasti J M 2011 β-actin specifically controls cell growth, migration, and the G-actin pool *Mol. Biol. Cell* **22** 4047–58

Burkel B M, Von Dassow G and Bement W M 2007 Versatile fluorescent probes for actin filaments based on the actin-binding domain of utrophin *Cell Motil. Cytoskelet.* **64** 822–32

Burnette D T, Manley S, Sengupta P, Sougrat R, Davidson M W, Kachar B and Lippincott-Schwartz J 2011 A role for actin arcs in the leading-edge advance of migrating cells *Nat. Cell Biol.* **13** 371–81

Burridge K and Guilluy G 2016 Focal adhesions, stress fibers and mechanical tension *Exp. Cell Res.* **343** 14–20

Burridge K and Chrzanowska-Wodnicka M 1996 Focal adhesions, contractility, and signaling *Annu. Rev. Cell Dev. Biol.* **12** 463–518

Burtnick L D, Koepf E K, Grimes J, Jones E Y and Stuart D I *et al* 1997 The crystal structure of plasma gelsolin: implications for actin severing, capping, and nucleation *Cell* **90** 661–70

Campbell P J, Yachida S and Mudie L J *et al* 2010 The patterns and dynamics of genomic instability in metastatic pancreatic cancer *Nature* **467** 1109–13

Campellone K G and Welch M D 2010 A nucleator arms race: cellular control of actin assembly *Nat. Rev. Mol. Cell Biol.* **11** 237–51

Campellone K G and Welch M D 2010 A nucleator arms race: cellular control of actin assembly *Nat. Rev. Mol. Cell Biol.* **11** 237–51

Carlier M F, Laurent V, Santolini J, Melki R, Didry D, Xia G X, Hong Y, Chua N H and Pantaloni D 1997 Actin depolymerizing factor (ADF/cofilin) enhances the rate of filament turnover: Implication in actin-based motility *J. Cell Biol.* **136** 1307–22

Carvalho A, Desai A and Oegema K 2009 Structural memory in the contractile ring makes the duration of cytokinesis independent of cell size *Cell* **137** 926–37

Castellino F, Ono S, Matsumura F and Luini A 1995 Essential role of caldesmon in the actin filament reorganization induced by glucocorticoids *J. Cell Biol.* **131** 1223–30

Castoria G, D'Amato L, Ciociola A, Giovannelli P, Giraldi T, Sepe L, Paolella G, Barone M V, Migliaccio A and Auricchio F 2011 Androgen-induced cell migration: role of androgen receptor/filamin A association *PLoS One* **6** e17218

Chaigne A, Campillo C and Gov N S *et al* 2013 A soft cortex is essential for asymmetric spindle positioning in mouse oocytes *Nat. Cell Biol.* **15** 958–66

Chan A Y, Bailly M, Zebda N, Segall J E and Condeelis J S 2000 Role of cofilin in epidermal growth factor-stimulated actin polymerization and lamellipod protrusion *J. Cell Biol.* **148** 531–42

Chan C, Beltzner C C and Pollard T D 2009 Cofilin dissociates Arp2/3 complex and branches from actin filaments *Curr. Biol.* **19** 537–45

Chang C Y, Leu J D and Lee Y J 2015 The actin depolymerizing factor (ADF)/cofilin signaling pathway and DNA damage responses in cancer *Int. J. Mol. Sci.* **16** 4095–120

Charras G T, Hu C K, Coughlin M and Mitchison T J 2006 Reassembly of contractile actin cortex in cell blebs *J. Cell Biol.* **175** 477–90

Chen Q, Nag S and Pollard T D 2012 Formins filter modified actin subunits during processive elongation *J. Struct. Biol.* **177** 32–9

Chen W T 1981 Mechanism of retraction of the trailing edge during fibroblast movement *J. Cell Biol.* **90** 187–200

Chesarone M A, Du Page A G and Goode B L 2010 Unleashing formins to remodel the actin and microtubule cytoskeletons *Nat. Rev. Mol. Cell Biol.* **11** 62–74

Chin Y R and Toker A 2010 The actin-bundling protein palladin is an Akt1-specific substrate that regulates breast cancer cell migration *Mol. Cell* **38** 333–44

Choi C K, Vicente-Manzanares M, Zareno J, Whitmore L A, Mogilner A and Horwitz A R 2008 Actin and α-actinin orchestrate the assembly and maturation of nascent adhesions in a myosin II motor-independent manner *Nat. Cell Biol.* **10** 1039–50

Choquet D, Felsenfeld D P and Sheetz M P 1997 Extracellular matrix rigidity causes strengthening of integrin-cytoskeleton linkages *Cell* **88** 39–48

Chrzanowska-Wodnicka M and Burridge K 1996 Rho-stimulated contractility drives the formation of stress fibers and focal adhesions *J. Cell Biol.* **133** 1403–15

Chung H, Kim B, Jung S H, Won K J, Jiang X, Lee C K, Lim S D, Yang S K, Song K H and Kim H S 2013 Does phosphorylation of cofilin affect the progression of human bladder cancer? *BMC Cancer* **13** 45

Clark A G, Dierkes K and Paluch E K 2013 Monitoring actin cortex thickness in live cells *Biophys. J.* **105** 570–80

Clausen M P and Lagerholm B C 2011 The probe rules in single particle tracking *Curr. Protein Peptide Sci.* **12** 699–713

Clausen M P and Lagerholm B C 2013 Visualization of plasma membrane compartmentalization by high-speed quantum dot tracking *Nano Lett.* **13** 2332–7

Clausen M P, Colin-York H, Schneider F, Eggeling C and Fritzsche M 2017 Dissecting the actin cortex density and membrane–cortex distance in living cells by super-resolution microscopy *J. Phys. D: Appl. Phys.* **50** 064002

Colin-York H, Shrestha D, Felce J H, Waithe D, Moeendarbary E, Davis S J, Eggeling C and Fritsche M 2016 Super-resolved traction force microscopy (STFM) *Nano Lett.* **16** 2633–8

Colombelli J, Besser A, Kress H, Reynaud E G, Girard P, Caussinus E, Haselmann U, Small J V, Schwarz U S and Stelzer E H 2009 Mechanosensing in actin stress fibers revealed by a close correlation between force and protein localization *J. Cell Sci.* **122** 1665–79

Coluccio L M and Tilney L G 1984 Phalloidin enhances actin assembly by prevententing monomer dissociation *J. Cell Biol* **99** 529–35

Cooper J A, Walker S B and Pollard T D 1983 Pyrene actin: documentation of the validity of a sensitive assay for actin polymerization *J. Muscle Res. Cell Motil.* **4** 253–62

Copeland J W and Treisman R 2002 The diaphanous-related formin mDia1 controls serum response factor activity through its effects on actin polymerization *Mol. Biol. Cell* **13** 4088–99

Courtemanche N, Pollard T D and Chen Q 2016 Avoiding artefacts when counting polymerized actin in live cells with LifeAct fused to fluorescent proteins *Nat. Cell Biol.* **18** 676–83

Cramer L P, Siebert M and Mitchison T J 1997 Identification of novel graded polarity actin filament bundles in locomoting heart fibroblasts: implications for the generation of motile force *J. Cell Biol.* **136** 1287–305

Cramer L P 1999 Role of actin-filament disassembly in lamellipodium protrusion in motile cells revealed using the drug jasplakinolide *Curr. Biol.* **9** 1095–105

D'Este E, Kamin D, Goettfert F, El-Hady A and Hell S W 2015 STED nanoscopy reveals the ubiquity of subcortical cytoskeleton periodicity in living neurons *Cell Rep.* **10** 1246–251

Dai J W and Sheetz M P 1999 Membrane tether formation from blebbing cells *Biophys J.* **77** 3363–70

Danninger C and Gimona M 2000 Live dynamics of GFP-calponin: isoform-specific modulation of the actin cytoskeleton and autoregulation by C-terminal sequences *J. Cell Sci.* **21** 3725–36

Davila M, Jhala D, Ghosh D, Grizzle W E and Chakrabarti R 2007 Expression of LIM kinase 1 is associated with reversible G1/S phase arrest, chromosomal instability and prostate cancer *Mol. Cancer* **8** 6–40

del Rio A, Perez-Jimenez R, Liu R, Roca-Cusachs P, Fernandez J M and Sheetz M P 2009 Stretching single talin rod molecules activates vinculin binding *Science* **323** 638–41

Delanote V, Vandekerckhove J and Gettemans J 2005 Plastins: versatile modulators of actin organization in (patho)physiological cellular processes *Acta Pharmacol. Sin.* **26** 769–79

DeMali K A and Burridge K 2003 Coupling membrane protrusion and cell adhesion *J. Cell Sci.* **116** 2389–97

DesMarais V, Ghosh M, Eddy R and Condeelis J 2005 Cofilin takes the lead *J. Cell Sci.* **118** 19–26

DesMarais V, Ichetovkin I, Condeelis J and Hitchcock-DeGregori S E 2002 Spatial regulation of actin dynamics: a tropomyosin-free, actin-rich compartment at the leading edge *J. Cell Sci.* **115** 4649–60

Di Modugno F, Bronzi G and Scanlan M J *et al* 2004 Human Mena protein, a serex-defined antigen overexpressed in breast cancer eliciting both humoral and CD8+ T-cell immune response *Int. J. Cancer* **109** 909–18

Dietrich C, Yang B, Fujiwara T, Kusumi A and Jacobson K 2002 Relationship of lipid rafts to transient confinement zones detected by single particle tracking *Biophys. J.* **82** 274–84

Dimchev G, Steffen A, Kage F, Dimchev V, Pernier J, Carlier M F and Rottner K 2017 Efficiency of lamellipodia protrusion is determined by the extent of cytosolic actin assembly *Mol. Biol. Cell* **28** 1311–25

Dixon R D, Arneman D K, Rachlin A S, Sundaresan N R, Costello M J, Campbell S L and Otey C A 2008 Palladin is an actin cross-linking protein that uses immunoglobulin-like domains to bind filamentous actin *J. Biol. Chem.* **283** 6222–31

Diz-Muñoz A, Krieg M, Bergert M, Ibarlucea-Benitez I, Muller D J, Paluch E and Heisenberg C P 2010 Control of directed cell migration in vivo by membrane-to-cortex attachment *PLoS Biol.* **8** e1000544

Dominguez R 2004 Actin-binding proteins—a unifying hypothesis *Trends Biochem. Sci.* **29** 572–8

Dominguez R 2010 Structural insights into de novo actin polymerization *Curr. Opin. Struct. Biol.* **20** 217–25

Dominguez R 2007 The beta-thymosin/WH2 fold: multifunctionality and structure *Ann. N. Y. Acad. Sci.* **1112** 86–94

Dos Remedios C G, Chhabra D, Kekic M, Dedova I V, Tsubakihara M, Berry D A and Nosworthy J 2003 Actin binding proteins: regulation of cytoskeletal microfilaments *Physiol. Rev.* **83** 433–73

Doyle T and Botstein D 1996 Movement of yeast cortical actin cytoskeleton visualized *in vivo Proc. Natl Acad. Sci. USA* **93** 3886–91

Draeger A, Gimona M, Stuckert A, Celis J E and Small J V 1991 Calponin. Developmental isoforms and a low molecular weight variant *FEBS Lett.* **291** 24–8

Drees F and Gertler B 2008 Ena/VASP: proteins at the tip of the nervous system *Curr. Opin. Neurobiol.* **18** 53–9

Du H Q, Chen L, Wang Y, Wang L J, Yan H, Liu H Y and Xiao H 2015 Increasing radiosensitivity with the downregulation of cofilin-1 in U251 human glioma cells *Mol. Med. Rep.* **11** 3354–60

Dubois F, Keller M and Calvayrac O *et al* 2016 RASSF1A suppresses the invasion and metastatic potential of human non-small cell lung cancer cells by inhibiting YAP activation through the GEF-H1/RhoB pathway *Cancer Res.* **76** 1627–40

Duleh S N and Welch M D 2012 Regulation of integrin trafficking, cell adhesion, and cell migration by WASH and the Arp2/3 complex *Cytoskeleton* **69** 1047–58

Dumpich M, Mannherz H G and Theiss C 2015 VEGF signaling regulates cofilin and the Arp2/3-complex within the axonal growth cone *Curr. Neurovasc. Res.* **12** 293–307

Dustin M L and Davis S J 2014 TCR signaling: the barrier within *Nat. Immunol.* **15** 136–7

Edidin M, Kuo S C and Sheetz M P 1991 Lateral movements of membrane glycoproteins restricted by dynamic cytoplasmic barriers *Science* **254** 1379–82

Edlund M, Lotano M A and Otey C A 2001 Dynamics of alpha-actinin in focal adhesions and stress fibers visualized with α-actinin-green fluorescent protein *Cell Motil. Cytoskelet.* **48** 190–200

Edwards K A, Demsky M, Montague R A, Weymouth N and Kiehart D P 1997 GFP-moesin illuminates actin cytoskeleton dynamics in living tissue and demonstrates cell shape changes during morphogenesis in *Drosophila Dev. Biol.* **191** 103–17

Edwards S D and Keep N H 2001 The 2.7 A crystal structure of the activated FERM domain of moesin: an analysis of structural changes on activation *Biochemistry* **40** 7061–8

Eggeling C *et al* 2009 Direct observation of the nanoscale dynamics of membrane lipids in a living cell *Nature* **457** 1159–62

Endlich N, Schordan E and Cohen C D *et al* 2009 Palladin is a dynamic actin-associated protein in podocytes *Kidney Int.* **75** 214–26

Estornes Y, Gay F, Gevrey J C, Navoizat S, Nejjari M, Scoazec J Y, Chayvialle J A, Saurin J C and Abello J 2007 Differential involvement of destrin and cofilin-1 in the control of invasive properties of Isreco1 human colon cancer cells *Int. J. Cancer* **121** 2162–71

Etienne-Manneville S 2011 Control of polarized cell morphology and motility by adherens junctions *Semin. Cell Dev. Biol.* **22** 850–7

Evans E A and Calderwood D A 2007 Forces and bond dynamics in cell adhesion *Science* **316** 1148–53

Fan L, Pellegrin S, Scott A and Mellor H 2010 The small GTPase Rif is an alternative trigger for the formation of actin stress fibers in epithelial cells *J. Cell Sci.* **123** 1247–52

Farsad K and De Camilli P 2003 Mechanisms of membrane deformation *Curr. Opin. Cell Biol.* **15** 372–81

Fehon R G, McClatchey A I and Bretscher A 2011 Organizing the cell cortex: the role of ERM proteins *Nat. Rev. Mol. Cell Biol.* **11** 276–87

Ferron F, Rebowski G, Lee S H and Dominguez R 2007 Structural basis for the recruitment of profilin–actin complexes during filament elongation by Ena/VASP *EMBO J.* **26** 4597–606

Fiolka R, Shao L, Rego E H, Davidson M W and Gustafsson M G L 2012 Time-lapse two-color 3D imaging of live cells with doubled resolution using structured illumination *Proc. Natl Acad. Sci. USA* **109** 5311–5

Fletcher D A and Mullins R D 2010 Cell mechanics and the cytoskeleton *Nature* **463** 485–92

Foran E, McWilliam P, Kelleher D, Croke D T and Long A 2006 The leukocyte protein L-plastin induces proliferation, invasion and loss of E-cadherin expression in colon cancer cells *Int. J. Cancer* **118** 2098–104

Foth B J, Goedecke M C and Soldati D 2006 New insights into myosin evolution and classification *Proc Natl Acad. Sci. USA* **103** 3681–6

Frame M C, Patel H, Serrels B, Lietha D and Eck M J 2010 The FERM domain: organizing the structure and function of FAK *Nat. Rev. Mol. Cell Biol.* **11** 802–14

Franke R P, Grafe M, Schnittler H, Seiffge D, Mittermayer C and Drenckhahn D 1984 Induction of human vascular endothelial stress fibres by fluid shear stress *Nature* **307** 648–9

Fratelli M, Demol H and Puype M *et al* 2002 Identification by redox proteomics of glutathionylated proteins in oxidatively stressed human T lymphocytes *Proc. Natl Acad. Sci. USA* **99** 3505–10

Friederich E, Pringault E, Arpin M and Louvard D 1990 From the structure to the function of villin, an actin-binding protein of the brush border *BioEssays* **12** 403–8

Friedl P and Wolf K 2010 Plasticity of cell migration: a multiscale tuning model *J. Cell Biol.* **188** 11–9

Friedland J C, Lee M H and Boettiger D 2009 Mechanically activated integrin switch controls $\alpha5\beta1$ function *Sci.* **323** 642–4

Fritzsche M, Fernandes R A, Colin-York H, Santos A M, Lee S F, Lagerholm B C, Davis S J and Eggeling C 2015 CalQuo: automated, simultaneous single-cell and population-level quanti cation of global intracellular Ca2+ responses *Sci. Rep.* **5** 16487

Fritzsche M, Erlenkamper C, Moeendarbary E, Charras G and Kruse K 2016 Actin kinetics shapes cortical network structure and mechanics *Sci. Adv.* **2** e1501337

Fritzsche M, Lewalle A, Duke T, Kruse K and Charras G 2013 Analysis of turnover dynamics of the submembranous actin cortex *Mol. Biol. Cell* **24** 757–67

Fritzsche M, Thorogate R and Charras G 2014 Quantitative analysis of ezrin turnover dynamics in the actin cortex *Biophys. J.* **106** 343–53

Fujii T, Iwane A H, Yanagida T and Namba K 2010 Direct visualization of secondary structures of F-actin by electron cryomicroscopy *Nature* **467** 724–8

Fujiwara T K, Iwasawa K and Kalay Z *et al* 2016 Confined diffusion of transmembrane proteins and lipids induced by the same actin meshwork lining the plasma membrane *Mol. Biol. Cell* **27** 1101–19

Fujiwara T, Ritchie K, Murakoshi H, Jacobson K and Kusumi A 2002 Phospholipids undergo hop diffusion in compartmentalized cell membrane *J. Cell Biol.* **157** 1071–81

Gaggioli C, Hooper S, Hidalgo-Carcedo C, Grosse R, Marshall J F, Harrington K and Sahai E 2007 Fibroblast-led collective invasion of carcinoma cells with differing roles for RhoGTPases in leading and following cells *Nat. Cell Biol.* **9** 1392–400

Galbraith C G, Yamada K M and Sheetz M P 2002 The relationship between force and focal complex development *J. Cell Biol.* **159** 695–705

Gardel M L, Sabass B, Ji L, Danuser G, Schwarz U S and Waterman C M 2008 Traction stress in focal adhesions correlates biphasically with actin retrograde flow speed *J. Cell Biol.* **183** 999–1005

Gau D M, Lesnock J L and Hood B L *et al* 2015 BRCA1 deficiency in ovarian cancer is associated with alteration in expression of several key regulators of cell motility – a proteomics study *Cell Cycle* **14** 1884–92

Geeves M A and Holmes K C 2005 The molecular mechanism of muscle contraction *Adv. Protein Chem.* **71** 161–93

Gertler F B, Niebuhr K, Reinhard M, Wehland J and Soriano P 1996 Mena, a relative of VASP and *Drosophila* enabled, is implicated in the control of microfilament dynamics *Cell* **87** 227–39

Ghosh M, Song X, Mouneimne G, Sidani M, Lawrence D S and Condeelis J S 2004 Cofilin promotes actin polymerization and defines the direction of cell motility *Science* **304** 743–6

Goicoechea S M, Bednarski B, Garcia-Mata R, Prentice-Dunn H, Kim H J and Otey C A 2009 Palladin contributes to invasive motility in human breast cancer cells *Oncogene* **28** 587–98

Goley E D and Welch M D 2006 The ARP2/3 complex: an actin nucleator comes of age *Nat. Rev. Mol. Cell Biol.* **7** 713–26

Gonzalez A M, Otey C, Edlund M and Jones J C 2001 Interactions of a hemidesmosome component and actinin family members *J. Cell Sci.* **114** 4197–206

Goode B L and Eck M J 2007 Mechanism and function of formins in the control of actin assembly *Annu. Rev. Biochem.* **76** 593–627

Gouin E, Welch M D and Cossart P 2005 Actin-based motility of intracellular pathogens *Curr. Opin. Microbiol.* **8** 35–45

Gowrishankar K, Ghosh S, Saha S, Rumamol C, Mayor S and Madan R 2012 Active remodeling of cortical actin regulates spatiotemporal organization of cell surface molecules *Cell* **149** 1353–67

Griffin B A, Adams S R and Tsien R Y 1998 Specific covalent labeling of recombinant protein molecules inside live cells *Science* **281** 269–72

Grikscheit K, Frank T, Wang Y and Grosse R 2015 Junctional actin assembly is mediated by Formin-like 2 downstream of Rac1 *J. Cell Biol.* **209** 367–76

Grimm H P, Verkhovsky A B, Mogilner A and Meister J J 2003 Analysis of actin dynamics at the leading edge of crawling cells: implications for the shape of keratocyte lamellipodia *Eur. Biophys. J.* **32** 563–77

Grone H J, Weber K, Helmchen U and Osborn M 1986 Villin-a marker of brush border differentiation and cellular origin in human renal cell carcinoma *Am. J. Pathol.* **124** 294–302

Guck J, Schinkinger S and Lincoln B *et al* 2005 Optical deformability as an inherent cell marker for testing malignant transformation and metastatic competence *Biophys. J.* **88** 3689–98

Guilluy C, Swaminathan V, Garcia-Mata R, O'Brien E T, Superfine R and Burridge K 2011 The Rho GEFs LARG and GEF-H1 regulate the mechanical response to force on integrins *Nat. Cell Biol.* **13** 722–27

Gunning P 2008 Emerging issues for tropomyosin structure, regulation, function and pathology *Adv. Exp. Med. Biol.* **644** 293–8

Guolla L, Bertrand M, Haase K and Pelling A E 2012 Force transduction and strain dynamics in actin stress fibres in response to nanonewton forces *J. Cell Sci.* **125** 603–13

Gupton S L, Anderson K L and Kole T P *et al* 2005 Cell migration without a lamellipodium: translation of actin dynamics into cell movement mediated by tropomyosin *J. Cell Biol.* **168** 619–31

Gurniak C B and Witke W 2007 HuGE, a novel GFP-actin-expressing mouse line for studying cytoskeletal dynamics *Eur. J. Cell Biol.* **86** 3–12

Guvakova M A, Adams J C and Boettiger D 2002 Functional role of alpha-actinin, PI 3-kinase and MEK1/2 in insulin-like growth factor I receptor kinase regulated motility of human breast carcinoma cells *J. Cell Sci.* **115** 4149–65

Hanahan D and Weinberg R A 2011 Hallmarks of cancer: the next generation *Cell* **144** 646–74

Hanakam F, Albrecht R, Eckerskorn C, Matzner M and Gerisch G 1996 Myristoylated and non-myristoylated forms of the pH sensor protein hisactophilin II: intracellular shuttling to plasma membrane and nucleus monitored in real time by a fusion with green fluorescent protein *EMBO J.* **15** 2935–43

Hanson J and Lowy J 1963 The structure of F-actin and the actin filaments isolated from muscle *J. Mol. Biol.* **6** 46–60

Hartman M A and Spudich J A 2012 The myosin superfamily at a glance *J. Cell Sci.* **125** 1627–32

Hashimoto Y, Skacel M and Adams J C 2005 Roles of fascin in human carcinoma motility and signaling: prospects for a novel biomarker? *Int J. Biochem. Cell Biol.* **37** 1787–804

Hashimoto Y, Kim D J and Adams J C 2011 The roles of fascins in health and disease *J. Pathol.* **224** 289–300

Hashimoto Y, Parsons M and Adams J C 2007 Dual actin-bundling and protein kinase C-binding activities of fascin regulate carcinoma cell migration downstream of Rac and contribute to metastasis *Mol. Biol. Cell* **18** 4591–602

Hatzakis N S, Bhatia V K, Larsen J, Madsen K L, Bolinger P Y, Kunding A H, Castillo J, Gether U, Hedegård P and Stamou D 2009 How curved membranes recruit amphipathic helices and protein anchoring motifs *Nat. Chem. Biol.* **5** 835–41

Hayakawa K, Tatsumi H and Sokabe M 2011 Actin filaments function as a tension sensor by tension-dependent binding of cofilin to the filament *J. Cell Biol.* **195** 721–7

Heasman S J and Ridley A J 2008 Mammalian Rho GTPases: new insights into their functions from *in vivo* studies *Nat. Rev. Mol. Cell Biol.* **9** 690–701

Heath J P 1983 Behaviour and structure of the leading lamella in moving fibroblasts. I. Occurrence and centripetal movement of arc-shaped microfilament bundles beneath the dorsal cell surface *J. Cell Sci.* **60** 331–54

Heinemann F, Vogel S K and Schwille P 2013 Lateral membrane diffusion modulated by a minimal actin cortex *Biophys. J.* **104** 1465–75

Herman I M 1993 Actin isoforms *Curr. Opin. Cell Biol.* **5** 48–55

Heuberger J and Birchmeier W 2010 Interplay of cadherin-mediated cell adhesion and canonical Wnt signaling *Cold Spring Harb. Perspect Biol.* **2** a002915

Hill C S, Wynne J and Treisman R 1995 The Rho family GTPases RhoA, Rac1, and CDC42Hs regulate transcriptional activation by SRF *Cell* **81** 1159–70

Hirata H, Tatsumi H and Sokabe M 2008 Mechanical forces facilitate actin polymerization at focal adhesions in a zyxin-dependent manner *J. Cell Sci.* **121** 2795–804

Hodge T and Cope M J 2000 A myosin family tree *J. Cell Sci.* **113** 3353–4

Hoffman B D, Grashoff C and Schwartz M A 2011 Dynamic molecular processes mediate cellular mechanotransduction *Nature* **475** 316–23

Holubcová Z, Howard G and Schuh M 2013 Vesicles modulate an actin network for asymmetric spindle positioning *Nat. Cell Biol.* **15** 937–47

Honda K, Yamada T, Endo R, Ino Y, Gotoh M, Tsuda H, Yamada Y, Chiba H and Hirohashi S 1998 Actinin-4, a novel actin-bundling protein associated with cell motility and cancer invasion *J. Cell Biol.* **140** 1383–93

Honda K, Yamada T, Hayashida Y, Idogawa M, Sato S, Hasegawa F, Ino Y, Ono M and Hirohashi S 2005 Actinin-4 increases cell motility and promotes lymph node metastasis of colorectal cancer *Gastroenterology* **128** 51–62

Honda K, Yamada T, Seike M, Hayashida Y, Idogawa M, Kondo T, Ino Y and Hirohashi S 2004 Alternative splice variant of actinin-4 in small cell lung cancer *Oncogene* **23** 5257–62

Honigmann A, Mueller V, Ta H, Schoenle A, Sezgin E, Hell S W and Eggeling C 2014a Scanning STED-FCS reveals spatiotemporal heterogeneity of lipid interaction in the plasma membrane of living cells *Nat. Commun.* **5** 5412

Honigmann A, Sadeghi S, Keller J, Hell S, Eggeling C and Vink R 2014b A lipid bound actin meshwork organizes liquid phase separation in model membranes *eLife* **3** e01671

Honore B, Madsen P, Andersen A H and Leffers H 1993 Cloning and expression of a novel human profilin variant, profilin II *FEBS Lett.* **330** 151–5

Hossain M M, Hwang D Y, Huang Q Q, Sasaki Y and Jin J P 2003 Developmentally regulated expression of calponin isoforms and the effect of h2-calponin on cell proliferation *Am. J. Physiol. Cell Physiol.* **284** 156–67

Hotulainen P and Lappalainen P 2006 Stress fibers are generated by two distinct actin assembly mechanisms in motile cells *J. Cell Biol.* **173** 383–94

Hou X, Katahira T, Ohashi K, Mizuno K, Sugiyama S and Nakamura H 2013 Coactosin accelerates cell dynamism by promoting actin polymerization *Dev. Biol.* **379** 53–63

Hu K, Ji L, Applegate K T, Danuser G and Waterman-Storer C M 2007 Differential transmission of actin motion within focal adhesions *Science* **315** 111–5

Huang L, Kuwahara I and Matsumoto K 2014 EWS represses cofilin 1 expression by inducing nuclear retention of cofilin 1 mRNA *Oncogene* **33** 2995–3003

Innocenti M, Zucconi A, Disanza A, Frittoli E, Areces L B, Steffen A, Stradal T E, Di Fiore P P, Carlier M F and Scita G 2004 Abi1 is essential for the formation and activation of a WAVE2 signalling complex *Nat. Cell Biol.* **6** 319–27

Insall R H and Machesky L M 2009 Actin dynamics at the leading edge: from simple machinery to complex networks *Dev. Cell* **17** 310–22

Ishikawa R, Yamashiro S and Matsumura F 1989a Differential modulation of actin-severing activity of gelsolin by multiple isoforms of cultured rat cell tropomyosin. Potentiation of protective ability of tropomyosins by 83-kDa nonmuscle caldesmon *J. Biol. Chem.* **264** 7490–7

Ishikawa R, Yamashiro S and Matsumura F 1989b Annealing of gelsolin-severed actin fragments by tropomyosin in the presence of Ca2+. Potentiation of the annealing process by caldesmon *J. Biol. Chem.* **264** 16764–70

Isogai T, van der Kammen R, Leyton-Puig D, Kedziora K M, Jalink K and Innocenti M 2015 Initiation of lamellipodia and ruffles involves cooperation between mDia1 and the Arp2/3 complex *J. Cell Sci.* **128** 3796–810

Ito T, Taniguchi H, Fukagai K, Okamuro S and Kobayashi 2015 A Inhibitory mechanism of FAT4 gene expression in response to actin dynamics during Src-induced carcinogenesis *PLoS One* **10** e0118336

Ivanovska J, Tregubova A and Mahadevan V *et al* 2013 Identification of DAPK as a scaffold protein for the LIMK/cofilin complex in TNF-induced apoptosis *Int. J. Biochem. Cell Biol.* **45** 1720–29

Iwasa J H and Mullins R D 2007 Spatial and temporal relationships between actin-filament nucleation, capping, and disassembly *Curr. Biol.* **17** 395–406

Jacobson K, Mouritsen O G and Anderson R G 2007 Lipid rafts: at a crossroad between cell biology and physics *Nat. Cell Biol.* **9** 7–14

Jacobson K, Sheets E D and Simson R 1995 Revisiting the fluid mosaic model of membranes *Science* **268** 1441–2

Janmey P A and Miller R T 2011 Mechanisms of mechanical signaling in development and disease *J. Cell Sci.* **124** 9–18

Janmey P A and Weitz D A 2004 Dealing with mechanics: mechanisms of force transduction in cells *Trends Biochem. Sci.* **29** 364–70

Jaqaman K, Kuwata H, Touret N, Collins R, Trimble W S, Danuser G and Grinstein S 2011 Cytoskeletal control of CD36 diffusion promotes its receptor and signaling function *Cell* **146** 593–606

Jiang N, Kham S K, Koh G S, Suang Lim J Y, Ariffin H, Chew F T and Yeoh A E 2011 Identification of prognostic protein biomarkers in childhood acute lymphoblastic leukemia *J. Proteom.* **74** 843–57

Jiang X, Gillen S, Esposito I, Giese N A, Michalski C W, Friess H and Kleeff J 2010 Reduced expression of the membrane skeleton protein β1-spectrin (SPTBN1) is associated with worsened prognosis in pancreatic cancer *Histol. Histopathol.* **25** 1497–506

Joanny J F, Kruse K, Prost J and Ramaswamy S 2013 The actin cortex as an active wetting layer *Eur. Phys. J.* E 36–52

Jockusch B M, Murk K and Rothkegel M 2007 The profile of profilins *Rev. Physiol. Biochem. Pharmacol.* **159** 131–49

Joo E, Surka M C and Trimble W S 2007 Mammalian SEPT2 is required for scaffolding nonmuscle myosin II and its kinases *Dev. Cell* **13** 677–90

Kabsch W, Mannherz H G, Suck D, Pai E F and Holmes K C 1990 Atomic structure of the actin: DNase I complex *Nature* **347** 37–44

Kage F, Winterhoff M and Dimchev V *et al* 2017 FMNL formins boost lamellipodial force generation *Nat. Commun.* **8**

Kamai T, Tsujii T, Arai K, Takagi K, Asami H, Ito Y and Oshima H 2003 Significant association of Rho/ROCK pathway with invasion and metastasis of bladder cancer *Clin Cancer Res.* **9** 2632–41

Karatzaferi C, Chinn M K and Cooke R 2004 The force exerted by a muscle cross-bridge depends directly on the strength of the actomyosin bond *Biophys. J.* **87** 2532–44

Kashiwada K, Nishida W and Hayashi K *et al* 1997 Coordinate expression of β-tropomyosin and caldesmon isoforms in association with phenotypic modulation of smooth muscle cells *J. Biol. Chem.* **272** 13404–5396

Kastrup L, Blom H, Eggeling C and Hell S W 2005 Fluorescence fluctuation spectroscopy in subdi raction focal volumes *Phys. Rev. Lett.* **94** 178104

Katayama E, Scott-Woo G and Ikebe M 1995 Effect of caldesmon on the assembly of smooth muscle myosin *J. Biol. Chem.* **270** 3919–25

Katoh K, Kano Y, Amano M, Kaibuchi K and Fujiwara K 2001a Stress fiber organization regulated by MLCK and Rho-kinase in cultured human fibroblasts *Am J. Physiol. Cell Physiol* **280** 1669–79

Katoh K, Kano Y, Amano M, Onishi H, Kaibuchi K and Fujiwara K 2001b Rho-kinase-mediated contraction of isolated stress fibers *J. Cell Biol.* **153** 569–84

Keppler A, Pick H, Arrivoli C, Vogel H and Johnsson K 2004 Labeling of fusion proteins with synthetic fluorophores in live cells *Proc. Natl Acad. Sci. USA* **101** 9955–9

Kerber M L and Cheney R E 2011 Myosin-X: a MyTH-FERM myosin at the tips of filopodia *J. Cell Sci.* **124** 3733–41

Ketschek A, Spillane M, Dun X P, Hardy H, Chilton J and Gallo G 2016 Drebrin coordinates the actin and microtubule cytoskeleton during the initiation of axon collateral branches *Dev. Neurobiol.* **76** 1092–110

Khatau S B, Hale C M, Stewart-Hutchinson P J, Patel M S, Stewart C L, Searson P C, Hodzic D and Wirtz D 2009 A perinuclear actin cap regulates nuclear shape *Proc. Natl Acad. Sci. USA* **106** 19017–22

Kikuchi S, Honda K and Tsuda H *et al* 2008 Expression and gene amplification of actinin-4 in invasive ductal carcinoma of the pancreas *Clin. Cancer Res.* **14** 5348–56

Kimura K, Ito M and Amano M *et al* 1996 Regulation of myosin phosphatase by Rho and Rho-associated kinase (Rho-kinase) *Science* **273** 245–8

Kinoshita M, Field C M, Coughlin M L, Straight A F and Mitchiso T J 2002 Self- and actin-templated assembly of Mammalian septins *Dev. Cell* **3** 791–802

Kitisin K, Ganesan N and Tan Y *et al* 2007 Disruption of transforming growth factor-β signaling through beta-spectrin ELF leads to hepatocellular cancer through cyclin D1 activation *Oncogene* **26** 7103–10

Klammt C and Lillemeier B F 2012 How membrane structures control T cell signaling *Front. Immunol.* **3** 291

Klamt F, Zdanov S, Levine R L, Pariser A, Zhang Y, Zhang B, Yu L R, Veenstra T D and Shacter E 2009 Oxidant-induced apoptosis is mediated by oxidation of the actin-regulatory protein cofilin *Nat. Cell Biol.* **11** 1241–6

Klann M T, Lapin A and Reuss M 2009 Stochastic simulation of signal transduction: impact of the cellular architecture on diffusion *Biophys. J.* **96** 5122–9

Klemke M, Wabnitz G H, Funke F, Funk B, Kirchgessner H and Samstag Y 2008 Oxidation of cofilin mediates T cell hyporesponsiveness under oxidative stress conditions *Immunity* **29** 404–13

Koestler S A, Rottner K, Lai F, Block J, Vinzenz M and Small J V 2009 F- and G-actin concentrations in lamellipodia of moving cells *PLoS One* **4** e4810

Koestler S A, Auinger S, Vinzenz M, Rottner K and Small J V 2008 Differentially oriented populations of actin filaments generated in lamellipodia collaborate in pushing and pausing at the cell front *Nat. Cell Biol.* **10** 306–13

Koestler S A, Steffen A and Nemethova M *et al* 2013 Arp2/3 complex is essential for actin network treadmilling as well as for targeting of capping protein and cofilin *Mol. Biol. Cell.* **24** 2861–75

Kong F, Garcia A J, Mould A P, Humphries M J and Zhu C 2009 Demonstration of catch bonds between an integrin and its ligand *J. Cell Biol.* **185** 1275–84

Kordowska J, Hetrick T, Adam L P and Wang C L 2006 Phosphorylated l-caldesmon is involved in disassembly of actin stress fibers and postmitotic spreading *Exp. Cell Res.* **312** 95–110

Kovar D R, Harris E S, Mahaffy R, Higgs H N and Pollard T D 2006 Control of the assembly of ATP- and ADP-actin by formins and profilin *Cell* **124** 423–35

Krause M and Gautreau A 2014 Steering cell migration: lamellipodium dynamics and the regulation of directional persistence *Nat. Rev. Mol. Cell Biol.* **15** 577–90

Krendel M, Kim S V, Willinger T, Wang T, Kashgarian M, Flavell R A and Mooseker M S 2009 Disruption of myosin 1e promotes podocyte injury *J. Am. Soc. Nephrol.* **20** 86–94

Krendel M and Mooseker M S 2005 Myosins: Tails (and heads) of functional diversity *Physiology* **20** 239–51

Kristó I, Bajusz I, Bajusz C, Borkúti P and Vilmos P 2016 Actin, actin-binding proteins, and actin-related proteins in the nucleus *Histochem. Cell Biol.* **145** 373–88

Kunda P, Pelling A E, Liu T and Baum B 2008 Moesin controls cortical rigidity, cell rounding, and spindle morphogenesis during mitosis *Curr. Biol.* **18** 91–101

Kuo J C, Han X, Hsiao C T, Yates J R and Waterman C M 2011 Analysis of the myosin-II-responsive focal adhesion proteome reveals a role for β-Pix in negative regulation of focal adhesion maturation *Nat. Cell Biol.* **13** 383–93

Kurita S, Watanabe Y, Gunji E, Ohashi K and Mizuno K 2008 Molecular dissection of the mechanisms of substrate recognition and F-actin-mediated activation of cofilin-phosphatase Slingshot-1 *J. Biol. Chem.* **283** 32542–52

Kursula P, Kursula I, Massimi M, Song Y H, Downer J, Stanley W A, Witke W and Wilmanns M 2008 High-resolution structural analysis of mammalian profilin 2a complex formation with two physiological ligands: the formin homology 1 domain of mDia1 and the proline-rich domain of VASP *J. Mol. Biol.* **375** 270–90

Kusumi A, Fujiwara T K, Chadda R, Xie M, Tsunoyama T A, Kalay Z, Kasai R and Suzuki K G 2012 Dynamic organizing principles of the plasma membrane that regulate signal transduction: commemorating the fortieth anniversary of Singer and Nicolson's fluid-mosaic model *Annu. Rev. Cell Dev. Biol.* **28** 215–50

Kusumi A, Nakada C, Ritchie K, Murase K, Suzuki K, Murakoshi H, Kasai R S, Kondo J and Fujiwara T 2005 Paradigm shift of the plasma membrane concept from the two-dimensional continuum fluid to the partitioned fluid: high-speed single-molecule tracking of membrane molecules *Annu Rev. Biophys.* **34** 351–78

Kuure S, Cebrian C, Machingo Q, Lu B C, Chi X, Hyink D, D'agati V, Gurniak C, Witke W and Costantini F 2010 Actin depolymerizing factors cofilin1 and destrin are required for ureteric bud branching morphogenesis *PLoS Genet.* 2010 **6** e1001176

Kwik J, Boyle S, Fooksman D, Margolis L, Sheetz M P and Edidin M 2003 Membrane cholesterol, lateral mobility, and the phosphatidylinositol 4,5-bisphosphate-dependent organization of cell actin *Proc. Natl Acad. Sci. USA* **100** 13964–9

Lai F P, Szczodrak M, Block J, Faix J, Breitsprecher D, Mannherz H G, Stradal T E, Dunn G A, Small J V and Rottner K 2008 Arp2/3 complex interactions and actin network turnover in lamellipodia *EMBO J.* **27** 982–92

Lambrechts A, Kwiatkowski A V, Lanier L M, Bear J E, Vandekerckhove J, Ampe C and Gertler FB 2000 cAMP-dependent protein kinase phosphorylation of EVL, a Mena/VASP relative, regulates its interaction with actin and SH3 domains *J. Biol. Chem.* **275** 36143–51

Lambrechts A, Verschelde J L, Jonckheere V, Goethals M, Vandekerckhove J and Ampe C 1997 The mammalian profilin isoforms display complementary affinities for PIP2 and proline-rich sequences *EMBO J.* **16** 484–94

Lane J, Martin T A, Watkins G, Mansel R E and Jiang W G 2008 The expression and prognostic value of ROCK I and ROCK II and their role in human breast cancer *Int. J. Oncol.* **33** 585–93

Lang A E, Schmidt G and Schlosser A *et al* 2010 *Photorhabdus luminescens* toxins ADP-ribosylate actin and RhoA to force actin clustering *Science* **327** 1139–42

Langhorst M F, Genisyuerek S and Stuermer C A O 2006 Accumulation of FlAsH/Lumio green in active mitochondria can be reversed by β-mercaptoethanol for specific staining of tetracysteine-tagged proteins *Histochem. Cell Biol.* **125** 743–7

Lappalainen P, Kessels M M, Cope M J and Drubin D G 1998 The ADF homology (ADF-H) domain: a highly exploited actin-binding module *Mol. Biol. Cell* **9** 1951–9

Lash L L, Wallar B J, Turner J D, Vroegop S M, Kilkuskie R E, Kitchen-Goosen S M, Xu H E and Alberts A S 2016 Small-molecule intramimics of formin autoinhibition: a new strategy to target the cytoskeletal remodeling machinery in cancer cells *Cancer Res.* **73** 6793–803

Lazarides E 1975 Tropomyosin antibody: the specific localization of tropomyosin in nonmuscle cells *J. Cell Biol.* **65** 549–61

Lazarides E and Burridge K 1975 α-actinin: immunofluorescent localization of a muscle structural protein in nonmuscle cell *Cell* **6** 289–98

Lazarides E and Weber K 1974 Actin antibody: the specific visualization of actin filaments in non-muscle cells *Proc. Natl Acad. Sci. USA* **71** 2268–72

Lee W M and Galbraith R M 1992 The extracellular actin-scavenger system and actin toxicity *N. Engl. J. Med.* **326** 1335–41

Leinweber B D, Leavis P C, Grabarek Z, Wang C L and Morgan K G 1999a Extracellular regulated kinase (ERK) interaction with actin and the calponin homology (CH) domain of actin-binding proteins *Biochem. J.* **1** 117–23

Leinweber B D, Tang J X, Stafford W F and Chalovich J M 1999b Calponin interaction with α-actinin-actin: evidence for a structural role for calponin *Biophys. J.* **77** 3208–217

Lenart P, Bacher C P and Daigle N *et al* 2005 A contractile nuclear actin network drives chromosome congression in oocytes *Nature* **436** 812–8

Leung T, Chen X Q, Manser E and Lim L 1996 The p160 RhoA-binding kinase ROK α is a member of a kinase family and is involved in the reorganization of the cytoskeleton *Mol. Cell Biol.* **16** 5313–27

Lewalle A, Fritzsche M, Wilson K, Thorogate R, Duke T and Charras G 2014 A phenomenological density-scaling approach to lamellipodial actin dynamics *Interface Focus* **4** 20140006

Li A, Dawson J C, Forero-Vargas M, Spence H J, Yu X, Konig I, Anderson K and Machesky L M 2010a The actin-bundling protein fascin stabilizes actin in invadopodia and potentiates protrusive invasion *Curr. Biol.* **20** 339–45

Li C, Yu S, Nakamura F, Pentikainen O T, Singh N, Yin S, Xin W and Sy M S 2010b Pro-prion binds filamin A, facilitating its interaction with integrin $\beta 1$, and contributes to melanomagenesis *J. Biol. Chem.* **285** 30328–39

Li G H, Arora P D, Chen Y, McCulloch C A and Liu P 2012 Multifunctional roles of gelsolin in health and diseases *Med. Res. Rev.* **32** 999–1025

Lin A Y, Prochniewicz E, James Z M, Svensson B and Thomas D D 2011 Large-scale opening of utrophin's tandem calponin homology (CH) domains upon actin binding by an induced-fit mechanism *Proc. Natl Acad. Sci. USA* **108** 12729–33

Lingwood D and Simons K 2010 Lipid rafts. As a membrane-organizing principle *Science* **327** 46–50

Liu Y, Belkina N V and Park C *et al* 2012 Constitutively active Ezrin increases membrane tension, slows migration, and impedes endothelial transmigration of lymphocytes *in vivo* in mice *Blood* **119** 445–53

Loomis P A, Zhen L, Sekerkova G, Changyaleket B, Mugnaini E and Bartles J R 2003 Espin cross-links cause the elongation of microvillus-type parallel actin bundles *in vivo J. Cell Biol.* **163** 1045–55

Loy C J, Sim K S and Yong E L 2003 Filamin-A fragment localizes to the nucleus to regulate androgen receptor and coactivator functions *Proc. Natl Acad. Sci. USA* **100** 4562–67

Lukinavicius G, Umezawa K and Olivier N *et al* 2013 A near-infrared fluorophore for live-cell super-resolution microscopy of cellular proteins *Nat. Chem.* **5** 132–9

Luxton G W, Gomes E R, Folker E S, Vintinner E and Gundersen G G 2010 Linear arrays of nuclear envelope proteins harness retrograde actin flow for nuclear movement *Science* **329** 956–9

Machaidze G, Sokoll A, Shimada A, Lustig A, Mazur A, Wittinghofer A, Aebi U and Mannherz H G 2010 Actin filament bundling and different nucleating effects of mouse Diaphanous-related formin FH2 domains on actin/ADF and actin/cofilin complexes *J. Mol. Biol.* **403** 529–45

Machesky L M and Li A 2010 Fascin: invasive filopodia promoting metastasis *Commun. Integr. Biol.* **3** 263–70

Madsen C D, Hooper S, Tozluoglu M, Bruckbauer A, Fletcher G, Erler J T, Bates P A, Thompson B and Sahai E 2015 STRIPAK components determine mode of cancer cell migration and metastasis *Nat. Cell Biol.* **17** 68–80

Maekawa M, Ishizaki T, Boku S, Watanabe N, Fujita A, Iwamatsu A, Obinata T, Ohashi K, Mizuno K and Narumiya S 1999 Signaling from Rho to the actin cytoskeleton through protein kinases ROCK and LIM-kinase *Science* **285** 895–8

Mahoney N M, Janmey P A and Almo S C 1997 Structure of the profilin-poly-L-proline complex involved in morphogenesis and cytoskeletal regulation *Nat. Struct. Biol.* **4** 953–60

Maimaiti Y, Tan J, Liu Z, Guo Y, Yan Y, Nie X, Huang B, Zhou J and Huang T 2017 Overexpression of cofilin correlates with poor survival in breast cancer: a tissue microarray analysis *Oncol. Lett.* **14** 2288–94

Mannherz H G, Goody R S, Konrad M and Nowak E 1980 The interaction of bovine pancreatic deoxyribonuclease I and skeletal muscle actin *Eur. J. Biochem.* **104** 367–79

Marston S, Pinter K and Bennett P 1992 Caldesmon binds to smooth muscle myosin and myosin rod and crosslinks thick filaments to actin filaments *J. Muscle Res. Cell Motil.* **13** 206–18

Martin S K, Kamelgarn M and Kyprianou N 2014 Cytoskeleton targeting value in prostate cancer treatment *Am. J. Clin. Exp. Urol.* **2** 15–26

Martin B R, Giepmans B N G, Adams S R and Tsien R Y 2005 Mammalian cell-based optimization of the biarsenical-binding tetracysteine motif for improved fluorescence and affinity *Nat. Biotechnol.* **23** 1308–14

Matthews B D, Overby D R, Mannix R and Ingber D E 2006 Cellular adaptation to mechanical stress: role of integrins, Rho, cytoskeletal tension and mechanosensitive ion channels *J. Cell Sci.* **119** 508–18

Mattila P K and Lappalainen P 2008 Filopodia: molecular architecture and cellular functions *Nat. Rev. Mol. Cell Biol.* **9** 446–54

Maul R S, Song Y, Amann K J, Gerbin S C, Pollard T D and Chang D D 2003 EPLIN regulates actin dynamics by cross-linking and stabilizing filaments *J. Cell Biol.* **160** 399–407

McConnell B V, Koto K and Gutierrez-Hartmann 2011 A nuclear and cytoplasmic LIMK1 enhances human breast cancer progression *Mol. Cancer* **10** 75

McConnell R E and Tyska M J 2007 Myosin-1a powers the sliding of apical membrane along microvillar actin bundles *J. Cell Biol.* **177** 671–81

McGough A, Pope B, Chiu W and Weeds A 1997 Cofilin changes the twist of F-actin: implications for actin filament dynamics and cellular function *J. Cell Biol.* **138** 771–81

McGrath J L, Tardy Y, Dewey C F Jr., Meister J J and Hartwig J H 1998 Simultaneous measurements of actin filament turnover, filament fraction, and monomer diffusion in endothelial cells *Biophys. J.* **75** 2070–78

McKillop D F and Geeves M A 1993 Regulation of the interaction between actin and myosin subfragment 1: evidence for three states of the thin filament *Biophys. J.* **65** 693–701

Melak M, Plessner M and Grosse R 2017 Actin visualization at a glance *J. Cell Sci.* **130** 525–30

Menez J, Le Maux Chansac B, Dorothee G, Vergnon I, Jalil A, Carlie M F, Chouaib S and Mami-Chouaib F 2004 Mutant α-actinin-4 promotes tumorigenicity and regulates cell motility of a human lung carcinoma *Oncogene* **23** 2630–39

Michael M and Yap A S 2013 The regulation and functional impact of actin assembly at cadherin cell–cell adhesions *Semin. Cell Dev. Biol.* **24** 298–307

Mierke C T 2013 The integrin $\alpha v\beta 3$ increases cellular stiffness and cytoskeletal remodeling dynamics to facilitate cancer cell invasion *New J. Phys.* **15** 015003

Mierke C T, Frey B, Fellner M, Herrmann M and Fabry B 2011 Integrin $\alpha 5\beta 1$ facilitates cancer cell invasion through enhanced contractile forces *J. Cell Sci.* **124** 369–83

Mierke C T, Zitterbart D P, Kollmannsberger P, Raupach C, Schlotzer-Schrehardt U, Goecke T W, Behrens J and Fabry B 2008 Breakdown of the endothelial barrier function in tumor cell transmigration *Biophys. J.* **94** 2832–46

Migocka-Patrzalek M, Makowiecka A, Nowak D, Mazur A J, Hofmann W A and Malicka-Błaszkiewicz M 2015 β- and γ-actins in the nucleus of human melanoma A375 cells *Histochem. Cell Biol.* **144** 417–28

Mikati M A, Breitsprecher D, Jansen S, Reisler E and Goode B L 2015 Coronin enhances actin filament severing by recruiting cofilin to filament sides and altering F-actin conformation *J. Mol. Biol.* **427** 3137–47

Miki H, Suetsugu S and Takenawa T 1998 WAVE, a novel WASP-family protein involved in actin reorganization induced by Rac *EMBO J.* **17** 6932–41

Minamiya Y, Nakagawa T, Saito H, Matsuzaki I, Taguchi K, Ito M and Ogawa J 2005 Increased expression of myosin light chain kinase mRNA is related to metastasis in non-small cell lung cancer *Tumour Biol.* **26** 153–7

Miralles F and Visa N 2006 Actin in transcription and transcription regulation *Curr. Opin. Cell Biol.* **18** 261–6

Miralles F, Posern G, Zaromytidou A I and Treisman R 2003 Actin dynamics control SRF activity by regulation of its coactivator MAL *Cell* **113** 329–42

Miralles F, Posern G, Zaromytidou A-I and Treisman R 2003 Actin dynamics control SRF activity by regulation of its coactivator MAL *Cell* **113** 329–42

Mitchison T J and Cramer L P 1996 Actin-based cell motility and cell locomotion *Cell* **84** 371–9

Mizoguchi T, Kawakami K and Itoh M 2016 Zebrafish lines expressing UAS-driven red probes for monitoring cytoskeletal dynamics *Genesis* **54** 483–9

Mizuno K 2013 Signaling mechanisms and functional roles of cofilin phosphorylation and dephosphorylation *Cell Signal* **25** 457–69

Moll R, Robine S, Dudouet B and Louvard D 1987 Villin: a cytoskeletal protein and a differentiation marker expressed in some human adenocarcinomas *Virchows Arch.* B **54** 155–69

Moon A and Drubin D G 1995 The ADF/cofilin proteins: stimulus-responsive modulators of actin dynamics *Mol. Biol. Cell* **6** 1423–31

Moriyama K, Iida K and Yahara I 1996 Phosphorylation of Ser-3 of cofilin regulates its essential function on actin *Genes Cells* **1** 73–86

Moseley J B, Okada K, Balcer H I, Kovar D R, Pollard T D and Goode B L 2006 Twinfilin is an actin-filament-severing protein and promotes rapid turnover of actin structures *in vivo J. Cell Sci.* **119** 1547–57

Mouneimne G, Soon L, DesMarais V, Sidani M, Song X, Yip S C, Ghosh M, Eddy R, Backer J M and Condeelis J 2004 Phospholipase C and cofilin are required for carcinoma cell directionality in response to EGF stimulation *J. Cell Biol.* **166** 697–708

Mounier N, Perriard J C, Gabbiani G and Chaponnier C 1997 Transfected muscle and non-muscle actins are differentially sorted by cultured smooth muscle and non-muscle cells *J. Cell Sci.* **110** 839–46

Mueller V *et al* 2011 STED nanoscopy reveals molecular details of cholesterol- and cytoskeleton-modulated lipid interactions in living cells *Biophys. J.* **101** 1651–60

Mulder J, Poland M, Gebbink M F, Calafat J, Moolenaar W H and Kranenburg O 2003 p116Rip is a novel filamentous actin-binding protein *J. Biol. Chem.* **278** 27216–23

Mullins R D, Heuser J A and Pollard T D 1998 The interaction of Arp2/3 complex with actin: nucleation, high affinity pointed end capping, and formation of branching networks of filaments *Proc. Natl Acad. Sci. USA* **95** 6181–6

Munsie L N, Caron N, Desmond C R and Truant R 2009 Lifeact cannot visualize some forms of stress-inducedt wisted F-actin *Nat. Methods* **6** 317

Murphy D A and Courtneidge S A 2011 The 'ins' and 'outs' of podosomes and invadopodia: characteristics, formation and function *Nat. Rev. Mol. Cell Biol.* **12** 413–26

Nagai S, Moreno O, Smith C A, Ivanchuk S, Romagnuolo R, Golbourn B, Weeks A, Seol H J and Rutka J T 2011 Role of the cofilin activity cycle in astrocytoma migration and invasion *Genes Cancer* **2** 859–69

Nagayama K, Yahiro Y and Matsumoto T 2011 Stress fibers stabilize the position of intranuclear DNA through mechanical connection with the nucleus in vascular smooth muscle cells *FEBS Lett.* **585** 3992–7

Nakashima K, Sato N, Nakagaki T, Abe H, Ono S and Obinata T 2005 Two mouse cofilin isoforms, muscle-type (MCF) and non-muscle type (NMCF), interact with F-actin with different efficiencies *J. Biochem* **138** 519–26

Nakatsuji H, Nishimura N, Yamamura R, Kanayama H O and Sasaki T 2008 Involvement of actinin-4 in the recruitment of JRAB/MICAL-L2 to cell–cell junctions and the formation of functional tight junctions *Mol. Cell. Biol.* **28** 3324–35

Narumiya S, Tanji M and Ishizaki T 2009 Rho signaling, ROCK and mDia1, in transformation, metastasis and invasion *Cancer Metas. Rev.* **28** 65–76

Naumanen P, Lappalainen P and Hotulainen P 2008 Mechanisms of actin stress fibre assembly *J. Microsc.* **231** 446–54

Nebl G, Meuer S C and Samstag Y 1996 Dephosphorylation of serine 3 regulates nuclear translocation of cofilin *J. Biol. Chem.* **271** 26276–80

Neely B A, Wilkins C E and Marlow L A *et al* 2016 Proteotranscriptomic analysis reveals stage specific changes in the molecular landscape of clear-cell renal cell carcinoma *PLoS One* **11** e0154074

Nemethova M, Auinger S and Small J V 2008 Building the actin cytoskeleton: filopodia contribute to the construction of contractile bundles in the lamella *J. Cell Biol.* **180** 1233–44

Neujahr R, Heizer C, Albrecht R, Ecke M, Schwartz J-M, Weber I and Gerisc G 1997 Three-dimensional patterns and redistribution of myosin II and actin in mitotic *Dictyostelium* cell *J. Cell Biol.* **139** 1793–804

Nevalainen E M, Braun A, Vartiainen M K, Serlachius M, Andersson L C, Moser M and Lappalainen P 2011 Twinfilin-2a is dispensable for mouse development *PLoS One* **6** e22894

Nishida E, Maekawa S and Sakai H 1984 Cofilin, a protein in porcine brain that binds to actin filaments and inhibits their interactions with myosin and tropomyosin *Biochemistry* **23** 5307–13

Nishita M, Aizawa H and Mizuno K 2002 Stromal cell-derived factor 1α activates LIM kinase 1 and induces cofilin phosphorylation for T-cell chemotaxis *Mol. Cell. Biol.* **22** 774–83

Nishita M, Tomizawa C, Yamamoto M, Horita Y, Ohashi K and Mizuno K 2005 Spatial and temporal regulation of cofilin activity by LIM kinase and Slingshot is critical for directional cell migration *J. Cell Biol.* **171** 349–59

Nodelman I M, Bowman G D, Lindberg U and Schutt C E 1999 X-ray structure determination of human profilin II: a comparative structural analysis of human profilins *J. Mol. Biol.* **294** 1271–85

Nolen B J *et al* 2009 Characterization of two classes of small molecule inhibitors of Arp2/3 complex *Nature* **460** 1031–34

Nomura K, Hayakawa K, Tatsumi H and Ono S 2016 Actin-interacting protein 1 promotes disassembly of actin-depolymerizing factor/cofilin-bound actin filaments in a pH-dependent manner *J. Biol. Chem.* **291** 5146–56

Nürnberg A, Kitzing T and Grosse R 2011 Nucleating actin for invasion *Nat. Rev. Cancer* **11** 177–87

O'Connell C B, Tyska M J and Mooseker M S 2007 Myosin at work: motor adaptations for a variety of cellular functions *Biochim. Biophys. Acta* **1773** 615–30

Obinata T, Nagaoka-Yasuda R, Ono S, Kusano K, Mohri K, Ohtaka Y, Yamashiro S, Okada K and Abe H 1997 Low molecular-weight G-actin binding proteins involved in the regulation of actin assembly during myofibrillogenesis *Cell Struct. Funct.* **22** 181–9

Oda T, Iwasa M, Aihara T, Maeda Y and Narita A 2009 The nature of the globular- to fibrous-actin transition *Nature* **457** 441–5

Odronitz F and Kollmar M 2007 Drawing the tree of eukaryotic life based on the analysis of 2,269 manually annotated myosins from 328 species *Genome Biol.* **8** R196

Ohashi K 2015 Roles of cofilin in development and its mechanisms of regulation *Dev. Growth Differ.* **57** 275–90

Olson E N and Nordheim A 2010 Linking actin dynamics and gene transcription to drive cellular motile functions *Nat. Rev. Mol. Cell Biol.* **11** 353–65

Ono S, Baillie D L and Benian G M 1999 UNC-60B, an ADF/cofilin family protein, is required for proper assembly of actin into myofibrils in *Caenorhabditis elegans* body wall muscle *J. Cell Biol.* **145** 491–502

Ono S and Ono K 2002 Tropomyosin inhibits ADF/cofilin-dependent actin filament dynamics *J. Cell Biol.* **156** 1065–76

Ostap E M 2008 Tropomyosins as discriminators of myosin function *Adv. Exp. Med. Biol.* **644** 273–82

Patil K S, Basak I, Pal R, Ho H P, Alves G, Chang E J, Larsen J P and Moller S G 2015 A proteomics approach to investigate miR-153-3p and miR-205-5p targets in neuroblastoma cells *PLoS One* **10** e0143969

Patil S B, Pawar M D and Bitar K N 2004 Direct association and translocation of PKC-α with calponin *Am. J. Physiol. Gastrointest. Liver Physiol.* **286** 954–63

Patrie K M, Drescher A J, Welihinda A, Mundel P and Margolis B 2002 Interaction of two actin-binding proteins, synaptopodin and α-actinin-4, with the tight junction protein MAGI-1 *J. Biol. Chem.* **277** 30183–90

Paunola E, Mattila P K and Lappalainen P 2002 WH2 domain: a small, versatile adapter for actin monomers *FEBS Lett.* **513** 92–7

Pawlak G and Helfman D M 2002 MEK mediates v-Src-induced disruption of the actin cytoskeleton via inactivation of the Rho-ROCK-LIM kinase pathway *J. Biol. Chem.* **277** 26927–33

Pearson M A, Reczek D, Bretscher A and Karplus P A 2000 Structure of the ERM protein moesin reveals the FERM domain fold masked by an extended actin binding tail domain *Cell* **101** 259–70

Pellegrin S P and Mellor H 2007 Actin stress fibres *J. Cell Sci.* **120** 3491–9

Peterburs P, Heering J, Link G, Pfizenmaier K, Olayioye M A and Hausser A 2009 Protein kinase D regulates cell migration by direct phosphorylation of the cofilin phosphatase slingshot 1 like *Cancer Res.* **69** 5634–8

Phang J M *et al* 2016 Structural characterization suggests models for monomeric and dimeric forms of full-length ezrin *Biochem. J.* **473** 2763–82

Phang L-K, Stanchi F and Gerhardt H 2013 Filopodia are dispensable for endothelial tip cell guidance *Development* **140** 4031–40

Pollard L W and Lord M 2014 Getting myosin-V on the right track: tropomyosin sorts transport in yeast *Bioarchitecture* **4** 35–8

Pollard T D and Borisy G G 2003 Cellular motility driven by assembly and disassembly of actin filaments *Cell* **112** 453–65

Pollard T D 2007 Regulation of actin filament assembly by Arp2/3 complex and formins *Annu. Rev. Biophys. Biomol. Struct.* **36** 451–77

Pollard T D and Cooper J A 2009 Actin, a central player in cell shape and movement *Science* **326** 1208–12

Pollard T D and Borisy G G 2003 Cellular motility driven by assembly and disassembly of actin filaments *Cell* **112** 453–65

Ponti A, Machacek M, Gupton S L, Waterman-Storer C M and Danuser G 2004 Two distinct actin networks drive the protrusion of migrating cells *Science* **305** 1782–6

Popowicz G M, Schleicher M, Noegel A A and Holak T A 2006 Filamins: promiscuous organizers of the cytoskeleton *Trends Biochem. Sci.* **31** 411–9

Pruyne D 2008 Tropomyosin function in yeast *Adv. Exp. Med. Biol.* **644** 168–86

Pula G and Krause M 2008 Role of Ena/VASP proteins in homeostasis and disease *Protein-Protein Interactions as New Drug Targets* (Handbook of Experimental Pharmacology vol 186) (Berlin: Springer) pp **186** 39–65

Reinhard M, Giehl K, Abel K, Haffner C, Jarchau T, Hoppe V, Jockusch B M and Walter U 1995 The proline-rich focal adhesion and microfilament protein VASP is a ligand for profilins *EMBO J.* **14** 1583–9

Ren J 1991 Relationship between development of microvilli on tumor cells and growth or metastatic potential of tumor cells *Hokkaido Igaku Zasshi* **66** 187–200

Ren J, Hamada J-I, Okada F, Takeichi N, Morikawa K, Hosokawa M and Kobayashi H 1990 Correlation between the presence of microvilli and the growth or metastatic potential of tumor cells *Cancer Sci.* **81** 920–6

Richards T A and Cavalier-Smith T 2005 Myosin domain evolution and the primary divergence of eukaryotes *Nature* **436** 1113–8

Ridley A J 2011 Life at the leading edge *Cell* **145** 1012–22

Riedl J *et al* 2008 Lifeact: a versatile marker to visualize F-actin *Nat. Methods* **5** 605–7

Riedl J, Flynn K C and Raducanu A *et al* 2010 Lifeact mice for studying F-actin dynamics *Nat. Methods* **7** 168–9

Riveline D, Zamir E, Balaban N Q, Schwarz U S, Ishizaki T, Narumiya S, Kam Z, Geiger B and Bershadsky A D 2001 Focal contacts as mechanosensors: externally applied local mechanical force induces growth of focal contacts by an mDia1-dependent and ROCK-independent mechanism *J. Cell Biol* **153** 1175–86

Rohr G and Mannherz H G 1978 Isolation and characterization of secretory actin. DNAase I complex from rat pancreatic juice *Eur. J. Biochem.* **89** 151–57

Romer L H, Birukov K G and Garcia J G N 2006 Focal adhesions: paradigm for a signaling nexus *Circul. Res.* **98** 606–16

Rottner K, Behrendt B, Small J V and Wehland J 1999 VASP dynamics during lamellipodia protrusion *Nat. Cell Biol.* **1** 321–22

Rotty J D, Wu C, Haynes E M, Suarez C, Winkelman J D, Johnson H E, Haugh J M, Kovar D R and Bear J E 2015 Profilin-1 serves as a gatekeeper for actin assembly by Arp2/3-dependent and -independent pathways *Dev. Cell* **32** 54–67

Roussos E T, Balsamo M and Alfor S K *et al* 2011b Mena invasive (MenaINV) promotes multicellular streaming motility and transendothelial migration in a mouse model of breast cancer *J. Cell Sci.* **124** 2120–31

Roussos E T, Condeelis J S and Patsialou A 2011a Chemotaxis in cancer *Nat. Rev. Cancer* **11** 573–87

Roussos E T, Goswami S and Balsamo M *et al* 2010 Mena invasive (Mena(INV)) and Mena11a isoforms play distinct roles in breast cancer cell cohesion and association with TMEM *Clin. Exp. Metas.* **28** 515–27

Roussos E T, Wang Y, Wyckoff J B, Sellers R S, Wang W, Li J, Pollard J W, Gertler F B and Condeelis J S 2011c Mena deficiency delays tumor progression and decreases metastasis in polyoma middle-T transgenic mouse mammary tumors *Breast Cancer Res.* **12** R101

Rouven B B, Pietuch A, Nehls S, Rother J and Janshoff A 2015 Ezrin is a major regulator of membrane tension in epithelial cells *Sci. Rep.* **5** 14700

Roux A, Cuvelier D, Nassoy P, Prost J, Bassereau P and Goud B 2005 Role of curvature and phase transition in lipid sorting and fussion of membrane tubules *EMBO J.* **24** 1537–45

Royer C and Lu X 2011 Epithelial cell polarity: a major gatekeeper against cancer? *Cell Death Differ.* **18** 1470–77

Rullo J, Becker H, Hyduk S J, Wong J C, Digby G, Arora P D, Cano A P, Hartwig J, McCulloch C A and Cybulsky M I 2012 Actin polymerization stabilizes $\alpha 4\beta 1$ integrin anchors that mediate monocyte adhesion *J. Cell Biol.* **197** 115–29

Saffman P G and Delbruck M 1975 Brownian motion in biological membranes *PNAS* **72** 3111–3

Saka S K, Honigmann A, Eggeling C, Hell S W, Lang T and Rizzoli S O 2014 Multi-protein assemblies underlie the mesoscale organization of the plasma membrane *Nat. Commun.* **5** 4509

Salbreux G, Charras G and Paluch E 2012 Actin cortex mechanics and cellular morphogenesis *Trends Cell Biol.* **22** 536–45

Saman P G and Delbruck M 1975 Brownian motion in biological membranes *PNAS* **72** 3111–3

Sampson E R, Yeh S Y, Chang H C, Tsai Y, Wang X, Ting H J and Chang C 2001 Identification and characterization of androgen receptor associated coregulators in prostate cancer cells *J. Biol. Regul. Homeost. Agents* **15** 123–9

Samstag Y, Dreizler E M, Ambach A, Sczakiel G and Meuer S C 1996 Inhibition of constitutive serine phosphatase activity in T lymphoma cells results in phosphorylation of pp19/cofilin and induces apoptosis *J. Immunol.* **156** 4167–73

Samstag Y, Eckerskorn C, Wesselborg S, Henning S, Wallich R and Meuer S C 1994 Costimulatory signals for human T-cell activation induce nuclear translocation of pp19/cofilin *Proc. Natl Acad. Sci. USA* **91** 4494–98

Samuel M S, Lopez J I and McGhee E J *et al* 2011 Actomyosin-mediated cellular tension drives increased tissue stiffness and β-catenin activation to induce epidermal hyperplasia and tumor growth *Cancer Cell* **19** 776–91

Sanders T A, Llagostera E and Barna M 2013 Specialized filopodia direct long-range transport of SHH during vertebrate tissue patterning *Nature* **497** 628–32

Sanz-Moreno V, Gadea G, Ahn J, Paterson H, Marra P, Pinner S, Sahai E and Marshall C J 2008 Rac activation and inactivation control plasticity of tumor cell movement *Cell* **135** 510–23

Sawada Y, Tamada M, Dubin-Thaler B J, Cherniavskaya O, Sakai R, Tanaka S and Sheetz M P 2006 Force sensing by mechanical extension of the Src family kinase substrate p130Cas *Cell* **127** 1015–26

Schachtner H, Li A, Stevenson D, Calaminus S D J, Thomas S G, Watson S P, Sixt M, Wedlich-Soldner R, Strathdee D and Machesky L M 2012 Tissue inducible Lifeact expression allows visualization of actin dynamics *in vivo* and *ex vivo* *Eur. J. Cell Biol.* **91** 923–9

Schell M J, Erneux C and Irvine R F 2001 Inositol 1,4,5-trisphosphate 3-kinase a associates with F-actin and dendritic spines via its N terminus *J. Biol. Chem.* **276** 37537–46

Schiller H B, Friedel C C, Boulegue C and Fassler R 2011 Quantitative proteomics of the integrin adhesome show a myosin II-dependent recruitment of LIM domain proteins *EMBO Rep.* **12** 259–66

Schmidt K and Nichols B J 2004 Functional interdependence between septin and actin cytoskeleton *BMC Cell Biol* **5** 43–56

Schoumacher M, Goldman R D, Louvard D and Vignjevic D M 2010 Actin, microtubules, and vimentin intermediate filaments cooperate for elongation of invadopodia *J. Cell Biol.* **189** 541–56

Schwarz U S, Balaban N Q, Riveline D, Bershadsky A, Geiger B and Safran S A 2002 Calculation of forces at focal adhesions from elastic substrate data: the effect of localized force and the need for regularization *Biophys. J.* **83** 1380–94

Scott J A, Shewan A M, den Elzen N R, Loureiro J J, Gertler F B and Yap A S 2006 Ena/VASP proteins can regulate distinct modes of actin organization at cadherin-adhesive contacts *Mol. Biol. Cell* **17** 1085–95

Searson P C, Hodzic D and Wirtz D 2009 A perinuclear actin cap regulates shape *Proc. Natl Acad. Sci. USA* **106** 19017–22

Sezgin E *et al* 2012 Partitioning, diffusion, and ligand binding of raft lipid analogs in model and cellular plasma membranes *Biochim. Biophys. Acta* **1818** 1777–84

Shemesh T, Verkhovsky A B, Svitkina T M, Bershadsky A D and Kozlov M M 2009 Role of focal adhesions and mechanical stresses in the formation and progression of the lamellipodium–lamella interface *Biophys. J.* **97** 1254–6124

Shishkin S, Eremina L, Pashintseva N, Kovalev L and Kovaleva M 2017 Cofilin-1 and other ADF/cofilin superfamily members in human malignant cells *Int J Mol Sci.* **18** 10

Shishkin S, Kovaleva M and Ivanov A *et al* 2011 Comparative proteomic study of proteins in prostate cancer and benign hyperplasia cells *J. Cancer Sci. Ther* **S1** 003

Shishkin S, Kovaleva M, Ivanov A, Eryomina L, Lisitskaya K, Toropugin I, Kovalev L, Okhritz V and Loran O 2016 Comparative proteomic study of proteins in prostate cancer and benign hyperplasia cells *J. Cancer Sci. Ther.* **2**

Shukla V K, Kabra A, Maheshwari D, Yadav R, Jain A, Tripathi S, Ono S, Kumar D and Arora A 2015 Solution structures and dynamics of ADF/cofilins UNC-60A and UNC-60B from *Caenorhabditis elegans Biochem. J.* **465** 63–78

Silacci P, Mazzolai L, Gauci C, Stergiopulos N, Yin H L and Hayoz D 2004 Gelsolin superfamily proteins: key regulators of cellular functions *Cell Mol. Life Sci.* **61** 2614–23

Simpson K J, Selfors L M, Bui J, Reynolds A, Leake D, Khvorova A and Brugge J S 2008 Identification of genes that regulate epithelial cell migration using an siRNA screening approach *Nat. Cell Biol.* **10** 1027–38

Sliogeryte K, Thorpe S D, Wang Z, Thompson C L, Gavara N and Knight M M 2016 Differential effects of LifeAct-GFP and actin-GFP on cell mechanics assessed using micropipette aspiration *J. Biomech.* **49** 310–7

Small J V, Stradal T, Vignal E and Rottner K 2002 The lamellipodium: where motility begins *Trends Cell Biol.* **12** 112–20

Small J V, Rottner K, Kaverina I and Anderson K I 1998 Assembling an actin cytoskeleton for cell attachment and movement *Biochim. Biophys. Acta.* **1404** 271–81

Smith M A, Blankman E, Gardel M L, Luettjohann L, Waterman C M and Beckerle M C 2010 A zyxin-mediated mechanism for actin stress fiber maintenance and repair *Dev. Cell* **19** 365–76
Smith M A, Blankman E, Gardel M L, Luettjohann L, Waterman C M and Beckerle M C 2010 A zyxin-mediated mechanism for actin stress fiber maintenance and repair *Dev. Cell* **19** 365–76
Snapp E 2010 Design and use of fluorescent fusion proteins in cell biology *Curr. Protoc. Cell Biol.* Unit–21.4

Somlyo A P and Somlyo A V 2000 Signal transduction by G-proteins, rho-kinase and protein phosphatase to smooth muscle and non-muscle myosin II *J. Physiol.* **2** 177–85

Song X, Chen X, Yamaguchi H, Mouneimne G, Condeelis J S and Eddy R J 2006 Initiation of cofilin activity in response to EGF is uncoupled from cofilin phosphorylation and dephosphorylation in carcinoma cells *J. Cell Sci.* **119** 2871–81

Sparrow N, Manetti M E, Bott M, Fabianac T, Petrilli A, Bates M L, Bunge M B, Lambert S and Fernandez-Valle C 2012 The actin-severing protein cofilin is downstream of neuregulin signaling and is essential for Schwann cell myelination *J. Neurosci.* **32** 5284–97

Spracklen A J, Fagan T N, Lovander K E and Tootle T L 2014 The pros and cons of common actin labeling tools for visualizing actin dynamics during *Drosophila* oogenesis *Dev. Biol.* **393** 209–26

Stanly T A *et al* 2016 Critical importance of appropriate fixation conditions for faithful imaging of receptor microclusters *Biol. Open* **5** 1343–50

Stark B C, Sladewski T E, Pollard L W and Lord M 2010 Tropomyosin and myosin-II cellular levels promote actomyosin ring assembly in fission yeast *Mol. Biol. Cell* **21** 989–1000

Steffen A, Koestler S A and Rottner K 2014 Requirements for and consequences of Rac-dependent protrusion *Eur. J. Cell Biol.* **93** 184–93

Steffen A, Rottner K, Ehinger J, Innocenti M, Scita G, Wehland J and Stradal T E 2004 Sra-1 and Nap1 link Rac to actin assembly driving lamellipodia formation *EMBO J.* **23** 749–59

Strasser P, Gimona M, Moessler H, Herzog M and Small J V 1993 Mammalian calponin. Identification and expression of genetic variants *FEBS Lett.* **330** 13–8

Suarez C, Carroll R T, Burke T A, Christensen J R, Bestul A J, Sees J A, James M L, Sirotkin V and Kovar D R 2015 Profilin regulates F-actin network homeostasis by favoring formin over Arp2/3 complex *Dev. Cell* **32** 43–53

Suh N, Yang X J, Tretiakova M S, Humphrey P A and Wang H L 2005 Value of CDX2, villin, and α-methylacyl coenzyme A racemase immunostains in the distinction between primary adenocarcinoma of the bladder and secondary colorectal adenocarcinoma *Mod. Pathol.* **18** 1217–22

Suraneni P, Rubinstein B, Unruh J R, Durnin M, Hanein D and Li R 2012 The Arp2/3 complex is required for lamellipodia extension and directional fibroblast cell migration *J. Cell Biol.* **197** 239–51

Surks H K, Riddick N and Ohtani K 2005 M-RIP targets myosin phosphatase to stress fibers to regulate myosin light chain phosphorylation in vascular smooth muscle cells *J. Biol. Chem.* **280** 42543–51

Svitkina T M, Bulanova E A, Chaga O Y, Vignjevic D M, Kojima S, Vasiliev J M and Borisy G G 2003 Mechanism of filopodia initiation by reorganization of a dendritic network *J. Cell Biol.* **160** 409–21

Swaminathan V, Mythreye K, O'Brien E T, Berchuck A, Blobe G C and Superfine R 2011 Mechanical stiffness grades metastatic potential in patient tumor cells and in cancer cell lines *Cancer Res.* **71** 5075–80

Sy M S, Li C, Yu S and Xin W 2010 The fatal attraction between pro-prion and filamin A: prion as a marker in human cancers *Biomark. Med.* **4** 453–64

Szent-Gyorgyi A G 2004 The early history of the biochemistry of muscle contraction *J. Gen. Physiol.* **123** 631–41

Tamada M, Sheetz M P and Sawada Y 2004 Activation of a signaling cascade by cytoskeleton stretch *Dev. Cell* **7** 709–18

Tanabe K, Yamazaki H and Inaguma Y *et al* 2014 Phosphorylation of drebrin by cyclin-dependent kinase 5 and its role in neuronal migration *PLoS One* **9** e92291

Tania N, Condeelis J and Edelstein-Keshet L 2013 Modeling the synergy of cofilin and Arp2/3 in lamellipodial protrusive activity *Biophys. J.* **105** 1946–55

Thenappan A, Li Y, Shetty K, Johnson L, Reddy E P and Mishra L 2009 New therapeutics targeting colon cancer stem cells *Curr. Colorectal Cancer Rep.* **5** 209

Thirion C, Stucka R, Mendel B, Gruhler A, Jaksch M, Nowak K J, Binz N, Laing N G and Lochmüller H 2001 Characterization of human muscle type cofilin (CFL2) in normal and regenerating muscle *Eur. J. Biochem* **268** 3473–82

Thorstensson R, Sterky C and Norberg R 1985 Preparation of highly purified F-actin-depolymerizing factor of human serum *Eur. J. Biochem* **147** 637–40

Tojkander S, Gateva G and Lappalainen P 2012 Actin stress fibers–assembly, dynamics and biological roles *J. Cell Sci.* **125** 1855–64

Tojkander S, Gateva G, Schevzov G, Hotulainen P, Naumanen P, Martin C, Gunning P W and Lappalainen P 2011 A molecular pathway for myosin II recruitment to stress fibers *Curr. Biol.* **21** 539–50

Tomasella A, Blangy A and Brancolini C 2014 A receptor-interacting protein 1 (RIP1)-independent necrotic death under the control of protein phosphatase PP2A that involves the reorganization of actin cytoskeleton and the action of cofilin-1 *J. Biol. Chem.* **289** 25699–710

Tominaga T, Sahai E, Chardin P, McCormick F, Courtneidge S A and Alberts A S 2000 Diaphanous-related formins bridge Rho GTPase and Src tyrosine kinase signaling *Mol. Cell* **5** 13–25

Totsukawa G, Yamakita Y, Yamashiro S, Hartshorne D J, Sasaki Y and Matsumura F 2000 Distinct roles of ROCK (Rho-kinase) and MLCK in spatial regulation of MLC phosphorylation for assembly of stress fibers and focal adhesions in 3T3 fibroblasts *J. Cell Biol.* **150** 797–806

Tsai C H, Lin L T and Wang C Y *et al* 2015 Over-expression of cofilin-1 suppressed growth and invasion of cancer cells is associated with up-regulation of let-7 microRNA *Biochim. Biophys. Acta.* **1852** 851–61

Tyska M J and Warshaw D M 2002 The myosin power stroke *Cell Motil Cytoskeleton* **51** 1–15

Uyeda T Q, Iwadate Y, Umeki N, Nagasaki A and Yumura S 2011 Stretching actin filaments within cells enhances their affinity for the myosin II motor domain *Plos One* **6** e26200

Vallenius T and Makela T P 2002 Clik1: a novel kinase targeted to actin stress fibers by the CLP-36 PDZ-LIM protei. *J. Cell Sci.* **115** 2067–73

Vallenius T, Luukko K and Makela T P 2000 CLP-36 PDZ-LIM protein associates with nonmuscle α-actinin-1 and α-actinin-4 *J. Biol. Chem.* **275** 11100–5

Vallenius T, Vaahtomeri K, Kovac B, Osiceanu A M, Viljanen M and Makela T P 2011 An association between NUAK2 and MRIP reveals a novel mechanism for regulation of actin stress fibers *J. Cell Sci.* **124** 384–93

Vallotton P and Small J V 2009 Shifting views on the leading role of the lamellipodium in cell migration: speckle tracking revisited *J. Cell Sci.* **122** 1955–8

Vandekerckhove J, Deboben A, Nassal M and Wieland T 1985 The phalloidin binding site of F-actin *EMBO J.* **4** 2815–8

Vartiainen M K, Mustonen T, Mattila P K, Ojala P J, Thesleff I, Partanen J and Lappalainen P 2002 The three mouse actin-depolymerizing factor/cofilins evolved to fulfill cell-type-specific requirements for actin dynamics *Mol. Biol. Cell* **13** 183–94

Vartiainen M K, Sarkkinen E M, Matilainen T, Salminen M and Lappalainen P 2003 Mammals have two twinfilin isoforms whose subcellular localizations and tissue distributions are differentially regulated *J. Biol. Chem.* **278** 34347–55

Vavylonis D, Kovar D R, O'shaughnessy B and Pollard T D 2006 Model of formin-associated actin filament elongation *Mol. Cell* **21** 455–66

Velkova A, Carvalho M A, Johnson J O, Tavtigian S V and Monteiro A N 2010 Identification of filamin A as a BRCA1-interacting protein required for efficient DNA repair *Cell Cycle* **9** 1421–33

Veniere S, Ampe C, Vandekerckhove J and Lambrechts A 2009 The interaction of proline-rich ligands with profilin probed with an enzyme-linked immunosorbent assay *J. Biomol. Screen* **14** 350–9

Verkhusha V V, Tsukita S and Oda H 1999 Actin dynamics in lamellipodia of migrating border cells in the *Drosophila* ovary revealed by a GFP-actin fusion protein *FEBS Lett.* **445** 395–401

Vicente-Manzanares M, Ma X, Adelstein R S and Horwitz A R 2009 Non-muscle myosin II takes centre stage in cell adhesion and migration *Nat. Rev. Mol. Cell. Biol.* **10** 778–90

Vicidomini G *et al* 2015 STED-FLCS: an advanced tool to reveal spatiotemporal heterogeneity of molecular membrane dynamics *Nano Lett.* **15** 5912–8

Vindin H and Gunning P 2013 Cytoskeletal tropomyosins: choreographers of actin filament functional diversity *J. Muscle Res. Cell Motil.* **34** 261–74

Visegrády B, Lorinczy D, Hild G, Somogyi B and Nyitrai M 2004 The effect of phalloidin and jasplakinolide on the flexibility and thermal stability of actin filaments *FEBS Lett.* **565** 163–6

Vitriol E A, McMillen L M, Kapustina M, Gomez S M, Vavylonis D and Zheng J Q 2015 Two functionally distinct sources of actin monomers supply the leading edge of lamellipodia *Cell Rep* **11** 433–45

Vitolo M I, Boggs A E, Whipple R A, Yoon J R, Thompson K, Matrone M A, Cho E H, Balzer E M and Martin S S 2013 Loss of PTEN induces microtentacles through PI3K-independent activation of cofilin *Oncogene* **32** 2200–10

Vitriol E A,, Wise A L, Berginski M E, Bamburg J R and Zheng J Q 2013 Instantaneous inactivation of cofilin reveals its function of F-actin disassembly in lamellipodia *Mol. Biol. Cell* **24** 2238–47

Wang C L and Coluccio L M 2010 New insights into the regulation of the actin cytoskeleton by tropomyosin *Int. Rev. Cell Mol. Biol.* **281** 91–128

Wang N, Butler J P and Ingber D E 1993 Mechanotransduction across the cell surface and through the cytoskeleton *Science* **260** 1124–7

Wang T, Cleary R A, Wang R and Tang D D 2014 Glia maturation factor-γ phosphorylation at Tyr-104 regulates actin dynamics and contraction in human airway smooth muscle *Am. J. Respir. Cell Mol. Biol.* **51** 652–9

Wang W, Eddy R and Condeelis J 2007 The cofilin pathway in breast cancer invasion and metastasis *Nat. Rev. Cancer* **7** 429–40

Wang W, Mouneimne G, Sidani M, Wyckoff J, Chen X, Makris A, Goswami S, Bresnick A R and Condeelis J S 2006 The activity status of cofilin is directly related to invasion, intravasation, and metastasis of mammary tumors *J. Cell Biol.* **173** 395–404

Wang Y, Kuramitsu Y, Kitagawa T, Baron B, Yoshino S, Maehara S, Maehara Y, Oka M and Nakamura K 2015 Cofilin-phosphatase slingshot-1L (SSH1L) is over-expressed in pancreatic cancer (PC) and contributes to tumor cell migration *Cancer Lett.* **360** 171–6

Wang K, Ash J F and Singer S J 1975 Filamin, a new high-molecular-weight protein found in smooth muscle and non-muscle cells *Proc. Natl. Acad. Sci. USA* **72** 1359–64

Wang Y L 1985 Exchange of actin subunits at the leading edge of living fibroblasts: possible role of treadmilling *J. Cell Biol.* **101** 597–602

Watanabe N, Madaule P, Reid T, Ishizaki T, Watanabe G, Kakizuka A, Saito Y, Nakao K, Jockusch B M and Narumiya S 1997 p140mDia, a mammalian homolog of *Drosophila* diaphanous, is a target protein for Rho small GTPase and is a ligand for profilin *EMBO* J **16** 3044–56

Watanabe N and Mitchison T J 2002 Single-molecule speckle analysis of actin filament turnover in lamellipodia *Science* **295** 1083–6

Wawrezinieck L, Rigneault H, Marguet D and Lenne P F 2005 Fluorescence correlation spectroscopy diffusion laws to probe the submicron cell membrane organization *Biophys. J.* **89** 4029–42

Weaver A M 2008 Cortactin in tumor invasiveness *Cancer Lett.* **265** 157–66

Weaver A M 2006 Invadopodia: specialized cell structures for cancer invasion *Clin. Exp. Metastasis* **23** 97–105

Weber K and Groeschel-Stewart U 1974 Antibody to myosin: the specific visualization of myosin-containing filaments in nonmuscle cells *Proc. Natl Acad. Sci. USA* **71** 4561–4

Wegner A and Isenberg G 1983 12-fold difference between the critical monomer concentrations of the two ends of actin filaments in physiological salt conditions *Proc. Natl. Acad. Sci. USA* **80** 4922–5

Wehland J, Osborn M and Weber K 1977 Phalloidin-induced actin polymerization in the cytoplasm of cultured cells interferes with cell locomotionand growth *Proc. Natl Acad. Sci. USA* **74** 5613–7

Wei R, Zhang Y, Shen L, Jiang W, Li C, Zhong M, Xie Y, Yang D, He L and Zhou Q 2012 Comparative proteomic and radiobiological analyses in human lung adenocarcinoma cells *Mol. Cell. Biochem.* **359** 151–9

Weins A, Schlondorff J S, Nakamura F, Denker B M, Hartwig J H, Stossel T P and Pollak M R 2007 Disease-associated mutant α-actinin-4 reveals a mechanism for regulating its F-actin-binding affinity *Proc. Natl Acad. Sci. USA* **104** 16080–5

Welsch T, Keleg S, Bergmann F, Bauer S, Hinz U and Schmidt J 2009 Actinin-4 expression in primary and metastasized pancreatic ductal adenocarcinoma *Pancreas* **38** 968–76

Westphal M, Jungbluth A, Heidecker M, Mühlbauer B, Heizer S J-M, Marriott G and Gerisch G 1997 Microfilament dynamics during cell movement and chemotaxis monitored using a GFP–actin fusion protein *Curr. Biol.* **7** 176–83

Wieser S, Moertelmaier M, Fuertbauer E, Stockinger H and Schutz G 2007 (Un)confined diffusion of CD59 in the plasma membrane determined by high-resolution single molecule microscopy *Biophys. J.* **92** 3719–28

Wieser S and Schutz G J 2008 Tracking single molecules in the live cell plasma membrane—do's and don't's *Methods* **46** 131–40

Winder S J and Ayscough K R 2005 Actin-binding proteins *J. Cell Sci.* **118** 651–4

Winder S J, Allen B G, Clement-Chomienne O and Walsh M P 1998 Regulation of smooth muscle actin–myosin interaction and force by calponin *Acta Physiol. Scand.* **164** 415–26

Winder S J, Hemmings L, Bolton S J, Maciver S K, Tinsley J M, Davies K E, Critchley D R and Kendrick-Jones J 1995 Calmodulin regulation of utrophin actin binding *Biochem. Soc. Trans.* **23** 397S

Witke W, Podtelejnikov A V, Di Nardo A, Sutherland J D, Gurniak C B, Dotti C and Mann M 1998 In mouse brain profilin I and profilin II associate with regulators of the endocytic pathway and actin assembly *EMBO J.* **17** 967–76

Witke W 2014 The role of profilin complexes in cell motility and other cellular processes *Trends Cell Biol.* **14** 461–9

Wolfenson H, Henis Y I, Geiger B and Bershadsk A D 2009 The heel and toe of the cell's foot: a multifaceted approach for understanding the structure and dynamics of focal adhesions *Cell Motil. Cytoskeleton* **66** 1017–29

Wollweber L, Stracke R and Gothe U 1981 The use of a simple method to avoid cell shrinkage during SEM preparation *J. Microsc.* **121** 185–9

Woolner S and Bement W M 2009 Unconventional myosins acting unconventionally *Trends Cell Biol.* **19** 245–52

Wu C, Asokan S B, Berginski M E, Haynes E M, Sharpless N E, Griffith J D, Gomez S M and Bear J E 2012 Arp2/3 is critical for lamellipodia and response to extracellular matrix cues but is dispensable for chemotaxis *Cell* **148** 973–87

Wu C, Haynes E, Asokan S B, Simon J M, Sharpless N E, Bladwin A S, Davis I J, Johnson G L and Bear J E 2013 Loss of Arp2/3 induces an NF-κB-dependent, nonautonomous effect on chemotactic signaling *J. Cell Biol.* **203** 907–16

Wu J-Q and Pollard T D 2005 Counting cytokinesis proteins globally and locally in fission yeast *Science* **310** 310–4

Wulf E, Deboben A, Bautz F A, Faulstich H and Wielan T 1979 Fluorescent phallotoxin, a tool for the visualization of cellular actin *Proc. Natl Acad. Sci. USA* **76** 4498–502

Wulfkuhle J D, Donina I E, Stark N H, Pope R K, Pestonjamasp K N, Niswonger M L and Luna E J 1999 Domain analysis of supervillin, an F-actin bundling plasma membrane protein with functional nuclear localization signals *J. Cell Sci.* **112** 2125–36

Wyckoff J B, Pinner S E, Gschmeissner S, Condeelis J S and Sahai E 2006 ROCK- and myosin-dependent matrix deformation enables protease-independent tumor-cell invasion in viv *Curr. Biol.* **16** 1515–23

Xu K, Zhong G S and Zhuang X W 2013 Actin, spectrin, and associated proteins form a periodic cytoskeletal structure in axons *Science* **339** 452–6

Xu K, Babcock H and Zhuang X 2012 Dual-objective STORM reveals three-dimensional lament organization in the actin cytoskeleton *Nat. Methods* **9** 185–8

Xu Y, Moseley J B and Sagot I *et al* 2004 Crystal structures of a formin homology-2 domain reveal a tethered dimer architecture *Cell* **116** 711–23

Yachida S, Jones S and Bozic I *et al* 2010 Distant metastasis occurs late during the genetic evolution of pancreatic cancer *Nature* **467** 1114–7

Yahara I, Aizawa H, Moriyama K, Iida K, Yonezawa N, Nishida E, Hatanaka H and Inagaki F 1996 A role of cofilin/destrin in reorganization of actin cytoskeleton in response to stresses and cell stimuli *Cell Struct. Funct.* **21** 421–4

Yamada S, Pokutta S, Drees F, Weis W I and Nelson W J 2005 Deconstructing the cadherin–catenin–actin complex *Cell* **123** 889–901

Yamada S, Yanamoto S, Yoshida H, Yoshitomi I, Kawasaki G, Mizuno A and Nemoto T K 2010 RNAi-mediated down-regulation of α-actinin-4 decreases invasion potential in oral squamous cell carcinoma *Int. J. Oral Maxillofac. Surg.* **39** 61–7

Yamamoto S, Tsuda H, Honda K, Kita T, Takano M, Tamai S, Inazawa J, Yamada T and Matsubara O 2007 Actinin-4 expression in ovarian cancer: a novel prognostic indicator independent of clinical stage and histological type *Mod. Pathol.* **20** 1278–85

Yamamoto S, Tsuda H, Honda K, Onozato K, Takano M, Tamai S, Imoto I, Inazawa J, Yamada T and Matsubara O 2009 Actinin-4 gene amplification in ovarian cancer: a candidate oncogene associated with poor patient prognosis and tumor chemoresistance *Mod. Pathol.* **22** 499–507

Yamashiro-Matsumura S and Matsumura F 1988 Characterization of 83-kilodalton nonmuscle caldesmon from cultured rat cells: stimulation of actin binding of nonmuscle tropomyosin and periodic localization along microfilaments like tropomyosin *J. Cell Biol.* **106** 1973–83

Yan H, Yang K, Xiao H, Zou Y J, Zhang W B and Liu H Y 2012 Over-expression of cofilin-1 and phosphoglycerate kinase 1 in astrocytomas involved in pathogenesis of radioresistance *CNS Neurosci. Ther.* **18** 729–36

Yan X D, Pan L Y, Yuan Y, Lang J H and Mao N 2007 Identification of platinum-resistance associated proteins through proteomic analysis of human ovarian cancer cells and their platinum-resistant subline *J. Proteome Res.* **6** 772–80

Yang C S, Hoelzle M, Disanza A, Scita G and Svitkina T 2009 Coordination of membrane and actin cytoskeleton dynamics during filopodia protrusion *PLos One* **4** e5678

Yang C and Svitkina T 2011 Filopodia initiation: focus on the Arp2/3 complex and formins *Cell Adhes. Migr* **5** 402–8

Yang C, Czech L, Gerboth S, Kojima S, Scita G and Svitkina T 2007 Novel roles of formin mDia2 in lamellipodia and filopodia formation in motile cells *PLoS Biol.* **5** e317

Yang H C and Pon L A 2002 Actin cable dynamics in budding yeast *Proc. Natl Acad. Sci. USA* **99** 751–6

Yang Z L, Miao X, Xiong L, Zou Q, Yuan Y, Li J, Liang L, Chen M and Chen S 2013 CFL1 and Arp3 are biomarkers for metastasis and poor prognosis of squamous cell/adenosquamous carcinomas and adenocarcinomas of gallbladder *Cancer Investig* **31** 132–9

Yang W, Thein S, Lim C Y, Ericksen R E, Sugii S, Xu F, Robinson R C, Kim J B and Han W 2014 Arp2/3 complex regulates adipogenesis by controlling cortical actin remodelling *e Biochem. J.* **464** 179–92

Yarmola E G, Somasundaram T, Boring T A, Spector I and Bubb M R 2000 Actin-latrunculin A structure and function. Differential modulation of actin-binding protein function by latrunculin *A. J. Biol. Chem.* **275** 28120–7

Ydenberg C A, Smith B A, Breitsprecher D, Gelles J and Goode B L 2011 Cease-fire at the leading edge: new perspectives on actin filament branching, debranching, and cross-linking *Cytoskeleton* **68** 596–602

Yeung K S and Paterson I 2002 Actin-binding marine macrolides: total synthesis and biological importance *Angew. Chem. Int. Edn Engl.* **41** 4632–4653

Yi K, Unruh J R, Deng M, Slaughter B D, Rubinstein B and Li R 2011 Dynamic maintenance of asymmetric meiotic spindle position through Arp2/3-complex-driven cytoplasmic streaming in mouse oocytes *Nat. Cell Biol.* **13** 1252–8

Yonezawa N, Homma Y, Yahara I, Sakai H and Nishida E 1991 A short sequence responsible for both phosphoinositide binding and actin binding activities of cofilin *J. Biol. Chem.* **266** 17218–21

Yonezawa N, Nishida E and Sakai H 1985 pH control of actin polymerization by cofilin *J. Biol. Chem.* **260** 14410–12

Yoshigi M, Hoffman L M, Jensen C C, Yost H J and Beckerle M C 2005 Mechanical force mobilizes zyxin from focal adhesions to actin filaments and regulates cytoskeletal reinforcement *J. Cell Biol.* **171** 209–15

Yoshioka K, Foletta V, Bernard O and Itoh K 2003 A role for LIM kinase in cancer invasion *Proc. Natl Acad. Sci. USA* **100** 7247–52

Yuan C B, Zhao R, Wan F J, Cai J H, Ji X P and Yu Y Y 2010 Significance of plasmic L-plastin levels in the diagnosis of colorectal cancer *Chinese J. Gastrointest. Surg.* **13** 687–90

Zagorska A, Deak M, Campbell D G, Banerjee S, Hirano M, Aizawa S, Prescott A R and Alessi D R 2010 New roles for the LKB1-NUAK pathway in controlling myosin phosphatase complexes and cell adhesion *Sci. Signal* **3** ra25

Zalli D, Neff L, Nagano K, Shin N Y, Witke W, Gori F and Baron R 2016 The actin-binding protein cofilin and its interaction with cortactin are required for podosome patterning in osteoclasts and bone resorption *in vivo* and *in vitro Bone Miner. Res.* **31** 1701–12

Zebda N, Bernard O, Bailly M, Welti S, Lawrence D S and Condeelis J S 2000 Phosphorylation of ADF/cofilin abolishes EGF-induced actin nucleation at the leading edge and subsequent lamellipod extension *J. Cell Biol.* **151** 1119–28

Zhang H H, Wang W, Feng L, Yang Y, Zheng J, Huang L and Chen D B 2015 S-nitrosylation of cofilin-1 serves as a novel pathway for VEGF-stimulated endothelial cell migration *J. Cell. Physiol.* **230** 406–17

Zhang H S, Zhao J W, Wang H, Zhang H Y, Ji Q Y, Meng L J, Xing F J, Yang S T and Wang Y 2014 LIM kinase 1 is required for insulin-dependent cell growth of osteosarcoma cell lines *Mol. Med. Rep.* **9** 103–8

Zhang X-F, Schaefer A W, Burnette D T, Schoonderwoert V T and Forscher P 2003 Rho-dependent contractile responses in the neuronal growth cone are independent of classical peripheral retrograde actin flow *Neuron* **40** 931–44

Zhang Y and Tong X 2010 Expression of the actin-binding proteins indicates that cofilin and fascin are related to breast tumour size *J. Int. Med. Res.* **38** 1042–8

Zhao H, Hakala M and Lappalainen P 2010 ADF/cofilin binds phosphoinositides in a multivalent manner to act as a PIP(2)-density sensor *Biophys. J.* **98** 2327–36

Zhao X H, Laschinger C, Arora P, Szaszi K, Kapus A and McCulloch C A 2007 Force activates smooth muscle α-actin promoter activity through the Rho signaling pathway *J. Cell Sci.* **120** 1801–9

Zhong C, Chrzanowska-Wodnicka M, Brown J, Shaub A, Belkin A M and Burridge K 1998 Rho-mediated contractility exposes a cryptic site in fibronectin and induces fibronectin matrix assembly *J. Cell Biol* **141** 539–51

Zhong Z, Yeow W S, Zou C, Wassell R, Wang C, Pestell R G, Quong J N and Quong A A 2010 Cyclin D1/cyclin-dependent kinase 4 interacts with filamin A and affects the migration and invasion potential of breast cancer cells *Cancer Res.* **70** 2114

Zhou G L, Zhang H and Field J 2014 Mammalian CAP (cyclase-associated protein) in the world of cell migration: roles in actin filament dynamics and beyond *Cell Adhes. Migr.* **8** 55–9

Zhou A X, Toylu A, Nallapalli R K, Nilsson G, Atabey N, Heldin C H, Boren J, Bergo M O and Akyurek L M 2011 Filamin a mediates HGF/c-MET signaling in tumor cell migration *Int. J. Cancer* **128** 839–46

Zhou K, Muroyama A, Underwood J, Leylek R, Ray S, Soderlingh S H and Lechler T 2013 Actin-related protein2/3 complex regulates tight junctions and terminal differentiation to promote epidermal barrier formation *PNAS* **110** E3820–E3829

Zicha D, Dobbie I M, Holt M R, Monypenny J, Soong D Y, Gray C and Dunn G A 2003 Rapid actin transport during cell prorusion *Science* **300** 142–5

IOP Publishing

Physics of Cancer, Volume 1 (Second Edition)
Interplay between tumor biology, inflammation and cell mechanics
Claudia Tanja Mierke

Chapter 8

Intermediate filaments and nuclear deformability during matrix invasion

Summary

Intermediate filaments belong to one of the three major classes of intracellular filaments and are hence an important component of the cellular cytoskeleton. The first role of intermediate filaments is their support of the mechanical properties of the cells. Intermediate filament proteins can assemble to produce polymers that build viscoelastic gels. Intermediate filament networks display a striking nonlinear elasticity with increasing stiffness that has been quantified by analyzing the shear modulus, which can even increase by an order of magnitude when the networks are probed by large stains. It is clear that intermediate filaments fulfill many different roles, regulating a broad variety of biological functions, such as the organization and assembly of microtubules and microfilaments, the regulation of nuclear structure and the connection of the chromatin to the nuclear lamina, and subsequently they facilitate the regulation of the activity of genes, the cell cycle and signal transduction pathways. This chapter introduces and discusses the role of intermediate filaments and nuclear deformability, including how they may affect cell matrix invasion through connective tissue. Special emphasis is placed on the role of intermediate filaments in providing cytoskeletal and nuclear stiffness and hence supporting cellular motility. As intermediate filaments are integrated into a broad cytoskeletal and nuclear network, the interaction between them and actin filaments in supporting cell motility is highlighted. In addition to cellular motility regulation, the strong involvement of intermediate filament proteins in the regulation of intracellular trafficking has been established, as they associate with key proteins of the vesicular membrane transport machinery and subsequently of the endocytic pathway.

Moreover, a number of intermediate filament proteins have been identified to function in the acquisition of tumorigenic properties. Finally, it is suggested that cancer cells may overcome the steric restrictions of dense 3D extracellular matrix

doi:10.1088/978-0-7503-1753-5ch8

fiber networks by undergoing mitosis and thereby restructuring their entire cytoskeleton including the intermediate filament network.

8.1 Structure and assembly of intermediate filaments

Intermediate filament proteins belong to a large family of proteins that make up one of the three major components of the cytoskeleton. The family of intermediate proteins includes a large number of proteins which are encoded in the human genome by 70 different genes. Moreover, multiple splice variants of the same gene increase the broad variety of proteins assembling intermediate filaments (Hesse *et al* 2001, Rogers *et al* 2004, 2005, Szeverenyi *et al* 2008). Intermediate filaments exhibit an intermediate size of 8–15 nm in diameter, compared to the other cytoskeletal components, such as microtubules and actin filaments (also termed microfilaments), whose diameters are approximately 25 nm and 5–8 nm, respectively (Ishikawa *et al* 1968, Goldmann and Follett 1969, 1970, Franke *et al* 1978, Fuchs and Weber 1994, Geisler and Weber 1982). Intermediate filaments are different from the other cytoskeletal proteins, as they are apolar, which prevents them from being used as tracks for motor proteins, as these proteins require polar tracks (Herrmann *et al* 2009).

Although there exist a large number of intermediate filaments, all the proteins of this family possess the same tripartite structural organization. In particular, intermediate filaments are built on a central α-helical rod domain that is flanked by two non-α-helical domains. The amino acids are organized in a heptad repeat pattern within the central rod domain. The broad variability in the primary structure of intermediate filaments is provided in the amino-terminal head and the carboxy-terminal tail domains. The *in silico* structural prediction of intermediate filaments proposes that the central domain is subdivided into four domains, interrupted by three linker domains (L1, L12, L2), where the heptad repeats assemble into four sub-helices (coil 1A, coil 1B, coil 2A, coil 2B). However, crystallographic analyses indicate that the segments coil 2A, L2 and coil 2B provide a continuous coiled domain (Geisler and Weber 1982, Quax-Jeuken *et al* 1983, Fuchs and Weber 1994, Parry *et al* 2007, Herrmann *et al* 2009). The structure of intermediate filaments is crucial for their self-assembly. Unlike the other cytoskeletal components, intermediate filament monomers do not possess any enzymatic activity, therefore their assembly is simply based on the direct association of several monomers. In particular, two α-helical rods associate in parallel and form a dimer, which then associates with another dimer in an antiparallel manner in order to build a tetramer. Tetramers rapidly associate laterally to assemble into unit length filaments (ULF), whose length is approximately 60 nm, and then they anneal longitudinally to each other in order to assemble longer filaments (Strelkov *et al* 2003). Their assembly into mature insoluble filaments and their disassembly into soluble components (such as from tetramers to monomers) are regulated by cycles of phosphorylation and dephosphorylation events (Izawa and Inagaki 2006).

Based on comparisons to the primary structure of intermediate filaments of higher vertebrates, they are grouped into six classes. Acidic and basic keratins belong to classes I and class II, respectively. The class III of intermediate filaments is the most heterogeneous collection as it contains desmin, glial fibrillary acidic protein (GFAP), peripherin and vimentin, whereas class IV encompasses α-internexin, nestin, synemin, syncoilin and the neurofilament (NF) triplet proteins, which are classified based on their molecular weight into NF-L (light), NF-M (medium) and NF-H (heavy). Lamins belong to class V and, unlike the intermediate filaments of classes I–IV, provide the cytoplasmic intermediate filaments, and they build the nuclear intermediate filaments. Class VI has been added recently and comprises two lens-specific proteins, the beaded filament structural proteins, Bfsp1 (also termed filensin) and Bfsp2 (also termed phakinin or CP49) (Herrmann *et al* 2009, Kornreich *et al* 2015). In more detail, keratins can only form heteropolymers in epithelial cells through the combination of I and II class proteins in a 1:1 ratio (Miller *et al* 1993), whereas type III intermediate filaments form mostly homopolymers, however, they can also associate with other type III or type IV intermediate filaments to form heteropolymers. Moreover, vimentin assembles into heteropolymeric filaments with desmin in vascular smooth muscle tissue (Quinlan and Franke 1982), with glial fibrillary acidic protein (GFAP) in glioma cells (Quinlan and Franke 1983) and with neurofilaments (Monteiro and Cleveland 1989).

During embryonic developmental processes and cell differentiation, the intermediate filament expression is highly regulated. For example, the expression of vimentin is widely distributed in mesenchymal cells, in leukocytes, in blood vessel endothelial cells and in distinct epithelial cells (Franke *et al* 1979, Schmid *et al* 1979, Lilienbaum *et al* 1986), whereas it can also be expressed in neurons during axonal regeneration together with peripherin (Shea *et al* 1993, Cochard and Paulin 1984). Moreover, intermediate filament expression is cell-type and tissue specific in adults (Lazarides 1982). Specifically, desmin is expressed in both cardiac and skeletal muscle cells (Schmid *et al* 1979), GFAP is located in astrocytes and in other glial cells (Jessen and Mirsky 1980, Chiu *et al* 1981) and peripherin is found to be present in peripheral neurons and in central neurons that project towards peripheral structures (Portier *et al* 1984, 1993, Errante *et al* 1998). Moreover, neurofilaments are expressed in mature neurons (Pannese 1962). In addition, the expression levels of intermediate filaments are altered under pathological conditions. For instance, overexpression of keratins has been connected to certain diseases (Toivola *et al* 2015) and hence alterations of neurofilament levels in distinct body compartments represent hallmarks of neurological disorders (Norgren *et al* 2003, Gaiottino *et al* 2013). In addition, GFAP and vimentin expression is increased in reactive astrocytes (Hol and Pekny 2015), whereas peripherin expression is enhanced after injury (Xiao *et al* 2006).

Intermediate filaments represent essential parts of the cellular architecture and functions. To fulfill a wide range of cellular function, the alterations in the head and tail domains of the intermediate filament proteins, their distinct expression and their posttranslational modifications are essential parts to determine the individual function of a certain intermediate filament. At the first glance, intermediate filaments have been revealed to support the mechanical strength of cells and maintain the entire structure of tissues by providing a rather static scaffold within the cytoplasm. Indeed, intermediate filaments contribute strongly to the cell's resistance to mechanical and nonmechanical stress and deformation. In particular, vimentin provides cellular stiffness in fibroblasts and chondrocytes (Wang and Stamenovic 2000, Haudenschild *et al* 2011), whereas it also maintains and contributes to cellular elastic properties and subsequently protects against compression by external stress application. In embryonic fibroblasts isolated from wild-type and vimentin knock-out mice, it has been found that vimentin filaments contribute to intracellular mechanics through their localization towards and stabilization of organelles inside the cell and hence increase cellular elastic behavior (Guo *et al* 2013, Mendez *et al* 2014). Keratins represent another example of intermediate filaments that are important for the integrity of the cellular structure of epithelial cells. Indeed, keratins are located mainly in epithelial cells for simple epithelia regulation, for instance, cellular stiffness in hepatic epithelial cells (Bordeleau *et al* 2012).

The traditional and established roles of intermediate filaments encompass the regulation of the nuclear shape, its nuclear structure and the activity of specific gene transcription. In more detail, lamins are localized within the nuclear lamina, which represents one of the three most prominent components of the nuclear envelope. The other two major components of the nuclear envelope are the nuclear membrane and the nuclear pore complexes. Lamins are highly critical for maintenance of the nuclear architecture and provide important mechanical anchorage functions for other structural proteins towards the nuclear envelope (Hutchison 2002). In addition, they distribute the nuclear pore complexes within the nuclear membrane and thereby regulate the transport of molecules and ions in and out of the nucleus (Hutchison 2002). Moreover, a strong physical connection between cytoplasmic intermediate filaments, actin filaments and microtubule and lamins is provided through the location of nuclear protein complexes at a specific position in the nuclear envelope (Bloom *et al* 1996, Tolstonog *et al* 2002, Stoffler *et al* 2003, Kiseleva *et al* 2004).

Several studies have consistently shown that correct lamina assembly is necessary for the establishment of DNA replication centers and have demonstrated that lamins are highly necessary for the elongation phase of the synthesis of DNA (Ellis *et al* 1997, Spann *et al* 1997). Lamins are also able to alter and stabilize heterochromatin domains (Chaturvedi and Parnaik 2010, Rochman *et al* 2010, Montes de Oca *et al* 2011, Solovei *et al* 2013, Zhang *et al* 2013) and additionally they affect the activity of the DNA transcription regulating RNA polymerase II (Spann *et al* 2002). Specifically, they facilitate DNA repair through the recruitment of the DNA damage response machinery (Manjiu *et al* 2006, Gonzalez-Suarez *et al* 2009, Gibbs-Seymour *et al* 2015, Ghosh *et al* 2015) and the phosphorylation of lamins is required for the breakdown of the nuclear envelope during mitosis (Peter *et al* 1990, Ward and Kirschner 1990, Beaudoin *et al* 2002, Kochin *et al* 2014). However, lamins are not the only intermediate filaments regulating nuclear activities, as also vimentin filaments, together with lamins, provide the controlled regulation of the nuclear shape (Sarria *et al* 1994, Coffinier *et al* 2011).

Intriguingly, the involvement of intermediate filaments in several other biological functions revealed a dynamic intermediate filament network with multiple cellular functions. All three major cytoskeletal filaments constantly communicate with each other such as intermediate filaments, microtubules and microfilaments and, moreover, they are connected to each other by proteins termed cytolinkers (Oriolo *et al* 2006, Conover and Gregorio 2011). Hence, the modulation of intermediate filaments causes an additional modification in the microtubule and microfilament networks (Oriolo *et al* 2006, Conover and Gregorio 2011). As expected, direct interaction between vimentin and actin filaments has been shown and rheological measurements have clearly revealed that these two cytoskeletal components contribute together to the overall cellular stiffness (Cary *et al* 1994, Esue *et al* 2006). In particular, vimentin alters the extracellular-signal-regulated kinases; (ERKs) signal transduction pathway (Perlson *et al* 2005, 2006, Kumar *et al* 2007, Virtakoivu *et al* 2015), whereas keratins are associated with the precise regulation of the cell cycle (Toivola *et al* 2001). Moreover, keratins and lamins fulfill a role in apoptosis (Rao *et al* 1996, Caulin *et al* 2000), whereas vimentin, keratins and nestin perform established roles in cell migration and cancer (Gilles *et al* 1999, Fortier *et al* 2013, Ju *et al* 2013, Narita *et al* 2014, Sutoh *et al* 2014).

8.2 Involvement of intermediate filaments in vesicular trafficking

Endocytosis represents an active process that is utilized by cells in order to engulf and phagocytize into vesicular structures, molecules, macromolecules, particles and fluids from the extracellular milieu. In contrast, cells can also exocytose substances or waste products in secretory vesicles. Moreover, transport in the endocytic and secretory pathways is facilitated by the formation of vesicles. In endocytosis, material from the extracellular milieu is phagocytized by a portion of the cell membrane that creates a vesicle around the substance to be internalized. This vesicle fuses in the next step with early endosomes, which are compartments possessing tubular extensions that are localized at the cell's periphery. Then the transport goods are sorted to late endosome in order to be delivered to lysosomes by fusion of the late endosome with the lysosome for degradation or recycling of the transport goods back to the cell membrane directly or through perinuclear recycling endosomes. The endocytic pathway is determined by progressive acidification of the endosomal organelles, the assembly of multivesicular bodies and the formation of late endosomes, the recruitment of lysosomal hydrolases from the trans-Golgi apparatus and subsequently the degradation of the transport goods. In contrast, in the secretory pathway the Golgi apparatus engulfs and thereby sorts out macromolecules into vesicles that are transported towards the cell membrane and fused to it in order to release the vesicle content outside the cell (Jahn 2004, Brooks 2009).

A complex molecular machinery precisely regulates the different steps of intracellular vesicular trafficking, such as the creation of the vesicle from the donor compartment, the selection of the transport goods, the overall movement of the vesicle, and the tethering, docking and fusion to the acceptor compartment's membrane. Rab proteins and small GTPases form the class of the Ras superfamily, and they represent key players of this machinery, that are localized to specific compartments in the endocytic and exocytic pathway and facilitate the regulation of the different steps of intracellular vesicular transport, which begins with the formation of the vesicle and ends with its fusion with the target compartment (Bhuin and Roy 2014). Each Rab protein is able to control the transport between defined compartments. In particular, Rab5 facilitates the first steps of endocytosis from the cell membrane to early endosomes and for early endosomes it governs biogenesis and maturation, regulates the cargo selection, drives the early endosome motility and fusion, and recruits multiple relevant molecules, which are required for endocytosis (Gorvel *et al* 1991, Bucci *et al* 1992, Nielsen *et al* 1999, Zeigerer *et al* 2012). Moreover, Rab4, Rab11 and Rab25 are all involved in the regulation of the recycling of endosomes to the cell membrane (Ullrich *et al* 1996, Casanova *et al* 1999, McCaffrey *et al* 2001). However, the role of intermediate filaments in vesicular trafficking has been revealed and several links between intermediate filaments and vesicular membrane transport machinery have been identified. Moreover, intermediate filaments play a role in endocytosis and exocytosis.

8.3 Intermediate filaments play a crucial role in cellular mechanical properties and cellular motility

Intermediate filaments provide a pronounced structural scaffold for many non-cellular materials, such as hair and nails. The mechanical properties of intracellular intermediate filaments are proposed to be required for the normal function of many soft tissues, and hence mutations in distinct intermediate filament proteins can lead to human diseases such as cardiomyopathies and skin-blistering disorders, which are associated with a failure of the affected tissues to withstand structurally mechanical stress. The structures of intermediate filament proteins and their method of assembling into filaments are highly different from those of the other cytoskeletal filaments such as F-actin filaments and microtubules and hence the mechanical properties of intermediate filaments also differ strongly from the other filaments of the cytoskeleton. The viscoelasticity of intermediate filament networks *in vitro*, and also their contribution to the entire viscoelasticity of living cells are increasingly being revealed using a wide variety of different biophysical techniques. These findings demonstrate how the unusual structures of intermediate filaments provide and support the normal function of a large number of different cell types.

Keratins (previously called cytokeratins) are intermediate filament-assembling proteins that provide cells with their mechanical properties and fulfill a variety of functions in epithelial cells. Keratin genes encode most of the intermediate filaments in the human genome and build the two largest sequence homology groups of this large multigene family, the type I and II keratin groups. They are highly differentiation-specific in their expression patterns, which suggests different cellular functions. In particular, mutations in most of them are often associated with specific tissue-fragility disorders and hence antibodies to keratins are important markers of tissue differentiation as well as helpful tools in diagnostic pathology. The first comprehensive keratin nomenclature used 2D isoelectric focusing and SDS-PAGE to map the keratin profiles of normal human epithelia, tumors and cultured cells (Moll *et al* 1982). In more detail, they grouped the basic-to-neutral type II keratins as K1–K8 and the acidic type I keratins as K9–K19 (figure 8.1) (Moll *et al* 1982).

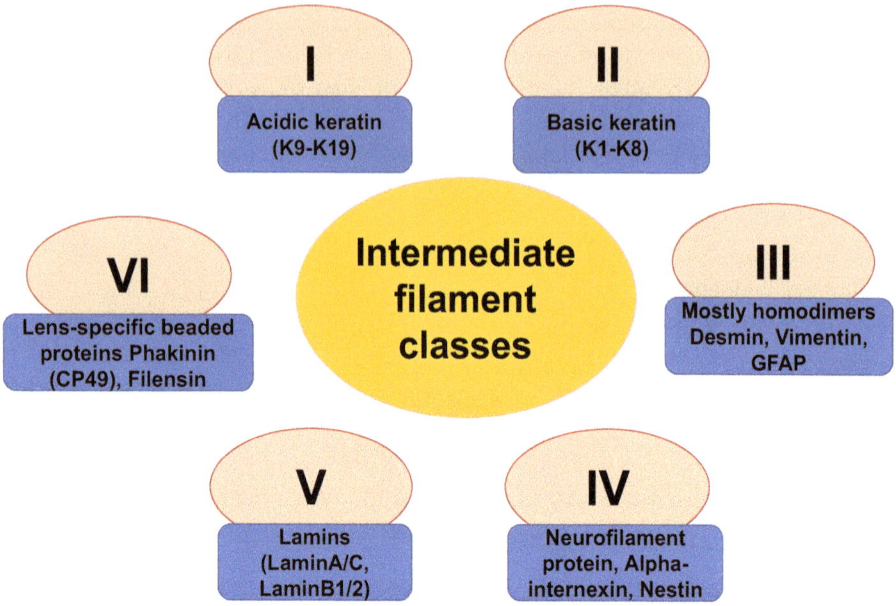

Figure 8.1. Keratin classes.

The current naming of keratins is not systematic and hence a reorganized and durable scheme is overdue. Genome analyses have shown that humans possess 54 functional keratin genes, 28 type I and 26 type II keratins, which form two clusters of 27 genes each on chromosomes 17q21.2 and 12q13.13, with the gene for the type I keratin K18 being located in the type II keratin gene domain (Hesse *et al* 2001, 2004, Rogers *et al* 2004, 2005). Knowledge of the extent of this mammalian gene family led to a revised nomenclature (Hesse *et al* 2004) based on an extended Moll system, K1–K8 and K9–K24 (Moll *et al* 1982, 1990, Chandler *et al* 1991, Zhang *et al* 2001, Sprecher *et al* 2002). Then, for both type I and II keratins, these four categories were arranged in the following numerical order:

1. Human epithelial keratins.
2. Human hair keratins.
3. Nonhuman epithelial/hair keratins.
4. Human keratin pseudogenes.

For historical reasons and because of the extensive number of existing publications, the Moll designation for the epithelial keratins K1–K8 and K9–K24 has been retained (Moll *et al* 1982, 1990, Chandler *et al* 1991, Zhang *et al* 2001, Sprecher *et al* 2002).

Many keratins are found in epithelia, which are exposed to multiple forms of stress. These keratin intermediate filaments are abundant in epithelia and build cytoskeletal networks providing cell-type-specific functions, such as cellular adhesion, migration and metabolism. There is a perpetual keratin filament turnover cycle that supports these cellular functions. In particular, this multistep process regulates the involvement of the cytoskeleton in motion, facilitating rapid protein biosynthesis, whereas the network is still intact during the remodeling process. However, the molecular mechanisms underlying the regulation of the keratin cycle are still elusive compared to the actin and microtubule networks. What regulatory role does the keratin cycle play in epithelial tissue function?

The protection of tissue and organs against microenvironmental alterations is a major functional role of epithelial tissues. In more detail, the renewal and repair of epithelia involves continuous cycles of cellular proliferation, migration and differentiation. Thus, the epithelial cytoskeleton is constantly remodeled in order to optimize epithelial functions. Keratin intermediate filaments are the most diverse and abundant cytoskeletal components of epithelial cells. More than 50 isotypes of keratin intermediate filaments have been expressed in epithelia (Schweizer *et al* 2006, Moll *et al* 2008, Bragulla and Homberger 2009). The regulatory functions of keratins in organelle trafficking, motility, translation, signaling, immune response and cell survival have been revealed, demonstrating that keratin intermediate filaments have the plasticity and network architecture required to regulate the epithelial function precisely (Toivola *et al* 2005, Kim *et al* 2006, 2007, Long *et al* 2006, Kim and Coulombe 2007, Magin *et al* 2007, Vijayaraj *et al* 2009, Depianto *et al* 2010, Ku *et al* 2010). How can keratins provide rigidity and strength, while still remaining dynamic and flexible in their structure? The molecular mechanisms that regulate keratin assembly, disassembly and network architecture are not yet well understood. What are the properties of a biosynthesis-independent multistep assembly and disassembly cycle of keratins that provides rapid network remodeling without the disruption of the whole network?

Time-lapse imaging of cultured monolayers of living cells expressing fluorescent keratins demonstrates that the keratin network is highly dynamic (Windoffer and Leube 1999, Yoon *et al* 2001, Windoffer *et al* 2004). These results suggest a perpetual cycle of keratin intermediate filament assembly and disassembly under standard 2D culture conditions (Koelsch *et al* 2010, Leube *et al* 2011). In particular, the cycle starts with the nucleation of keratin particles at the cell periphery, which is in most cases closely associated with lamellipodial focal adhesions. The next step is the elongation of newly assembled keratin particles during actin-dependent translocation to the peripheral 'old' keratin network. After the integration of precursor keratin particles into the existing network, keratin intermediate filaments are transported to the nucleus and they form bundles. In some cases, these novel bundles disassemble into soluble oligomers, which rapidly diffuse through the cytoplasm and are then available for another cycle of nucleation in the cell periphery near the cell membrane. In addition, other bundles mature into a stable network; they surround the nucleus and are anchored to desmosomes (cell–cell adhesions) or hemidesmosomes (cell–matrix adhesions). Taken together, cycling of the keratin bundles keeps the epithelial cytoskeleton in motion without losing structural integrity.

8.3.1 The assembly of the keratin intermediate filament network

In cultured epithelial cells, the assembly of keratin intermediate filaments (called nucleation) starts in the cell periphery in the close neighborhood of focal adhesions (Windoffer *et al* 2006). The focal adhesions are known to connect actin bundles to the cell membrane and subsequently to the extracellular matrix microenvironment (Petit and Thiery 2000, Geiger *et al* 2001, Carragher and Frame 2004), which thereby induces alterations in the microtubule network architecture (Krylyshkina *et al* 2003, Small and Kaverina 2003). Through the keratin intermediate filament nucleation at the cell periphery close to focal adhesions, the coordinated restructuring of the entire cytoskeleton is achieved. This spatial coordination of the restructuring is of special relevance for migrating cells. In line with this, a pronounced increase of keratin particle assembly has been reported in the lamellipodia of migrating cells (Woell *et al* 2005, Koelsch *et al* 2010, Rolli *et al* 2010). An appropriate nucleus for the initiation of the intermediate filament assembly has been identified *in vitro*: it should have an approximately 60 nm long unit length filament (ULF) containing 32 monomers (Herrmann *et al* 1999, Herrmann *et al* 2002). However, in living cells it is still not understood whether the particles that can assemble keratin intermediate filaments (termed keratin intermediate filament precursors) (Windoffer *et al* 2004) are the same as ULFs, because the resolution of a standard light microscope is lower than a single ULF and hence it is not possible to distinguish between single ULFs.

As there is no keratin intermediate filament precursor polarity, both ends are equally suited to support the elongation by oligomer addition. *In vitro* observations of vimentin intermediate filaments indicate that single and multiple ULFs are added at either end without any differences (Kirmse *et al* 2007). In accordance with these observations, live-cell imaging of keratins detects continuous particle elongation and fusion of larger particles (Windoffer *et al* 2004, 2006, Woell *et al* 2005). As long as keratins retain their free ends, they still elongate. When particles approach the keratin intermediate filament network, they integrate via their ends and thus add another novel branch to the filament network (Windoffer *et al* 2004, 2006, Woell *et al* 2005). However, in mutant keratins, which cause blistering skin diseases in humans, these elongated filaments cannot be assembled; instead, only short-lived spheroidal granules are built close to focal adhesions (Werner *et al* 2004, Windoffer *et al* 2006).

Keratin networks are heterogeneous, because they are assembled by two to ten different isotypes. However, the regulatory modes and mechanisms governing their organization and distribution are still elusive, and they seem to depend partly on cellular polarity. It has been suggested that the intracellular distribution of keratin isotypes is related to their primary sequence, but this has not been confirmed by experimental data. In particular, in the intestinal epithelia of mice the isotypes K20 and K8 are codistributed throughout the cell, whereas in the umbrella cells of the bladder K20 is located in the apical domain, indicating cell-type-specific regulatory mechanisms (Magin *et al* 2006). Another phenomenological feature of the network organization is the coexistence of individual filaments and bundles (inter-filament assemblies). The bundling of intermediate filaments increases the mechanical stability and reduces the turnover (Flitney *et al* 2009, Lee and Coulombe 2009, Kim *et al* 2010), which are two prerequisites for elastic and durable cytoskeletal scaffolding. However, this is not always observed, as in the absence of plectin, which acts as a cytoskeletal crosslinker, the bundling is enhanced, but plectin seems to be dysfunctional due to reduced cellular tension (Osmanagic-Myers *et al* 2006). In cultured cells, the bundling is reflected by an increase to the keratin intermediate filament diameter toward the nucleus, which is evoked by lateral association of keratin intermediate filaments (Windoffer *et al* 2004, Lee and Coulombe 2009, Koelsch *et al* 2010). Several factors influence the bundling: (i) intermediate-filament-associated proteins (Krieg *et al* 1997, Xu *et al* 2000, Makino *et al* 2001, Listwan and Rothnagel 2004, Long *et al* 2006, Osmanagic-Myers *et al* 2006, Boczonadi *et al* 2007, Ishikawa *et al* 2010); (ii) intrinsic and isotype-specific properties of keratin intermediate filaments (Eichner *et al* 1986, Blessing *et al* 1993, Hofmann *et al* 2000), which bundle spontaneously *in vitro* (named self-organization) (Lee and Coulombe 2009, Kim *et al* 2010); and (iii) phosphorylation, which is thought to coincide with bundling after mechanical and chemical stress application (Strnad *et al* 2001, Flitney *et al* 2009).

8.3.2 The disassembly of keratin intermediate filaments

As keratin intermediate filament assembly is highly favored over disassembly, there may be mechanisms to remove assembled filaments and especially dense filament bundles (figure 8.2), which would otherwise interfere with cellular functions. Suitable mechanisms to regulate this balance are the degradation of keratin intermediate filament polypeptides and the disassembly of keratin intermediate filaments into

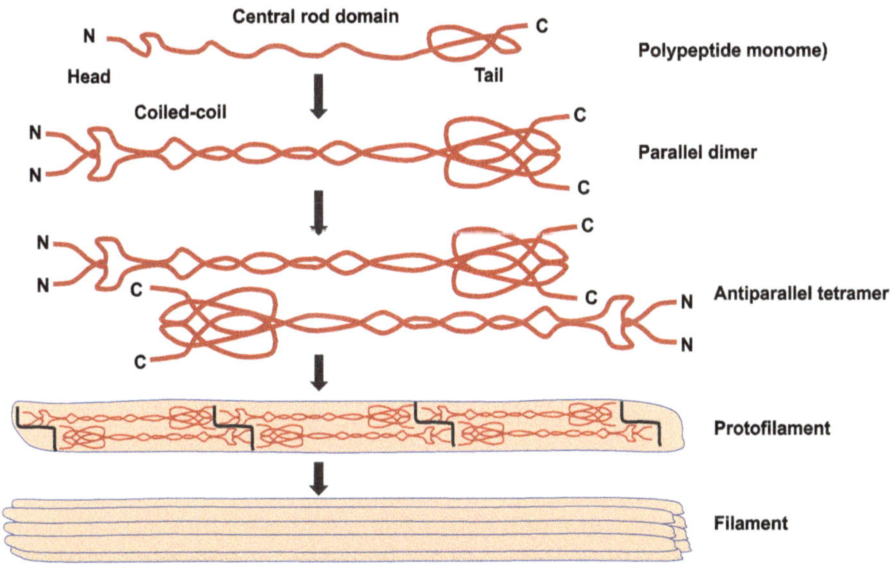

Figure 8.2. Assembly of keratins.

reusable subunits. One fact that supports the first mechanism is the ubiquitination of keratins and their subsequent proteasomal degradation (Ku and Omary 2000, Loeffek *et al* 2010, Rogel *et al* 2010). This mechanism is elevated in stress and pathology as a consequence of increased network restructuring (Zatloukal *et al* 2007, Jaitovich *et al* 2008, Na *et al* 2010). This process has been used to reduce aggregates typical of keratinopathies by the addition of chemical chaperones and chaperone-associated ubiquitin ligases into tissue (Lee *et al* 2008, Chamcheu *et al* 2010, Loeffek *et al* 2010). However, the second mechanism seems to be the most prominent mode in rapidly proliferating cultured cells, as time-lapse fluorescence recordings have demonstrated that keratin intermediate filament assembly occurs independently of and in the absence of protein biosynthesis (Windoffer *et al* 2004, Koelsch *et al* 2010). The detection of single inward-moving keratin intermediate filament bundles further highlights that they dissolve over time without the appearance of distinct fragments, which leads to the suggestion that the released subunits are nonfilamentous. The release of soluble subunits seems to be similar to the lateral subunit exchange, which has been observed for intermediate filaments at the equilibrium state (Eriksson *et al* 2009). Although how the disassembly is regulated remains elusive, the involvement of phosphorylation seems to be most likely, because inhibition of p38 MAPK or PKCζ activities leads to enhanced network stability and, unexpectedly, to increased kinase activity, which results in increased keratin intermediate filament network turnover (Woell *et al* 2007, Sivaramakrishnan *et al* 2009). Moreover, the nonfilamentous keratin pool is elevated during mitosis and in different stress conditions, such as elevated network remodeling, which occur simultaneously with increased keratin phosphorylation (Chou *et al* 1993, Liao and Omary 1996, Omary *et al* 1998, Strnad *et al* 2002, Ridge *et al* 2005). Interestingly, sumoylation has also been implicated in keratin network dynamics (Snider *et al* 2011).

What are the properties of the soluble keratin fraction?
Heterotypic, nonfilamentous keratins build up the biochemically defined soluble pool of tetramers and/or small oligomeric keratin assemblies (Soellner *et al* 1985, Chou *et al* 1993, Bachant and Klymkowsky 1996). In order to inhibit immediate assembly after biosynthesis or directly after filament disassembly, the soluble, nonassembled state needs to be stabilized. This stabilization may be performed by protein modification, association with chaperones such as Hsp70 and Hsc70, interaction with intermediate filament-associated proteins (IFAPs) or binding to 14-3-3 proteins (Liao and Omary 1996, Wiche 1998, Planko *et al* 2007, Mashukova *et al* 2009). 14-3-3 proteins bind to phosphorylated target proteins and are then able to alter their conformation (Kjarland *et al* 2006, Díaz-Moreno *et al* 2009). Thus, Ser phosphorylation of keratin subunits along the head domain occurs immediately after biosynthesis or disassembly in order to inhibit the assembly at nonpermissive sites in the cytoplasm. In particular, hyperphosphorylation by Cdk1, Plk1, Rho-kinase and Aurora B is important for the local breakdown of several intermediate filament classes during mitosis and is hence essential for the efficient segregation of intermediate filament networks into the two daughter cells (Izawa and Inagaki

2006). Due to the small size of the disassembled subunits and their solubility in the aqueous cytoplasm, it is suggested that they can be distributed rapidly within the cytoplasmic space by diffusion. Indeed, a fast diffusing pool has been detected by fluorescence recovery after photobleaching (Koelsch *et al* 2010). The nucleation continues even in the presence of substances that disrupt actin filaments and microtubules. This suggests that diffusive delivery of keratins to peripheral nucleation sites is sufficient (Woell *et al* 2005, Koelsch *et al* 2009).

8.3.3 Regulation of the keratin cycle in space and time

A continuous transport of filamentous keratins to the nucleus is necessary for keratin cycling (figure 8.3). The keratin isotype, cell-type-specific properties and other factors such as filament-associated proteins seem to regulate this transport process. In particular, growing keratin particles move along actin stress fibers with a velocity of approximately 300 nm min^{-1} (Woell *et al* 2005, Koelsch *et al* 2009). This movement is supposed to be directly coupled to lamellar actin treadmilling through plectin-facilitated linkage (Litjens *et al* 2003, Rezniczek *et al* 2004). In addition, keratin particles are also transported along microtubules (Yoon *et al* 2001, Liovic *et al* 2003, Woell *et al* 2005, Windoffer *et al* 2006).

The molecular mechanism of the subsequent inward-directed movement of the keratin network (Windoffer and Leube 1999, Yoon *et al* 2001, Koelsch *et al* 2010) is not yet understood. The intrinsic elasticity of the filaments (Kreplak *et al* 2008), in combination with their nuclear anchorage, may be responsible through the interaction between plectin and the cytoplasmic nuclear membrane protein nesprin-3 (Wilhelmsen *et al* 2005). As an alternative, actin filaments and/or microtubules are also supposed to be involved in this movement. In more detail, it has been observed that energy depletion inhibits keratin intermediate filament motility, which suggests an energy-requiring and motor-protein-driven active movement process (Hollenbeck *et al* 1989, Strnad *et al* 2001, Yoon *et al* 2001).

Focal adhesion-dependent nucleation is *in vitro* another major cycle determinant. It is known that the cytoskeletal crosslinker plectin (particularly isoform 1f) is in close proximity to focal adhesions and is able to bind to keratins (Nikolic *et al* 1996,

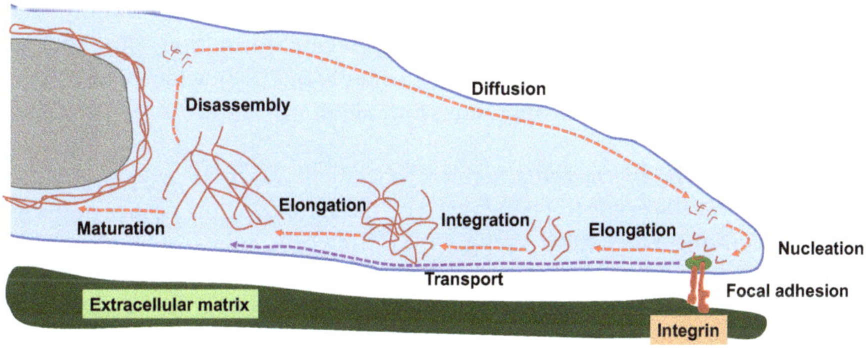

Figure 8.3. The keratin cycle.

Steinboeck *et al* 2000, Litjens *et al* 2003; Rezniczek *et al* 2003). Other candidate proteins that may regulate keratin dynamics are integrins, vinculin, metavinculin, talin and zyxin. All these focal adhesion proteins can bind to intermediate filaments (Kreis *et al* 2005, Ivaska *et al* 2007, Kostan *et al* 2009, Sun *et al* 2008a, 2008b, 2010). In addition to these structural components, focal adhesion-dependent signal transduction seems to be involved in keratin nucleation. Among additional factors, PKCs and MAPK have been discussed (Omary *et al* 1992, Ridge *et al* 2005, Osmanagic-Myers *et al* 2006, Akita *et al* 2007, Woell *et al* 2007, Bordeleau *et al* 2008, 2010, Sivaramakrishnan *et al* 2009). However, the spatiotemporal interaction of these components with the keratin assembly at focal adhesions is not well understood.

Cell shape alterations have been seen frequently in motile and dividing cells and are additionally suggested to enhance cycling, whereas stably anchored and non-moving cells in mature tissues are not supposed to cycle extensively. In order to account for the gradual transition between the moving and nonmoving cells, regulators need to be present to provide graded and locally restricted responses. In line with this, cycling of wild-type and mutant keratins is slowed down by p38 MAPK inhibitory drugs, which are supposed to affect keratin phosphorylation (Woell *et al* 2007). Indeed, this seems to be a possible mechanism for the regulation of keratin cycling through stress-induced signaling. In addition, the compartmentalization of kinases and phosphatases, such as focal adhesion kinase and protein kinase C, can further help to regulate the localized alteration of the network configuration precisely. Members of the plakin family, which link cytoskeletal networks, and members of the plakophilin family, which are localized to adhesion sites, are both good candidates to regulate these organizational functions, as they possess keratin binding sites and can affect kinase/phosphatase activity (Osmanagic-Myers *et al* 2006, Bass-Zubek *et al* 2009, Kostan *et al* 2009, Bordeleau *et al* 2010). Moreover, the shear stress increased PKCζ-dependent phosphorylation of K18-S33 leads to an increased exchange rate of the keratin intermediate filament network (Sivaramakrishnan *et al* 2009). In particular, the relevance of signaling-dependent keratin phosphorylation for dynamic network organization is most prominent in the case of a disease: autoantibodies from the skin-blistering disease (*Pemphigus vulgaris*) have been demonstrated to facilitate p38 MAPK-dependent keratin retraction (Berkowitz *et al* 2005). Moreover, toxic liver injury induced by the antifungal drug griseofulvin mediates increased keratin phosphorylation, increases soluble keratins and leads to enhanced aggregate formation (Ku *et al* 1996, Stumptner *et al* 2001, Toivola *et al* 2004, Fortier *et al* 2010), which can be inhibited *in vitro* by p38 MAPK inhibitors (Nan *et al* 2006).

How can the keratin intermediate filaments exit from the turnover cycle?

The plasticity of cells needs to be balanced by the stabilizing properties of the keratin network promoting mechanical strength for resting cells. In cultured interphase cells, the desmosome- and hemidesmosome-anchored filaments as well as the perinuclear cage-like structures are stabilized under this condition and no longer dynamic. Moreover, desmosome-anchored filaments are more tensile compared to the rest of the keratin intermediate filament network and are able to withstand disruption by the tyrosine phosphatase inhibitor vanadate (Strnad *et al* 2002). In a stable tissue, cycling is less important, because mechanical functions are more important.

Anchorage-dependent mechanosensing seems to affect keratin intermediate filament stability by altered interaction with regulatory proteins such as desmoplakin, BPAG1, plectin, periplakin and epiplakin facilitating keratin intermediate filament dynamics (Bornslaeger *et al* 1996, Wan *et al* 2004, Osmanagic-Myers *et al* 2006, Boczonadi *et al* 2007, Spazierer *et al* 2008, Ishikawa *et al* 2010). For the perinuclear network, attachment to the nuclear envelope through the plectin–nesprin-3a connection (Wilhelmsen *et al* 2005) provides filament stabilization. In the case of junction-associated keratin intermediate filaments, special keratins such as K80 (Langbein *et al* 2010) and keratin-binding proteins such as desmoplakin, plakophilin, plectin and BPAG1 seem to be involved in the keratin intermediate filament stabilization (figure 8.4) (Guo *et al* 1995, Eger *et al* 1997, Holthoefer *et al* 2007, Kostan *et al* 2009, Green *et al* 2010). All these results demonstrate that anchorage protects keratin intermediate filaments against disassembly, whereas the filament turnover has not been analyzed in detail at the single bundle or filament level. In more detail, chemical and/or biophysical keratin modification seems to be the result of the coupling of keratin intermediate filaments to the mechanotransducive systems and hence may support keratin intermediate filament stabilization, similar to the phosphorylation of the tail domain of the intermediate filament glial fibrillary acidic protein (GFAP), which impairs the turnover (Takemura *et al* 2002). Finally, this property can be seen as maturation, which has been observed for the other cytoskeletal filaments, such as the association with proteins and the detyrosination of tubulin (Bulinski and Gundersen 1991, Arce *et al* 2008, Konishi and Setou 2009, Ikegami and Setou 2010).

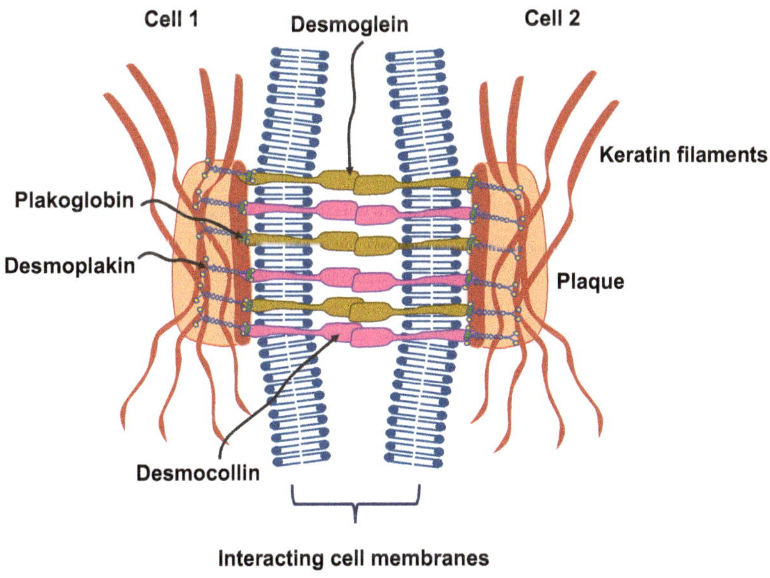

Figure 8.4. Stabilization of keratins at cell junctions.

What are the advantages of cytoskeletal cycling?
The cycling of keratins is more efficient than degradation and *de novo* biosynthesis. The cycling process has also been recognized for other cytoskeletal components, such as for the actin system, which is characterized by filament treadmilling and retrograde flow in migrating cells (Small and Resch 2005, Schaus *et al* 2007, Michalski and Carlsson 2010). Another example of the cycling process is the dynamic instability of microtubules, for example the switching between growing and shrinking (Mitchison and Kirschner 1984, Gardner *et al* 2008). Time-dependent, cyclic alterations of the total amount of cellular components increase the diversification of functional states and enhance the probability for the cell to adapt to microenvironmental alterations through immediate responses, ensuring survival and function (Wolf *et al* 2005, Vogel and Sheetz 2009).

The cycling of keratins is a mechanism for probing and sensing the cell periphery for the occurrence of new cell adhesions. The presence/absence of keratins alters the stability of desmosomes and hemidesmosomes (Long *et al* 2006). In the case of focal adhesions, the cycle is accelerated by enhanced nucleation and thus promotes network growth toward the leading edge. It has been suggested that in migrating cells, this cycle participates in the complex assembly and disassembly mechanisms needed to move the cell body (Proux-Gillardeaux *et al* 2005). Indeed, upon filament attachment to hemidesmosomes and desmosomes, the cycle decreases and hence favors mechanical stability. This mechanism may additionally support *in vivo* distribution patterns, such as localization in the terminal web, a dense filamentous network below the apical surface of the polarized epithelial cells of the gut (Oriolo *et al* 2007). Collectively, keratin cycling may be regarded as a continuous sensing of the immediate extracellular surroundings until new physical adhesion contacts with other cells and/or the extracellular matrix can be established and stabilized through desmosomes and hemidesmosomes. Without *de novo* protein biosynthesis, the cell has a variety of ways to respond to microenvironmental alterations within a small time frame. Thus, it is the dynamical remodeling of keratin intermediate filament networks that guarantees the adaptability and subsequently the function of moving cells.

Another function of keratin cycling is to maintain an intact network during epithelial differentiation, which promotes gradual polypeptide exchange without the disruption of filaments. Different admixtures of keratins are found in basal versus suprabasal epidermal keratinocytes and thus, basal type keratins can be detected in cells that possess no corresponding mRNAs (Lersch and Fuchs 1988, Reichelt *et al* 2001).

How does the keratin cycling process alter the epithelial functions?
Keratin intermediate filaments have been seen as a rather static component of the cytoskeleton, providing the mechanical strength of the epithelia. Moreover, this property is essential for resting cells, providing the whole epithelium with mechanical strength. Disruption of the keratin intermediate filament network integrity and, consequently, of epithelial rigidity induces skin-blistering disease (Coulombe *et al* 2009). Indeed, keratin intermediate filaments fulfill an important role in dynamic

processes such as wound healing and cancer metastasis (Paladini *et al* 1996, Mazzalupo *et al* 2003, Knösel *et al* 2006, Ptitsyn *et al* 2008, Karantza 2011). The keratin cycle can run at different length scales, ranging from diffusible filament precursors to macromolecular network components, involving the entire cytoplasm and hence it seems to be a key promoting factor.

Importantly, the keratin cycling regulates epithelial motility, migration and vesicle trafficking, which is known to be important for the epithelial stress response. The regulation of the keratin cycling is connected to keratin modification, such as phosphorylation, and is hence a target of signaling pathways. A hypothesis is that the predisposition of transgenic mice carrying phosphorylation-deficient K8 and K18 mutants to liver disease (Ku and Omary 2006) and, conversely, the cytoprotective effects of glycosylated keratins (Ku *et al* 2010), are associated with differences in keratin dynamics. The extent to which keratins impact on protein biosynthesis through 14-3-3 proteins or glucose transporters is determined through the keratin dynamics (Kim *et al* 2006, Vijayaraj *et al* 2009). The activity of 14-3-3 proteins depends on targeted protein phosphorylation and their nucleo-cytoplasmic distribution (Mackintosh 2004). The expression of K17 (and possibly additional type I keratins) recruits 14-3-3 to the cytoplasm, where it activates the mammalian target of the rapamycin (mTOR) pathway (Kim *et al* 2006). However, the presence of keratins provides the correct localization and function for the glucose transporter GLUT, whereas in the absence of keratins increased levels of glucose can be taken up and an AMPK-facilitated down-regulation of mTORC1 occurs (Vijayaraj *et al* 2009). Taken together, the degree of translational stimulation through elongation factor 4A (eIF4A)–plakophilin-1 seems to be guided through the interaction of keratins with plakophilin-1 (Wolf *et al* 2010). Do distinct keratin assembly forms, such as filamentous or nonfilamentous keratin, possess different specific regulatory functions? In more detail, is it true that filamentous subdomains fulfill specific regulatory roles, while nonfilamentous keratins function solely in a context-specific manner?

Another prediction is that the accumulation of several cycle intermediates leads to functional consequences for the behavior of the epithelial cell. In line with this, cells lacking keratins or producing mutant keratins, which are characterized by reduced keratin intermediate filaments and increased soluble keratins, are able to migrate even faster in 2D scratch assays (Morley *et al* 2003, Long *et al* 2006, Seltmann *et al* 2013).

Does the cycling process exist in other intermediate filament systems?

Other intermediate filament networks have been shown to utilize similar subunit exchange mechanisms to those found for keratins (Tsuruta and Jones 2003, Mignot *et al* 2007, Burgstaller *et al* 2010). There is evidence for intense cross-talk between vimentin intermediate filaments and focal adhesions (Seifert *et al* 1992, Gonzalez *et al* 2001, Tsuruta and Jones 2003, Spurny *et al* 2008, Bhattacharya *et al* 2009, Burgstaller *et al* 2010) and in addition it has been demonstrated that plectin 1f is involved in the recruitment of growing vimentin particles to focal adhesions (Burgstaller *et al* 2010). Particle elongation occurs through end-to-end annealing for vimentin and neurofilament intermediate filament proteins (Colakoğlu and Brown 2009). Moreover, the cytoplasmic transport of filament precursor particles has been reported for many intermediate filament types (Brown 2003, Helfand *et al* 2004, Woell *et al* 2005, Barry *et al* 2007, Koelsch *et al* 2009). However, differences have been observed: although actin filaments play an important role for keratin intermediate filament particles, microtubules seem to be more important for other intermediate filaments, such as vimentin, peripherin and neurofilaments (Roy *et al* 2000, Wang *et al* 2000, Yoon *et al* 2001, Helfand *et al* 2003). In addition, bundling has been found for neuronal type IV intermediate filaments. In more detail, highly phosphorylated intermediate filaments are tightly packed and located at the center of the axoplasm, whereas less phosphorylated and hence more dynamic intermediate filaments are found to be localized in the peripheral axoplasm (Sihag *et al* 2007, Kushkuley *et al* 2009).

The cycling of the intermediate filament cytoskeleton is also important for many cell types, including neurons for growth cone probing (Chan *et al* 2003), mesenchymal cells and astrocytes for motility (Eckes *et al* 1998, Lepekhin *et al* 2001, Nieminen *et al* 2006, Pan *et al* 2008) and endothelial cells for adaptation to fluid shear stress (Helmke *et al* 2001). In particular, intermediate filament cycling seems to be essential for the regeneration occurring upon neuronal axon injury or in diverse stress situations, such as in reactive gliosis (Pekny and Lane 2007).

The strongest evidence for the importance of intermediate filament cycling comes from observations in *C. elegans*, where the inhibition of sumoylation leads to a reduction in the intermediate filament network turnover associated with enhanced intermediate filament bundling and aggregation (Kaminsky *et al* 2009). However, it has also been demonstrated that intermediate filament phosphorylation is linked to network organization, and these are in turn both coupled to hemidesmosome-facilitated mechanotransduction processes (Zhang *et al* 2011).

Finally, several observations support the view that the plasticity of the epithelial keratin intermediate filament network not only relies on alterations in biosynthesis and degradation, but is solely provided by the cycles of assembly and disassembly. Although this concept is supported by microscopic observations in cultured living cells, the underlying molecular mechanisms still need to be confirmed *in vivo* and correlated with observations on *in vitro* keratin intermediate filament assembly (Herrmann *et al* 2002) as well as alternative concepts of intermediate filament network dynamics (Ngai *et al* 1990, Miller *et al* 1991, Chang *et al* 2006). Some questions remain unanswered. What is the driving force of the cycle? How do keratin-associated proteins such as plectin, 14-3-3, Akt-1, Hsp70 and others affect the cycle? How do microenvironmental factors such as mechanical force, cytokines and microbes modulate the keratin cycle and what effects does the cycle have on the cell? How can keratin isotypes affect the keratin intermediate filament cycling? What impact does keratin modification have on the cycling process? Which cellular processes are associated with the keratin cycling and how is this achieved? Is the cycling necessary for the stress protection facilitated by keratins? Is the cycling concept universal and hence observed in other intermediate filaments?

8.4 Viscoelasticity of purified intermediate filaments *in vitro*

Purified intermediate filament proteins can be assembled *in vitro* to produce polymers closely related to those assembled in living cells and subsequently these filaments form viscoelastic gels. The crosslinks tightly connecting the intermediate filaments in the network are provided by specific bonds between polypeptides extending from the filament surface and ionic interactions facilitated by divalent cations. In more detail, intermediate filament networks display a striking nonlinear elasticity with increasing stiffness that has been quantified by determining the shear modulus. In particular, the shear modulus increases even an order of magnitude, when the networks are deformed to large strains, which resemble those strains that soft tissues are probed *in vivo*. Individual intermediate filaments can be elongated by stretching to more than two or three times their resting length without breaking or rupture. Many different rheometric methods have been employed to quantify the viscoelasticity of intermediate filament networks over a broad range of timescales and strain magnitudes. Moreover, the mechanical roles of different classes of intermediate filament have also been investigated by using an even wider range of microrheological methods *in vitro* in mesenchymal and epithelial cells. Indeed, these experiments have demonstrated the effects on cellular mechanical properties when intermediate filaments are genetically or pharmacologically altered or even disrupted or when normal or mutant intermediate filament proteins are expressed in these cells. Consistently with *in vitro* rheology findings, the mechanical role of intermediate filaments is more prominently displayed by the cells when they are probed by larger and more frequent deformations. In more detail, the mechanical properties of individual intermediate filaments of different types have been analyzed directly by applying forces to them and then imaging their deflection. As an alternative method, when forces are applied to the intermediate filaments, they have been inferred from images by assuming that the polymer contours are solely deformed by thermal energy.

The mechanical properties of intermediate filament networks constituted *in vitro* either as homogeneous networks or as composite network, in which they are copolymerized with F-actin, have been analyzed by a large number of rheologic methods. The unique mechanical properties of intermediate filaments seem to be based on two major structural differences between intermediate filaments and the other cytoskeletal polymers such as F-actin and microtubules. Intermediate filaments are much more flexible than microtubules and actin filaments. Moreover, this flexibility differs largely from other cytoskeletal polymers by orders of magnitude and can be quantified by the persistence length l_p, which is a measure of the distance over which a filament exhibits an approximately straight configuration. In particular, l_p is defined by the expression $\langle \cos \theta(s) \rangle = e^{-s/l_p}$, where $\langle \cos\theta(s) \rangle$ represents an ensemble average of the angle θ formed by two tangents drawn at distances, s, along the contour of the filament. Moreover, the persistence length is related to the elastic bending constant of the filament K by the expression $K = \lambda_p / k_B T$, where $k_B T$ represents the thermal energy. In addition, the great flexibility of intermediate filaments seems to be related to the greater degree of disorder and open hydrated space within these intermediate filaments compared to actin filaments or tubulin polymers. How precisely the subunit packing and higher-order structure of intermediate filaments contributes to their flexibility and resistance to breakage is not yet clear, but many different measurement approaches reveal that intermediate filaments can potentially contribute the mechanical phenotype of the cells and tissues that cannot be provided by the other polymer types.

Several distinct features unique to the intermediate filament network mechanical properties emerge from all of these studies, however, some issues correlating with the magnitude of the intermediate filament network stiffness and even the nature of inter-filament connections still require further investigation. Unlike other elements of the cytoskeleton, individual intermediate filaments as well as the associated intermediate filament networks are able to withstand large deformations that would rupture F-actin filaments or microtubules (Janmey *et al* 1991, Kreplak *et al* 2005, Guzman *et al* 2006). Apart from not getting ruptured at large strain, intermediate filament networks increase their elastic moduli, and hence the incremental stiffness of vimentin, neurofilament, and other intermediate filament types can exhibit ten times larger values at 100% strain that in the limit of a low strain (Janmey *et al* 1991, Leterrier *et al* 1996, Schopferer *et al* 2009, Bertaud *et al* 2010, Lin *et al* 2010, Pawelzyk *et al* 2014). The elastic moduli depend on the intermediate filament network protein connections, which are different from that of other biopolymer gels. Instead, the shear moduli of fibrin and actin networks scale with at least the square of the protein concentration, whereas the shear moduli of vimentin and desmin networks increase much more gradually with power law exponents that are as low as 0.5 (Janmey *et al* 1991, Schopferer *et al* 2009, Lin *et al* 2010) relating the elastic modulus to the concentration. However, it is not yet clear, why there is such a discrepancy between intermediate filament and other biopolymer gels in their physical behavior.

The molecular mechanisms that connect intermediate filaments to each other and the manner in which they build mechanically resistant networks are also not yet well understood. Specific crosslinking proteins do not seem to be required for the network formation, and additionally several bonds between the intermediate filament subunit C-terminal extensions and the sides of other filaments have been revealed (Bousquet *et al* 2001, Esue *et al* 2006, Pawelzyk *et al* 2014). However, complementary attractive interactions between neurofilament sidearms are also supposed to connect intermediate filaments to each other (Gou *et al* 1998). The most common procedure to assemble intermediate filament networks *in vitro* is to add divalent cations, usually Mg^{2+}, at several millimolar concentrations. Although the mechanisms by which divalent ions can crosslink intermediate filaments are not yet clearly understood, it has been hypothesized that they involve either specific metal-binding bonds (Lin *et al* 2010) or polyelectrolyte effects, which rely on the high surface charge of all intermediate filaments (Huisman *et al* 2011, Janmey *et al* 2014). However, the identifying of the molecular mechanisms that are involved in the crosslinking of intermediate filaments and the formation of bundles remains elusive and hence represents a major challenge to define this system in the same great detail that is available for the network formation by other biopolymers.

8.5 Functional role of intermediate filaments in mechanotransduction processes

The process of cellular mechanotransduction represents an integration of multiple mechanical cues that are derived from sensing, transmission of force and the transduction of a mechanical signal into a biochemical response. There exists extensive evidence that cell–cell, cell–extracellular matrix and flow forces are actively sensed by cells through their junctional protein complexes in different cellular situations (Riveline *et al* 2001, Weber *et al* 2012, Conway *et al* 2013, Sanghvi-Shah and Weber 2017). The different mechanical forces affect the structure, assembly, adhesive strength, function, and signaling of these adhesive complexes, which in turn has effects on the cytoskeleton. Cytoplasmic intermediate filaments function as an elastic and conductive network for force transmission and propagation of mechanical stimuli within and between cells through adhesion complexes. Indeed, cytoplasmic intermediate filaments serve as modulators of specific signal transduction pathways in a variety of biological conditions. Abundant availability, overall cytoplasmic presence and subcellular reorganization of intermediate filaments depend on cellular conditions, which supports cytoplasmic intermediate filaments to function in various signaling pathways in several ways. Based on this view, cytoplasmic intermediate filaments act to transduce mechanical stimuli during development while integrating an adaptable physical environment through cell signal transduction processes.

Fluid flow shear stress fulfills important roles in the development of the vasculature system. Intriguingly, fluid flow shear stress represents an important mechanical stimulus in tissues, which is not often primarily associated with exposure to fluid flow shear stresses, such as bone and glandular epithelia. Indeed, fluid flow shear stress experiments have revealed an important role of intermediate filaments in mechanotransduction pathways. In particular, cytoplasmic intermediate filaments can change their network organization usually by mechanisms such as conformational alterations, changes in their assembly, interacting proteins and others. Mechanical forces such as shear stress can cause a rapid reorganization of vimentin and keratin intermediate filament networks in multiple cell types, which leads to the suggestion that they fulfill a role in the spatial redistribution of intracellular forces (Helmke *et al* 2000, 2001, Yoon *et al* 2001, Sivaramakrishnan *et al* 2008). Shear stress elevates the stiffness of the keratin intermediate filament network in the peripheral region of the cytoplasm (Sivaramakrishnan *et al* 2008). In line with this, shearing also dramatically transforms the keratin intermediate filaments into more 'wavy' tonofibril bundles, which is a process that is induced by K8 and K18 phosphorylation on the serine residues 73 and 33, respectively (Flitney *et al* 2009, Sivaramakrishnan *et al* 2009). The phosphorylation driven reorganization of intermediate filaments is facilitated by a variety of protein kinases including PKCδ and PKCζ (Ridge *et al* 2005, Sivaramakrishnan *et al* 2009). The phosphorylation of keratins in the regulatory head domain (K18 pSer33) recruits binding of 14-3-3 that provides a dynamic exchange and the remodeling of the entire network (Sivaramakrishnan *et al* 2009). In order to impair hyperphosphorylation evoked disruption, keratin intermediate filaments can recruit epiplakin, which perhaps serves as a chaperone during filament reorganization (Spazierer *et al* 2008). In addition to the local deformation of the intermediate filaments, there is increased association of vimentin intermediate filaments with β3-integrin focal contacts, further stabilizing cell–matrix adhesions (Tsuruta and Jones 2003). Similarly, due to shear stress, endothelial cells initiate a transition from cell–cell adhesion loading on VE-cadherin to interaction of platelet endothelial cell adhesion molecule-1 (PECAM-1) with vimentin in order to stabilize cell–cell junctions (Conway *et al* 2013). Due to this behavior, mechanical loads may be transduced from one cytoskeletal network to another. Indeed, the keratin intermediate filaments display less motion, when actomyosin-dependent rigidity is enhanced, which seems to be a consequence of stress generated by the actomyosin cytoskeleton transmission of prestress the keratin intermediate filament network (Nolting and Koster 2013).

Experimentally evoked physical forces, induced by optical tweezers and fibronectin beads on epithelial cells, can promote the alteration of both the K8/K18 intermediate filaments and the actin filament network through Rho–ROCK pathway signaling (Bordeleau *et al* 2012). Tensile forces reinforce stress fibers by coordination between the Solo protein a RhoA GEF and the K8/K18 intermediate filament network (Fujiwara *et al* 2016). The actin stress fiber assembly and contractility are also altered by vimentin filament driven regulation of RhoA and GEF-H1 (RhoA GEF protein; Jiu *et al* 2017). The decoupling the intermediate filaments from the mechanotransduction pathway has even demonstrated unrecognized roles of intermediate filaments in this process. For instance, cells with impaired vimentin expression exhibit decreased mechanical resistance to the effects of flow (Tsuruta and Jones 2003). Moreover, mutant keratin intermediate filament network cannot withstand mechanical stress (Ma *et al* 2001), and hence show a pronounced reorganization of the filaments into discrete aggregates (Russell *et al* 2004). In the absence of vimentin intermediate filaments or their displaced anchorage, which is based on the impairment or loss of plectin, cells show compromised activation of FAK and its downstream targets Src, ERK1/2, and p38 and subsequently the migration of the cells is impaired. In more detail, exploiting stress conditions in the absence of plectin induces prominent fragmentation of the intermediate filament network (Gregor *et al* 2014). In agreement with these results, cytoplasmic intermediate filaments sense tension through the upstream mechanosensors and hence initiate rearrangements that act as stress buffers. How can cytoplasmic intermediate filaments sense the tension? This is not yet fully understood and requires further investigation. A highly speculative possibility seems to be that cytoplasmic intermediate filaments alter their conformation or assembly in response to stress in order to uncover cryptic sites crucial for sensing tension. For instance, the vimentin Cys327 site is blocked under tension (Johnson *et al* 2007, Pérez-Sala *et al* 2015). Cytoplasmic intermediate filaments of all types display plasticity in their structural folding, which seem to provide both elasticity and the possible cryptic unmasking of several sites (possible additional binding sites). These conformational alterations in the intermediate filament structure within the polymerized filament may have pronounced impacts on cellular signal transduction processes and hence provide a linkage between the maintenance of the physical architecture and biochemical signal transduction pathways.

8.6 The role of intermediate and actin filament interactions

Cellular mechanical activity generated by the connection between the extracellular matrix and the actin cytoskeleton seems to be essential for the regulation of cell adhesion, spreading and migration during normal and cancerous development. In particular, keratins are the intermediate filament proteins of epithelial cells, expressed as pairs in a lineage/differentiation manner. The hallmarks of all relatively simple epithelia are intermediate filaments that contain only keratins 8/18 (K8/K18). In addition, in these epithelia the intermediate filaments K8/K18 play a key regulatory role in adhesion and migration, through facilitating integrin interactions with the extracellular matrix, actin adaptor proteins and signaling molecules at focal adhesions (figure 8.5). Using a K8 knockdown in rat H4 hepatoma cells and their K8/K18-containing wild-type counterparts seeded on fibronectin-coated substrata of different rigidities, it has been shown that the K8/K18 intermediate filament-lacking cells are unable to spread and hence display an altered actin fiber organization when seeded on a low-rigidity substrate. In addition, a concomitant reduction of local cellular stiffness, especially at focal adhesions generated by fibronectin-coated microbeads attached to the dorsal cell surface, has been detected (Bordeleau *et al* 2012). Furthermore, this K8/K18 intermediate filament modulation of cellular stiffness and actin fiber organization is mediated by RhoA–ROCK signaling (Bordeleau *et al* 2012). Taken together, these results reveal that K8/K18 intermediate filaments contribute to the cellular stiffness–extracellular matrix rigidity interaction through alteration of their Rho-dependent actin organization and cytoskeletal dynamics.

It has been shown that the ability of cells to sense and adapt to mechanical properties from the extracellular matrix is crucial for many biological processes,

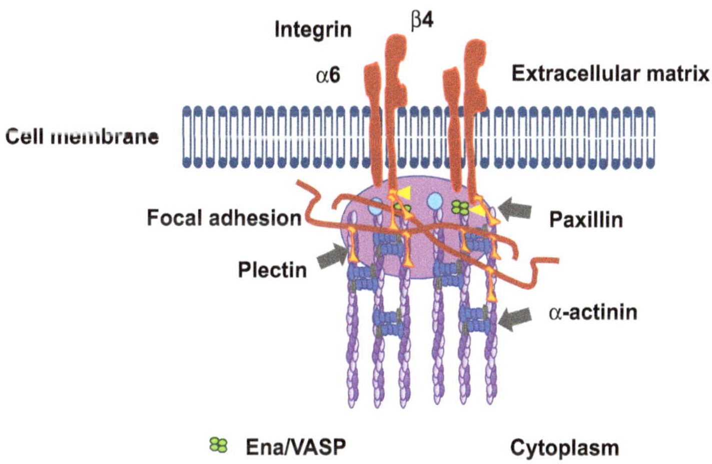

Figure 8.5. Intermediate filament and actin filament interaction.

such as the involvement of mechanical force in supporting embryonic development (Butcher *et al* 2009). In particular, embryonic stem cells progressively stiffen when they undergo differentiation and hence adapt their individual stiffness to the rigidity of the underlying extracellular matrix (Pajerowski *et al* 2007). Similarly, there is evidence for the involvement of increased extracellular matrix rigidity in promoting the occurrence and development of primary tumors, and subsequently in supporting cancer metastasis by enhancing the migration of certain escaping cancer cells (Lopez *et al* 2008). In more detail, aggressive and invasive cancer cells in suspension (where they are independent of interaction with the extracellular matrix) are more compliant than less aggressive cancer cells, which in turn are more compliant than healthy cells (Guck *et al* 2005). However, when these tumorigenic cancer cells are adhered on a rigid extracellular matrix substratum, they display increased contractility (Paszek *et al* 2005). Alterations in matrix rigidity can cause great differences in cell behavior and thus highlight how important changes in the mechanical properties of the microenvironment are for the cells to adapt and counterbalance the extracellular matrix constraints. Indeed, an extracellular matrix-produced stress is perceived and integrated intracellularly through the involvement of integrin receptors, which act as mechanotransducers by inducing signaling cascades in order to facilitate cellular responses such as cell migration and contractility through modulation of the actin cytoskeleton at focal adhesions (Matthews *et al* 2006, Butcher *et al* 2009). Cell contractility and the associated internal stiffness can be measured by tracking the force-induced displacement of fibronectin-coated beads that are bound to integrin receptors and subsequently to focal adhesions generated at the dorsal cell surface (Bausch *et al* 1999, Matthews *et al* 2004). These measurements at the cellular level have shown that a depolymerization of the actin cytoskeleton reduces cell stiffness, which demonstrates that this cytoskeletal network can respond to the mechanical force applied at focal adhesions (Matthews *et al* 2004, Bordeleau *et al* 2011). There is a molecular-based balance between internal stiffness and extracellular force exerted at focal adhesions, which is maintained by alteration of the fibrillar actin contractility (McBeath *et al* 2004, Paszek *et al* 2005, Matthews *et al* 2006) In particular, this is mediated by activation of Rho and the effector ROCK, which is in turn a regulator of the myosin light chain (Asparuhova *et al* 2009, Amano *et al* 2010). This Rho-dependent contractility suggests prominent actin cytoskeleton involvement in the interaction between cell stiffness and extracellular matrix rigidity.

Keratins belong to the largest family of cytoskeletal proteins and are grouped into type I (K9–28) and type II (K1–K8 and K71–K80) subfamilies (Amano *et al* 2010). Keratin intermediate filaments are heteropolymers that include at least one type I and one type II keratin. Due to cellular differentiation and special cell lineage, these keratins are coordinately expressed as specific pairs. Intermediate filaments from all simple epithelial cells contain K8/K18 and most of them also express two to three other keratins (Omary and Ku 1997, Pekny and Lane 2007). In more detail, K8 and K18 are the ancestral genes for the multiple specialized type II and type I keratin classes, respectively, and are the first cytoplasmic intermediate filament genes expressed in the embryo when stem cells differentiate into different cell lineages

(Oshima *et al* 1996, Coulombe and Wong 2004). In cancer disease, there is indeed some evidence that the persistence of K8/K18 intermediate filaments is a hallmark of invasive squamous cell carcinoma, where such perturbed K8/K18 expression contributes to cell invasiveness based on actin-dependent motility (Alam *et al* 2011). In addition, point mutations in K8 and K18 genes cause intermediate filament disorganization and consequently a predisposition to liver cirrhosis (Omary *et al* 2009), while in turn cirrhosis enhances hepatic tissue stiffness. This extracellular matrix-linked mechanical alteration is mostly associated with the occurrence of the hepatocellular carcinoma (Kuo *et al* 2010, Jung *et al* 2011). Keratin intermediate filaments assemble an elastic and flexible cytoskeletal network, which is responsible for the ability of epithelial cells to withstand mechanical stress (Coulombe and Wong 2004). Moreover, it can be hypothesized that K8/K18 intermediate filaments need to be included as the factor responsible for providing the interaction between actin-mediated cell stiffness and extracellular matrix rigidity in simple epithelial cells.

This hypothesis has been addressed by using monolayer cultures of K8-knockdown H4-II-E-C3 (shK8b) rat hepatoma cells and their K8/K18-containing wild-type counterparts (H4ev) (Bordeleau *et al* 2010). The intermediate filaments of hepatocytes are made up solely of K8/K18 (Omary and Ku 1997, Oshima 2002), which means that a loss of K8 leads to the degradation of K18, leading to the epithelial cells lacking K8/K18 intermediate filaments (Galarneau *et al* 2007). H4ev cells and shK8b cells have been cultured on fibronectin-coated substrates of diverse rigidity (Wang and Pelham 1998). After their adhesion, a laser-tweezers-driven force was exerted on a fibronectin-coated microbead attached to integrins (Matthews *et al* 2004, Bordeleau *et al* 2008). Then the cell stiffness at focal adhesions was determined by measuring the bead displacement (Matthews *et al* 2004). In addition, the involvement of Rho–ROCK was investigated in shK8b cells and compared to H4ev cells regarding cell stiffness and fibrillar actin organization. The results revealed an intervention of K8/K18 intermediate filaments in the cell stiffness–extracellular matrix rigidity interaction through alteration of the Rho-dependent actin organization and dynamics.

Intermediate filaments consist of a heterogenic group of evolutionarily conserved proteins and are specified in a tissue-, cell-type- and context-dependent manner. Intermediate filaments feature in multiple cellular processes, such as the maintenance of cell and tissue integrity and the response and adaptation to various external stresses. Inherited mutations in intermediate filament coding sequences can lead to a broad array of clinical disorders. As a consequence, the expression, assembly and organization of intermediate filaments are precisely regulated (Chung *et al* 2013). Indeed, cell migration is a cell-based phenomenon in which intermediate filaments act as effectors and regulators. In the following the focus is on vimentin and keratin and how intermediate filaments provide the cell's mechanical properties, cytoarchitecture assembly and adhesion, and how they regulate the pathways, which impact cellular motility.

Ten-nanometer-wide intermediate filaments were first described in muscle (Ishikawa *et al* 1968). They are composed of the most diverse and heterogeneous group of proteins among intracellular cytoskeletal fibers. There are approximately

70 genes that encode for intermediate filament-forming proteins in the human genome. Among them are 54 genes encoding for keratin proteins that occur in epithelia (Schweizer *et al* 2006, Pan *et al* 2013). Intermediate filaments can be subdivided into six major subtypes based on their gene substructure or sequence homology regarding their signature within the central rod domain. All intermediate filaments proteins self-assemble into approximately 10 nm wide filaments by forming obligatory or facultative heteropolymers, along with a defining tripartite domain structure with a central α-helical rod domain with long-range, coiled-coil forming heptad repeats, flanked by variable end domains located at their N- and C-termini. Intermediate filament proteins display pronounced heterogeneity, because their molecular mass ranges from 40 kDa (type I keratin 19) to 240 kDa (type IV nestin). However, focusing on individual keratins, their primary structure is evolutionarily well conserved. Intermediate filament systems are present across nearly all multicellular eukaryotes (Erber *et al* 1998). The evidence suggests that they appeared as nuclear proteins similar to lamins in lower eukaryotes such as *Dictyostelium* (Batsios *et al* 2012). The presence of the intermediate filament-like crescentin in *Caulobacter crescenti* (Ausmees *et al* 2003) raises the question of whether intermediate filaments might have existed earlier, in prokaryotes.

Another important signature feature of the intermediate filament superfamily of genes and proteins is the tissue-type-, differentiation-program- and context-dependent nature of their regulation. Accordingly, the list of functions fulfilled by intermediate filaments in their natural biological setting is growing longer as intermediate filaments and their associated proteins have been linked to cell motility (Toivola *et al* 2010, Burke and Stewart 2013, Pan *et al* 2013). Due to their status as abundant fibrous elements inside cells, intermediate filaments regulate cellular migration on the mechanical and cytoarchitectural levels. Intermediate filaments also impact on migration as regulators, because they can interact with and regulate various cellular effectors, such as signaling molecules (Pan *et al* 2013).

There are intermediate filament proteins such as vimentin that consistently stimulate cell migration and invasion independently of the setting, whereas others, such as various keratins, exert a more variable, fine-tuned and complex impact on these processes. Beyond the type of intermediate filament protein, additional determinants such as the expression level, associated proteins, intracellular organization and covalent modifications such as phosphorylation, act together to regulate cell migration. In addition, the cellular and biological features are crucially important. The nature and impact of various intermediate filaments during migration under normal and disease settings reflect their pervasive integration in a microenvironment-dependent manner within the broader background of the cell.

Which basic attributes of intermediate filaments are relevant for their properties and function *in vivo*? Similar to F-actin and microtubules, intermediate filaments depend on an array of partner proteins for their assembly, organization, function and regulation. Moreover, plakin family proteins are 'cytoskeletal organizers' that link intermediate filaments, microtubules and actin at several strategic locations within the cell's interior (Leung *et al* 2001). Beyond their signature plakin domain, plakin family members are large and possess a modular substructure that helps them act as key organizers of the cytoskeleton (Leung *et al* 2002). Plakin proteins facilitate intermediate filament attachment to the cytoplasmic so-called plaque domain present in cell–cell desmosome adhesions and cell–matrix hemidesmosome adhesions to the other two main cytoskeletal protein filaments, such as F-actin and microtubules, as well as to the surface of the nucleus (Leung *et al* 2001, Leung *et al* 2002, Desai *et al* 2009).

Intermediate filament proteins are modified by posttranslational modifications such as phosphorylation, O-glycosylation, ubiquitination, sumoylation and acetylation (figure 8.6) (Izawa and Inagaki 2006, Hyder *et al* 2008, Zencheck *et al* 2012). In particular, these modifications are site-specific within the intermediate filament protein backbone, reversible and regulate nearly all aspects of their assembly, organization, biochemical and mechanical properties, and their function (Omary *et al* 2004, Rogel *et al* 2010, Pan *et al* 2013). Associated proteins and posttranslational modifications of intermediate filaments help to define the polymerization status and their intracellular organization. Actively migrating, polarized cells possess a specific intermediate filament system reorganized around the nucleus or at their rear, trailing end (Paladini *et al* 1996, Helfand *et al* 2011, Weber *et al* 2012) and when mutant intermediate filament proteins are expressed (Morley *et al* 2003).

What is the interplay between intermediate filaments, cell adhesion and other cytoskeletal components?

Desmosomes are composed of transmembrane cadherins, armadillo proteins such as plakoglobin and plakophilins, and also plakin proteins such as desmoplakin, which anchors desmosomal plaques to intermediate filaments on the intracellular site (Desai *et al* 2009). Thus, desmosomes maintain tissue integrity even under mechanical stress conditions (Desai *et al* 2009) that start at an early stage during mouse embryogenesis (Gallicano *et al* 1998). In particular, pro-migratory stimuli such as epidermal growth factor (EGF) support the assembly and functional state of

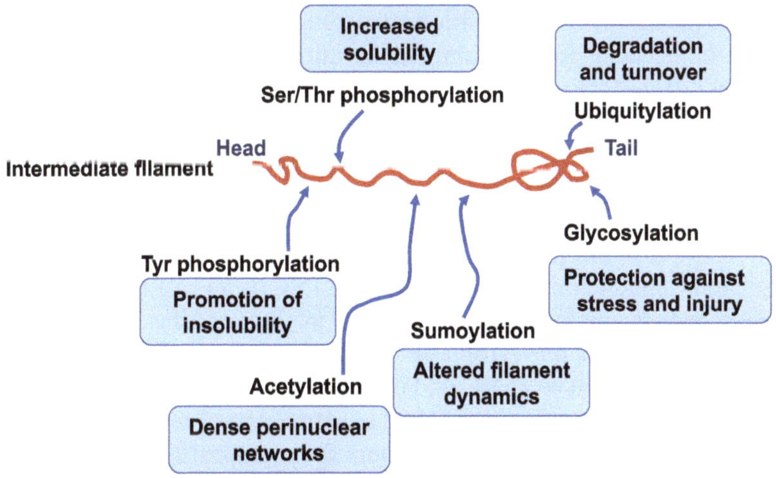

Figure 8.6. Posttranscriptional modification of intermediate filaments.

desmosomes, hemidesmosomes and the whole network architecture of the intermediate filaments (Keski-Oja *et al* 1981, Baribault *et al* 1989, Chung *et al* 2012, Felkl *et al* 2012). However, the induction of cell migration is typically linked to weaker desmosome-facilitated cell–cell adhesion (Kitajima 2013). Moreover, increased turnover of desmosomes and their decreased colocalization with keratins have been revealed in migrating oral squamous cell carcinoma cells (Roberts *et al* 2011).

Moreover, cell migration is a function of dynamic interactions between extracellular matrix components and the cell's cortex. The transmembrane, adhesion-facilitating entity in hemidesmosomes is the $\alpha6\beta4$ integrin heterodimer, which binds to extracellular laminin (Giancotti and Ruoslahti 1999). Intracellularly, the integrin connection to intermediate filaments is provided by plakin proteins such as the bullous pemphigoid antigens 1 and 2 (BPAG1 and BPAG2), as well as plectin (Leung *et al* 2002). In the complete absence of keratins, the hemidesmosome components are still present, but they are scattered in skin keratinocytes. However, these cells lacking keratins adhere faster to the extracellular matrix microenvironment and show increased migration compared to wild-type cells (Seltmann *et al* 2013). Re-expression of the K5–K14 keratin pair alone, which is typically expressed in progenitor basal keratinocytes, in such keratin-free skin keratinocytes reverses this phenotype, even when the K5–K14 expression is at a sub-physiological level. In contrast, keratinocytes with no expression for BPAG1 show a normal density of hemidesmosomes at the cell–matrix interface, but lack a cytoplasmic plaque as well as attachment to keratin intermediate filaments and display a delayed wound healing response (Guo *et al* 1995). The knockdown of actinin-4, which can bind to actin, leads to a loss of directionality during migration of keratinocytes. In more detail, this behavior is correlated with a mislocalization of $\alpha6\beta4$ integrin as well as BPAG1e and defects in cellular polarity and lamellipodial dynamics (Hamill *et al* 2013). It has been suggested that the p90 ribosomal protein S6 kinase (RSK) is involved in hemidesmosome remodeling (Frijns *et al* 2010, Faure *et al* 2012), and also in the regulation of wound-inducible K17 (Pan *et al* 2011). What is the influence of K17 in complete keratin-null and/or actinin-4 knockdown keratinocytes? It has also been shown that the knockdown of K8 in cultured hepatoma cells decreases cell migration in a 2D scratch-wound assay (Bordeleau *et al* 2010), reduces cell spreading, alters Rho-dependent actin fiber organization and reduces local stiffness at focal adhesions, suggesting that an interaction between K8/K18 intermediate filaments and Rho-facilitated actin dynamics occurs through plectin, RACK1 and Src (Bordeleau *et al* 2012).

Linker of nucleoskeleton and cytoskeleton (LINC) is a protein complex located at the nuclear membrane and it participates in linking the nuclear lamina to cytoskeletal proteins on the cytoplasmic side (Mellad *et al* 2011). In particular, Nesprin-3 is a component of LINC and can associate with plectin (Wilhelmsen *et al* 2005). However, the disruption of LINC via expression of mutated nesprin reduces the transmission of intracellular forces, alters the organization of F-actin as well as vimentin intermediate filaments and causes reduced migration and polarization in mouse embryonic fibroblasts (Lombardi *et al* 2011). Similarly, depletion of

Nesprin-3 in human aortic endothelial cells impairs the organization of vimentin intermediate filaments and reduces cell migration (Morgan *et al* 2011). In line with this, depletion of other major intermediate filaments, such as nestin, vimentin and glial fibrillary acidic protein in astrocytes, changes the localization and rotation of the nucleus during astrocyte migration (Dupin *et al* 2011) and subsequently reduces their migration (Lepekhin *et al* 2001). Cell migration has been shown to depend on dynamic alterations of the position and shape of the nucleus (Friedl *et al* 2011) and additionally it has been demonstrated that intermediate filaments contribute to nuclear architecture in skin keratinocytes (Pan *et al* 2013) and migrating cells (Friedl *et al* 2011).

How is the keratinocyte migration regulated?
Although there is an interaction between intermediate filaments and their associated proteins in supporting cell motility, it is obvious that there are converging migration phenotypes which are still exhibited by many genetic null mutants in mouse skin keratinocytes. Genetic loss of epiplakin, a plakin family member, in mice results in increased skin keratinocyte migration (Goto *et al* 2006), associated with the loss of keratin intermediate filament bundling in postwounding (Ishikawa *et al* 2010). In line with this, increased migration occurs in mouse keratinocytes that are genetically null for plectin (Osmanagic-Myers *et al* 2006), plakoglobin (Yin *et al* 2005), plakophilin (Yin *et al* 2005) and keratin 6 (K6a/K6b) (Yin *et al* 2005, Rotty and Coulombe 2012). Moreover, the loss of K6a/K6b, plectin or plakoglobin leads to Src family kinase activation and changes to the F-actin reorganization (Goto *et al* 2006, Osmanagic-Myers *et al* 2006, Todorovic *et al* 2010). A plectin deficiency phenotype demonstrated that the intermediate filament is supposed to indirectly regulate the organization and stability of microtubules through an interaction with the plectin1c isoform. In addition, this then has an associated impact on focal adhesion dynamics and the directional migration of keratinocytes (Valencia *et al* 2013).

Finally, the Src kinase is known to regulate leading edge protrusion through Rac and Cdc42 signaling and thus facilitate focal adhesion dynamics and building of invadopodia. In addition, Src is able to induce epithelial-to-mesenchymal transitions (EMT) directly.

Intermediate filaments, the epithelial-to-mesenchymal transition, tumor growth, invasiveness and metastasis
Invasion of the local microenvironment (such as the connective tissue stroma) by cancer cells is a critical initial step in determining in cancer metastasis. In addition, epithelial cancer cell invasion is mediated by the EMT. It is called this because epithelial cells lose their polarity and other characteristics, such as E-cadherin and keratin expression, and can adopt a fibroblast-like morphology, which includes vimentin expression and leads to aggressive migratory properties. One of the key differences between the epithelial and mesenchymal phenotypes is based on the tight cell–cell and cell–matrix contacts provided by epithelial cells compared to the weak contacts of mesenchymal cells (Thiery *et al* 2009, Nakamura and Tokura 2011). The process of EMT seems to be an important mechanism to account for the increased

motility and invasiveness of epithelial-derived cancer cells (Kalluri and Weinberg 2009, Yilmaz and Christofori 2009, Mendez *et al* 2010).

Because epithelial cells can undergo EMT, their intermediate filament system switches from being keratin-dominated to vimentin-dominated, which is another characteristic feature of mesenchymal cells. Indeed, many cancer cell lines display both a keratin-based and a vimentin-based intermediate filament network and these possess distinct intracellular organization and regulatory mechanisms (Gilles *et al* 1999). Cell-culture-based (Chu *et al* 1993, 1996) and xenograft-assays-based *in vivo* studies (Hendrix *et al* 1997) revealed that vimentin has a pro-migratory potential, which is facilitated by the interplay between vimentin and keratins 8/18 and the regulatory role of focal adhesions.

Src has been shown to interact directly with keratin intermediate filaments in a K6-dependent fashion through a novel, nonphosphotyrosine-dependent contact involving Src's SH2 domain using K6a/K6b null keratinocytes (Rotty and Coulombe 2012). Also, Src's localization in the membrane, where it is transiently inactive, is reduced in K6a/K6b null keratinocytes (Rotty and Coulombe 2012). Whether such findings also occur in epiplakin, plakophilin, plakoglobin and/or plectin null keratinocytes is still elusive (Bordeleau *et al* 2012). However, the keratin-containing multiprotein partnership, its mechanisms and the effectors through which it is so precisely regulated during keratinocyte migration, are not yet fully understood. The apparent paradox between the wound-inducible character of K6 and its negative influence for pure cell migration has also been found for several other cytoskeletal proteins. How is the optimal speed and mode of cellular migration achieved under certain conditions? It seems that this has to be determined for each case separately (Weber *et al* 2012).

Keratin 6 and K16 are often up-regulated in various types of carcinomas, suggesting that they may serve as useful diagnostic markers (Karantza 2011). K6's impact on keratinocyte migration (and possibly also its interaction with Src) may help to explain clinical correlations (Depianto *et al* 2010). For example, the lack of K6 expression correlates with an aggressive behavior for endometrial carcinomas (Stefansson *et al* 2006), whereas decreased K6 expression is coupled with reoccurrence of K8/K18 expression and consequently correlates with the acquisition of malignancy in mouse skin subjected to chemical carcinogenesis (Larcher *et al* 1992). These findings suggest that the functional significance of inducing or modifying K6 (and possibly K16, plakoglobin, plectin and others) may be a natural strategy to withstand dedifferentiation- and malignancy-promoting signaling and cellular processes, such as EMT (Rotty and Coulombe 2012). However, the link between keratins and cancer has not yet been clearly revealed and hence cannot be regarded as simple (Karantza 2011). Higher levels of K16 lead to poorer survival rates among breast cancer patients with metastatic reoccurrence (Joosse *et al* 2012), while higher levels of keratins 5, 6 and 17 have been associated with a poor prognosis in breast cancer (van de Rijn *et al* 2002, Abd El-Rehim *et al* 2004, de Silva Ruland *et al* 2011). Whether a keratin or a conglomerate of intermediate filament proteins and binding partners promotes or inhibits cell migration and displays cancer cell properties is determined by the overall microenvironemtal background within the cell, such as associated proteins, posttranslational modifications and biological setting.

How do intermediate filaments affect cellular mechanics and migration behavior? Cells develop a polarized cytoarchitecture upon the initiation of cell migration, as their front and rear contain different molecular components and possess different functional properties (Laemmermann and Sixt 2009, Friedl and Wolf 2010). As a cell senses microenvironmental properties, signaling events, actin polymerization and myosin motor function can be spatially regulated in order to generate membrane protrusions at the leading edge and retractive forces at the trailing edge. Mechanical signals facilitate polarized cell protrusions and directional migration and hence there is evidence that intermediate filaments regulate cell migration from the cellular mechanics point of view (Eckes *et al* 1998, Gilles *et al* 1999, Seltmann *et al* 2013).

In addition, mechanotransduction processes involving intermediate filaments are important for the attachment of epithelial cells to their microenvironment, such as the extracellular matrix. A mechanotransduction pathway has been revealed in *C. elegans* that includes hemidesmosome-like elements regulating intermediate filaments (Zhang *et al* 2011). In particular, it has been observed that muscle contraction mechanically alters the epidermis and activates p21-activated kinase (PAK). Then PAK phosphorylates intermediate filament proteins, which induces hemidesmosome biogenesis. Thus, hemidesmosomes act as mechanosensors, which, when under tension, trigger intracellular signaling pathways that promote epithelial morphogenesis. Cell–cell junctions are involved in local traction force generation, providing long-range gradients of intra- and inter-cellular tension during collective cell migration (Trepat and Fredberg 2011). When investigating *Xenopus gastrulation*, the application of a punctual mechanical force to single *Xenopus mesendoderm* cells through magnetic tweezers and cadherin-coated beads connected to the cells promoted polarized protrusions at the opposite end of the force (and the cell) and subsequently induced persistent directional cell migration (Weber *et al* 2012). It has been suggested that this localized tension increases induces plakoglobin-dependent redistribution of the keratin intermediate filaments at the cell's rear. Indeed, these events based on keratin and plakoglobin are needed for force-induced, polarized cell protrusions and normal mesendoderm polarity, as well as organization *in vivo*.

8.7 The role of vimentin as a promotor of cell migration during cancer progression

Vimentin and desmin are the most prominent filaments in the class III intermediate filament subgroup. The investigation of the mechanical influence of type III intermediate filaments on cell mechanical properties has revealed divergent results, as the results of mechanical measurements are strongly dependent on the type of probing the cell and the magnitude of the exerted strain on the cell during the biophysical experiments. In particular, the most commonly utilized biophysical method to characterize the cortical stiffness of cells is by the indention of the cells' surface with a conical tip or a colloidal bead, which has been glued to an AFM cantilever. Moreover, the AFM method enables us the identify stiffness differences between different cell types, which can even be determined at various degrees of deformation and often be combined with simultaneous imaging of the cells. The calculation of the absolute magnitude of elastic moduli using AFM or other microrheological techniques is difficult, as various assumptions about contact geometry, the homogeneity of the material and conservation of the volume need to be performed (Charrier and Janmey 2016).

Vimentin is expressed significantly throughout embryogenesis, whereas it is restricted to mesenchymal cell types in the adult setting, such as fibroblasts, bone marrow-derived blood cells and endothelial cells (Hansson *et al* 1984, Dellagi *et al* 1985). Vimentin can reoccur in adults, as it is drastically up-regulated after injury to diverse tissues, such as muscle, the central nervous system and various connective tissues, and during EMT. Vimentin exerts pleiotropic and context-dependent roles in cells (Satelli and Li 2011) and, in particular, has a marked impact on cell migration in several physiologically normal settings (Ivaska *et al* 2007). For instance, vimentin is essential for lymphocyte adhesion to the endothelial cell lining of blood and lymphoid vessels and their subsequent transmigration through the endothelium into the connective tissue (Nieminen *et al* 2006), for fibroblast or individual breast cancer cell motility (Mendez *et al* 2010) and for the *in vitro* wound healing and closure processes of alveolar epithelial cells (Rogel *et al* 2011).

Vimentin has been detected at unusually high levels in many types of epithelial cancers (Satelli and Li 2011). Vimentin expression is necessary for the invasive phenotypes of prostate cancer cells (Singh *et al* 2003, Wei *et al* 2008), soft tissue sarcoma cells and breast cancer cells, in *in vitro* assays (Zhu *et al* 2011). In line with this, inhibition of vimentin expression in a squamous carcinoma cell model not only decreases motility (McInroy and Maatta 2007), it also promotes a more pronounced epithelial phenotype, as supported by the up-regulation of K13, K14 and K15 (Paccione *et al* 2008) and alteration of the cellular shape (Mendez *et al* 2010). Conversely, vimentin overexpression has been shown to increase prostate cancer cell invasion (Zhao *et al* 2008) and invadopodia elongation (Schoumacher *et al* 2010).

Numerous studies support the idea that vimentin and the process of cellular migration mutually regulate one another. The tumor suppressor adenomatous polyposis coli (APC), which is frequently mutated or lost in colorectal cancer, directly binds to and thus regulates vimentin organization (Sakamoto *et al* 2013). In addition, in migrating astrocytes APC is required for vimentin intermediate filament alignment with the microtubule network (Sakamoto *et al* 2013). A C-terminal APC truncation mutant can bind to and disorganize vimentin, whereas the keratin intermediate filament organization remains unaltered when expressed in human SW480 colon cancer cells. It has been suggested that the loss of APC in cancer cells that have undergone EMT alters vimentin intermediate filament organization and impacts on motility and invasiveness. In cultured breast epithelial cells, over-expression of oncogenic H-Ras-V12G or the transcription factor Slug, which both favor cell migration and EMT, induces vimentin expression. In turn, vimentin expression is necessary for H-Ras-V12G- and Slug-induced migration and expression of the receptor tyrosine kinase Axl, whereas epithelial markers such as K6 are suppressed (Vuoriluoto *et al* 2011). Overexpression of Axl is able to rescue the decreased migration phenotype of a breast cancer cell line expressing vimentin siRNA, suggesting that vimentin acts through interaction with Axl.

An RNAi screen has identified regulators of vimentin expression such as the mitochondrial enzyme methylenetetrahydrofolate dehydrogenase 2 (MTHFD2), which is required for vimentin expression and network organization (Lehtinen *et al* 2013). Similar to vimentin, the siRNA-mediated knockdown of MTHFD eliminates the migration of breast cancer cells and their invasion into the extracellular matrix, suggesting their interdependence in cell motility. Moreover, vimentin expression is also regulated by miRNAs. In more detail, overexpression of mir-138, which is down-regulated in several primary tumors, leads to decreased vimentin expression and reduced cell migration and invasion in renal cell carcinoma cell lines (Yamasaki *et al* 2012). Similarly, mir-30a represses vimentin expression and thus consequently cell migration and invasion in breast cancer cell lines (Cheng *et al* 2012). As some cancer cells show decreased expression of mir-138 and mir-30a (Cheng *et al* 2012, Yamasaki *et al* 2012), these findings may provide evidence regarding how vimentin expression is up-regulated in EMT and cancer.

8.7.1 Vimentin knock-out cells

Several biomechanical studies have been carried out on mesenchymal cells with a disrupted or even no vimentin network (Klymkowsky 1981, Holwell *et al* 1997). The first identification of the role of vimentin in providing cellular mechanical properties utilized a rotational force magnetic twisting cytometer (Wang and Ingber 1994) to investigate primary fibroblasts from vimentin knock-out mice (Eckes *et al* 1998). In particular, vimentin null cells displayed a lower cortical rigidity than wild-type cells and impaired migratory capacities that leads to the hypothesis that vimentin plays a role in the stabilization of actin and microtubules networks within cells. Another subsequent study of fibroblasts isolated from vimentin knock-out mice even quantified the mechanical contribution of vimentin on the cells (Wang and Stamenovic 2000). Using the same magnetic twisting device, it has been demonstrated that the mechanical alteration of those fibroblasts is based on the lack of vimentin and in addition, a softer cortex and the decreased ability to stiffen, is solely detectable only when cells are exposed to large strain. Using AFM, the stiffness mapping revealed that the cortical stiffness of vimentin null cells measured at small strains, is not altered compared to wild-type cells, however, in the absence of vimentin it is observed that the cells are more strongly and repeatedly deformable. Hence, a loss of vimentin also elevates the amount of viscous loss during a cycle of cell deformation (Mendez *et al* 2014). Another study using fibroblasts from knock-out vimentin mice that are probed by optical magnetic twisting cytometry, revealed that the cortical rigidity of cells is not affected by the loss of vimentin, whereas the strain intensity applied to the cells during the measurement was not reported (Guo *et al* 2013). However, the same study reported that the interior cytoplasmic rigidity of fibroblasts null for vimentin using active bead microrheometry is decreased by a factor of two, which evokes an elevated velocity of vesicular trafficking, and hence it can be proposed that vimentin filaments are crucial for the stabilization of the position of organelles in the cytoplasm (Guo *et al* 2013).

8.7.2 Disruption of vimentin by drugs

To determine the contribution of vimentin network of cells to their mechanical properties, pharmacological drugs specifically targeting vimentin or the enzymes can be used that alter vimentin's phosphorylation and inducing the disruption, aggregation or depolymerization of the entire vimentin network. In particular, withaferin A treatment causes the disruption of the vimentin network and finally, leads to its aggregation (Thaiparambil *et al* 2011). In addition, incubation of suspended natural killer cells with withaferin A leads to a global cell softening of about 20%, when probed at large strains using a microfluidic optical stretcher (Gladilin *et al* 2014). Calyculin A is a pharmacological drug that targets vimentin phosphatases and hence causes the disruption of the vimentin network (Eriksson *et al* 1992, Eriksson *et al* 1998). In T lymphocytes, the vimentin network is primarily organized as a cortical cage that preserves the mechanical integrity of the cell since the collapse of this network is evoked by addition of calyculin A, which then enhances cell deformability by about 40% and thereby softens the cell, as it has been quantified by high g-force centrifugation onto adhesive substrates and a subsequently morphological analysis of the remaining cells (Brown *et al* 2001). Although calyculin A is an inhibitor of vimentin protein phosphatases, it can also inhibit other cellular phosphatases such as the myosin light chain phosphatase and hence alter the actomyosin contractility of the cell. Moreover, acrylamide can also be used to depolymerize and aggregate vimentin intermediate filaments (Eckert 1985). The characterization of acrylamide-treated primary human articular chondrocytes revealed that the impaired vimentin network integrity causes a three-fold softening of the entire cells as evaluated by exerting large strains to cells that are embedded in alginate gels (Haudenschild *et al* 2011). In this study, a low 4 mM acrylamide concentration is used to restrict the effect on other cytoskeletal components, whereas this concentration is still high enough to alter the nuclear lamina architecture and mitochondrial homeostasis (Hay and De Boni 1991). Moreover, at higher concentrations acrylamide can even disrupt actin and microtubules networks (Sager 1989) and thus an effect of acrylamide on other cellular elements such as actin or microtubules can no longer be excluded.

In addition, the overexpression of an oncogene protein (simian virus 40 large T antigen) is another approach to cause a perinuclear reorganization of the vimentin network in human fibroblasts. Moreover, this spatial reorganization of the network evokes a two-fold increase of the Young's modulus of the cytoplasm, when quantified using a colloidal probe force-mode AFM (Ducker *et al* 1991), in the region where the vimentin density is elevated (Rathje *et al* 2014).

8.7.3 The mutant desmin disrupts vimentin

Another possible mechanism to disrupt the entire vimentin network is through the expression of wild-type or mutated desmin in cells that originally express solely vimentin. In particular, desmin is able to copolymerize with vimentin and hence dominant negative desmin mutants can cause a collapse of the former highly functional vimentin network. The expression of the L345P desmin mutant in rat fibroblasts leads to a perinuclear aggregation of the vimentin network that is associated with a local stiffening of the cytoplasm in certain areas, which are probed by AFM (Plodinec *et al* 2011). At least three prominent desmin mutants affect the elastic modulus of fibroblasts. Expression of GFP-fused wild-type desmin causes a small softening of the cell, which is possibly provided by a destabilization of the entire intermediate filament network through the GFP-desmin fusion protein. In contrast, the desA213V point mutant of desmin, which is still able to polymerize filaments, stiffens the cell possibly through an increase in the total amount of intermediate filaments but also potentially by changing the mechanical properties of the intermediate filaments into which this mutant is incorporated. The nonfilament forming desmin-mutant desL345P evokes a complex effect on cellular stiffness, as it leads to a breakdown of the endogenous vimentin network around the nucleus and subsequently it strongly stiffens the perinuclear region and thereby causes the rest of the cell to be slightly less stiff than under normal conditions. However, desmin fulfills a negligible role in the mechanical properties of the cellular cortex, however, it is still a major determinant of the overall cellular stiffness (Charrier *et al* 2018). In particular, higher forces are needed to cause high indentation depths in desmin-mutated cells than in wild-type cells, where desmin intermediate filaments are typically located. Moreover, a heat-shock treatment enhanced the proportion of desmin-mutant cells with aggregates and hence leads to the production of a secondary peak in the distribution of Young's moduli (Even *et al* 2017).

8.8 The impact of keratins 8/18 on epithelial cell migration

Type II keratin 8 has also been shown to be involved in cell migration and cancer metastasis (Raul *et al* 2004, Bordeleau *et al* 2010). Similar to vimentin, K8 is expressed broadly during development, but is then restricted to simple epithelial lineages, such as in the liver, gut, kidney and lungs, in adults (Moll *et al* 1982). Moreover, K8 expression is induced or enhanced in special cancer types, such as breast, lung and pancreatic cancers, and specific tumor-derived cell lines (Moll *et al* 1982). Dissimilar to vimentin, K8's impact on cancer cell migration and invasion is inhibitory. Thus, the balance between vimentin and K8/K18 expression is supposed to be a key determinant of the migratory properties and invasiveness of various types of cancer cells *in vitro* and *in vivo*.

In line with this, it has been reported that treatment of pancreatic cancer Panc-1 cells with sphingosylphosphorylcholine (SPC), which is a bioactive lipid, leads to keratin phosphorylation, promotes a dramatic reorganization of the keratins' intermediate filaments to the perinuclear region, reduces cellular elasticity and induces pronounced cell migration (Beil *et al* 2003). These observations are supported by other studies. One study showed that SPC treatment facilitates the expression of transglutaminase-2 expression in Panc-1 cells, which precedes JNK kinase activation and phosphorylation of K8 at Ser 431 (Park *et al* 2011). Another study reported that SPC can activate ERK kinase upstream of keratin intermediate filament reorganization and provides phosphorylation of K8 and K18 at Ser 431 and Ser 52, respectively, in pancreatic and gastric cancer cells (Busch *et al* 2012). However, do these events contribute to the so-called mechanical softening of the cytoplasm in SPC-treated Panc-1 cells (Beil *et al* 2003), which contributes to their enhanced motile behavior?

The expression and/or site-specific phosphorylation of K8 (and of its partner K18) alters the migratory behavior and invasiveness of various types of cancer cells. An inhibitory influence for K8 on cell migration was observed by studies in which pancreatic cancer cells (Busch *et al* 2012) and a weakly invasive subclone of MDA-MB-468 breast cancer cells were subjected to K8 knockdown (Iyer *et al* 2013), a highly invasive subclone of MDA-MB-435 breast cancer cells was transfected to overexpress K8 (Iyer *et al* 2013) and KLE endometrial cancer cells and also HepG2 hepatocellular cancer cells expressed no K8/K18 due to silencing (Fortier *et al* 2013). By contrast, other studies reported that the loss of K8 phosphorylation at either Ser 73 or Ser 431 led to elevated migration and enhanced metastatic potential for oral squamous cell carcinoma cells (Alam *et al* 2011) and colorectal cancer cells (Mizuuchi *et al* 2009). In addition, K8-dependent stimulation of cell migration led to reduced collective migration of hepatoma cells after K8 silencing (Bordeleau *et al* 2010). Taken together, the silencing of the desmosomal plaque protein plakophilin 3 enhances the migration and metastasis of human colon carcinoma cells (Kundu *et al* 2008), while another study speculated that this result is caused by increased levels of K8 protein and phosphatase PRL-3, which evokes K8 dephosphorylation (Khapare *et al* 2012).

Keratin-dependent activation of Akt signaling seems to be important during tumorigenesis. Indeed, lactotransferrin has anti-tumor activity and is even down-regulated in cancer (Bezault *et al* 1994). In particular, interaction with lactotransferrin blocks the binding of K18 to 14-3-3σ and hence suppresses K18-induced Akt activation and its impact on cancer cell proliferation and invasion (Deng *et al* 2012). However, others have shown that K17 interacts with 14-3-3σ and regulates the Akt-mTOR signaling pathway (Kim *et al* 2006), whereas vimentin interacts with and becomes activated by Akt to promote cancer cell invasion (Zhu *et al* 2011). Although not related to migration so far, O-linked N-acetylglucosamine modification of K18 induces Akt activity and consequently protects against liver injury (Ku *et al* 2010).

In addition to vimentin and keratin, the increased expression of nestin, which is a class IV intermediate filament protein and a marker of stem/progenitor cells (Lendahl *et al* 1990), has been found in multiple types of tumors (Florenes *et al* 1994, Parry *et al* 2008). Moreover, nestin regulates cellular motility and metastatic properties, whereas it does not affect the growth of prostate (Kleeberger *et al* 2007) and pancreatic cancer cells (Matsuda *et al* 2011). Based on its connection to stem/progenitor cells, it may be possible that nestin expression can be useful to identify cancer stem cells (Matsuda *et al* 2012). However, this is still a hypothesis that needs to be tested in future studies.

Taken together, the complex role of intermediate filaments during cell migration has been detected. In particular, intermediate filaments affect cell migration because (i) they are intrinsic determinants of cellular micromechanical properties and (ii) they contribute to the regulation of certain migratory pathways. Finally, the impact of intermediate filaments during cell migration in normal and disease settings reflects their full integration into the cytoskeletal framework of migratory and invasive cells.

8.9 Interaction between filamin A and vimentin in cellular motility

The role of filamin A and vimentin in cell migration

Cell adhesion and spreading are regulated by diverse and complex interactions between the cytoskeleton and extracellular matrix proteins. Thus, it has been suggested that the interaction of the intermediate filament protein vimentin with the actin crosslinking protein filamin A may regulate the spreading in HEK-293 and 3T3 cells (Kim *et al* 2010). In more detail, filamin A and vimentin-expressing cells spread on collagen and exhibit numerous cell extensions containing filamin A and vimentin. However, when the cells are treated with small interfering RNA (siRNA) in order to knockdown filamin A or vimentin, they spread badly. Both of these filamin A and vimentin knockdown cell populations have exhibited a less than 50% decrease in cell adhesion, cell surface expression and $\beta1$ integrin activation. In more detail, the knockdown of filamin A decreases vimentin phosphorylation and blocks recruitment of vimentin to cell extensions, whereas the knockdown of filamin A and/or vimentin diminishes the exertion of cell protrusions. Decreased vimentin phosphorylation, impaired cell spreading and reduced $\beta1$ integrin surface expression and decreased activation occur in cells that have been cultured with the protein kinase C inhibitor bisindolylmaleimide. In addition, cell spreading has been shown to be reduced by siRNA-facilitated knockdown of protein kinase C-ε. Immunoprecipitation of cell lysates and pull-down assays with purified proteins has revealed a direct binding interaction between filamin A and vimentin. Filamin A is additionally associated with the protein kinase C-ε, which has been observed to increase in cell extensions. Taken together, these data propose that filamin A associates with vimentin and protein kinase C-ε, thereby inducing vimentin phosphorylation, which is crucial for $\beta1$ integrin activation and cell spreading on collagen substrates.

The adhesion and spreading of fibroblasts is critical for development, tissue remodeling and wound healing. Cell spreading is characterized by the extension of cell membrane protrusions, including filopodia and lamellipodia (Albrecht-Buehler and Lancaster 1976, Guillou *et al* 2008, Johnston *et al* 2008, Ladwein and Rottner 2008). The exertion of protrusions during cell spreading on extracellular matrix proteins such as collagen and fibronectin is regulated by complex cell–matrix signaling pathways involving $\beta1$ integrins, small GTPases and actin-binding proteins, which facilitate actomyosin remodeling (Price *et al* 1998, Zimerman *et al* 2004, Li *et al* 2005). Whereas the actin cytoskeleton is crucially involved in cell adhesion and spreading (Serrels *et al* 2007, Vicente-Manzanares *et al* 2009), the intermediate filament cytoskeleton and, in particular, vimentin seems to be additionally involved in cell adhesion and spreading (Eckes *et al* 1998, Ivaska *et al* 2002, 2007, Tsuruta and Jones 2003, Fortin *et al* 2010). As actin and vimentin cytoskeletons colocalize at the leading edge of spreading cells (Correia *et al* 1999), the two systems may also be functionally correlated during cell adhesion and spreading. However, the mechanisms through which the two cytoskeletal components act together in order to promote cell spreading are not yet well understood. Numerous proteins can bind to both actin and intermediate filaments and thus may functionally coordinate the two different cytoskeletal systems. Candidate proteins for the interaction are plectin (Seifert *et al* 1992, Svitkina *et al* 1996), BPAG1 (Yang *et al* 1996), fimbrin (Correia *et al* 1999) and filamin A (Brown and Binder 1992). Filamin A is an actin crosslinking protein inducing gelation, membrane stability, adhesion and motility (Brotschi *et al* 1978, Glogauer *et al* 1998, Stossel *et al* 2001, Feng and Walsh 2004, Popowicz *et al* 2006). In addition, filamin A is appropriately located in order to regulate cell–matrix interactions, as it binds to actin and regulates the properties of $\beta 1$ integrins (Loo *et al* 1998, Stossel *et al* 2001, Popowicz *et al* 2006). Moreover, filamin A is concentrated in focal adhesions (Glogauer *et al* 1998, Critchley 2000, Campbell 2008) and hence regulates cell adhesion through the control of the cell surface expression (Meyer *et al* 1998) and activation (Kim *et al* 2008) of $\beta 1$ integrins.

Vimentin has several functions in common with filamin A, such as the regulation of cell adhesion and motility (Eckes *et al* 1998, Tsuruta and Jones 2003, Nieminen *et al* 2006, McInroy and Maatta 2007, Whipple *et al* 2008), the clustering of $\beta 1$ integrin in focal adhesions (Kreis *et al* 2005) and the mechanical stabilization of cells (Eckes *et al* 1998). In more detail, the expression of filamin A and vimentin is elevated during shear stress in osteoblasts (Jackson *et al* 2008), indicating that they protect against external mechanical stimulation. As with filamin A, many functions of vimentin are facilitated upon its phosphorylation (Eriksson *et al* 2004, Izawa and Inagaki 2006, Chou *et al* 2007, Sihag *et al* 2007). In particular, the phosphorylation of vimentin by PKC is necessary for the recycling of $\beta 1$ integrins to the cell membrane and, consequently, for cell migration (Ivaska *et al* 2002, Fortin *et al* 2010). In addition, filamin A can bind PKC (Glogauer *et al* 1998, Tigges *et al* 2003, Feng and Walsh 2004), which then facilitates the phosphorylation of mediates vimentin (Ando *et al* 1989, Ogawara *et al* 1995, Yasui *et al* 2001, Iwasza and Inagaki 2006).

In summary, it has been demonstrated that filamin A and vimentin coregulate the activation and cell surface expression of $\beta 1$ integrin in order to provide cell spreading in fibroblasts. Indeed, PKC-facilitated phosphorylation of vimentin is involved in $\beta 1$ integrin activation as well as cell spreading and consequently this phosphorylation is dependent on filamin A, which can bind directly to vimentin.

Does filamin A interact with vimentin during cell spreading?

Filamin A is an important determinant for cell spreading (Kim *et al* 2008). In addition to revealing novel filamin A binding proteins involved in filamin A-mediated cell spreading, an isotope-coded affinity tag analysis was performed on filamin A immunoprecipitates. Twenty-one peptides have been identified with >99% certainty and at least six different proteins were predicted to be differentially expressed in suspended and spreading cells. Among these six proteins was vimentin. Starting from the previously identified roles of filamin A and vimentin in cell adhesion and migration (Eckes *et al* 1998, Bellanger *et al* 2000, Vadlamudi *et al* 2002, Tsuruta and Jones 2003, Nieminen *et al* 2006), subsequent investigations have focused on the roles of filamin A and vimentin in the onset of cell spreading.

The coregulation of filamin A and vimentin in cell spreading

The extrusion of filopodia and/or lamellipodia by cells is an important marker of early cell spreading (figure 8.7) (Price *et al* 1998, Zimerman *et al* 2004, Li *et al* 2005). It has been shown whether filamin A and/or vimentin are present in the earliest build extensions of spreading cells. Indeed, immunofluorescence revealed that in control HEK cells spreading on collagen, filamin A and vimentin colocalized within these extensions. As true filopodial and lamellipodial extensions consist of cortactin, vinculin and paxillin in addition to actin filaments (Nobes and Hall 1995, Artym *et al* 2006, Perrin *et al* 2006, Webb *et al* 2007), the presence of these proteins in cell extensions could be confirmed using immunofluorescence. To investigate the individual roles of filamin A and vimentin in cell spreading, RNA interference was performed to diminish filamin A and vimentin expression in HEK cells. To evaluate the combined roles of filamin A and vimentin in cell spreading, siRNA was used for vimentin in cells stably transfected with a filamin A shRNA, thereby creating a double knockdown cell line. As the initial extension of filopodia and lamellipodia normally occurs within 30 min of after plating (Defilippi *et al* 1999, Ren *et al* 1999, Arthur and Burridge 2001), time points between 15 and 120 min were assessed for

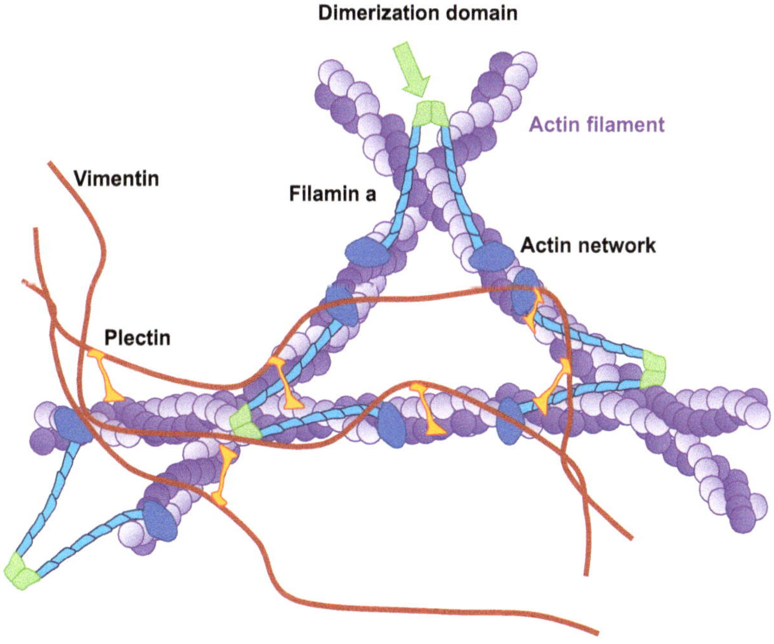

Figure 8.7. Interaction of vimentin and filamin A.

differences in cell spreading between control and filamin/vimentin-deficient cells. Unexpectedly, knockdown of either filamin A or vimentin expression leads to a pronounced reduction in the number of cell extensions. Concurrent knockdown of filamin A and vimentin did not reveal further reductions in the number of cell extensions compared with each individual knock-downs of filamin A or vimentin, suggesting that the role of vimentin in mediating cell spreading is not additive to the role of filamin A. Similar results have been obtained for the reductions in cell extension numbers upon silencing of filamin A and/or vimentin in mouse 3T3 fibroblasts. In more detail, filamin A expression was then rescued by introducing a filamin cDNA into filamin knockdown cells. The vimentin expression was restored in vimentin-knockdown cells after four days following transient and hence reversible siRNA knockdown. In addition, the formation of cell extensions by filamin- and vimentin-deficient HEK-293 cells was rescued after the restoration of filamin or vimentin expression.

What is the difference between filamin A and vimentin during actin assembly in spreading cells?

The single and combined effects of filamin A and vimentin have been analyzed on the generation of actin filament free barbed ends in spreading cells. Filamin A can regulate the GTPase activity of Rac and Cdc42 (Ohta *et al* 1999, Bellanger *et al* 2000, Vadlamudi *et al* 2002), which then control cell spreading through determining the generation of free barbed ends, which are required for actin filament assembly (Winokur and Hartwig 1995, Sun *et al* 2007). Indeed, the incorporation of rhodamine-labeled actin monomers into growing actin filaments was pronouncedly decreased in filamin A-deficient cells. No difference was found in the number of free barbed ends in vimentin-knockdown cells compared to their wild-type counterparts. However, incorporation of actin monomers into actin filaments was hardly detectable after concurrent knockdown of filamin A and vimentin. Finally, these data indicate that vimentin contributes to the generation of free barbed ends in the absence of filamin A, whereas vimentin is not able to regulate actin assembly independently.

As filamin A plays a role in binding and regulating the activity of Rac and Cdc42 (Bellanger *et al* 2000, Stossel *et al* 2001, Vadlamudi *et al* 2002), it has been investigated whether vimentin supports the association between filamin A and Rac and/or Cdc42. Thus, filamin A was immunoprecipitated from spreading control and vimentin-deficient cells and then the immunoprecipitates were immunoblotted for Rac and Cdc42. As expected, the amount of filamin A-bound Rac and Cdc42 did not differ significantly between the control and vimentin-deficient cells. In particular, the levels of active, GTP-bound Rac and Cdc42 were similar in the control and vimentin-deficient cells. Finally, the data suggest that vimentin may regulate cell spreading independently of the filamin A–small GTPase-actin assembly pathway.

Do filamin A and vimentin facilitate $\beta 1$ integrin expression and activation?

As cell spreading is $\beta 1$ integrin-dependent (Price *et al* 1998, Jovic *et al* 2007, Frame and Norman 2008) and filamin A-null cells express fewer cell surface $\beta 1$ integrins

(Meyer *et al* 1998), it has been suggested that filamin A and vimentin coregulate cell surface $\beta 1$ integrin expression. To address the individual and collective effects of filamin A and vimentin knockdown on $\beta 1$ integrin expression and activation, HEK cells were spread on collagen for 15 min to allow initial adhesion (Zimerman *et al* 2004). Then, cells were suspended and incubated with the 4B4 antibody in order to detect cell surface $\beta 1$ integrins. In addition, active, ligand-bound $\beta 1$ integrins were labeled with the 12G10 antibody. Flow cytometric analysis revealed that cell surface $\beta 1$ integrin was reduced to 50% upon knockdown of filamin A and/or vimentin. This phenomenon was rescued after reexpression of filamin or vimentin, which is still consistent with the cell spreading results. In particular, 12G10 staining intensity was reduced four-fold after filamin A and/or vimentin-knockdown. Taken together, these findings showed that filamin and vimentin are important determinants of $\beta 1$ integrin expression, while they are especially important for regulating $\beta 1$ integrin activation.

Which proteins are associated with nascent cell adhesions? To investigate this collagen-coated magnetite beads were incubated with cells and isolated and bead-associated proteins were immunoblotted (Glogauer *et al* 1998, Glogauer *et al* 1998, Zhao *et al* 2007). There was indeed more bead-associated vinculin and paxillin in control cells compared to filamin A- or vimentin-deficient cells, suggesting that filamin A and vimentin are both necessary for the generation of stable adhesions. Moreover, these data suggest that filamin A and vimentin regulate $\beta 1$ integrin activation, cell adhesion and the early steps of cell spreading, possibly using the same mechanism.

The role of vimentin phosphorylation for $\beta 1$ integrin expression, activity and cell spreading

As expected, the phosphorylation of vimentin facilitated by PKC-ε is a major regulatory step for $\beta 1$ integrin trafficking and cell migration (Ivaska *et al* 2002, Fortin *et al* 2010). In particular, the role of PKC-mediated phosphorylation of vimentin in early cell spreading was analyzed. Cells were treated for 90 min with bisindolylmaleimide (BIM) (1 μM), which inhibits PKC-ε (Ivaska *et al* 2002, Ivaska *et al* 2002) and then plated on collagen-coated planar substrates. In order to verify the specificity of BIM as a PKC inhibitor, all experiments were additionally performed with cells pretreated with calphostin-C (100 nM). In BIM- and calphostin-C-treated cells undergoing early spreading, vimentin was decreasingly phosphorylated at serines-6, -38, and -50, which are PKC phosphorylation sites (Ando *et al* 1989, Ogawara *et al* 1995, Yasui *et al* 2001, Izawa and Inagaki 2006). Indeed, BIM-treated and calphostin-C-treated cells both displayed a significant reduction in the number of cell extensions compared to untreated controls. In more detail, the BIM- and calphostin-C-induced effects on cell spreading were comparable to siRNA knockdown of vimentin in magnitude. In order to validate the use of BIM as an inhibitor of PKC-ε, the phenotype of BIM-treated and calphostin-C-treated cells was compared to the phenotype of PKC-ε-deficient cells obtained by siRNA knockdown. In line with this, PKC-ε-knockdown cells displayed reduced numbers of cell extensions compared to those detected in vimentin-knockdown, BIM-treated and calphostin-C-treated cells. Although vimentin is not a direct substrate for PKC-ε (Ivaska *et al* 2002), its phosphorylation is PKC-ε dependent. Thus, these findings suggest that the phosphorylation of vimentin regulated by PKC-ε may be a key regulatory step for initial cell adhesion and spreading.

PKC-ε-facilitated vimentin phosphorylation is a critical for cell surface $\beta 1$ integrin expression (Ivaska *et al* 2002). In flow cytometric analysis, nonpermeabilized cells performing initial spreading displayed less active and total cell surface $\beta 1$ integrin after knockdown of filamin A and/or vimentin, BIM treatment, calphostin-C treatment or PKC-ε knockdown. Thus, these data indicate that the phosphorylation of vimentin by PKC seems to be crucial for cell surface $\beta 1$ integrin expression and activation.

Does filamin A bind directly to vimentin in spreading cells under in vitro conditions? As filamin A can bind PKC (Glogauer *et al* 1998, Tigges *et al* 2003, Feng and Walsh 2004), it has been hypothesized that filamin A facilitates vimentin phosphorylation by interacting with it. In order to investigate the filamin A–vimentin interaction, filamin A was immunoprecipitated from the lysates of spreading HEK-293 cells and subsequently blotted for vimentin. Consistently, a small increase in vimentin–filamin A association in response to spreading has been seen, and a small increase of filamin A-associated vimentin in spreading versus suspended cells has been found by immunoprecipitation. These data suggest that filamin A–vimentin interactions are mostly constitutive and depend only moderately on cell adhesion.

Dot-blot analysis revealed that purified filamin A and vimentin seem to interact directly *in vitro*. In addition, the question of whether there is direct interaction was explored using purified GST-tagged vimentin bound to glutathione beads, with the beads being incubated with purified FLAG-tagged filamin A. A direct filamin A–vimentin interaction was detected by immunoblotting the bead-associated proteins for filamin A. Conversely, purified vimentin associated with purified FLAG-filamin A bound to beads was detected by immunoblotting for vimentin. As filamin A has been shown to bind to vimentin directly, it has been suggested that filamin A regulates the phosphorylation of vimentin by PKC.

The role of filamin A in vimentin phosphorylation and reorganization

Indeed, the phosphorylation by PKC is needed for the assembly and disassembly of vimentin filaments and distribution in the cytoplasm (Eriksson *et al* 2004, Izawa and Inagaki 2006, Sihag *et al* 2007), and for processes that affect cell adhesion and motility through possible regulation of the recycling and trafficking of $\beta 1$ integrins to the cell membrane (Ivaska *et al* 2002, Fortin *et al* 2010). The spatial colocalization of vimentin and PKC-ε has been determined through immunofluorescence and it has been confirmed that vimentin and $\beta 1$ integrins colocalize in spreading control cells. However, this colocalization can be disrupted after knockdown of filamin A or treatment with the cell permeable and reversible inhibitor of PKC BIM to block PKC. Moreover, the cytoplasmic distribution of vimentin was pronouncedly altered among filamin A-expressing, filamin A-deficient and BIM-treated cells. In particular, control cells displayed a well-defined vimentin filament network throughout the cytoplasm, in which vimentin was located in cell extensions. In contrast, the location of vimentin in filamin A-deficient or BIM-treated cells appeared as a single focal accumulation in the cytoplasm. These data support the hypothesis that filamin A expression and PKC activity are both required for the reorganization and redistribution of vimentin.

What is the role of filamin A in PKC-driven phosphorylation of vimentin in respect of the association between PKC and filamin A? Coimmunoprecipitation revealed that filamin A associates with phosphorylated vimentin as well as with phosphorylated PKC-ε, which leads to it being catalytically active through its phosphorylation at serine-729 (Cenni *et al* 2002). Indeed, using coimmunofluorescence, the spatial colocalization of filamin A and PKC-ε can be found after 30 min of cell spreading on collagen-coated planar substrates.

The effect of filamin A knockdown on PKC-mediated phosphorylation of vimentin was analyzed using phospho-specific antibodies on serine-6, serine-33, serine-38 and serine-50, all of which are known to be phosphorylated by PKC (Ando *et al* 1989, Ogawara *et al* 1995, Yasui *et al* 2001, Izawa and Inagaki 2006). Knockdown of filamin A eliminated vimentin phosphorylation at serines-6, -38 and -50 in spreading cells, whereas there was no detectable alteration of phosphorylation at serine-33. In more detail, filamin A-deficient cells revealed lower levels of the phosphorylated, catalytically active form of PKC-ε. These findings emphasize that filamin A is crucial for PKC-mediated phosphorylation of vimentin. However, phosphorylation of vimentin at serines-6, -38 and -50 can be restored after pretreatment of filamin-knockdown cells with the PKC activator, bryostatin. In addition, the impaired assembly of cell extensions and reduced cell surface $\beta 1$ integrin expression found in filamin-deficient cells is reversed after bryostatin treatment.

The major finding is that in spreading fibroblasts, filamin A can bind to vimentin and hence act as a scaffold for PKC-provided phosphorylation of vimentin. Moreover, these functions regulate the activation of $\beta 1$ integrins and their trafficking to the cell surface. The direct interaction between filamin and vimentin leads to a novel mechanism by which the actin and intermediate filament cytoskeletons coregulate cell adhesion and spreading. Although actin and intermediate filament cytoskeletons have distinct roles in regulating cell adhesion, spreading and motility (Ivaska *et al* 2007, Serrels *et al* 2007, Vicente-Manzanares *et al* 2009), the functions of these two cytoskeletal systems may be interdependent and coregulated (Green *et al* 1987, Chang and Goldman 2004, Kasas *et al* 2005, Arocena 2006). Proteins that bind both actin and intermediate filaments such as plectin (Seifert *et al* 1992, Svitkina *et al* 1996), BPAG1 (Yang *et al* 1996) and fimbrin (Correia *et al* 1999) can combine the functions of these different cytoskeletal systems. In summary, these findings provide a new mechanism in which the actin-binding protein filamin A serves as a linker protein between vimentin and actin filaments.

Filamin A colocalizes with vimentin (Brown and Binder 1992) and filamin A and vimentin can bind directly *in vitro*. In particular, the two proteins have been reported to colocalize at the extensions of spreading cells, which is indeed consistent with the colocalization of filamin A and vimentin in pseudopodia of cancer cells (Jia *et al* 2005). As there is direct interaction between these proteins and their roles in regulating cell motility (Eckes *et al* 1998, 36, Stossel *et al* 2001, McInroy and Maatta 2007), it seems to be likely that filamin A and vimentin also coregulate early events in cell spreading.

Filamin A and vimentin coregulate cell spreading

Cell spreading and migration require the extrusion of cell extensions such as filopodia and lamellipodia (Guillou *et al* 2008, Johnston *et al* 2008, Ladwein and Rottner 2008). Indeed, the formation of cell extensions can be evaluated by recording the numbers, and they can even be validated as authentic filopodia/lamellipodia through immunostaining for vinculin, paxillin and cortactin. The assembly of these actin-rich cell extensions is facilitated by $\beta 1$ integrins, small GTPases and actin-binding proteins (Zimerman *et al* 2004, Li *et al* 2005), such as filamin A which regulates $\beta 1$ integrin activity and small GTPases by binding to the actin filaments (Bellanger *et al* 2000, Stossel *et al* 2001, Vadlamudi *et al* 2002). Knockdown of filamin A and/or vimentin inhibits the extrusion of cell extensions, which is consistent with hindered migration in vimentin-deficient cells (Eckes *et al* 1998, McInroy and Maatta 2007). In addition, knockdown of filamin A and vimentin blocked cell spreading compared to control cells, in which either filamin A or vimentin was knocked down, suggesting that filamin A and vimentin coregulate early processes in cell spreading.

Actin-free barbed ends have been reported to decrease in number after knockdown of filamin A, whereas a knockdown of vimentin has no effect. Surprisingly, combined knockdown of filamin A and vimentin further decreases free barbed end formation, suggesting that vimentin may also support actin assembly in the absence of filamin A, partially superseding the role of filamin A. In particular, vimentin associates with PAK, which is a protein that facilitates actin-free barbed end formation via the LIM kinase–cofilin pathway (Chan *et al* 2002, Goto *et al* 2002). Indeed, mutation of vimentin's PAK-binding site eliminated PAK activation (Li *et al* 2005), indicating that vimentin is involved in PAK activation and actin-free barbed end formation. However, vimentin does not modulate free barbed end formation independently of filamin A and in addition knockdown of vimentin does not interfere with binding between filamin A and Rac or Cdc42, the small GTPases regulating filopodia and lamellipodia formation, respectively (Nobes and Hall 1995). These findings show that filamin A–vimentin interactions do not affect actin assembly directly, while they seem to regulate $\beta 1$ integrin function as an upstream target protein in the cell-spreading process.

The coregulation of $\beta 1$ integrin expression and activation by filamin A and vimentin

In the early stages of cell spreading and migration, cell surface $\beta 1$ integrins are endocytosed and still need to be transported back to the cell membrane (Skalski and Coppolino 2005, Dunphy *et al* 2006, Jovic *et al* 2007). As the location, availability and activation of cell surface integrins are crucial regulatory factors in cell adhesion and spreading (Guillou *et al* 2008), it has been supposed filamin A and vimentin regulate $\beta 1$ integrin activation in spreading. Moreover, $\beta 1$ integrin activation assays have shown that initial $\beta 1$ integrin–ligand binding is dependent on the presence of both filamin A and vimentin. In line with cell spreading, ligand-bound $\beta 1$ integrins decrease equivalently after knockdown of filamin A, vimentin, or both proteins, indicating that they coregulate $\beta 1$ integrin activation. In particular, focal adhesion proteins have revealed that filamin A and vimentin are both required for the targeting and localization of vinculin and paxillin to focal adhesions. However, other studies have shown the importance of vimentin for cell adhesion (Eckes *et al* 1998, Tsuruta and Jones 2003, Nieminen *et al* 2006), thus filamin A has been identified as a novel regulator of cell adhesion. Whereas some reports suggest that cell surface expression of $\beta 1$ integrins is regulated by filamin A (Meyer *et al* 1998) or vimentin (Nieminen *et al* 2006, Rizki *et al* 2007), other results show that filamin A and vimentin coregulate $\beta 1$ integrin trafficking, $\beta 1$ integrin activation, cell adhesion and spreading.

The role of vimentin phosphorylation by PKC in delivering $\beta 1$ integrin activity and cell spreading

PKC-ε-facilitated vimentin phosphorylation, $\beta 1$ integrin recycling and directed cell migration seem to be linked processes (Ivaska *et al* 2005), while the regulatory role of vimentin phosphorylation during the onset of cell spreading is not yet well understood. In particular, it has been reported that vimentin-knockdown (as well as inhibition of PKC) impairs the formation of cell extensions and reduces the $\beta 1$ integrin–ligand binding and expression on the cell surface. Moreover, cell extensions in the early stages of spreading have been found to be enriched with filamin A, vimentin and PKC-ε. Thus, filamin A, vimentin and phospho-vimentin seem to be involved in the same pathway for regulating cell spreading. As $\beta 1$ integrins associate with filamin A (Loo *et al* 1998, Kiema *et al* 2006) and with vimentin (Ivaska *et al* 2005, Kreis *et al* 2005), while filamin A additionally associates with PKC (Glogauer *et al* 1998, Tigges *et al* 2003), it has been proposed that filamin A regulates the activity of $\beta 1$ integrins and cell spreading through facilitating the phosphorylation of vimentin, which is performed by PKC.

The role of filamin A in PKC-facilitated vimentin phosphorylation

There is a close spatial relationship between filamin A and PKC, which is in line with the finding that filamin A is phosphorylated by PKC (Glogauer *et al* 1998, Tigges *et al* 2003). In more detail, the catalytically active, phosphorylated form of PKC-ε can be coimmunoprecipitated with filamin A, which is important since the PKC-ε isoform is essential for integrin recycling (Ivaska *et al* 2005). Moreover, knockdown of filamin A impairs vimentin phosphorylation at serines-6, -38 and -50, which are known PKC targets for phosphorylation (Ando *et al* 1989, Ogawara *et al* 1995, Izawa and Inagaki 2006). While these findings do not indicate a direct interaction between filamin A and PKC-ε, they strongly suggest a role for filamin A in promoting vimentin phosphorylation by PKC. Moreover, knockdown of filamin A prevents the translocation of vimentin to the cell periphery. The same result is observed after inhibition of PKC by the PKC inhibitor, BIM. These results support the model in which phosphorylation by PKC-ε is necessary for vimentin reorganization and for $\beta 1$ integrin trafficking to the cell membrane (Ivaska *et al* 2005).

In addition, filamin A regulates this crucial vimentin phosphorylation step. Taken together, filamin A, vimentin and PKC regulate $\beta 1$ integrin activity, cell adhesion and the exertion of early cell extensions caused by the binding of filamin A to vimentin.

8.10 The role of nuclear intermediate filaments in cell invasion

Mutations in nuclear lamins or other proteins of the nuclear envelope are the cause of a group of phenotypically diverse genetic disorders called laminopathies. These laminopathies have diverse symptoms, ranging from muscular dystrophy through neuropathy to premature aging syndromes. The precise disease mechanisms are not yet well understood. However, there has been substantial progress in our understanding of the biological nature of the nuclear structure and thus also of laminopathies. In particular, the dysfunction of the nuclear envelope is based on altered nuclear activity, impaired structural dynamics and signal transduction. As a result of these observations, small molecules are used as effective therapeutic substances.

Since the discovery of the constituents of the nuclear lamina in 1978 (Gerace *et al* 1978), nuclear lamins have been discussed in respect to and proposed for diverse roles in almost everything that takes place in the nucleus. These early studies were based on biochemistry and cell biology, with the goal of revealing the basic principles that regulate nuclear organization. The nuclear envelope became a medical research field in the mid-1990s, as mutations in emerin were found in patients with Emery–Dreifuss muscular dystrophy (EDMD) (Bione *et al* 1994). The LMNA gene that encodes all A-type nuclear lamins has also been linked to EDMD (Bonne *et al* 1999). Therefore, extensive connections between nuclear structure and human diseases have been identified. With approximately 15 diseases, including dystrophic and progeroid syndromes, being associated with LMNA mutations and mutations in the genes encoding associated nuclear envelope proteins, the following questions have been raised. Why do alterations in nuclear envelope proteins lead directly to a disease? What are the precise mechanisms underlying the pathology of the disease? Do A-type lamins play a role in aging? Although many discoveries have been made in this field, many aspects are still not well known.

Nuclear lamins

The nuclear envelope consists of two membranes: (i) the outer nuclear membrane, which is continuously connected to the endoplasmic reticulum, and (ii) the inner nuclear membrane, which is associated with the nuclear lamina. The nuclear pore complexes are holes in the nuclear envelope that facilitate transport between the cytoplasm and nucleus. The nuclear lamina consists of nuclear lamins, which have been identified as lamins A, B and C (Gerace *et al* 1978). These proteins belong to the only class of intermediate filament proteins in the nucleus and build the associated filamentous structures that underlie the nuclear envelope and interact with neighboring proteins (Gerace and Huber 2012). Lamins A and C are classed as

A-type lamins and are encoded by the LMNA gene through alternative splicing. In addition, three different lamin B family members (B-type lamins) are encoded by only two genes (lamin B1 by LMNB1 and lamins B2 and B3 by LMNB2).

In more detail, A- and B-type lamins have totally different properties, possibly due to their different isoelectric points, which enable B-type lamins to be associated with the nuclear envelope during mitosis, whereas A-type lamins are soluble. Moreover, the expression patterns are different; B-type lamins are expressed in most or all cell types and A-type lamins are expressed only during cell differentiation in many developmental lineages (Roeber *et al* 1989). At the cellular level, it has been proposed that both A-type and B-type lamins have structural roles in the nucleus, and also roles in a range of other activities, such as coordination of transcription and replication. While the specific functions of A-type lamins are still elusive, a number of discoveries point to key interactions between lamins and cell proliferation, differentiation and stress response pathways. In more detail, both A- and B-type lamins undergo posttranslational processing based on a C-terminal CaaX motif that leads to a series of modifications (Weber *et al* 1989). Only lamin C avoids this, as it lacks the C terminus due to alternative splicing of the LMNA transcript. As a first step, the cysteine residue is farnesylated. Next, proteolytic processing evokes the cleavage after the cysteine residue, followed by a carboxymethylation of the new C-terminal residue. Many membrane-associated proteins such as Ras undergo this processing event. In more detail, in the case of lamin A, isoprenylation is a transient event, as a second proteolytic event facilitated by the zinc metalloproteinase Zmpste24 leads to excision of another 15 amino acids. Thus, mature lamin A lacks the modified cysteine. This process is clearly important for pathologic states, because laminopathies are linked to altered processing of lamin A and loss-of-function mutations in ZMPSTE24.

The purpose of farnesylation of lamin A is still unclear. It has been thought that the transient farnesylation event was needed, through association of the hydrophobic farnesyl group with the nuclear envelope, to provide initial recruitment of lamin A to the nuclear periphery (Hennekes and Nigg 1994). However, after assembly into filaments, farnesylation of lamin A may no longer be essential. Consistent with this, the nucleus has been reported to be the site of both lamin A carboxymethylation and proteolytic cleavage by ZMPSTE24. However, several studies using mice and/or cells have been engineered to express mutant forms of lamin A, indicating that farnesylation is not required for recruitment (Davies *et al* 2010). When only a nonfarnesylated version of lamin A is expressed, normal localization of the lamin A variant to the nuclear periphery has been detected (Davies *et al* 2010, Lee *et al* 2010), although mice generated in this way will develop cardiomyopathy. In addition, mice expressing only lamin C (not farnesylated) or a mature (preprocessed) lamin A are unexpectedly normal and have the correct localization of the respective protein to the nuclear periphery. Although these studies do not preclude a more subtile role for lamin A processing in filament assembly or envelope association, they raise questions about the importance of these events in the mouse.

Since the identification of diseases caused by mutations of genes encoding nuclear lamina proteins, studies have focused on understanding the molecular mechanisms causing these specific phenotypes. In particular, it seems to be crucial to understand how the nuclear lamina interacts with structural proteins such as chromatin, transcription factors and other signaling partners, which will reveal the mechanistic links to disease. The overall picture of how lamins regulate all of these pathways and how this regulation leads to disease is becoming clearer, but it is not yet sharp. Knowing the mechanisms by which mutations in lamins cause these rare diseases will help to reveal common conditions and help to provide a good laminopathies model for muscle diseases and cardiomyopathy. As mutations in the nuclear lamina lead to rapid aging-like diseases, defining the precise role of the nuclear lamina in controlling human aging will be of great interest.

8.11 The role of cell division in cellular motility

As discussed above, there is an interaction between intermediate filaments and actin filaments and they work together to provide cellular functions such as cell motility. An interaction between the three major components of the cytoskeleton was proposed a long time ago, and it is becoming likely that there is such an interaction between intermediate filaments, actin filaments and microtubules. The latter cytoskeletal components play an essential role in effecting cell division by assembling the mitotic spindle. However, microtubules can interfere with the actomyosin cytoskeleton within the cells in order to regulate cell motility.

References and further reading

Abd El-Rehim D M, Pinder S E, Paish C E, Bell J, Blamey R W, Robertson J F, Nicholson R I and Ellis I O 2004 Expression of luminal and basal cytokeratins in human breast carcinoma *J. Pathol.* **203** 661–71

Akhmanova A and Steinmetz M O 2008 Tracking the ends: a dynamic protein network controls the fate of microtubule tips *Nat. Rev. Mol. Cell Biol.* **9** 309–22

Akita Y, Kawasaki H, Imajoh-Ohmi S, Fukuda H, Ohno S, Hirano H, Ono Y and Yonekawa H 2007 Protein kinase C epsilon phosphorylates keratin 8 at Ser8 and Ser23 in GH4C1 cells stimulated by thyrotropin-releasing hormone *FEBS J.* **274** 3270–85

Alam H *et al* 2011 Loss of keratin 8 phosphorylation leads to increased tumor progression and correlates with clinico-pathological parameters of OSCC patients *PLoS One* **6** e27767

Alam H, Kundu S T, Dalal S N and Vaidya M M 2011 Loss of keratins 8 and 18 leads to alterations in $\alpha6\beta4$-integrin-mediated signalling and decreased neoplastic progression in an oral-tumour-derived cell line *J. Cell Sci.* **124** 2096–106

Albrecht-Buehler G and Lancaster R M 1976 A quantitative description of the extension and retraction of surface protrusions in spreading 3T3 mouse fibroblasts *J. Cell Biol.* **71** 370–82

Amano M, Nakayama M and Kaibuchi K 2010 Rho-kinase/ROCK: A key regulator of the cytoskeleton and cell polarity *Cytoskeleton* **67** 545–54

Ando S, Tanabe K, Gonda Y, Sato C and Inagaki M 1989 Domain- and sequence-specific phosphorylation of vimentin induces disassembly of the filament structure *Biochemistry* **28** 2974–9

Arce C A, Casale C H and Barra H S 2008 Submembraneous microtubule cytoskeleton: regulation of ATPases by interaction with acetylated tubulin *FEBS J.* **275** 4664–74

Arocena M 2006 Effect of acrylamide on the cytoskeleton and apoptosis of bovine lens epithelial cells *Cell Biol. Int.* **30** 1007–12

Arthur W T and Burridge K 2001 RhoA inactivation by p190RhoGAP regulates cell spreading and migration by promoting membrane protrusion and polarity *Mol. Biol. Cell* **12** 2711–20

Artym V V, Zhang Y, Seillier-Moiseiwitsch F, Yamada K M and Mueller S C 2006 Dynamic interactions of cortactin and membrane type 1 matrix metalloproteinase at invadopodia: defining the stages of invadopodia formation and function *Cancer Res.* **66** 3034–43

Asparuhova M B, Gelman L and Chiquet M 2009 Role of the actin cytoskeleton in tuning cellular responses to external mechanical stress *Scand. J. Med. Sci. Sports* **19** 490–9

Ausmees N, Kuhn J R and Jacobs-Wagner C 2003 The bacterial cytoskeleton: an intermediate filament-like function in cell shape *Cell* **115** 705–13

Bachant J B and Klymkowsky M W 1996 A nontetrameric species is the major soluble form of keratin in *Xenopus* oocytes and rabbit reticulocyte lysates *J. Cell Biol.* **132** 153–65

Baribault H, Blouin R, Bourgon L and Marceau N 1989 Epidermal growth factor-induced selective phosphorylation of cultured rat hepatocyte 55-kD cytokeratin before filament reorganization and DNA synthesis *J. Cell Biol.* **109** 1665–76

Barry D M, Millecamps S, Julien J P and Garcia M L 2007 New movements in neurofilament transport, turnover and disease *Exp. Cell Res.* **313** 2110–20

Bass-Zubek A E, Godsel L M, Delmar M and Green K J 2009 Plakophilins: multifunctional scaffolds for adhesion and signaling *Curr. Opin. Cell Biol.* **21** 708–16

Batsios P, Peter T, Baumann O, Stick R, Meyer I and Graf R 2012 A lamin in lower eukaryotes? *Nucleus* **3** 237–43

Bausch A R, Moller W and Sackmann E 1999 Measurement of local viscoelasticity and forces in living cells by magnetic tweezers *Biophys. J.* **76** 573–9

Beaudouin J, Gerlich D, Daigle N, Eils R and Ellenberg J 2002 Nuclear envelope breakdown proceeds by microtubule-induced tearing of the lamina *Cell* **108** 83–96

Beil M, Micoulet A and von Wichert G *et al* 2003 Sphingosylphosphorylcholine regulates keratin network architecture and visco-elastic properties of human cancer cells *Nat. Cell Biol.* **5** 803–11

Bellanger J M, Astier C, Sardet C, Ohta Y, Stossel T P and Debant A 2000 The Rac1- and RhoG-specific GEF domain of Trio targets filamin to remodel cytoskeletal actin *Nat. Cell Biol.* **2** 888–92

Berkowitz P, Hu P, Liu Z, Diaz L A, Enghild J J, Chua M P and Rubenstein D S 2005 Desmosome signaling. Inhibition of p38MAPK prevents pemphigus vulgaris IgG-induced cytoskeleton reorganization *J. Biol. Chem.* **280** 23778–84

Bertaud J, Qin Z and Buehler M J 2010 Intermediate filament-deficient cells are mechanically softer at large deformation: a multi-scale simulation study *Acta Biomater.* **6** 2457–66

Bezault J, Bhimani R, Wiprovnick J and Furmanski P 1994 Human lactoferrin inhibits growth of solid tumors and development of experimental metastases in mice *Cancer Res.* **54** 2310–2

Bhattacharya R, Gonzalez A M, Debiase P J, Trejo H E, Goldman R D, Flitney F W and Jones J C 2009 Recruitment of vimentin to the cell surface by $\beta 3$ integrin and plectin mediates adhesion strength *J. Cell Sci.* **122** 1390–400

Bhuin T and Roy J K 2014 Rab proteins: The key regulators of intracellular vesicle transport *Exp. Cell Res.* **328** 1–19

Bione S, Maestrini E, Rivella S, Mancini M, Regis S, Romeo G and Toniolo D 1994 Identification of a novel X-linked gene responsible for Emery–Dreifuss muscular dystrophy *Nat. Genet.* **8** 323–7

Blessing M, Ruether U and Franke W W 1993 Ectopic synthesis of epidermal cytokeratins in pancreatic islet cells of transgenic mice interferes with cytoskeletal order and insulin production *J. Cell Biol.* **120** 743–55

Bloom S, Lockard V G and Bloom M 1996 Intermediate filament-mediated stretch-induced changes in chromatin: a hypothesis for growth initiation in cardiac myocytes *J. Mol. Cell. Cardiol.* **28** 2123–7

Boczonadi V, McInroy L and Määttä A 2007 Cytolinker cross-talk: periplakin N-terminus interacts with plectin to regulate keratin organisation and epithelial migration *Exp. Cell Res.* **313** 3579–91

Bonne G *et al* 1999 Mutations in the gene encoding lamin A/C cause autosomal dominant Emery--Dreifuss muscular dystrophy *Nat. Genet.* **21** 285–8

Bordeleau F, Bessard J, Marceau N and Sheng Y 2011 Measuring integrated cellular mechanical stress response at focal adhesions by optical tweezers *J. Biomed. Opt.* **16** 095005

Bordeleau F, Bessard J, Sheng Y and Marceau N 2008 Keratin contribution to cellular mechanical stress response at focal adhesions as assayed by laser tweezers *Biochem. Cell Biol.* **86** 352–9

Bordeleau F, Galarneau L, Gilbert S, Loranger A and Marceau N 2010 Keratin 8/18 modulation of protein kinase C-mediated integrin-dependent adhesion and migration of liver epithelial cells *Mol. Biol Cell.* **21** 1698–713

Bordeleau F, Myrand Lapierre M E, Sheng Y and Marceau N 2012 Keratin 8/18 regulation of cell stiffness-extracellular matrix interplay through modulation of Rho-mediated actin cytoskeleton dynamics *PLoS One* **7** e38780

Bordeleau F, Lapierre M E, Sheng Y and Marceau N 2012 Keratin 8/18 regulation of cell stiffness-extracellular matrix interplay through modulation of rho-mediated actin cytoskeleton dynamics *PLoS One* **7** e38780

Bornslaeger E A, Corcoran C M, Stappenbeck T S and Green K J 1996 Breaking the connection: displacement of the desmosomal plaque protein desmoplakin from cell–cell interfaces disrupts anchorage of intermediate filament bundles and alters intercellular junction assembly *J. Cell Biol.* **134** 985–1001

Bousquet O, Ma L L, Yamada S, Gu C H, Idei T, Takahashi K and Coulombe P A 2001 The nonhelical tail domain of keratin 14 promotes filament bundling and enhances the mechanical properties of keratin intermediate filaments *in vitro J. Cell Biol.* **155** 747–53

Bragulla H H and Homberger D G 2009 Structure and functions of keratin proteins in simple, stratified, keratinized and cornified epithelia *J. Anat.* **214** 516–59

Brooks D A 2009 The endosomal network *Int. J. Clin. Pharmacol. Ther.* **47** S9–17

Brotschi E A, Hartwig J H and Stossel T P 1978 The gelation of actin by actin-binding protein *J. Biol Chem.* **253** 8988–93

Brown A 2003 Axonal transport of membranous and nonmembranous cargoes: a unified perspective *J. Cell Biol.* **160** 817–21

Brown K D and Binder L I 1992 Identification of the intermediate filament-associated protein gyronemin as filamin. Implications for a novel mechanism of cytoskeletal interaction *J. Cell Sci.* **102** 19–30

Brown M J, Hallam J A, Colucci-Guyon E and Shaw S 2001 Rigidity of circulating lymphocytes is primarily conferred by vimentin intermediate filaments *J. Immunol.* **166** 6640–6

Bucci C, Parton R G, Mather I H, Stunnenberg H, Simons K, Hoflack B and Zerial M 1992 The small GTPase Rab5 functions as a regulatory factor in the early endocytic pathway *Cell* **70** 715–28

Bulinski J C and Gundersen G G 1991 Stabilization of post-translational modification of microtubules during cellular morphogenesis *Bioessays* **13** 285–93

Burgstaller G, Gregor M, Winter L and Wiche G 2010 Keeping the vimentin network under control: cell–matrix adhesion-associated plectin 1f affects cell shape and polarity of fibroblasts *Mol. Biol. Cell.* **21** 3362–75

Burke B and Stewart C L 2013 The nuclear lamins: flexibility in function *Nat. Rev. Mol. Cell Biol.* **14** 13–24

Busch T, Armacki M, Eiseler T, Joodi G, Temme C, Jansen J, von Wichert G, Omary M B, Spatz J and Seufferlein T 2012 Keratin 8 phosphorylation regulates keratin reorganization and migration of epithelial tumor cells *J. Cell Sci.* **125** 2148–59

Butcher D T, Alliston T and Weaver V M 2009 A tense situation: forcing tumor progression *Nat. Rev. Cancer* **9** 108–22

Campbell I D 2008 Studies of focal adhesion assembly. *Biochem Soc Trans.* **36** 263–6

Campellone K G and Welch M D 2010 A nucleator arms race: cellular control of actin assembly *Nat. Rev. Mol. Cell Biol.* **11** 237–51

Carragher N O and Frame M C 2004 Focal adhesion and actin dynamics: a place where kinases and proteases meet to promote invasion *Trends Cell Biol.* **14** 241–9

Cary R B, Klymkowsky M W, Evans R M, Domingo A, Dent J A and Backhus L E 1994 Vimentin's tail interacts with actin-containing structures *in vivo J. Cell Sci.* **107** 1609–22

Casanova J E, Wang X, Kumar R, Bhartur S G, Navarre J, Woodrum J E, Altschuler Y, Ray G S and Goldenring J R 1999 Association of Rab25 and Rab11a with the apical recycling system of polarized Madin–Darby canine kidney cells *Mol. Biol. Cell* **10** 47–61

Caulin C, Ware C F, Magin T M and Oshima R G 2000 Keratin-dependent, epithelial resistance to tumor necrosis factor-induced apoptosis *J. Cell Biol.* **149** 17–22

Cenni V, Doppler H, Sonnenburg E D, Maraldi N, Newton A C and Toker A 2002 Regulation of novel protein kinase C epsilon by phosphorylation *Biochem. J.* **363** 537–45

Chamcheu J C, Virtanen M, Navsaria H, Bowden P E, Vahlquist A and Törmä H 2010 Epidermolysis bullosa simplex due to KRT5 mutations: mutation-related differences in cellular fragility and the protective effects of trimethylamine N-oxide in cultured primary keratinocytes *Br. J. Dermatol.* **162** 980–9

Chan M W, Arora P D and McCulloch C A 2007 Cyclosporin inhibition of collagen remodelling is mediated by gelsolin *Am. J. Physiol. Cell Physiol.* **293** C1049–58

Chan W K, Yabe J T, Pimenta A F, Ortiz D and Shea T B 2003 Growth cones contain a dynamic population of neurofilament subunits *Cell Motil. Cytoskelet.* **54** 195–207

Chan W, Kozma R, Yasui Y, Inagaki M, Leung T, Manser E and Lim L 2002 Vimentin intermediate filament reorganization by Cdc42: involvement of PAK and p70 S6 kinase *Eur. J. Cell Biol.* **81** 692–701

Chandler J S, Calnek D and Quaroni A 1991 Identification and characterization of rat intestinal keratins. Molecular cloning of cDNAs encoding cytokeratins 8, 19 and a new 49 kDa type I cytokeratin (cytokeratin 21) expressed by differentiated intestinal epithelial cells *J. Biol. Chem.* **266** 11932–8

Chang L and Goldman R D 2004 Intermediate filaments mediate cytoskeletal crosstalk *Nat. Rev. Mol. Cell Biol.* **5** 601–13

Chang L, Shav-Tal Y, Trcek T, Singer R H and Goldman R D 2006 Assembling an intermediate filament network by dynamic cotranslation *J. Cell Biol.* **172** 747–58

Charrier E E and Janmey J A 2016 Mechanical properties of intermediate filament proteins *Methods Enzymol* **568** 35–57

Charrier E E, Montel L, Asnacios A, Delort F, Vicart P, Gallet F, Batonnet-Pichon S and Hénon S 2018 The desmin network is a determinant of the cytoplasmic stiffness of myoblasts *Biol. Cell.* **110** 77–90

Chaturvedi P and Parnaik V K 2010 Lamin A rod domain mutants target heterochromatin protein 1α and β for proteasomal degradation by activation of F-box protein, fbxw10 *PLoS One* **5** e10620

Chen S C, Kennedy B K and Lampe P D 2013 Phosphorylation of connexin43 on S279/282 may contribute to laminopathy-associated conduction defects *Exp. Cell Res* **319** 888–96

Cheng C W, Wang H W, Chang C W, Chu H W, Chen C Y, Yu J C, Chao J I, Liu H F, Ding S L and Shen C Y 2012 MicroRNA-30a inhibits cell migration and invasion by down-regulating vimentin expression and is a potential prognostic marker in breast cancer *Breast Cancer Res. Treat* **134** 1081–93

Chhabra E S and Higgs H N 2007 The many faces of actin: matching assembly factors with cellular structures *Nat. Cell Biol.* **9** 1110–21

Chiu F C, Norton W T and Fields K L 1981 The cytoskeleton of primary astrocytes in culture contains actin, glial fibrillary acidic protein, and the fibroblast-type filament protein, vimentin *J. Neurochem.* **37** 147–55

Chong S A, Lee W, Arora P D, Laschinger C, Young E W, Simmons C A, Manolson M, Sodek J and McCulloch C A 2007 Methylglyoxal inhibits the binding step of collagen phagocytosis *J. Biol. Chem.* **282** 8510–20

Chou C F, Riopel C L, Rott L S and Omary M B 1993 A significant soluble keratin fraction in 'simple' epithelial cells. Lack of an apparent phosphorylation and glycosylation role in keratin solubility *J. Cell Sci.* **105** 433–44

Chou Y H, Flitney F W, Chang L, Mendez M, Grin B and Goldman R D 2007 The motility and dynamic properties of intermediate filaments and their constituent proteins *Exp. Cell Res.* **313** 2236–43

Chu Y W, Runyan R B, Oshima R G and Hendrix M J 1993 Expression of complete keratin filaments in mouse L cells augments cell migration and invasion *Proc. Natl Acad. Sci. USA* **90** 4261–5

Chu Y W, Seftor E A, Romer L H and Hendrix M J 1996 Experimental coexpression of vimentin and keratin intermediate filaments in human melanoma cells augments motility *Am. J. Pathol.* **148** 63–9

Chung B M, Murray C I, Van Eyk J E and Coulombe P A 2012 Identification of novel interaction between annexin A2 and keratin 17: evidence for reciprocal regulation *J. Biol. Chem* **287** 7573–81

Chung B-M, Rotty J D and Coulombe P A 2013 Networking galore: Intermediate filaments and cell migration *Curr. Opin. Cell Biol.* **25** 600–12

Chung B-M, Rotty J D and Coulombe P A 2014 Networking galore: intermediate filaments and cell migration *Curr. Opin. Cell Biol.* **25** 600–12

Cochard P and Paulin D 1984 Initial expression of neurofilaments and vimentin in the central and peripheral nervous system of the mouse embryo *in vivo J. Neurosci.* **4** 2080–94

Coffinier C, Jung H J and Nobumori C *et al* 2011 Deficiencies in lamin B1 and lamin B2 cause neurodevelopmental defects and distinct nuclear shape abnormalities in neurons *Mol. Biol. Cell* **22** 4683–93

Colakoğlu G and Brown A 2009 Intermediate filaments exchange subunits along their length and elongate by end-to-end annealing *J. Cell Biol.* **185** 769–77

Conover G M and Gregorio C C 2011 The desmin coil 1b mutation K190A impairs nebulin Z-disc assembly and destabilizes actin thin filaments *J. Cell Sci.* **124** 3464–76

Correia I, Chu D, Chou Y-H, Goldman R D and Matsudaira P 1999 Integrating the actin and vimentin cytoskeletons: adhesion-dependent formation of fimbrin–vimentin complexes in macrophages *J. Cell Biol.* **146** 831–42

Conway D E, Breckenridge M T, Hinde E, Gratton E, Chen C S and Schwartz M A 2013 Fluid shear stress on endothelial cells modulates mechanical tension across VE-cadherin and PECAM-1 *Curr. Biol.* **23** 1024–30

Coulombe P A and Wong P 2004 Cytoplasmic intermediate filaments revealed as dynamic and multipurpose scaffolds *Nat. Cell Biol.* **6** 699–706

Coulombe P A, Kerns M L and Fuchs E 2009 Epidermolysis bullosa simplex: a paradigm for disorders of tissue fragility *J. Clin. Invest.* **119** 1784–93

Critchley D R 2000 Focal adhesions—the cytoskeletal connection *Curr. Opin. Cell Biol.* **12** 133–9

Davies B S J *et al* 2010 An accumulation of non-farnesylated prelamin A causes cardiomyopathy but not progeria *Hum. Mol. Genet.* **19** 2682–94

de Silva Rudland S, Platt-Higgins A, Winstanley J H, Jones N J, Barraclough R, West C, Carroll J and Rudland P S 2011 Statistical association of basal cell keratins with metastasis-inducing proteins in a prognostically unfavorable group of sporadic breast cancers *Am. J. Pathol.* **179** 1061–72

Defilippi P, Olivo C, Venturino M, Dolce L, Silengo L and Tarone G 1999 Actin cytoskeleton organization in response to integrin-mediated adhesion Micro Res *Technol* **47** 67–78

Dellagi K, Tabilio A, Portier M M, Vainchenker W, Castaigne S, Guichard J, Breton-Gorius J and Brouet J C 1985 Expression of vimentin intermediate filament cytoskeleton in acute nonlymphoblastic leukemias *Blood* **65** 1444–52

Deng M, Zhang W and Tang H *et al* 2013 Lactotransferrin acts as a tumor suppressor in nasopharyngeal carcinoma by repressing AKT through multiple mechanisms *Oncogene* **32** 4273–83

Depianto D, Kerns M L, Dlugosz A A and Coulombe P A 2010 Keratin 17 promotes epithelial proliferation and tumor growth by polarizing the immune response in skin *Nat. Genet.* **42** 910–4

Desai B V, Harmon R M and Green K J 2009 Desmosomes at a glance *J Cell Sci* **122** 4401–7

Díaz-Moreno I, Hollingworth D, Frenkiel T A, Kelly G, Martin S, Howell S, García-Mayoral M, Gherzi R, Briata P and Ramos A 2009 Phosphorylation-mediated unfolding of a KH domain regulates KSRP localization via 14-3-3 binding *Nat. Struct. Mol. Biol.* **16** 238–46

Ducker W A, Senden T J and Pashley R M 1991 Direct measurement of colloidal forces using an atomic force microscope *Nature* **353** 239–41

Dunphy J L, Moravec R, Ly K, Lasell T K, Melancon P and Casanova J E 2006 The Arf6 GEF GEP100/BRAG2 regulates cell adhesion by controlling endocytosis of beta1 integrins *Curr. Biol.* **16** 315–20

Dupin I, Sakamoto Y and Etienne-Manneville S 2011 Cytoplasmic intermediate filaments mediate actin-driven positioning of the nucleus *J. Cell Sci.* **124** 865–72

Eckert B 1985 Alteration of intermediate filament distribution in PtK1 cells by acrylamide *Eur. J. Cell Biol.* **37** 169–74

Eckes B, Dogic D, Colucci-Guyon E, Wang N, Maniotis A, Ingber D and Krieg T 1998 Impaired mechanical stability, migration and contractile capacity in vimentin-deficient fibroblasts *J. Cell Sci.* **111** 1897–907

Eger A, Stockinger A, Wiche G and Foisner R 1997 Polarization-dependent association of plectin with desmoplakin and the lateral submembrane skeleton in MDCK cells *J. Cell Sci.* **110** 1307–16

Eichner R, Sun T T and Aebi U 1986 The role of keratin subfamilies and keratin pairs in the formation of human epidermal intermediate filaments *J. Cell Biol.* **102** 1767–77

Ellis D J, Jenkins H, Whitfield W G and Hutchinson C J 1997 GST-lamin fusion proteins act as dominant negative mutants in *Xenopus* egg extract and reveal the function of the lamina in DNA replication *J. Cell Sci.* **110** 2507–18

Erber A, Riemer D, Bovenschulte M and Weber K 1998 Molecular phylogeny of metazoan intermediate filament proteins *J. Mol Evol.* **47** 751–62

Eriksson J E, Dechat T, Grin B, Helfand B, Mendez M, Pallari H M and Goldman R D 2009 Introducing intermediate filaments: from discovery to disease *J. Clin. Invest.* **119** 1763–71

Eriksson J E, He T, Trejo-Skalli A V, Harmala-Brasken A S, Hellman J, Chou Y H and Goldman R D 2004 Specific *in vivo* phosphorylation sites determine the assembly dynamics of vimentin intermediate filaments *J. Cell Sci.* **117** 919–32

Eriksson J E, Brautigan D L, Vallee R, Olmsted J, Fujiki H and Goldman R D 1992 Cytoskeletal integrity in interphase cells requires protein phosphatase activity *Proc. Natl Acad. Sci. USA* **89** 11093–7

Eriksson J E, Toivola D M, Sahlgren C, Mikhailov A and Harmala-Brasken A S 1998 Strategies to assess phosphoprotein phosphatase and protein kinase-mediated regulation of the cytoskeleton *Methods Enzymol* **298** 542–69

Errante L, Tang D, Gardon M, Sekerkova G, Mugnaini E and Shaw G 1998 The intermediate filament protein peripherin is a marker for cerebellar climbing fibres *J. Neurocytol.* **27** 69–84

Esue O, Carson A A, Tseng Y and Wirtz D 2006 A direct interaction between actin and vimentin filaments mediated by the tail domain of vimentin *J. Biol. Chem.* **281** 30393–9

Even C, Abramovici G, Delort F, Rigato A F, Bailleux V, de Sousa Moreira A, Vicart P, Rico F, Batonnet-Pichon S and Briki F 2017 Mutation in the core structure of desmin intermediate filaments affects myoblast elasticity *Biophys J.* **113**(3) 627–36

Faure E, Garrouste F, Parat F, Monferran S, Leloup L, Pommier G, Kovacic H and Lehmann M 2012 P2Y2 receptor inhibits EGF-induced MAPK pathway to stabilize keratinocyte hemidesmosomes *J. Cell Sci.* **125** 4264–77

Felkl M, Tomas K, Smid M, Mattes J, Windoffer R and Leube R E 2012 Monitoring the cytoskeletal EGF response in live gastric carcinoma cells *PLoS One* **7** e45280

Feng Y and Walsh C A 2004 The many faces of filamin: a versatile molecular scaffold for cell motility and signalling *Nat. Cell Biol.* **6** 1034–1038

Flitney E W, Kuczmarski E R, Adam S A and Goldman R D 2009 Insights into the mechanical properties of epithelial cells: the effects of shear stress on the assembly and remodeling of keratin intermediate filaments *FASEB J.* **23** 2110–9

Florenes V A, Holm R, Myklebost O, Lendahl U and Fodstad O 1994 Expression of the neuroectodermal intermediate filament nestin in human melanomas *Cancer Res.* **54** 354–6

Fortier A M, Asselin E and Cadrin M 2013 Keratin 8 and 18 loss in epithelial cancer cells increases collective cell migration and cisplatin sensitivity through Claudin1 up-regulation *J. Biol. Chem.* **288** 11555–71

Fortier A M, Riopel K, Désaulniers M and Cadrin M 2010 Novel insights into changes in biochemical properties of keratins 8 and 18 in griseofulvin-induced toxic liver injury *Exp. Mol. Pathol.* **89** 117–25

Fortier A M, Asselin E and Cadrin M 2013 Keratin 8 and 18 loss in epithelial cancer cells increases collective cell migration and cisplatin sensitivity through claudin1 up-regulation *J. Biol. Chem.* **288** 11555–71

Fortin S, Le Mercier M, Camby I, Spiegl-Kreinecker S, Berger W, Lefranc F and Kiss R 2010 Galectin-1 is implicated in the protein kinase C epsilon/vimentin-controlled trafficking of integrin-β1 in glioblastoma cells *Brain Pathol* **20** 39–49

Frame M and Norman J 2008 A tal(in) of cell spreading *Nat. Cell Biol.* **10** 1017–9

Franke W W, Schiller D L, Hatzfeld M and Winter S 1983 Protein complexes of intermediate-sized filaments: melting of cytokeratin complexes in urea reveals different polypeptide separation characteristics *Proc. Natl Acad. Sci. USA* **80** 7113–7

Franke W W, Schmid E, Osborn M and Weber K 1978 Different intermediate-sized filaments distinguished by immunofluorescence microscopy *Proc. Natl Acad. Sci. USA* **75** 5034–8

Franke W W, Schmid E, Osborn M and Weber K 1979 Intermediate-sized filaments of human endothelial cells *J. Cell Biol.* **81** 570–80

Friedl P and Wolf K 2010 Plasticity of cell migration: a multiscale tuning model *J. Cell Biol.* **188** 11–9

Friedl P, Wolf K and Lammerding J 2011 Nuclear mechanics during cell migration *Curr. Opin. Cell Biol.* **23** 55–64

Frijns E, Sachs N, Kreft M, Wilhelmsen K and Sonnenberg A 2010 EGF-induced MAPK signaling inhibits hemidesmosome formation through phosphorylation of the integrin *J. Biol. Chem.* **285** 37650–62

Fuchs E and Weber K 1994 Intermediate filaments: structure, dynamics, function, and disease *Annu. Rev. Biochem.* **63** 345–82

Fujiwara S, Ohashi K, Mashiko T, Kondo H and Mizuno K 2016 Interplay between Solo and keratin filaments is crucial for mechanical force-induced stress fiber reinforcement *Mol. Biol. Cell* **58** 7250–7

Gaiottino J, Norgren N and Dobson R *et al* 2013 Increased neurofilament light chain blood levels in neurodegenerative neurological diseases *PLoS One* **8** e75091

Galarneau L, Loranger A, Gilbert S and Marceau N 2007 Keratins modulate hepatic cell adhesion, size and G1/S transition *Exp. Cell Res.* **313** 179–94

Gallicano G I, Kouklis P, Bauer C, Yin M, Vasioukhin V, Degenstein L and Fuchs E 1998 Desmoplakin is required early in development for assembly of desmosomes and cytoskeletal linkage *J. Cell Biol.* **143** 2009–22

Gardner M K, Hunt A J, Goodson H V and Odde D J 2008 Microtubule assembly dynamics: new insights at the nanoscale *Curr. Opin. Cell Biol.* **20** 64–70

Geiger B, Bershadsky A, Pankov R and Yamada K M 2001 Transmembrane crosstalk between the extracellular matrix-cytoskeleton crosstalk *Nat. Rev. Mol. Cell Biol.* **2** 793–805

Geisler N and Weber K 1981 Comparison of the proteins of two immunologically distinct intermediate-sized filaments by amino acid sequence analysis: desmin and vimentin *Proc. Natl Acad. Sci. USA* **78** 4120–3

Geisler N and Weber K 1982 The amino acid sequence of chicken muscle desmin provides a common structural model for intermediate filaments proteins *EMBO J.* **1** 1649–56

Gerace L, Blum A and Blobel G 1978 Immunocytochemical localization of the major polypeptides of the nuclear pore complex-lamina fraction interphase and mitotic distribution *J. Cell Biol.* **79** 546–66

Gerace L and Huber M D 2012 Nuclear lamina at the crossroads of the cytoplasm and nucleus *J. Struct. Biol.* **177** 24–31

Ghosh S, Liu B, Wang Y, Hao Q and Zhou Z 2015 Lamin a is an endogenous SIRT6 activator and promotes SIRT6-mediated DNA repair *Cell Rep.* **13** 1396–406

Giancotti F G and Ruoslahti E 1999 Integrin signaling *Science* **285** 1028–32

Gibbs-Seymour I, Markiewicz E, Bekker-Jensen S, Mailand N and Hutchison C J 2015 Lamin A/C-dependent interaction with 53BP1 promotes cellular responses to DNA damage *Aging Cell* **14** 162–9

Gilles C, Polette M, Zahm J M, Tournier J M, Volders L, Foidart J M and Birembaut P 1999 Vimentin contributes to human mammary epithelial cell migration *J. Cell Sci.* **112** 4615–25

Gladilin E, Gonzalez P and Eils R 2014 Dissecting the contribution of actin and vimentin intermediate filaments to mechanical phenotype of suspended cells using high-throughput deformability measurements and computational modeling *J. Biomech.* **47** 2598–605

Glogauer M, Arora P, Chou D, Janmey P A, Downey G P and McCulloch C A 1998 The role of actin-binding protein 280 in integrin-dependent mechanoprotection *J. Biol. Chem.* **273** 1689–98

Glogauer M, Arora P, Yao G, Sokholov I, Ferrier J and McCulloch C A 1997 Calcium ions and tyrosine phosphorylation interact coordinately with actin to regulate cytoprotective responses to stretching *J. Cell Sci.* **110** 11–21

Goldman R D and Follett E A 1970 Birefringent filamentous organelle in BHK-21 cells and its possible role in cell spreading and motility *Science* **169** 286–8

Goldman R D and Follett E A 1969 The structure of the major cell processes of isolated BHK21 fibroblasts *Exp. Cell. Res.* **57** 263–76

Gonzalez A M, Otey C, Edlund M and Jones J C 2001 Interactions of a hemidesmosome component and actinin family members *J. Cell Sci.* **114** 4197–206

Gonzalez-Suarez I, Redwood A B and Perkins S M *et al* 2009 Novel roles for A-type lamins in telomere biology and the DNA damage response pathway *EMBO J.* **28** 2414–27

Gorvel J P, Chavrier P, Zerial M and Gruenberg J 1991 Rab5 controls early endosome fusion *in vitro Cell* **64** 915–25

Goto H, Tanabe K, Manser E, Lim L, Yasui Y and Inagaki M 2002 Phosphorylation and reorganization of vimentin by p21-activated kinase (PAK) *Genes Cells* **7** 91–7

Goto M, Sumiyoshi H, Sakai T, Fassler R, Ohashi S, Adachi E, Yoshioka H and Fujiwara S 2006 Elimination of epiplakin by gene targeting results in acceleration of keratinocyte migration in mice *Mol. Cell Biol.* **26** 548–58

Gou J P, Gotow T, Janmey P A and Leterrier J F 1998 Regulation of neurofilament interactions in vitro by natural and synthetic polypeptides sharing Lys-Ser-Pro sequences with the heavy neurofilament subunit NF-H: neurofilament crossbridging by antiparallel sidearm overlapping *Med. Biol. Eng. Comput.* **36** 371–87

Green K J, Geiger B, Jones J C, Talian J C and Goldman R D 1987 The relationship between intermediate filaments and microfilaments before and during the formation of desmosomes and adherens-type junctions in mouse epidermal keratinocytes *J. Cell Biol.* **104** 1389–402

Green K J, Getsios S, Troyanovsky S and Godsel L M 2010 Intercellular junction assembly, dynamics, and homeostasis *Cold Spring Harb. Perspect. Biol.* **2** a000125

Gregor M, Osmanagic-Myers S, Burgstaller G, Wolfram M, Fischer I, Walko G, Resch G P, Jörgl A, Herrmann H and Wiche G 2014 Mechanosensing through focal adhesion-anchored intermediate filaments *FASEB J.* **28** 715–29

Guck J *et al* 2005 Optical deformability as an inherent cell marker for testing malignant transformation and metastatic competence *Biophys. J.* **88** 3689–98

Guillou H, Depraz-Depland A, Planus E, Vianay B, Chaussy J, Grichine A, Albiges-Rizo C and Block M R 2008 Lamellipodia nucleation by filopodia depends on integrin occupancy and downstream Rac1 signaling *Exp. Cell Res.* **314** 478–88

Guo L, Degenstein L, Dowling J, Yu Q C, Wollmann R, Perman B and Fuchs E 1995 Gene targeting of BPAG1: abnormalities in mechanical strength and cell migration in stratified epithelia and neurologic degeneration *Cell* **81** 233–43

Guo L, Degenstein L, Dowling J, Yu Q-C, Wollmann R, Perman B and Fuchs E 1995 Gene targeting of BPAG1: abnormalities in mechanical strength and cell migration in stratified epithelia and neurologic degeneration *Cell* **81** 233–43

Guo M, Ehrlicher A J, Mahammad S, Fabich H, Jensen M H, Moore J R and Weitz D A 2013 The role of vimentin intermediate filaments in cortical and cytoplasmic mechanics *Biophys. J.* **105** 1562–8

Guzman C, Jeney S, Kreplak L, Kasas S, Kulik A J, Aebi U and Forro L 2006 Exploring the mechanical properties of single vimentin intermediate filaments by atomic force microscopy *J. Mol. Biol.* **360** 623–30

Hamill K J, Hopkinson S B, Skalli O and Jones J C 2013 Actinin-4 in keratinocytes regulates motility via an effect on lamellipodia stability and matrix adhesions *FASEB J.* **27** 546–56

Hansson G K, Starkebaum G A, Benditt E P and Schwartz S M 1984 Fc-mediated binding of IgG to vimentin-type intermediate filaments in vascular endothelial cells *Proc. Natl Acad. Sci. USA* **81** 3103–7

Haudenschild D R, Chen J F, Pang N N, Steklov N, Grogan S P, Lotz M K and D'Lima D D 2011 Vimentin contributes to changes in chondrocyte stiffness in osteoarthritis *J. Orthopaedic Res.* **29** 20–5

Hay M and De Boni U 1991 Chromatin motion in neuronal interphase nuclei: changes induced by disruption of intermediate filaments *Cell Motil. Cytoskelet.* **18** 63–75

Helfand B T, Chang L and Goldman R D 2004 Intermediate filaments are dynamic and motile elements of cellular architecture *J. Cell Sci.* **117** 133–41

Helfand B T, Chang L and Goldman R D 2003 The dynamic and motile properties of intermediate filaments *Annu. Rev. Cell Dev. Biol.* **19** 445–67

Helfand B T *et al* 2011 Vimentin organization modulates the formation of lamellipodia *Mol. Biol. Cell* **22** 1274–89

Helmke B P, Thakker D B, Goldman R D and Davies P F 2001 Spatiotemporal analysis of flow-induced intermediate filament displacement in living endothelial cells *Biophys. J.* **80** 184–94

Helmke B P, Goldman R D and Davies P F 2000 Rapid displacement of vimentin intermediate filaments in living endothelial cells exposed to flow *Circ. Res.* **86** 745–52

Hendrix M J, Seftor E A, Seftor R E and Trevor K T 1997 Experimental co-expression of vimentin and keratin intermediate filaments in human breast cancer cells results in phenotypic interconversion and increased invasive behavior *Am. J. Pathol* **150** 483–95
Hennekes H and Nigg E A 1994 The role of isoprenylation in membrane attachment of nuclear lamins. A single point mutation prevents proteolytic cleavage of the lamin A precursor and confers membrane binding properties *J. Cell Sci.* **107** 1019–29
Herrmann H, Baer H, Kreplak L, Strelkov S V and Aebi U 2007 Intermediate filaments: from cell architecture to nanomechanics *Nat. Rev. Mol. Cell Biol.* **8** 562–73
Herrmann H, Haener M, Brettel M, Ku N O and Aebi U 1999 Characterization of distinct early assembly units of different intermediate filament proteins *J. Mol. Biol.* **286** 1403–20
Herrmann H, Wedig T, Porter R M, Lane E B and Aebi U 2002 Characterization of early assembly intermediates of recombinant human keratins *J. Struct. Biol.* **137** 82–96
Herrmann H, Strelkov S V, Burkhard P and Aebi U 2009 Intermediate filaments: primary determinants of cell architecture and plasticity *J. Clin. Investig.* **119** 1772–83
Hesse M, Magin T M and Weber K 2001 Genes for intermediate filament proteins and the draft sequence of the human genome: novel keratin genes and a surprisingly high number of pseudogenes related to keratins 8 and 18 *J. Cell Sci.* **114** 2569–75
Hesse M, Zimek A, Weber K and Magin T M 2004 Comprehensive analysis of keratin gene clusters in humans and rodents *Eur. J. Cell Biol.* **83** 19–26
Hesse M, Magin T M and Weber K 2001 Genes for intermediate filament proteins and the draft sequence of the human genome: novel keratin genes and a surprisingly high number of pseudogenes related to keratin genes 8 and 18 *J. Cell Sci.* **114** 2569–75
Hofmann I, Mertens C, Brettel M, Nimmrich V, Schnölzer M and Herrmann H 2000 Interaction of plakophilins with desmoplakin and intermediate filament proteins: an *in vitro* analysis *J. Cell Sci.* **113** 2471–83
Hol E M and Pekny M 2015 Glial fibrillary acidic protein (GFAP) and the astrocyte intermediate filament system in diseases of the central nervous system *Curr. Opin. Cell Biol.* **32** 121–30
Hollenbeck P J, Bershadsky A D, Pletjushkina O T, Tint I S and Vasiliev J M 1989 Intermediate filament collapse is an ATP-dependent and actin-dependent process *J. Cell Sci.* **92** 621–31
Holthoefer B, Windoffer R, Troyanovsky S and Leube R E 2007 Structure and function of desmosomes *Int. Rev. Cytol.* **264** 65–163
Holwell T A, Schweitzer S C and Evans R M 1997 Tetracycline regulated expression of vimentin in fibroblasts derived from vimentin null mice *J. Cell Sci.* **110** 1947–56
Huisman E M, Wen Q, Wang Y H, Cruz K, Kitenbergs G, Erglis K and Janmey P A 2011 Gelation of semiflexible polyelectrolytes by multivalent counterions *Soft Matter* **7** 7257–61
Hutchison C J 2002 Lamins: building blocks or regulators of gene expression? *Nat. Rev. Mol. Cell Biol.* **3** 848–58
Hyder C L, Pallari H M, Kochin V and Eriksson J E 2008 Providing cellular signposts--post-translational modifications of intermediate filaments *FEBS Lett.* **582** 2140–8
Ikegami K and Setou M 2010 Unique post-translational modifications in specialized microtubule architecture *Cell Struct. Funct.* **35** 15–22
Ishikawa H, Bischoff R and Holtzer H 1968 Mitosis and intermediate-sized filaments in developing skeletal muscle *J. Cell Biol.* **38** 538–55
Ishikawa K *et al* 2010 Epiplakin accelerates the lateral organization of keratin filaments during wound healing *J. Dermatol. Sci.* **60** 95–104

Ishikawa K *et al* 2010 Epiplakin accelerates the lateral organization of keratin filaments during wound healing *J. Dermatol Sci.* **60** 95–104

Ishikawa H, Bischoff R and Holtzer H 1968 Mitosis and intermediate-sized filaments in developing skeletal muscle *J. Cell Biol.* **38** 538–55

Ivaska J, Pallari H M, Nevo J and Eriksson J E 2007 Novel functions of vimentin in cell adhesion, migration, and signaling *Exp. Cell Res.* **313** 2050–62

Ivaska J, Vuoriluoto K, Huovinen T, Izawa I, Inagaki M and Parker P J 2005 PKCepsilon-mediated phosphorylation of vimentin controls integrin recycling and motility *EMBO J.* **24** 3834–45

Ivaska J, Whelan R D, Watson R and Parker P J 2002 PKC epsilon controls the traffic of β1 integrins in motile cells *EMBO J.* **21** 3608–19

Iyer S V, Dange P P, Alam H, Sawant S S, Ingle A D, Borges A M, Shirsat N V, Dalal S N and Vaidya M M 2013 Understanding the role of keratins 8 and 18 in neoplastic potential of breast cancer derived cell lines *PLoS One* **8** e53532

Izawa I and Inagaki M 2006 Regulatory mechanisms and functions of intermediate filaments: a study using site- and phosphorylation state-specific antibodies *Cancer Sci.* **97** 167–74

Jackson W M, Jaasma M J, Tang R Y and Keaveny T M 2008 Mechanical loading by fluid shear is sufficient to alter the cytoskeletal composition of osteoblastic cells *Am. J. Physiol. Cell Physiol.* **295** C1007–C1015

Jahn R 2004 Principles of exocytosis and membrane fusion *Ann. N. Y. Acad. Sci.* **1014** 170–78

Jaitovich A, Mehta S, Na N, Ciechanover A, Goldman R D and Ridge K M 2008 Ubiquitin-proteasome-mediated degradation of keratin intermediate filaments in mechanically stimulated A549 cells *J. Biol. Chem.* **283** 25348–55

Janmey P A, Euteneuer U, Traub P and Schliwa M 1991 Viscoelastic properties of vimentin compared with other filamentous biopolymer networks *J. Cell Biol.* **113** 155–60

Janmey P A, Slochower D R, Wang Y H, Wen Q and Cebers A 2014 Polyelectrolyte properties of filamentous biopolymers and their consequences in biological fluids *Soft Matter* **10** 1439–49

Jessen K R and Mirsky R 1980 Glial cells in the enteric nervous system contain glial fibrillary acidic protein *Nature* **286** 736–7

Jia Z, Barbier L, Stuart H, Amraei M, Pelech S, Dennis J W, Metalnikov P, O'Donnell P and Nabi I R 2005 Tumor cell pseudopodial protrusions. Localized signaling domains coordinating cytoskeleton remodeling, cell adhesion, glycolysis, RNA translocation, and protein translation *J. Biol. Chem.* **280** 30564–73

Jiu Y, Peränen J, Schaible N, Cheng F and Eriksson J E 2017 Vimentin intermediate filaments control actin stress fiber assembly through GEF-H1 and RhoA *J. Cell Sci.* **130** 892–902

Johansen L D *et al* 2008 IKAP localizes to membrane ruffles with filamin A and regulates actin cytoskeleton organization and cell migration *J. Cell Sci.* **121** 854–64

Johnson C P, Tang H-Y, Carag C, Speicher D W and Discher D E 2007 Forced unfolding of proteins within cells *Science* **317** 663–6

Johnston S A, Bramble J P, Yeung C L, Mendes P M and Machesky L M 2008 Arp2/3 complex activity in filopodia of spreading cells *BMC Cell Biol.* **9** 65

Joosse S A, Hannemann J, Spotter J, Bauche A, Andreas A, Muller V and Pantel K 2012 Changes in keratin expression during metastatic progression of breast cancer: impact on the detection of circulating tumor cells *Clin. Cancer Res.* **18** 993–1003

Jovic M, Naslavsky N, Rapaport D, Horowitz M and Caplan S 2007 EHD1 regulates β1 integrin endosomal transport: effects on focal adhesions, cell spreading and migration *J. Cell Sci.* **120** 802–14

Ju J H, Yang W, Lee K M, Oh S, Nam K, Shim S, Shin S Y, Gye M C, Chu I S and Shin I 2013 Regulation of cell proliferation and migration by keratin 19-induced nuclear import of early growth response-1 in breast cancer cells *Clin. Cancer Res.* **19** 4335–46

Jung K S, Kim S U, Ahn S H, Park Y N, Kim do Y, Park J Y, Chon C Y, Choi E H and Han K H 2011 Risk assessment of hepatitis B virus-related hepatocellular carcinoma development using liver stiffness measurement (FibroScan) *Hepatology* **53** 885–94

Kalluri R and Weinberg R A 2009 The basics of epithelial–mesenchymal transition *J. Clin. Investig.* **119** 1420–8

Kaminsky R, Denison C, Bening-Abu-Shach U, Chisholm A D, Gygi S P and Broday L 2009 SUMO regulates the assembly and function of a cytoplasmic intermediate filament protein in *C. elegans Dev. Cell* **17** 724–35

Karantza V 2011 Keratins in health and cancer: more than mere epithelial cell markers *Oncogene* **30** 127–38

Kasas S *et al* 2005 Superficial and deep changes of cellular mechanical properties following cytoskeleton disassembly *Cell Motil. Cytoskelet.* **62** 124–32

Keski-Oja J, Lehto V P and Virtanen I 1981 Keratin filaments of mouse epithelial cells are rapidly affected by epidermal growth factor *J. Cell Biol.* **90** 537–41

Khapare N, Kundu S T and Sehgal L *et al* 2012 Plakophilin3 loss leads to an increase in PRL3 levels promoting K8 dephosphorylation, which is required for transformation and metastasis *PLoS One* **7** e38561

Kiema T, Lad Y, Jiang P, Oxley C L, Baldassarre M, Wegener K L, Campbell I D, Ylanne J and Calderwood D A 2006 The molecular basis of filamin binding to integrins and competition with talin *Mol. Cell* **21** 337–47

Kim H, Nakamura F, Lee W, Shifrin Y, Arora P and McCulloch C A 2010 Filamin A is required for vimentin-mediated cell adhesion and spreading *Am. J. Physiol. Cell Physiol.* **298** C221-36

Kim H, Sengupta A, Glogauer M and McCulloch C A 2008 Filamin A regulates cell spreading and survival via β1 integrins *Exp. Cell Res.* **314** 834–46

Kim J S, Lee C H and Coulombe P A 2010 Modeling the self-organization property of keratin intermediate filaments *Biophys. J.* **99** 2748–56

Kim S and Coulombe P A 2007 Intermediate filament scaffolds fulfill mechanical, organizational and signaling functions in the cytoplasm *Genes Dev.* **21** 1581–97

Kim S, Kellner J, Lee C H and Coulombe P A 2007 Interaction between the keratin cytoskeleton and eEF1Bgamma affects protein synthesis in epithelial cells *Nat. Struct. Mol. Biol.* **14** 982–3

Kim S, Wong P and Coulombe P A 2006 A keratin cytoskeletal protein regulates protein synthesis and epithelial cell growth *Nature* **441** 362–5

Kirmse R, Portet S, Muecke N, Aebi U, Herrmann H and Langowski J 2007 A quantitative kinetic model for the *in vitro* assembly of intermediate filaments from tetrameric vimentin *J. Biol. Chem.* **282** 18563–72

Kiseleva E, Allen T D, Rutherford S, Bucci M, Wente S R and Goldberg M W 2004 Yeast nuclear pore complexes have a cytoplasmic ring and internal filaments *J. Struct. Biol.* **145** 272–88

Kitajima Y 2013 New insights into desmosome regulation and pemphigus blistering as a desmosome-remodeling disease Kaohsiung *J. Med Sci.* **29** 1–13

Kjarland E, Keen T J and Kleppe R 2006 Does isoform diversity explain functional differences in the 14-3-3 protein family? *Curr. Pharm. Biotechnol.* **7** 217–23

Kleeberger W, Bova G S, Nielsen M E, Herawi M, Chuang A Y, Epstein J I and Berman D M 2007 Roles for the stem cell associated intermediate filament Nestin in prostate cancer migration and metastasis *Cancer Res.* **67** 9199–206

Klymkowsky M W 1981 Intermediate filaments in 3T3 cells collapse after intracellular injection of a monoclonal anti-intermediate filament antibody *Nature* **291** 249–51

Knösel T, Emde V, Schluens K, Schlag P M, Dietel M and Petersen I 2006 Cytokeratin profiles identify diagnostic signatures in colorectal cancer using multiplex analysis of tissue microarrays *Cell. Oncol.* **28** 167–75

Knowles G C, McKeown M, Sodek J and McCulloch C A 1991 Mechanism of collagen phagocytosis by human gingival fibroblasts: importance of collagen structure in cell recognition and internalization *J. Cell Sci.* **98** 551–8

Kochin V, Shimi T, Torvaldson E, Adam S A, Goldman A, Pack C G, Melo-Cardenas J, Imanishi S Y, Goldman R D and Eriksson J E 2014 Interphase phosphorylation of lamin A *J. Cell Sci.* **127** 2683–96

Koelsch A, Windoffer R and Leube R E 2009 Actin-dependent dynamics of keratin filament precursors *Cell Motil. Cytoskelet.* **66** 976–85

Koelsch A, Windoffer R, Würflinger T, Aach T and Leube R E 2010 The keratin-filament cycle of assembly and disassembly *J. Cell Sci.* **123** 2266–72

Konishi Y and Setou M 2009 Tubulin tyrosination navigates the kinesin-1 motor domain to axons *Nat. Neurosci.* **12** 559–67

Kornreich M, Avinery R, Malka-Gibor E, Laser-Azogui A and Beck R 2015 Order and disorder in intermediate filament proteins *FEBS Lett* **589** 2464–76

Kostan J, Gregor M, Walko G and Wiche G 2009 Plectin isoform-dependent regulation of keratin-integrin $\alpha6\beta4$ anchorage via Ca^{2+}/calmodulin *J. Biol. Chem.* **284** 18525–36

Kreis S, Schoenfeld H J, Melchior C, Steiner B and Kieffer N 2005 The intermediate filament protein vimentin binds specifically to a recombinant integrin $\alpha2/\beta1$ cytoplasmic tail complex and co-localizes with native $\alpha2/\beta1$ in endothelial cell focal adhesions *Exp. Cell Res.* **305** 110–21

Kreplak L, Bar H, Leterrier J F, Herrmann H and Aebi U 2005 Exploring the mechanical behavior of single intermediate filaments *J. Mol. Biol.* **354** 569–77

Kreplak L, Herrmann H and Aebi U 2008 Tensile properties of single desmin intermediate filaments *Biophys. J.* **94** 2790–9

Krieg P, Schuppler M, Koesters R, Mincheva A, Lichter P and Marks F 1997 Repetin (Rptn), a new member of the 'fused gene' subgroup within the S100 gene family encoding a murine epidermal differentiation protein *Genomics* **43** 339–48

Krylyshkina O, Anderson K I, Kaverina I, Upmann I, Manstein D J, Small J V and Toomre D K 2003 Nanometer targeting of microtubules to focal adhesions *J. Cell Biol.* **161** 853–9

Ku N O and Omary M B 2000 Keratins turn over by ubiquitination in a phosphorylation-modulated fashion *J. Cell Biol.* **149** 547–52

Ku N O and Omary M B 2006 A disease- and phosphorylation-related non-mechanical function for keratin 8 *J. Cell Biol.* **174** 115–25

Ku N O, Michie S A, Soetikno R M, Resurreccion E Z, Broome R L, Oshima R G and Omary M B 1996 Susceptibility to hepatotoxicity in transgenic mice that express a dominant-negative human keratin 18 mutant *J. Clin. Investig.* **98** 1034–46

Ku N O, Toivola D M, Strnad P and Omary M B 2010 Cytoskeletal keratin glycosylation protects epithelial tissue from injury *Nat. Cell Biol.* **12** 876–85

Kumar N, Robidoux J, Daniel K W, Guzman G, Floering L M and Collins S 2007 Requirement of vimentin filament assembly for β3-adrenergic receptor activation of ERK MAP kinase and lipolysis *J. Biol. Chem.* **282** 9244–50

Kundu S T, Gosavi P, Khapare N, Patel R, Hosing A S, Maru G B, Ingle A, Decaprio J A and Dalal S N 2008 Plakophilin3 down-regulation leads to a decrease in cell adhesion and promotes metastasis *Int J. Cancer* **123** 2303–14

Kuo Y H, Lu S N, Hung C H, Kee K M, Chen C H, Hu T H, Lee C M, Changchien C S and Wang J H 2010 Liver stiffness measurement in the risk assessment of hepatocellular carcinoma for patients with chronic hepatitis *Hepatol. Int.* **4** 700–06

Kushkuley J, Chan W K, Lee S, Eyer J, Leterrier J F, Letournel F and Shea T B 2009 Neurofilament cross-bridging competes with kinesin-dependent association of neurofilaments with microtubules *J. Cell Sci.* **122** 3579–86

Ladwein M and Rottner K 2008 On the Rho'd: the regulation of membrane protrusions by Rho-GTPases *FEBS Lett.* **582** 2066–74

Laemmermann T and Sixt M 2009 Mechanical modes of 'amoeboid' cell migration *Curr. Opin. Cell Biol.* **21** 636–44

Langbein L, Eckhart L, Rogers M A, Praetzel-Wunder S and Schweizer J 2010 Against the rules: human keratin K80: two functional alternative splice variants, K80 and K80.1, with special cellular localization in a wide range of epithelia *J. Biol. Chem.* **285** 36909–21

Larcher F, Bauluz C, Diaz-Guerra M, Quintanilla M, Conti C J, Ballestin C and Jorcano J L 1992 Aberrant expression of the simple epithelial type II keratin 8 by mouse skin carcinomas but not papillomas *Mol. Carcinog.* **6** 112–21

Lazarides E 1982 Intermediate filaments: a chemically heterogeneous, developmentally regulated class of proteins *Annu. Rev. Biochem.* **51** 219–50

Lee C H and Coulombe P A 2009 Self-organization of keratin intermediate filaments into cross-linked networks *J. Cell Biol.* **186** 409–21

Lee D, Santos D, Al-Rawi H, McNeill A M and Rugg E L 2008 The chemical chaperone trimethylamine N-oxide ameliorates the effects of mutant keratins in cultured cells *Br. J. Dermatol.* **159** 252–5

Lee R, Chang S Y, Trinh H, Tu Y, White A C, Davies B S, Bergo M O, Fong L G, Lowry W E and Young S G 2010 Genetic studies on the functional relevance of the protein prenyl-transferases in skin keratinocytes *Hum. Mol. Genet.* **19** 1603–17

Lee W, Sodek J and McCulloch C A 1996 Role of integrins in regulation of collagen phagocytosis by human fibroblasts *J. Cell Physiol.* **168** 695–704

Lehtinen L, Ketola K, Makela R, Mpindi J P, Viitala M, Kallioniemi O and Iljin K 2013 High-throughput RNAi screening for novel modulators of vimentin expression identifies MTHFD2 as a regulator of breast cancer cell migration and invasion *Oncotarget* **4** 48–63

Lendahl U, Zimmerman L B and McKay R D 1990 CNS stem cells express a new class of intermediate filament protein *Cell* **60** 585–95

Lepekhin E A, Eliasson C, Berthold C H, Berezin V, Bock E and Pekny M 2001 Intermediate filaments regulate astrocyte motility *J. Neurochem.* **79** 617–25

Lersch R and Fuchs E 1988 Sequence and expression of a type II keratin, K5, in human epidermal cells *Mol. Cell. Biol.* **8** 486–93

Leterrier J F and Eyer J 1987 Properties of highly viscous gels formed by neurofilaments in vitro. A possible consequence of a specific inter-filament cross-bridging *Biochem. J.* **245** 93–101
Leterrier J F, Käs J, Hartwig J, Vegners Rand and Janmey P A 1996 Mechanical effects of neurofilament cross-bridges *J. Biol. Chem.* **271** 15687–94
Leube R E, Moch M, Kölsch A and Windoffer R 2011 'Panta rhei': perpetual cycling of the keratin cytoskeleton *BioArchitecture* **1** 39–44
Leung C L, Green K J and Liem R K 2002 Plakins: a family of versatile cytolinker proteins *Trends Cell Biol.* **12** 37–45
Leung C L, Liem R K, Parry D A and Green K J 2001 The plakin family *J. Cell Sci.* **114** 3409–10
Li Q F, Spinelli A M, Wang R, Anfinogenova Y, Singer H A and Tang D D 2006 Critical role of vimentin phosphorylation at Ser-56 by p21-activated kinase in vimentin cytoskeleton signaling *J. Biol. Chem.* **281** 34716–24
Li S, Guan J L and Chien S 2005 Biochemistry and biomechanics of cell motility *Annu. Rev. Biomed. Eng.* **7** 105–50
Liao J and Omary M B 1996 14-3-3 proteins associate with phosphorylated simple epithelial keratins during cell cycle progression and act as a solubility cofactor *J. Cell Biol.* **133** 345–57
Lilienbaum A, Legagneux V, Portier M M, Dellagi K and Paulin D 1986 Vimentin gene: expression in human lymphocytes and in Burkitt's lymphoma cells *EMBO J.* **5** 2809–14
Lin Y C, Broedersz C P, Rowat A C, Wedig T, Herrmann H, MacKintosh F C and Weitz D A 2010 Divalent cations crosslink vimentin intermediate filament tail domains to regulate network mechanics *J. Mol. Biol.* **399** 637–44
Liovic M, Mogensen M M, Prescott A R and Lane E B 2003 Observation of keratin particles showing fast bidirectional movement colocalized with microtubules *J. Cell Sci.* **116** 1417–27
Listwan P and Rothnagel J A 2004 Keratin bundling proteins *Methods Cell Biol.* **78** 817–27
Litjens S H, Koster J, Kuikman I, van Wilpe S, de Pereda J M and Sonnenberg A 2003 Specificity of binding of the plectin actin-binding domain to *β*4 integrin *Mol. Biol. Cell* **14** 4039–50
Loeffek S, Wöll S, Hoehfeld J, Leube R E, Has C, Bruckner-Tuderman L and Magin T M 2010 The ubiquitin ligase CHIP/STUB1 targets mutant keratins for degradation *Hum. Mutat.* **31** 466–76
Lombardi M L, Jaalouk D E, Shanahan C M, Burke B, Roux K J and Lammerding J 2011 The interaction between nesprins and sun proteins at the nuclear envelope is critical for force transmission between the nucleus and cytoskeleton *J. Biol. Chem.* **286** 26743–53
Long H A, Boczonadi V, McInroy L, Goldber M and Määttä A 2006 Periplakin-dependent re-organisation of keratin cytoskeleton and loss of collective migration in keratin-8-down-regulated epithelial sheets *J. Cell Sci.* **119** 5147–59
Loo D T, Kanner S B and Aruffo A 1998 Filamin binds to the cytoplasmic domain of the *β*1-integrin. Identification of amino acids responsible for this interaction *J. Biol. Chem.* **273** 23304–12
Lopez J I, Mouw J K and Weaver V M 2008 Biomechanical regulation of cell orientation and fate *Oncogene* **27** 6981–93
Ma L, Yamada S, Wirtz D and Coulombe P A 2001 A 'hot-spot' mutation alters the mechanical properties of keratin filament networks *Nat. Cell Biol.* **3** 503–06
Mackintosh C 2004 Dynamic interactions between 14-3-3 proteins and phosphoproteins regulate diverse cellular processes *Biochem. J.* **381** 329–42
Magin T M, Reichelt J and Chen J 2006 The role of keratins in epithelial homeostasis *Skin Barrier* ed P M Elias and K R Feingold (New York: Taylor and Francis) pp 141–170

Magin T M, Vijayaraj P and Leube R E 2007 Structural and regulatory functions of keratins *Exp. Cell Res.* **313** 2021–32

Makino T, Takaishi M, Morohashi M and Huh N H 2001 Hornerin, a novel profilaggrin-like protein and differentiation-specific marker isolated from mouse skin *J. Biol. Chem.* **276** 47445–52

Manjiu K, Muralikrishna B and Parnaik V K 2006 Expression of disease-causing lamin A mutants impairs the formation of DNA repair foci *J. Cell Sci.* **199** 2704–14

Mashukova A, Oriolo A S, Wald F A, Casanova M L, Kröger C, Magin T M, Omary M B and Salas P J 2009 Rescue of atypical protein kinase C in epithelia by the cytoskeleton and Hsp70 family chaperones *J. Cell Sci.* **122** 2491–503

Matsuda Y, Kure S and Ishiwata T 2012 Nestin and other putative cancer stem cell markers in pancreatic cancer *Med. Mol. Morphol.* **45** 59–65

Matsuda Y, Naito Z, Kawahara K, Nakazawa N, Korc M and Ishiwata T 2011 Nestin is a novel target for suppressing pancreatic cancer cell migration, invasion and metastasis *Cancer Biol. Ther.* **11** 512–23

Matthews B D, Overby D R, Alenghat F J, Karavitis J, Numaguchi Y, Allen P G and Ingber D E 2004 Mechanical properties of individual focal adhesions probed with a magnetic microneedle *Biochem. Biophys. Res. Commun.* **313** 758–64

Matthews B D, Overby D R, Mannix R and Ingber D E 2006 Cellular adaptation to mechanical stress: role of integrins, Rho, cytoskeletal tension and mechanosensitive ion channels *J. Cell Sci.* **119** 508–18

Mazzalupo S, Wong P, Martin P and Coulombe P A 2003 Role for keratins 6 and 17 during wound closure in embryonic mouse skin *Dev. Dyn.* **226** 356–65

McBeath R, Pirone D M, Nelson C M, Bhadriraju K and Chen C S 2004 Cell shape, cytoskeletal tension, and RhoA regulate stem cell lineage commitment *Dev. Cell.* **6** 483–95

McCaffrey M W, Bielli A, Cantalupo G, Mora S, Roberti V, Santillo M, Drummond F and Bucci C 2001 Rab4 affects both recycling and degradative endosomal trafficking *FEBS Lett.* **495** 21–30

McInroy L and Maatta A 2007 Down-regulation of vimentin expression inhibits carcinoma cell migration and adhesion *Biochem. Biophys. Res. Commun.* **360** 109–14

Mellad J A, Warren D T and Shanahan C M 2011 Nesprins LINC the nucleus and cytoskeleton *Curr. Opin. Cell Biol.* **23** 47–54

Mendez M G, Kojima S and Goldman R D 2010 Vimentin induces changes in cell shape, motility, and adhesion during the epithelial to mesenchymal transition *FASEB J.* **24** 1838–51

Mendez M G, Restle D and Janmey P A 2014 Vimentin enhances cell elastic behavior and protects against compressive stress *Biophys. J.* **107** 314–23

Meng X, Yuan Y, Maestas A and Shen Z 2004 Recovery from DNA damage-induced G2 arrest requires actin-binding protein filamin-A/actin-binding protein 280 *J. Biol. Chem.* **279** 6098—105

Meyer S C, Sanan D A and Fox J E 1998 Role of actin-binding protein in insertion of adhesion receptors into the membrane *J. Biol. Chem.* **273** 3013–20

Michalski P J and Carlsson A E 2010 The effects of filament aging and annealing on a model lamellipodium undergoing disassembly by severing *Phys. Biol.* **7** 026004

Mignot C, Delarasse C and Escaich S *et al* 2007 Dynamics of mutated GFAP aggregates revealed by real-time imaging of an astrocyte model of Alexander disease *Exp. Cell Res.* **313** 2766–79

Miller R K, Vikstrom K and Goldman R D 1991 Keratin incorporation into intermediate filament networks is a rapid process *J. Cell Biol.* **113** 843–55

Miller R K, Khuon S and Goldman R D 1993 Dynamics of keratin assembly: exogenous type I keratin rapidly associates with type II keratin *in vivo J. Cell Biol.* **122** 123–35

Mitchison T and Kirschner M 1984 Dynamic instability of microtubule growth *Nature* **312** 237–42

Mizuuchi E, Semba S, Kodama Y and Yokozaki H 2009 Down-modulation of keratin 8 phosphorylation levels by PRL-3 contributes to colorectal carcinoma progression *Int J. Cancer.* **124** 1802–10

Moll R, Divo M and Langbein L 2008 The human keratins: biology and pathology *Histochem. Cell Biol.* **129** 705–33

Moll R, Franke W W, Schiller D L, Geiger B and Krepler R 1982 The catalog of human cytokeratins: patterns of expression in normal epithelia, tumors and cultured cells *Cell* **31** 11–24

Moll R, Schiller D L and Franke W W 1990 Identification of protein IT of the intestinal cytoskeleton as a novel type I cytokeratin with unusual properties and expression patterns *J. Cell Biol.* **111** 567–80

Monteiro M J and Cleveland D W 1989 Expression of NF-L and NF-M in fibroblasts reveals coassembly of neurofilament and vimentin subunits *J. Cell Biol.* **108** 579–83

Montes de Oca R, Andreassen P R and Wilson K L 2011 Barrier-to-autointegration factor influences specific histone modifications *Nucleus* **2** 580–90

Morgan J T, Pfeiffer E R, Thirkill T L, Kumar P, Peng G, Fridolfsson H N, Douglas G C, Starr D A and Barakat A I 2011 Nesprin-3 regulates endothelial cell morphology, perinuclear cytoskeletal architecture, and flow-induced polarization *Mol. Biol. Cell* **22** 4324–34

Morley S M, D'Alessandro M and Sexton C *et al* 2003 Generation and characterization of epidermolysis bullosa simplex cell lines: scratch assays show faster migration with disruptive keratin mutations *Br. J. Dermatol.* **149** 46–58

Morley S M *et al* 2003 Generation and characterization of epidermolysis bullosa simplex cell lines: scratch assays show faster migration with disruptive keratin mutations *Br. J. Dermatol.* **149** 46–58

Muchir A, Bonne G, van der Kooi A J, van Meegen M, Baas F, Bolhuis P A, de Visser M and Schwartz K 2000 Identification of mutations in the gene encoding lamins A/C in autosomal dominant limb girdle muscular dystrophy with atrioventricular conduction disturbances (LGMD1B) *Hum. Mol. Genet.* **9** 1453–9

Muchir A, Shan J, Bonne G, Lehnart S E and Worman H J 2009 Inhibition of extracellular signal-regulated kinase signaling to prevent cardiomyopathy caused by mutation in the gene encoding A-type lamins *Hum. Mol. Genet.* **18** 241–7

Muchir A, Wu W and Worman H J 2009 Reduced expression of A-type lamins and emerin activates extracellular signal-regulated kinase in cultured cells *Biochim. Biophys. Acta* **1792** 75–81

Na N, Chandel N S, Litvan J and Ridge K M 2010 Mitochondrial reactive oxygen species are required for hypoxia-induced degradation of keratin intermediate filaments *FASEB J.* **24** 799–809

Nakamura F, Osborn T M, Hartemink C A, Hartwig J H and Stossel T P 2007 Structural basis of filamin A functions *J. Cell Biol.* **179** 1011–25

Nakamura M and Tokura Y 2011 Epithelial–mesenchymal transition in the skin *J. Dermatol. Sci.* **61** 7–13

Nan L, Dedes J, French B A, Bardag-Gorce F, Li J, Wu Y and French S W 2006 Mallory body (cytokeratin aggresomes) formation is prevented *in vitro* by p38 inhibitor *Exp. Mol. Pathol.* **80** 228–40

Narita K, Matsuda Y, Seike M, Naito Z, Gemma A and Ishiwata T 2014 Nestin regulates proliferation, migration, invasion and stemness of lung adenocarcinoma *Int. J. Oncol.* **44** 1118–30

Ngai J, Coleman T R and Lazarides E 1990 Localization of newly synthesized vimentin subunits reveals a novel mechanism of intermediate filament assembly *Cell* **60** 415–27

Nielsen E, Severin F, Backer J M, Hyman A A and Zerial M 1999 Rab5 regulates motility of early endosomes on microtubules *Nat. Cell Biol.* **1** 376–82

Nieminen M, Henttinen T, Merinen M, Marttila-Ichihara F, Eriksson J E and Jalkanen S 2006 Vimentin function in lymphocyte adhesion and transcellular migration *Nat. Cell Biol.* **8** 156–62

Nikolic B, Mac Nulty E, Mir B and Wiche G 1996 Basic amino acid residue cluster within nuclear targeting sequence motif is essential for cytoplasmic plectin–vimentin network junctions *J. Cell Biol.* **134** 1455–67

Nobes C D and Hall A 1995 Rho, rac, and cdc42 GTPases regulate the assembly of multimolecular focal complexes associated with actin stress fibers, lamellipodia, and filopodia *Cell* **81** 53–62

Nolting J F and Koster S 2013 Influence of microfluidic shear on keratin networks in living cells *N. J. Phys.* **15** 045025

Norgren N, Rosengren L and Stigbrand T 2003 Elevated neurofilament levels in neurological diseases *Brain Res.* **987** 25–31

Ogawara M *et al* 1995 Differential targeting of protein kinase C and CaM kinase II signalings to vimentin *J. Cell Biol.* **131** 1055–66

Ohta Y, Suzuki N, Nakamura S, Hartwig J H and Stossel T P 1999 The small GTPase RalA targets filamin to induce filopodia *Proc. Natl Acad. Sci. USA* **96** 2122–8

Omary M B and Ku N O 1997 Intermediate filament proteins of the liver: emerging disease association and functions *Hepatology* **25** 1043–8

Omary M B, Baxter G T, Chou C F, Riopel C L, Lin W Y and Strulovici B 1992 PKC epsilon-related kinase associates with and phosphorylates cytokeratin 8 and 18 *J. Cell Biol.* **117** 583–93

Omary M B, Coulombe P A and McLean W H 2004 Intermediate filament proteins and their associated diseases *N. Engl. J. Med.* **351** 2087–100

Omary M B, Ku N O, Liao J and Price D 1998 Keratin modifications and solubility properties in epithelial cells and in vitro *Subcell. Biochem.* **31** 105–40

Omary M B, Ku N O, Strnad P and Hanada S 2009 Toward unraveling the complexity of simple epithelial keratins in human disease *J. Clin. Investig.* **119** 1794–805

Oriolo A S, Wald F A, Ramsauer V P and Salas P J 2007 Intermediate filaments: a role in epithelial polarity *Exp. Cell Res.* **313** 2255–64

Oriolo A S, Wald F A, Canessa G and Salas P J 2006 GCP6 binds to intermediate filaments: A novel function of keratins in the organization of microtubules in epithelial cells *Mol. Biol. Cell* **18** 781–94

Oshima R G 2002 Apoptosis and keratin intermediate filaments *Cell Death Differ.* **9** 486–92

Oshima R G, Baribault H and Caulin C 1996 Oncogenic regulation and function of keratins 8 and 18 *Cancer Metas. Rev.* **15** 445–71

Osmanagic-Myers S, Gregor M, Walko G, Burgstaller G, Reipert S and Wiche G 2006 Plectin-controlled keratin cytoarchitecture affects MAP kinases involved in cellular stress response and migration *J. Cell Biol.* **174** 557–68
Osmanagic-Myers S, Gregor M, Walko G, Burgstaller G, Reipert S and Wiche G 2006 Plectin-controlled keratin cytoarchitecture affects MAP kinases involved in cellular stress response and migration *J. Cell Biol.* **174** 557–68
Paccione R J, Miyazaki H, Patel V, Waseem A, Gutkind J S, Zehner Z E and Yeudall W A 2008 Keratin down-regulation in vimentin-positive cancer cells is reversible by vimentin RNA interference, which inhibits growth and motility *Mol. Cancer Ther.* **7** 2894–903

Pajerowski J D, Dahl K N, Zhong F L, Sammak P J and Discher D E 2007 Physical plasticity of the nucleus in stem cell differentiation *Proc. Natl Acad. Sci. USA* **104** 15619–24

Paladini R D, Takahashi K, Bravo N S and Coulombe P A 1996 Onset of re-epithelialization after skin injury correlates with a reorganization of keratin filaments in wound edge keratinocytes: defining a potential role for keratin 16 *J. Cell Biol.* **132** 381–97
Paladini R D, Takahashi K, Bravo N S and Coulombe P A 1996 Onset of re-epithelialization after skin injury correlates with a reorganization of keratin filaments in wound edge keratinocytes: defining a potential role for keratin 16 *J. Cell Biol.* **132** 381–97
Pan X, Hobbs R P and Coulombe P A 2013 The expanding significance of keratin intermediate filaments in normal and diseased epithelia *Curr. Opin. Cell Biol.* **25** 47–56

Pan X, Kane L A, Van Eyk J E and Coulombe P A 2011 Type I keratin 17 protein is phosphorylated on serine 44 by p90 ribosomal protein S6 kinase 1 (RSK1) in a growth- and stress-dependent fashion *J. Biol. Chem.* **286** 42403–13

Pan Y, Jing R, Pitre A, Williams B J and Skalli O 2008 Intermediate filament protein synemin contributes to the migratory properties of astrocytoma cells by influencing the dynamics of the actin cytoskeleton *FASEB J.* **22** 3196–206

Pannese E 1962 Detection of neurofilaments in the perikaryon of hypertrophic nerve cells *J. Cell Biol.* **13** 457–9

Park M K, Lee H J, Shin J, Noh M, Kim S Y and Lee C H 2011 Novel participation of transglutaminase-2 through c-Jun N-terminal kinase activation in sphingosylphosphorylcholine-induced keratin reorganization of PANC-1 cells *Biochim. Biophys. Acta* **1811** 1021–9

Parry S, Savage K, Marchio C and Reis-Filho J S 2008 Nestin is expressed in basal-like and triple negative breast cancers *J. Clin. Pathol.* **61** 1045–50

Parry D A, Strelkov S V, Burkard P, Aebi U and Herrmann H 2007 Towards a molecular description of intermediate filament structure and assembly *Exp. Cell Res.* **313** 2204–16

Paszek M J *et al* 2005 Tensional homeostasis and the malignant phenotype *Cancer Cell* **8** 241–54

Pawelzyk P, Mucke N, Herrmann H and Willenbacher N 2014 Attractive interactions among intermediate filaments determine network mechanics *in vitro Plos One* **9** e109450

Pekny M and Lane E B 2007 Intermediate filaments and stress *Exp. Cell Res.* **313** 2244–54

Pérez-Sala D, Oeste C L, Martínez A E, Carrasco M J, Garzón B and Cañada F J 2015 Vimentin filament organization and stress sensing depend on its single cysteine residue and zinc binding *Nat. Commun.* **6** 7287

Perlson E, Hanz S, Ben-Yaakov K, Segal-Ruder Y, Seger R and Fainzilber M 2005 Vimentin-dependent spatial translocation of an activated MAP kinase in injured nerve *Neuron* **45** 715–26

Perlson E, Michaelevski I, Kowalsman N, Ben-Yaakov K, Shaked M, Seger R, Eisenstein M and Fainzilber M 2006 Vimentin binding to phosphorylated ERK sterically hinders enzymatic dephosphorylation of the kinase *J. Mol. Biol.* **364** 938–44

Perrin B J, Amann K J and Huttenlocher A 2006 Proteolysis of cortactin by calpain regulates membrane protrusion during cell migration *Mol. Biol. Cell* **17** 239–50

Peter M, Nakagawa J, Dorée M, Labbé J C and Nigg E A 1990 *In vitro* disassembly of the nuclear lamina and M phase-specific phosphorylation of lamins by cdc2 kinase *Cell* **61** 591–602

Petit V and Thiery J P 2000 Focal adhesions: structure and dynamics *Biol. Cell* **92** 477–94

Planko L, Boehse K, Hoehfeld J, Betz R C, Hanneken S, Eigelshoven S, Kruse R, Noethen M M and Magin T M 2007 Identification of a keratin-associated protein with a putative role in vesicle transport *Eur. J. Cell Biol.* **86** 827–39

Plodinec M, Loparic M, Suetterlin R, Herrmann H, Aebi U and Schoenenberger C A 2011 The nanomechanical properties of rat fibroblasts are modulated by interfering with the vimentin intermediate filament system *J. Struct. Biol.* **174** 476–84

Popowicz G M, Schleicher M, Noegel A A and Holak T A 2006 Filamins: promiscuous organizers of the cytoskeleton *Trends Biochem. Sci.* **31** 411–9

Portier M M, de Néchaud B and Gros F 1984 Peripherin, a new member of the intermediate filament protein family *Dev. Neurosci.* **6** 335–44

Portier M M, Escurat M, Landon F, Djabali K and Bousquet O 1993 Peripherin and neurofilaments: expression and role during neural development *C. R. Acad. Sci. III* **316** 1124–40

Price L S, Leng J, Schwartz M A and Bokoch G M 1998 Activation of Rac and Cdc42 by integrins mediates cell spreading *Mol. Biol. Cell* **9** 1863–71

Proux-Gillardeaux V, Gavard J, Irinopoulou T, Mège R M and Galli T 2005 Tetanus neurotoxin-mediated cleavage of cellubrevin impairs epithelial cell migration and integrin-dependent cell adhesion *Proc. Natl Acad. Sci. USA* **102** 6362–7

Ptitsyn A A, Weil M M and Thamm D H 2008 Systems biology approach to identification of biomarkers for metastatic progression in cancer *BMC Bioinformatics* **9** S8

Quax-Jeuken Y E, Quax W J and Bloemendal H 1983 Primary and secondary structure of hamster vimentin predicted from the nucleotide sequence *Proc. Natl Acad. Sci. USA* **80** 3548–52

Quinlan R A and Franke W W 1982 Heteropolymer filaments of vimentin and desmin in vascular smooth muscle tissue and cultured baby hamster kidney cells demonstrated by chemical crosslinking *Proc. Natl Acad. Sci. USA* **79** 3452–6

Quinlan R A and Franke W W 1983 Molecular interactions in intermediate-sized filaments revealed by chemical cross-linking. Heteropolymers of vimentin and glial filament protein in cultured human glioma cells *Eur. J. Biochem.* **132** 477–84

Rao L, Perez D and White E 1996 Lamin proteolysis facilitates nuclear events during apoptosis *J. Cell Biol.* **135** 1441–55

Rathje L S Z, Nordgren N, Pettersson T, Ronnlund D, Widengren J, Aspenstrom P and Gad A K B 2014 Oncogenes induce a vimentin filament collapse mediated by HDAC6 that is linked to cell stiffness *Proc. Natl Acad. Sci. USA* **111** 1515–20

Raul U, Sawant S, Dange P, Kalraiya R, Ingle A and Vaidya M 2004 Implications of cytokeratin 8/18 filament formation in stratified epithelial cells: induction of transformed phenotype *Int J. Cancer* **111** 662–8

Reichelt J, Büssow H, Grund C and Magin T M 2001 Formation of a normal epidermis supported by increased stability of keratins 5 and 14 in keratin 10 null mice *Mol. Biol. Cell* **12** 1557–68

Ren X D, Kiosses W B and Schwartz M A 1999 Regulation of the small GTP-binding protein Rho by cell adhesion and the cytoskeleton *EMBO J.* **18** 578–85

Rezniczek G A, Abrahamsberg C, Fuchs P, Spazierer D and Wiche G 2003 Plectin 5'-transcript diversity: short alternative sequences determine stability of gene products, initiation of translation and subcellular localization of isoforms *Hum. Mol. Genet.* **12** 3181–94

Rezniczek G A, Janda L and Wiche G 2004 Plectin *Methods Cell Biol.* **78** 721–55

Ridge K M, Linz L, Flitney F W, Kuczmarski E R, Chou Y H, Omary M B, Sznajder J I and Goldman R D 2005 Keratin 8 phosphorylation by protein kinase C delta regulates shear stress-mediated disassembly of keratin intermediate filaments in alveolar epithelial cells *J. Biol. Chem.* **280** 30400–05

Ridge K M, Linz L, Flitney F W, Kuczmarski E R, Chou Y H, Omary M B, Sznajder J I and Goldman R D 2005 Keratin 8 phosphorylation by protein kinase c δ regulates shear stress-mediated disassembly of keratin intermediate filaments in alveolar epithelial cells *J. Biol. Chem.* **280** 30400–05

Riveline D, Zamir E, Balaban N Q, Schwarz U S, Ishizaki T, Narumiya S, Kam Z, Geiger B and Bershadsky A D 2001 Focal contacts as mechanosensors: externally applied local mechanical force induces growth of focal contacts by an mDia1-dependent and ROCK-independent mechanism *J. Cell Biol.* **153** 1175–85

Rizki A, Mott J D and Bissell M J 2007 Polo-like kinase 1 is involved in invasion through extracellular matrix *Cancer Res.* **67** 11106–10

Roberts B J, Pashaj A, Johnson K R and Wahl J K 2011 Desmosome dynamics in migrating epithelial cells requires the actin cytoskeleton *Exp. Cell Res.* **317** 2814–22

Rochman M, Malicet C and Bustin M 2010 Hmgn5/nsbp1: a new member of the HMGN protein family that affects chromatin structure and function *Biochim. Biophys. Acta* **1799** 86–92

Roeber R A, Weber K and Osborn M 1989 Differential timing of nuclear lamin A/C expression in the various organs of the mouse embryo and the young animal: a developmental study *Development* **105** 365–78

Rogel M R, Jaitovich A and Ridge K M 2010 The role of the ubiquitin proteasome pathway in keratin intermediate filament protein degradation *Proc. Am. Thorac. Soc* **7** 71–6

Rogel M R, Soni P N, Troken J R, Sitikov A, Trejo H E and Ridge K M 2011 Vimentin is sufficient and required for wound repair and remodeling in alveolar epithelial cells *FASEB J.* **25** 3873–83

Rogers M A, Edler L, Winter H, Langbein L, Beckman I and Schweizer J 2005 Characterization of new members of the human type II keratin gene family and a general evaluation of the keratin gene domain on chromosome 12q13.13 *J. Investig. Dermatol.* **124** 536–44

Rogers M A, Winter H, Langbein L, Bleiler R and Schweizer J 2004 The human type I keratin gene family: characterization of new hair follicle specific members and evaluation of the chromosome 17q21.2 gene domain *Differentiation* **72** 527–40

Rogers M A, Edler L, Winter H, Langbein L, Beckmann I and Schweizer J 2005 Characterization of new members of the human type II keratin gene family and a general evaluation of the keratin gene domain on chromosome 12q13.13 J *Investig. Dermatol* **124** 536–44

Rolli C G, Seufferlein T, Kemkemer R and Spatz J P 2010 Impact of tumor cell cytoskeleton organization on invasiveness and migration: a microchannel-based approach *PLoS One* **5** e8726

Rotty J D and Coulombe P A 2012 A wound-induced keratin inhibits Src activity during keratinocyte migration and tissue repair *J. Cell Biol.* **197** 381–9

Roy S, Coffee P, Smith G, Liem R K, Brady S T and Black M M 2000 Neurofilaments are transported rapidly but intermittently in axons: implications for slow axonal transport *J. Neurosci.* **20** 6849–61

Russell D, Andrews P D, James J and Lane E B 2004 Mechanical stress induces profound remodelling of keratin filaments and cell junctions in epidermolysis bullosa simplex keratinocytes *J. Cell Sci.* **117** 5233–43

Sager P R 1989 Cytoskeletal effects of acrylamide and 2,5-hexanedione: selective aggregation of vimentin filaments *Toxicol. Appl. Pharmacol.* **97** 141–55

Sakamoto Y, Boeda B and Etienne-Manneville S 2013 APC binds intermediate filaments and is required for their reorganization during cell migration *J. Cell Biol.* **200** 249–58

Sanghvi-Shah R and Weber G F 2017 Intermediate filaments at the junction of mechanotransduction, migration, and development *Front. Cell Dev. Biol.* **5** 81

Sarria A J, Lieber J G, Nordeen S K and Evans R M 1994 The presence or absence of a vimentin-type intermediate filament network affects the shape of the nucleus in human SW-13 cells *J. Cell Sci.* **107** 1593–607

Satelli A and Li S 2011 Vimentin in cancer and its potential as a molecular target for cancer therapy *Cell. Mol. Life Sci.* **68** 3033–46

Schaus T E, Taylor E W and Borisy G G 2007 Self-organization of actin filament orientation in the dendritic-nucleation/array-treadmilling model *Proc. Natl Acad. Sci. USA* **104** 7086–91

Schmid E, Tapscott S, Bennett G S, Croop J, Fellini S A, Holtzer H and Franke W W 1979 Differential location of different types of intermediate-sized filaments in various tissues of the chicken embryo *Differentiation* **15** 27–40

Schopferer M, Bar H, Hochstein B, Sharma S, Mucke N, Herrmann H and Willenbacher N 2009 Desmin and vimentin intermediate filament networks: their viscoelastic properties investigated by mechanical rheometry *J. Mol. Biol.* **388** 133–43

Schoumacher M, Goldman R D, Louvard D and Vignjevic D M 2010 Actin, microtubules, and vimentin intermediate filaments cooperate for elongation of invadopodia *J. Cell Biol.* **189** 541–56

Schweizer J *et al* 2006 New consensus nomenclature for mammalian keratins *J. Cell Biol.* **174** 169–74

Seifert G J, Lawson D and Wiche G 1992 Immunolocalization of the intermediate filament-associated protein plectin at focal contacts and actin stress fibers *Eur. J. Cell Biol.* **59** 138–47

Seltmann K, Roth W, Kroeger C, Loschke F, Lederer M, Huettelmaier S and Magin T M 2013 Keratins mediate localization of hemidesmosomes and repress cell motility *J. Invest Dermatol.* **133** 181–90

Serrels B, Serrels A, Brunton V G, Holt M, McLean G W, Gray C H, Jones G E and Frame M C 2007 Focal adhesion kinase controls actin assembly via a FERM-mediated interaction with the Arp2/3 complex *Nat. Cell Biol.* **9** 1046–56

Shea T B, Beermann M L and Fischer I 1993 Transient requirement for vimentin in neuritogenesis: intracellular delivery of anti-vimentin antibodies and antisense oligonucleotides inhibit neurite initiation but not elongation of existing neurites in neuroblastoma *J. Neurosci. Res.* **36** 66–76

Sihag R K, Inagaki M, Yamaguchi T, Shea T B and Pant H C 2007 Role of phosphorylation on the structural dynamics and function of types III and IV intermediate filaments *Exp. Cell Res.* **313** 2098–109

Singh S, Sadacharan S, Su S, Belldegrun A, Persad S and Singh G 2003 Overexpression of vimentin: role in the invasive phenotype in an androgen-independent model of prostate cancer *Cancer Res.* **63** 2306–11

Sivaramakrishnan S, Schneider J L, Sitikov A, Goldman R D and Ridge K M 2009 Shear stress induced reorganization of the keratin intermediate filament network requires phosphorylation by protein kinase C zeta *Mol. Biol. Cell.* **20** 2755–65

Sivaramakrishnan S, DeGiulio J V, Lorand L, Goldman R D and Ridge K M 2008 Micromechanical properties of keratin intermediate filament networks *Proc. Natl Acad. Sci. USA* **105** 889–94

Skalski M and Coppolino M G 2005 SNARE-mediated trafficking of $\alpha5\beta1$ integrin is required for spreading in CHO cells *Biochem. Biophys. Res. Commun.* **335** 1199–210

Small J V and Kaverina I 2003 Microtubules meet substrate adhesions to arrange cell polarity *Curr. Opin. Cell Biol.* **15** 40–7

Small J V and Resch G P 2005 The comings and goings of actin: coupling protrusion and retraction in cell motility *Curr. Opin. Cell Biol.* **17** 517–23

Snider N T, Weerasinghe S V, Iñiguez-Lluhí J A, Herrmann H and Omary M B 2011 Keratin hypersumoylation alters filament dynamics and is a marker for human liver disease and keratin mutation *J. Biol. Chem.* **286** 2273–84

Soellner P, Quinlan R A and Franke W W 1985 Identification of a distinct soluble subunit of an intermediate filament protein: tetrameric vimentin from living cells *Proc. Natl Acad. Sci. USA* **82** 7929–33

Solovei I, Wang A S and Thanisch K *et al* 2013 LBR and lamina A/C sequentially tether peripheral heterochromatin and inversely regulate differentiation *Cell* **152** 584–98

Spann T P, Goldman A E, Wang C, Huang S and Goldman R D 2002 Alteration of nuclear lamin organization inhibits rna polymerase ii-dependent transcription *J. Cell Biol.* **156** 603–08

Spann T P, Moir R D, Goldman A E, Stick R and Goldman R D 1997 Disruption of nuclear lamin organization alters the distribution of replication factors and inhibits DNA synthesis *J. Cell Biol.* **136** 1201–12

Spazierer D, Raberger J, Gross K, Fuchs P and Wiche G 2008 Stress-induced recruitment of epiplakin to keratin networks increases their resistance to hyperphosphorylation-induced disruption *J. Cell Sci.* **121** 825–33

Sprecher E, Itin P, Whittock N V, McGrath A, Meyer R, DiGiovanna J J, Bale S J, Uitto J and Richard G 2002 Refined mapping of Naegeli–Franceschetti–Jadasohn syndrome to a 6 cM interval on chromosome 17q11.2-q21 and investigation of candidate genes *J. Investig. Dermatol.* **119** 692–8

Spurny R, Gregor M, Castañón M J and Wiche G 2008 Plectin deficiency affects precursor formation and dynamics of vimentin networks *Exp. Cell Res.* **314** 3570–80

Stefansson I M, Salvesen H B and Akslen L A 2006 Loss of p63 and cytokeratin 5/6 expression is associated with more aggressive tumors in endometrial carcinoma patients *Int. J. Cancer* **118** 1227–33

Steinboeck F A, Nikolic B, Coulombe P A, Fuchs E, Traub P and Wiche G 2000 Dose-dependent linkage, assembly inhibition and disassembly of vimentin and cytokeratin 5/14 filaments through plectin's intermediate filament-binding domain *J. Cell Sci.* **113** 483–91

Stoffler D, Feja B, Fahrenkrog B, Walz J, Typke D and Aebi U 2003 Cryo-electron tomography provides novel insights into nuclear pore architecture: implications for nucleocytoplasmic transport *J. Mol. Biol.* **328** 243–51

Stossel T P, Condeelis J, Cooley L, Hartwig J H, Noegel A, Schleicher M and Shapiro S S 2001 Filamins as integrators of cell mechanics and signalling *Nat. Rev. Mol. Cell Biol.* **2** 138–45

Strelkov S V, Hermann H and Aebi U 2003 Molecular architecture of intermediate filaments *Bioessays* **25** 243–51

Strnad P, Windoffer R and Leube R E 2001 *In vivo* detection of cytokeratin filament network breakdown in cells treated with the phosphatase inhibitor okadaic acid *Cell Tissue Res.* **306** 277–93

Strnad P, Windoffer R and Leube R E 2002 Induction of rapid and reversible cytokeratin filament network remodeling by inhibition of tyrosine phosphatases *J. Cell Sci.* **115** 4133–48

Stumptner C, Fuchsbichler A, Lehner M, Zatloukal K and Denk H 2001 Sequence of events in the assembly of Mallory body components in mouse liver: clues to the pathogenesis and significance of Mallory body formation *J. Hepatol.* **34** 665–75

Sun C X, Magalhaes M A and Glogauer M 2007 Rac1 and Rac2 differentially regulate actin free barbed end formation downstream of the fMLP receptor *J. Cell Biol.* **179** 239–45

Sun L P, Wang L, Wang H, Zhang Y H and Pu J L 2010 Connexin 43 remodeling induced by LMNA gene mutation Glu82Lys in familial dilated cardiomyopathy with atrial ventricular block *Chin. Med. J. (Engl.)* **123** 1058–62

Sun N, Critchley D R, Paulin D, Li Z and Robson R M 2008a Human α-synemin interacts directly with vinculin and metavinculin *Biochem. J.* **409** 657–67

Sun N, Critchley D R, Paulin D, Li Z and Robson R M 2008b Identification of a repeated domain within mammalian α-synemin that interacts directly with talin *Exp. Cell Res.* **314** 1839–49

Sun N, Huiatt T W, Paulin D, Li Z and Robson R M 2010 Synemin interacts with the LIM domain protein zyxin and is essential for cell adhesion and migration *Exp. Cell Res.* **316** 491–505

Sutoh Y M, Hatakeyama S, Habuchi T, Inoue T, Nakamura T, Funyu T, Wiche G, Ohyama C and Tsuboi S 2014 Vimentin intermediate filament and plectin provide a scaffold for invadopodia, facilitating cancer cell invasion and extravasation for metastasis *Eur. J. Cell Biol.* **93** 157–69

Svitkina T M, Verkhovsky A B and Borisy G G 1996 Plectin sidearms mediate interaction of intermediate filaments with microtubules and other components of the cytoskeleton *J. Cell Biol.* **135** 991–1007

Szeverenyi I, Cassidy A J and Chung C W *et al* 2008 The human intermediate filament database: comprehensive information on a gene family involved in many human diseases *Hum. Mutat.* **29** 351–60

Takemura M, Gomi H, Colucci-Guyon E and Itohara S 2002 Protective role of phosphorylation in turnover of glial fibrillary acidic protein in mice *J. Neurosci.* **22** 6972–9

Thaiparambil J T, Bender L, Ganesh T, Kline E, Patel P, Liu Y and Marcus A I 2011 Withaferin A inhibits breast cancer invasion and metastasis at sub-cytotoxic doses by inducing vimentin disassembly and serine 56 phosphorylation *Int J. Cancer* **129** 2744–55

Thiery J P, Acloque H, Huang R Y and Nieto M A 2009 Epithelial–mesenchymal transitions in development and disease *Cell* **139** 871–90

Tigges U, Koch B, Wissing J, Jockusch B M and Ziegler W H 2003 The F-actin cross-linking and focal adhesion protein filamin A is a ligand and *in vivo* substrate for protein kinase C alpha *J. Biol. Chem.* **278** 23561–9

Todorovic V, Desai B V, Patterson M J, Amargo E V, Dubash A D, Yin T, Jones J C and Green K J 2010 Plakoglobin regulates cell motility through Rho- and fibronectin-dependent Src signaling *J. Cell Sci.* **123** 3576–86

Toivola D M, Ku N O, Resurreccion E Z, Nelson D R, Wright T L and Omary M B 2004 Keratin 8 and 18 hyperphosphorylation is a marker of progression of human liver disease *Hepatology* **40** 459–66

Toivola D M, Strnad P, Habtezion A and Omary M B 2010 Intermediate filaments take the heat as stress proteins *Trends Cell Biol.* **20** 79–91

Toivola D M, Tao G Z, Habtezion A, Liao J and Omary M B 2005 Cellular integrity plus: organelle-related and protein-targeting functions of intermediate filaments *Trends Cell Biol.* **15** 608–17

Toivola D M, Boor P, Alam C and Strnad P 2015 Keratins in health and disease *Curr. Opin. Cell Biol.* **32** 73–81

Toivola D M, Nieminen M I, Hesse M, He T, Baribault H, Magin T M, Omary M B and Eriksson J E 2001 Disturbances in hepatic cell-cycle regulation in mice with assembly-deficient keratins 8/18 *Hepatology* **34** 1174–83

Tolstonog G V, Sabasch M and Traub P 2002 Cytoplasmatic intermediate filaments are stably associated with nuclear matrices and potentially modulate their DNA-binding function *DNA Cell Biol.* **21** 213–39

Trepat X and Fredberg J J 2011 Plithotaxis and emergent dynamics in collective cellular migration *Trends Cell Biol.* **21** 638–46

Tsuruta D and Jones J C 2003 The vimentin cytoskeleton regulates focal contact size and adhesion of endothelial cells subjected to shear stress *J. Cell Sci.* **116** 4977–84

Tsuruta D and Jones J C 2003 The vimentin cytoskeleton regulates focal contact size and adhesion of endothelial cells subjected to shear stress *J. Cell Sci.* **116** 4977–84

Ullrich O, Reinsch S, Urbe S, Zerial M and Parton R G 1996 Rab11 regulates recycling through the pericentriolar recycling endosome *J. Cell Biol.* **135** 913–24

Vadlamudi R K, Li F, Adam L, Nguyen D, Ohta Y, Stossel T P and Kumar R 2002 Filamin is essential in actin cytoskeletal assembly mediated by p21-activated kinase 1 *Nat. Cell Biol.* **4** 681–90

Valencia R G, Walko G, Janda L, Novacek J, Mihailovska E, Reipert S, Andra-Marobela K and Wiche G 2013 Intermediate filament-associated cytolinker plectin 1c destabilizes microtubules in keratinocytes *Mol. Biol. Cell.* **24** 768–84

van de Rijn M, Perou C M and Tibshirani R *et al* 2002 Expression of cytokeratins 17 and 5 identifies a group of breast carcinomas with poor clinical outcome *Am. J. Pathol.* **161** 1991–6

Vicente-Manzanares M, Choi C K and Horwitz A R 2009 Integrins in cell migration—the actin connection *J. Cell Sci* **122** 199–206

Vijayaraj P, Kroeger C, Reuter U, Windoffer R, Leube R E and Magin T M 2009 Keratins regulate protein biosynthesis through localization of GLUT1 and -3 upstream of AMP kinase and Raptor *J. Cell Biol.* **187** 175–84

Virtakoivu R, Mai A and Mattila E *et al* 2015 Vimentin-ERK signaling uncouples Slug gene regulatory function *Cancer Res.* **75** 2349–62

Vogel V and Sheetz M P 2009 Cell fate regulation by coupling mechanical cycles to biochemical signaling pathways *Curr. Opin. Cell Biol.* **21** 38–46

Vuoriluoto K, Haugen H, Kiviluoto S, Mpindi J P, Nevo J, Gjerdrum C, Tiron C, Lorens J B and Ivaska J 2011 Vimentin regulates EMT induction by Slug and oncogenic H-Ras and migration by governing Axl expression in breast cancer *Oncogene* **30** 1436–48

Wade R H 2009 On and around microtubules: an overview *Mol. Biotechnol.* **43** 177–91

Wan H *et al* 2004 Striate palmoplantar keratoderma arising from desmoplakin and desmoglein 1 mutations is associated with contrasting perturbations of desmosomes and the keratin filament network *Br. J. Dermatol.* **150** 878–91

Wang L, Ho C L, Sun D, Liem R K and Brown A 2000 Rapid movement of axonal neurofilaments interrupted by prolonged pauses *Nat. Cell Biol.* **2** 137–41

Wang N and Ingber D E 1994 Control of cytoskeletal mechanics by extracellular matrix, cell shape, and mechanical tension *Biophys. J.* **66** 2181–9

Wang N and Stamenovic D 2000 Contribution of intermediate filaments to cell stiffness, stiffening, and growth *Am. J. Physiol. Cell Physiol.* **279** C188–C194

Wang Y L and Pelham R J Jr 1998 Preparation of a flexible, porous polyacrylamide substrate for mechanical studies of cultured cells *Methods Enzymol.* **298** 489–96

Wang N and Stamenovic D 2000 Contribution of intermediate filaments to cell stiffness, stiffening, and growth *Am. J. Physiol.* **279** C188–C194

Ward G E and Kirschner M W 1990 Identification of cell-cycle regulated phosphorylation sites on nuclear lamin C *Cell* **61** 561–77

Webb B A, Jia L, Eves R and Mak A S 2007 Dissecting the functional domain requirements of cortactin in invadopodia formation *Eur. J. Cell Biol.* **86** 189–206

Weber G F, Bjerke M A and DeSimone D W 2012 A mechanoresponsive cadherin-keratin complex directs polarized protrusive behavior and collective cell migration *Dev. Cell* **22** 104–15

Weber K, Plessmann U and Traub P 1989 Maturation of nuclear lamin A involves a specific carboxy-terminal trimming, which removes the polyisoprenylation site from the precursor; implications for the structure of the nuclear lamina *FEBS Lett.* **257** 411–4

Weber G F, Bjerke M A and DeSimone D W 2012 A mechanoresponsive cadherin–keratin complex directs polarized protrusive behavior and collective cell migration *Dev. Cell* **22** 104–15

Wei J *et al* 2008 Overexpression of vimentin contributes to prostate cancer invasion and metastasis via Src regulation *Anticancer Res.* **28** 327–34

Werner N S, Windoffer R, Strnad P, Grund C, Leube R E and Magin T M 2004 Epidermolysis bullosa simplex-type mutations alter the dynamics of the keratin cytoskeleton and reveal a contribution of actin to the transport of keratin subunits *Mol. Biol. Cell.* **15** 990–1002

Whipple R A, Balzer E M, Cho E H, Matrone M A, Yoon J R and Martin S S 2008 Vimentin filaments support extension of tubulin-based microtentacles in detached breast tumor cells *Cancer Res.* **68** 5678–88

Wiche G 1998 Role of plectin in cytoskeleton organization and dynamics *J. Cell Sci.* **111** 2477–86

Wilhelmsen K, Litjens S H, Kuikman I, Tshimbalanga N, Janssen H, van den Bout I, Raymond K and Sonnenberg A 2005 Nesprin-3, a novel outer nuclear membrane protein, associates with the cytoskeletal linker protein plectin *J. Cell Biol.* **171** 799–810

Windoffer R and Leube R E 1999 Detection of cytokeratin dynamics by time-lapse fluorescence microscopy in living cells *J. Cell Sci.* **112** 4521–34

Windoffer R, Koelsch A, Woell S and Leube R E 2006 Focal adhesions are hotspots for keratin filament precursor formation *J. Cell Biol.* **173** 341–8

Windoffer R, Woell S, Strnad P and Leube R E 2004 Identification of novel principles of keratin filament network turnover in living cells *Mol. Biol. Cell* **15** 2436–48

Winokur R and Hartwig J H 1995 Mechanism of shape change in chilled human platelets *Blood* **85** 1796–804

Woell S, Windoffer R and Leube R E 2007 p38 MAPK-dependent shaping of the keratin cytoskeleton in cultured cells *J. Cell Biol.* **177** 795–807

Woell S, Windoffer R and Leube R E 2005 Dissection of keratin dynamics: different contributions of the actin and microtubule systems *Eur. J. Cell Biol.* **84** 311–28

Wolf A, Krause-Gruszczynska M, Birkenmeier O, Ostareck-Lederer A, Huettelmaier S and Hatzfeld M 2010 Plakophilin 1 stimulates translation by promoting eIF4A1 activity *J. Cell Biol.* **188** 463–71

Wolf D M, Vazirani V V and Arkin A P 2005 Diversity in times of adversity: probabilistic strategies in microbial survival games *J. Theor. Biol.* **234** 227–53

Wong P and Coulombe P A 2003 Loss of keratin 6 (K6) proteins reveals a function for intermediate filaments during wound repair *J. Cell Biol.* **163** 327–37

Xiao S, McLean J and Robertson J 2006 Neuronal intermediate filaments and ALS: a new look at an old question *Biochim. Biophys. Acta* **1762** 1001–12

Xu Z, Wang M R, Xu X, Cai Y, Han Y L, Wu K M, Wang J, Chen B S, Wang X Q and Wu M 2000 Novel human esophagus-specific gene c1orf10: cDNA cloning, gene structure, and frequent loss of expression in esophageal cancer *Genomics* **69** 322–30

Yamasaki T, Seki N, Yamada Y, Yoshino H, Hidaka H, Chiyomaru T, Nohata N, Kinoshita T, Nakagawa M and Enokida H 2012 Tumor suppressive microRNA138 contributes to cell migration and invasion through its targeting of vimentin in renal cell carcinoma *Int. J. Oncol.* **41** 805–17

Yang S H, Qiao X, Fong L G and Young S G 2008 Treatment with a farnesyltransferase inhibitor improves survival in mice with a Hutchinson–Gilford progeria syndrome mutation *Biochim. Biophys. Acta* **1781** 36–9

Yang Y, Dowling J, Yu Q C, Kouklis P, Cleveland D W and Fuchs E 1996 An essential cytoskeletal linker protein connecting actin microfilaments to intermediate filaments *Cell* **86** 655–65

Yasui Y, Goto H, Matsui S, Manser E, Lim L, Nagata K and Inagaki M 2001 Protein kinases required for segregation of vimentin filaments in mitotic process *Oncogene* **20** 2868–76

Yilmaz M and Christofori G 2009 EMT, the cytoskeleton and cancer cell invasion *Cancer Metas. Rev.* **28** 15–33

Yin T, Getsios S, Caldelari R, Kowalczyk A P, Muller E J, Jones J C and Green K J 2005 Plakoglobin suppresses keratinocyte motility through both cell–cell adhesion-dependent and -independent mechanisms *Proc. Natl Acad. Sci. USA* **102** 5420–5

Yoon K H, Yoon M, Moir R D, Khuon S, Flitney F W and Goldman R D 2001 Insights into the dynamic properties of keratin intermediate filaments in living epithelial cells *J. Cell Biol.* **153** 503–16

Zatloukal K, French S W, Stumptner C, Strnad P, Harada M, Toivola D M, Cadrin M and Omary M B 2007 From Mallory to Mallory–Denk bodies: what, how and why? *Exp. Cell Res.* **313** 2033–49

Zeigerer A, Gilleron J and Bogorad R L *et al* 2012 Rab5 is necessary for the biogenesis of the endolysosomal system in vivo *Nature* **485** 465–70

Zencheck W D, Xiao H and Weiss L M 2012 Lysine post-translational modifications and the cytoskeleton *Essays Biochem.* **52** 135–45

Zhang H, Landmann F, Zahreddine H, Rodriguez D, Koch M and Labouesse M 2011 A tension-induced mechanotransduction pathway promotes epithelial morphogenesis *Nature* **471** 99–103

Zhang J S, Wang L, Huang H, Nelson M and Smith D I 2001 Keratin 23 (K23), a novel acidic keratin, is highly induced by histone deacetylase inhibitors during differentiation of pancreatic cancer cells *Genes Chromosomes Cancer* **30** 123–35

Zhang S, Schones D E, Malicet C, Rochman M, Zhou M, Foisner R and Bustin M 2013 High mobility group protein N5 (HMGN5) and lamina-associated polypeptide 2α (LAP2α) interact and reciprocally affect their genome-wide chromatin organization *J. Biol. Chem.* **288** 18104–9

Zhao X H, Laschinger C, Arora P, Szaszi K, Kapus A and McCulloch C A 2007 Force activates smooth muscle α-actin promoter activity through the Rho signaling pathway *J. Cell Sci.* **120** 1801–9

Zhao Y, Yan Q, Long X, Chen X and Wang Y 2008 Vimentin affects the mobility and invasiveness of prostate cancer cells *Cell Biochem. Funct.* **26** 571–7

Zhu Q S *et al* 2011 Vimentin is a novel AKT1 target mediating motility and invasion *Oncogene* **30** 457–70

Zimerman B, Volberg T and Geiger B 2004 Early molecular events in the assembly of the focal adhesion-stress fiber complex during fibroblast spreading *Cell Motil. Cytoskelet.* **58** 143–59

www.ingramcontent.com/pod-product-compliance
Ingram Content Group UK Ltd.
Pitfield, Milton Keynes, MK11 3LW, UK
UKHW051910100926
8556IPUK00005B/839

9 780750 317511